Chaotic Oscillators

CHAOTIC OSCILLATORS

THEORY AND APPLICATIONS

Edited by

Tomasz Kapitaniak

Division of Control and Dynamics
Technical University of Łódź
Poland

Published by

World Scientific Publishing Co. Pte. Ltd.
P O Box 128, Farrer Road, Singapore 9128
USA office: Suite 1B, 1060 Main Street, River Edge, NJ 07661
UK office: 73 Lynton Mead, Totteridge, London N20 8DH

Library of Congress Cataloging-in-Publication Data
Chaotic oscillators : theory and applications / edited by Tomasz
   Kapitaniak.
      p.   cm.
   Includes bibliographical references.
   ISBN 9810206534 -- ISBN 9810206542 (pbk)
   1. Chaotic behavior in systems.   I. Kapitaniak, Tomasz.
Q172.5.C45C45   1991
003'.7--dc20   91-11836
                                                                      CIP

Copyright © 1992 by World Scientific Publishing Co. Pte. Ltd.

All rights reserved. This book, or parts thereof, may not be reproduced in any form or by any means, electronic or mechanical, including photocopying, recording or any information storage and retrieval system now known or to be invented, without written permission from the Publisher.

Printed in Singapore by Utopia Press.

INTRODUCTION

There is great interest in chaotic systems currently which has begun after Lorenz work on the model of atmospheric convection. Edward Lorenz in his famous work 'Deterministic nonperiodic flow', which was published in 1963, presents a system of three differential equations which are deterministic but show very irregular behavior. It should be mentioned here that Poincaré in 1889 considered the possibility of such systems and many of the modern ideas and developments in chaos theory can be traced back to his classical work on celestial mechanics. One hundred years ago Poincaré published a memoir (1890) on the stability of the solar system where he focused on a simplified model situation: the restricted three-body problem. For this work he was awarded a mathematical price offered by King Oscar II of Sweden. Poincaré was the first to point out that unstable motion of apparently simple systems can be extraordinarily complicated, but unfortunately the methods of his days could not solve the equations of motion of the solar system and other similar dynamical systems which he considered. He developed the new branch of mathematics, topology, which has turned out to be a powerful tool for describing chaotic systems. Later, Birkhoff in 1927 proved what Poincaré had conjectured: the existence in the restricted planar three-body problem of infinitely many periodic orbits. These important results of Poincaré and Birkhoff are not readily applicable to problems in applied science, so the recent explosive growth of chaotic dynamics and its applications follows from Lorenz work. The remarkable fact that determinism implies neither regular behavior nor predictability has got a major impact on many fields of science, engineering, and mathematics. A great number of other deterministic equations showing chaotic behavior have been obtained, both as simple, analytical theoretical mathematical systems and as models of real physical, biological, chemical systems. These include both systems of nonlinear ordinary differential equation and maps.

A great number of systems with chaotic behavior have been found (for example: May (1976), Feigenbaum (1978, 1980), Greborgi et al. (1982), Jeffries and Perez (1983), Yamaguchi and Sakai (1983), Collet and Eckmann (1980), Nauenberg and Rudnick (1981)). Scientists from many different branches of science including: mathematics, physics, chemistry, biology, mechanical, electrical and civil engineering are working on chaotic experimental systems and models, and hence developing a new way of modelling the phenomena in the real world.

After a decade of extensive research in chaotic dynamics the success is well documented in the great number of papers and several books as shown in References.

The books can be divided into five groups:
(a) collections of papers or invited reviews,
(b) conference proceedings,
(c) monographs,
(d) texts,
(e) popular work.

The research in physics and fluid dynamics is well documented in two collections of papers (Hao (1985, 1990) and Cvitanovitch (1986)).

Collections of invited reviews, mainly by physicists, can be found in the series Directions in Chaos (ed. Hao (1988, 1989)). The reviews by mathematicians, physicists and biologists can be found in Holden (1986).

In the past few years a great number of conferences and workshops devoted to chaotic dynamics had been organized. After some of them the proceedings were published (for example: Salam and Levi (1988), Helleman (1980), Kawakami (1990), Christiansen and Parmentier (1989)). In most of them the papers by researchers from various branches of science can be found.

Most of the monographs on chaos are written by mathematicians: Guckenheimer and Holmes (1983), Sparrow (1982), Steeb and Louw (1986), Lichtenberg and Lieberman (1982), Rulle (1990) or physicists: Berge et al. (1984), Schuster (1984), Kaneko (1986), Hao (1989), Lee (1990). When they consider oscillatory examples (for example Guckenheimer and Holmes (1983)) these are discussed from mathematical point of view. The books which specially consider the chaotic dynamics in engineering problems are: Thompson and Steward (1986), Moon (1987), El Naschie (1990) and Kapitaniak (1991), where a number of examples of applications are presented. The examples of applications of the methods of chaotic dynamics in noisy mechanical systems can be found in Kapitaniak (1990), who considers the effect of random noise on chaotic behavior. In El Naschie (1990) and Kapitaniak (1990) besides the usual chaotic behavior in time the chaotic behavior in space (spatial chaos) is also considered.

As understanding of the nonlinear phenomena requires solid mathematical background, textbooks for undergraduate as well as graduate students of applied mathematics such as: Arrowsmith and Place (1990) and Wiggins (1990) are available. An introduction to chaotic dynamics with all elementary information on nonlinear phenomena can be found in Baker and Gollub (1990). The texts which cover nonlinear problems in fluid dynamics and plasma physics are: Ottino (1989) and Infeld and Rowlands (1990).

A few popular books like Gleick (1987) and Steward (1989) are written to give the flavor of chaos, but they do not allow the reader a significant measure of participation in the subject.

Up to now there have not been reprint collections which documented the achievements of the research on chaotic behavior of nonlinear oscillators. The aim of this selection is to try to fulfill this gap. In this collection we have presented the papers on chaotic behavior of nonlinear oscillators. First we will give some fundamental papers about quantifying chaotic systems and later present a number of chaotic oscillators with applications in physics and engineering.

In each section we start with papers which provide fundamental information about the problem and later more advanced papers are given. Prior knowledge of basic ideas of nonlinear dynamics (Baker and Gollub (1990), Kapitaniak (1991)) might be an advantage in studying problems presented in this selection.

This book could be very useful for anyone with fundamental knowledge of nonlinear oscillators who likes to start a research in this subject as well as the source of references for researchers who are already working in the field.

Contents

Introduction

I. Chaos before Chaos

I.1. Frequency Demultiplication (*Nature* **120**, (1927) 363-364)
B. Van der Pol and J. Van der Mark — 4

I.2. On Nonlinear Differential Equations of the Second Order: I. The Equation $\ddot{y} - k(1 - y^2)\dot{y} + y = b\lambda k\cos(\lambda t + a)$, k Large (*J. London Math. Soc.* **20**, (1945) 180-189)
M.L. Cartwright and J.E. Littlewood — 6

I.3. Nonlinear Vibrations of a Buckled Beam under Harmonic Excitation (*ASME J. Appl. Mech.* **38**, (1971) 467-476)
W.-Y. Tseng and J. Dugundji — 16

I.4. Random Phenomena Resulting from Nonlinearity in the System Described by Duffing's Equation (*Int. J. Non-Linear Mech.* **20**, (1985) 481-491)
Y. Ueda — 26

II. Description and Quantification of Chaotic Behavior

II.1. Strange Attractors (*Math. Intelligencer* **2**, (1980) 126-137)
D. Ruelle — 40

II.2. Geometry from a Time Series (*Phys. Rev. Lett.* **45**, (1980) 712-716)
N.H. Packard, J.P. Crutchfield, J.D. Farmer and R.S. Shaw — 52

II.3. Deterministic Representation of Chaos in Classical Dynamics (*Phys. Lett.* **107A**, (1985) 125-128)
M. Zak — 57

II.4. Determining Lyapunov Exponents from a Time Series (*Physica* **16D**, (1985) 285-317)
A. Wolf, J.B. Swift, H.L. Swinney and J.A. Vastano — 61

III. Analytical Methods

III.1. A Partial Differential Equation with Infinitely Many Periodic Orbits: Chaotic Oscillations of a Forced Beam (*Arch. Rat. Mech. Anal.* **76**, (1981) 135-165)
P. Holmes and J. Marsden ... 98

III.2. Second Order Averaging and Bifurcations to Subharmonics in Duffing's Equation (*J. Sound Vib.* **78**, (1981) 161-174)
C. Holmes and P. Holmes ... 129

III.3. Subharmonic and Homoclinic Bifurcations in a Parametrically Forced Pendulum (*Physica* **16D**, (1985) 1-13)
B.P. Koch and R.W. Leven ... 143

III.4. Power Spectra of Strange Attractors near Homoclinic Orbits (*Phys. Rev. Lett.* **58**, (1987) 1699-1702)
V. Brunsden and P. Holmes ... 156

III.5. The Mel'nikov's Technique for Highly Dissipative Systems (*SIAM J. Appl. Math.* **47**, (1987) 232-243)
F. Salam ... 160

III.6. The Variational Approach to the Theory of Subharmonic Bifurcations (*Physica* **26D**, (1987) 251-276)
N. Takimoto and H. Yamashida ... 172

III.7. The Refined Approximate Criterion for Chaos in a Two-State Mechanical Oscillator (*Ingenieur-Archiv* **58**, (1988) 354-366)
W. Szemplińska-Stupnicka ... 198

III.8. Analytical Condition for Chaotic Behaviour of the Duffing Oscillator (*Phys. Lett.* **A144**, (1990) 322-324)
T. Kapitaniak ... 211

IV. Classical Nonlinear Oscillators: Duffing, Van der Pol and Pendulum

IV.1. Universal Scaling Property in Bifurcation Structure of Duffing's and of Generalized Duffing's Equations (*Phys. Rev.* **A28**, (1983) 1654-1658)
S. Sato, M. Sano and Y. Sawada ... 219

IV.2.	Superstructure in the Bifurcation Set of the Duffing's Equation, $\ddot{x} + d\dot{x} + x + x^3 = f \cos(\omega t)$ (*Phys. Lett.* **A107**, (1985) 351-355) U. Parlitz and W. Lauterborn	224
IV.3.	On the Understanding of Chaos in Duffing's Equation Including a Comparison with Experiment (*ASME J. Appl. Mech.* **53**, (1986) 5-9) E. Dowell and C. Pezeshki	229
IV.4.	Chaotically Transitional Phenomena in the Forced Negative-Resistance Oscillator (*IEEE Trans. Circuits Syst.* **CAS-28**, (1981) 217-224) Y. Ueda and N. Akamatsu	234
IV.5.	Period-Doubling Cascades and Devil's Staircases of the Driven Van der Pol Oscillator (*Phys. Rev.* **A36**, (1987) 1428-1434) U. Parlitz and W. Lauterborn	242
IV.6.	Chaotic States and Routes to Chaos in the Forced Pendulum (*Phys. Rev.* **A26**, (1982) 3483-3496) D. D'Humieres, M.R. Beasley, B.A. Huberman and A. Libchaber	249
IV.7.	Josephson Junctions and Circle Maps (*Solid State Commun.* **51**, (1984) 231-234) P. Bak, T. Bohr, M.H. Jensen and P. Christiansen	263
IV.8.	Resonance and Symmetry Breaking for the Pendulum (*Physica* **31D**, (1988) 252-268) J. Miles	267

V. Other Oscillatory Systems

V.1.	Experiments on Chaotic Motions of a Forced Nonlinear Oscillator: Strange Attractors (*ASME J. Appl. Mech.* **47**, (1980) 638-644) F.C. Moon	289

V.2. Chaos after Period-Doubling Bifurcations in the Resonance of an Impact Oscillator (*Phys. Lett.* **A91**, (1982) 5-8)
J.M.T. Thompson and R. Ghaffari ... 296

V.3. Devil's Attractors and Chaos of a Driven Impact Oscillator (*Phys. Lett.* **A107**, (1985) 343-346)
H.M. Isomäki, J. von Boehm and J. Räty ... 300

V.4. A Periodically Forced Piecewise Linear Oscillator (*J. Sound Vib.* **90**, (1983) 129-155)
S.W. Shaw and P. Holmes ... 304

V.5. Complex Dynamics of Compliant Off-Shore Structures (*Proc. R. Soc. London* **A387**, (1983) 407-427)
J.M.T. Thompson ... 331

V.6. Chaos Generated by the Cutting Process (*Phys. Lett.* **A117**, (1986) 384-386)
I. Grabec ... 352

V.7. The Stability of Modes at Rest in a Chaotic System (*J. Sound Vib.* **138**, (1990) 421-431)
S.-R. Hsieh and S.W. Shaw ... 355

V.8. Chaotic Motion of an Elastic-Plastic Beam (*ASME J. Appl. Mech.* **55**, (1988) 185-189)
B. Poddar, F.C. Moon and S. Mukherjee ... 366

V.9. Chua's Circuit Family (*Proc. IEEE* **75**, (1987) 1022-1032)
S. Wu ... 371

V.10. Chaos via Torus Breakdown (*IEEE Trans. Circuits Syst.* **CAS-34**, (1987) 240-253)
T. Matsumoto, L.O. Chua and R. Tokunaga ... 382

VI. Chaos in Noisy Systems

VI.1. Fluctuations and the Onset of Chaos (*Phys. Lett.* **A77**, (1980) 407-410)
J.P. Crutchfield and B.A. Huberman ... 401

VI.2. Noise versus Chaos in Acousto-optic Bistability
(*Phys. Rev.* **A30**, (1984) 336-342)
R. Vallée, C. Delisle and J. Chrostowski   405

VI.3. Generalized Multistability and Noise-Induced Jumps in a Nonlinear Dynamical System (*Phys. Rev.* **A32**, (1985) 402-408)
F.T. Arecchi, R. Badii and A. Politi   412

VI.4. Chaotic Distribution of Non-Linear Systems Perturbed by Random Noise (*Phys. Lett.* **A116**, (1986) 251-254)
T. Kapitaniak   419

VI.5. Influence of Perturbations on Period-Doubling Bifurcation
(*Phys. Rev.* **A36**, (1987) 2413-2417)
H. Svensmark and M. Samuelsen   423

VI.6. Chaos in a Noisy Mechanical System with Stress Relaxation
(*J. Sound Vib.* **123**, (1988) 391-396)
T. Kapitaniak   428

VI.7. Homoclinic Chaos in Systems Perturbed by Weak Langevin Noise
(*Phys. Rev.* **A41**, (1990) 668-681)
A.R. Bulsara, W.C. Schieve and E.W. Jacobs   434

VII. Strange Nonchaotic Attractors

VII.1. Quasiperiodically Forced Dynamical Systems with Strange Nonchaotic Attractors (*Physica* **26D**, (1987) 277-294)
F.J. Romeiras, A. Bondeson, E. Ott, T.M. Antonsen, Jr. and C. Grebogi   452

VII.2. Dimensions of Strange Nonchaotic Attractors
(*Phys. Lett.* **A137**, (1989) 167-172)
M. Ding, C. Grebogi and E. Ott   470

VII.3. Route to Chaos via Strange Non-chaotic Attractors
(*J. Phys.* **A23**, (1990) L383-L387)
T. Kapitaniak, E. Ponce and J. Wojewoda   476

VIII. Spatial Chaos

VIII.1. Chaos as a Limit in a Boundary Value Problem
(Z. Naturforsch. **A39**, (1984) 1200-1203)
C. Kahlert and O.E. Rössler 484

VIII.2. On the Connection between Statical and Dynamical Chaos
(Z. Naturforsch. **A44**, (1989) 645-650)
M.S. El Naschie and S. Al Athel 488

VIII.3. Spatially Complex Equilibria of Buckled Rods
(Arch. Rat. Mech. Anal. **101**, (1988) 319-348)
A. Mielke and P. Holmes 494

VIII.4. Spatial Chaos and Localization Phenomena in Nonlinear
Elasticity (Phys. Lett. **A126**, (1988) 491-496)
J.M.T. Thompson and L.N. Virgin 524

VIII.5. Soliton Chaos Models for Mechanical and Biological
Elastic Chains (Phys. Lett. **A147**, (1990) 275-281)
M.S. El Naschie and T. Kapitaniak 530

VIII.6. Spatiotemporal Dynamics in a Dispersively Coupled Chain of
Nonlinear Oscillators (Phys. Rev. **A39**, (1989) 4835-4842)
D.K. Umberger, C. Grebogi, E. Ott and B. Afeyan 537

IX. Fractal Basin Boundaries

IX.1. Fractal Basin Boundaries (Physica **17D**, (1985) 125-153)
S.W. McDonald, C. Grebogi, E. Ott and J.A. Yorke 548

IX.2. Fractal Basin Boundaries and Homoclinic Orbit for
Periodic Motion in a Two-Well Potential
(Phys. Rev. Lett. **55**, (1985) 1439-1442)
F.C. Moon and G.-H. Li 577

IX.3. Fractal Basin Boundaries of an Impacting Particle
(Phys. Lett. **A126**, (1988) 484-490)
H.M. Isomäki, J. von Boehm and R. Räty 581

IX.4. Integrity Measures Quantifying the Erosion of Smooth and Fractal Basins of Attraction (*J. Sound Vib.* **135**, (1989) 453-475)

M.S. Soliman and J.M.T. Thompson   588

References   611

I. Chaos Before Chaos

In the eighties the study of chaotic dynamics became very popular within the applied science. In this section we have given some early examples of chaotic behavior although the word 'chaos' is not used in them.

Paper I.1 is an article published by Van der Pol in 1927 where the first experimental observation of 'chaos' is reported.

Mathematical model showing possibility of strange behavior of Van der Pol's equation is presented in the second paper.

It has been suspected that chaotic oscillations in solids were observed long before the eighties but were either ignored or could not be explained. The experiments of Tseng and Dugundji (Paper I.3) on oscillations of a buckled beam are such an example.

Paper I.4 by Ueda on strange behavior of the system described by Duffing's equation is one of the first reports on chaotic oscillators and it stimulated further research in this field.

Frequency Demultiplication.

It is a well-known fact that when a sinusoidal E.M.F. (of the form $E_0 \sin \omega t$) is available, it is a relatively simple matter to design an electrical system such that alternating currents or potential differences will occur in the system, having a frequency which is a whole multiple of the applied E.M.F., e.g. 2ω, 3ω, etc. For example, when the E.M.F. $E_0 \sin \omega t$ is applied to a diode-rectifier, the current in the anode circuit will include a component of double frequency, i.e. 2ω. This is therefore one method of frequency multiplication. Several other methods could easily be mentioned.

Now we found it is also possible to design an electrical system such that when the above-mentioned

E.M.F., $E_0 \sin \omega t$, is applied to it, currents and potential differences occur in the system the frequencies of which are whole submultiples of the frequency of the applied E.M.F., e.g. $\omega/2$, $\omega/3$, $\omega/4$ up to $\omega/40$.

To this end one can make use of the remarkable synchronising properties of relaxation-oscillations,

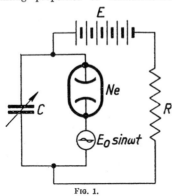

FIG. 1.

i.e. oscillations the time period of which is determined by the approximate expression $T = \pi/2\,CR$, a relaxation time (Balth. van der Pol, "On Relaxation Oscillations," *Phil. Mag.*, p. 978, 1926; also *Zeitschr. f. hochfreq. Technik*, 29, 114; 1927).

Let Ne in Fig. 1 represent a neon glow lamp, R a resistance of the order of a few megohms, C a variable condenser of approximately maximum 3500 cm. capacity and E a battery of say 200 volts. In the absence of the E.M.F. $E_0 \sin \omega t$, this system will oscillate with a time period $T = aCR$ where a is a number of the order unity. With the E.M.F. $E_0 \sin \omega t$ present, where E_0 may be of the order of 10 volts (considerably lower voltages also give the same result) it is found that the system is only capable of oscillating with *discrete frequencies*, these being determined by *whole submultiples of the applied frequency*. For example, with $E_0 = 0$, give C a small value such that the natural relaxation frequency of the system is 1000 periods per second. Next apply the alternating voltage $E_0 \sin \omega t$, where ω may be made $2\pi \times 1000$ sec.$^{-1}$, then the system will go on oscillating with a frequency 1000 sec.$^{-1}$. When now the applied $E_0 \sin \omega t$ is left as before but C is gradually increased to a much greater value, it will be found that the system continues to oscillate with a frequency 1000 sec.$^{-1}$. If C is next increased still further, the frequency of the oscillations in the system (as detected, for example, with a telephone coupled loosely in some way to the system) suddenly drops to 1000/2 sec.$^{-1}$, to maintain this value over a certain range of the capacity value. If C is increased still more, the frequency suddenly jumps to 1000/3 sec.$^{-1}$, and so on up to 1000/40 sec.$^{-1}$. In some recent experiments it was found possible to obtain a frequency demultiplication up to the ratio 1 : 1/200. Often an irregular noise is heard in the telephone receivers before the frequency jumps to the next lower value. However, this is a subsidiary phenomenon, the main effect being the regular frequency demultiplication. It may be noted that while the production of harmonics, as with frequency multiplication, furnishes us with tones determining the musical major scale, the phenomenon of frequency-division renders the musical minor scale audible. In fact, with a properly chosen 'fundamental' ω, the turning of the condenser in the region of the third to the sixth subharmonic strongly reminds one of the tunes of a bagpipe.

In conclusion, we give in Fig. 2 the measured time periods (which are thus found to be a series of discrete subharmonics) as a function of the setting of the condenser C. The dotted line in the figure gives the frequency with which the system oscillates in the absence of the applied alternating E.M.F. The shaded parts correspond to those settings of the condenser where an irregular noise is heard. In the actual experiment the resistance R was, for ease of adjustment, replaced by a diode. The experiment, however, succeeds just as well with an ohmic resistance R. Obviously the same experiment succeeds with all systems capable of producing relaxation-oscillations such as described in the papers quoted.

BALTH. VAN DER POL.
J. VAN DER MARK.

Natuurkundig Laboratorium der
N. V. Philips' Gloeilampenfabrieken,
Eindhoven, Aug. 5.

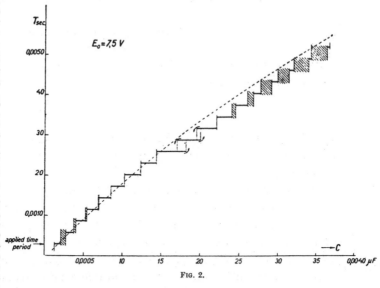

FIG. 2.

ON NON-LINEAR DIFFERENTIAL EQUATIONS OF THE SECOND ORDER:
I. THE EQUATION $\ddot{y} - k(1-y^2)\dot{y} + y = b\lambda k \cos(\lambda t + a)$, k LARGE

M. L. CARTWRIGHT *and* J. E. LITTLEWOOD†

1. In the present short preliminary survey we confine ourselves, to fix ideas, to equations of the form

$$\ddot{y} + f(y)\dot{y} + g(y) = p(t),$$

where f, g are real and analytic for real y, p is real and analytic and has period ϖ in (real) t, and $\underline{\lim} f > 0$ as $y \to \pm \infty$. A specially important case is that of $g(y) \equiv y$.

There is some general theory of such equations. A trajectory (or "motion") with initial conditions $y(t_0) = \xi$, $\dot{y}(t_0) = \eta$ (ξ, η real) at some fixed $t = t_0$ is said to have the point $P = (\xi, \eta)$ as "representative point". If ξ', η' are the values of y, $\dot{y}$ at $t = t_0 + \varpi$, the transformation T from P to $P' = (\xi', \eta') = TP = T(\xi, \eta)$ is (1, 1) and continuous (in fact analytic).

With the condition $\underline{\lim} f > 0$ and suitable conditions on g (fulfilled for $g = y$), every trajectory is bounded as $t \to \infty$, and T transforms a large

---

† Received 12 December, 1945; read 13 December, 1945.

Reprinted with permission from J. London Mathematical Society,
Vol. 20 (1945) pp. 180–189

circle in the P space into a domain in its interior. Further, the vector V, or $P \to TP$, makes exactly one revolution as P moves positively round the circumference of a large circle. Hence Poincaré's "fixed point" theorem holds, and Birkhoff's "index number" proof of it is valid†. There is at least one "fixed point" (f.p. for short) of the transformation T, and there corresponds *a periodic motion (p.m. for short) of period* ϖ; this need not, however, be stable‡.

A periodic trajectory or motion (p.m.), of least period $n\varpi$, with $n > 1$, is called a *subharmonic of order* n. We are interested in the class K of "limiting trajectories"§, the class whose representative points are the limit points of $T^m(\xi, \eta)$ as $m \to \infty$ for some ξ, η; this class as a whole is invariant under T, and its trajectories are all bounded in $-\infty < t < +\infty$. The simplest possible case is that in which K is a single point; there is then a stable p.m. of order 1 and every trajectory converges to it. The next simplest case is that of a finite number of p.m., to some one of which every trajectory converges. In the general topological theory, however, other possibilities, indeed very "bad" ones, have to be contemplated, and it is found very difficult in any given case to rule them out—for the best of reasons, as we shall see.

2. The two simplest cases (in respect of the functions f and g) are:

(i) $f > c > 0$ for all y, together with suitable conditions on g, valid when $g = y$;

(ii) *small* departure from linearity, $f = \epsilon F$, $g = y + \epsilon G$, with fixed F, G and small ϵ∥.

After these, since f is not to be of constant sign, and since it is positive for large $|y|$, it must have two zeros at least; if we add symmetry about $y = 0$ and take the simplest possible g and p, our search leads to an equation

---

† N. Levinson, *Journal of Math. and Physics*, 22 (1943), 41–48, and *Annals of Math.*, 45 (1945), 723–727.

‡ A f.p. P_0 (and its corresponding p.m.) is called stable if $T^m P$ converges to P_0 as $m \to \infty$ for all P of a neighbourhood of P_0. It is called totally unstable if the reversed trajectories are stable : this is equivalent to the statement that P's near and not identical with P_0 ultimately recede from P_0 under iteration of T. f.p. and p.m. that are neither stable nor totally unstable we shall call non-stable. There is no point in classifying them here, since the ones that occur in our work are extremely singular, or multiple.

§ "Recurrent motions" of Birkhoff, *Dynamical systems* (New York, 1927), 198.

∥ $f = a + \epsilon F$ for $a > 0$ is ruled out as a special case of (i).

that can be normalized as

(E) $$\ddot{y} - k(1-y^2)\dot{y} + y = b\lambda k \cos(\lambda t + \alpha)\dagger.$$

k must not be small, or we are in case (ii); the next possibility of simplification is to suppose it large, which gives us the equation of the title.

3. If $b > \tfrac{2}{3}$ and $k > k_0(b, \lambda)\ddagger$, (E) shows the simplest possible behaviour: there is a stable p.m. of order 1, period $\varpi = 2\pi/\lambda$, to which every trajectory converges.

If, however, $b < \tfrac{2}{3}$, and k is large enough, (E) shows a rich variety of behaviour, some of it very bizarre§.

We have to exclude certain intervals of b; in order that these should be a small proportion of the whole interval $(0, \tfrac{2}{3})$, we need to introduce an arbitrarily small positive δ. There then exist $\epsilon_\delta = \epsilon(\lambda, \delta)$, small with δ, and $k_0 = k_0(\lambda, \delta)$, with the following properties. If $k \geqslant k_0$, there is a set of excluded intervals in $(0, \tfrac{2}{3})$, including among them $(0, \delta)$ and $(\tfrac{2}{3}-\delta, \tfrac{2}{3})$, of total length ϵ_δ at most. The remainder of $(0, \tfrac{2}{3})$ is also a set of intervals, $\mathcal{B}$, say; this varies with k, but has length at least $\tfrac{2}{3} - \epsilon_\delta$. $\mathcal{B}$ divides into two parts (roughly equal), $\mathcal{B}_1$ and $\mathcal{B}_2$.

When b belongs to an interval I_1 of $\mathcal{B}_1$, (E) has a set of stable sub-harmonics of order $2n+1$‖, and most¶ trajectories converge each to some one of these. n is constant in I_1 and is of order $(\tfrac{2}{3}-b)k$.

When b belongs to an interval I_2 of $\mathcal{B}_2$, (E) has a set of stable sub-harmonics of order $2n+1$, and another of order $2n-1$; most trajectories converge each to some member of one of the two sets. It possesses a further set Σ, infinite in number, of p.m. of a great variety of "structures" (described in more detail later). It possesses further a set X, of the power

---

† See B. van der Pol, *Proc. Institute of Radio Engineers*, 22 (1934), 1051–1086, § IX, 1080–1082. Some graphical solutions are given by D. L. Herr, *Proc. Institute of Radio Engineers*, 27 (1939), 396–402.

‡ This is the simplest instance of a point that should be emphasized. We never assert that behaviour is more and more nearly such and such as k increases, always that it is *exactly* such and such so soon as k exceeds a certain k_0 [here $k_0(b, \lambda)$]. In fact k is not "large", but only "large enough".

§ Our faith in our results was at one time sustained only by the experimental evidence that stable sub-harmonics of two distinct orders did occur. [See B. van der Pol and J. van der Mark, *Nature*, 120 (1927), 363.] It is this that leads to the startling consequences; the consequences themselves relate to non-stable motions (which the experiments naturally did not reveal).

‖ Any p.m. of order m shifted a period ϖ is another one of the same order, and so gives rise to a "set", m in number.

¶ The general sense of "most" is fairly obvious: to define it precisely would occupy too much space.

of the continuum, of non-periodic limiting trajectories, of the type described† as "discontinuous recurrent". If we denote the sets of representative points P in the (ξ, η) plane also by Σ and X, then every point of Σ is a limit point of points of Σ and also a limit point of points of X. A point of Σ is thus non-stable, and is clearly a highly singular, or multiple‡ f.p. The number n (which is of the order of k) is constant in I_2. Moreover the set K and its subsets Σ, X remain topologically equivalent throughout I_2. Thus a point of Σ remains "infinitely multiple" for all b of the interval I_2, contrary to the natural expectation that multiplicity would be confined to isolated values of b.

For b of an I_1 (of $\mathcal{B}_1$) there is a set of non-stable subharmonics of order $2n+1$.

To complete the account of f.p. we observe finally that for all b of $(\delta, \frac{2}{3}-\delta)$ there is a single f.p. of order 1, and it is totally unstable. We shall call its representative point P_u, and denote by K_0 the set K less the point P_u.

As b increases (from δ to $\frac{2}{3}-\delta$), jumping the excluded intervals, the number n decreases (down to $O(\delta k)$). We have nothing to say about the *transitions* from one stable period $(2n+1)\varpi$ to two stable periods $(2n\pm1)\varpi$ and *vice versa*; these take place in the excluded intervals§.

4. It follows from the famous "last geometrical theorem" of Poincaré∥ *inter alia*, that if a transformation T, which is $(1, 1)$, continuous and area-preserving in the annulus between two curves, has f.p. of order n_1 on one curve and f.p. of a different order n_2 on the other, such that the points go round the curves once in n_1 and n_2 transformations respectively, then it has f.p. in the annulus of every order N such that m/N lies between $1/n_1$ and $1/n_2$ for some integer m; if $n_1 = 2n+1$ and $n_2 = 2n-1$, it has f.p. of orders $2n$, $4n\pm1$, $6n\pm1$, $8n\pm1$, $8n\pm3$, ... It seems generally taken for granted that an *annulus* is essential to such behaviour. But in our case of stable periods $(2n\pm1)\varpi$ there is no annulus: K_0 is a connected set of zero area separating P_u from ∞, and the stable f.p. of K_0 are limit points of $T^m P$ (as $m \to \infty$) both for P near P_u and for P near ∞. There would seem, moreover, to be a much richer "fine structure" of

---

† G. D. Birkhoff, *Acta Math.*, 43 (1922), 1–119.

‡ *Cf.* L. Bieberbach, *Differentialgleichungen* (Berlin, 1930), 72–74, and J. Hadamard, *Bull. de la Soc. Math. de France*, 26 (1901).

§ We may, however, mention that as b increases through an interval I_2 the shorter period $(2n-1)\varpi$ extends its sphere of influence at the expense of the longer (and there consistent behaviour in an I_1).

∥ See G. D. Birkhoff, *Dynamical systems* (New York, 1927), 165.

non-stables than is provided by the annulus†. For example, there exist 4 distinct kinds of set of least period $2n\varpi$, 6 of least period $4n\varpi$, and a considerable number of least period $6n\varpi$. The annulus may have as few as 2 of least period $2n\varpi$, and none of least periods $4n\varpi$, $6n\varpi$.

5. We proceed to sketch the methods of the work. At the end we shall be in a position to explain more precisely the "structure" of the non-stable p.m.

We are concerned only with the case $\delta < b < \frac{2}{3}-\delta$; the case $b > \frac{2}{3}$ has been disposed of, and that of small b constitutes a separate problem. By A we denote an absolute constant, by $A(x, y, ...)$ a constant depending only on the parameters shown, by L a constant $A(\lambda)$, by B a constant $A(b, \lambda)$‡, by D a constant $A(b, \lambda, \delta)$: all of them to be positive§. The constant of an O is to be of type L. We shall sometimes allow ourselves the licence of using $o(1)$ without precise definition. The lower bound k_0, which is ultimately a D, is to be rechosen at any moment when the run of the argument requires it.

We use the symbol Q for the current point (t, y) of a trajectory, ϕ for the phase $\lambda t + \alpha$, η for $y-1$, y_1 for $\int y\,dt$. Let $F(y) = y - \frac{1}{3}y^3 - \frac{2}{3} = -\eta^2 - \frac{1}{3}\eta^3$. Then we have the identity‖

$$(1) \qquad \dot{y} - \dot{y}_0 = k[F(y)]_{t_0}^{t} + bk(\sin\phi - \sin\phi_0) + [y_1]_{t_0}^{t}.$$

6. It is easy to show that every trajectory eventually satisfies $|y| < L$, $|\dot{y}| < Lk$, and crosses $y = 0$ infinitely often. We have now the two following key results.

LEMMA 1. *Let d be any positive constant¶. Then there is a $k_0(b, \lambda, d)$ such that when $k \geqslant k_0$ the following properties hold.*

(i) *Suppose that an eventual trajectory (for which consequently $|y| < L$) has a piece XYZ ($t_X < t_Y < t_Z$) lying entirely above $y = 1-dk^{-\frac{1}{2}}$, and that*

---

† K_0 seems to be an example of "bad" topological behaviour comparable with Birkhoff's "bad curve". See *Bull. de la Soc. Math. de France*, 60 (1932), 1–26.

‡ These are relatively rare, since b has an A for upper bound; they arise from the possibility that b is small.

§ It is convenient to think of them as large, and to use them in the denominator where lower bounds are in question.

‖ In which, the origin $t = 0$ being at our disposal, we generally take $t_0 = 0$, $\dot{y}_0 = \dot{y}(0)$.

¶ We choose it in different ways at different times: it is always at worst a D.

XY lasts a time at least $3\pi/\lambda$†. Then for any Q of YZ

(2) $\qquad |\dot{y}| < A(\lambda, d), \quad |\ddot{y}| < A(\lambda, d) k^{\frac{1}{2}}, \quad |\dddot{y}| < A(\lambda, d) k, \quad \ldots,$

(3) $\qquad \dot{y}(1-y^2) + b\lambda \cos\phi = O\big(A(\lambda, d) k^{-\frac{1}{2}}\big),$

(4) $\qquad F(y) = C - b(1-\sin\phi) - y_1/k + O\big(A(\lambda, d) k^{-1}\big),$

where C is a constant (in YZ).

(ii) *At a Q that has been preceded by a piece of an eventual trajectory lying above $y = 1+d$ and lasting a time $2\log k/(dk)$ at least, we have*‡

(5) $\qquad |\dot{y}| < A(\lambda, d), \quad |\ddot{y}| < A(\lambda, d), \quad \ldots,$

$\big($with various consequences, e.g. (3) with k^{-1} in place of $k^{-\frac{1}{2}}$ on the right$\big)$.

LEMMA 2. *Consider the equation*

$$\frac{dz}{dx} = z^2 - x^2 + 1 + \alpha - 2\beta x \quad [x \geqslant 0, \; z(0) = 0].$$

There is, for any real β, a critical value $\alpha_0(\beta) > -1$ of α; if $\alpha < \alpha_0$, z, initially positive, changes sign at $x = x_0(\alpha, \beta) > 0$; if $\alpha > \alpha_0$, $z \to \infty$ at a finite asymptote $x = x_0(\alpha, \beta)$. Also $\alpha_0(\beta)$ increases with β and $\alpha_0(0) = 0$, so that $\alpha_0(\beta)$ has the same sign as β. When $\beta < 0$, $\alpha_0(\beta) + \beta^2 < 0$.

7. We shall now follow the course of a trajectory Γ starting with a vertex§ $\dot{y}_0 = 0$, $|y_0 - 2| < D_0 k^{-1}$, $|\phi - \tfrac{1}{2}\pi| < D_0 k^{-\frac{1}{2}}$; D_0 is ultimately chosen to be a particular D. In the first place we deduce from Lemma 1 that, if $k > k_0(\lambda, D_0)$, Γ does not reach $y = 1$ till a time at least k/B has elapsed; during this time we have $|\dot{y}| < L$, and y satisfies (4), with error term $O\big(A(\lambda, D_0) k^{-1}\big)$, and (from initial values) the constant C has a value $\tfrac{4}{3} + O(D_0 k^{-1})$. During any part of this motion of length $O(1)$, y_1/k varies by $O(k^{-1})$; up to the point where it first reaches $y = 1$, Γ consists of successive waves, of length ϖ, with equation, to error $O(k^{-1})$,

$$F(y) = C_m - b(1 - \sin\phi),$$

the successive C_m decreasing by the increments of y_1/k.

---

† Some proviso of this kind is essential for the truth of the results.

‡ It is convenient to use the same d in (i) and (ii). The general sense of Lemma 1 is that spending time in the regions $|y| > 1$ has a strongly "settling" effect on the trajectory. The region $|y| < 1$ is unsettling, and the trajectory needs a little time to recover after being in it.

§ We must begin *somewhere*. It is as a matter of fact the case that every trajectory except that with period ϖ passes sooner or later through such a vertex.

It is important to observe that the upper bound L of $|\dot{y}|$ is independent of D_0, though k_0 is not.

Next, let Γ arrive at $y = 1$ for the first time at U, with
$$\phi_U = -\tfrac{1}{2}\pi - \omega \quad (|\omega| \leqslant \pi), \quad v = -\dot{y}_U,$$
and let $V = v + bk(1 - \cos\omega)$. By Lemma 1, (2) and (3), with $d = 0$, we have $v < L$, $|\sin\omega| < Bk^{-\frac{1}{2}}$, and so† $|\omega| < Bk^{-\frac{1}{2}}$; hence

(6) $\qquad\qquad v < L, \quad V < L, \quad |\omega| < Bk^{-\frac{1}{2}}.$

We prove next that v and ω, and so all three of v, V, ω, are approximately "linked"‡. If $v^* = \lambda(\tfrac{1}{2}b)^{\frac{1}{2}}$, we have in fact

(7) $\qquad\qquad V = v^*\{1 + \beta^2 + a_0(\beta)\} + o(1), \quad \beta = \lambda^{-1}(v^* k)^{\frac{1}{2}} \sin\omega,$

where a_0 is the function of Lemma 2.

The idea of the proof of this is easily intelligible. We consider the reversed motion (r.m.) from U, over a time of slightly larger order than $k^{-\frac{1}{2}}$. If in equation (1) we change the sign of t and then write
$$y = 1 + v^{*\frac{1}{2}} k^{-\frac{1}{2}} z, \quad t = v^{*-\frac{1}{2}} k^{-\frac{1}{2}} x,$$
and reject terms with small coefficients, we obtain the equation of Lemma 2, with β as above, and $a = v/v^* - 1$. The two types of behaviour in Lemma 2 correspond to z changing sign and going to ∞ respectively, for finite x, and finite x corresponds to t of order $k^{-\frac{1}{2}}$. Since the r.m. does not behave in either of these ways, we expect a and β to be approximately linked by $a = a_0(\beta)$, which corresponds to (7).

8. There are now three possibilities about v: (i) $v > v^* + \delta$; (ii) $v < v^* - \delta$; (iii) $|v - v^*| \leqslant \delta$. The third we describe as a "gap" case. The direct motion (d.m.) from U with the sign of z changed has for its z, x equation that of the r.m. with the sign of β changed. In case (i) this involves (for the new a, β) $a > a_0(\beta)$, and $z \to \infty$ for $x \to x_0$. Corresponding to this the (t, y) trajectory does in fact acquire a large (negative) velocity.

In case (ii) the d.m. has $a < a_0(\beta)$ and its z changes sign at a finite x_0. The corresponding behaviour of the (t, y) trajectory is to make a "dip" below $y = 1$, of depth at most $D_1 k^{-\frac{1}{2}}$, and emerge upwards after time at most $Dk^{-\frac{1}{2}}$ (D_1 and D do not depend on D_0 but on δ). Suppose its next

---

† ω is easily seen not to be near $\pm\pi$.
‡ The linkage is not exact (there is some "play", varying with the past history of Γ), but it is convenient sometimes to speak as if it were.

return to $y=1$ is at U_1. Between U and U_1, Γ *conforms to Lemma* 1 *with* $d = D_1$, provided that k exceeds a k_0 rechosen to depend on D_1; the equation of Γ between U, U_1 is given by (4) with $C = b + o(1)$. From this it follows that U_1 is approximately $U + \varpi$. There is linkage of v_1, ω_1, V_1 at U_1; further, we have

$$(8) \qquad V_1 - V = [y_1]_U^{U_1} = M + o(1),$$

where M is an $A(\lambda, b)$.

The three alternatives concerning v_1 and the gap $v^* \pm \delta$ arise again at U_1; either v_1 is in the gap; or $v_1 > v^* + \delta$ and the velocity y becomes (negatively) large; or $v_1 < v^* - \delta$ and there is another dip, of depth $D_2 k^{-\frac{1}{2}}$ at most, Γ conforms to Lemma 1 with $d = \max(D_1, D_2)$, and, if k_0 is rechosen to depend on D_2, there is a fresh arrival at $U_2 = U_1 + \varpi + o(1)$. And so on.

Now $v_n > v^* + \delta$ is approximately equivalent to $V_n > v^* + \delta$, and the latter inequality must, by (8), hold after $n = 1 + [v^*/M + o(1)] < L$ dips at most. In this case the final D_n and d_n are D's, and the final k_0 is a $k_0(\lambda, \delta)$.

To sum up, with a change in the meaning of U†, there is a $k_0(\lambda, \delta, D_0)$; provided $k \geq k_0$, Γ arrives (from above), after possible dips, at U on $y = 1$, where either $|v - v^*| \leq \delta$, or $v^* + \delta < v < L$. Suppose the latter. Γ now descends in time $O(Dk^{-\frac{1}{2}})$ ($\dot{y}$ attaining order k) to a vertex (inverted) with $\dot{y} = 0$, $|y + 2| < A_1(\lambda, \delta) k^{-1}$, $|\phi - \frac{1}{2}\pi| < A_2(\lambda, \delta) k^{-\frac{1}{2}}$. If we now choose $D_0 = \max\{A_1(\lambda, \delta), A_2(\lambda, \delta)\}$ (this does not involve a vicious circle), Γ has arrived at an (inverted) vertex satisfying the same conditions as the starting point. It repeats (in an inverted form‡) its former behaviour, arriving after (inverted) dips at U' on $y = -1$ with

$$\dot{y}_{U'} = v', \quad \phi_{U'} = \tfrac{1}{2}\pi - \omega', \quad V' = v' + bk(1 - \cos\omega'),$$

v', ω', V' linked, and either v' in the gap $v^* \pm \delta$ or else $v^* + \delta < v' < L$.

9. We have the identity

$$(9) \qquad V' + V = -(\tfrac{4}{3} - 2b)k + \int_U^{U'} y\, dt.$$

We must next study the relations between two Γ's of the kind considered, Γ_1, Γ_2, from U_1, U_2 to U_1', U_2', respectively; U_1, U_2 being near, as also U_1', U_2'; $v_1, v_2 > v^* + \delta$; finally $v_1', v_2' > v^* - \delta$ (so possibly in the gap).

---

† The new U is the old U_n.
‡ Inverted behaviour can be inferred from direct by changing ϕ into $\phi + \pi$.

Let ζ denote a ("small") number of the form $\exp(-k/D)$. Then we have

LEMMA 3. *Either* $U_1 U_1'$, $U_2 U_2'$ *have y's differing everywhere by $O(\zeta)$, or else they do not intersect.*

The proof of this involves a long analysis of the behaviour of the difference of ordinates $y_1 - y_2$†. The non-intersection has two vital consequences: (i) $\omega_1' - \omega_2'$ and $\omega_1 - \omega_2$ have (simply as a matter of geometry) opposite signs; (ii) $y_1 - y_2$ is of constant sign. It is necessary to prove further that (iii) $|y_1 - y_2| > |V_1 - V_2|/(Dk) + O(\zeta)$ over a range of t of length at least k/D. It then follows from (9) that, with error $O(\zeta)$,

(10) $\quad \operatorname{sgn}(V_1' - V_2') = -\operatorname{sgn}(V_1 - V_2), \quad |V_1' - V_2'| < (1 - D^{-1})|V_1 - V_2|.$

Similar relations hold for $V_1'' - V_2''$ and so on, provided we do not arrive at a $v^{(n)}$ in the gap $v^* \pm \delta$. These results evidently open the way towards a proof of "quasi-convergence" of Γ_1, Γ_2 [convergence except for error $o(1)$, here $O(\zeta)$] as $t \to \infty$. The final step to strict convergence involves a separate argument.

Whether there are two sets or one set of stable p.m. depends roughly on whether there is or is not a $\bar{v}$, not in the gap, such that $\bar{v}' = v^*$. To discuss this, and to control the interference by "gaps", we need to know the effect on V', for a given $V > v^* + \delta$, of an increase δb of order $1/(k \log k)$, say. (This requires a long and unexpectedly subtle analysis; the *result* is the expected one, that V' varies smoothly with b.) There is no interference by gaps provided that $(v^* + \delta)'$ is neither near v^* nor near $\bar{v}$ if that exists: it is these conditions that require us to exclude intervals of b.

10. We are now in a position to discuss in more detail the structure of the non-stable p.m. when b is in $\mathcal{B}_2$. Consider a "stream" of trajectories ("settled") through the "gap" on $y = 1$ (whose ends have t's linked with $v^* \pm \delta$). A trajectory through the $v^* + \delta$ end ends on $y = -1$ (for the last time, after possible inverted dips) at U_1' say; that through the $v^* - \delta$ end ends at U_2', approximately 2ϖ later‡. A "continuity" argument shows that a continuous stream exists between these extreme trajectories: the narrow original stream, "scattered" by the central part

---

† It depends incidentally on using a number of Γ's "intermediate" between Γ_1, Γ_2 (whose existence has to be *proved*). Extremes of a mesh do not intersect if consecutives do not; hence it is enough to prove results for $\omega_1 - \omega_2$ conveniently small, *e.g.* less than k^{-10}.

‡ The trajectories cross near $y = -2$ at time ϖ approximately after the gap.

of the gap, spreads out on arrival at $y = -1$ into a "delta" extending from U_1' to U_2'. Now this delta contains *two* gaps on $y = -1$ of the $v^* \pm \delta$ type, and one of the $\bar{v} \pm \delta$ type. The sub-streams through the first two are scattered similarly to the original one, that through the third has a sub-stream through the gap (on $y = 1$) to which $\bar{v}$ leads. Given any structure of gaps built up by repetitions of these processes to ∞, and backwards to $-\infty$, there exists a continuous stream through any finite section $-T < t < T$ of it; and by a limit argument there exist trajectories (probably one only) through all the gaps of any possible structure. These provide the trajectories X of §3; and when the structure is periodic they provide a quasi-periodic trajectory having the structure. The final step to the existence of a strictly periodic trajectory having the structure requires a (difficult) index-number argument. This last is the only topological argument we have occasion to use.

Girton College, Cambridge.

Trinity College, Cambridge.

W.-Y. TSENG
Senior Engineer,
Northern Research and
Engineering Company,
Cambridge, Mass.

J. DUGUNDJI
Professor,
Department of Aeronautics
and Astronautics,
Massachusetts Institute of Technology,
Cambridge, Mass.

Nonlinear Vibrations of a Buckled Beam Under Harmonic Excitation[1]

A buckled beam with fixed ends, excited by the harmonic motion of its supporting base, was investigated analytically and experimentally. Using Galerkin's method the governing partial differential equation reduced to a modified Duffing equation, which was solved by the harmonic balance method. Besides the solution of simple harmonic motion (SHM), other branch solutions involving superharmonic motion (SPHM) were found experimentally and analytically. The stability of the steady-state SHM and SPHM solutions were analyzed by solving a variational Hill-type equation. The importance of the second mode on these results was examined by a similar stability analysis. The Runge-Kutta numerical integration method was used to investigate the snap-through problem. Intermittent, as well as continuous, snap-through behavior was obtained. The theoretical results agreed well with the experiments.

Introduction

THE dynamic behavior of a buckled beam with fixed ends, excited by the harmonic motion of its supporting base, is considered here. This is an extension of the straight beam case examined previously by the same authors [1],[2] and relates to the general problem of the dynamic stability of thin arch and shell structures.

The thin arch problem has been studied both theoretically and experimentally by a number of authors in the past few years [2–12]. Humphreys [4] examined a circular arch under impulse-step and rectangular pulse loading by using the analog computer. Lock [5] determined the critical step-pressure loads of an arch by the numerical integration of the equations of motions and by an infinitesimal stability analysis. Some authors [6, 7] have applied an energy criterion for the snap-through problem of an arch, but this is less successful. Mettler [8] applied the method of averaging to investigate the stability and the vibration of a sine arch under harmonic excitation. In this investigation, he found the jump phenomenon of the simple harmonic motion (SHM), superharmonic motion (SPHM) and subharmonic motion (SBHM) resonances. His observation that the jump phenomenon is the kinetic snap-through in analogy to the static snap-through seems inaccurate since this jump to a higher branch solution may not necessarily cause the arch to snap-through. To obtain information on snap-through, one should include the dynamic overshoot effect due to transient response. Generally, the method of averaging gives little information about the snap-through. Also, it is noted that little experimental work exists on the problem of an arch under harmonic excitation.

The present investigation will include:

1 The steady-state solutions of SHM and SPHM.
2 Snap-through analysis.
3 The effects of the initial static deflection.

Both analytical and experimental work are considered here. The present article is a partial condensation of a longer report by one of the authors [13].

Formulation and General Solution

The governing differential equation of a buckled beam (the beam originally flat, which has been compressed past the critical buckling load, P_{cr}, to a static deflection position W_0) with fixed ends and excited by the base motion W_B is (see Fig. 1)

$$EI \frac{\partial^4}{\partial x^4}(W + W_0) - \frac{\partial}{\partial x}\left[N_x \frac{\partial}{\partial x}(W + W_0)\right]$$
$$= -m\left(\frac{\partial^2 W}{\partial t^2} + \frac{\partial^2 W_B}{\partial t^2}\right) - C\frac{\partial W}{\partial t} \quad (1)$$

---

[1] This research was sponsored by the Air Force Office of Scientific Research under AFOSR Contract No. F44620-69-C-0091 and is part of a PhD thesis by the first author, Wu-Yang Tseng.
[2] Numbers in brackets designate References at end of paper.
Contributed by the Applied Mechanics Division and presented at the Winter Annual Meeting, New York, N. Y., November 29–December 3, 1970, of THE AMERICAN SOCIETY OF MECHANICAL ENGINEERS.
Discussion on this paper should be addressed to the Editorial Department, ASME, United Engineering Center, 345 East 47th Street, New York, N. Y. 10017, and will be accepted until July 20, 1971. Discussion received after the closing date will be returned. Manuscript received by ASME Applied Mechanics Division, January 12, 1970. Paper No. 70-WA/APM-48.

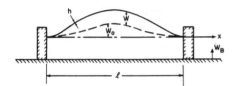

Fig. 1 Basic configuration

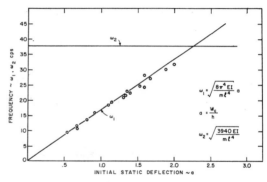

Fig. 2 Linear natural frequency

$$N_x = -P_0 + \frac{EA}{2l} \int_0^l \left[\frac{\partial}{\partial x}(W + W_0)\right]^2 dx \quad (2)$$

Boundary conditions are $W = \partial W/\partial x = 0$, at $x = 0$ and l, where E = Young's modulus; I = moment of inertia; N_x = total tension force on beam; W = beam displacements; W_0 = initial static deflection; m = mass/unit length; W_B = base displacement; c = damping coefficient; A = beam cross section; l = length of beam, x = longitudinal axis; t = time in seconds; P_0 is a fictitious compressive force on the beam defined by

$$P_0 = P_{cr} + \frac{AE}{2l} \int_0^l \left(\frac{dW_0}{dx}\right)^2 dx \quad (3)$$

Since P_{cr} is the fundamental buckling load of a clamped-clamped beam, i.e., $P_{cr} = 4\pi^2 EI/l^2$, the initial static deflection W_0 of the buckled beam is

$$W_0 = \frac{ah}{2}\left(1 - \cos\frac{2\pi x}{l}\right) \quad (4)$$

where h is the beam thickness and "a" is the ratio W_c/h, with $W_c \doteq W_0(l/2)$.

For the present range of initial static deflections, i.e., $a < 2$, the solution of (1) may be written as

$$W(x, t) = \sum_{n=1}^{2} \phi_n(x)\tilde{q}_n(t) \quad (5)$$

where $\tilde{q}_1$ and $\tilde{q}_2$ are generalized coordinates and ϕ_1 and ϕ_2 are the first and second buckling modes of the beam, respectively,

$$\phi_1 = \frac{ah}{2}(1 - \cos 2\pi\xi) \quad (6a)$$

$$\phi_2 = ah[\beta(k\xi - \sin k\xi) + \cos k\xi - 1] \quad (6b)$$

where $k = 8.986$, $\beta = 2/k$ and $\xi = x/l$. Applying Galerkin's method and assuming harmonic excitation of the base, $W_B = A_{F_0} \sin \omega_F t$ leads to the equations

$$\frac{d^2\tilde{q}_1}{dt^2} + \frac{c}{m}\frac{d\tilde{q}_1}{dt} + \omega_1^2\tilde{q}_1 + \frac{3}{2}\omega_1^2\tilde{q}_1^2 + \frac{1}{2}\omega_1^2\tilde{q}_1^3$$
$$+ 2.263\omega_1^2(\tilde{q}_1 + 1)\tilde{q}_2^2 = \frac{4}{3}\omega_F^2 A_F \sin \omega_F t \quad (7)$$

$$\frac{d^2\tilde{q}_2}{dt^2} + \frac{c}{m}\frac{d\tilde{q}_2}{dt} + \frac{5.05}{a^2}\omega_1^2\tilde{q}_2 + 6.356\omega_1^2\tilde{q}_2^3$$
$$+ 1.404\omega_1^2(\tilde{q}_1^2 + 2\tilde{q}_1)\tilde{q}_2 = 0 \quad (8)$$

where $A_F = A_{F_0}/ah$ and ω_1 is the natural frequency of the first mode in infinitesimal amplitude and has the value

$$\omega_1^2 = \frac{8\pi^4 EI}{ml^4} a^2 \quad (9)$$

It is noticed from (9) that ω_1 is linearly proportional to "a" and equal to zero when $a = 0$, which corresponds to classical buckling theory. This was the purpose of choosing the first and second buckling modes.[3] The natural frequency of the second mode in infinitesimal amplitude is found from (8) to be

$$\omega_2^2 = \frac{5.05}{a^2}\omega_1^2 = 3940 \frac{EI}{ml^4} \quad (10)$$

The numerical results of (9) and (10) are shown in Fig. 2. It is interesting to note that $\omega_1 = \omega_2$ at $a = 2.25$.

Now let $\omega_F t = n\tau$ and $\tilde{q}_1 = q_1 - 1$, $\tilde{q}_2 = q_2$, then (7) and (8) become simply

$$\frac{d^2 q_1}{d\tau^2} + 2n\zeta\sqrt{\alpha}\frac{dq_1}{d\tau} + n^2 K_1\alpha q_1 + n^2 K_2\alpha q_1^3 + n^2 K_4\alpha q_1 q_2^2$$
$$= n^2 K_3 A_F \sin n\tau \quad (11)$$

$$\frac{d^2 q_2}{d\tau^2} + 2n\zeta\sqrt{\alpha}\frac{dq_2}{d\tau} + \frac{n^2 K_5\alpha}{a^2} q_2$$
$$+ n^2 K_6\alpha q_2^3 + n^2 K_7\alpha(q_1^2 - 1)q_2 = 0 \quad (12)$$

where $K_1 = -0.5$, $K_2 = 0.5$, $K_3 = 1.333$, $K_4 = 2.263$, $K_5 = 5.05$, $K_6 = 6.356$, $K_7 = 1.404$, $\alpha = (\omega_1/\omega_F)^2 = 1/\Omega^2$, $\zeta = c/(2m\omega_1)$ and n is any integer. It is readily seen that $q_2 = 0$ is a solution of (11) and (12). The governing differential equation for q_1, when $q_2 = 0$, becomes

$$\frac{d^2 q_1}{d\tau^2} + 2n\zeta\sqrt{\alpha}\frac{dq_1}{d\tau} + n^2 K_1\alpha q_1 + n^2 K_2\alpha q_1^3$$
$$= n^2 K_3 A_F \sin n\tau \quad (13)$$

This equation is the same form as Duffing's equation [1], except here K_1 is negative. The general solution of (13) before snapping through is the same as in reference [1]; i.e., the solution of (13) can be approximated as

$$q_1 = y_0 + \sum_{k=1}^{3}(x_k \sin k\tau + y_k \cos k\tau) \quad (14)$$

Substituting (14) into (13) and using the method of harmonic balance for the constant and the first three harmonics will give seven nonlinear coupled algebraic equations [1, 13]. For simplicity and ease of explanation, only the no-damping case ($\zeta = 0$) will be considered. For this case, $y_1 = x_2 = y_3 = 0$ and $x_1 = r_1$, $y_2 = r_2$, $x_3 = r_3$. The seven equations, for $n = 1$, then reduce to the following four equations:

---

[3] The choice of the first and second vibration modes of the beam would give a slight difference in coefficients of (7) and (8), but would not make $\omega_1 = 0$ at $a = 0$.

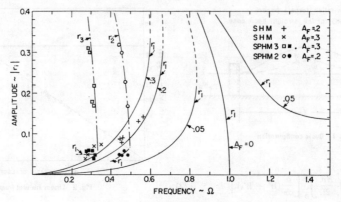

Fig. 3 Overall steady-state solutions

$$A_0 y_0 + \tfrac{3}{4} K_2 \alpha (2 r_1 r_2 r_3 - r_1^2 r_2) = 0 \quad (15a)$$

$$A_1 r_1 + 3 K_2 \alpha (y_0 r_2 r_3 - y_0 r_1 r_2 - \tfrac{1}{4} r_1^2 r_3 - \tfrac{1}{4} r_2^2 r_3) = K_3 A_F \quad (15b)$$

$$A_2 r_2 + 3 K_2 \alpha (y_0 r_1 r_3 - \tfrac{1}{2} y_0 r_1^2 - \tfrac{1}{2} r_1 r_2 r_3) = 0 \quad (15c)$$

$$A_3 r_3 + 3 K_2 \alpha (y_0 r_1 r_2 - \tfrac{1}{4} r_1 r_2^2 - \tfrac{1}{12} r_1^3) = 0 \quad (15d)$$

where

$$A_0 = \alpha \{ K_1 + K_2 [y_0^2 + \tfrac{3}{2}(r_1^2 + r_2^2 + r_3^2)] \}$$

$$A_1 = K_1 \alpha - 1 + \tfrac{3}{4} K_2 \alpha (4 y_0^2 + r_1^2 + 2 r_2^2 + 2 r_3^2)$$

$$A_2 = K_1 \alpha - 4 + \tfrac{3}{4} K_2 \alpha (4 y_0^2 + 2 r_1^2 + r_2^2 + 2 r_3^2)$$

$$A_3 = K_1 \alpha - 9 + \tfrac{3}{4} K_2 \alpha (4 y_0^2 + 2 r_1^2 + 2 r_2^2 + r_3^2)$$

Equation (13) is valid only when $q_2 = 0$. This means that q_2 does not become parametrically excited by the first-mode oscillations of q_1, i.e., any infinitesimal disturbance in q_2 will eventually die out. The study of the unstable regions for q_2 will be discussed later. The specific cases to be studied will deal with the $q_2 = 0$ solution, i.e., (13)–(15).

Specific Cases[4]

SHM Solution. For the SHM case, i.e., $|r_1| \gg |r_2|$ and $|r_3|$, further simplification of (15) can be made by neglecting r_2 and r_3 components and discarding the associated equations (15c) and (15d), which leads to the equations

$$A_0 y_0 = 0 \quad (16)$$

$$A_1 r_1 = K_3 A_F \quad (17)$$

Since $y_0 \neq 0$, (16) and (17) yield

$$y_0^2 = \frac{-K_1}{K_2} - \frac{3}{2} r_1^2 \quad (18)$$

$$\left(-2 K_1 \alpha - 1 - \frac{15 K_2 \alpha}{4} r_1^2 \right) r_1 = K_3 A_F \quad (19)$$

The nonlinear natural frequency relation can be obtained from (19) by setting $A_F = 0$ and $r_1 \neq 0$, which gives the "backbone curve" as

$$r_1^2 = -\frac{4}{15 K_2 \alpha}(2 K_1 \alpha + 1) = -\frac{4}{15 K_2}(2 K_1 + \Omega^2) \quad (20)$$

where $\Omega = \omega/\omega_1$, is the nonlinear natural frequency.

---
[4] The effects of SPHM order 3/2 is less important than those of SPHM order 2 and 3 and will not be discussed here.

The numerical results of (19) and (20) are shown in Fig. 3 by the solid curves. From the figure, one can see that (19) is a soft-spring-type solution and there exist jump points which are the basic information to investigate dynamic snap-through criteria to be discussed later.

The SHM solution, (18) and (19), gives a good approximation to the complete solution with small forcing term, $A_F < 0.2$ and Ω not close to $1/2$ and $1/3$. For the $A_F > 0.2$ case, the r_2 component becomes significant, although $|r_2|$ is still less than $|r_1|$. Also the solution of (18) and (19) is a good approximation for the small ζ case, say $\zeta < 0.01$.

SPHM Order 3. For the SPHM order 3 solution, one assumes $|r_3| \gg |r_1| \gg |r_2|$. Then, neglecting the r_2 component and discarding the associated equation (15c) leads to the equations

$$A_0 y_0 = 0 \quad (21a)$$

$$A_1 r_1 - \tfrac{3}{4} K_2 \alpha r_1^2 r_3 = K_3 A_F \quad (21b)$$

$$A_3 r_3 - \frac{K_2 \alpha}{4} r_1^3 = 0 \quad (21c)$$

Since $y_0 \neq 0$, (21a) gives

$$y_0^2 = \frac{-K_1}{K_2} - \frac{3}{2}(r_1^2 + r_3^2) \quad (22)$$

Also (21c) can be rewritten as

$$\left[K_1 \alpha - 9 + \frac{3}{4} K_2 \alpha (4 y_0^2 + r_3^2) \right] r_3 = \frac{K_2 \alpha}{4}(r_1 - 6 r_3) r_1^2 \quad (23)$$

Since $|r_3| \gg |r_1|$, the bold-faced terms can be neglected, which yields

$$A_3 \approx 0 \quad (24)$$

Placing (22) into (24) gives

$$r_3^2 = \frac{-4(2 K_1 \alpha + 9)}{15 K_2 \alpha} - \frac{4}{5} r_1^2 \quad (25)$$

It is seen from (25) that SPHM order 3 solution exists only at $\Omega < 1/3$. By the assumption $|r_3| \gg |r_1|$, (25) can be further approximated as

$$r_3^2 \approx \frac{-4(2 K_1 \alpha + 9)}{15 K_2 \alpha} \quad (26)$$

(Note: for stable solution, r_3 is positive). Eliminating y_0 and r_3 components in (21b) by using (22) and (26) gives

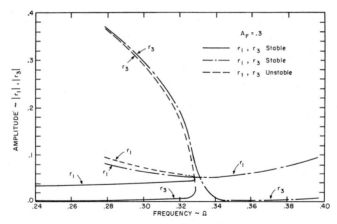

Fig. 4 Detail of transition of SPHM 3

$$\left\{6.2 - 0.4K_1\alpha - K_2\alpha \left[3.75r_1^2 + \sqrt{\frac{-2.4(2K_1\alpha + 9)}{K_2\alpha}} r_1\right]\right\} r_1 = K_3 A_F \quad (27)$$

For the small A_F case, say $A_F < 0.2$, (22), (26), and (27) give almost identical results for y_0, r_1 and r_3 as from (21a, b, and c). The numerical results of (22), (26) and (27) for r_1 and r_3 corresponding to SPHM order 3 are shown in Fig. 3. They appear to bifurcate from the main SHM solution near where the "shifted backbone curve order 3;" i.e., (26) intersects the SHM solution.

The details of the actual transition between these two solutions, $|r_1| > |r_3|$ and $|r_3| > |r_1|$, can be found by solving (15) numerically using a Newton iteration method, and are shown in Fig. 4. [Note: (15) gives a very small r_2 component while not affecting y_0, r_1, and r_3 of (21). This negligible r_2 component is not shown in the figure.]

For the $\zeta \neq 0$ case the complete seven equations have to be solved. In order to obtain the solution of SPHM order 3, the critical damping ratio, ζ, must be very small [1, 13]. Similarly (22), (26), and (27) give a good approximation for the $\zeta < 0.001$ case.

SPHM Order 2. For the SPHM order 2 solution, one assumes $|r_2| \gg |r_1| \gg |r_3|$. Then, neglecting the r_3 component and discarding the associated equation (15d), leads to the equations

$$A_0 y_0 - \tfrac{3}{4} K_2 \alpha r_2 r_1^2 = 0 \quad (28a)$$

$$A_1 r_1 - 3K_2\alpha y_0 r_1 r_2 = K_3 A_F \quad (28b)$$

$$A_2 r_2 - \tfrac{3}{2} K_2 \alpha y_0 r_1^2 = 0 \quad (28c)$$

Equations (28a and c) can be rearranged as

$$K_1 \alpha y_0 + K_2\alpha (y_0^3 + \tfrac{3}{2} r_1^2 y_0 + \tfrac{3}{2} r_2^2 y_0 - \tfrac{3}{4} r_1^2 r_2) = 0 \quad (29)$$

$$(K_1\alpha - 4)r_2 + \tfrac{3}{4}K_2\alpha(4y_0^2 r_2 + 2r_1^2 r_2 + r_2^3 - 2y_0 r_1^2) = 0 \quad (30)$$

For $|y_0| > |r_2| \gg |r_1|$, the bold-faced terms in (29) and (30) will drop out leaving the following two simple equations as

$$y_0^2 = -\frac{K_1}{K_2} - \frac{3}{2} r_2^2 \quad (31)$$

$$K_1\alpha - 4 + \tfrac{3}{4}K_2\alpha(4y_0^2 + r_2^2) = 0 \quad (32)$$

Solving (31) and (32) will give

$$y_0^2 = -\frac{K_1}{K_2} + \frac{2(2K_1\alpha + 4)}{5K_2\alpha} \quad (33)$$

$$r_2^2 = \frac{-4(2K_1\alpha + 4)}{15 K_2\alpha} \quad (34)$$

Equation (34) is the "shifted backbone curve order 2," which shows that the SPHM order 2 solution exists only for $\Omega < 0.5$. Substituting (33) and (34) into (28b) yields a simple equation in r_1 as (Note: for stable solution r_2 is negative)

$$\left\{2.2 - 0.4K_1\alpha + 4K_2\alpha\left[\frac{-K_1}{K_2} + \frac{2(2K_1\alpha + 4)}{5K_2\alpha}\right]^{1/2}\right. \\ \left. \times \left[\frac{-4(2K_1\alpha + 4)}{15K_2\alpha}\right]^{1/2} + \tfrac{3}{4}K_2\alpha r_1^2\right\} r_1 = K_3 A_F \quad (35)$$

For the $A_F < 0.2$ case, equations (33)–(35) give a close result as from (28).

The numerical results of (33)–(35) are shown in Fig. 3. Bifurcation from SHM solution is also seen. Near $\Omega = 0.5$, the complete solution of (15) showed a transition to the SPHM order 2 solution similar to that given in Fig. 4 for the SPHM order 3 case. This solution of (15) now gives a very small r_3 component, while not affecting y_0, r_1 and r_2 of (28). Also, equations (33)–(35) give a good approximation for small damping $\zeta < 0.001$ present.

Exact Solution of Free Vibration. The free-vibration solution equation can be easily obtained from (13) and $\zeta = 0$, $A_F = 0$, and $n^2\alpha = \omega_1^2$, which gives

$$\frac{d^2 q}{dt^2} + K_1 \omega_1^2 q_1 + K_2 \omega_1^2 q_1^3 = 0 \quad (36)$$

With initial conditions taken as $q_1 = q_u$, $dq_1/dt = 0$ at $t = 0$, the period, T, of a complete cycle is obtained from (36) as the elliptic integral,

$$T = 2 \int_{q_l}^{q_u} \frac{dq_1}{\omega_1 \{(q_u^2 - q_1^2)[K_1 + 0.5K_2(q_u^2 + q_1^2)]\}^{1/2}} \quad (37)$$

The lower bound q_l of the integral is determined as the position at which the value of the square root in (37) becomes zero, i.e., the position at which $dq_1/dt = 0$, $q_1 = q_l$. For the case of small vibrations without snap-through, the lower bound, q_l, is taken to be the root of the equation

$$K_1 + 0.5K_2(q_u^2 + q_l^2) = 0 \quad (38)$$

or

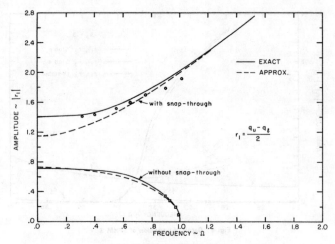

Fig. 5 Nonlinear natural frequency

$$q_l = \pm\sqrt{-\frac{2K_1}{K_2} - q_u^2} \quad (39)$$

(Note: q_l is taken to have the same sign as q_u.) Since the quantity in the square root must be positive, free vibrations without snap-through exist if and only if

$$|q_u| < \sqrt{-\frac{2K_1}{K_2}} = \sqrt{2} \quad (40)$$

For $|q_u| > 2$, free vibrations with snap-through will take place and the lower bound q_l in (37) becomes $q_l = -q_u$ which is the root of $q_u^2 - q_l^2 = 0$ in (37).

The numerical results of the exact free-vibration solutions are shown in Fig. 5, plotted as $|r_1| = |0.5(q_u - q_l)|$ versus $\Omega = \omega/\omega_1$ where $\omega = 2\pi/T$. Fig. 5 also shows the approximate solution for free vibrations without snap-through as obtained from the "backbone curve" of (20). The approximate solution of free vibrations with snap-through can be obtained from (15) by setting $A_F = y_0 = r_2 = 0$ and neglecting r_2 component and its associated equation (15d) which gives

$$r_1^2 = \frac{4}{3K_2}(\Omega^2 - K_1) \quad (41)$$

These results are also shown in Fig. 5. It is seen that approximate solution is good for $|r_1| > 1.8$.

A related free vibrations analysis is given by Eisley [14].

Stability Conditions

Stability of the First-Mode Solution. To investigate stability a small variation η from the periodic state of equilibrium is considered. Substituting $q_1 = q_0 + \eta$, where q_0 is the steady-state solution (14) which satisfies (13) into (13) and keeping only first-order terms of η gives

$$\frac{d^2\eta}{d\tau^2} + 2n\zeta\sqrt{\alpha}\,\frac{d\eta}{d\tau} + (n^2K_1\alpha + 3n^2K_2\alpha q_0^2)\eta = 0 \quad (42)$$

Then, placing the steady-state solution q_0 of (14) into (42) and using new variable $\bar{\tau} = \tau/2$, will result in the extended form of the Mathieu-Hill equation

$$\frac{d^2\eta}{d\bar{\tau}^2} + 4n\zeta\sqrt{\alpha}\,\frac{d\eta}{d\bar{\tau}}$$

$$+ \left[\theta_0 + 2\sum_{\nu=1}^{6}(\theta_{s\nu}\sin 2\nu\bar{\tau} + \theta_{c\nu}\cos 2\nu\bar{\tau})\right]\eta = 0 \quad (43)$$

where θ_0, $\theta_{s\nu}$, and $\theta_{ic\nu}$ are functions of $(y_0, x_1, y_1, x_2, y_2, x_3, y_3, \alpha, K_1$ and $K_2)$.

Following the same procedures as in Hayashi [15], except here more terms are involved for better accuracy [1, 13]. The dimension of the characteristic determinant is taken as 7×7 for the even stability regions and 6×6 for the odd stability regions. For stable solutions the two determinants

$$\begin{aligned}\Delta_{\text{even}}(0) &= |7 \times 7| \\ \Delta_{\text{odd}}(0) &= |6 \times 6|\end{aligned} \quad (44)$$

must both be greater than zero.

The stability investigation showed that SHM solutions were stable except in the usual overhanging portion near $\Omega = 1$ and in the vicinity of $\Omega = 1/m$, where $m = 3/2, 2, 3\ldots$. For the small $\zeta < 0.001$ case, the SPHM solutions, which bifurcated from the unstable SHM vicinities, were found to be stable up to large amplitudes. The unstable solutions studied by this analysis are plotted with dash lines in the previous mentioned figures.

Stability of the Second-Mode Solution. As mentioned in the section, "Formulation and General Solution," to insure the only existence of the first-mode solution q_1, the second-mode solution q_2 in (12) must not be parametrically excited by the first-mode oscillations. To investigate the stability of q_2 in (12), one may employ the method used by Lock [5], in which q_2 is taken as an infinitesimal variational variable. Substituting the steady-state solution q_1 which satisfies (13) into (12) and neglecting higher-order terms, q_2^2, one obtains the same equation as (43) with the same time variable $\bar{\tau}$, except here one has now

$$\begin{aligned}K_1 &= \frac{5.05}{a^2} - 1.404 \\ K_2 &= 0.468\end{aligned} \quad (45)$$

Then, the stability criterion for q_2 is readily obtained by substituting the new variables of (45) into (43) and (44).

Numerical evaluations of these stability determinants, (44), were carried out for the SHM, SPHM order 3 and SPHM order 2

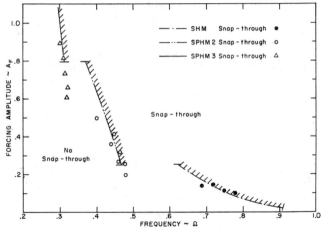

Fig. 6 Snap-through regions

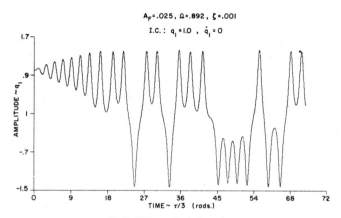

Fig. 7 SHM snap-through response

solutions. These were evaluated for the range $\Omega < 1.2$, $|r_i| < 0.4$ and the initial static deflection variable $a = 0.5-5.0$. These numerical results showed that unstable solution of q_2 appeared only when $a > 2.0$. For all experimental and theoretical cases considered here, "a" was less than 2.0 and, hence, the second mode did not play any role here. It is of interest to recall that, at $a = 2.25$, the natural frequency of the first mode ω_1 equals that of the second mode ω_2; see Fig. 2.

Further details of these first and second-mode stability investigations are given in reference [13].

Snap-Through Analysis

Numerical Analysis of Snapping Phenomenon. From the previous stability analysis, one can find two stable steady-state solutions for the buckled beam at frequencies just below $\Omega = 1/k$, where $k = 1, 2, 3, \ldots$ etc. In reality the upper branch solution may not be obtained due to the dynamic overshoot in the transient response, which causes the beam to snap-through during its attempt to achieve this upper branch solution. This snapping phenomenon generally occurs near jump points. To study this phenomenon, one may solve (13) for the transient response directly by numerical methods and observe when snap-through occurs. Accordingly, the Runge-Kutta numerical integration method was employed using a time increment $\Delta \tau = 0.05$ rad. The calculations were performed in the vicinity of jump points, a damping ratio of about $\zeta = 0.001$ was considered, and the following initial condition (IC), corresponding to the beam at rest, was employed:

$$q_1 = 1.0, \quad \frac{dq_1}{d\tau} = 0; \quad \text{at} \quad \tau = 0 \qquad (46)$$

The snap-through regions are shown in Fig. 6, plotted as forcing amplitude A_F versus frequency Ω. The snap-through boundary gives the frequency Ω for which a given forcing amplitude A_F will first encounter snap-through. It is noted that, in addition to SHM snap-through, there also occur snap-throughs of SPHM order 2 and 3.

Some typical snap-through responses are shown in Figs. 7 and 8. For the SHM snap-through, the predominant SHM component before snap-through is noted. Also, when $A_F < 0.02$ approximately, no SHM snap-through occurs and the upper branch solution can be obtained. For the SPHM order 3 snap-through, the development of a significant SPHM order 3 component is clearly evident before snap-through.

The presence of small damping doesn't affect the SHM snap-

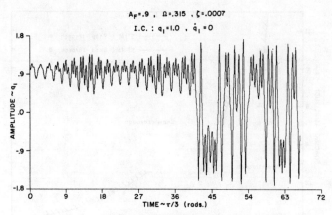

Fig. 8 SPHM 3 snap-through response

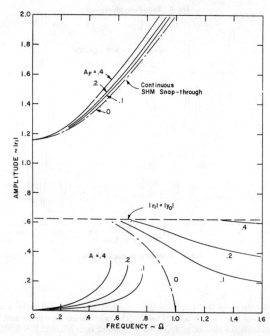

Fig. 9 Overall behavior of SHM solution

through, but it may inhibit the SPHM order 3 snap-through in the example shown in Fig. 8.

Continuous SHM Snap-Through. The preceding discussion related to the first onset of snap-through, and the response was characterized by an intermittent snapping; see Figs. 7 and 8. This intermittent snapping also persists generally at higher forcing amplitudes A_F and higher frequencies Ω. However, under certain conditions, a well-defined continuous snap-through behavior may exist.

An approximate solution for continuous SHM snap-through can be obtained from (15) by setting $y_0 = r_2 = 0$ and neglecting the r_3 component and its associated (15d). This gives

$$(K_1\alpha - 1 + \tfrac{3}{4}K_2\alpha r_1^2)r_1 = K_3 A_F \qquad (47)$$

Numerical results for this continuous SHM snap-through solution (47) are shown in Fig. 9, together with the previous without snap-through solution (19). Only the stable branches of each solution are shown. It is noted that the condition $|r_1| = |y_0|$ acts as a cutoff to the without snap-through solution, since above this $|r_1|$, the one-sided vibrations spill over to the other side and cause snap-through.

Some continuous SHM snap-through responses were also obtained as before using the Runge-Kutta numerical integration method and the initial conditions of (46). A typical such response is given in reference [13] for the $A_F = 0.1$, $\Omega = 0.8$ and $\zeta = 0.001$ case.

Experiment

A spring steel beam with 18-in. length and 0.021 in. $\times$ 0.5 in. cross section was compressed to buckle with the nondimensional initial static deflection "a" approximately equal to 1.5. The buckled beam then was rigidly clamped at both ends and mounted

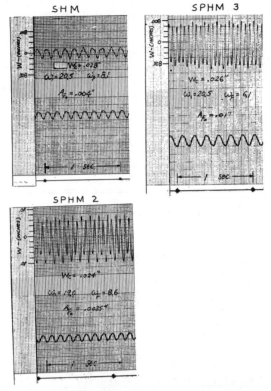

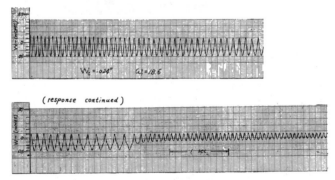

Fig. 10 Experimental SHM, SPHM response

Fig. 11 Experimental free-vibration response

on a shake table. Checking the symmetric property about the straight beam position, it was found that the initial static deflection "a" was equal on either side and so was the natural frequency, ω_1, on either side. The critical damping ratio ζ was found to be approximately 0.0007 from a transient decay test. The table was oscillated over a frequency range from 2–50 cps. The amplitude and frequency of the shake table were measured using strain gages and the response of the beam was recorded by a capacitor probe at the midpoint of the beam. A two-channel Sanborn recorder was used to record the responses of the shake table and the beam. Further details of the test setup are given in reference [13].

The experimental points for linear natural frequency versus initial static deflection "a" are shown in Fig. 2. No experimental points with $a > 2.0$ were obtained due to the second mode becoming unstable. The experimental points for nonlinear natural frequency without and with snap-through are plotted in Fig. 5, where now $\Omega = \omega/\omega_1$.

Some experimentally obtained points for the steady-state SHM and SPHM responses are plotted in Fig. 3. Typical records of these steady-state SHM and SPHM responses are given in Fig. 10, where the top trace shows the beam response and the lower trace shows the base motion. Fig. 11 shows an experimental record of free-vibration response of the beam. The transition from the hard spring behavior with snap-through to the soft spring behavior without snap-through is readily apparent

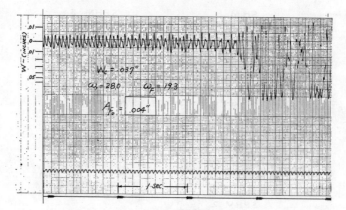

Fig. 12 Experimental SHM snap-through

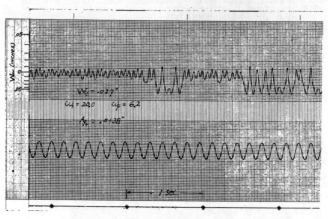

Fig. 13 Experimental SPHM 3 snap-through

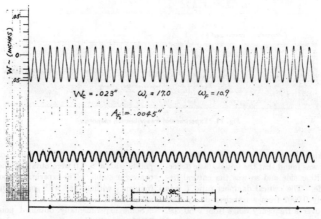

Fig. 14 Experimental continuous SHM snap-through

(recall Fig. 5). Some experimental points for the SHM and SPHM snap-through boundaries are shown in Fig. 6. Typical records of the SHM snap-through, SPHM order 3 snap-through, and continuous SHM snap-through are given in Figs. 12–14, respectively.

Conclusions

1 The buckled beam displays a soft-spring-type behavior at small amplitudes before snapping-through, and a hard-spring-type behavior after snapping-through.

2 For small amplitudes without snap-through, there exist in addition to the conventional SHM solution, other solutions of SPHM order $k = 2, 3, \ldots$. These solutions give SPHM component amplitudes near the "shifted backbone curve order k," i.e., replacing Ω by $k\Omega$ in the basic backbone curve $r_1^2 = -4(2K_1 + \Omega^2)/15K_2$. For small A_F and very small damping, all these additional solutions are dominated by the SPHM components and tend to have the system oscillate near its own natural frequency ω_1. For large A_F these additional solutions involve substantial forcing frequency components as well as the SPHM components.

3 These solutions of SPHM are stable. They seem to bifurcate from the SHM solution near where the shifted backbone curve intersects the SHM solution. For small A_F, these bifurcation points would be near $\Omega = 1/k$. The SHM solution itself is unstable near these bifurcation points.

4 Multiple solutions may exist at a fixed Ω depending on the initial conditions.

5 One-mode approximation is good for initial static deflection, $a < 2.0$ and $\Omega < 1$. Second mode must be included for $a > 2.0$ case, since it may be parametrically excited by the first-mode oscillations at $\Omega < 1$.

6 Linear natural frequency ω_1 is linearly proportional to the initial static deflection, a, with $\omega_1 = 0$ at $a = 0$ and $\omega_1 = \omega_2$ at $a = 2.25$.

7 The steady-state solution is valid for the maximum response amplitude to be less than 0.4 approximately.

8 Dynamic overshoot in the transient period will cause the beam to snap-through. The snapping phenomenon generally occurs at jump points of SHM and SPHM solutions.

9 The snap-through behavior is usually intermittent. Under certain conditions a well-defined continuous SHM snap-through exists.

10 The theoretical analysis agrees well with the experiment.

11 The present investigation has attempted to show the importance of superharmonics and snap-through in understanding the vibration behavior of buckled beams. The techniques and the results here can be extended to plates, curved panels, and shells in order to better understand their vibration behavior.

References

1 Tseng, W. Y., and Dugundji, J., "Nonlinear Vibrations of a Beam Under Harmonic Excitation," JOURNAL OF APPLIED MECHANICS Vol. 37, No. 2, TRANS. ASME, Vol. 92, Series E, June 1970, pp. 292–297.

2 Bolotin, V. V., *Dynamic Stability of Elastic Systems*, Holden-Day Inc., San Francisco, Calif., 1964.

3 Herrmann, G., "Dynamic Stability of Structures," *Proceedings of the International Conference*, Northwestern University, Evanston, Ill., October 18–20, 1965; Pergamon Press, 1967.

4 Humphreys, J. S., "Dynamic Snap Buckling of Shallow Arches," *AIAA Journal*, May 1966, pp. 878–886.

5 Lock, M. H., "Snapping of a Shallow Sinusoidal Arch Under a Step Pressure Load," *AIAA Journal*, July 1966, pp. 1249–1256.

6 Humphreys, J. S., "The Adequacy of Energy Criterion for Dynamic Buckling," *AIAA Journal*, May 1966, pp. 921–923.

7 Gjelsvik, A., and Bodner, S. R., "The Energy Criterion and Snap-Buckling of Arches," *Journal of Engineering Mechanics Division*, ASCE, Oct. 1962, pp. 89–134.

8 Mettler, E., "Stability and Vibration Problems of Mechanical Systems Under Harmonic Excitation," *Dynamics Stability of Structures, Proceedings of the International Conference*, Northwestern University, Evanston, Ill., Oct. 1965; Pergamon Press, 1967, pp. 169–188.

9 Anderson, D. L., and Lindberg, H. E., "Dynamic Pulse Buckling of Cylindrical Shells Under Transient Lateral Pressures," *AIAA Journal*, Apr. 1968, pp. 589–598.

10 Navaratna, D. R., Pian, T. H. H., and Witmer, E. A., "Stability Analysis of Shells of Revolution by the Finite Element Method," *AIAA Journal*, Feb. 1968, pp. 355–361.

11 Budiansky, B., and Roth, R. S., "Axisymmetric Dynamic Buckling of Clamped Shallow Spherical Shells," *Collected Papers on Instability of Shell Structures—1962*, NASA Langley Research Center TND-1510, Dec. 1962.

12 Goodier, J. N., and McIvor, I. J., "The Elastic Cylindrical Shell Under Nearly Uniform Radial Impulse," JOURNAL OF APPLIED MECHANICS, Vol. 3, TRANS. ASME, Vol. 86, Series E, 1964, pp. 259–266.

13 Tseng, W. Y., "Nonlinear Vibrations of Straight and Buckled Beams Under Harmonic Excitation," M.I.T., Aeroelastic and Structures Research Lab. Report TR 159-1, Air Force Office of Scientific Research, AFOSR 69-2157 TR, Nov. 1969; also written under same title as PhD thesis, M.I.T., Department of Aeronautics and Astronautics, Nov. 1969.

14 Eisley, J. G., "Large Amplitude Vibration of Buckled Beams and Rectangular Plates," *AIAA Journal*, Dec. 1964, pp. 2207–2209.

15 Hayashi, C., *Nonlinear Oscillations in Physical Systems*, McGraw-Hill, New York, 1964.

RANDOM PHENOMENA RESULTING FROM NON-LINEARITY IN THE SYSTEM DESCRIBED BY DUFFING'S EQUATION†

YOSHISUKE UEDA

Department of Electrical Engineering, Kyoto University, Kyoto 606, Japan

1. INTRODUCTION

In physical phenomena, uncertainties lie between causes and effects. When uncertain factors are small, their effects may be neglected in most physical systems and the phenomena under consideration are treated as deterministic ones. Whereas in non-linear systems on some conditions, however small uncertain factors may be, they sometimes cause global changes of state variables even in the steady states. These kind of phenomena originate from global structures of the solutions for non-linear equations describing physical systems. In such systems, steady motions may be observed exhibiting stochastic properties.

This paper deals with random oscillations which occur in a series-resonance circuit containing a saturable inductor under the impression of a sinusoidal voltage. Firstly, a differential equation is derived from the electrical circuit under discussion. The discrete dynamical system is introduced by using the solutions of the differential equation. The terminology used in the following descriptions is explained briefly. Secondly, experimental results concerning random oscillations are obtained by using analog and digital computers. Finally, the experimental results are examined and the problems arising from them are summarized.

The phenomenon treated in this paper should be called turbulence in electric circuits. A series of results obtained in this paper disclose an important feature of non-linear phenomena not only in electrical circuits but also in general physical systems.

2. STEADY OSCILLATIONS AND ATTRACTORS

2.1. Differential equation

A series-resonance circuit containing a saturable inductor is shown in Fig. 1. With the notation of the figure, the equation for the circuit is written as

$$\left. \begin{array}{l} n\dfrac{d\phi}{dt} + Ri_R = E\sin\omega t \\[2mm] Ri_R = \dfrac{1}{C}\int i_c \, dt, \qquad i = i_R + i_c \end{array} \right\} \qquad (1)$$

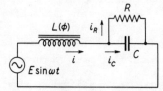

Fig. 1. Series-resonance circuit with non-linear inductance.

---

† This paper was translated by the author from his article in Japanese published in the *Transactions of the Institute of Electrical Engineers of Japan*, Vol. A98, March 1978, with kind permission of the Institute.

Reprinted with permission from Int. J. Non-Linear Mechanics,
Vol. 20, pp. 481–491. Copyright (1985), Pergamon Press PLC.

where n is the number of turns of the inductor coil, and ϕ is the magnetic flux in the core. Let us consider the case in which the saturation curve of the core is expressed by

$$i = a\phi^3 \tag{2}$$

It is to be noted that the effect of hysteresis is neglected in equation (2). Here we introduce the dimensionless variable x, defined by

$$\phi = \Phi_n x \tag{3}$$

where Φ_n is an appropriate base quantity of the flux and is fixed by the relation

$$n\omega^2 C\Phi_n = a\Phi_n^3 \tag{4}$$

Then, eliminating i_R and i_C in equation (1) and using equations (2), (3) and (4), we obtain the well-known Duffing's equation

where

$$\left. \begin{aligned} \frac{d^2x}{d\tau^2} + k\frac{dx}{d\tau} + x^3 &= B\cos\tau \\ \\ \tau = \omega t - \tan^{-1}k, \quad k &= \frac{1}{\omega CR} \\ \\ B &= \frac{E}{n\omega\Phi_n}\sqrt{1+k^2}. \end{aligned} \right\} \tag{5}$$

2.2. Discrete dynamical system

Equation (5) is rewritten in simultaneous form as

$$\frac{dx}{d\tau} = y, \quad \frac{dy}{d\tau} = -ky - x^3 + B\cos\tau \tag{6}$$

A discrete dynamical system on the xy plane is introduced by using the solutions of equation (6). In order to see this, first consider the solution $(x(\tau, x_0, y_0), y(\tau, x_0, y_0))$ of equation (6), which, when $\tau = 0$, is at the point $p_0 = (x_0, y_0)$ of the xy plane. Let $p_1 = (x_1, y_1)$ denote the point specified by $x_1 = x(2\pi, x_0, y_0)$, $y_1 = y(2\pi, x_0, y_0)$; then either a C^∞-diffeomorphism f_λ

where

$$\left. \begin{aligned} f_\lambda : R^2 &\to R^2 \\ p_0 &\mapsto p_1 \\ \\ \lambda = (k, B) &\in \Lambda \end{aligned} \right\} \tag{7}$$

of the xy plane into itself or a discrete dynamical system on R^2 is defined. For the theory of dynamical systems, see refs. [1–6].

For the circuit with dissipation ($k > 0$), f_λ is of class D or a dissipative system for large displacements and is a contractive mapping of the xy plane into itself. This implies that an orbit $\mathrm{Orb}(p) = \{f_\lambda^n(p) \mid n \in Z\}$ of the discrete dynamical system (7) which starts from an arbitrary point $p \in R^2$ is positively stable in the sense of Lagrange and that there exists positively asymptotically stable, f_λ-invariant, maximum compact set $\Delta(f_\lambda)$ with zero area. Therefore, investigation of the steady oscillations leads to examining the maximum compact set $\Delta(f_\lambda)$ on R^2 and the behavior of its neighboring orbits.

2.3. Steady oscillations and attractors

Since there exist uncertain factors, such as noise, in actual electric circuits, changes of voltages and/or currents are represented approximately by the solutions of the differential equations of the circuit. Accordingly, when the representative point of the circuit moves along an asymptotically stable solution, effects of noise may be neglected and deterministic phenomenon occurs. But stochastic phenomenon is caused when the representative point wanders, under the influence of noise, in the neighborhood of infinite solutions.

In the following, let us define the attractor as asymptotically stable, f_λ-invariant, compact set on the xy plane which exhibits a steady oscillation sustained in the actual circuit of Fig. 1. An attractor exhibiting a periodic oscillation is either a fixed point or a periodic group of the discrete dynamical system. An attractor exhibiting a random oscillation is considered to be an f_λ-invariant compact set containing infinite minimal sets.

3. EXPERIMENTS ON THE RANDOM OSCILLATIONS

In this section, experimental results obtained by using analog and digital computers are shown. Simulation and/or calculation errors are unavoidable in the computer solutions for the differential equation. Therefore, random quantities are not introduced intentionally but these errors are regarded as uncertainties acting on the system. These errors seem to be sufficiently small compared with noises in the actual circuit.

In the electric circuit as shown in Fig. 1, random oscillations can be observed within some intervals of the applied voltage and the circuit's constants. In fact, they occur in the ranges $B = 9.9-13.3$ for $k = 0.1$ and $k = 0-0.31$ for $B = 12.0$. In the following, experimental results are given for the steady oscillations, which occur in the computer-simulated systems for the representative values of the system parameters

$$\lambda = (k, B) = (0.1, 12.0) \tag{8}$$

3.1. Waveforms

Figure 2 shows two waveforms of the steady states sustained in the analog computer-simulated system. Figure 2(a) shows a periodic oscillation containing remarkable higher harmonic components. The upper waveform is the applied voltage $B \cos \tau$ and the lower one is the normalized magnetic flux $x(\tau)$. Figure 2(b) shows a random oscillation. This waveform is not reproducible in every analog computer experiment. In the digital simulation, different waveforms are observed depending on the integration method and the step size. Therefore, the waveform of Fig. 2(b) is a realization of the random process $\{X(\tau)\}$.

3.2. Phase-plane analysis

Figure 3 shows a long-term orbit (a realization) of a random oscillation. The movement of images under iterations of f_λ is not uniquely determined even for the same initial point, but

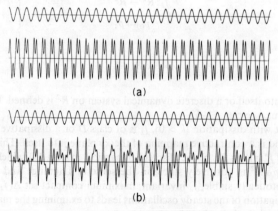

Fig. 2. Waveforms of the steady oscillations in the system prescribed by equation (8).

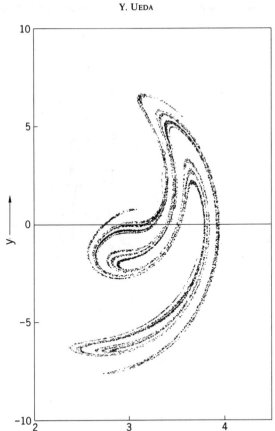

Fig. 3. Observed orbit corresponding to the random oscillation.

the general aspect (location, shape and size) of the orbit is reproducible, and further it seems stable in the Poisson sense. Therefore, a set of points as shown in Fig. 3 should be regarded as an outline of an attractor M representing the random oscillation.

Figure 4 shows fixed points $^1D^1$ (directly unstable), $^1I^1$ and $^2I^1$ (inversely unstable), and outlines of unstable $W^u(^1D^1)$ and stable $W^s(^1D^1)$ manifolds of $^1D^1$. If the unstable manifolds (thick lines in the figure) are prolonged, they tend to the orbit of Fig. 3. In other words, the closure of unstable manifolds of $^1D^1$, i.e. $ClW^u(^1D^1)$, is regarded as the attractor M for the random oscillation.

Figure 5 shows the global phase-plane structure of the diffeomorphism f_λ. The stable manifolds (thick lines) of the saddle $^2D^1$ (directly unstable fixed point) are the boundary of the two domains of attraction, and the sink S^1 is the attractor corresponding to the periodic oscillation of Fig. 2(a).

3.3. Spectral analysis

In this section, spectral analysis of the random oscillation is shown. To this end, we regard the random process $\{X(\tau)\}$ as the periodic random process $\{X_T(\tau)\}$ with a sufficiently long period T, where T is a multiple of 2π. That is, let $x_T(\tau)$ be a periodic function with period T, which coincides with a realization $x(\tau)$ of $\{X(\tau)\}$ in the interval $(-T/2, T/2]$. Then a realization $x_T(\tau)$ is expanded into Fourier series as

$$x_T(\tau) = \frac{a_0}{2} + \sum_{m=1}^{\infty} (a_m \cos m\omega_0 \tau + b_m \sin m\omega_0 \tau), \qquad \omega_0 = \frac{2\pi}{T} \qquad (9)$$

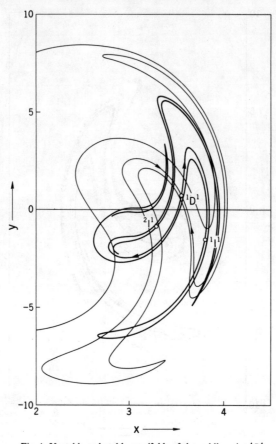

Fig. 4. Unstable and stable manifolds of the saddle point $^1D^1$.

where
$$a_m = \frac{2}{T} \int_{-T/2}^{T/2} x_T(\tau) \cos m \omega_0 \tau \, d\tau$$
$$b_m = \frac{2}{T} \int_{-T/2}^{T/2} x_T(\tau) \sin m \omega_0 \tau \, d\tau \qquad (10)$$
$$m = 0, 1, 2, \ldots$$

Fourier's coefficients a_m and b_m are random variables because $x_T(\tau)$ is a realization of the random process $\{X_T(\tau)\}$. From these coefficients, the mean value $m_X(\tau)$ and the average power spectrum $\Phi_X(\omega)$ of the random process $\{X(\tau)\}$ can be estimated as follows.

$$m_X(\tau) = \langle X(\tau) \rangle = \lim_{T \to \infty} \langle X_T(\tau) \rangle$$
$$\doteqdot \langle X_T(\tau) \rangle = \left\langle \frac{a_0}{2} \right\rangle + \sum_{m=1}^{\infty} [\langle a_m \rangle \cos m \omega_0 \tau + \langle b_m \rangle \sin m \omega_0 \tau] \qquad (11)$$

$$\Phi_X(\omega) = \lim_{T \to \infty} \left\langle \frac{1}{T} \left| \int_{-T/2}^{T/2} x_T(\tau) e^{-i\omega\tau} d\tau \right|^2 \right\rangle$$
$$\doteqdot \Phi_X(m\omega_0) = \frac{2\pi}{\omega_0} \left\langle \frac{1}{4}(a_m^2 + b_m^2) \right\rangle \qquad (12)$$
$$\omega_0 = \frac{2\pi}{T}$$

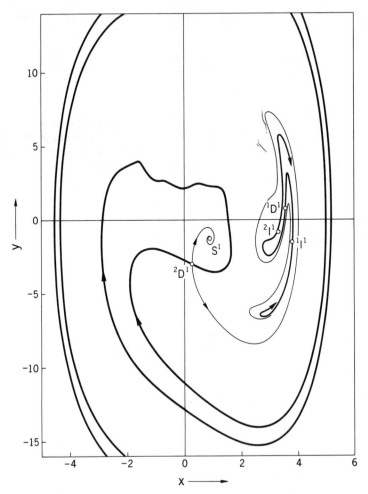

Fig. 5. Phase-plane structure of the diffeomorphism f_λ, $\lambda = (0.1, 12.0)$.

The ensemble average is calculated by regarding successive waveforms at the intervals $((n - 1/2)T, (n + 1/2)T]$ $(n = 0, 1, 2, \ldots, N_s)$ as sample processes of $\{X_T(\tau)\}$.

Let us give the results thus estimated for the specified values of the system parameters given by equation (8). The mean value of $\{X(\tau)\}$ is given by

$$\begin{aligned} m_X(\tau) = {} & 1.72 \cos \tau + 0.22 \sin \tau \\ & + 1.21 \cos 3\tau - 0.26 \sin 3\tau \\ & + 0.25 \cos 5\tau - 0.06 \sin 5\tau \\ & + 0.07 \cos 7\tau - 0.02 \sin 7\tau \\ & + 0.02 \cos 9\tau - 0.01 \sin 9\tau \end{aligned} \quad (13)$$

The mean value is found to be a periodic function. This indicates that the random process $\{X(\tau)\}$ is a non-stationary random process. Figure 6 shows the average power spectrum estimated by using equation (12). In the figure, line spectra at $\omega = 1, 3, 5, \ldots$ indicate the periodic components of the mean value as given by equation (13), and numerical values attached to line spectra represent the power concentrated on those frequencies. In every computer experiment, the general aspect (location, shape and size) of the average power spectrum is reproducible. The average power of the random process $\{X(\tau)\}$ is given by

$$\lim_{T\to\infty} \frac{1}{T}\int_{-T/2}^{T/2} \langle X_T^2(\tau)\rangle \, d\tau \doteq \frac{1}{T}\int_{-T/2}^{T/2} \langle X_T^2(\tau)\rangle \, d\tau = 3.08 \qquad (14)$$

3.4. Spectral decomposition of the power

It is easily seen that, due to the non-linearity of the inductor, the results of Fig. 6 and equation (14) in the preceding section do not have dimension of the electric power. Therefore, let us here examine the situation of spectral dispersion of electric power supplied by the source of single frequency. The terminal voltage of the capacitor

$$v(\tau) = \frac{B}{\sqrt{1+k^2}} \sin(\tau + \tan^{-1} k) - y(\tau) \qquad (15)$$

is also a random process $\{V(\tau)\}$, and the mean value is given by

$$\begin{aligned}
m_V(\tau) = &\; 0.97 \cos\tau + 13.60 \sin\tau \\
&+ 0.77 \cos 3\tau + 3.63 \sin 3\tau \\
&+ 0.28 \cos 5\tau + 1.27 \sin 5\tau \\
&+ 0.16 \cos 7\tau + 0.51 \sin 7\tau \\
&+ 0.06 \cos 9\tau + 0.15 \sin 9\tau \\
&+ 0.02 \cos 11\tau + 0.04 \sin 11\tau \\
&+ 0.00 \cos 13\tau + 0.01 \sin 13\tau
\end{aligned} \qquad (16)$$

Figure 7 shows an average power spectrum of $\{V(\tau)\}$.† The average power of the random process is given by

$$\lim_{T\to\infty} \frac{1}{T}\int_{-T/2}^{T/2} \langle V_T^2(\tau)\rangle \, d\tau \doteq \frac{1}{T}\int_{-T/2}^{T/2} \langle V_T^2(\tau)\rangle \, d\tau = 104 \qquad (17)$$

From these results, spectral decomposition of the electric power dissipated in the shunt resistance with the capacitor $kv^2(\tau)$ is calculated. That is, the power supplied by the single frequency is decomposed and dissipated in the resistor at the following rate of frequencies:

fundamental component	89.3%
third harmonic component	6.6%
random component	3.1%
fifth harmonic component	0.8%

The results of this section are obtained by using analog and digital computers. In the analog computer experiments, various types of non-linear elements and a number of operating speeds (time scales) are used. In the digital computer experiments, integration

---

† In the present discussion, voltage, current and time are normalized by $n\omega\Phi_n$, $a\Phi_n^3$ and $1/\omega$, respectively. In this case, the source voltage is given by

$$\frac{B}{\sqrt{1+k^2}} \sin(\tau + \tan^{-1} k) = 1.19 \cos\tau + 11.88 \sin\tau$$

and the mean value of the current by

$$\langle X^3(\tau)\rangle = 13.70 \cos\tau + 0.39 \sin\tau + \cdots$$

Therefore, the power supplied by the source turns out to be 10.4.

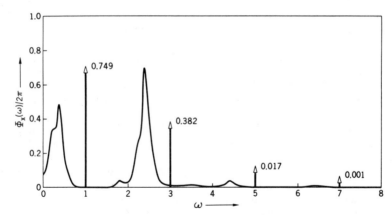

Fig. 6. Average power spectrum of the random process $\{X(\tau)\}$.

methods of Runge–Kutta–Gill and of Hamming with various step sizes are used, and further experiments have also been executed for both single and double precisions. It is confirmed that these results agree well with each other not only qualitatively but also quantitatively.

Fourier's transformation, equation (10), has been carried out by applying the discrete FFT algorithm

$$\left. \begin{array}{l} a_m = \dfrac{1}{N} \displaystyle\sum_{k=-(N-1)}^{N} x\left(k\dfrac{T}{2N}\right) \cos\left(mk\dfrac{\pi}{N}\right), \\ \qquad m = 0, 1, 2, \ldots, N \\[1em] b_m = \dfrac{1}{N} \displaystyle\sum_{k=-(N-1)}^{N} x\left(k\dfrac{T}{2N}\right) \sin\left(mk\dfrac{\pi}{N}\right), \\ \qquad m = 1, 2, \ldots, N-1 \end{array} \right\} \quad (18)$$

for $2N$ sampled values $x(\tau_k)$ at $\tau_k = kT/2N$ ($k = -(N-1), \ldots, N$). In the calculations, taking the errors related to the approximations of continuous variables by discrete variables and of infinite interval by finite interval into account, the values $T = 2\pi \times 2^{10}$, $2N = 2^{15}$ are used, and the ensemble average is estimated for 100 ($N_s = 99$) realizations.

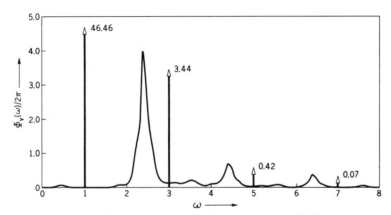

Fig. 7. Average power spectrum of the random process $\{V(\tau)\}$.

4. DISCUSSION OF THE EXPERIMENTAL RESULTS

In the present section, the experimental results given in the preceding section are discussed and the problems arising from them are summarized.

(1) The random oscillation is not a special one which appears only for particular values of the system parameters, but can be observed in a rather wide range of values. From Figs. 3 and 4 the attractor M of the random oscillation specified by the parameter values of equation (8) is regarded as a closure of unstable manifolds of the directory unstable fixed point ${}^1D^1$ of f_λ, i.e. $M = ClW^u({}^1D^1)$. An appearance of the attractor seems to change continuously when the parameters are varied in the neighborhood of $\lambda = (0.1, 12.0)$. The movement of images in the attractor M under iterations of f_λ is not reproducible, but seems to be stable in the Poisson sense.

(2) The stable manifolds $W^s({}^1D^1)$ intersect the unstable manifolds $W^u({}^1D^1)$ forming a homoclinic cycle. As we see in Fig. 4, most intersections (doubly asymptotic points) are transversal, but, as is seen in the neighborhood of the point $(2.8, -2.0)$ of Fig. 4, it is expected that there exist some tangential points (doubly asymptotic points of the special type) on the prolonged manifolds. This fact suggests that the structure of the attractor may be unstable in the sense of Andronov–Pontryagin.

Existence of homoclinic points indicates that the attractor M contains infinite periodic groups. As shown by experiments, every periodic group is unstable. This fact implies that, even if sinks exist, their domains of attraction are so narrow that they are subject to the perturbations by the uncertainties acting on the system.

(3) The observed orbit $\{X(2n\pi), Y(2n\pi)\}$ ($n \in Z^+$) of the discrete dynamical system, in other words, stroboscopic sequence of the computer solution with the same period as that of the periodic forcing seems to be a 2-dimensional stationary sequence taking values in the attractor.

(4) The random process $\{X(\tau)\}$ can be regarded as a sample process of the periodic non-stationary random process. Therefore, the mean value $m_X(\tau)$ is a periodic function with period 2π and the correlation function of the random component $\{R(\tau)\} = \{X(\tau) - m_X(\tau)\}$ is invariant under the periodic translations: $\tau \to \tau + 2n\pi$ ($n \in Z$) [7]–[9].

In the computer experiments, an outline of the average power spectrum of the random process is reproducible. This indicates that the average power spectrum of the process is characterized by the structure of solutions regardless of the nature of uncertain factors. In other words, the simulation and/or calculation errors are not amplified into a random process but only bring about randomness in the phenomenon.

(5) From the above facts, the genesis and the properties of the random oscillation are summarized as follows. "The representative point of the actual system (which is not prescribed by the solution of the differential equation in the mathematical sense) continues to transit randomly among the infinitely many solutions due to the perturbations by uncertain factors of the system. The average power spectrum of the random oscillation depends practically not on the nature of uncertain factors but on the structure of the solutions emanating from the attractor".

In succession, the attractor representing random oscillations should be defined appropriately by "the asymptotically stable, compact, f_λ-invariant set which contains infinitely minimal sets connected to one another by the influence of uncertain factors in the actual system".

Because of those aforementioned, we have called this type of oscillation "the randomly transitional oscillation" [10]. The difference between randomly transitional oscillations and almost periodic oscillations is explained as follows. The attractor of the former is composed of infinite minimal sets, whereas that of the latter is made up of a single minimal set.

The problems arising from the above matters are summarized as follows.

(6) Let $\Omega(f_\lambda)$ be a set of non-wandering points in the domain of attraction for the attractor $M = ClW^u({}^1D^1)$. Is $\Omega(f_\lambda)$ identical with M, or a proper subset of M? Does $\Omega(f_\lambda)$ contain minimal sets different from periodic groups? Does $\Omega(f_\lambda)$ contain minimal sets different from periodic groups? How is $\Omega(f_\lambda)$ decomposed?

(7) From the above item (1), the attractor M seems to be structurally stable in some sense. What is the concept of structural stability? That is, what kinds of space (of differential equations) and topology are used for the discussion of structural stability?

(8) How does the transition probability of the stroboscopic sequence $\{X(2n\pi), Y(2n\pi)\}$ ($n \in Z^+$) and the stochastic properties of $\{X(\tau)\}$ depend on the nature of uncertain factors of the system?

As mentioned in item (4), an average power spectrum scarcely depends on the simulation and/or calculation errors but is determined from the structure of the solutions emanating from the attractor. This fact seems applicable to general electric circuits provided that the uncertain factors are sufficiently small and have no special characteristics. For the case in which this conjecture does not hold, namely, for the case in which some kind of resonance may be expected, what kind of relationship is expected between the nature of the noise and the structure of the solutions passing the attractor? Under the influence of random noise having the characteristics above, the phenomenon must be discussed by introducing random parameters into the differential equations describing the electric circuit.

5. CONCLUSION

In the present paper, random phenomena resulting from non-linearity have been studied in the series-resonance circuit containing a saturable inductor. As a result of this investigation, a part of the genesis and of the stochastic properties of the random oscillation has been first clarified. This phenomenon should be called turbulence in electric circuits.

Although an example of randomly transitional phenomena has been studied in detail for the system described by Duffing's equation, this kind of phenomena have been observed in another non-linear system. Hence they may be regarded as general steady phenomena in non-linear systems [10]. Further, it seems interesting to examine the phenomena in reference to the turbulence in fluid dynamics [11].

The unsolved problems (6)–(8) pointed out in the preceding section relate closely to the global structure of solutions of differential equations both in time and in space. They also relate to uncertain factors of actual systems. They are really fundamental and difficult problems. It is hoped that these problems will deserve attention as material for further study.

Acknowledgements—The author wishes to express his sincere thanks to Professor Michiyoshi Kuwahara and Professor Chikasa Uenosono, both of Kyoto University for their thoughtful consideration and encouragement. He is likewise grateful to Professor Ken-ich Shiraiwa of Nagoya University and Professor Hisanao Ogura of the Kyoto Institute of Technology for their many useful comments and generous advice. The author also appreciates the assistance he received from Associate Professors Hiroshi Kawakami and Norio Akamatsu, both of Tokushima University, Miss Keiko Tamaki and Miss Yuriko Yamamoto, both of Kyoto University.

(The manuscript was received 30 June 1977, and the revised one 30 September 1977 by the Inst. Elect. Engrs. of Japan.)

REFERENCES

1. K. Shiraiwa, *Theory of Dynamical Systems*. Iwanami Shoten (1974).
2. N. Levinson, *Annls Math.* **45**, 723 (1944); **49**, 738 (1948).
3. V. V. Nemytskii and V. V. Stepanov, *Qualitative Theory of Differential Equations*. Princeton University Press, Princeton (1960).
4. G. D. Birkhoff, *Collected Mathematical Papers*. Dover, New York (1968).
5. S. Smale, *Bull. Am. Math. Soc.* **73**, 747 (1967).
6. Z. Nitecki, *Differentiable Dynamics*. MIT Press, Cambridge, Mass. (1971).
7. R. L. Stratonovich, *Topics in the Theory of Random Noise*, Gordon and Breach, New York (1963).
8. H. Ogura, *Trans. Inst. Elect. Commun Engrs, Japan.* **53-C**, 133 (March 1970).
9. H. Ogura, *IEEE Trans. Inf. Theory*, **IT-17**, 143 (1971).
10. Y. Ueda *et al.*, *Trans. Inst. Elect. Commun Engrs, Japan.* **56-A**, 218 (April 1973).
11. J. E. Marsden and M. McCracken, *The Hopf Bifurcation and Its Applications*. Springer, Berlin (1976).

POSTSCRIPT

I deem it a great honour to be given the opportunity to translate my article into English and I would like to express my thanks to the members of the editorial board. In the following I am writing down some comments and fond memories of days past when I was preparing the manuscript with tremendous difficulty.

It was on 27 November 1961 when I met with chaotic motions in an analog computer simulating a forced self-oscillatory system. Since then my interest has been held by the phenomenon, and I have been fascinated by the problem "What are steady states in non-linear systems?". After nearly ten years, I understood "randomly transitional phenomenon", I published my findings in the *Transactions of the Institute of Electronics and Communication Engineers of Japan*, Vol. 56, April 1973 [10]. My paper then received a number of unfavorable criticisms from some of my colleagues: such as, "Your results are of no importance because you have not examined the effects of simulation and/or calculation errors at all", "Your paper is of little importance because it is merely an experimental result", "Your result is no more than an almost periodic oscillation. Don't form a selfish concept of steady states", and so forth. Professor Hiromu Momota of the Institute of Plasma Physics was the first to appreciate the worth of my work. He said "Your results give an important feature relating to stochastic phenomena" on

3 March 1974. Through his good offices I joined the Collaborating Research Program at the Institute of Plasma Physics of Nagoya University. These events gave me such unforgettable impressions that I continued the research with tenacity. At this moment I yearn for those days with great appreciation for their criticisms and encouragements.

By the middle of the 1970s, I had obtained many data of strange attractors for some systems of differential equations, but I had no idea to what journals and/or conferences I might submit these results. I was then lucky enough to meet with Professor David Ruelle who was visiting Japan in the early summer of 1978. He advised me to submit my results to the *Journal of Statistical Physics* [P1]. Further, he named the strange attractor of Fig. 3 "Japanese Attractor" and introduced it to the whole world [P2–P5]. At that time chaotic behavior in deterministic systems began to come under the spotlight in various fields of natural sciences. I fortunately had several opportunities to present my accumulated results [P6–P11]. It is worth while mentioning that, due to the efforts of Professor David Ruelle and Professor Jean-Michel Kantor, the Japanese Attractor will be displayed at the National Museum of Sciences, Techniques and Industries which will open in Paris, 1986. In these circumstances this paper is a commemorative for me and I sincerely appreciate their kindness on these matters.

As the reader will notice in this translation and also in ref. [P1] I was rather nervous of using the term "strange attractor", because I had no understanding of its mathematical definition in those days. Although I do not think I fully understand the definition of it even today, I begin to use the term "strange attractor" without hesitation because it seems to agree with reality. However, it seems to me that the term "chaos", though it is short and simple, is a little bit exaggerated. In the universe one does have a lot more complicated, mysterious and incomprehensible phenomena! I should be interested in readers' views of my opinion.

REFERENCES TO POSTSCRIPT

P1. Y. Ueda, *J. statist. Phys.* **20**, 181 (1979).
P2. D. Ruelle, *La Recherche*, **11**, 132 (1980).
P3. D. Ruelle, *The Mathematical Intelligencer* **2**, 126 (1980).
P4. D. Ruelle, *Mathematics Calendar*. Springer, Berlin (November 1981).
P5. D. Ruelle, *Czech. J. Phys.* **A32**, 99 (1982).
P6. Y. Ueda, New approaches to non-linear problems in dynamics, *SIAM J. appl. Math.* 311 (1980).
P7. Y. Ueda, *Annls N.Y. Acad. Sci.* **357**, 422 (1980).
P8. Y. Ueda and N. Akamatsu, *IEEE Trans Circuits and Systems*, **28**, 217 (1981).
P9. H. Ogura *et al.*, *Prog. theoret. Phys.* **66**, 2280 (1981).
P10. Y. Ueda, *Proc. 24th Midwest Symposium on Circuits and Systems*, p. 549. University of New Mexico (1981).
P11. Y. Ueda and H. Ohta, *Chaos and Statistical Methods*, p. 161. Springer, Berlin (1984).

II. Description and Quantification of Chaotic Behavior

This section starts with the popular Paper II.1 which introduces fundamental properties of strange attractors basing on the simplest examples.

In many physical systems equations of motion are unknown and all the information about the system has to be obtained from time series. Paper II.2 describes the method of reconstructing an attractor from time series basing on delay coordinates and embedding theorem.

Deterministic representations of chaos in classical dynamics are dealt with in Paper II.3.

Lyapunov exponents describing sensitivity of the system to the initial conditions are a fundamental feature to describe chaos. The methods of estimating Lyapunov exponents both from equations of motion and time series are presented in Paper II.4. Also the Fortran codings for both methods are given.

Strange Attractors*

David Ruelle

Introduction: Deterministic Systems with a Touch of Fantasy

Systems with an irregular, non periodic, "chaotic" time evolution are frequently encountered in physics, chemistry, and biology. Think for example of the smoke rising in still air from a cigarette. Oscillations appear at a certain height in the smoke column, and they are so complicated as to apparently defy understanding. Although the time evolution obeys strict deterministic laws, the system seems to behave according to its own free will. Physicists, chemists, biologists, and also mathematicians have tried to understand this situation. We shall see how they have been helped by the concept of *strange attractor,* and by the use of modern computers.

A strange attractor consists of a infinity of points, in the plane as shown on Figure 1A, or in m-dimensional space. These points correspond to the states of a chaotic system. Strange attractors are relatively abstract mathematical objects, but computers give them some life, and draw pictures of them. (See the illustrations, and note that the computer may mark only a finite number of points.) It may well be that the reader has access to a computer, and can reproduce some of the "experiments" described below.

The Description of Time Evolution: Dynamical Systems

We specify the state of a physical, chemical, or biological system by parameters $x_1, x_2, \ldots, x_m$. A chemical system for example would be described by the concentrations of various reactants. The parameters vary with time, and we denote by

$$x_1(t), x_2(t), \ldots, x_m(t)$$

their values at time t. For simplicity we shall consider first only integer values of t (time expressed in seconds, or in years). We shall come back later to the case of continuously varying time.

How do we determine the time evolution of the system, in other words its *dynamics*? We shall admit that the parameters specifying the system at time $t + 1$ are given functions of the parameters at time t. We may thus write

$$\left.\begin{aligned}
x_1(t+1) &= F_1(x_1(t), x_2(t), \ldots, x_m(t)) \\
x_2(t+1) &= F_2(x_1(t), x_2(t), \ldots, x_m(t)) \\
&\cdots \\
x_m(t+1) &= F_m(x_1(t), x_2(t), \ldots, x_m(t))
\end{aligned}\right\} \quad (1)$$

We assume that the functions $F_1, F_2, \ldots, F_m$ are continuous and have continuous derivatives. This "technical" differentiability condition will be satisfied in our examples. We shall see later why it is important.

Given *initial values* $x_1(0), x_2(0), \ldots, x_m(0)$ for the parameters we can, using (1), compute $x_1(t), x_2(t), \ldots, x_m(t)$ successively for all positive integer times t. Thus, knowing the state of the system at time zero one may compute its state at time t. We say that the functions $F_1, F_2, \ldots, F_m$ determine a discrete time *dynamical system*. It is a *differentiable* dynamical system because we have assumed that the functions $F_1, F_2, \ldots, F_m$ have continuous derivatives.

An Example: The Hénon Attractor

Let us now examine a concrete case. Let $m = 2$, and write x, y instead of x_1, x_2. We are given

$$F_1(x, y) = y + 1 - ax^2$$
$$F_2(x, y) = bx$$

with $a = 1.4$ and $b = 0.3$. The relations (1) thus take the form

$$\left.\begin{aligned}
x(t+1) &= y(t) + 1 - ax(t)^2 \\
y(t+1) &= bx(t)
\end{aligned}\right\} \quad (2)$$

Given $x(0), y(0)$ we may compute $x(t)$ and $y(t)$ for $t = 1, 2, \ldots, 10,000$ for instance, keeping everywhere sixteen significant figures. Done by hand this calculation would take many months and, since its interest is not obvious, nobody undertook it. For a digital computer on

---

* Translated by the author from his French article published in *La Recherche* N° 108, Février 1980, with kind permission of La Recherche.

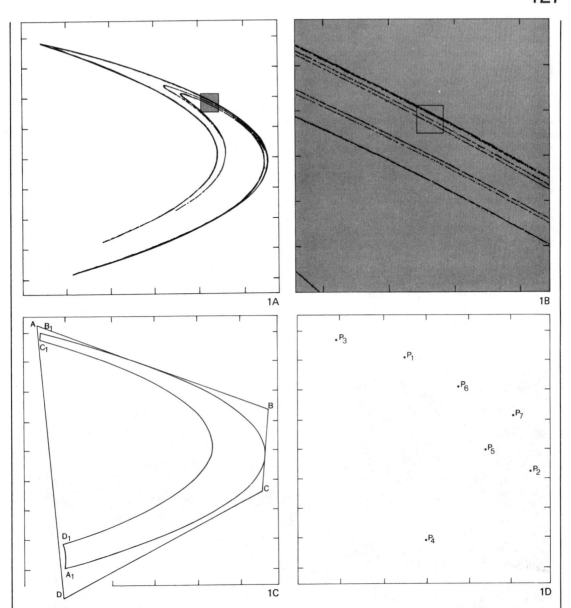

Figure 1. *The Hénon attractor.* A computer has been asked to mark points of coordinates $x(t), y(t)$ for t going from 1 to 10,000. The point $(x(0), y(0))$ is given, and the following points are determined by

$$x(t+1) = y(t) + 1 - ax(t)^2, \quad y(t) = bx(t)$$

with $a = 1.4$ and $b = 0.3$. Figure 1A shows the result. The 10,000 points distribute themselves on a complex system of lines: the *Hénon attractor*. It is an example of a *strange attractor*. Magnification of the little square in Figure 1A yields 1B, and magnification of the little square in 1B would again yield a similar picture. Each new magnification resolves lines into more lines. The Hénon attractor is associated with a map of the plane which sends the point (x, y) to $(F_1(x, y), F_2(x, y))$, with $F_1(x, y) = y + 1 - ax^2$, $F_2(x, y) = bx$. In particular, the quadrilateral $ABCD$ of Figure 1C is mapped inside itself into $A_1 B_1 C_1 D_1$. Notice that F_1, F_2 are polynomials, and therefore have continuous derivatives

$$\partial F_1/\partial x = -2ax \qquad \partial F_1/\partial y = 1$$
$$\partial F_2/\partial x = b \qquad \partial F_2/\partial y = 0$$

One can see that the surface of $A_1 B_1 C_1 D_1$ is equal to three tenths of the surface of $ABCD$ (the factor $b = 0.3$ is given, up to sign, by the determinant of the above derivatives). In Figure 1D one has kept $b = 0.3$ but taken $a = 1.3$. The strange attractor disappears, and is replaced by the seven points of a periodic attractor.

the other hand, this boring and repetitive task is not a problem. Michel Hénon, of the observatory in Nice, did the first calculations with an HP-65 programmable pocket computer. He then went on to a more powerful machine (IBM 7040). That computer had a plotter, which marked on a sheet of paper the points with coordinates $x(t)$, $y(t)$, for t ranging from 1 to 10,000. Figure 1A shows the picture obtained. Unexpectedly, the ten thousand points lie on a system of lines with complex structure. If the little square of Figure 1A is magnified, Figure 1B is obtained. If the square of Figure 1B were magnified, one would obtain again a similar picture, and so on, each magnification revealing lines which were not previously visible [1].

What happens if the initial point $(x(0), y(0))$ is changed? Well, for a "bad" choice $(x(t), y(t))$ will go to infinity (and in particular, leave the sheet of paper). For a "good" choice, $(x(1), y(1)), (x(2), y(2)), \ldots$, will rapidly get close to the "noodle" of Figure 1A, and the general aspect of this picture will be reproduced after a few thousand points have been marked.

Our "noodle" is the *Hénon attractor*. It is an example of a *strange attractor*. Let me mention, among other curiosities, that the attractor may suddenly disappear when the parameters a, b in (2) are changed. Taking for instance $a = 1.3$ and $b = 0.3$ one sees the points $(x(t), y(t))$ approaching, when t increases, a set of seven points $P_1, \ldots, P_7$ (Figure 1D). Instead of a strange attractor we now have a *periodic attractor* (of period 7).

In trying to understand the Hénon attractor, it is helpful to consider the map F of the plane to itself defined by (2). If X has coordinates x and y, $F(X)$ has coordinates

$$F_1(x,y) = y + 1 - ax^2, \quad F_2(x,y) = bx$$

Call X_t the point with coordinates $x(t)$, $y(t)$. Then $X_1 = F(X_0)$, $X_2 = F(F(X_0))$, etc ..., X_t is obtained from X_0 by applying t times the map F. Figure 1C shows a quadrilateral $ABCD$, and its *image* $A_1 B_1 C_1 D_1$ by F. This image is by definition the set of points $F(X)$ with X in the quadrilateral $ABCD$. Hénon has chosen the quadrilateral $ABCD$ in such a manner that it contains the image $A_1 B_1 C_1 D_1$. Figure 1C shows that the quadrilateral is "folded in two" by the map F. If the initial point X_0 is in $ABCD$, then X_1 is in the image $A_1 B_1 C_1 D_1$, and thus again in $ABCD$. All the points $X_1, X_2, \ldots, X_t, \ldots$ are therefore in the quadrilateral $ABCD$, and the Hénon attractor is also contained in that quadrilateral.

 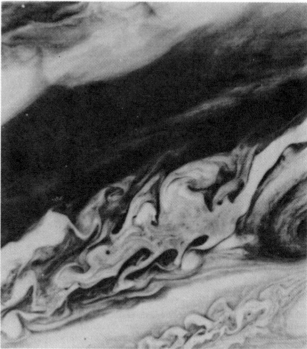

Smoke rising from a cigarette. – The atmosphere of Jupiter. Two of the many examples of systems whose evolution through time involves oscillations which can be described by strange attractors.   (Clichés E. Rousseau & IPS)

Another Example: The Solenoid

We shall now examine an attractor in three dimensions, i.e., we shall take $n = 3$ in the formulae (1). Instead of writing explicit expressions for the functions F_1, F_2, F_3, we describe geometrically the map F of three-dimensional space to itself which they define. (This map F sends the point with coordinates x_1, x_2, x_3 to the point with coordinates $F_1(x_1, x_2, x_3), F_2(x_1, x_2, x_3), F_3(x_1, x_2, x_3)$). We suppose that F takes a ring A (the solid torus of Figure 2A), stretches it, makes it thinner, folds it, and places it in the manner drawn in Figure 2B. This figure shows both A, and its image $F(A)$ by the map F. The image $F(A)$ winds twice around the central hole of the ring A.

Starting from a point X_0 in the ring A, we write $X_1 = F(X_0), X_2 = F(X_1), \ldots$ Figure 2C shows the five thousand points $X_{51}, X_{52}, \ldots, X_{5050}$ (together with the set $F(A)$). A new strange attractor appears. Since the point X_0 is arbitrary in A, it is not in general on the attractor, but $X_1, X_2, X_3, \ldots$ get progressively closer to it. This is why we have marked the points starting at X_{51}. It is fascinating to observe the plotter (of the HP 9830A) draw the picture. About once per second a click is produced and a point is marked, in an apparently random manner. It takes a fairly long time before one can guess the final form of the attractor.

The attractor of Figure 2C has been called a *solenoid*. Indeed the picture is suggestive of electric wires around an axis. To understand this structure, note that the solenoid is contained not only in the ring A of Figure 2A, but also in its image $F(A)$ drawn in Figure 2C, and also in $F(F(A))$, $F(F(F(A))), \ldots$ The image $F(A)$ is the inside of a tube which winds twice around the central hole of A, $F(F(A))$ is in a thinner tube which winds four times around the hole, $F(F(F(A)))$ is still thinner and winds around eight times, etc. . . . The solenoid is thus contained in very thin tubes winding around many times, and this explains how it looks.

Sensitive Dependence with Respect to Initial Conditions: How Errors Grow with Time

Remember that the parameters $x_1(t), x_2(t), \ldots, x_m(t)$ are supposed to describe a physical, chemical, or biological

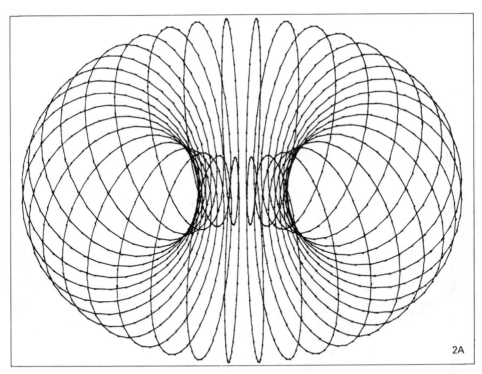

Figure 2. *The solenoid.* Figure 2A is a perspective view of a ring A in three dimensional space. A map F stretches A, makes it thinner, folds it, and places the image $F(A)$ inside A so that $F(A)$ turns twice around the central hole of A, as shown in Figure 2B. Figure 2C shows $F(A)$ again, and also 5,000 points successively defined by $X_{t+1} = F(X_t)$ starting from some initial point X_0. The 5,000 points produce a wiry structure. It is a new strange attractor, called *solenoid*.

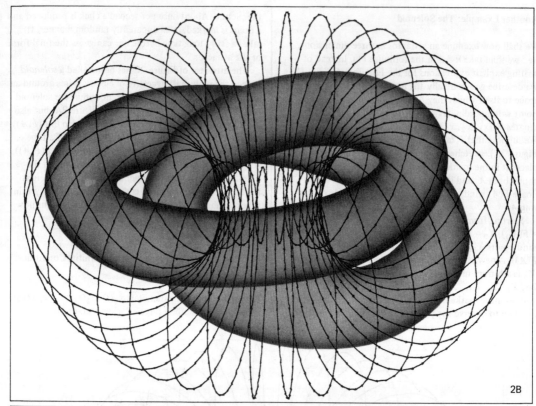

2B

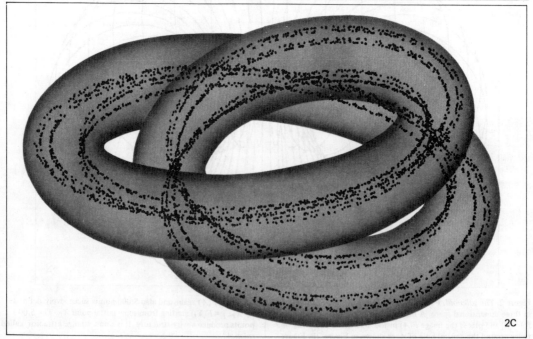

2C

system at time t. We assume that the system has a deterministic time-evolution defined by the equations (1). With what precision can we predict the evolution if the choice of the initial values $x_1(0), x_2(0), \ldots, x_m(0)$ is slightly in error, as is always the case for experimental data? How will the error increase (or decrease) with increasing t? The answer will of course depend on the given functions $F_1, F_2, \ldots, F_m$, and on the initial values $x_1(0), x_2(0), \ldots, x_m(0)$. For the two strange attractors which we have examined (the Hénon attractor and the solenoid) a small error (or uncertainty) on the initial values gives an error (or uncertainty) at time t, which increases rapidly with t.

Let us verify this assertion for the Hénon attractor. We know that there is, around the attractor, a quadrilateral $ABCD$ such that the map F folds the quadrilateral in two. As Figure 1C shows, the folding in two is accompanied by stretching. Thus if X_t and X'_t correspond to initial data X_0 and X'_0 close to each other, the distance $d(X_t, X'_t)$ generally increases with t. At least this is the case as long as this distance remains small; when the distance from X_t to X'_t becomes of the order of the total size of the attractor it cannot increase any more. Numerically one finds

$$d(X_t, X'_t) \sim d(X_0, X'_0) . a^t \qquad (3)$$

with $a \approx 1.52$. Since $a > 1$, the factor a^t increases rapidly (exponentially) with t. Therefore *the error $d(X_t, X'_t)$ increases exponentially with time*. The rate of exponential increase is determined by a (or by its logarithm $\lambda = \ln a$ called characteristic exponent, here $\lambda \approx 0.42$).

We may argue similarly for the solenoid. The map F stretches a tube containing the solenoid and, because of this stretching, formula (3) remains valid, with a different choice of $a > 1$.

The exponential increase of errors described by formula (3) is expressed by saying that the dynamical system under consideration has *sensitive dependence on initial condition*.

Notice that to give a precise meaning to (3) we have to take $d(X_0, X'_0)$ "infinitesimal". The assumption that $F_1, F_2, \ldots, F_m$ have continuous derivatives is used here. Notice also that, for given X_0, there may be exceptional X'_0 for which the error does not grow as indicated by (3) (it may for instance decrease).

A Little Bit of Mathematics: A Definition of Strange Attractors

Let us come back to the general dynamical system described by the equations (1). We call F the map of m dimensional space to itself which sends X with coordinates $x_1, \ldots, x_m$ to $F(X)$ with coordinates $F_1(x_1, \ldots, x_m), \ldots, F_m(x_1, \ldots, x_m)$. We shall say that a bounded set A in m-dimensional space is a *strange attractor* for the map F if there is a set U with the following properties:

(a) U is an m-dimensional *neighborhood* of A, i.e., for each point X of A, there is a little ball centered at X and entirely contained in U. In particular A is contained in U.

(b) For every initial point X_0 in U, the point X_t with coordinates $x_1(t), \ldots, x_m(t)$ remains in U for positive t; it becomes and stays as close as one wants to A for t large enough. This means that A is *attracting*.

(c) There is sensitive dependence on initial condition when X_0 is in U. This makes A a *strange* attractor.

In the case of the Hénon attractor one can take for U the quadrilateral $ABCD$ (Figure 1C), in the case of the solenoid one can take for U the solid torus A (Figure 2).

The above definition allows the practical determination of strange attractors in computer studies, but it is not quite complete mathematically. It is desirable also to impose the following condition.

(d) One can choose a point X_0 in A such that, arbitrarily close to each other point Y in A, there is a point X_t for some positive t. This *indecomposability condition* implies that A cannot be split into two different attractors.

It would also be necessary to make the notion of sensitive dependence on initial condition more precise. This however, leads to questions which are not too well understood. It must be said that the mathematical theory of strange attractors is difficult and, in part, still in its infancy. The solenoid is well understood, thanks to the work of Steve Smale [2] of Berkeley. By contrast, it has not been *proved* that Figures 1A and 1B do not just show a periodic orbit of very long period. The fact that the Hénon attractor exists as a strange attractor is for the time being a *belief* based on computer calculations! Perhaps our definition of strange attractors will have to be changed to adapt to more general situations. Do not take it too seriously.

It seems that the phrase "strange attractor" first appeared in print in a paper by Floris Takens (of Groningen) and myself [3]. I asked Floris Takens if he had created this remarkably successful expression. Here is his answer: "Did you ever ask God whether he created this damned universe? . . . I don't remember anything . . . I often create without remembering it . . ." The creation of strange attractors thus seems to be surrounded by clouds and thunder. Anyway, the name is beautiful, and well suited to these astonishing objects, of which we understand so little.

Besides strange attractors, we should remember that there are also non strange attractors. For instance *attracting fixed points*. The point A is an attracting fixed point if X_t gets arbitrarily close to A when t increases, provided X_0 is in a neighborhood U of A. In that case of course errors decrease when t increases, and there is no sensitive dependence on initial conditions. Attracting fixed points belong to the *periodic attractors*, which we have already met (Figure 1D). A periodic attractor has a finite number of points.

Attracting fixed points have been known for a long time. They describe an asymptotically stationary situation, i.e., for large t, X_t practically no longer depends on t. In the same manner the periodic attractors describe an asymptotically periodic situation. Scientists had got used to the notion that the asymptotic behavior of natural phenomena should be stationary, or perhaps periodic. Only recently did interest arise in the "chaotic" behavior, with sensitive dependence on initial condition, which occurs in many natural phenomena.

Strange Attractors in Nature

To describe the systems which they encounter, physicists, chemists, and biologists use equations of the type (1), or differential equations in the case of continuous time. One should not underestimate the amount of idealization implied by such a description. Certain parameters are selected as variables $x_1, \ldots, x_m$, others are ignored, and various simplifications are made. Idealization is a basic ingredient of all natural sciences, and a serious scientist must show that the natural system which he considers obeys deterministic laws of the type (1) with a good approximation. He may then look for strange attractors, either by the direct study of experimental results, or by computer simulation. In this manner, the "chaos" which occurs in certain phenomena becomes understandable, and it may be hoped that this understanding will lead to practical applications.

The study of "chaotic" or "turbulent" time evolutions in natural phenomena is now only at its beginnings. Progress is slow, due in part to experimental difficulties, in part to the insufficient development of the theory. In the absence of a satisfactory mathematical theory, computers play an important role in the interpretation of data.

We shall now discuss some examples of chaotic phenomena, and in particular the problem of fluid turbulence. In order to do this we shall have to use a continuous time t rather than a discrete time.

The Lorenz Attractor, and Meteorological Predictions

In order to define differentiable dynamical systems with continuous time we replace the equations (1) by differential equations

$$\left. \begin{aligned} \frac{d}{dt} x_1(t) &= G_1(x_1(t), \ldots, x_m(t)) \\ &\ldots \\ \frac{d}{dt} x_m(t) &= G_m(x_1(t), \ldots, x_m(t)) \end{aligned} \right\} \quad (4)$$

If $G_1, \ldots, G_m$ satisfy certain conditions (existence of continuous derivatives, etc.) the equations (4) uniquely determine the functions $x_1(t), \ldots, x_m(t)$ of time t when the initial data $x_1(0), \ldots, x_m(0)$ are known. The equations (4) thus define a deterministic evolution with continuous time, just as the equations (1) defined a deterministic evolution with discrete time.

Let us take for example $m = 3$, and write $x_1(t) = x$, $x_2(t) = y$, $x_3(t) = z$. We consider the differential equations

$$\left. \begin{aligned} \frac{dx}{dt} &= -\sigma x + \sigma y \\ \frac{dy}{dt} &= -xy + rx - y \\ \frac{dz}{dt} &= xy - bz \end{aligned} \right\} \quad (5)$$

with $\sigma = 10$, $b = 8/3$, and $r = 28$. Figure 3 shows the trajectory of the point (x, y, z) corresponding to the solution of these equations with initial condition $(0, 0, 0)$. It appears that we have here again a strange attractor, and one can show that there is indeed sensitive dependence on initial condition.

The attractor of Figure 3 is the *Lorenz attractor*, named after Edward Lorenz, professor in the Meteorology department of the Massachusetts Institute of Technology. The equations (5) were indeed first written and studied by Lorenz [4]. These equations give an approximate description of a horizontal fluid layer heated from below. The warmer fluid formed at the bottom is lighter. It tends to rise, creating convection currents. If the heating is sufficiently intense, the convection takes place in an irregular, turbulent manner. This phenomenon takes place for instance in the earth atmosphere, and since it has sensitive dependence on initial condition, it is understandable that meteorologists cannot predict the state of the atmosphere with precision a long time in advance. The work of Ed Lorenz thus gives some theoretical excuse to the well-known unreliability of weather forecasts.

Fluid Turbulence: One of the Great Unsolved Problems of Theoretical Physics

Turbulence is a phenomenon easily produced by opening the tap over the bath tub or the kitchen sink. The nature

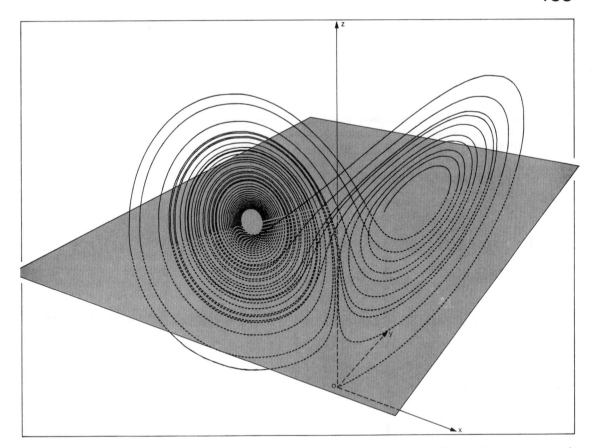

Figure 3. *The Lorenz attractor.* This beautiful figure has been obtained by Oscar Lanford, of Berkeley. It illustrates a new strange attractor, the *Lorenz attractor,* which is approached by the solutions of the Lorenz system of equations:

$$\frac{dx}{dt} = -10x + 10y, \quad \frac{dy}{dt} = -xz + 28x - y, \quad \frac{dz}{dt} = xy - \frac{8}{3}z.$$

Lanford has chosen the solution which starts from the origin $(0, 0, 0)$ at time $t = 0$. It makes one loop to the right, then a few loops to the left, then to the right, and so on in irregular manner. One follows the solution here for fifty loops. The part below the plane $z = 27$ is drawn as a dotted line. If one would take, instead of $(0, 0, 0)$, a nearby initial condition, the new solution would soon deviate from the old one, and the numbers of loops to the left and to the right would no longer be the same. There is *sensitive dependence with respect to initial conditions*. The Lorenz equations are suggested by a problem of atmospheric convection. Edward Lorenz has used the sensitive dependence on initial condition observed with the above equations to justify the imprecision of weather forecasting.

of turbulence remains however rather mysterious and controversial.

One may in principle describe the time evolution of a viscous fluid by equations of the form (4). The number m will have to be taken infinite, because the state of the fluid at a given instant of time requires an infinite number of variables for its description. We admit that there are no further problems, and write $X(t)$ and G instead of $x_1(t), x_2(t), \ldots$, and $G_1, G_2, \ldots$. The equations (4) can then be written in compact form as

$$\frac{d}{dt} X(t) = G_\mu(X(t)) \tag{6}$$

We have introduced a parameter μ in (6) to indicate the intensity of external action on the fluid. (If there is no external action, viscosity brings the fluid to rest, and there is no turbulence). In the example of the tap, μ might give the degree of opening of the tap. In the convection equations (5) of Lorenz, μ is replaced by r, which is proportional to the temperature difference between the top and the bottom of the fluid layer. In many hydrodynamical problems, the role of μ is taken by a parameter called *Reynolds number*.

If $\mu = 0$, i.e., if there is no external action, the fluid tends to a state of rest $X(t) = X_0$. This state corresponds to an attracting fixed point X_0 for our dynamical system.

For small μ one observes again a steady state $X(t) = X_\mu$. As μ is further increased, one often sees periodic oscillations in the fluid. This means that asymptotically

$$X(t) = f(\omega t)$$

where f is a function of period 2π and ω the frequency of the oscillations. This situation corresponds to a periodic attractor for continuous time, i.e., a circle or "attracting limit cycle". For sufficiently large μ, the fluid motion becomes irregular, chaotic: turbulence has set in.

When I became interested in turbulence, around 1970, Lorenz' paper of 1963 was not known to physicists and mathematicians. The most popular theory of turbulence was that of Lev D. Landau of Moscow [5]. According to this theory, the time evolution of a turbulent fluid is asymptotically given by

$$X(t) = f_k(\omega_1 t, \omega_2 t, \ldots, \omega_k t) \tag{7}$$

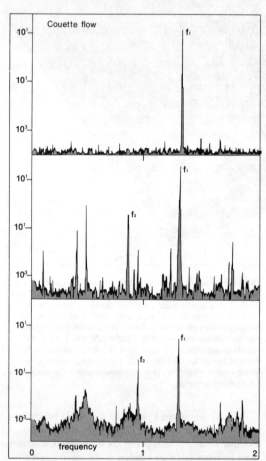 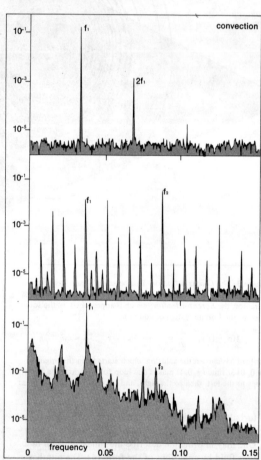

Figure 4. *Frequency spectra*. A frequency analysis of the time dependence of a phenomenon is possible, whether this dependence is periodic or not. One obtains thus a "frequency spectrum" giving the square of the amplitude associated with each frequency. The spectra on the left of the figure have been measured by R. Fenstermacher for the Couette flow (the interval between two coaxial circular cylinders is filled with fluid, and the inner cylinder is rotated at constant speed). The spectra on the right have been measured by S. Benson for a convective flow (a liquid layer is heated from below, the hot liquid formed below is lighter and rises, producing convection currents).

The different spectra shown correspond to different speeds of rotation (Couette) or different intensities of heating (convection). The spectra at the top contain isolated peaks corresponding to a certain frequency and its harmonics: the system is *periodic*. The spectra in the middle row exhibit several independent frequencies: the system is *quasi periodic*. The spectra at the bottom show some wide peaks on a background of *continuous spectrum*, this suggests that a strange attractor is present. Notice that the frequency spectra are shown with a logarithmic vertical scale.

where f_k is a periodic function of period 2π in each of its arguments, and $\omega_1, \omega_2, \ldots, \omega_k$ are independent frequencies. A function of t of the form (7) is called quasiperiodic. (One can see that the corresponding quasiperiodic attractor is a k-dimensional torus). A quasiperiodic function has a non periodic, irregular aspect, suggestive of turbulence. However a small change in initial conditions simply replaces $\omega_1 t, \ldots, \omega_k t$ by $\omega_1 t + \alpha_1, \ldots, \omega_k t + \alpha_k$ with small $\alpha_1, \ldots, \alpha_k$. There is thus no sensitive dependence on initial conditions.

It was tempting to appeal to strange attractors rather than quasiperiodic attractors to interpret turbulence. A mathematical argument against quasiperiodic attractors is their fragility. My attention had been drawn on this fragility, or absence of "structural stability" by the seminars of René Thom at the Institut des Hautes Etudes Scientifiques (Bures-sur-Yvette). By a small perturbation of (6) one can destroy a quasiperiodic attractor and, if $k \geqslant 3$, obtain a strange attractor. I had published this result with Floris Takens [3] in 1971, and we had on this occasion proposed the idea that turbulence is described by strange attractors. While structural stability may not be as important an aspect of things as we thought at the time, the connection between strange attractors and turbulence was a lucky idea.

It remained to be seen if strange attractors would give a better description of turbulence than quasiperiodic attractors. There is no direct experimental test of sensitive dependence on initial condition in hydrodynamics. One may however do a frequency analysis of the fluid velocity at a point, considered as a function of time. The function giving the square of the amplitude versus the frequency is called *frequency spectrum* (see Figure 4). For a quasiperiodic function the frequency spectrum is formed of discrete peaks at the frequencies $\omega_1, \ldots, \omega_k$ and their linear combinations with integer coefficients. By contrast if the time evolution is governed by a strange attractor one may obtain a continuous frequency spectrum.

It was known that the frequency spectrum of a turbulent fluid is continuous, but this fact was attributed to the accumulation of a large number of independent frequencies simulating, in the limit, a continuous spectrum. Recently (1974–75), delicate experiments performed by Guenter Ahlers at Bell Labs (Murray Hill, NJ), Jerry Gollub

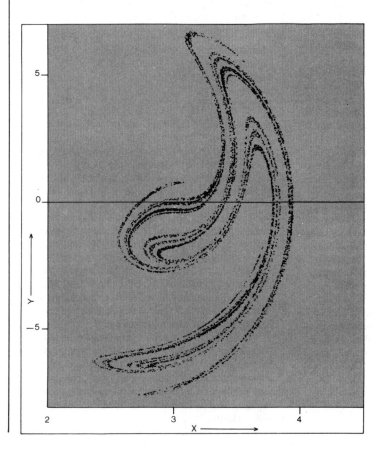

Figure 5. *A Japanese attractor.* This picture shows a strange attractor invented by Yoshisuke Ueda, of Kyoto University. It is obtained by solving the differential equation

$$\frac{d^2x}{dt^2} + k\frac{dx}{dt} + x^3 = B\cos t$$

for $k = 0.1$ and $B = 12$, and marking the points with coordinates $x(2n\pi), \dfrac{d}{dt}x(2n\pi)$ for integer n (discrete time $t = 2n\pi$). Depending on the initial conditions one obtains either the above attractor, or a single point (attracting fixed point). Y. Ueda has studied strange attractors numerically for a number of years on analog and digital computers. Esthetically, his pictures are probably the finest obtained to this date.

and Harry Swinney at City College (New York) [6], and others, have shown that things happen differently. When one increases the parameter μ describing the system, the transition to the continuous spectrum characteristic of turbulence is rapid. There is no progressive accumulation of many independent discrete frequencies. So it seems that the onset of turbulence may well correspond to the appearance of strange attractors.

Other Chaotic Phenomena: Turbulence Everywhere

It should here be mentioned that frictionless mechanical systems (conservative systems) give rise neither to strange attractors, nor in fact to attractors at all. Actually, a theorem of mechanics, Liouville's theorem, asserts that time evolution preserves volumes in phase space. This prevents the volume contraction which occurs near an attractor. On the other hand, conservative systems often show sensitive dependence on initial condition.

The physico-chemical systems which give rise to strange attractors are the *dissipative systems,* i.e., those for which a "noble" form of energy (for instance mechanical, electrical, or chemical energy) is changed into heat [7]. These systems actually exhibit an interesting behavior only if they are constantly fed some noble energy, otherwise they go to rest.

One knows chemical reactions which are periodic in time (see inset). I asked in 1971 a chemist, specialist of these periodic reactions, if he thought that one would find chemical reactions with chaotic time dependence. He answered that if an experimentalist obtained a chaotic record in the study of a chemical reaction, he would throw away the record, saying that the experiment was unsuccessful. Things, fortunately, have changed, and we now have several examples of non periodic chemical reactions.

The magnetism of the earth perhaps gives an example of a strange attractor. It is known that the earth magnetic field reverses itself at irregular intervals. This phenomenon occurred at least sixteen times in the last four million years. Geophysicists have written "dynamo equations" with chaotic solutions which describe irregular changes of direction of a magnetic field. There is however as yet no quantitatively satisfactory theory.

Ecologists have studied non periodic models in population dynamics. If m species have, in the year $t + 1$, populations $x_1(t + 1), \ldots, x_m(t + 1)$ determined by the equations (1) in terms of the populations in the year t, one may expect strange attractors to occur. In fact, already for $m = 1$, the equation

$$x(t + 1) = Rx(t)(1 - x(t))$$

gives rise to nonperiodic behavior [8].

One imagines easily that strange attractors may play a role in economics, where periodic processes (economic cycles) are well-known. In fact, let us suppose that the macroeconomical evolution equations contain a parameter μ describing, say, the level of technological development. By analogy with hydrodynamics we would guess that for small μ the economy is in a steady state and that, as μ increases, periodic or quasiperiodic cycles may develop. For high μ chaotic behavior with sensitive dependence on initial condition would be present. This discussion is some

---

A Periodic Chemical Phenomenon: The Belousov-Zhabotinski Reaction

For about twenty years now, an oscillating reaction has been known to chemists. The oscillations have a period of the order of one minute, and continue for perhaps an hour, until the reagents are exhausted. If reagents are added continuously, while reaction products are removed, the oscillations proceed periodically forever. The reaction is, roughly speaking, the oxydization of malonate by bromate, catalyzed by Cerium. The experiment is fairly easy to realize: here is the recipe.

Malonic acid	0.3 M
Cerous nitrate	0.005–0.01 M
Sulfuric acid	3.0 M
Sodium bromate	0.05–0.01 M
Ferroin	a little

M means "molar", for instance sulfuric acid occurs at the concentration of 3 moles per liter. Ferroin is an oxidation reduction indicator (obtained by mixing in water a small amount of o-phenanthroline and ferrous sulfate). In practice one prepares one solution with part of the reagents (in water), and another solution with the rest of the reagents. The oscillating reaction starts when the two solutions are mixed. Perhaps the mathematical reader should be warned that diluting sulfuric acid produces heat and requires caution (see a chemistry text). The sulfuric solution should be allowed to cool before being mixed with the other solution, otherwise the oscillations will not be seen. During the reaction, the ferroin turns from blue to purple to red, making the oscillations visible. At the same time the Cerium ion changes from pale yellow to colorless, so that all kinds of hues are produced.

The Belousov-Zhabotinski reaction, which we just described, caused astonishment and some disbelief among chemists when it was discovered. Other periodic chemical reactions have now been discovered, in particular in systems of biological origin. One speculates on the physiological significance of these reactions, but little is really known with certainty.

what metaphorical, but its conclusions are suggestive, and a more detailed analysis may be useful.

To conclude this list of examples, let me mention a dynamical system of vital interest to everyone of us: the heart. The normal cardiac regime is periodic, but there are many nonperiodic pathologies (like ventricular fibrillation) which lead to the steady state of death. It seems that great medical benefit might be derived from computer studies of a realistic mathematical model which would reproduce the various cardiac dynamical regimes.

The application of the ideas which we have discussed often poses serious methodological problems. How does one maintain constant experimental conditions, and how does one make precise measurements? In any case, the recognition of the role of strange attractors in many problems is a great conceptual progress. The nonperiodic fluctuations of a dynamical system do not necessary indicate an experiment spoilt by mysterious random forces; they often point to a strange attractor, which one may try to understand [9].

I have not spoken of the esthetic appeal of strange attractors. These systems of curves, these clouds of points suggest sometimes fireworks or galaxies, sometimes strange and disquieting vegetal proliferations. A realm lies there of forms to explore, and harmonies to discover.

References

1. M. Hénon. A two-dimensional mapping with a strange attractor. Commun. Math. Phys. *50*, 69–77 (1976). See also S. D. Feit. Characteristic exponents and strange attractors. Commun. Math. Phys. *61*, 249–260 (1978). J. H. Curry. On the Hénon transformation. Commun. Math. Phys. *68*, 129–140 (1979).
2. S. Smale. Differentiable dynamical systems. Bull. Amer. Math. Soc. *73*, 747–817 (1967).
3. D. Ruelle, F. Takens. On the nature of turbulence. Commun. Math. Phys. *20*, 167–192 (1971); *23*, 343–344 (1971). See also S. Newhouse, D. Ruelle, F. Takens. Occurrence of strange Axiom A attractors near quasiperiodic flows on T^m, $m \geq 3$. Commun. Math. Phys. *64*, 35–40 (1978).
4. E. N. Lorenz. Deterministic nonperiodic flow. J. Atmos. Sci. *20*, 130–141 (1963).
5. L. D. Landau, E. M. Lifshitz. *Fluid mechanics.* Pergamon, Oxford, 1959.
6. H. L. Swinney, J. P. Gollub. The transition to turbulence. Physics Today *31*, No. 8, 41–49 (1978).
7. I. Prigogine. *Introduction to thermodynamics of irreversible processes.* Wiley, New York, 1962. [Does not cover chaotic time evolutions. I prefer this little book on the physics of dissipative systems to later and more ambitions works of the same author].
8. R. May. Simple mathematical models with very complicated dynamics. Nature *261*, 459–467 (1976).
9. A wealth of information is contained in the proceedings of two conferences organized by the New York Academy of Sciences. *Bifurcation theory and applications in scientific disciplines.* Ann. N. Y. Acad. Sci. 316 (1979). *Nonlinear dynamics.* Ann. N. Y. Acad. Sci. (To appear).

David Ruelle
Institut des Hautes Etudes Scientifiques
F-91440 Bures-sur-Yvette, France

Geometry from a Time Series

N. H. Packard, J. P. Crutchfield, J. D. Farmer, and R. S. Shaw

Dynamical Systems Collective, Physics Department, University of California, Santa Cruz, California 95064
(Received 13 November 1979)

It is shown how the existence of low-dimensional chaotic dynamical systems describing turbulent fluid flow might be determined experimentally. Techniques are outlined for reconstructing phase-space pictures from the observation of a single coordinate of any dissipative dynamical system, and for determining the dimensionality of the system's attractor. These techniques are applied to a well-known simple three-dimensional chaotic dynamical system.

PACS numbers: 47.25.-c

Lorenz originally demonstrated that very simple low-dimensional systems could display "chaotic" or "turbulent" behavior.[1] Attractors which display such behavior were termed "strange attractors" by Ruelle and Takens,[2] who then went on to conjecture that these strange attractors are the cause of turbulent behavior in fluid flow. The experiments of Gollub and Swinney have strengthened the conjecture,[3] but the question still remains: How can we discern the nature of the strange attractor underlying turbulence from observing the actual fluid flow?

Data obtained by experimentalists examining turbulent fluid flow often take the form of a "time series," which is to say, a series of values sampled at regular intervals. We address here the problem of using such a time series to reconstruct a finite-dimensional phase-space picture of the sampled system's time evolution. From this picture we can then obtain the asymptotic properties of the system, such as the positive Liapunov characteristic exponents, which are a measure of how chaotic the system is,[4-6] and topological characteristics such as the attractor's topological dimension. We illustrate these reconstruction methods by applying them to a time series obtained from sampling one coordinate of a three-dimensional chaotic dynamical system first studied by Rossler,[7] and then comparing the resulting values of the Liapunov exponents to those obtained by a different method.

The dynamical system of interest is a set of three ordinary differential equations:

$$\dot{x} = -(y+z),$$
$$\dot{y} = x + 0.2y, \qquad (1)$$
$$\dot{z} = 0.4 + xz - 5.7z.$$

These equations have a chaotic attractor which is illustrated in Fig. 1, which was obtained from an analog computer simulation.

The heuristic idea behind the reconstruction method is that to specify the state of a three-dimensional system at any given time, the measurement of *any* three independent quantities should be sufficient, where "independent" is not yet formally defined, but will become operationally defined. We conjecture that any such sets of three independent quantities which uniquely and smoothly label the states of the attractor are diffeomorphically equivalent. The three quantities typically used are the values of each state-space coordinate, $x(t)$, $y(t)$, and $z(t)$. We have found that beginning with a time series obtained by sampling a single coordinate of Eq. (1), one can obtain a variety of three independent quantities which appear to yield a faithful phase-space representation of the dynamics in the original x, y, z space. One possible set of three such quantities is the value of the coordinate with its values at two previous times,[8] e.g., $x(t)$, $x(t-\tau)$, and $x(t-2\tau)$. Another set obtained by making the time delays small, and taking differences is $x(t)$, $\dot{x}(t)$, and $\ddot{x}(t)$. Figure 2 shows a reconstruction of the $(x, \dot{x})$ picture from the time series taken from sampling the x coordinate of Eq. (1). Comparison of Figs. 1 and 2 certainly indicates that topological characteristics and geometrical form of the attractor remain intact when viewed in the $(x, \dot{x})$ coordinates. For an experimentalist observing some chaotic phenomenon, such as turbulent fluid flow, the construction of phase-space coordinates might not be as simple as the case illustrated above. In many cases the experimentalist has no *a priori* knowledge of how many dimensions a dynamical description would require, nor the quantities appropriate to the construction of such a description. So far there is no universally applicable method of phase-space construction, though the nature of the phenomenon might suggest possible alternatives. In a study of fluid turbulence, for example, the experimentalist might try using the velocity of the fluid in different directions, at different points in space, and at different times.

After having obtained a phase-space picture like that shown in Fig. 2, if the attractor is of sufficiently simple topology, one can use methods which have been previously developed[1,4] to con-

FIG. 1. (x,y) projection of Rossler (Ref. 7).

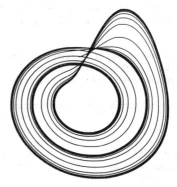

FIG. 2. $(x, \dot{x})$ reconstruction from the time series.

struct a one-dimensional return map, and then from the return map one can obtain the positive characteristic exponent of tne attractor. Roughly speaking, the procedure consists of making a cut along the attractor, coordinatizing it with the unit interval $(0,1)$, and accumulating a return map by observing where successive passes of the trajectory through the cut occur. The result is a return map of the form $x(n+1) = f(x(n))$, and the positive characteristic exponent is found by computing

$$\lambda = \lim_{N \to \infty} \frac{1}{N} \sum_{i=1}^{N} \ln \left| \frac{df}{dx} \right|_{x_i}$$

or alternately, by computing

$$\lambda = \int_0^1 P(x') \ln \left| \frac{df}{dx} \right|_{x'} dx'$$

if one knows (or has accumulated empirically) the equilibrium probability distribution $P(x)$. See Shaw[4] for more complete discussions of the computation of the characteristic exponents with use of this method.

Equations (1) are sufficiently simple that one can explicitly obtain a new set of three differential equations describing the dynamics of the state space comprised of a coordinate along with its first and second derivatives. Table I contains a comparison of the characteristic exponents for the original system [Eq. (1)], the transformed $(y, \dot{y}, \ddot{y})$ system, and the $(x, \dot{x})$ reconstructed return map, and shows very good agreement. The first two entries were obtained using the method of neighboring trajectories,[5,9] and the third entry was obtained using the return map method outlined above. The former method requires explicit knowledge of the dynamical equations, while the latter method depends on the dynamical system's attractor having sufficiently simple topology.

When trying to apply these reconstruction techniques to actual turbulence data, one of the first questions will be exactly what dimension the system's attractor is. Note that the topological dimension of the attractor is directly related to the number of nonnegative characteristic exponents (see Bennetin, Galgani, and Strelcyn,[5] Shimada and Nagashima,[9] and Crutchfield[10]). A spectrum of all negative characteristic exponents implies a pointlike zero-dimensional attractor; one zero characteristic exponent with all others negative implies a one dimensional attractor; one positive and one zero characteristic exponent corresponds to the observation of folded-sheet-like structures making up the attractor; two positive characteristic exponents correspond to volumelike structures; and so on. The case of two zero characteristic exponents corresponds to a two-torus (two dimensional), but this should be distinguishable from a sheetlike chaotic attractor by the observation of two sharp incommensurate frequencies in the power spectrum. The dimension referred to above is the topological dimension of the attractor; we must realize that the Cantor-set structure typical of these objects implies a nonintegral fractal dimension[11] which can be expressed in terms of the characteristic exponents.[12] However, at any finite degree of resolution the observed topological dimension will be some integer value, though nonintegral dimension might be obtained by varying the resolution of the observation to see scaling in the structure of the attractor.

We now outline a procedure for determining the dimension of a smooth dynamical system. We begin with the idea that the "dimension" of a system being observed corresponds to the number of independent quantities needed to specify the state of the system at any given instant. Thus the observed dimension of an attractor is the number of independent quantities needed to specify a point on the attractor. For an attractor in an n-dimension phase space, we can discover the number of independent quantities needed for such a specification by slicing the attractor with $(n-1)$-dimension hypersheets defined by one coordinate being constant. The topological dimension of the attractor corresponds to the minimum number of sheets which, when intersected with each other and the attractor, will yield a countable number of points.[11]

If one chooses as phase-space coordinates the value of some variable along with time-delayed values of the same variable, this slicing of phase space can be accomplished by constructing conditional probability distributions. We define the

TABLE I. Comparison of characteristic exponents from original (x,y,z) system, transformed $(y, \dot{y}, \ddot{y})$ system, and construction of return map from $(x, \dot{x}, \ddot{x})$ system.

	Characteristic exponent value
(x,y,z) system [Eq. (1)]	0.0677 ± 0.0005
$(y, \dot{y}, \ddot{y})$ system [Eq. (2)]	0.0680 ± 0.0005
$(x, \dot{x})$ return map reconstruction	0.0677 ± 0.0001

kth-order conditional probability distribution of a coordinate x, $P(x|x_1,x_2,\ldots;\tau)$, as the probability of observing the value x given that x_1 was observed time τ before, x_2 was observed time 2τ before, and so on. If we take τ to be small, the k conditions are equivalent to specification of the value of x at some time along with the value of all its derivatives up to order $k-1$. In fact, we must have $\tau \ll I/\Lambda$, where I is the degree of accuracy with which one can specify a state, and where Λ is the sum of all the positive-characteristic exponents, otherwise the information generating properties of the flow would randomize the samples with respect to each other.[4] In practice, it is easy to choose τ one or two orders of magnitude smaller than I/Λ. The dimensionality of the attractor is the number of conditions needed to yield an extremely sharp conditional probability distribution, in which case the system is determined by the conditions. These conditional probability distributions have been accumulated for the system given by Eq. (1), illustrated by the sequence in Fig. 3. We observe that the second-order conditional probability distribution is extremely sharp, implying that the attractor is two dimensional (sheets), which is indeed the observed structure. Other methods for determining the dimension of attractors will be reported elsewhere.[13]

The presence of observational noise in an experiment would be manifested in the increased width of the sharpest peaks obtainable in the sequence of probability distributions. For a noise level of δ, the width of the nth (sharp) probability distribution should be $\sim n\delta$. Thus for high-dimensional attractors, low-noise data is of paramount importance.

We have outlined techniques for reconstructing a phase-space picture from observing a single coordinate of any dynamical system. For systems which have only one positive-characteristic exponent along with sufficiently simple topology, we can obtain its value. We have also outlined a procedure for determining the dimensionality of an attractor from the observation of a single coordinate. All these techniques should be directly applicable to time series obtained from observing turbulence, as well as any other physical system, to construct a finite-dimensional phase-space picture of the system's attractor, provided such a low-dimensional structure exists. These ideas have recently been utilized by Roux et al.[14] to construct a phase-space picture of the chaotic attractor underlying chemical turbulence in the Belousof-Zhabotinsky reaction.

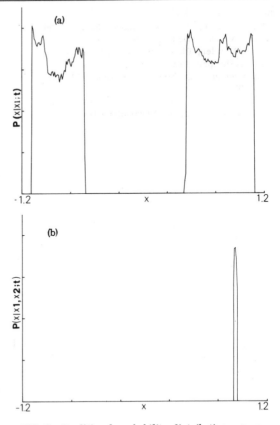

FIG. 3. Conditional-probability-distribution sequence for Eq. (1): (a) x vs $P(x|x_1;t)$; (b) x vs $P(x|x_1,x_2;t)$; where $x_1 = 0$, $x_2 = 0.495$, and $t = 0.2$ time units.

We have benefitted from many stimulating conversations with R. Abraham, W. Burke, J. Guckenheimer, and T. Jacobson. We also thank F. Bridges for the use of his microcomputer. This work was supported in part by the National Science Foundation under Grant No. 44315 0-21299 and in part by the John and Fanny Hertz Foundation.

---

[1]E. N. Lorenz, J. Atmos. Sci. **20**, 130 (1963).

[2]D. Ruelle and E. Takens, Commun. Math. Phys. **50**, 69 (1976).

[3]J. P. Gollub and H. L. Swinney, Phys. Rev. Lett. **35**, 927 (1975).

[4]R. S. Shaw, Ph.D. thesis, University of California at Santa Cruz, 1978 (to be published).

[5]G. Bennentin, L. Galgani, and J. M. Strelcyn, Phys.

Rev. A **14**, 2338 (1976).

[6]Ya. B. Piesin, Dokl. Akad. Nauk SSSR **226**, 196 (1976) [Sov. Math. Dokl. **17**, 196 (1976)].

[7]O. E. Rossler, Phys. Lett. **57A**, 397 (1976).

[8]D. Ruelle, private communication.

[9]Shimada and Nagashima, Prog. Theor. Phys. **61**, 1605 (1979).

[10]J. P. Crutchfield, Senior thesis, University of California at Santa Cruz, 1979 (unpublished).

[11]B. Mandelbrot, *Fractals: Form, Chance, and Dimension* (Freeman, San Francisco, 1977).

[12]H. Mori, Prog. Theor. Phys. **63**, 1044 (1980).

[13]H. Froehling, J. P. Crutchfield, J. D. Framer, N. H. Packard, R. S. Shaw, and L. Wennerberg, "On Determining the Dimension of Chaotic Flows" (to be published).

[14]J. C. Roux, A. Rossi, S. Bachelart, and C. Vidal, Phys. Lett. **77A**, 391 (1980).

DETERMINISTIC REPRESENTATION OF CHAOS IN CLASSICAL DYNAMICS

Michail ZAK

Applied Mechanics Technology Section, Jet Propulsion Laboratory,
California Institute of Technology, Pasadena, CA 91109, USA

Received 24 May 1984
Revised manuscript received 18 September 1984

Chaos in an Anosov-type mechanical system is eliminated by referring the governing equations to a specially selected rapidly oscillating (non-inertial) frame of reference in which the stabilization effect is caused by inertia forces. The result is generalized to any orbitally unstable mechanical system.

1. Introduction. Until recently, the random motions in classical dynamics were considered as a result of random input. However, in recent years simple physical and mathematical simulation experiments, substantiated by some theoretical results, have demonstrated the existence of another type of random motions which are generated by fully deterministic systems without any random input.

The main properties of these systems are the following: (a) they are necessarily nonlinear, and (b) they are supersensitive to changes of initial conditions.

Formal utilization of the corresponding mathematical models leads to solutions possessing random properties, and therefore, the behavior of the original deterministic systems becomes unpredictable.

In this note a new interpretation of this phenomenon is discussed. This interpretation is based upon the concept that some of the mathematical properties (including stability) of the momentum equations can be changed by referring motions to non-inertial frames of reference.

The main idea of such an approach is that the transport motion of the frame of reference adopts all the "non-smooth" components of the motion so that the relative motion remains "smooth". It is expected that the scale of the transport motion is much smaller than the scale characterizing the examined motion. The stabilization effect in this frame of reference is caused by rapidly oscillating inertia forces which change the potential energy of the system in the new frame of reference [1].

2. Criteria of chaos. Formally, chaotic behavior is caused by instability of trajectories (orbital instability). This special type of instability may be accompanied by a classical (Lyapunov) instability. Indeed, if the velocity of a particle is decomposed as $\mathbf{v} = v\boldsymbol{\tau}$ ($\boldsymbol{\tau}$ is the unit vector along the trajectory), then the classical and orbital instabilities are identified with instabilities of v and $\boldsymbol{\tau}$, respectively. In other words, the classical instability is characterized by large deflections of the total kinetic energy, while the orbital instability is associated only with redistributions of this energy between different coordinates. That is why the orbital instability may not lead to classical attractors and chaos can emerge.

The best illustration to this statement is shown by Anosov systems [2], where orbital instability leads to the stochasticity of the corresponding motions.

We will start with the inertial motion of a particle M on a smooth surface S with constant negative gaussian curvature (fig. 1a)

$$G = G_0 \times \text{const.} < 0. \qquad (1)$$

Remembering that the trajectories of inertial motions must be geodesics of S [3], we will compare two different trajectories assuming that initially they are parallel and that the distance between them, ϵ_0, is very small.

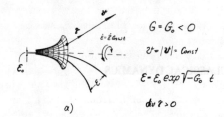

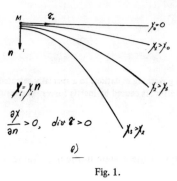

Fig. 1.

As shown in differential geometry [3], the distance between such geodesics will exponentially increase:

$$\epsilon = \epsilon_0 \exp[(-G_0)^{1/2} t] . \quad (2)$$

Hence, no matter how small is the initial distance ϵ_0, the current distance ϵ tends to infinity. That is why the motion, once recorded, cannot be reproduced again, and consequently, it attains stochastic features.

For non-inertial motions the trajectories of the particle will not be geodesic, while the rate of their deviation from geodesics is characterized by the geodesic curvature χ (fig. 1b):

$$\chi = F_n/v^2 , \quad (3)$$

in which v is the velocity of the particle, F_n is the normal (to the trajectory) component of the force field F.

Following ref. [3], it can be shown that in the case of a potential force field the condition of orbital instability (1) and the equation of a disturbed motion (2) can be replaced by

$$G_0 + (1/mv^2) \partial^2\Pi/\partial\epsilon^2 < 0, \quad F = \mathrm{grad}\,\Pi , \quad (4)$$

$$\epsilon = \epsilon_0 \exp\{[-G_0 + (1/mv^2) \partial^2\Pi/\partial\epsilon^2]^{1/2} t\} . \quad (5)$$

Thus, the orbital instability is eliminated if

$$\partial^2\Pi/\partial\epsilon^2 \geqslant -mv^2 G_0 . \quad (6)$$

126

Such a force potential Π can be induced, for instance, by a rapidly oscillating force field [1]:

$$F_n = F \sin \omega t, \quad \omega \to \infty , \quad (7)$$

where F_n is normal to geodesics.
Then

$$\Pi = \tfrac{1}{2} m\dot{\tilde{\epsilon}}^2 , \quad (8)$$

in which

$$\dot{\epsilon} = \dot{\tilde{\epsilon}} \cos \omega t \quad (9)$$

is the velocity fluctuations normal to the geodesics.

3. New frame of reference. Returning to the inertial motion (2) let us refer it to a frame of reference which rotates about the axis of symmetry of the pseudosphere S (fig. 1) with the oscillating transport velocity (9). Then the momentum equation of the particle attains an additional inertia force (7) which induces the potential field (8).

Selecting the transport velocity (9) such that

$$\dot{\tilde{\epsilon}} = -mv^2 G_0 \epsilon \quad (G_0 < 0) \quad (10)$$

one provides the condition of the orbital stability of the relative motion in the new frame of reference:

$$\partial^2\Pi/\partial\epsilon^2 = -mv^2 G_0 . \quad (11)$$

The equation of the relative motion now is written in the following form (see eq. (2)):

$$\epsilon = \epsilon_0 = \mathrm{const} . \quad (12)$$

Thus, in the new frame of reference an initial error does not grow — it remains constant.

Returning to the original (inertial) system one obtains the resultant velocity by summing the relative and transport velocities:

$$v_\tau = v_0 , \quad (13)$$

$$v_\epsilon = -mv_0^2 G_0 \epsilon_0 \cos \omega t \quad (\omega \to \infty) , \quad (14)$$

in which v_τ and v_ϵ are the velocity components parallel and normal to the undisturbed (geodesic) trajectory, respectively.

The equations of the disturbed motion in the original frame of reference are

$$\sigma = v_0 t , \quad (15)$$

$$\epsilon = \epsilon_0 + (1/\omega) mv_0^2 G_0 \epsilon_0 \sin \omega t \quad (\omega \to \infty) , \quad (16)$$

in which σ is the arc coordinate along the undisturbed (geodesic) trajectory.

As follows from eqs. (13)–(16) the motion in the original frame of reference is stable in the sense that the current deviations of displacements and velocities do not exceed their initial values. However, the displacement–time function (16) is not differentiable because its derivative (14) is multivalued. Indeed, for any arbitrarily small interval Δt there always exists such a large frequency $\omega > \Delta t/2\pi$ that within this interval the velocity (14) runs through all its values. In other words, one arrived at stability in the class of non-differentiable functions. (The mathematical meaning of this result will be discussed below.)

Thus, chaotic motion of a particle on a smooth pseudosphere is represented by the "mean" motion (15) along the undisturbed geodesic trajectory (with the constant velocity (13)) and the fluctuation motion (16) normal to this trajectory. The "amplitude" of these fluctuations is vanishingly small, but the velocity "amplitude" is finite. It is worth emphasizing that this amplitude is proportional to the gaussian curvature of the surface S, i.e., to the degree of the orbital instability. Therefore, it can be associated with the measure of the uncertainty in the description of the motion.

It is worth mentioning that both "mean" and "fluctuation" components representing the originally chaotic motion are stable. That is why they are not sensitive to initial uncertainties and are fully reproducible. In other words, such a representation of the originally chaotic motion is deterministic.

One should notice that the condition $\omega \to \infty$ is a mathematical idealization. Practically, ω is finite:

$$\omega \gg 1/T, \qquad (17)$$

where T is a time scale over which changes of the parameters of the motion are negligible. In the same sense the concepts of differentiability and multivaluedness have to be understood. Indeed, the multivaluedness of the functions (14) and (16) means that the time interval between two different values of these functions is smaller than the scale of observation T of the examined motion, and therefore, these values can be associated with "almost" the same argument.

4. Some generalizations. The result formulated above can be generalized to any orbitally unstable mechanical system.

First, the same stabilization effect (4) occurs even if the undisturbed motion of the particle M on the surface S is non-inertial and the gaussian curvature of this surface is not constant (but negative).

Second, any finite-degree-of-freedom mechanical system can be represented by a unit-mass particle in a specially selected configuration space [3]: if a mechanical system has N generalized coordinates q^i and is characterized by the kinetic energy

$$W = \tfrac{1}{2} a_{ij} q^i q^j, \qquad (18)$$

then the configuration space is introduced as an N-dimensional space with the following metric tensor:

$$g_{ij} = a_{ij}, \qquad (19)$$

while the curvature of this space (which may be non-euclidean) is expressed by the Gauss formula [3]. Therefore, all the results obtained for the motion of a particle in actual space can be formally applied to the motion of a mechanical system using the concept of configuration space.

5. Discussion. Thus, chaos in solutions to the governing equations of orbitally unstable mechanical systems can be eliminated by appropriate selection of the frame of reference.

What is the mathematical meaning of this procedure? First, it is necessary to distinguish the chaos as a characteristic of solutions to certain mathematical equations from the chaos as a characteristic of motions of some mechanical systems. The first type of chaos cannot be eliminated since it is an inherent property of the mathematical equation. However, chaos as a characteristic of mechanical motions can be eliminated by changing the way of their description, i.e., by referring these equations to such a frame of reference which provides the best "view" of the motion.

Second, it is a well established fact that the concept of stability is related to a certain class of functions, or a type of space: the same solution can be stable in one space and unstable in another, depending on the definition of the "distance" between two solutions. Indeed, if the distance between the solutions in (16) is defined as

$$\rho = \sum_{k=0}^{n} \max |\epsilon_2^{(k)}(t) - \epsilon_1^{(k)}(t)|, \qquad (20)$$

then the solution (16) is stable for $n = 0, 1$, but it is

unstable for $n = 2, 3, ...$, since its derivatives $\epsilon^{(2)}$, $\epsilon^{(3)}$, ..., etc., are unbounded. In other words, the concept of stability as well as chaos is an attribute of a mathematical model rather than of a physical phenomenon.

Hence, from a formal mathematical point of view, the occurrence of chaos in description of mechanical motions means only that these motions cannot be properly described by smooth functions if the scale of observation is finite. A similar strategy can be applied for the deterministic representation of chaos in continua [4–6].

The work described in this paper was carried out by the Jet Propulsion Laboratory, California Institute of Technology, under contract with the National Aeronautics and Space Administration. The author takes this occasion to express his sincere gratitude to Professor M. Mayer (UCI) and Dr. M. Weinberg (JPL) for a number of valuable remarks.

References

[1] L.D. Landau, and E.M. Lifshitz, Course of theoretical physics, Vol. 1 (Pergamon, London, 1965) pp. 93–95.
[2] V.I. Arnold, Mathematical methods in classical mechanics, (Springer, Berlin, 1978) pp. 314–316.
[3] J.L. Synge, Philos. Trans. R. Soc. A226 (1926) 31.
[4] M. Zak, Acta Mech. 43 (1982) 97.
[5] M. Zak, Solid Mech. Arch. 7 (1982) 467; 8 (1983) 1.
[6] M. Zak, Acta Mech. 72 (1984) 119.

DETERMINING LYAPUNOV EXPONENTS FROM A TIME SERIES

Alan WOLF†, Jack B. SWIFT, Harry L. SWINNEY and John A. VASTANO

Department of Physics, University of Texas, Austin, Texas 78712, USA

Received 18 October 1984

We present the first algorithms that allow the estimation of non-negative Lyapunov exponents from an *experimental* time series. Lyapunov exponents, which provide a qualitative and quantitative characterization of dynamical behavior, are related to the exponentially fast divergence or convergence of nearby orbits in phase space. A system with one or more positive Lyapunov exponents is defined to be chaotic. Our method is rooted conceptually in a previously developed technique that could only be applied to analytically defined model systems: we monitor the *long-term* growth rate of *small* volume elements in an attractor. The method is tested on model systems with known Lyapunov spectra, and applied to data for the Belousov–Zhabotinskii reaction and Couette–Taylor flow.

Contents
1. Introduction
2. The Lyapunov spectrum defined
3. Calculation of Lyapunov spectra from differential equations
4. An approach to spectral estimation for experimental data
5. Spectral algorithm implementation*
6. Implementation details*
7. Data requirements and noise*
8. Results
9. Conclusions

*Appendices**
A. Lyapunov spectrum program for systems of differential equations
B. Fixed evolution time program for λ_1

1. Introduction

Convincing evidence for deterministic chaos has come from a variety of recent experiments [1–6] on dissipative nonlinear systems; therefore, the question of detecting and quantifying chaos has become an important one. Here we consider the spectrum of Lyapunov exponents [7–10], which has proven to be the most useful dynamical diagnostic for chaotic systems. Lyapunov exponents are the average exponential rates of divergence or convergence of nearby orbits in phase space. Since nearby orbits correspond to nearly identical states, exponential orbital divergence means that systems whose initial differences we may not be able to resolve will soon behave quite differently – predictive ability is rapidly lost. Any system containing at least one positive Lyapunov exponent is defined to be chaotic, with the magnitude of the exponent reflecting the time scale on which system dynamics become unpredictable [10].

For systems whose equations of motion are explicitly known there is a straightforward technique [8, 9] for computing a complete Lyapunov spectrum. This method cannot be applied directly to experimental data for reasons that will be discussed later. We will describe a technique which for the first time yields estimates of the non-negative Lyapunov exponents from finite amounts of experimental data.

A less general procedure [6, 11–14] for estimating only the dominant Lyapunov exponent in experimental systems has been used for some time. This technique is limited to systems where a well-defined one-dimensional (1-D) map can be recovered. The technique is numerically unstable and the literature contains several examples of its improper application to experimental data. A discussion of the 1-D map calculation may be found

†Present address: The Cooper Union, School of Engineering, N.Y., NY 10003, USA.
*The reader may wish to skip the starred sections at a first reading.

in ref. 13. In ref. 2 we presented an unusually robust 1-D map exponent calculation for experimental data obtained from a chemical reaction.

Experimental data inevitably contain external noise due to environmental fluctuations and limited experimental resolution. In the limit of an infinite amount of noise-free data our approach would yield Lyapunov exponents by definition. Our ability to obtain good spectral estimates from experimental data depends on the quantity and quality of the data as well as on the complexity of the dynamical system. We have tested our method on model dynamical systems with known spectra and applied it to experimental data for chemical [2, 13] and hydrodynamic [3] strange attractors.

Although the work of characterizing chaotic data is still in its infancy, there have been many approaches to quantifying chaos, e.g., fractal power spectra [15], entropy [16–18, 3], and fractal dimension [proposed in ref. 19, used in ref. 3–5, 20, 21]. We have tested many of these algorithms on both model and experimental data, and despite the claims of their proponents we have found that these approaches often fail to characterize chaotic data. In particular, parameter independence, the amount of data required, and the stability of results with respect to external noise have rarely been examined thoroughly.

The spectrum of Lyapunov exponents will be defined and discussed in section 2. This section includes table I which summarizes the model systems that are used in this paper. Section 3 is a review of the calculation of the complete spectrum of exponents for systems in which the defining differential equations are known. Appendix A contains Fortran code for this calculation, which to our knowledge has not been published elsewhere. In section 4, an outline of our approach to estimating the non-negative portion of the Lyapunov exponent spectrum is presented. In section 5 we describe the algorithms for estimating the two largest exponents. A Fortran program for determining the largest is contained in appendix B. Our algorithm requires input parameters whose selection is discussed in section 6. Section 7 concerns sources of error in the calculations and the quality and quantity of data required for accurate exponent estimation. Our method is applied to model systems and experimental data in section 8, and the conclusions are given in section 9.

2. The Lyapunov spectrum defined

We now define [8, 9] the spectrum of Lyapunov exponents in the manner most relevant to spectral calculations. Given a continuous dynamical system in an n-dimensional phase space, we monitor the long-term evolution of an *infinitesimal* n-sphere of initial conditions; the sphere will become an n-ellipsoid due to the locally deforming nature of the flow. The i th one-dimensional Lyapunov exponent is then defined in terms of the length of the ellipsoidal principal axis $p_i(t)$:

$$\lambda_i = \lim_{t \to \infty} \frac{1}{t} \log_2 \frac{p_i(t)}{p_i(0)}, \qquad (1)$$

where the λ_i are ordered from largest to smallest†. Thus the Lyapunov exponents are related to the expanding or contracting nature of different directions in phase space. Since the orientation of the ellipsoid changes continuously as it evolves, the directions associated with a given exponent vary in a complicated way through the attractor. One cannot, therefore, speak of a well-defined direction associated with a given exponent.

Notice that the linear extent of the ellipsoid grows as $2^{\lambda_1 t}$, the area defined by the first two principal axes grows as $2^{(\lambda_1 + \lambda_2)t}$, the volume defined by the first three principal axes grows as $2^{(\lambda_1 + \lambda_2 + \lambda_3)t}$, and so on. This property yields another definition of the spectrum of exponents:

---

†While the existence of this limit has been questioned [8, 9, 22], the fact is that the orbital divergence of *any* data set may be quantified. Even if the limit does not exist for the underlying system, or cannot be approached due to having finite amounts of noisy data, Lyapunov exponent estimates could still provide a useful characterization of a given data set. (See section 7.1.)

the sum of the first j exponents is defined by the long term exponential growth rate of a j-volume element. This alternate definition will provide the basis of our spectral technique for experimental data.

Any continuous time-dependent dynamical system without a fixed point will have at least one zero exponent [22], corresponding to the slowly changing magnitude of a principal axis tangent to the flow. Axes that are on the average expanding (contracting) correspond to positive (negative) exponents. The sum of the Lyapunov exponents is the time-averaged divergence of the phase space velocity; hence any dissipative dynamical system will have at least one negative exponent, the sum of all of the exponents is negative, and the post-transient motion of trajectories will occur on a zero volume limit set, an attractor.

The exponential expansion indicated by a positive Lyapunov exponent is incompatible with motion on a bounded attractor unless some sort of *folding* process merges widely separated trajectories. Each positive exponent reflects a "direction" in which the system experiences the repeated stretching and folding that decorrelates nearby states on the attractor. Therefore, the long-term behavior of an initial condition that is specified with *any* uncertainty cannot be predicted; this is chaos. An attractor for a dissipative system with one or more positive Lyapunov exponents is said to be "strange" or "chaotic".

The signs of the Lyapunov exponents provide a qualitative picture of a system's dynamics. One-dimensional maps are characterized by a single Lyapunov exponent which is positive for chaos, zero for a marginally stable orbit, and negative for a periodic orbit. In a three-dimensional continuous dissipative dynamical system the only possible spectra, and the attractors they describe, are as follows: $(+, 0, -)$, a strange attractor; $(0, 0, -)$, a two-torus; $(0, -, -)$, a limit cycle; and $(-, -, -)$, a fixed point. Fig. 1 illustrates the expanding, "slower than exponential," and contracting character of the flow for a three-dimensional system, the Lorenz model [23]. (All of the model systems that we will discuss are defined in table I.) Since Lyapunov exponents involve long-time averaged behavior, the short segments of the trajectories shown in the figure cannot be expected to accurately characterize the positive, zero, and negative exponents; nevertheless, the three distinct types of behavior are clear. In a continuous four-dimensional dissipative system there are three possible types of strange attractors: their Lyapunov spectra are $(+, +, 0, -)$, $(+, 0, 0, -)$, and $(+, 0, -, -)$. An example of the first type is Rossler's hyperchaos attractor [24] (see table I). For a given system a change in parameters will generally change the Lyapunov spectrum and may also change both the type of spectrum and type of attractor.

The magnitudes of the Lyapunov exponents *quantify* an attractor's dynamics in information theoretic terms. The exponents measure the rate at which system processes create or destroy information [10]; thus the exponents are expressed in bits of information/s or bits/orbit for a continuous system and bits/iteration for a discrete system. For example, in the Lorenz attractor the positive exponent has a magnitude of 2.16 bits/s (for the parameter values shown in table I). Hence if an initial point were specified with an accuracy of one part per million (20 bits), the future behavior could not be predicted after about 9 s [20 bits/(2.16 bits/s)], corresponding to about 20 orbits. After this time the small initial uncertainty will essentially cover the entire attractor, reflecting 20 bits of new information that can be gained from an additional measurement of the system. This new information arises from scales smaller than our initial uncertainty and results in an inability to specify the state of the system except to say that it is somewhere on the attractor. This process is sometimes called an information gain – reflecting new information from the heat bath, and sometimes is called an information loss – bits shifted out of a phase space variable "register" when bits from the heat bath are shifted in.

The average rate at which information contained in transients is lost can be determined from

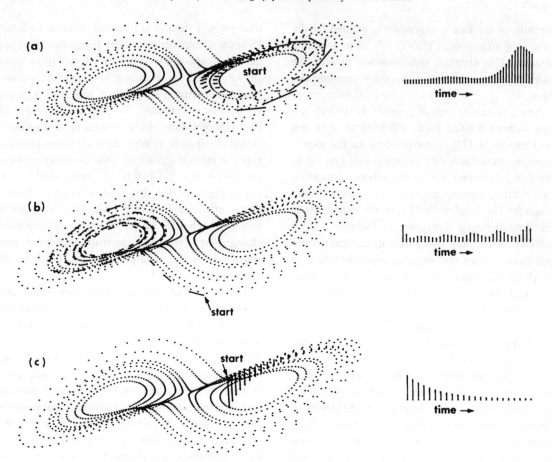

Fig. 1. The short term evolution of the separation vector between three carefully chosen pairs of nearby points is shown for the Lorenz attractor. a) An expanding direction ($\lambda_1 > 0$); b) a "slower than exponential" direction ($\lambda_2 = 0$); c) a contracting direction ($\lambda_3 < 0$).

the negative exponents. The asymptotic decay of a perturbation to the attractor is governed by the least negative exponent, which should therefore be the easiest of the negative exponents to estimate†.

For the Lorenz attractor the negative exponent is so large that a perturbed orbit typically becomes indistinguishable from the attractor, by "eye", in less than one mean orbital period (see fig. 1).

---

†We have been quite successful with an algorithm for determining the dominant (smallest magnitude) negative exponent from pseudo-experimental data (a single time series extracted from the solution of a model system and treated as an experimental observable) for systems that are nearly integer-dimensional. Unfortunately, our approach, which involves measuring the mean decay rate of many induced perturbations of the dynamical system, is unlikely to work on many experimental systems. There are several fundamental problems with the calculation of negative exponents from experimental data, but of greatest importance is that *post-transient* data may not contain resolvable negative exponent information and *perturbed* data must reflect properties of the unperturbed system, that is, perturbations must only change the state of the system (current values of the dynamical variables). The response of a physical system to a non-delta function perturbation is difficult to interpret, as an orbit separating from the attractor may reflect either a locally repelling region of the attractor (a positive contribution to the negative exponent) or the finite rise time of the perturbation.

Table I
The model systems considered in this paper and their Lyapunov spectra and dimensions as computed from the equations of motion

System	Parameter values	Lyapunov spectrum (bits/s)[†]	Lyapunov dimension[‡]
Hénon: [25] $$X_{n+1} = 1 - aX_n^2 + Y_n$$ $$Y_{n+1} = bX_n$$	$a = 1.4$ $b = 0.3$	$\lambda_1 = 0.603$ $\lambda_2 = -2.34$ (bits/iter.)	1.26
Rossler–chaos: [26] $$\dot{X} = -(Y + Z)$$ $$\dot{Y} = X + aY$$ $$\dot{Z} = b + Z(X - c)$$	$a = 0.15$ $b = 0.20$ $c = 10.0$	$\lambda_1 = 0.13$ $\lambda_2 = 0.00$ $\lambda_3 = -14.1$	2.01
Lorenz: [23] $$\dot{X} = \sigma(Y - X)$$ $$\dot{Y} = X(R - Z) - Y$$ $$\dot{Z} = XY - bZ$$	$\sigma = 16.0$ $R = 45.92$ $b = 4.0$	$\lambda_1 = 2.16$ $\lambda_2 = 0.00$ $\lambda_3 = -32.4$	2.07
Rossler–hyperchaos: [24] $$\dot{X} = -(Y + Z)$$ $$\dot{Y} = X + aY + W$$ $$\dot{Z} = b + XZ$$ $$\dot{W} = cW - dZ$$	$a = 0.25$ $b = 3.0$ $c = 0.05$ $d = 0.5$	$\lambda_1 = 0.16$ $\lambda_2 = 0.03$ $\lambda_3 = 0.00$ $\lambda_4 = -39.0$	3.005
Mackey–Glass: [27] $$\dot{X} = \frac{aX(t+s)}{1+[X(t+s)]^c} - bX(t)$$	$a = 0.2$ $b = 0.1$ $c = 10.0$ $s = 31.8$	$\lambda_1 = 6.30\text{E}{-}3$ $\lambda_2 = 2.62\text{E}{-}3$ $|\lambda_3| < 8.0\text{E}{-}6$ $\lambda_4 = -1.39\text{E}{-}2$	3.64

[†] A mean orbital period is well defined for Rossler chaos (6.07 seconds) and for hyperchaos (5.16 seconds) for the parameter values used here. For the Lorenz attractor a characteristic time (see footnote – section 3) is about 0.5 seconds. Spectra were computed for each system with the code in appendix A.
[‡] As defined in eq. (2).

The Lyapunov spectrum is closely related to the fractional dimension of the associated strange attractor. There are a number [19] of different fractional-dimension-like quantities, including the fractal dimension, information dimension, and the correlation exponent; the difference between them is often small. It has been conjectured by Kaplan and Yorke [28, 29] that the information dimension d_f is related to the Lyapunov spectrum by the equation

$$d_f = j + \frac{\sum_{i=1}^{j} \lambda_i}{|\lambda_{j+1}|}, \quad (2)$$

where j is defined by the condition that

$$\sum_{i=1}^{j} \lambda_i > 0 \quad \text{and} \quad \sum_{i=1}^{j+1} \lambda_i < 0. \quad (3)$$

The conjectured relation between d_f (a *static* property of an attracting set) and the Lyapunov exponents appears to be satisfied for some model systems [30]. The calculation of dimension from this equation requires knowledge of all but the most negative Lyapunov exponents.

3. Calculation of Lyapunov spectra from differential equations

Our algorithms for computing a non-negative Lyapunov spectrum from experimental data are inspired by the technique developed independently by Bennetin et al. [8] and by Shimada and Nagashima [9] for determining a complete spectrum from a set of differential equations. Therefore, we describe their calculation (for brevity, the ODE approach) in some detail.

We recall that Lyapunov exponents are defined by the long-term evolution of the axes of an infinitesimal sphere of states. This procedure could be implemented by defining the principal axes with initial conditions whose separations are as small as computer limitations allow and evolving these with the nonlinear equations of motion. One problem with this approach is that in a chaotic system we cannot guarantee the condition of small separations for times on the order of hundreds of orbital periods†, needed for convergence of the spectrum.

This problem may be avoided with the use of a phase space plus tangent space approach. A "fiducial" trajectory (the center of the sphere) is defined by the action of the nonlinear equations of motion on some initial condition. Trajectories of points on the surface of the sphere are defined by the action of the linearized equations of motion on points infinitesimally separated from the fiducial trajectory. In particular, the principal axes are defined by the evolution via the linearized equations of an initially orthonormal vector frame anchored to the fiducial trajectory. By definition, *principal axes defined by the linear system are always infinitesimal relative to the attractor*. Even in the linear system, principal axis vectors diverge in magnitude, but this is a problem only because computers have a limited dynamic range for storing numbers. This divergence is easily circumvented. What has been avoided is the serious problem of principal axes finding the global "fold" when we really only want them to probe the local "stretch."

To implement this procedure the fiducial trajectory is created by integrating the nonlinear equations of motion for some post-transient initial condition. Simultaneously, the linearized equations of motion are integrated for n different initial conditions defining an arbitrarily oriented frame of n orthonormal vectors. We have already pointed out that each vector will diverge in magnitude, but there is an additional singularity – in a chaotic system, each vector tends to fall along the local direction of most rapid growth. Due to the finite precision of computer calculations, the collapse toward a common direction causes the tangent space orientation of all axis vectors to become indistinguishable. These two problems can be overcome by the repeated use of the Gram-Schmidt reorthonormalization (GSR) procedure on the vector frame:

Let the linearized equations of motion act on the initial frame of orthonormal vectors to give a set of vectors $\{v_1, \ldots, v_n\}$. (The desire of each vector to align itself along the λ_1 direction, and the orientation-preserving properties of GSR mean that the initial labeling of the vectors may be done arbitrarily.) Then GSR provides the following orthonormal set $\{v'_1, \ldots, v'_n\}$:

$$v'_1 = \frac{v_1}{\|v_1\|},$$

$$v'_2 = \frac{v_2 - \langle v_2, v'_1 \rangle v'_1}{\|v_2 - \langle v_2, v'_1 \rangle v'_1\|},$$

$$\vdots$$

$$v'_n = \frac{v_n - \langle v_n, v'_{n-1} \rangle v'_{n-1} - \cdots - \langle v_n, v'_1 \rangle v'_1}{\|v_n - \langle v_n, v'_{n-1} \rangle v'_{n-1} - \cdots - \langle v_n, v'_1 \rangle v'_1\|},$$

(4)

---

†Should the mean orbital period not be well-defined, a characteristic time can be either the mean time between intersections of a Poincaré section or the time corresponding to a dominant power spectral feature.

where $\langle\,,\,\rangle$ signifies the inner product. The frequency of reorthonormalization is not critical, so long as neither the magnitude nor the orientation divergences have exceeded computer limitations. As a rule of thumb, GSR is performed on the order of once per orbital period.

We see that GSR never affects the direction of the first vector in a system, so this vector tends to seek out the direction in tangent space which is most rapidly growing (components along other directions are either growing less rapidly or are shrinking). The second vector has its component along the direction of the first vector removed, and is then normalized. Because we are changing its direction, vector v_2 is not free to seek out the most rapidly growing direction. Because of the manner in which we are changing it, it also is not free to seek out the second most rapidly growing direction†. Note however that the vectors v'_1 and v'_2 span the same two-dimensional subspace as the vectors v_1 and v_2. *In spite of repeated vector replacements, the space these vectors define continually seeks out the two-dimensional subspace that is most rapidly growing.* The area defined by these vectors is proportional to $2^{(\lambda_1+\lambda_2)t}$ [8]. The length of vector v_1 is proportional to $2^{\lambda_1 t}$ so that monitoring length and area growth allows us to determine both exponents. In practice, as v'_1 and v'_2 are orthogonal, we may determine λ_2 directly from the mean rate of growth of the projection of vector v_2 on vector v'_2. In general, the subspace spanned by the first k vectors is unaffected by GSR so that the long-term evolution of the k-volume defined by these vectors is proportional to 2^μ where $\mu = \sum_{i=1}^{k}\lambda_i t$. Projection of the evolved vectors onto the new orthonormal frame correctly updates the rates of growth of each of the first k-principal axes in turn, providing estimates of the k largest Lyapunov exponents. Thus GSR allows the integration of the vector frame for as long as is required for spectral convergence.

Fortran code for the ODE procedure appears in appendix A. We illustrate the use of this procedure for the Rossler attractor [26]. The spectral calculation requires the integration of the 3 equations of motion and 9 linearized equations for on the order of 100 orbits of model time (a few cpu minutes on a VAX 11/780) to obtain each exponent to within a few percent of its asymptotic value. In practice we consider the asymptotic value to be attained when the mandatory zero exponent(s) are a few orders of magnitude smaller than the smallest positive exponent. The convergence rate of zero and positive exponents is about the same, and is much slower than the convergence rate of negative exponents. Negative exponents arise from the nearly uniform attractiveness of the attractor which can often be well estimated from a few passes around an attractor, non-negative exponents arise from a once-per-orbit stretch and fold process that must be sampled on the order of hundreds of times (or more) for reasonable convergence.

The method we have described for finding Lyapunov exponents is perhaps more easily understood for a discrete dynamical system. Here we consider the Hénon map [25] (see table I). The linearization of this map is

$$\begin{pmatrix}\delta X_{n+1}\\ \delta Y_{n+1}\end{pmatrix} = J_n \begin{pmatrix}\delta X_n\\ \delta Y_n\end{pmatrix}, \quad (5)$$

where

$$J_n = \begin{bmatrix} -2.8X_n & 1 \\ 0.3 & 0 \end{bmatrix} \quad (6)$$

and X_n is the $(n-1)$st iterate of an arbitrary initial condition X_1.

An orthonormal frame of principal axis vectors such as $((0,1),(1,0))$ is evolved by applying the product Jacobian to each vector. For either vector

---

†This is clear when we consider that we may obtain different directions of vector v_2 at some specified time if we exercise our freedom to choose the intermediate times at which GSR is performed. That is, beginning with a specified v_1 and v_2 at time t_i, we may perform replacements at times t_{i+1} and t_{i+2}, obtaining the vectors v'_1, v'_2 and then v''_1, v''_2 or we may propagate directly to time t_{i+2}, obtaining v_1^*, v_2^*. v''_2 and v_2^* are *not* parallel; therefore, the details of propagation and replacement determine the orientation of v_2.

the operation may be written in two different ways. For example, for the vector $(0,1)$ we have

$$\begin{pmatrix} \delta X_n \\ \delta Y_n \end{pmatrix} = J_{n-1} \left(J_{n-2} \cdots J_1 \begin{pmatrix} 0 \\ 1 \end{pmatrix} \right), \qquad (7)$$

or, by regrouping the terms,

$$\begin{pmatrix} \delta X_n \\ \delta Y_n \end{pmatrix} = [J_{n-1} J_{n-2} \cdots J_1] \begin{pmatrix} 0 \\ 1 \end{pmatrix}. \qquad (8)$$

In eq. (7) the latest Jacobi matrix multiplies each *current* axis vector, which is the initial vector multiplied by all previous Jacobi matrices. The magnitude of each current axis vector diverges, and the angular separation between the two vectors goes to zero. Fig. 2 shows that divergent behavior is visible within a few iterations. GSR corresponds to the replacement of each current axis vector. Lyapunov exponents are computed

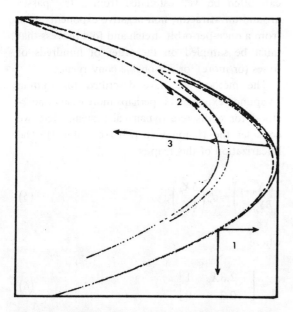

Fig. 2. The action of the product Jacobian on an initially orthonormal vector frame is illustrated for the Hénon map: (1) initial frame; (2) first iterate; and (3) second iterate. By the second iteration the divergence in vector magnitude and the angular collapse of the frame are quite apparent. Initial conditions were chosen so that the angular collapse of the vectors was uncommonly *slow*.

from the growth rate of the length of the first vector and the growth rate of the area defined by both vectors.

In eq. (8) the *product* Jacobian acts on each of the *initial* axis vectors. The columns of the product matrix converge to large multiples of the eigenvector of the biggest eigenvalue, so that elements of the matrix diverge and the matrix becomes singular. Here GSR corresponds to factoring out a large scalar multiplier of the matrix to prevent the magnitude divergence, and doing row reduction with pivoting to retain the linear independence of the columns. Lyapunov exponents are computed from the eigenvalues of the long-time product matrix†.

We emphasize that Lyapunov exponents are not local quantities in either the spatial or temporal sense. Each exponent arises from the average, with respect to the dynamical motion, of the local deformation of various phase space directions. Each is determined by the *long-time* evolution of a *single* volume element. Attempts to estimate exponents by averaging local contraction and expansion rates of phase space are likely to fail at the point where these contributions to the exponents are combined. In fig. 3a we show vector v_1' at each renormalization step for the Lorenz attractor over the course of several hundred orbits [32]. The apparent multivaluedness of the most rapidly growing direction (in some regions of the attractor) shows that this direction is not simply a function of position on the attractor. While this direction is often nearly parallel to the flow on the Lorenz attractor (see fig. 3b) it is usually nearly transverse to the flow for the Rossler attractor. We conclude that exponent calculation by averaging *local* divergence estimates is a dangerous procedure.

†We are aware of an attempt to estimate Lyapunov spectra from experimental data through direct estimation of local Jacobian matrices and formation of the long time product matrix [31]. This calculation is essentially the same as ours (we avoid matrix notation by diagonalizing the system at each step) and has the same problems of sensitivity to external noise, and to the amount and resolution of data required for accurate estimates.

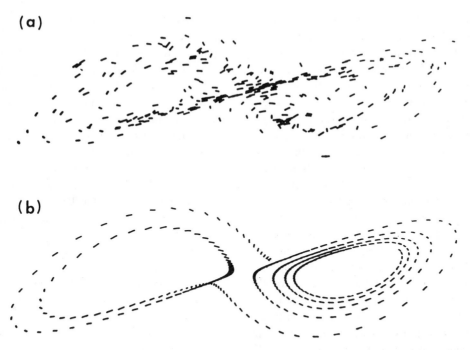

Fig. 3. A modification to the ODE spectral code (see appendix A) allows us to plot the running direction of greatest growth (vector v'_1) in the Lorenz attractor. In (a), infrequent renormalizations confirm that this direction is not single-valued on the attractor. In (b), frequent renormalizations show us that this direction is usually nearly parallel to the flow. In the Rossler attractor, this direction is usually nearly orthogonal to the flow.

4. An approach to spectral estimation for experimental data

Experimental data typically consist of discrete measurements of a single observable. The well-known technique of phase space reconstruction with delay coordinates [2, 33, 34] makes it possible to obtain from such a time series an attractor whose Lyapunov spectrum is identical to that of the original attractor. We have designed a method, conceptually similar to the ODE approach, which can be used to estimate non-negative Lyapunov exponents from a reconstructed attractor. To understand our method it is useful to summarize what we have discussed thus far about exponent calculation.

Lyapunov exponents may be defined by the *phase space* evolution of a sphere of states. Attempts to apply this definition numerically to equations of motion fail since computer limitations do not allow the initial sphere to be constructed sufficiently small. In the ODE approach one avoids this problem by working in the *tangent space* of a fiducial trajectory so as to obtain always infinitesimal principal axis vectors. The remaining divergences are easily eliminated with Gram–Schmidt reorthonormalization.

The ODE approach is not directly applicable to experimental data as the linear system is not available. All is not lost provided that the linear approximation holds on the smallest length scales defined by our data. Our approach involves working in a reconstructed attractor, examining orbital divergence on length scales that are always as small as possible, using an approximate GSR procedure in the reconstructed *phase space* as necessary. To simplify the ensuing discussion we will assume that the systems under consideration possess at least one positive exponent.

To estimate λ_1 we in effect monitor the long-term evolution of a single pair of nearby orbits. Our reconstructed attractor, though defined by a single trajectory, can provide points that may be considered to lie on different trajectories. We choose points whose temporal separation in the original time series is at least one mean orbital period, because a pair of points with a much smaller temporal separation is characterized by a zero Lyapunov exponent. Two data points may be considered to define the early state of the first principal axis so long as their spatial separation is small. When their separation becomes large we would like to perform GSR on the vector they define (simply normalization for this single vector), which would involve replacing the non-fiducial data point with a point closer to the fiducial point, in the same direction as the original vector. With finite amounts of data, we cannot hope to find a replacement point which falls exactly along a specified line segment in the reconstructed phase space, but we can look for a point that comes close. *In effect, through a simple replacement procedure that attempts to preserve orientation and minimize the size of replacement vectors, we have monitored the long-term behavior of a single principal axis vector*. Each replacement vector may be evolved until a problem arises, and so on. This leads us to an estimate of λ_1. (See fig. 4a.)

The use of a finite amount of experimental data does not allow us to probe the desired infinitesimal length scales of an attractor. These scales are also inaccessible due to the presence of noise on finite length scales and sometimes because the chaos-producing structure of the attractor is of negligible spatial extent. A discussion of these points is deferred until section 7.1.

An estimate of the sum of the two largest exponents $\lambda_1 + \lambda_2$ is similarly obtained. In the ODE procedure this involves the long-term evolution of a fiducial trajectory and a pair of tangent space vectors. In our procedure a triple of points is evolved in the reconstructed attractor. Before the area element defined by the triple becomes comparable to the extent of the attractor we mimic GSR by keeping the fiducial point, replacing the remainder of the triple with points that define a smaller area element and that best preserve the element's phase space orientation. Renormalizations are necessary solely because vectors grow too large, *not* because vectors will collapse to indistinguishable directions in phase space (this is unlikely with the limited amounts of data usually available in experiments). The exponential growth rate of area elements provides an estimate of $\lambda_1 + \lambda_2$. (See fig. 4b.)

Our approach can be extended to as many non-negative exponents as we care to estimate: $k + 1$ points in the reconstructed attractor define a k-volume element whose long-term evolution is possible through a data replacement procedure that attempts to preserve phase space orientation and probe only the small scale structure of the attractor. The growth rate of a k-volume element provides an estimate of the sum of the first k Lyapunov exponents.

In principle we might attempt the estimation of negative exponents by going to higher-dimensional volume elements, but information about contracting phase space directions is often impossible to resolve. In a system where fractal structure can be resolved, there is the difficulty that the volume elements involving negative exponent directions collapse exponentially fast, and are therefore numerically unstable for experimental data (see section 7.1).

5. Spectral algorithm implementation

We have implemented several versions of our algorithms including simple "fixed evolution time" programs for λ_1 and $\lambda_1 + \lambda_2$, "variable evolution time" programs for $\lambda_1 + \lambda_2$, and "interactive" programs that are used on a graphics machine†.

†The interactive program avoids the profusion of input parameters required for our increasingly sophisticated expo-

In appendix B we include Fortran code and documentation for the λ_1 fixed evolution time program. This program is not sophisticated, but it is concise, easily understood, and useful for learning about our technique. We do not include the fixed evolution time code for $\lambda_1 + \lambda_2$ (though it is briefly discussed at the end of appendix B) or our other programs, but we will supply them to interested parties. We can also provide a highly efficient data base management algorithm that can be used in any of our programs to eliminate the expensive process of exhaustive search for nearest neighbors. We now discuss the fixed evolution time program for λ_1 and the variable evolution time program for $\lambda_1 + \lambda_2$ in some detail.

5.1. Fixed evolution time program for λ_1

Given the time series $x(t)$, an m-dimensional phase portrait is reconstructed with delay coordinates [2, 33, 34], i.e., a point on the attractor is given by $\{x(t), x(t+\tau), \ldots, x(t+[m-1]\tau)\}$ where τ is the almost arbitrarily chosen delay time. We locate the nearest neighbor (in the Euclidean sense) to the initial point $\{x(t_0), \ldots, x(t_0 + [m-1]\tau)\}$ and denote the distance between these two points $L(t_0)$. At a later time t_1, the initial length will have evolved to length $L'(t_1)$. The length element is propagated through the attractor for a time short enough so that only small scale attractor structure is likely to be examined. If the evolution time is too large we may see L' shrink as the two trajectories which define it pass through a folding region of the attractor. This would lead to an underestimation of λ_1. We now look for a new data point that satisfies two criteria reasonably well: its separation, $L(t_1)$, from the evolved fiducial point is small, and the angular separation between the evolved and replacement elements is small (see fig. 4a). If an adequate replacement point cannot be found, we retain the points that were being used. This procedure is repeated until the fiducial trajectory has traversed the entire data file, at which point we estimate

$$\lambda_1 = \frac{1}{t_M - t_0} \sum_{k=1}^{M} \log_2 \frac{L'(t_k)}{L(t_{k-1})}, \qquad (9)$$

where M is the total number of replacement steps. In the fixed evolution time program the time step $\Delta = t_{k+1} - t_k$ (EVOLV in the Fortran program) between replacements is held constant. In the limit of an infinite amount of noise-free data our procedure always provides replacement vectors of infinitesimal magnitude with no orientation error, and λ_1 is obtained by definition. In sections 6 and 7 we discuss the severity of errors of orientation and finite vector size for finite amounts of noisy experimental data.

5.2. Variable evolution time program for $\lambda_1 + \lambda_2$

The algorithm for estimating $\lambda_1 + \lambda_2$ is similar in spirit to the preceeding algorithm, but is more complicated in implementation. A trio of data points is chosen, consisting of the initial fiducial point and its two nearest neighbors. The area $A(t_0)$ defined by these points is monitored until a replacement step is both *desirable* and *possible* – the evolution time is variable. This mandates the use of several additional input parameters: a minimum number of evolution steps between replacements (JUMPMN), the number of steps to evolve backwards (HOPBAK) when a replacement site proves inadequate, and a maximum length or area before replacement is attempted.

---

nent programs. This program allows the operator to observe: the attractor, a length or area element evolving over a range of times, the best replacement points available over a range of times, and so forth. Each of these is seen in a two or three-dimensional projection (depending on the graphical output device) with terminal output providing supplementary information about vector magnitudes and angles in the dimension of the attractor reconstruction. Using this information the operator chooses appropriate evolution times and replacement points. The program is currently written for a Vector General 3405 but may easily be modified for use on other graphics machines. A 16mm movie summarizing our algorithm and showing the operation of the program on the Lorenz attractor has been made by one of the authors (A.W.).

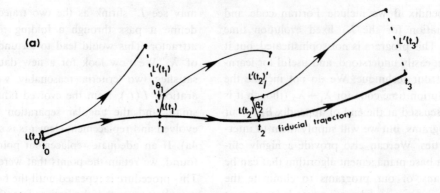

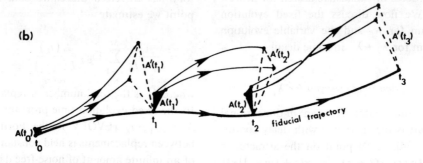

Fig. 4. A schematic representation of the evolution and replacement procedure used to estimate Lyapunov exponents from experimental data. a) The largest Lyapunov exponent is computed from the growth of length elements. When the length of the vector between two points becomes large, a new point is chosen near the reference trajectory, minimizing both the replacement length L and the orientation change ϑ. b) A similar procedure is followed to calculate the sum of the two largest Lyapunov exponents from the growth of area elements. When an area element becomes too large or too skewed, two new points are chosen near the reference trajectory, minimizing the replacement area A and the change in phase space orientation between the original and replacement area elements.

Evolution continues until a "problem" arises. In our implementation the problem list includes: a principal axis vector grows too large or too rapidly, the area grows too rapidly, and the skewness of the area element exceeds a threshold value. Whenever any of these criteria are met, the triple is evolved backwards HOPBAK steps and a replacement is attempted. If replacement fails, we will pull the triple back another HOPBAK steps, and try again. This process is repeated, if necessary, until the triple is getting uncomfortably close to the previous replacement site. At this point we take the best available replacement point, and jump forward at least JUMPMN steps to start the next evolution. At the first replacement time, t_1, the two points not on the fiducial trajectory are replaced with two new points to obtain a smaller area $A(t_1)$ whose orientation in phase space is most nearly the same as that of the evolved area $A'(t_1)$. Determining the set of replacement points that best preserves area orientation presents no fundamental difficulties.

Propagation and replacement steps are repeated (see fig. 4b) until the fiducial trajectory has traversed the entire data file at which point we estimate

$$\lambda_1 + \lambda_2 = \frac{1}{t_M - t_0} \sum_{k=1}^{M} \log_2 \frac{A'(t_k)}{A(t_{k-1})}, \qquad (10)$$

where t_k is the time of the kth replacement step.

It is often possible to verify our results for λ_1 through the use of the $\lambda_1 + \lambda_2$ calculation. For attractors that are very nearly two dimensional there is no need to worry about preserving orientation when we replace triples of points. These elements may rotate and deform within the plane of the attractor, but replacement triples always lie within this same plane. Since λ_2 for these attractors is zero, area evolution provides a direct estimate for λ_1. With experimental data that appear to define an approximately two-dimensional attractor, an independent calculation of d_f from its definition (feasible for attractors of dimension less than three [35]) may justify this approach to estimating λ_1.

6. Implementation details

6.1. Selection of embedding dimension and delay time

In principle, when using delay coordinates to reconstruct an attractor, an embedding [34] of the original attractor is obtained for any sufficiently large m and almost any choice of time delay τ, but in practice accurate exponent estimation requires some care in choosing these two parameters. We should obtain an embedding if m is chosen to be greater than twice the dimension of the underlying attractor [34]. However, we find that attractors reconstructed using smaller values of m often yield reliable Lyapunov exponents. For example, in reconstructing the Lorenz attractor from its x-coordinate time series an embedding dimension of 3 is adequate for accurate exponent estimation, well below the sufficient dimension of 7 given by ref. [34]†. When attractor reconstruction is performed in a space whose dimension is too low, "catastrophes" that interleave distinct parts of the attractor are likely to result. For example, points on separate lobes of the Lorenz attractor may be coincident in a *two*-dimensional reconstruction of the attractor. When this occurs, replacement elements may contain points whose separation in the original attractor is very large; such elements are liable to grow at a dramatic rate in our reconstructed attractor in the short term, providing an enormous contribution to the estimated exponent. As these elements tend to blow up almost immediately, they are also quite troublesome to replace‡.

If m is chosen too large we can expect, among other problems, that noise in the data will tend to decrease the density of points defining the attractor, making it harder to find replacement points. Noise is an infinite dimensional process that, unlike the deterministic component of the data, fills each available phase space dimension in a reconstructed attractor (see section 7.2). Increasing m past what is minimally required has the effect of unnecessarily increasing the level of contamination of the data.

Another problem is seen in a three-dimensional reconstruction of the Hénon attractor. The reconstructed attractor looks much like the original attractor sitting on a two-dimensional sheet, with this sheet showing a simple twist in three-space. We expect that this behavior is typical; when m is increased, surface curvature increases*. Increasing m therefore makes it increasingly difficult to satisfy orientation constraints at replacement time, as the attractor is not sufficiently flat on the smallest length scales filled out by the fixed quantity of data. It is advisable to check the stationarity of

---

†We have found that it is often possible to ignore several components of evolving vectors in computing their average exponential rate of growth: keeping two or more components of the vector often suffices for this purpose. As our discussion of "catastrophes" will soon make clear, the search for replacement points most often requires that all of the delay coordinates be used.

‡If two points lie at opposite ends of an attractor, it is possible that their separation vector lies entirely outside of the attractor so that no orientation preserving replacement can be found. If this goes undetected, the current pair of points is likely to be retained for an orbital period or longer, until these points are accidentally thrown close together.

*A simple study for the Hénon system showed that for reconstructions of increasing dimension the mean distance between the points defining the attractor rapidly converged to an attractor independent value. The fold put in each new phase space direction by the reconstruction process tended to make the concept of "nearby point in phase space" meaningless for this finite data set.

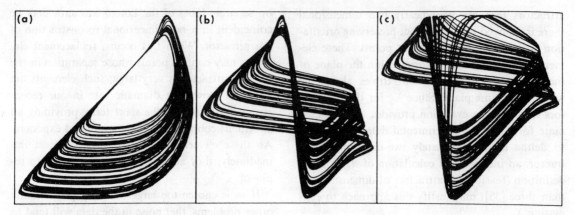

Fig. 5. The strange attractor in the Belousov–Zhabotinskii reaction is reconstructed by the use of delay coordinates from the bromide ion concentration time series [2]. The delays shown are a) $\frac{1}{12}$; b) $\frac{1}{2}$; and c) $\frac{3}{4}$ of a mean orbital period. Notice how the folding region of the attractor evolves from a featureless "pencil" to a large scale twist.

results with m to ensure robust exponent estimates.

Choice of delay time is also governed by the necessity of avoiding catastrophes. In our data [2] for the Belousov–Zhabotinskii chemical reaction (see fig. 5) we see a dramatic difference in the reconstructed attractors for the choices $\tau = 1/12$, $\tau = 1/2$ and $\tau = 3/4$ of the mean orbital period. In the first case we obtain a "pencil-like" region which obscures the folding region of the attractor. This structure opens up *and grows larger* relative to the total extent of the attractor for the larger values of τ, which is clearly desirable for our algorithms. We choose τ neither so small that the attractor stretches out along the line $x = y = z = \ldots$, nor so large that $m\tau$ is much larger than the orbital period. A check of the stationarity of exponent estimates with τ is again recommended.

6.2. Evolution times between replacements

Decisions about propagation times and replacement steps in these calculations depend on additional input parameters, or in the case of the interactive program, on the operator's judgement. (The stationarity of λ_1 values over ranges of all algorithm parameters is illustrated for the Rossler attractor in figs. 6a–6d.) Accurate exponent calculation therefore requires the consideration of the following interrelated points: the desirability of maximizing evolution times, the tradeoff between minimizing replacement vector size and minimizing the concomitant orientation error, and the manner in which orientation errors can be expected to accumulate. We now discuss these points in turn.

Maximizing the propagation time of volume elements is highly desirable as it both reduces the frequency with which orientation errors are made and reduces the cost of the calculation considerably (element propagation involves much less computation than element replacement). In our variable evolution time program this is not much of a problem, as replacements are performed only when deemed necessary (though the program has been made conservative in such judgments). In the interactive algorithm this is even less of a problem, as an experienced operator can often process a large file with a very small number of replacements. The problem is severe, however, in our fixed evolution time program, which is otherwise desirable for its extreme simplicity. In this program replacements are attempted at fixed time steps, independent of the behavior of the volume element.

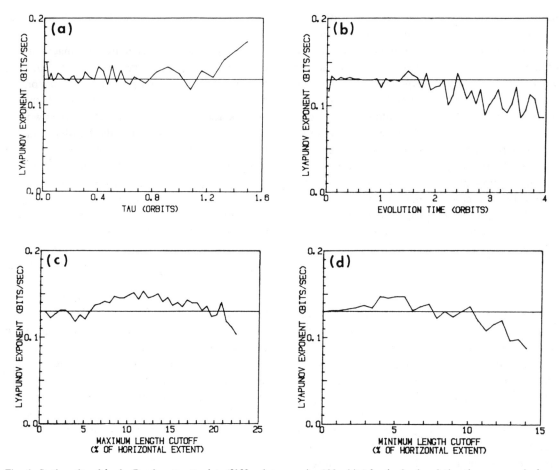

Fig. 6. Stationarity of λ_1 for Rossler attractor data (8192 points spanning 135 orbits) for the fixed evolution time program is shown for the input parameters: a) Tau (delay time); b) evolution time between replacement steps; c) maximum length of replacement vector length allowed; and d) minimum length of replacement vector allowed. The correct value of the positive exponent is 0.13 bits/s and is shown by the horizontal line in these figures.

Our numerical results on noise-free model systems have produced the expected results: too frequent replacements cause a dramatic loss of phase space orientation, and too infrequent replacements allow volume elements to grow overly large and exhibit folding. For the Rossler, Lorenz, and the Belousov–Zhabotinskii attractors, each of which has a once-per-orbit chaos generating mechanism, we find that varying the evolution time in the range $\frac{1}{2}$ to $1\frac{1}{2}$ orbits almost always provides stable exponent estimates. In systems where the mechanism for chaos is unknown, one must check for exponent stability over a wide range of evolution times. For such systems it is perhaps wise to employ only the variable evolution time program or the interactive program.

There are other criteria that may affect replacement times for variable evolution time programs such as avoiding regions of high phase space velocity, where the density of replacement points is likely to be small. Such features are easily integrated into our programs.

In the Lorenz attractor, the separatrix between the two lobes of the attractor is not a good place to find a replacement element. An element chosen here is likely to contain points that will almost immediately fly to opposite lobes, providing an enormous contribution to an exponent estimate. This effect is certainly related to the chaotic nature of the attractor, but is not directly related to the values of the Lyapunov exponents. This has the same effect as the catastrophes that can arise from too low a value of embedding dimension as discussed in section 6.1. While we are not aware of any foolproof approach to detecting troublesome regions of attractors it may be possible for an exponent program to avoid catastrophic replacements. For example, we may monitor the *future* behavior of potential replacement points and reject those whose separation from the fiducial trajectory is atypical of their neighbors.

6.3. *Shorter lengths versus orientation errors*

With a given set of potential replacement points some compromise will be necessary between the goals of minimizing the length of replacement vectors and minimizing changes in phase space orientation. On the one hand, short vectors may in general be propagated further in time, resulting in less frequent orientation errors. On the other hand, we may wish to minimize orientation errors directly. We must also consider that short vectors are likely to contain relatively large amounts of noise.

In the fixed evolution time program the search for replacements involves looking at successively larger length scales for a minimal orientation change. In the variable evolution time program, points satisfying minimum length and orientation standards are assigned scores based on a linear weighting (with heuristically chosen weighting factors) of their lengths and orientation changes. We have also performed numerical studies by searching successively larger angular displacements while attempting to satisfy a minimum length criterion. Fortunately, we find that these different approaches perform about equally well. Attempts to solve the tradeoff problem analytically have suggested "optimal" choices of initial vector magnitude, but due to the system dependent nature of these calculations, we cannot be confident that our results are of general validity.

The problem of considering the magnitude of evolved or replacement vectors is complicated by the fact that at a given point in an attractor, the orientation of a vector can determine whether or not it is too large. If we consider a system with an underlying 1-D map such as the Rossler attractor, it is the magnitude of the vector's component transverse to the attractor that is relevant. In this case our algorithm is closely related to obtaining the Lyapunov exponent of the map through a determination of its local slope profile [13]. The transverse vector component plays the role of the chord whose image under the map provides a slope estimate. This chord should obviously be no longer than the smallest resolvable structure in the 1-D map, a highly system-dependent quantity. Since the underlying maps of commonly studied model and physical systems have not had much detailed structure on small length scales (consider the logistic equation, cusp maps, and the Belousov–Zhabotinskii map [2]) we have somewhat arbitrarily decided to consider 5–10% of the transverse attractor extent as the maximum acceptable magnitude of a vector's transverse component.

6.4. *The accumulation of orientation errors*

The problem of the accumulation of orientation errors is reasonably well understood. Consider for simplicity a very nearly two-dimensional system with a $(+, 0, -)$ spectrum, such as the Lorenz attractor. Post-transient data traverse the subspace characterized by the positive and zero exponents. Length propagation with replacement on the attractor is clearly susceptible to orientation error that will mix contributions from the positive and zero exponents in some complex, system dependent manner. Now consider the nth replacement step (see fig. 4a) with an orientation change within the plane of the attractor of ϑ_n. Further, let the angle the replacement vector makes with respect to the vector v'_1 be $\bar{\vartheta}_n$. We make the crucial assumption that vectors are propagated for a time t that is long enough that growth along directions v'_1 and v'_2 are reasonably well characterized by the exponents λ_1 and λ_2 respectively. Then for the new replacement vector

$$L(t_n) = L(v'_1 \cos \bar{\vartheta}_n + v'_2 \sin \bar{\vartheta}_n) \tag{11}$$

and at the next replacement

$$L'(t_{n+1}) = L(v'_1 (\cos \bar{\vartheta}_n) 2^{\lambda_1 t_r} + v'_2 (\sin \bar{\vartheta}_n) 2^{\lambda_2 t_r}), \tag{12}$$

where t_r is the time between successive replacement steps ($t_{n+1} - t_n$). The contribution to eq. (9) from this evolution is then

$$\tfrac{1}{2} \log_2 \left[\cos^2 \bar{\vartheta}_n 2^{2\lambda_1 t_r} + \sin^2 \bar{\vartheta}_n 2^{2\lambda_2 t_r} \right] \tag{13}$$

and the angle the next replacement vector $L(t_{n+1})$ makes with v'_1 is

$$\bar{\vartheta}_{n+1} = \arctan(b \cdot \tan \bar{\vartheta}_n) + \vartheta_{n+1}, \tag{14}$$

where

$$b = 2^{(\lambda_2 - \lambda_1) t_r}. \tag{15}$$

If we assume all angles are small compared to unity and set $\bar{\vartheta}_0 = \vartheta_0$, eq. (14) implies that

$$\bar{\vartheta}_n = \sum_{m=0}^{n} \vartheta_{n-m} b^m. \tag{16}$$

If the orientation changes have zero mean and are uncorrelated from replacement to replacement then an average over the changes gives

$$\langle \bar{\vartheta}_n^2 \rangle = \frac{\vartheta_M^2 (1 - b^{2n+2})}{1 - b^2}, \tag{17}$$

where ϑ_M is an angular change on replacement on the order of the ANGLMX parameter in the fixed evolution time program of appendix B. From eqs. (9), (13), and (17) we find the fractional error in λ_1 to be

$$\frac{\Delta \lambda_1}{\lambda_1} = \frac{-\vartheta_M^2}{2(\ln 2) N_t \lambda_1 t_r} \left[N_t - \frac{b^2 (1 - b^{2N_t})}{1 - b^2} \right], \tag{18}$$

where N_t is the total number of replacement steps. If there is no degeneracy, i.e., $b^2 \ll 1$, eqs. (17) and (18) show that orientation errors do not accumulate, i.e, there is no N_t dependence, and our fractional error in λ_1 is

$$\frac{\Delta \lambda_1}{\lambda_1} = \frac{-\vartheta_M^2}{2(\ln 2) \lambda_1 t_r}. \tag{19}$$

For the Lorenz attractor, b^2 has a value of about 0.33 for an evolution time of one orbit, so an orientation error of about 19 degrees results in a 10% error in λ_1. If we can manage to evolve the vector for two orbits, the permissible initial orientation error is about 27 degrees, and so on. We see that a given orientation error at replacement time shrinks to a value negligible compared to the next orientation error, provided that propagation times are long enough. Orientation errors do not accumulate because there is no memory of previous errors.

This calculation may be generalized to an attractor with an arbitrary Lyapunov spectrum and a similar result is obtained. The ease of estimating the ith exponent depends on how small the quantity $2^{(\lambda_{i+1} - \lambda_i) t_r}$ is. Problems arise when successive exponents are very close or identical. Hyperchaos, with a spectrum of $[0.16, 0.03, 0.00, \simeq -40]$ bits/s and an orbital period of about 5.16 s, has an easily determinable first exponent, but distinguishing λ_2 from λ_3 is more difficult.

7. Data requirements and noise

7.1. Probing small length scales

As we have already pointed out, the infinitesimal length scales on which the definition of Lyapunov exponents rely are inaccessible in ex-

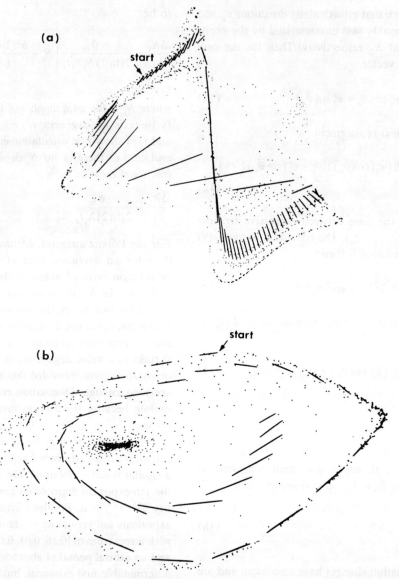

Fig. 7. Experimental data for two different Belousov–Zhabotinskii systems shows chaos on large and small length scales respectively. In the Texas attractor [2] a), the separation between a single pair of points is shown for one orbital period. In the French attractor [36]; b), the separation between a pair of points is shown for two periods. Estimation of Lyapunov exponents is quite difficult for the latter system.

perimental data. There are three somewhat related reasons why this is so: (1) a finite amount of attractor data can only define finite length scales; (2) the stretching and folding that is the chaotic element of a flow may occur on a scale small compared to the extent of the attractor; and (3) noise defines a length scale below which separations are meaningless. We discuss each of these problems in turn and then consider whether exponent estimation is possible in spite of them.

The finiteness of a data set means that the fixed evolution time program undoubtedly allows principal axis vectors to grow far too large on occasion and also to completely lose the proper phase space orientation, yet we almost invariably obtain accurate exponent estimates for noise-free model systems defined by small data sets. This is because *on the time scale of several orbital periods*, orbital divergences may be moderately well characterized by Lyapunov exponents in sufficiently chaotic systems. Averaging many such segments together in our algorithms is therefore likely to mask *infrequent* large errors.

The problem of "chaos on a small length scale" is a system dependent one. Consider a system such as the Rossler attractor in which chaos generation occurs on a relatively large length scale. Here it is quite easy to distinguish between true exponential divergence of nearby orbits and a temporary divergence due to local changes in the attractor's shape. If, however, the Rossler attractor were "crossed" with a periodic motion of sufficiently large amplitude, we would lose the ability to detect the mechanism for chaos as it would manifest itself only on length scales that we must regard as suspiciously small. For such a system it is difficult to conceive of any means of recovering exponents from experimental data.

We have observed this problem to some degree for the Couette–Taylor system, which makes a transition from motion on a 2-torus to chaos. In such a system chaos can arise from small stretches and folds on the torus. When we use the interactive program to monitor the evolution of lengths in the Couette–Taylor attractors, we seem to observe this effect; that is, the separation vectors do not exhibit dramatic growth but instead oscillate in magnitude. Such an oscillation could indicate a stretching and folding so that we might wish to attempt a replacement, or it could simply indicate a variation of the attractor's shape, which should be ignored. In figs. 7a and 7b we show experimental data for attractors of the large scale [2] and small scale [36] varieties, both arising from the Belousov–Zhabotinskii system. Exponent estimation in the latter case is quite difficult. The presence of external noise on length scales as small as the chaos generation mechanism will of course further complicate exponent calculations.

Even though infinitesimal length scales are not accessible, Lyapunov exponent estimation may still be quite feasible for many experimental systems. The same problem arises in calculations of the fractal dimension of strange attractors. Fractal structure does not exist in nature, where it is truncated on atomic scales, nor does it exist in any computer representation of a dynamical system, where finite precision truncates scaling. In these calculations we hope that as the smallest *accessible* length scales are approached, scaling converges to the zero length scale limit. Similarly, provided that chaos production is nearly the same on infinitesimal and the smallest accessible length scales, our calculations on the small scales may provide accurate results. A successful calculation requires that one has enough data to approach the appropriate length scales, ignores anything on the length scale of the noise, and has an attractor with a macroscopic stretching/folding mechanism.

7.2. *Noise*

The effects of noise in our algorithms fall into two categories which we have named the "statistical" and the "catastrophic". The former category deals with such problems as point-to-point jitter that cause us to estimate volumes inaccurately; this was the motivation for avoiding highly skewed replacement elements. Catastrophes can arise even in the absence of noise either from too low an embedding dimension (section 6.1), or from too little data compounded with high attractor complexity (section 6.2). In the presence of noise, catastrophes occur because noise drives a faraway data point into the replacement "arena." Noise in physical systems can be broken into two categories: measurement noise, i.e., simple lack of resolution, and dynamical noise, i.e., fluctuations in the state of the system or its parameters which enter into the dynamics. In the former case it is clear that the system possesses well defined exponents that are potentially recoverable. Strictly speaking, in the latter case Lyapunov exponents are not well defined, but some work [37] has suggested that a system may be characterized by numbers that are the Lyapunov exponents for the noise-free system averaged over the range of noise-induced states.

Our first study of the effects of noise on our algorithms involved adding dynamical noise to the Hénon attractor, that is, a small uniformly distributed random number was added to each coordinate as the map was being iterated. These data were then processed with the fixed evolution time program. For noise of sufficiently large amplitude, λ_1 could not be accurately determined. Specifically, the average initial separation between replacement points grew with the noise level (noise causing diffusion of the 1.26-dimensional attractor into the two-dimensional phase space) and the large final separations were not much affected by the noise. The result was an underestimate of the positive exponent. A nearly identical result was obtained for the addition of measurement noise (addition of a random term to each element of the time series, after the entire series has been generated) to the Hénon attractor.

It is ironic that measurement noise is not a problem unless large amounts of data are available to define the attractor. Noise is only detectable when the point density is high enough to provide replacements near the noise length scale. This suggests a simple approach to the noise problem, simply avoiding principal axis vectors whose magnitude is smaller than some threshold value we select. If this value is chosen to be somewhat larger than the noise level, the fractional error in determining initial vector magnitudes may be reduced to an acceptable level. Avoiding noisy length scales is not a trivial matter, as noise may not be of constant amplitude throughout an attractor and the noise length scale may be difficult to determine. Again, this approach can only work if scales larger than the noise contain accurate information about orbital divergence rates in the zero length scale limit. In fig. 6d we confirm that a straightforward small distance cutoff works in the case of the Rossler attractor by showing stationarity of the estimated exponent over a broad range of cutoff values.

7.3. Low pass filtering

Another approach to reducing the effects of noise, closely related to the use of a small distance cutoff, involves low pass filtering of the data before beginning exponent estimation. Rather severe filtering may be possible for systems with a once-per-orbit chaos producing mechanism – the filter cutoff approaching (orbital period)$^{-1}$. Filtering can be expected to distort shapes, eliminate small scale structure, and scramble phase, but we do not expect the divergent nature of the attractor to be lost.

A demonstration of the use of filtering for the Belousov–Zhabotinskii attractor is shown in fig. 8. Filtering dramatically altered the shape of the reconstructed attractor, but the estimated values of λ_1 differed by at most a few percent for reasonable cutoff frequencies. A similar calculation for the Rossler attractor indicated that the low-frequency cutoff could be moved all the way down to the attractor's sole large spectral feature before the exponent estimate showed any noticeable effect. Results for the Lorenz attractor, with its much more complicated spectral profile, are not quite as impressive. An analytical proof of the low pass filtering invariance of Lyapunov exponents (with conditions on the cutoff frequency relative to the orbital period and the replacement frequency) has proved elusive. Of course, low pass filtering fails to help with exponent estimation if there is substantial contamination of the data at frequencies lower than the filter cutoff. In a simple study of multi-periodic data with added white noise [3] the estimated exponent returned (very nearly) to zero for a sufficient amount of filtering. It thus appears that in some cases external noise can be distinguished from chaos by this procedure.

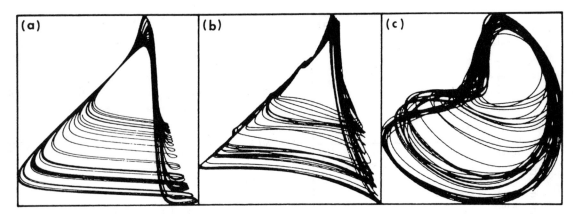

Fig. 8. a) Unfiltered experimental data for the Belousov–Zhabotinskii reaction [2]; b) the same data, low-pass filtered with a filter cutoff of 0.046 Hz, compared to the mean orbital frequency of 0.009 Hz. Our estimate of λ_1 for these data was only 5% lower than the estimate from the unfiltered data. Replacement frequencies in the region of stationarity for these results were approximately 0.005 Hz. c) the data are over-filtered at 0.023 Hz. λ_1 differs by only 20% from the exponent estimate for unfiltered data.

7.4. Data requirements

We now address the important questions of the quality and quantity of experimental data required for accurate exponent calculation. The former question is more easily disposed of – resolution requirements are so highly system dependent that we cannot make any general statements about them. In one study with the fixed evolution time program, the largest exponent was repeatedly computed for Rossler and Lorenz attractor data, the resolution of which was decreased one bit at a time from 16 bits. In each case the estimates were reasonably good for data with as few as 5 bits resolution. In fig. 9 we show the results of bit chopping for these systems as well as for Belousov–Zhabotinskii data. As a conservative rule of thumb we suggest a minimum of 8 meaningful bits of precision be used for exponent calculations. We strongly suggest that the resolution of experimental data be artificially lowered as we did for the model systems. If a plot of λ_1 versus resolution does not show an initial plateau for at least one or two bits, the initial data are suspect for the purpose of exponent calculations.

The amount of data required to calculate Lyapunov exponents depends on three distinct factors: the number of points necessary to provide an adequate number of replacement points, the number of orbits of data necessary to probe stretching (but not folding) within the attractor, and the number of data points per orbit that allow for proper attractor reconstruction with delay coordinates.

We first estimate how many points are required to "fill out" the structure of an attractor to provide replacement points. A simple minded estimate of this factor depends on the following factors: the fractal dimension of the attractor, the number of non-negative exponents in the system, the number of exponents we are attempting to compute (the dimension of each volume element), and the geometric requirements for acceptable replacement points. A more accurate calculation of this number will depend on such detailed information as the attractor's fractal structure and its probability density, which are not typically available for experimental data and the effective noise level in the system (which depends on both the actual level of contamination and the dimension of the reconstructed attractor). We assume that our data are uniformly distributed over a d-dimensional attractor of extent L and ignore noise-induced diffusion of the data. Thus, the density of

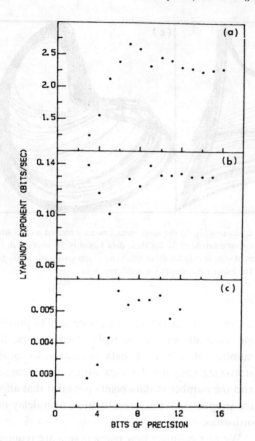

Fig. 9. The results of bit chopping (simulated measurement noise) for the a) Lorenz; b) Rossler; and c) Belousov–Zhabotinskii systems. For each system at least 5 bits of precision in the data are required for accurate exponent estimates.

points, ρ, is

$$\rho = \frac{N}{L^d}. \qquad (20)$$

The mean number of replacement points located in a region of length ε (SCALMX) and angular size ϑ (ANGLMX) is given by

$$N_r = V_d(\varepsilon, \vartheta)\rho \qquad (21)$$

where V_d, the volume of a d-dimensional search cone, is proportional to $\varepsilon^d \vartheta^{d-1}$, with d the (nearest integer) dimension of the attractor. N_r may be set to 1 for λ_1 estimation. Combining these expres-

sions and solving for N, we obtain

$$N(\lambda_1) \propto \frac{1}{\vartheta^{d-1}} \left[\frac{L}{\varepsilon}\right]^d. \qquad (22)$$

A nearly identical calculation for the number of points required for area replacement results in

$$N(\lambda_1 + \lambda_2) \propto \frac{1}{\vartheta^{d-2}} \left[\frac{L}{\varepsilon}\right]^d, \qquad (23)$$

and in general,

$$N\left(\sum_{i=1}^{j} \lambda_i\right) \propto \frac{1}{\vartheta^{d-j}} \left[\frac{L}{\varepsilon}\right]^d. \qquad (24)$$

We have ignored several prefactors that might modify these estimates by at most an order of magnitude. Also, the variance of the density of points is so high that the data requirement should probably be substantially increased to ensure that replacements are almost always available when needed, not just on the average. Perhaps surprisingly, the number of points required for estimating $\lambda_1 + \lambda_2$ is not significantly larger than $N(\lambda_1)$ (our estimate actually predicts it to be smaller). While we must double the number of points in the attractor to have a good chance of finding a pair of replacement points rather than a single one, the search volume for area replacement is actually larger (a larger solid angle of a potential replacement sphere is acceptable) than the search volume for length replacement. For area evolution there are pairs of points that define highly skewed replacement elements, but these are sufficiently unlikely that we can ignore their effect on $N(\lambda_1 + \lambda_2)$. For calculation of exponents past λ_2, the required point density does not change significantly. In our numerical work, in a best case scenario $L/\varepsilon \approx 5$, and the maximum value of ϑ is about 0.6 radians. In a worst case calculation L/ε is about 10, and ϑ is about 0.3 radians. From these values eq. (24) predicts to first order that between $\approx 10^d$ and $\approx 30^d$ points are necessary to fill out a d-dimensional attractor, independent of how many non-negative exponents we are calculating.

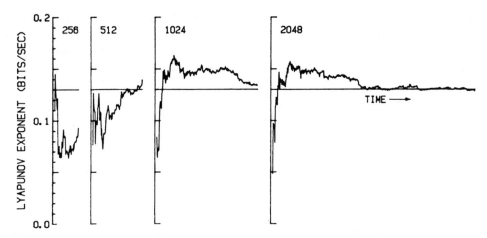

Fig. 10. The temporal convergence of λ_1 is shown as a function of the number of data points defining the Rossler attractor. The sampling rate is held constant; the longer time series contain more orbits. Note that lengthening the time series not only allows more time for convergence but also provides more replacement candidates at each replacement step. (Here the embedding dimension was 3, the embedding delay (τ) was a sixth of an orbit, and the evolution time-step was three-quarters of an orbit.)

The next factor we consider is the number of orbits of data required. The analysis is simplest if chaos arises through a once per orbit stretching and folding mechanism, which may be represented by a discrete mapping in one or more dimensions (as, for example, in the Lorenz, Rossler, hyperchaos, and Belousov–Zhabotinskii attractors). Exponent convergence requires that volume elements be operated on many times by the mapping until the element has sampled the slope profile of the map, suitably weighted by the map's probability density. The Lorenz and Rossler attractors have simple underlying 1-D maps; on the order of 10 to 100 map points are required for adequate slope profiles [13]. For these attractors, we expect between 10 and 100 orbits worth of data will be required for estimating λ_1 or for confirming that $\lambda_2 = 0$. No additional orbits are required for area propagation in a system defined by a 1-D map. If the system had an underlying 2-D map as hyperchaos does, we might expect, depending on the complexity of the map, that roughly the square of this number of orbits would be required. This would provide the same sampling resolution for the slope profile of the map (see ref. 13) in each dimension. In general, for a system defined by an n-D map, the number of orbits of data required to estimate any non-negative exponent is given by a constant, C, raised to a power which is the number of positive exponents. The number of positive exponents is approximately the dimension of the attractor minus one, thus the required number of orbits is about C^{d-1}. C is a system dependent quantity depending on the amount of structure in the map, perhaps in the range 10 to 100.

The last and simplest point we consider is the required number of points per orbit, P. There is no benefit to choosing P any larger than is absolutely necessary. We might try to choose P so small that in an evolution of a single step, the average replacement vector would grow to as large a separation as we care to allow. In the Lorenz attractor, for example, we might decide to allow the average replacement vector to grow by a factor of 32 in a single time step, so that we would have one data point per 6 orbits. The problem is that with data this sparse we are unlikely to obtain a good reconstruction of our attractor. Often, the relationship $m\tau \approx 1$ is used in reconstructions, where m is the embedding dimension and τ is the delay in units of orbital periods. We assume that reconstruction is performed in an approximately

d-dimensional space, and the delay corresponds to a single sample time, so that $\tau = 1/P$. We then obtain a requirement of about d points per orbit†.

When the number of points per orbit is multiplied by the number of orbits, we obtain a required number of points ranging from $d \times 10^{d-1}$ to $d \times 100^{d-1}$, which we can compare to the point density requirement of between 10^d and 30^d points. Since all three requirements must be met, the larger of these two quantities determines the amount of data required. In each of these two ranges of values the complexity of the underlying map (if any) determines which end of the range is appropriate. Therefore we may conclude that for up to about a 10-dimensional system the required number of data points ranges from 10^d to 30^d. We compute this range for several systems: Hénon attractor ($d = 1.26$), 30–100 points; Rossler attractor ($d = 2.01$), 100–1000 points; hyperchaos ($d = 3.005$), 1000–30000 points; delay differential equation ($d = 3.64$), 4000–200 000 points. We see that the amount of data required to estimate nonnegative exponents rises exponentially fast with the dimension of the attractor, the identical problem with calculations of fractal dimension by all algorithms of the distance scaling variety [35]. Fig. 10 shows the convergence of our λ_1 estimate as the number of points used is increased for the Rossler attractor. It is important to note that while it may take 32 000 points to define an attractor, it is generally not necessary to evolve completely through the data before the exponent estimate converges. For example, see fig. 10.

8. Results

We now present our results for the various model and experimental systems on which our algorithms have been tested. We emphasize that no explicit use was made of the differential equations defining the model systems, except to produce a dynamical observable (the x-coordinate time series) which was then treated as experimental data. For the equations that define each system see table I. The quoted uncertainty values for each system were calculated either from the known values of the exponents or from the variation of our results with changes in parameters.

Hénon attractor

For the Hénon map, we obtained the positive exponent to within 5% with only 128 points defining the attractor.

Rossler attractor

For the Rossler attractor, we found the first exponent with a 5% error using 1024 points. The second exponent was measured as less than 6% of the first with 2048 points defining the attractor. Six points per mean orbital period were used to define the attractor.

Lorenz attractor

The Lorenz system was the most difficult test of the fixed evolution time program because its ill-defined orbital period made it difficult to avoid catastrophic replacements near the separatrix. In using the interactive program the operators simply

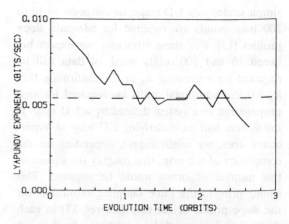

Fig. 11. Stationarity of λ_1 with evolution time is shown for Belousov–Zhabotinskii data [2] (compare to fig. 6b).

---

†We note that d points per orbit is a very small number compared to the sampling rate required for λ_1 estimation with an underlying 1-D map. Construction of a map requires that orbital intersections with the Poincaré section be determined with high accuracy, often necessitating 100 or more points per orbit. Our technique thus allows a factor of about 10 times more orbits for a given size data file.

avoided that region at replacement time. The interactive runs determined the positive exponent to within 3%, and measured the second exponent as less than 5% of the first, using 8192 points. The fixed evolution time code measured the first exponent to within 5% and found the second exponent to be less than 10% of the first, using 8192 points in both cases.

Hyperchaos

For this system we obtained the largest exponent to within 10% using 8192 points and the sum of the two positive exponents to within 15% using 16384 points.

Delay differential equation

Using 65536 points, we computed the largest of the two positive exponents to within 10% and found the sum of the first two exponents to within 20%.

Belousov–Zhabotinskii reaction

In fig. 11 we show the result of our algorithm used on a time series of 65536 points spanning 400 orbital periods. The system was in a chaotic regime near a period-three window. The exponent calculated by the algorithm is stable over a range of parameter values. We also calculated the exponent using 1-D map analysis [2] as a comparison. Our algorithm gives a result in the plateau region of 0.0054 ± 0.0005 bits/s, while the 1-D map estimation yields a result of 0.0049 ± 0.0010 bits/s. Thus the estimates are in agreement.

Couette–Taylor

For the Couette–Taylor experiment we computed the largest Lyapunov exponent as a function of Reynolds number from data sets (at each Reynolds number) consisting of 32768 points spanning about 200 mean orbital periods. Our results are given in fig. 12. Earlier studies of power spectra and phase portraits had indicated that the onset of chaos occurred at $R/R_c \simeq 12$, where R_c marks the transition to Taylor vortex flow. This onset of chaos is confirmed and quantified by the calculation of λ_1.

9. Conclusions

The algorithms we have presented can detect and quantify chaos in experimental data by accurately estimating the first few non-negative Lyapunov exponents. Moreover, our numerical studies have shown that deterministic chaos can be distinguished in some cases from external noise (as in the Belousov–Zhabotinskii attractor) and topological complexity (as in the Lorenz attractor). However, this requires a reasonable quantity of accurate data, and the attractor must not be of very high dimension.

As with other diagnostics used in chaotic dynamical systems, the calculation of Lyapunov exponents is still in its infancy, but we believe that the approach to exponent estimation that we have described here is workable. We encourage experimentation with our algorithms.

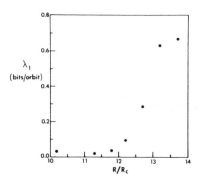

Fig. 12. The largest Lyapunov exponent for experimental Couette–Taylor data is shown as a function of Reynolds number. The shape of this curve (but not its absolute magnitude, see reference [3]) was independently verified by the calculation of the metric entropy h_μ – which is equal to λ_1 if there is a single positive exponent [38].

Acknowledgements

We thank J. Doyne Farmer for helpful discussions and for calculating the Lyapunov spectrum of the delay differential equation; Mark Haye for assistance in programming the Vector General; and Shannon Spires for programming support, and for producing the Lyapunov exponent film. This research is supported by the Department of Energy, Office of Basic Energy Sciences contract DE-AS05-84ER13147. H. Swinney acknowledges the support of a Guggenheim Fellowship and J. Vastano acknowledges the support of an Exxon Fellowship.

Appendix A

Lyapunov spectrum program for systems of differential equations

This program computes the complete Lyapunov spectrum of a nonlinear system whose equations of motion and their linearizations are supplied by the user in the routine FCN. The program is set up here for the three variable Lorenz system but is easily modified for any other system of equations.

```
      PROGRAM ODE
C
C     N = # OF NONLINEAR EQTNS., NN = TOTAL # OF EQTNS.
C
      PARAMETER N=3
      PARAMETER NN=12
      EXTERNAL FCN
C
      DIMENSION Y(NN),ZNORM(N),GSC(N),CUM(N),C(24),W(NN,9)
C
C     INITIAL CONDITIONS FOR NONLINEAR SYSTEM
C
      Y(1) = 10.0
      Y(2) = 1.0
      Y(3) = 0.0
C
C     INITIAL CONDITIONS FOR LINEAR SYSTEM (ORTHONORMAL FRAME)
C
      DO 10 I = N+1,NN
         Y(I) = 0.0
   10 CONTINUE
      DO 20 I = 1,N
         Y((N+1)*I) = 1.0
         CUM(I) = 0.0
   20 CONTINUE
C
C     INTEGRATION TOLERANCE, # OF INTEGRATION STEPS,
C     TIME PER STEP, AND I/O RATE
C
      TYPE*, 'TOL,NSTEP,STPSZE,IO?'
      ACCEPT*,TOL,NSTEP,STPSZE,IO
C
C     INITIALIZATION FOR INTEGRATOR
C
      NEQ = NN
      X = 0.0
      IND = 1
```

```
C
        DO 100 I = 1,NSTEP
            XEND = STPSZE*FLOAT(I)
C
C     CALL ANY ODE INTEGRATOR - THIS IS AN IMSL ROUTINE
C
            CALL DVERK (NEQ,FCN,X,Y,XEND,TOL,IND,C,NEQ,W,IER)
C
C     CONSTRUCT A NEW ORTHONORMAL BASIS BY GRAM-SCHMIDT METHOD
C
C     NORMALIZE FIRST VECTOR
C
            ZNORM(1) = 0.0
            DO 30 J = 1,N
                ZNORM(1) = ZNORM(1)+Y(N*J+1)**2
    30      CONTINUE
            ZNORM(1) = SQRT(ZNORM(1))
            DO 40 J = 1,N

                Y(N*J+1) = Y(N*J+1)/ZNORM(1)
    40      CONTINUE
C
C     GENERATE THE NEW ORTHONORMAL SET OF VECTORS.
C
            DO 80 J = 2,N
C
C     GENERATE J-1 GSR COEFFICIENTS.
C
                DO 50 K = 1,(J-1)
                    GSC(K) = 0.0
                    DO 50 L = 1,N
                        GSC(K) = GSC(K)+Y(N*L+J)*Y(N*L+K)
    50          CONTINUE
C
C     CONSTRUCT A NEW VECTOR.
C
                DO 60 K = 1,N
                    DO 60 L = 1,(J-1)
                        Y(N*K+J) = Y(N*K+J)-GSC(L)*Y(N*K+L)
    60          CONTINUE
C
C     CALCULATE THE VECTOR'S NORM
C
                ZNORM(J) = 0.0
                DO 70 K = 1,N
                    ZNORM(J) = ZNORM(J)+Y(N*K+J)**2
    70          CONTINUE
                ZNORM(J) = SQRT(ZNORM(J))
C
C     NORMALIZE THE NEW VECTOR.
C
                DO 80 K = 1,N
                    Y(N*K+J) = Y(N*K+J)/ZNORM(J)
    80      CONTINUE
```

```
C
C      UPDATE RUNNING VECTOR MAGNITUDES
C
       DO 90 K = 1,N
          CUM(K) = CUM(K)+ALOG(ZNORM(K))/ALOG(2.)
 90    CONTINUE
C
C      NORMALIZE EXPONENT AND PRINT EVERY IO ITERATIONS
C
       IF (MOD(I,IO).EQ.0) TYPE*,X,(CUM(K)/X,K = 1,N)
C
 100   CONTINUE
C
       CALL EXIT
       END
C
       SUBROUTINE FCN (N,X,Y,YPRIME)
C
C      USER DEFINED ROUTINE CALLED BY IMSL INTEGRATOR.
C
       DIMENSION Y(12),YPRIME(12)
C
C      LORENZ EQUATIONS OF MOTION
C
       YPRIME(1) = 16.*(Y(2)-Y(1))
       YPRIME(2) = -Y(1)*Y(3)+45.92*Y(1)-Y(2)
       YPRIME(3) = Y(1)*Y(2)-4.*Y(3)
C
C      3 COPIES OF LINEARIZED EQUATIONS OF MOTION.
C
       DO 10 I = 0,2
          YPRIME(4+I) = 16.*(Y(7+I)-Y(4+I))
          YPRIME(7+I) = (45.92-Y(3))*Y(4+I)-Y(7+I)-Y(1)*Y(10+I)
          YPRIME(10+I) = Y(2)*Y(4+I)+Y(1)*Y(7+I)-4.*Y(10+I)
 10    CONTINUE
C
       RETURN
       END
```

See section 3 and refs. 8, 9 for a discussion of the ODE algorithm.

Appendix B

Fixed evolution time program for λ_1

A time series (of length NPT) is read from a data file, along with the parameters necessary to reconstruct the attractor, namely, the dimension of the phase space reconstruction (DIM), the reconstruction time delay (TAU), and the time between the data samples (DT), required only for normalization of the exponent. Three other input parameters are required: length scales that we consider to be too large (SCALMX) and too small (SCALMN) and a constant propagation time (EVOLV) between replacement attempts. SCALMX is our estimate of the length scale on which the local structure of the attractor is no longer being probed; SCALMN is the length scale on which noise is expected to appear. We also supply a maximum angular error to be accepted at replacement time (ANGLMX), but we do not consider this a free parameter as its selection is not likely to have much effect on exponent estimates. It is usually fixed at 0.2 or 0.3 radians.

The calculation is initiated by carrying out an exhaustive search of the data file to locate the nearest neighbor to the first point (the fiducial point), omitting points closer than SCALMN†. The main program loop, which carries out repeated cycles of propagating and replacing the first principal axis vector is now entered. The current pair of points is propagated EVOLV steps through the attractor and its final separation is computed. The log of the ratio of final to initial separation of this pair updates a running average rate of orbital divergence. A replacement step is then attempted. The distance of each delay coordinate point to the evolved fiducial point is then determined. Points closer than SCALMX but further away than SCALMN are examined to see if the change in angular orientation is less than ANGLMX radians. If more than one candidate point is found, the point defining the smallest angular change is used for replacement. If no points satisfy these criteria, we loosen the larger distance criterion to accept replacement points as far as twice SCALMX away. If necessary the large distance criterion is relaxed several more times, at which point we tighten this constraint and relax the angular acceptance criterion. Continued failure will eventually result in our keeping the pair of points we had started out with, as this pair results in no change whatsoever in phase space orientation. We now go back to the top of the main loop where the new points are propagated. This process is repeated until the fiducial trajectory reaches the end of the data file, by which time we hope to see stationary behavior of λ_1. See section 6 for a discussion of how to choose the input parameters.

The fixed evolution time code for $\lambda_1 + \lambda_2$ estimation is too long to present here (350 lines of Fortran) but we discuss its structure briefly. This program begins by reading in a dynamical observable and many of the same input parameters as the code for λ_1 estimation. A number of parameters are also required to evaluate the quality of replacement triples: the maximum allowed triple "skewness", the maximum angular deviation of each replacement vector from the plane defined by the last triple, and weighting factors for the relative importance of skewness, size of replacement vectors, and angular errors in choosing replacement vectors.

The structure of this program is very similar to the program for λ_1: locate the two nearest neighbors of the first delay coordinate point, determine the initial area defined by this triple, enter the main program loop for repeated evolution and replacement. Triples are evolved EVOLV steps through the attractor and replacement is performed. Triple replacement is a more complicated process than pair replacement, which involved minimizing a single angular separation and a length. Our approach to triple replacement is a two step process; first we find a small set of points which, together with the fiducial trajectory, define small separation vectors and lie close to the required two-dimensional subspace. We then determine which of all of the possible pairs of these points will make the best replacement triple. In the first step, the qualifications of up to 20 potential replacement points are saved. Separation and orientation requirements of replacement points are varied so that a moderate number of candidates is almost always obtained. In the second step every possible pair of these points is assigned a score based on how close the replacement triple is to the old two-dimensional subspace and how numerically stable the orientation of the triple is believed to be. (It is possible that the individual replacement *vectors* lie very close to the old two-dimensional subspace but that the replacement *area element* is nearly orthogonal to the same subspace!) The relative importance of replacement lengths, skewness, orientation changes, etcetera, are weighted by the user chosen factors. The highest scoring pair of points is used in the replacement triple. The calculations in this program involve dot products and the Gram-Schmidt process and so are independent of the dimension of the reconstructed attractor – no complicated geometry is required in the coding.

---

†Such an exhaustive search is very inefficient for large arrays; then more efficient algorithms should be employed. See, for example, ref. 39.

```
      PROGRAM FET1
      INTEGER DIM,TAU,EVOLV
      DIMENSION X(16384),PT1(12),PT2(12)
C
C     **DEFINE DELAY COORDINATES WITH A STATEMENT FUNCTION**
C     **Z(I,J)=JTH COMPONENT OF ITH RECONSTRUCTED ATTRACTOR POINT**
C
      Z(I,J) = X(I+(J-1)*TAU)
C
      OPEN (UNIT=1,FILE='INPUT.',TYPE='OLD')
C
      TYPE*, 'NPT,DIM,TAU,DT,SCALMX,SCALMN,EVOLV ?'
      ACCEPT*,NPT,DIM,TAU,DT,SCALMX,SCALMN,EVOLV
C
C     **IND POINTS TO FIDUCIAL TRAJECTORY**
C     **IND2 POINTS TO SECOND TRAJECTORY**
C     **SUM HOLDS RUNNING EXPONENT ESTIMATE SANS 1/TIME**
C     **ITS IS TOTAL NUMBER OF PROPAGATION STEPS**
C
      IND = 1
      SUM = 0.0
      ITS = 0
C
C     **READ IN TIME SERIES**
C
      DO 10 I = 1,NPT
         READ (1,*) X(I)
   10 CONTINUE
C
C     **CALCULATE USEFUL SIZE OF DATAFILE**
C
      NPT = NPT - DIM*TAU - EVOLV
C
C     **FIND NEAREST NEIGHBOR TO FIRST DATA POINT**
C
      DI = 1.E38
C
C     **DONT TAKE POINT TOO CLOSE TO FIDUCIAL POINT**
C
      DO 30 I = 11,NPT
C
C     **COMPUTE SEPARATION BETWEEN FIDUCIAL POINT AND CANDIDATE**
C
         D = 0.0
         DO 20 J = 1,DIM
            D = D+(Z(IND,J)-Z(I,J))**2
   20    CONTINUE
         D = SQRT(D)
C
C     **STORE THE BEST POINT SO FAR BUT NO CLOSER THAN NOISE SCALE**
C
         IF (D.GT.DI.OR.D.LT.SCALMN) GO TO 30
         DI = D
         IND2 = I
   30 CONTINUE
```

```
C
C        **GET COORDINATES OF EVOLVED POINTS**
C
      40 DO 50 J = 1,DIM
            PT1(J) = Z(IND+EVOLV,J)
            PT2(J) = Z(IND2+EVOLV,J)
      50 CONTINUE
C
C        **COMPUTE FINAL SEPARATION BETWEEN PAIR, UPDATE EXPONENT**
C
         DF = 0.0
         DO 60 J = 1,DIM
            DF = DF+(PT1(J)-PT2(J))**2
      60 CONTINUE
         DF = SQRT(DF)
         ITS = ITS+1
         SUM = SUM+ALOG(DF/DI)/(FLOAT(EVOLV)*DT*ALOG(2.))
         ZLYAP = SUM/FLOAT(ITS)
         TYPE*,ZLYAP,EVOLV*ITS,DI,DF
C
C        **LOOK FOR REPLACEMENT POINT**
C        **ZMULT IS MULTIPLIER OF SCALMX WHEN GO TO LONGER DISTANCES**
C
         INDOLD = IND2
         ZMULT = 1.0
         ANGLMX = 0.3
      70 THMIN = 3.14
C
C        **SEARCH OVER ALL POINTS**
C
         DO 100 I = 1,NPT
C
C        **DONT TAKE POINTS TOO CLOSE IN TIME TO FIDUCIAL POINT**
C
            III = IABS(I-(IND+EVOLV))
            IF (III.LT.10) GO TO 100
C
C        **COMPUTE DISTANCE BETWEEN FIDUCIAL POINT AND CANDIDATE**
C
            DNEW = 0.0
            DO 80 J = 1,DIM
               DNEW = DNEW+(PT1(J)-Z(I,J))**2
      80    CONTINUE
            DNEW = SQRT(DNEW)
C
C        **LOOK FURTHER AWAY THAN NOISE SCALE, CLOSER THAN ZMULT*SCALMX**
C
            IF (DNEW.GT.ZMULT*SCALMX.OR.DNEW.LT.SCALMN) GO TO 100
C
C        **FIND ANGULAR CHANGE OLD TO NEW VECTOR**
C
            DOT = 0.0
            DO 90 J = 1,DIM
               DOT = DOT+(PT1(J)-Z(I,J))*(PT1(J)-PT2(J))
      90    CONTINUE
```

```
              CTH = ABS(DOT/(DNEW*DF))
              IF (CTH.GT.1.0) CTH = 1.0
              TH = ACOS(CTH)
      C
      C     **SAVE POINT WITH SMALLEST ANGULAR CHANGE SO FAR**
      C
              IF (TH.GT.THMIN) GO TO 100
              THMIN = TH
              DII = DNEW
              IND2 = I
        100 CONTINUE
              IF (THMIN.LT.ANGLMX) GO TO 110
      C
      C     **CANT FIND A REPLACEMENT - LOOK AT LONGER DISTANCES**
      C
              ZMULT = ZMULT+1.
              IF (ZMULT.LE.5.) GO TO 70
      C
      C     **NO REPLACEMENT AT 5*SCALE, DOUBLE SEARCH ANGLE, RESET DISTANCE**
      C
              ZMULT = 1.0
              ANGLMX = 2.*ANGLMX
              IF (ANGLMX.LT.3.14) GO TO 70
              IND2 = INDOLD + EVOLV
              DII = DF
        110 CONTINUE
              IND = IND+EVOLV
      C
      C     **LEAVE PROGRAM WHEN FIDUCIAL TRAJECTORY HITS END OF FILE**
      C
              IF (IND.GE.NPT) GO TO 120
              DI = DII
              GO TO 40
        120 CALL EXIT
              END
```

References

[1] See the references in: H.L. Swinney, "Observations of Order and Chaos in Nonlinear Systems," Physica 7D (1983) 3 and in: N.B. Abraham, J.P. Gollub and H.L. Swinney, "Testing Nonlinear Dynamics," Physica 11D (1984) 252.

[2] J.-C. Roux, R.H. Simoyi and H.L. Swinney, "Observation of a Strange Attractor," Physica 8D (1983) 257.

[3] A. Brandstater, J. Swift, H.L. Swinney, A. Wolf, J.D. Farmer, E. Jen and J.P. Crutchfield, "Low-Dimensional Chaos in a Hydrodynamic System," Phys. Rev. Lett 51 (1983) 1442.

[4] B. Malraison, P. Atten, P. Berge and M. Dubois, "Turbulence-Dimension of Strange Attractors: An Experimental Determination for the Chaotic Regime of Two Convective Systems," J. Physique Lettres 44 (1983) L-897.

[5] J. Guckenheimer and G. Buzyna, "Dimension Measurements for Geostrophic Turbulence," Phys. Rev. Lett. 51 (1983) 1438.

[6] J.P. Gollub, E.J. Romer and J.E. Socolar, "Trajectory divergence for coupled relaxation oscillators: measurements and models," J. Stat. Phys. 23 (1980) 321.

[7] V.I. Oseledec, "A Multiplicative Ergodic Theorem. Lyapunov Characteristic Numbers for Dynamical Systems," Trans. Moscow Math. Soc. 19 (1968) 197.

[8] G. Benettin, L. Galgani, A. Giorgilli and J.-M. Strelcyn,

"Lyapunov Characteristic Exponents for Smooth Dynamical Systems and for Hamiltonian Systems; A Method for Computing All of Them," Meccanica 15 (1980) 9.

[9] I. Shimada and T. Nagashima, "A Numerical Approach to Ergodic Problem of Dissipative Dynamical Systems," Prog. Theor. Phys. 61 (1979) 1605.

[10] R. Shaw, "Strange Attractors, Chaotic Behavior, and Information Flow," Z. Naturforsch. 36A (1981) 80.

[11] J.L. Hudson and J.C. Mankin, "Chaos in the Belousov–Zhabotinskii Reaction," J. Chem. Phys. 74 (1981) 6171.

[12] H. Nagashima, "Experiment on Chaotic Response of Forced Belousov–Zhabotinskii Reaction," J. Phys. Soc. Japan 51 (1982) 21.

[13] A. Wolf and J. Swift, "Progress in Computing Lyapunov Exponents from Experimental Data," in: Statistical Physics and Chaos in Fusion Plasmas, C.W. Horton Jr. and L.E. Reichl, eds. (Wiley, New York, 1984).

[14] J. Wright, "Method for Calculating a Lyapunov Exponent," Phys. Rev. A29 (1984) 2923.

[15] S. Blacher and J. Perdang, "Power of Chaos," Physica 3D (1981) 512.

[16] J.P. Crutchfield and N.H. Packard, "Symbolic Dynamics of Noisy Chaos," Physica 7D (1983) 201.

[17] P. Grassberger and I. Procaccia, "Estimation of the Kolmogorov Entropy from a Chaotic Signal," Phys. Rev. A28 (1983) 2591.

[18] R. Shaw, The Dripping Faucet, (Aerial Press, Santa Cruz, California, 1984).

[19] J.D. Farmer, E. Ott and J.A. Yorke, "The Dimension of Chaotic Attractors," Physica 7D (1983) 153.

[20] S. Ciliberto and J.P. Gollub, "Chaotic Mode Competition in Parametrically Forced Surface Waves"—preprint.

[21] P. Grassberger and I. Procaccia, "Characterization of Strange Attractors," Phys. Rev. Lett. 50 (1983) 346.

[22] H. Haken, "At Least One Lyapunov Exponent Vanishes if the Trajectory of an Attractor Does Not Contain a Fixed Point," Phys. Lett. 94A (1983) 71.

[23] E.N. Lorenz, "Deterministic Nonperiodic Flow," J. Atmos. Sci. 20 (1983) 130.

[24] O.E. Rossler, "An Equation for Hyperchaos," Phys. Lett. 71A (1979) 155.

[25] M. Hénon, "A Two-Dimensional Mapping with a Strange Attractor," Comm. Math. Phys. 50 (1976) 69.

[26] O.E. Rossler, "An Equation for Continuous Chaos," Phys. Lett. 57A (1976) 397.

[27] M.C. Mackey and L. Glass, "Oscillation and Chaos in Physiological Control Systems," Science 197 (1977) 287.

[28] J. Kaplan and J. Yorke, "Chaotic behavior of multidimensional difference equations," in: Functional Differential Equations and the Approximation of Fixed Points, Lecture Notes in Mathematics, vol. 730, H.O. Peitgen and H.O. Walther, eds. (Springer, Berlin), p. 228; P. Frederickson, J. Kaplan, E. Yorke and J. Yorke, "The Lyapunov Dimension of Strange Attractors," J. Diff. Eqs. 49 (1983) 185.

[29] L.-S. Young, "Dimension, Entropy, and Lyapunov Exponents," Ergodic Theory and Dynamical Systems 2 (1982) 109. F. Ledrappier, "Some Relations Between Dimension and Lyapunov Exponents," Comm. Math. Phys. 81 (1981) 229.

[30] D.A. Russell, J.D. Hanson and E. Ott, "Dimension of Strange Attractors," Phys. Rev. Lett. 45 (1980) 1175.

[31] D. Ruelle, private communication.

[32] After R. Shaw, unpublished.

[33] N.H. Packard, J.P. Crutchfield, J.D. Farmer and R.S. Shaw, "Geometry from a Time Series," Phys. Rev. Lett. 45 (1980) 712.

[34] F. Takens, "Detecting Strange Attractors in Turbulence," in Lecture Notes in Mathematics, vol. 898, D.A. Rand and L.-S. Young, eds. (Springer, Berlin, 1981) p. 366.

[35] H.S. Greenside, A. Wolf, J. Swift and T. Pignataro, "The Impracticality of a Box-Counting Algorithm for Calculating the Dimensionality of Strange Attractors," Phys. Rev. A25 (1982) 3453.

[36] J.-C. Roux and A. Rossi, "Quasiperiodicity in chemical dynamics," in: Nonequilibrium Dynamics in Chemical Systems, C. Vidal and A. Pacault, eds. (Springer, Berlin, 1984), p. 141.

[37] J.P. Crutchfield, J.D. Farmer and B.A. Huberman, "Fluctuations and Simple Chaotic Dynamics," Phys. Rep. 92 (1982) 45.

[38] R. Bowen and D. Ruelle, "The Ergodic Theory of Axiom-A Flows," Inv. Math. 29 (1975) 181. D. Ruelle, "Applications conservant une mesure absolument continué par rapport à dx sur [0,1]," Comm. Math. Phys. 55 (1977) 47.

[39] D.E. Knuth, The Art of Computer Programming, vol. 3 – Sorting and Searching, (Addison–Wesley, Reading, Mass., 1975).

III. Analytical Methods

Unfortunately there are no general analytical methods which allow to predict chaotic behavior, but there are some which are useful in these investigations.

Paper III.1 describes Melnikov method which allows to predict transversal intersection of stable and unstable manifolds of a hyperbolic fixed point. Such an intersection does not mean the existence of a stable strange attractor but is a necessary condition for such an attractor.

Bifurcations to subharmonics play an important role in the transition to chaos in oscillators. The methods which allow their investigation are presented in Paper III.2 (second order averaging) and Paper III.6 (variational methods).

Paper III.3 describes the application of Melnikov method to the investigation of subharmonic and homoclinic bifurcations of a parametrically forced pendulum, while Paper III.4 shows its application to predict the power spectra of near homoclinic motion.

Melnikov method is generally applicable to near Hamiltonian systems with small damping. Its generalization to the systems with big damping is presented in Paper III.5.

Application of classical methods of nonlinear oscillations such as for example: harmonic balance to predict the domain of chaotic behavior in parameter space is discussed in Paper III.7. The combination of harmonic balance analysis and Feigenbaum's universal properties of period-doubling allowed to obtain conditions for chaos given in Paper III.8.

A Partial Differential Equation with Infinitely many Periodic Orbits: Chaotic Oscillations of a Forced Beam

PHILIP HOLMES & JERROLD MARSDEN

Communicated by D. D. JOSEPH

Abstract

This paper delineates a class of time-periodically perturbed evolution equations in a Banach space whose associated Poincaré map contains a Smale horseshoe. This implies that such systems possess periodic orbits with arbitrarily high period. The method uses techniques originally due to MELNIKOV and applies to systems of the form $\dot{x} = f_0(x) + \varepsilon f_1(x, t)$, where $\dot{x} = f_0(x)$ is Hamiltonian and has a homoclinic orbit. We give an example from structural mechanics: sinusoidally forced vibrations of a buckled beam.

§ 1. Introduction: A Physical Model

In this paper we give sufficient conditions on T-periodically forced evolution equations in a Banach space for the existence of a Smale horseshoe for the time-T map of the dynamics. This implies the existence of infinitely many periodic orbits of arbitrarily high period and suggests the existence of a strange attractor. The results here are an extension to infinite dimensions of some of those in HOLMES [1979a, b, 1980a] and CHOW, HALE & MALLET-PARET [1980].

The techniques used are invariant manifolds, nonlinear semigroups and an extension to infinite dimensions of MELNIKOV's method [1963] for planar ordinary differential equations. The results are applied to the equations of a nonlinear, periodically forced, buckled beam. As the external force is increased, we show that a global bifurcation occurs, resulting in the transversal intersection of stable and unstable manifolds. This leads to all the complex dynamics of a horseshoe (SMALE [1963]).

The study of chaotic motion in dynamical systems is now a burgeoning industry. The mechanism given here is just one of many that can lead to chaotic dynamics. For a different mechanism occuring in reaction-diffusion equations, see GUCKENHEIMER [1979].

A physical model will help motivate the analysis. Consider a beam that is buckled by an external load Γ, so there are two stable and one unstable equilibrium states (see Figure 1). The whole structure is then shaken with a transverse periodic displacement, $f \cos \omega t$. The beam moves due to its inertia. In a related experiment (see TSENG & DUGUNDJI [1971] and MOON & HOLMES [1979], and remarks below), one observes periodic motion about either of the two stable equilibria for small f, but as f is increased, the motion becomes aperiodic or chaotic. The mathematical problem is to provide theorems that help to explain this behavior.

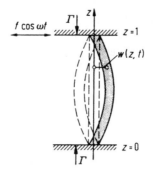

Fig. 1. The forced, buckled beam

There are a number of specific models that can be used to describe the beam in Figure 1. One of these is the following partial differential equation for the transverse deflection $w(z, t)$ of the centerline of the beam:

$$\ddot{w} + w'''' + \Gamma w'' - \varkappa \left(\int_0^1 [w']^2 \, d\zeta \right) w'' = \varepsilon(f \cos \omega t - \delta \dot{w}), \tag{1}$$

where $\dot{} = \partial/\partial t$, $' = \partial/\partial z$, Γ = external load, $\varkappa$ = stiffness due to "membrane" effects, δ = damping, and ε is a parameter used to measure the relative size of f and δ. Amongst many possible boundary conditions we shall choose $w = w'' = 0$ at $z = 0, 1$; i.e., simply supported, or hinged ends. With these boundary conditions, the eigenvalues of the linearized, unforced equations, i.e., complex numbers λ such that

$$\lambda^2 w + w'''' + \Gamma w'' = 0$$

for some non-zero w satisfying $w = w'' = 0$ at $z = 0, 1$, form a countable set

$$\lambda_j = \pm \pi j \sqrt{\Gamma - \pi^2 j^2}, \quad j = 1, 2, \ldots$$

Thus, if $\Gamma < \pi^2$, all eigenvalues are imaginary and the trivial solution $w = 0$ is formally stable; for positive damping it is Liapunov stable. We shall henceforth assume that

$$\pi^2 < \Gamma < 4\pi^2,$$

in which case the solution $w = 0$ is unstable with one positive and one negative eigenvalue and the nonlinear equation (1) with $\varepsilon = 0$, $\varkappa > 0$ has two nontrivial stable buckled equilibrium states.

A simplified model for the dynamics of (1) is obtained by seeking lowest mode solutions of the form

$$w(z, t) = x(t) \sin(\pi z).$$

Substitution into (1) and taking the inner product with the basis function $\sin(\pi z)$, gives a Duffing type equation for the modal displacement $x(t)$:

$$\ddot{x} - \beta x + \alpha x^3 = \varepsilon(\gamma \cos \omega t - \delta \dot{x}), \qquad (2)$$

where $\beta = \pi^2(\Gamma - \pi^2) > 0$, $\alpha = \varkappa \pi^4/2$ and $\gamma = 4f/\pi$. Equation (2) was studied at length in earlier papers (see Holmes [1979a, 1979b] and Holmes & Marsden [1979]). This work uses Melnikov's method; see Melnikov [1963], Arnold [1964], and Holmes [1980a]. Closely related results are obtained by Chow, Hale & Mallet-Paret [1980]. This method allows one to estimate the separation between stable and unstable manifolds and to determine when they intersect transversally. The method given in the above references applies to periodically perturbed two-dimensional flows such as the dynamics of equation (2). In this paper we extend these ideas to infinite dimensional evolution equations on Banach spaces and apply the method to the evolution equation (1).

Tseng & Dugundji [1971] studied the one and two mode Galerkin approximations of (1) and found "chaotic snap-through" motions in numerical integrations. Such motions were also found experimentally but were not studied in detail. Subsequently, Moon & Holmes [1979] found similar motions in experiments with an elastic, ferromagnetic beam and showed that a single-mode Galerkin approximation could indeed admit infinite sets of periodic motions of arbitrarily high period (Holmes [1979b]).

It is known that the time t-maps of the Euler and Navier-Stokes equations written in Lagrangian coordinates are smooth. Thus the methods of this paper apply to these equations, in principle. On regions with no boundary, one can regard the Navier-Stokes equations with forcing as a perturbation of a Hamiltonian system (the Euler equations); see Ebin & Marsden [1970]. Thus, if one knew that a homoclinic orbit existed for the Euler equations, then the methods of this paper would produce infinitely many periodic orbits with arbitrarily high period, indicative of turbulence. No specific examples of this are known to us. One possibility, however, is periodically forced surface waves. See Gollub [1980].

An unforced sine-Gordon equation possesses heteroclinic orbits, as was shown by Levi, Hoppensteadt & Miranker [1978]. Methods of this paper were used by Holmes [1980b] to show that this system, with weak periodic forcing and damping and defined on a finite spatial domain, contains horseshoes. The methods should also be useful for travelling wave problems on infinite domains, such as the Korteweg-de Vries equation.

Acknowledgements. We are grateful to the following people for their interest in this work and their comments: Haim Brezis, Jack Carr, Paul Chernoff, Shui-Nee Chow, John Guckenheimer, Jack Hale, Mark Levi, Richard McGehee, Sheldon Newhouse, David Rand and Jeffrey Rauch.

The work of PHILIP HOLMES was supported in part by the U.S. National Science Foundation Grant ENG 78-02891 and that of JERROLD MARSDEN by the U.S. National Science Foundation Grant MCS 78-06718, U.S. Army Research Office Grant DAAG 29-79C-0086, and a Killam visiting fellowship at the University of Calgary.

§ 2. Abstract Hypotheses

We consider an evolution equation in a Banach space X of the form

$$\dot{x} = f_0(x) + \varepsilon f_1(x, t) \tag{3}$$

where f_1 is periodic of period T in t. Our hypotheses on (3) are as follows:

(H1). (a) *Assume $f_0(x) = Ax + B(x)$ where A is an (unbounded) linear operator which generates a C^0 one parameter group of transformations on X and where $B: X \to X$ is C^∞. Assume that $B(0) = 0$ and $DB(0) = 0$.*

(b) *Assume $f_1: X \times S^1 \to X$ is C^∞ where $S^1 = \mathbb{R}/(T)$, the circle of length T.*

Assumption 1 implies that the associated suspended autonomous system on $X \times S^1$,

$$\begin{aligned}\dot{x} &= f_0(x) + \varepsilon f_1(x, \theta), \\ \dot{\theta} &= 1,\end{aligned} \tag{4}$$

has a smooth local flow, F_t^ε. This means that $F_t^\varepsilon: X \times S^1 \to X \times S^1$ is a smooth map defined for small $|t|$ which is jointly continuous in all variables ε, t, $x \in X$, $\theta \in S^1$ and for x_0 in the domain of A, $t \mapsto F_t^\varepsilon(x_0, \theta_0)$ is the unique solution of (4) with initial condition x_0, θ_0.

This implication results from a theorem of SEGAL [1962]. For a simplified proof, see HOLMES & MARSDEN [1978, Prop. 2.5] and for generalizations, see MARSDEN & MCCRACKEN [1976].

The final part of assumption 1 follows:

(c) *Assume that F_t^ε is defined for all $t \in \mathbb{R}$ for $\varepsilon > 0$ sufficiently small.*

To verify this in examples, one must obtain an *a priori* bound on the X-norm of solutions of (4) to ensure they do not escape to infinity in a finite time. This is sufficient by the local existence theory alluded to above. In examples of concern to us, (c) will be verified using straightforward energy estimates. See HOLMES & MARSDEN [1978] for related examples.

Our second assumption is that the unperturbed system is Hamiltonian. This means that X carries a skew symmetric continuous bilinear map $\Omega: X \times X \to \mathbb{R}$ which is weakly non-degenerate (*i.e.*, $\Omega(u, v) = 0$ for all v implies $u = 0$) called the *symplectic form* and there is a smooth function $H_0: X \to \mathbb{R}$ such that

$$\Omega(f_0(x), u) = dH_0(x) \cdot u$$

for all x in D_A, the domain of A. Consult ABRAHAM & MARSDEN [1978] and CHERNOFF & MARSDEN [1974] for details about Hamiltonian systems. For example, these assumptions imply that the unperturbed system conserves energy:

$$H_0(F_t^0(x)) = H_0(x).$$

(For $\varepsilon = 0$ we drop the dependence on θ.) We summarize this condition and further restrictions as follows:

(H2). (a) *Assume that the unperturbed system $\dot{x} = f_0(x)$ is Hamiltonian with energy $H_0: X \to \mathbb{R}$.*

(b) *Assume these is a symplectic 2-manifold $\Sigma \subset X$ invariant under the flow F_t^0 and that on Σ the fixed point $p_0 = 0$ has a homoclinic orbit $x_0(t)$, i.e.,*

$$\dot{x}_0(t) = f_0(x_0(t))$$

and

$$\lim_{t \to +\infty} x_0(t) = \lim_{t \to -\infty} x_0(t) = 0.$$

Remarks on Assumption 2.

(a) For a non-Hamiltonian two-dimensional version, see HOLMES [1980a] and CHOW, HALE & MALLET-PARET [1980]. Non-Hamiltonian infinite dimensional analogues could probably be developed by using the methods of this paper.

(b) The condition that Σ be symplectic means that Ω restricted to vectors tangent to Σ defines a non-degenerate bilinear form. We also note that by a general theorem of CHERNOFF & MARSDEN [1974], the restriction of F_t^0 to Σ is generated by a smooth vector field on Σ; *i.e.*, the dynamics within Σ is governed by *ordinary* differential equations. The situation described in assumption 2 is illustrated in Figure 2(a).

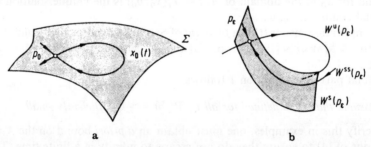

Fig. 2a. Phase portrait on Σ for $\varepsilon = 0$; b. Perturbation of invariant manifolds; $\varepsilon > 0$.

(c) Assumption 2 can be replaced by a similar assumption on the existence of heteroclinic orbits connecting two saddle points and the existence of transverse heteroclinic orbits can then be proven using the methods below. For details in the two-dimensional case, see HOLMES [1980a]. Theorems guaranteeing the existence of saddle connections may be found in CONLEY & SMOLLER [1974] and KOPELL & HOWARD [1979].

(d) To apply the techniques that follow, one must be able to calculate $x_0(t)$ either explicitly or numerically. In our examples, we find it analytically; for numerical methods, see HASSARD [1980].

Next we introduce a non-resonance hypothesis.

(H3) (a) *Assume that the forcing term $f_1(x, t)$ in (3) has the form*

$$f_1(x, t) = A_1 x + f(t) + g(x, t) \tag{5}$$

where $A_1: X \to X$ is a bounded linear operator, f is periodic with period T, $g(x, t)$ is t-periodic with period T and satisfies $g(0, t) = 0$, $D_x g(0, t) = 0$, so g admits the estimate

$$\|g(x, t)\| \leq (\text{Const}) \|x\|^2 \tag{6}$$

for x in a neighborhood of 0.

(b) *Suppose that the "linearized" system*

$$\dot{x}_L = A x_L + \varepsilon A_1 x_L + \varepsilon f(t) \tag{7}$$

has a T-periodic solution $x_L(t, \varepsilon)$ such that $x_L(t, \varepsilon) = O(\varepsilon)$.

Remarks on (H3)

1. For finite dimensional systems, (H3) can be replaced by the assumption that 1 does not lie in the spectrum of e^{TA}; i.e. none of the eigenvalues of A resonate with the forcing frequency.

2. For the beam problem, with $f(t) = f(z) \cos \omega t$, (b) means that $\omega \neq \pm \lambda_n$, $n = 1, 2, \ldots$, where $i \lambda_n$ are the purely imaginary eigenvalues of A. This is seen by solving the component forced linear oscillator equations. As we shall see, more delicate non-resonance requirements would be necessary for general (smooth) T-periodic perturbations, not of the form (5).

3. For the beam problem we can take $g = 0$. We have included it in the abstract theory for use in other examples such as the sine-Gordon equation.

Next, we need an assumption that A_1 contributes positive damping and that $p_0 = 0$ is a saddle.

(H4) (a) *For $\varepsilon = 0$, e^{TA} has a spectrum consisting in two simple real eigenvalues $e^{\pm \lambda T}$, $\lambda \neq 0$, with the rest of the spectrum on the unit circle.*

(b) *For $\varepsilon > 0$, $e^{T(A + \varepsilon A_1)}$ has a spectrum consisting in two simple real eigenvalues $e^{T \lambda_\varepsilon^\pm}$ (varying continuously in ε from perturbation theory; cf. KATO [1977]) with the rest of the spectrum, σ_R^ε, inside the unit circle $|z| = 1$ and obeying the estimates*

$$C_2 \varepsilon \leq \text{distance } (\sigma_R^\varepsilon, |z| = 1) \leq C_1 \varepsilon \tag{8}$$

for C_1, C_2 positive constants.

Remarks on (H4). 1. In general it can be awkward to estimate the spectrum of e^{TA} in terms of the spectrum of A. Some information is contained in HILLE & PHILLIPS [1957] and CARR [1980]. See also CARR & MALHARDEEN [1980], VIDAV [1970], SHIZUTA [1979] and RAUCH [1979]. For the beam problem with $\varepsilon = 0$ it is sufficient to use these two facts or a direct calculation:

(a) if A is skew adjoint, then $\sigma(e^{tA}) = $ closure of $e^{t\sigma(A)}$;

(b) if $X = X_1 \oplus X_2$, where X_2 is finite dimensional (the eigenspace of the real eigenvalues in the beam problem) and B_1 is skew adjoint on X_1 and $B_2: X_2 \to X_2$

is a (bounded) linear operator, then

$$\sigma(e^{t(B_1 \oplus B_2)}) = \text{closure } (e^{\sigma t B_1}) \cup e^{t\sigma(B_2)}.$$

For $\varepsilon > 0$ the abstract theorems are not very helpful. In the beam example the eigenfunctions of $A + \varepsilon A_1$ can be computed explicitly and form a basis for X, so the estimates (8) can be done directly; in fact σ_R^ε consists of a circle a distance $O(\varepsilon)$ inside the unit circle; see Appendix A.

2. The estimate dist $(\sigma_R^\varepsilon, |z| = 1) \geq C_2 \varepsilon$ guarantees that

$$L_\varepsilon = \text{Id} - e^{T(A + \varepsilon A_1)}$$

is invertible and

$$\|L_\varepsilon^{-1}\| \leq \text{const}/\varepsilon. \tag{9}$$

3. The estimate dist $(\sigma_R^\varepsilon, |z| = 1) \leq C_1 \varepsilon$ guarantees that the eigenvalue $\exp(T\lambda_\varepsilon^-)$ will be the closest to the origin for ε small. This is needed below for the existence of an invariant manifold corresponding to λ_ε^-.

Finally, we need an extra hypothesis on the nonlinear term. We have already assumed B vanishes at least quadratically, as does g. Now we assume B vanishes cubically.

(H5) $B(0) = 0$, $DB(0) = 0$ and $D^2 B(0) = 0$.

This means that in a neighborhood of 0,

$$\|B(x)\| \leq \text{Const } \|x\|^3.$$

(Actually $B(x) = o(\|x\|^2)$ would do).

Remarks on (H5). 1. The necessity of having B vanish cubically is due to the possibility of the spectrum of A accumulating at zero. If this can be excluded for other reasons, (H5) can be dropped and (H4) simplified. There is a similar phenomenon for ordinary differential equations noted by JACK HALE. Namely, if the linear system

$$\frac{d}{dt}\begin{Bmatrix} x \\ \dot{x} \\ y \end{Bmatrix} = \begin{Bmatrix} \dot{x} \\ x \\ -\varepsilon y \end{Bmatrix}$$

is perturbed by nonlinear terms plus forcing, to guarantee that the trivial solution (0, 0, 0) perturbs to a periodic solution as in lemma 1 below, one needs the nonlinear terms to be $o(|x| + |\dot{x}| + |y|)^3$.

2. For nonlinear wave equations, positivity of energy may force $D^2 B(0) = 0$.

Consider the suspended system (4) with its flow $F_t^\varepsilon: X \times S^1 \to X \times S^1$. Let $P^\varepsilon: X \to X$ be defined by

$$P^\varepsilon(x) = \pi_1 \cdot (F_T^\varepsilon(x, 0))$$

where $\pi_1: X \times S^1 \to X$ is the projection onto the first factor. The map P^ε is just the Poincaré map for the flow F_t^ε. Note that $P^0(p_0) = p_0$, and that fixed points of P^ε correspond to periodic orbits of F_t^ε.

Lemma 1. *For $\varepsilon > 0$ small, there is a unique fixed point p_ε of P^ε near $p_0 = 0$; moreover $p_\varepsilon - p_0 = O(\varepsilon)$, i.e. there is a constant K such that $\|p_\varepsilon\| \leq K\varepsilon$ for all (small) ε.*

For ordinary differential equations, lemma 1 is a standard fact about persistence of fixed points, assuming 1 does not lie in the spectrum of e^{TA} (i.e., p_0 is hyperbolic). For general partial differential equations, the validity of lemma 1 can be a delicate matter. In our context of smooth perturbations of linear systems with assumptions (H1)–(H5), the result is proved in Appendix A, along with the following.

Lemma 2. *For $\varepsilon > 0$ sufficiently small, the spectrum of $DP^\varepsilon(p_\varepsilon)$ lies strictly inside the unit circle with the exception of the single real eigenvalue $e^{T\lambda_\varepsilon^+} > 1$.*

In lemma 1 we saw the fixed point p_0 perturbs to another fixed point p_ε for the perturbed system. The same is true for the invariant manifolds; see Figure 2(b):

Lemma 3. *Corresponding to the eigenvalues $e^{T\lambda_\varepsilon^\pm}$ there are unique invariant manifolds $W^{ss}(p_\varepsilon)$ (the strong stable manifold) and $W^u(p_\varepsilon)$ (the unstable manifold) of p_ε for the map P^ε such that*
  i. *$W^{ss}(p_\varepsilon)$ and $W^u(p_\varepsilon)$ are tangent to the eigenspaces of $e^{T\lambda_\varepsilon^\pm}$ respectively at p_ε;*
  ii. *they are invariant under P^ε;*
  iii. *if $x \in W^{ss}(p_\varepsilon)$, then*

$$\lim_{n \to \infty} (P^\varepsilon)^n (x) = p_\varepsilon,$$

and if $x \in W^u(p_\varepsilon)$ then

$$\lim_{n \to -\infty} (P^\varepsilon)^n (n) = p_\varepsilon.$$

  iv. *For any finite t^*, $W^{ss}(p_\varepsilon)$ is C^r close as $\varepsilon \to 0$ to the homoclinic orbit $x_0(t)$, $t^* \leq t < \infty$ and for any finite t_*, $W^u(p_\varepsilon)$ is C^r close to $x_0(t)$, $-\infty < t \leq t_*$ as $\varepsilon \to 0$. Here, r is any fixed integer, $0 \leq r < \infty$.*

This lemma follows from the invariant manifold theorems (HIRSCH, PUGH & SHUB [1977]) and the smoothness of the flow of equations (4), discussed in Appendix A.

The Poincaré map P^ε was associated with the section $X \times \{0\}$ in $X \times S^1$. Equally well, we can take the section $X \times \{t_0\}$ to get Poincaré maps $P^\varepsilon_{t_0}$. By definition,

$$P^\varepsilon_{t_0}(x) = \pi_1(F^\varepsilon_T(x, t_0)).$$

[The Poincaré maps on different sections are related as follows: let $U^\varepsilon_{t,s}: X \to X$ be the evolution operators defined by $U^\varepsilon_{t,s}(x) = \pi_1(F^\varepsilon_{t-s}(x, s))$. Then $U^\varepsilon_{t,s} = U^\varepsilon_{t,r} \circ U^\varepsilon_{r,s}$ and $P^\varepsilon_{t_0} = U_{T+t_0, T+s_0} \circ P^\varepsilon_{s_0} \circ U^{-1}_{t_0, s_0}$.]

There is an analogue of Lemmas 1, 2 and 3 for $P^\varepsilon_{t_0}$. Let $p_\varepsilon(t_0)$ denote its unique fixed point and $W^{ss}_\varepsilon(p_\varepsilon(t_0))$ and $W^u_\varepsilon(p_\varepsilon(t_0))$ be its strong stable and unstable manifolds. Lemma 2 implies that the stable manifold $W^s(p_\varepsilon)$ of p_ε has codimension 1 in X. The same is then true of $W^s(p_\varepsilon(t_0))$ as well.

Let $\gamma_\varepsilon(t)$ denote the periodic orbit of the (suspended) system (4) with $\gamma_\varepsilon(0) = (p_\varepsilon, 0)$. We have

$$\gamma_\varepsilon(t) = (p_\varepsilon(t), t). \tag{10}$$

The invariant manifolds for the periodic orbit γ_ε are denoted $W_\varepsilon^{ss}(\gamma_\varepsilon)$, $W_\varepsilon^s(\gamma_\varepsilon)$ and $W_\varepsilon^u(\gamma_\varepsilon)$. We have

$$W_\varepsilon^s(p_\varepsilon(t_0)) = W_\varepsilon^s(\gamma_\varepsilon) \cap (X \times \{t_0\}),$$
$$W_\varepsilon^{ss}(p_\varepsilon(t_0)) = W_\varepsilon^{ss}(\gamma_\varepsilon) \cap (X \times \{t_0\}),$$

and

$$W_\varepsilon^u(p_\varepsilon(t_0)) = W_\varepsilon^u(\gamma_\varepsilon) \cap (X \times \{t_0\}).$$

See Figure 3.

We wish to study the structure of $W_\varepsilon^u(p_\varepsilon(t_0))$ and $W_\varepsilon^s(p_\varepsilon(t_0))$ and their intersections. To do this, we first study the perturbation of solution curves in $W_\varepsilon^{ss}(\gamma_\varepsilon)$, $W_\varepsilon^s(\gamma_\varepsilon)$ and $W_\varepsilon^u(\gamma_\varepsilon)$.

Choose a point, say $x_0(0)$ on the homoclinic orbit for the unperturbed system. Choose a codimension 1 hyperplane H transverse to the homoclinic orbit at $x_0(0)$. Since $W_\varepsilon^{ss}(p_\varepsilon(t_0))$ is C^r close to $x_0(0)$, it intersects H in a unique point, say $x_\varepsilon^s(t_0, t_0)$. Define $(x_\varepsilon^s(t, t_0), t)$ to be the unique integral curve of the suspended system (4) with initial condition $x_\varepsilon^s(t_0, t_0)$. Define $x_\varepsilon^u(t, t_0)$ in a similar way.

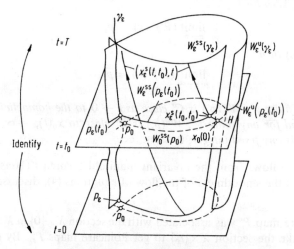

Fig. 3. The perturbed manifolds

The initial conditions $x_\varepsilon^s(t_0, t_0)$ and $x_\varepsilon^u(t_0, t_0)$ are not conveniently computable. This difficulty is, however, unimportant and is taken care of by the boundary conditions at $t \to \pm \infty$. We have

$$x_\varepsilon^s(t_0, t_0) = x_0(0) + \varepsilon v^s + O(\varepsilon^2)$$
and
$$x_\varepsilon^u(t_0, t_0) = x_0(0) + \varepsilon v^u + O(\varepsilon^2)$$
(11)

by construction, where $\|O(\varepsilon^2)\| \leq \text{Constant} \cdot \varepsilon^2$ and v^s and v^u are fixed vectors. Notice that

$$(P_{t_0}^\varepsilon)^n x_\varepsilon^s(t_0, t_0) = x_\varepsilon^s(t_0 + nT, t_0) \to p_\varepsilon(t_0) \text{ as } n \to \infty.$$

Since $x_\varepsilon^s(t, t_0)$ is an integral curve of a perturbation, we can write

$$x_\varepsilon^s(t, t_0) = x_0(t - t_0) + \varepsilon x_1^s(t, t_0) + O(\varepsilon^2), \tag{12}$$

where $x_1^s(t, t_0)$ is the solution of the first variation equation

$$\frac{d}{dt} x_1^s(t, t_0) = Df_0(x_0(t - t_0)) \cdot x_1^s(t, t_0) + f_1(x_0(t - t_0), t), \tag{13}$$

with

$$x_1^s(t_0, t_0) = v^s.$$

This linearization procedure is justified by the proof of smoothness of the time t-maps; see Appendix A. There is a similar formula for $x_\varepsilon^u(t, t_0)$. Notice that when $\varepsilon \to 0$, the curve $x_\varepsilon^s(t, t_0)$ approaches the homoclinic orbit $x(t - t_0)$ with a phase shift t_0. In (12), $O(\varepsilon^2)$ means a term bounded by a constant $\times \varepsilon^2$, on each finite time interval. For $x_\varepsilon^s(t, t_0)$, the error $O(\varepsilon^2)$ is uniform as $t \to +\infty$ since $(x_\varepsilon^s(t, t_0), t)$ converges to the periodic orbit $\gamma_\varepsilon(t)$, by construction. Similarly the error $O(\varepsilon^2)$ in the corresponding equation for $x_\varepsilon^u(t, t_0)$ is uniform as $t \to -\infty$.

§ 3. The Melnikov Function

Recall that $\Omega: X \times X \to \mathbb{R}$ denotes the symplectic form on X relative to which f_0 is Hamiltonian. Define the *Melnikov function* by

$$\Delta_\varepsilon(t, t_0) = \Omega(f_0(x_0(t - t_0)), x_\varepsilon^s(t, t_0) - x_\varepsilon^u(t, t_0)) \tag{14}$$

and set

$$\Delta_\varepsilon(t_0) = \Delta_\varepsilon(t_0, t_0).$$

Lemma 4. *If ε is sufficiently small and $\Delta_\varepsilon(t_0)$ has a simple zero at some t_0 and maxima and minima that are at least $O(\varepsilon)$, then $W_\varepsilon^u(p_\varepsilon(t_0))$ and $W_\varepsilon^s(p_\varepsilon(t_0))$ intersect transversally near $x_0(0)$.*

Proof. First note that, by lemma 2, $W_\varepsilon^s(p_\varepsilon)$ has codimension 1. As $\varepsilon \to 0$, the perturbation theory of invariant manifolds shows that $W_\varepsilon^s(p_\varepsilon) \xrightarrow{C^r} W^{sc}(p_0)$, where $W^{sc}(p_0)$ is the center-stable manifold for F_T, the time T map for the unperturbed system.

Let $d_\varepsilon(t, t_0) = x_\varepsilon^s(t, t_0) - x_\varepsilon^u(t, t_0)$ and $d_\varepsilon(t_0) = d_\varepsilon(t_0, t_0)$.

Let $T_{x_0(0)}\Sigma$ be the tangent space to Σ at $x_0(0)$ and let

$$(T_{x_0(0)}\Sigma)^\Omega = \{v \in X \mid \Omega(v, w) = 0 \text{ for all } w \in T_{x_0(0)}\Sigma\}$$

be its Ω-orthogonal complement. Because Σ is finite (2)-dimensional and symplectic,

$$X = T_{x_0(0)}\Sigma \oplus (T_{x_0(0)}\Sigma)^\Omega. \tag{15}$$

Let $P: X \to T_{x_0(0)}\Sigma$ be the projection onto the first factor, associated with the decomposition (15). Set

$$d_\varepsilon^\Sigma(t_0) = P(d_\varepsilon(t_0)) \tag{16}$$

(see Fig. 4).

By definition,

$$\Delta_\varepsilon(t_0) = \Omega(f_0(x_0(0)), d_\varepsilon(t_0)) = \Omega(f_0(x_0(0)), d_\varepsilon^\Sigma(t_0)).$$

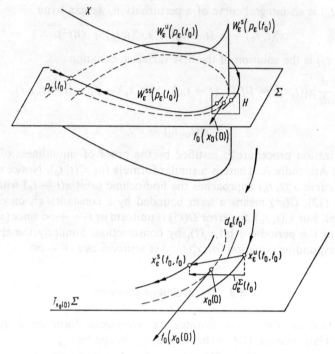

Fig. 4. Geometry for the Melnikov function

If $\Delta_\varepsilon(t_0)$ has a simple zero, then as a function of t_0, $d_\varepsilon^\Sigma(t_0)$ changes orientation relative to $f_0(x_0(0))$ within Σ as t_0 changes. Thus, as $W_0^{sc}(p_0)$ is codimension 1, $d_\varepsilon^\Sigma(t_0)$ also changes orientation relative to it.

The tangent space of $W_\varepsilon^s(p_\varepsilon(t_0))$ near $x_0(t)$ is ε-close to that of $W_0^{sc}(p_0)$; and the tangent space of $W_\varepsilon^u(p_\varepsilon(t_0))$ near $x_0(0)$ is ε-close to the vector $f_0(x_0(0))$, uniformly for $0 \leq t_0 \leq T$. This follows from the perturbation theory of invariant manifolds. Let $v_\varepsilon(t_0)$ denote the vector from $x_\varepsilon^u(t_0, t_0)$ to the nearest point on $W_\varepsilon^s(p_\varepsilon(t_0))$ (it is easy to see that there is a unique such point in an ε-neighborhood of $x_0(0)$). By (11) and the tangent space estimates just discussed, it follows that

$$v_\varepsilon(t_0) = d_\varepsilon^\Sigma(t_0) + O(\varepsilon^2). \tag{17}$$

Thus if $d_\varepsilon^\Sigma(t_0)$ passes through zero, changing orientation relative to $f_0(x_0(0))$ with an amplitude $O(\varepsilon)$, then by (17), $v_\varepsilon(t_0)$ must do the same. It follows that $W_\varepsilon^u(p_\varepsilon(t_0))$ and $W_\varepsilon^s(p_\varepsilon(t_0))$ then intersect transversally near the t_0 at which $\Delta_\varepsilon(t_0)$ has its zero. ∎

The next lemma gives a formula for $\Delta_\varepsilon(t_0)$ that uses the Hamiltonian nature of f_0. This formula is needed in examples to check effectively that $\Delta_\varepsilon(t_0)$ has simple zeros.

Lemma 5. *The following formula holds:*

$$\Delta_\varepsilon(t_0) = -\varepsilon \int_{-\infty}^{\infty} \Omega(f_0(x_0(t-t_0)), f_1(x_0(t-t_0)), t)\, dt + O(\varepsilon^2). \tag{18}$$

Proof. By (12) we can write $\Delta_\varepsilon(t, t_0) = \Delta_\varepsilon^+(t, t_0) - \Delta_\varepsilon^-(t, t_0) + O(\varepsilon^2)$, where

$$\Delta_\varepsilon^+(t, t_0) = \Omega(f_0(x_0(t - t_0)), \varepsilon x_1^s(t, t_0))$$

and

$$\Delta_\varepsilon^-(t, t_0) = \Omega(f_0(x_0(t - t_0)), \varepsilon x_1^u(t, t_0)).$$

Using (13), we get

$$\frac{d}{dt}\Delta_\varepsilon^+(t, t_0) = \Omega(Df_0(x_0(t, t_0)) \cdot f_0(x_0(t - t_0)), \varepsilon x_1^s(t, t_0))$$

$$+ \Omega(f_0(x_0(t - t_0)), \varepsilon\{Df_0(x_0(t - t_0)) \cdot x_1^s(t, t_0) + f_1(x_0(t - t_0), t)\}).$$

Since f_0 is Hamiltonian, Df_0 is Ω-skew. Therefore the terms involving x_1^s drop out, leaving

$$\frac{d}{dt}\Delta_\varepsilon^+(t, t_0) = \Omega(f_0(x_0(t - t_0)), \varepsilon f_1(x_0(t - t_0), t)).$$

Integrating, we have

$$-\Delta_\varepsilon^+(t_0, t_0) = \varepsilon \int_{t_0}^{\infty} \Omega(f_0(x_0(t - t_0)), f_1(x_0(t - t_0), t)) \, dt, \quad (19)$$

since

$$\Delta_\varepsilon^+(\infty, t_0) = \Omega(f_0(p_0), \varepsilon f_1(p_0, \infty)) = 0, \text{ because } f_0(p_0) = 0.$$

Similarly, we obtain

$$\Delta_\varepsilon^-(t_0, t_0) = \varepsilon \int_{-\infty}^{t_0} \Omega(f_0(x_0(t - t_0)), f_1(x_0(t - t_0), t)) \, dt$$

and adding gives the stated formula. ∎

The expression $\int_{-\infty}^{\infty} \Omega(f_0(x_0(t - t_0)), f_1(x_0(t - t_0), t)) \, dt$ is an "averaged bracket" over the orbit $x_0(t)$; if f_1 is Hamiltonian (time dependent), this is just an integrated Poisson bracket over the orbit $x_0(t)$; cf. ARNOLD [1964]. The power of MELNIKOV's method rests in the fact that this formula renders the leading term of $\Delta_\varepsilon(t_0)$ computable.

We summarize the situation as follows:

Theorem 1. *Let hypotheses 1–5 hold. Let*

$$M(t_0) = \int_{-\infty}^{\infty} \Omega(f_0(x_0(t - t_0)), f_1(x_0(t - t_0), t)) \, dt. \quad (20)$$

Suppose that $M(t_0)$ has a simple zero as a function of t_0. Then for $\varepsilon > 0$ sufficiently small, the stable manifold $W_\varepsilon^s(p_\varepsilon(t_0))$ of p_ε for $P_{t_0}^\varepsilon$ and the unstable manifold $W_\varepsilon^u(p_\varepsilon(t_0))$ intersect transversally. (We shall also call M the Melnikov function).

In Section 5 we discuss consequences of this result. Before doing so we discuss some examples related to the physical model of the beam.

§ 4. Examples

1. (See HOLMES [1979 b]): Consider the forced Duffing equation

$$\ddot{x} - \beta x + \alpha x^3 = \varepsilon(\gamma \cos \omega t - \delta \dot{x}).$$

Here the unperturbed ($\varepsilon = 0$) system is $\ddot{x} - \beta x + \alpha x^3 = 0$, i.e.,

$$\frac{d}{dt}\begin{pmatrix} x \\ \dot{x} \end{pmatrix} = \begin{pmatrix} \dot{x} \\ \beta x - \alpha x^3 \end{pmatrix} \tag{21}$$

which is Hamiltonian in $X = \mathbb{R}^2 = \Sigma$ with

$$H(x, \dot{x}) = \frac{\dot{x}^2}{2} - \frac{\beta x^2}{2} + \frac{\alpha x^4}{4}. \tag{22}$$

The flow of this system is the "figure eight" pattern shown in Figure 5(a). The homoclinic orbit is given by

$$(x_0(t), \dot{x}_0(t)) = \left(\sqrt{\frac{2\beta}{\alpha}} \operatorname{sech}(\sqrt{\beta} t), -\beta \sqrt{\frac{2}{\alpha}} \operatorname{sech} \sqrt{\beta} t \tanh \sqrt{\beta} t \right). \tag{23}$$

We have based the solution at $(x_0(0), \dot{x}_0(0)) = (\sqrt{2\beta/\alpha}, 0)$. The symplectic form is $\Omega((x, \dot{x}), (y, \dot{y})) = x\dot{y} - \dot{x}y$, so by (20) the Melnikov function is

$$M(t_0) = \int_{-\infty}^{\infty} \Omega\left(\begin{pmatrix} \dot{x} \\ \beta x - \alpha x^3 \end{pmatrix}, \begin{pmatrix} 0 \\ \gamma \cos \omega t - \delta \dot{x} \end{pmatrix}\right) dt = \int_{-\infty}^{\infty} \dot{x}(\gamma \cos \omega t - \delta \dot{x}) \, dt, \tag{24}$$

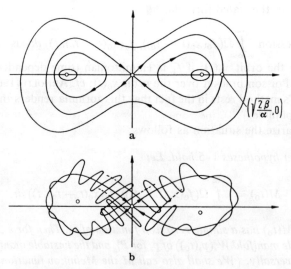

Fig. 5a and b. The single mode model.
a Unperturbed case; $\varepsilon = 0$; **b** Perturbed Case; $\varepsilon >$ small, $\gamma > \gamma_c$

where x and $\dot{x}$ are given in (23) (with t replaced by $t - t_0$). The integral (24) may be evaluated by standard methods (HOLMES [1979b]) to yield

$$M(t_0) = -\frac{4\delta\beta^{\frac{3}{2}}}{3\alpha} + 2\gamma\omega\sqrt{\frac{2}{\alpha}}\frac{\sin(\omega t_0)}{\cosh\left(\frac{\pi\omega}{2\sqrt{\beta}}\right)}. \tag{25}$$

Thus, if

$$\gamma > \gamma_c = \frac{2\delta\beta^{\frac{3}{2}}}{3\omega\sqrt{2\alpha}}\cosh\left(\frac{\pi\omega}{2\sqrt{\beta}}\right), \tag{26}$$

then M has simple zeros and so by Theorem 1 the stable and unstable manifolds intersect transversely for ε sufficiently small. See Fig. 5(b).

2. The two mode Galerkin approximation of the beam equation (1) is given as follows (*cf.* TSENG & DUGUNDJI [1971], HOLMES [1979a]):

$$\frac{d}{dt}\begin{pmatrix}x\\ \dot{x}\\ y\\ \dot{y}\end{pmatrix} = \begin{pmatrix}\dot{x}\\ \beta_1 x - \alpha(x^2 + 4y^2)x\\ \dot{y}\\ -\beta_2 x - 4\alpha(x^2 + 4y^2)y\end{pmatrix} + \varepsilon\begin{pmatrix}0\\ \gamma_1\cos\omega t - \delta_1\dot{x}\\ 0\\ \gamma_2\cos\omega t - \delta_2\dot{y}\end{pmatrix} \tag{27}$$

where $\beta_1 = \pi\sqrt{\Gamma - \pi^2} > 0$, $\beta_2 = \pi\sqrt{4\pi^2 - \Gamma} > 0$ and $\alpha = \frac{\varkappa\pi^4}{2}$.

Here the plane Σ, spanned by the vectors $(1, 0, 0, 0)^T$ and $(0, 1, 0, 0)^T$, is a symplectic 2 manifold and the unperturbed homoclinic orbit is given by

$$(x_0(t_0), \dot{x}_0(t_0), y_0(t_0), \dot{y}_0(t_0))$$
$$= \left(\sqrt{\frac{2\beta}{\alpha}}\operatorname{sech}(\sqrt{\beta}t), -\beta\sqrt{\frac{2}{\alpha}}\operatorname{sech}\sqrt{\beta}t\tanh\sqrt{\beta}t, 0, 0\right). \tag{28}$$

The Melnikov function is found to be

$$M(t_0) = \int_{-\infty}^{\infty}\dot{x}[(\gamma_1\cos\omega t - \delta_1\dot{x}) + \dot{y}(\gamma_2\cos\omega t - \delta_2\dot{y})]\,dt, \tag{29}$$

and since $\dot{y}_0 = 0$ by (28), the computation of (29) reduces to that of Example 1. The non-resonance condition (H3) for this example becomes $\omega^2 \neq \beta_2$.

3. The partial differential equation of the beam:

$$\ddot{w} + w'''' + \Gamma w'' - \varkappa\left(\int_0^1 [w']^2\,d\zeta\right)w'' = \varepsilon(f\cos\omega t - \delta\dot{w})$$

with boundary conditions

$$w = w'' = 0 \text{ at } z = 0, 1.$$

The basic space is $X = H_0^2 \times L^2$ where H_0^2 denotes the set of all H^2 functions on $[0, 1]$ satisfying the boundary conditions $w = 0$ at $z = 0, 1$. For $x = (w, \dot{w}) \in X$, the X-norm is the "energy" norm $\|x\|^2 = |w''|^2 + |\dot{w}|^2$ where $|\cdot|$ denotes the

L_2 norm. Write the equation as

$$\frac{dx}{dt} = f_0(x) + \varepsilon f_1(x, t),$$

where

$$f_0(x) = Ax + B(x) \text{ and } f_1(x, t) = \begin{pmatrix} 0 \\ f \cos \omega t - \delta \dot{w} \end{pmatrix} = A_1 x + f(t).$$

Here A and A_1 are the linear operators

$$A \begin{pmatrix} w \\ \dot{w} \end{pmatrix} = \begin{pmatrix} \dot{w} \\ -w'''' - \Gamma w'' \end{pmatrix}, A_1 \begin{bmatrix} w \\ \dot{w} \end{bmatrix} = \begin{bmatrix} 0 \\ -\delta \dot{w} \end{bmatrix},$$

with domains $D(A_1) = X$ (A_1 is bounded) and

$$D(A) = \{(w, \dot{w}) \in H^4 \times H^2 \mid w = 0, w'' = 0 \text{ and } \dot{w} = 0 \text{ at } z = 0, 1\},$$

and B is the nonlinear mapping of X to X given by

$$B(x) = \begin{pmatrix} 0 \\ \varkappa \left(\int_0^1 [w']^2 \, d\zeta \, w'' \right) \end{pmatrix}.$$

In the forcing term $f(t) = f \cos \omega t$, f can be a function of z, i.e., a spatially distributed load. Let $\bar{f}$ denote the mean of f. We expand f in a Fourier series with period twice the beam length:

$$f(z) = \bar{f} + \sum_{n=1}^{\infty} \{\alpha_n \sin(n\pi z) + \beta_n \cos(n\pi z)\}.$$

The coefficients $\bar{f}$ and α_1 will be important in calculations that follow.

The theorems of HOLMES & MARSDEN [1978] show that A is a generator and that B and f_1 are smooth maps. This, together with the energy estimates, shows that the equations generate a global flow $F_t^\varepsilon: X \times S^1 \to X \times S^1$ consisting of C^∞ maps for each ε and t. See appendix A for details. If x_0 lies in the domain of the (unbounded) operator A, then $F_t^\varepsilon(x_0, s)$ is t-differentiable and the equation is literally satisfied. Thus Hypothesis 1 holds.

For $\varepsilon = 0$ the equation is readily verified to be Hamiltonian using the symplectic form

$$\Omega((w_1, \dot{w}_1), (w_2, \dot{w}_2)) = \int_0^1 \{\dot{w}_2(z) w_1(z) - \dot{w}_1(z) w_2(z)\} \, dz$$

and

$$H(w, \dot{w}) = \frac{1}{2} |\dot{w}|^2 - \frac{\Gamma}{2} |w'|^2 + \frac{1}{2} |w''|^2 + \frac{\varkappa}{4} |w'|^4.$$

The invariant symplectic 2-manifold Σ is the plane in X spanned by the functions $(a \sin \pi z, b \sin \pi z)$ and the homoclinic loop is given by

$$w_0(z, t) = \frac{2}{\pi} \sqrt{\frac{\Gamma - \pi^2}{\varkappa}} \sin(\pi z) \operatorname{sech}(t\pi \sqrt{\Gamma - \pi^2}).$$

Hypothesis 2 therefore holds. For $\varepsilon = 0$ one finds by direct calculation that the spectrum of $Df_0(p_0)$, where $p_0 = (0, 0)$, is discrete with two real eigenvalues

$$\pm \lambda = \pm \pi \sqrt{\Gamma - \pi^2}$$

and the remainder pure imaginary (since $\Gamma < 4\pi^2$) at

$$\lambda_n = \pm n\pi \sqrt{\Gamma - n^2\pi^2}, \quad n = 2, 3, \ldots$$

(*cf.* HOLMES [1979a]). Hypothesis (H3a) clearly holds with $g \equiv 0$. Condition (H3b) holds by direct calculation using Fourier series and the following nonresonance assumption:

$$j^2\pi^2(j^2\pi^2 - \Gamma) \neq \omega^2, \quad j = 2, 3, 4, \ldots \quad (30)$$

Expanding $w(z, t)$ in the eigenfunctions $\sin j\pi z$ of the linearized problem and using Galerkin's method we obtain an infinite set of second order ordinary differential equations for the modal coefficient $a_j(t)$:

$$\ddot{a}_j + \varepsilon \delta \dot{a}_j + j^2\pi^2(j^2\pi^2 - \Gamma) a_j = \varepsilon \gamma_j \cos \omega t, \quad j = 1, 2, \ldots, \quad (31)$$

where

$$w(z, t) = \sum_{j=1}^{\infty} a_j(t) \sin(j\pi z) \text{ and } \gamma_j = \int_0^1 f(z) \sin(j\pi z) \, dz.$$

It is easy to check that if (30) holds, then (31) has a unique periodic solution $w_L(z, t) = w_L(z, t + 2\pi/w)$ of $O(\varepsilon)$. Moreover, the eigenvalues of the perturbed operator $e^{T(A + \varepsilon A_1)}$ and the unperturbed operator e^{TA} can be calculated directly from (31), with the γ_j set to zero (*cf.* HOLMES [1979a]). One obtains a countable set of eigenvalues for the flow given by

$$\lambda_n^\varepsilon = \frac{1}{2}\left[-\varepsilon \delta \pm \sqrt{\varepsilon^2 \delta^2 + n^2\pi^2(\Gamma - n^2\pi^2)}\right], \quad n = 1, 2, \ldots \quad (32)$$

Exponentiation of (32) reveals that hypothesis (H4) holds (here $T = 2\pi/\omega$). Finally, it is clear that (H5) also holds for this example.

The Melnikov function (20) is given by

$$M(t_0) = \int_{-\infty}^{\infty} \Omega \left(\begin{matrix} \dot{w} \\ -w'''' + \varkappa |w'|^2 w'' - \Gamma w'' \end{matrix}, \begin{matrix} 0 \\ f \cos \omega t - \delta \dot{w} \end{matrix}\right) dt$$

$$= \int_{-\infty}^{\infty} \left(\int_0^1 f \cos \omega t \, \dot{w}(z, t - t_0) - \delta \dot{w}(z, t - t_0) \, \dot{w}(z, t - t_0) \, dz\right) dt.$$

Substituting the expressions for $w, \dot{w}$ along the homoclinic orbit, we evaluate the integral as in Example 1. One gets

$$-M(t_0) = \frac{2\omega}{\pi} \sqrt{\frac{\Gamma - \pi^2}{\varkappa}} \left(\frac{\alpha_1}{2} + \frac{2\bar{f}}{\pi}\right) \frac{\sin(\omega t_0)}{\cosh\left(\frac{\omega}{2\sqrt{\Gamma - \pi^2}}\right)} + \frac{4\delta(\Gamma - \pi^2)^{\frac{3}{2}}}{3\pi\varkappa}$$

Thus, if

$$\left|\frac{\alpha_1}{2} + \frac{2\bar{f}}{\pi}\right| > \frac{2\delta(\Gamma - \pi^2)}{3\omega\sqrt{\varkappa}} \left[\cosh\left(\frac{\omega}{2\sqrt{\Gamma - \pi^2}}\right)\right],$$

then the hypotheses of Theorem 1 hold and so the stable and unstable manifolds intersect transversally. Note that in the spatial integral evaluated, in the expression for Ω, only the components $\bar{f}$ and α_1 of $f(z)$ survive, due to orthogonality of the other Fourier components with the solution

$$\dot{w}_0(t) = \frac{2}{\sqrt{\varkappa}} (\Gamma - \pi^2) \sin(\pi z) \operatorname{sech}(t\pi\sqrt{\Gamma - \pi^2}) \tanh(t\pi\sqrt{\Gamma - \pi^2}).$$

It should be realized that, while the formal calculations of $M(t_0)$ in the second and third examples are similar to that of $M(t_0)$ in the two dimensional example given first, the full power of Theorem 1 is necessary since in the four and infinite dimensional cases, the perturbed manifolds $W^{ss}_\varepsilon(p_\varepsilon(t_0))$ and $W^u_\varepsilon(p_\varepsilon(t_0))$ do not lie in Σ.

We close this section with a comment on the bifurcations in which the transversal intersections are created. Since the Melnikov function $M(t_0)$ has nondegenerate maxima and minima in all three examples, it can be shown that, near the parameter values at which $M(t_0) = M'(t_0) = 0$, but $M''(t_0) \neq 0$, the stable and unstable manifolds $W^s_\varepsilon(p_\varepsilon(t_0))$, $W^u_\varepsilon(p_\varepsilon(t_0))$ have quadratic tangencies. The mechanism, described by NEWHOUSE [1974, 1979], then implies that $P^\varepsilon_{t_0}$ can have infinitely many stable periodic orbits of arbitrarily high periods near the bifurcation point, at least in the first two (finite dimensional) examples. In practice it may be difficult to distinguish these long period stable periodic points with their small basins, from transient chaos, noise and from "true" chaos itself. We note that transient chaos has been observed in experimental work (MOON & HOLMES [1979]).

§ 5. Consequences of Transversal Intersection

If the hypotheses of Theorem 1 hold, we obtain a Poincaré map $P^\varepsilon_{t_0}: X \to X$ having the following property: there is a hyperbolic saddle point p_ε which has a 1 dimensional unstable manifold intersecting a codimension 1 stable manifold transversally. For $X = \mathbb{R}^2$, this situation implies that the dynamics contains a horseshoe (see SMALE [1967]). For instance, one can conclude the existence of infinitely many periodic points with arbitrarily high period. See Figure 6. Together with global attractivity due to positive damping, this suggests the presence of a strange attractor (cf. HOLMES [1979a, b]).

A particularly noteworthy method for analysis in such cases has been given by CONLEY & MOSER; see MOSER [1973], Chapter III. The attractive feature of their method is that it reduces the proof to one of finding *explicit estimates* on what $P^\varepsilon_{t_0}$ does to horizontal and vertical strips near the saddle point. This enables one to generalize the argument to dimension ≥ 2 and to Banach spaces X. In particular, it applies to the beam example. Specifically, we prove the following result in Appendix B.

Theorem 2. *If the diffeomorphism $P^\varepsilon_{t_0}: X \to X$ possesses a hyperbolic saddle point p_ε and an associated transverse homoclinic point $q \in W^u_\varepsilon(p_\varepsilon) \cap W^s_\varepsilon(p_\varepsilon)$, with $W^u_\varepsilon(p_\varepsilon)$*

of dimension 1 and $W^s_\varepsilon(p_\varepsilon)$ of codimension 1, then some power of $P^\varepsilon_{t_0}$ possesses an invariant zero dimensional hyperbolic set Λ homeomorphic to a Cantor set on which a power of $P^\varepsilon_{t_0}$ is conjugate to a shift on two symbols.

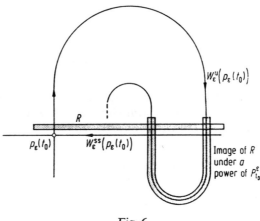

Fig. 6

As in the finite dimensional case, this implies

Corollary 1. *A power of $P^\varepsilon_{t_0}$ restricted to Λ possesses a dense set of periodic points there are points of arbitrarily high period and there is a non-periodic orbit dense in Λ.*

The hyperbolicity of Λ under a power of $P^\varepsilon_{t_0}$ and the theorem on structural stability of ROBBIN [1971] implies that the situation of Theorem 2 persists under perturbations:

Corollary 2. *If $\overline{P}: X \to X$ is a diffeomorphism that is sufficiently close to $P^\varepsilon_{t_0}$ in C^1 norm, then a power of $\overline{P}$ has an invariant set $\overline{\Lambda}$ and there is a homeomorphism $h: \overline{\Lambda} \to \Lambda$ such that $(P^\varepsilon_{t_0})^N \circ h = h \circ \overline{P}^N$ for a suitable power N.*

Thus, the complex dynamics of $(P^\varepsilon_{t_0})^N$ near Λ cannot be removed by making small changes in lower order (bounded) terms in the governing equations (3).

Although the dynamics near Λ is complex, we do not assert that Λ is a strange attractor. In fact, Λ is unstable in the sense that its generalized unstable manifold (or outset), $W^u(\Lambda)$ is non-empty (it is one dimensional) and thus points starting near Λ may wander, remaining near Λ for a relatively long time, but eventually leaving a neighborhood of Λ and approaching an attractor. This kind of behavior has been referred to as *transient chaos* (or pre-turbulence). In two dimensions (Example 1 of § 4), Λ can coexist with two simple sinks of period one or with a strange attractor, depending on the parameter values (see HOLMES [1979b]). As noted earlier, there is experimental evidence for transient chaos in the magnetic cantilever problem, in addition to the evidence for sustained non-periodic notions (HOLMES & MOON [1979]).

Appendix A. Perturbations of Fixed Points and Invariant Manifolds for Partial Differential Equations

We begin by recalling the local existence result.

Proposition A.1. *Let X be a Banach space and U_t a linear semigroup on X with generator A and domain $D(A)$. Let $B: X \to X$ be C^k, $k \geq 1$. Let $G = A + B$ on $D(A)$. Then*

$$\frac{dx}{dt} = G(x); \quad x_0 = x(0) \in X \qquad \text{(A-1)}$$

defines a unique local semiflow $F_t(x_0)$: If $x_0 \in D(A)$, then $F_t(x_0) \in D(A) = D(G)$, is X-differentiable and satisfies (A-1) *with initial condition x_0, $F_t(x_0)$ is the unique such solution and moreover, F_t extends to a C^k map of an open set in X to X for each $t \geq 0$.*

The basic idea is to use Picard iteration on the corresponding integral equation

$$x(t) = U_t x_0 + \int_0^t U_{t-s} B(x(s)) \, ds, \qquad \text{(A-2)}$$

where $U_t = e^{tA}$. One sets $F_t(x_0) = x(t)$. For details, see HOLMES & MARSDEN [1978], Prop. 2.5.

Suppose $G = A + B$ and $\bar{G} = A + \bar{B}$ both satisfy the conditions of proposition A.1. and F_t and $\bar{F}_t$ are the corresponding semiflows.

Proposition A.2. *We have the following estimates on $F_t - \bar{F}_t$: fix $T > 0$ and suppose F_t and $\bar{F}_t$ map the bounded set $S \subset X$ into the ball B_R of radius R and that on B_R we have*

(i) $$\sup_{x \in B_R} \| B(x) - \bar{B}(x) \| < C,$$

(ii) $$\| B(x) - B(y) \| \leq K \| x - y \|$$

and assume

(iii) $$\| e^{tA} \| \leq M e^{|t|\beta} \text{ for } M > 0, \beta > 0.$$

Let $\bar{M} = M e^{T\beta}$. Then

$$\| F_t(t) - \bar{F}_t(x) \| \leq \bar{M} C T e^{K|t|\bar{M}}. \qquad \text{(A-3)}$$

Furthermore, assume that for $x, y \in B_R$,

(iv) $$\| [DB(x) - DB(y)] \cdot v \| \leq K_1 \|v\| \|x - y\|,$$

(v) $$\| [DB(x) - D\bar{B}(x)] \cdot v \| \leq C_1 \|v\|,$$

(vi) $$\| DB(x) \cdot v \| \leq M_2 \|v\|$$

and (vii) *for $x \in S$, $\| D\bar{F}_t(x) \cdot u \| \leq M_3 \|u\|$ for $|t| \leq T$. Then*

$$\| [DF_t(x) - D\bar{F}_t(x)] \cdot u \| \leq T M_3 \bar{M} (C_1 + K_1 \bar{M} C T e^{kT\bar{M}}) e^{M_2 |t| \bar{M}} \|u\|. \qquad \text{(A-4)}$$

In particular, if B and $\bar{B}$ are close in C^1 norm on bounded sets then $DF_t(x)$ and $D\bar{F}_t(x)$ are close in the *operator* norm; *cf.* Prop. 2.13 of HOLMES & MARSDEN [1978].

Proof. Let $x(t) = F_t(x)$ and $\bar{x}(t) = \bar{F}_t(x)$. From (A-2) we have

$$x(t) - \bar{x}(t) = \int_0^t U_{t-s}[B(x(s)) - \bar{B}(\bar{x}(s))]\,ds$$

$$= \int_0^t U_{t-s}[B(x(s)) - B(\bar{x}(s))]\,ds$$

$$+ \int_0^t U_{t-s}[B(\bar{x}(s)) - \bar{B}(\bar{x}(s))]\,ds.$$

Thus from (i), (ii) and (iii),

$$\|x(t) - \bar{x}(t)\| \leq MK \int_0^t \|x(s) - \bar{x}(s)\|\,ds + \bar{M}CT.$$

Estimate (A-3) thus follows. To prove (A-4) we recall (HOLMES & MARSDEN [1978, Eq. 14]) that $DF_t(x)$ satisfies the first variation equation:

$$DF_t(x)\,u = U_t \cdot u + \int_0^t U_{t-s} DB(F_s(x)) \cdot [DF_s(x) \cdot u]\,ds.$$

(Note that from this one can choose $M_3 = \bar{M}e^{\bar{M}M_2T}$.) Therefore

$DF_t(x) \cdot u - D\bar{F}_t(x) \cdot u$

$$= \int_0^t U_{t-s}[DB(F_s(x)) \cdot (DF_s(x) \cdot u) - D\bar{B}(\bar{F}_s(x)) \cdot (D\bar{F}_s(x) \cdot u)]\,ds$$

$$= \int_0^t U_{t-s}[DB(F_s(x)) \cdot \{DF_s(x) \cdot u - D\bar{F}_s(x) \cdot u\}\,ds]$$

$$+ \int_0^t U_{t-s}[\{DB(F_s(x)) - DB(\bar{F}_s(x))\} \cdot \{D\bar{F}_s(x) \cdot u\}]\,ds$$

$$+ \int_0^t U_{t-s}\{DB(\bar{F}_s(x)) - D\bar{B}(\bar{F}_s(x))\} \cdot \{D\bar{F}_s(x) \cdot u\}\,ds.$$

Thus $\|DF_t(x) \cdot u - D\bar{F}_t(x) \cdot u\| \leq \bar{M}M_2 \int_0^t \|DF(x) \cdot u - D\bar{F}_s(x) \cdot u\|\,ds +$
$+ \bar{M}K_1\bar{M}CT^2 e^{KT\bar{M}} M_3 \|u\| + \bar{M}C_1TM_3\|u\|$,
from which (A-4) follows. ∎

Condition (ii) holds if DB is bounded on B_R and (iii) is automatic for any C^0 semigroup and serves only to define the constants. Condition (iv) holds if D^2B is bounded on B_R, (vi) just says DB is bounded on B_R, and we have already noted that one can choose $M_3 = \bar{M}e^{\bar{M}M_2T}$ to obtain (vii).

Proposition A.3. *Under assumptions* 1(a), (b) *and* (c) *of* §2, *the bounded linear operators* $DP^\varepsilon(p_0): X \to X$ *converge in norm as* $\varepsilon \to 0$ *to* $DP^0(p_0)$.

Proof. As $\varepsilon \to 0$, $\varepsilon f_1(x, t) \to 0$ locally in x (uniformly in t) along with its derivative. Thus, by proposition A.2 $DF_t^\varepsilon(x) \to DF_t^0(x)$ as $\varepsilon \to 0$, where the convergence is in norm. Since $P^\varepsilon(x) = \pi_1 \cdot (F_T^\varepsilon(x, 0))$, the result follows. ∎

Remarks. (a) Norm convergence of evolution operators in general is not to be expected, even for linear operators. It is true here because the unbounded part is fixed and the perturbation is bounded; *cf.* the Trotter-Kato theorem, KATO [1977, p. 502]. (b) These estimates generalize to higher derivatives in the obvious way.

We now prove Lemmas 1 and 2 of Section 2.

Proof of Lemma 1. By (H3b) we have an $x_L(t, \varepsilon)$ satisfying

$$x_L(t, \varepsilon) = e^{t(A+\varepsilon A_1)} x_L(0, \varepsilon) + \int_0^t \varepsilon e^{(t-s)(A+\varepsilon A_1)} f(s) \, ds \tag{A-5}$$

and

$$x_L(T, \varepsilon) = x_L(0, \varepsilon), \quad x_L(t, \varepsilon) = O(\varepsilon).$$

We seek a curve $x(t, \varepsilon)$ such that

$$x(t, \varepsilon) = e^{t(A+\varepsilon A_1)} x(0, \varepsilon) + \int_0^t \varepsilon e^{(t-s)(A+\varepsilon A_1)} f(s) \, ds$$

$$+ \int_0^t e^{(t-s)(A+\varepsilon A_1)} [B(x(s, \varepsilon)) + \varepsilon g(x(s, \varepsilon), s)] \, ds \tag{A-6}$$

and

$$x(T, \varepsilon) = x(0, \varepsilon), \quad x(T, \varepsilon) = O(\varepsilon).$$

We first claim that for ε sufficiently small, $\|x(0, \varepsilon)\| \leq (\text{Const}) \, \varepsilon$ implies that $\|x(t, \varepsilon)\| \leq (\text{Const}) \, \varepsilon$ for $0 \leq t \leq T$. To obtain this, subtract (A-5) and (A-6):

$$x(t, \varepsilon) - x_L(t, \varepsilon) = e^{t(A+\varepsilon A_1)} [x(0, \varepsilon) - x_L(0, \varepsilon)]$$

$$+ \int_0^t e^{(t-s)(A+\varepsilon A_1)} [B(x(s, \varepsilon)) + \varepsilon g(x(s, \varepsilon), s)] \, ds. \tag{A-7}$$

Thus

$$\|x(t, \varepsilon) - x_L(t, \varepsilon)\| \leq (\text{Const}) \, \varepsilon + (\text{Const}) \int_0^t \{\|x(s, \varepsilon)\|^3 + \varepsilon \|x(s, \varepsilon)\|^2\} \, ds. \tag{A-8}$$

The estimates on B and g are valid since for ε small enough the solutions will remain in a neighborhood of 0 for $0 \leq t \leq T$. From (A-8) we obtain an estimate of the form

$$\|x(t, \varepsilon)\| \leq (\text{Const}) \, \varepsilon + \text{Const} \int_0^t \|x(s, \varepsilon)\| \, ds$$

by using $\|x_L(t, \varepsilon)\| \leq (\text{Const}) \, \varepsilon$ and dropping the cube and square. Thus, by Gronwall's inequality,

$$\|x(t, \varepsilon)\| \leq (\text{Const}) \, \varepsilon \tag{A-9}$$

as desired.

Next, let B_ε be the ball of radius ε about $x_L(0, \varepsilon) = x_L(T, \varepsilon)$. Consider the map

$$P^\varepsilon : B_\varepsilon \to X; \quad x(0, \varepsilon) \mapsto x(T, \varepsilon).$$

We seek a fixed point of P^ε.

From (A-7) note that $x(0, \varepsilon)$ is a fixed point of P^ε if and only if it is a fixed point of the map
$$\mathscr{F}_\varepsilon : B_\varepsilon \to X,$$
$$\mathscr{F}_\varepsilon(x(0, \varepsilon)) = x_L(0, \varepsilon) + L_\varepsilon^{-1} \int_0^T e^{(T-s)(A+\varepsilon A_1)} [B(x(s, \varepsilon)) + \varepsilon g(x(s, \varepsilon), s)] \, ds, \quad \text{(A-10)}$$
where
$$L_\varepsilon = (Id - e^{T(a+\varepsilon A_1)}).$$

Claim 1. *For ε small, $\mathscr{F}_\varepsilon$ maps B_ε to itself.*

Proof. From (9), (A-9) and (H-5),
$$\|\mathscr{F}_\varepsilon(x(0, \varepsilon)) - x_L(0, \varepsilon)\| \leq \frac{\text{Const}}{\varepsilon} \int_0^T (\text{Const}) [\|x(s, \varepsilon)\|^3 + \varepsilon \|x(s, \varepsilon)\|^2] \, ds$$
$$\leq (\text{Const}) \, \varepsilon^2.$$

This is less than ε for ε sufficiently small.

Claim 2. *$\mathscr{F}_\varepsilon$ is a contraction; i.e. has Lipschitz constant <1.*

Proof. Indeed, the derivative of $\mathscr{F}_\varepsilon$ is
$$D\mathscr{F}_\varepsilon(x(0, \varepsilon)) = L_\varepsilon^{-1} \int_0^1 e^{(T-s)(A+\varepsilon A_1)} [DB(x(s, \varepsilon)) \circ DF^\varepsilon_{s,0}(x(0, \varepsilon))$$
$$+ \varepsilon D_x g(x(s, \varepsilon), s) \circ DF^\varepsilon_{s,0}(x(0, \varepsilon), s)] \, ds.$$

Estimating as above,
$$\| D\mathscr{F}_\varepsilon(x(0, \varepsilon))\| \leq \frac{(\text{Const})}{\varepsilon} \cdot \int_0^T (\text{const}) \cdot \varepsilon^2 \, ds \leq (\text{Const}) \, \varepsilon$$

so if ε is small enough, this is less than 1. ∎

Thus $\mathscr{F}_\varepsilon$ has a unique fixed point in B_ε, so lemma 1 is proved. ∎

Proof of Lemma 2. The Poincaré map P^ε is the map $x(0, \varepsilon) \mapsto x(T, \varepsilon)$ determined by equation (A-6). Thus $DP^\varepsilon(x(0, \varepsilon)) = DP^\varepsilon(p_\varepsilon)$ maps $v(0)$ to $v(T)$, determined by
$$v(T) = e^{(A+\varepsilon A_1)T} v(0) + \int_0^T e^{(A+\varepsilon A_1)(T-s)} [DB(x(s, \varepsilon)) \cdot DF^\varepsilon_{s,0}(x(0, \varepsilon)) \cdot v(s)$$
$$+ \varepsilon Dg(x(s, \varepsilon)) \cdot DF^\varepsilon_{s,0}(x(0, \varepsilon)) \cdot v(s)] \, ds.$$

By lemma 1, $x(s, \varepsilon) = O(\varepsilon)$ and B is cubic, so $DB(x(s, \varepsilon)) = O(\varepsilon^2)$; also $\varepsilon Dg(x(s, \varepsilon)) = O(\varepsilon^2)$. Thus for ε small, $DP^\varepsilon(p_\varepsilon) - e^{(A+\varepsilon A_1)T} = O(\varepsilon^2)$ where the $O(\varepsilon^2)$ estimate is in norm. Now $e^{(A+\varepsilon A_1)T}$ has spectrum shifted toward the origin by an amount $O(\varepsilon)$ and so it follows by perturbation of spectra that $DP^\varepsilon(p_\varepsilon)$ has its spectrum shifted by $O(\varepsilon)$ in the same direction as well. (See KATO [1977], Chapter 4, §3.) ∎

Finally we make some remarks on why the flow is global for the beam example. First of all, the fact that A generates a group in X follows the same pattern as the proposition 2.4 of HOLMES & MARSDEN [1978], so is omitted. Secondly, B and f_1 are smooth maps since multiplication $H^1 \times H^1 \to H^1$ is continuous and bilinear. Moreover it is clear that B and f_1 have bounded derivatives on bounded sets. Thus hypotheses 1(a) and (b) hold. To prove (c) we use energy functionals. To begin we consider the unforced case.

Proposition A.4. *Consider equation* (1) *with* $f = 0$ *and* $\delta > 0$. *Then its flow* F_t^ε *on* X *is globally defined. If* $\Gamma < \pi^2$, *and* $\varepsilon > 0$, *then* $(0, 0)$ *is stable: i.e. for any* $x \in X$,
$$\lim_{t \to +\infty} F_t^\varepsilon(x) = (0, 0).$$

Proof. Consider the energy
$$H(w, \dot{w}) = \frac{1}{2}|\dot{w}|^2 - \frac{\Gamma}{2}|w'|^2 + \frac{1}{2}|w''|^2 + \frac{\varkappa}{4}|w'|^4$$
where $|w|^2 = \int_0^1 |w(\zeta, t)|^2 \, d\zeta$ is the square of the L^2 norm. We compute:
$$\frac{d}{dt} H(w, \dot{w}) = -\varepsilon \delta |\dot{w}|^2 \leq 0.$$

Also, we have the elementary estimate
$$H(w, \dot{w}) \geq \begin{cases} \frac{1}{2} \|(w, \dot{w})\|_X^2 & \text{if } |w'|^2 \geq \Gamma/\varkappa, \\ -\Gamma^2/2\varkappa & \text{if } |w'|^2 \leq \Gamma/\varkappa. \end{cases}$$

In particular, $\|(w, \dot{w})\|_X$ is *a priori* bounded for all t, so the flow is global in time from the local existence theory.

Notice that H strictly decreases along any trajectory if $\varepsilon \delta > 0$. (Thus there can be no closed orbits.) Along any trajectory $H(x(t))$ decreases, so it converges to a limit, say H_∞, as $t \to +\infty$. From
$$H(x(s)) - H(x(t)) = -\int_s^t \frac{d}{dt} H(x(t)) \, dt = \varepsilon \delta \int_s^t |\dot{w}(\tau)|^2 \, d\tau,$$
we see that $\dot{w}$ satisfies a Cauchy condition, so it converges in L^2. If $\Gamma < \pi^2$, then the estimate $|w''|^2 \geq \pi^2 |w'|^2$ shows that
$$H(w, \dot{w}) \geq \frac{1}{2}|\dot{w}|^2 + \frac{(\pi^2 - \Gamma)}{2}|w'|^2 \geq 0.$$

Now since H is strictly decreasing, and $H \geq 0$, $\dot{w}$ converges to 0 in L^2. Also H must converge to 0 so from the above inequality, $w' \to 0$ in L^2. From the original expression for H, $w \to 0$ in H^2. ∎

For $\Gamma > \pi^2$, $F_t^\varepsilon(x)$ will generally converge to one of the stable fixed points.

Proposition A.5. *The flow of* (1) *on* $X \times S^1$ *is global in time for any* $\varepsilon > 0$, $\delta > 0$.

Proof. The same energy function has
$$\frac{d}{dt} H(w, \dot{w}) = -\varepsilon \delta |\dot{w}|^2 + (\varepsilon \cos \omega t)\left(\int_0^1 f(\zeta) \dot{w}(\zeta, b) \, d\zeta\right),$$
which decreases for $|\dot{w}|^2$ large (by the Schwartz inequality) and so $\|(w, \dot{w})\|_X^2$ is again bounded uniformly for all time. ∎

Appendix B. The Birkhoff-Smale Homoclinic Theorem in Infinite Dimensions

The goal of this appendix is to outline the proof of theorem 2 for a C^∞ diffeomorphism $P: X \to X$ where X is a Banach space. Let $p \in X$ be a fixed point of P that is hyperbolic; specifically assume that $\sigma(DP(p))$ is the union of two compact sets not meeting the unit circle. In fact, assume the piece exterior to the unit circle is a single point. Let $W^u(p)$ and $W^s(p)$ be the corresponding unstable and stable manifolds, so $W^u(p)$ is one dimensional and $W^s(p)$ has codimension one. Assume $q \in W^u(p) \cap W^s(p)$ is a transverse homoclinic point. We wish to show that there is an integer N such that P^N has an invariant set Λ such that P^N restricted to Λ is conjugate to a shift on two symbols.

We shall only outline the argument since many of the details are similar to the two dimensional case. The plan is that due to CONLEY & MOSER. The reader wishing to reconstruct all the details should consult MOSER [1973], Chapter III. (Some notes of DAVID RAND were also helpful.)

Abstract Shoes

Let Z be a Banach space and let $Q \subset Z \times \mathbb{R}$ denote the unit box:

$$Q = \{(x, y) \in Z \times \mathbb{R} \mid \|x\| \leq 1, 0 \leq y \leq 1\}.$$

Fix a number μ, $0 < \mu < 1$. A map $u: B = \{x \in Z \mid \|x\| \leq 1\} \to \mathbb{R}$ is called *horizontal* (or μ-horizontal) if $0 \leq u(x) \leq 1$ for $x \in B$ and if $|u(x_1) - u(x_2)| \leq \mu \|x_1 - x_2\|$ for $x_1, x_2 \in B$. Let $\mathcal{H}$ (or $\mathcal{H}_\mu$) denote the set of such maps. For u_1 and u_2 in $\mathcal{H}$ satisfying $0 \leq u_1(x) < u_2(x) \leq 1$ let

$$H = \{(x, y) \in Q \mid u_1(x) < y < u_2(x)\}$$

and call such a set a *horizontal strip*. Its *diameter* is

$$d(H) = \sup_{x \in B} (u_2(x) - u_1(x)).$$

A *vertical* (or μ-vertical) curve is a map

$$v: [0, 1] \to B$$

such that $\|v(y_1) - v(y_2)\| \leq \mu |y_1 - y_2|$ for $y_1, y_2 \in [0, 1]$. The set of all such curves is denoted $\mathcal{V}$ (or $\mathcal{V}_\mu$). A *vertical strip* is the closure of an open set V in Q bounded by open subsets of $B \times \{0\}$ and $B \times \{1\}$ and by vertical curves joining their respective boundaries. We set

$$d(V) = \sup_{\alpha,\beta,y} \|v_\alpha(y) - v_\beta(y)\|,$$

where $\{v_\alpha\}$ denotes the vertical curves comprising the sides of V as described above. See Figure 7.

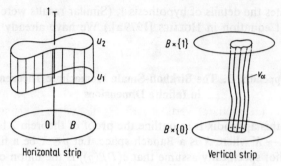

Fig. 7

Lemma B.1. *If $H^1 \supset H^2 \supset \ldots$ is a nested sequence of μ-horizontal strips and if $d(H^k) \to 0$ as $k \to \infty$, then $\bigcap_{k=1}^{\infty} H^k$ is the graph of a μ-horizontal map.*

Proof. Let u_1^k and u_2^k be the two functions defining H^k. Then $u_1^k(x) - u_2^k(x) \to 0$. But by the nesting assumption, $u_1^k(x)$ is increasing and $u_2^k(x)$ is decreasing, so they converge to, say, $u(x)$. Letting $k \to \infty$ in $|u_2^k(x_1) - u_2^k(x_2)| \leq \mu \|x_1 - x_2\|$ gives the μ-horizontality of u. ∎

There is an analogous lemma for vertical strips.

Lemma B.2. *A μ-horizontal graph and a μ-vertical curve intersect in exactly one point.*

Proof. Let u and v describe the μ-horizontal graph and μ-vertical curve respectively. Let (x, y) lie on their graphs: Then $x = v(y)$ and $y = u(x)$, so $x - v(u(x)) = 0$, and $y - u(v(y)) = 0$. Let $g(y) = y - u(v(y))$ so $g: [0, 1] \to \mathbb{R}$. Now if $0 \leq y_1 < y_2 \leq a$, then

$$|u(v(y_2)) - u(v(y_1))| \leq \mu |v(y_2) - v(y_1)| \leq \mu^2 |y_2 - y_1|$$

and so as $\mu^2 < 1$,

$$g(y_2) - g(y_1) = y_2 - y_1 + [u(v(y_2)) - u(v(y_1))] > 0.$$

Thus g is strictly increasing. However, $g(0) \leq 0$ and $g(1) \geq 0$ so g has exactly one zero. ∎

By lemma B.2, there is a well-defined map

$$\chi: \mathscr{H}_\mu \times \mathscr{V}_\mu \to Q.$$

Define norms on $\mathscr{H}_\mu \times \mathscr{V}_\mu$ and Q by

$$\|(u, v)\| = \sup_{x \in B} |u(x)| + \sup_{y \in [0,1]} \|v(y)\|,$$

$$\|(x, y)\| = \|x\| + |y|.$$

Lemma B.3. *χ is Lipschitz continuous with Lipschitz constant $(1 - \mu)^{-1}$.*

This is straightforward (see MOSER [1973], p. 71).

Definition B.4. Fix numbers μ and ν satisfying $0 < \mu < 1$ and $0 < \nu < 1$. Let H_1 and H_2 be two disjoint μ-horizontal strips and V_1 and V_2 be two disjoint μ-vertical strips. Let
$$\phi: H_1 \cup H_2 \to V_1 \cup V_2$$
be a homeomorphism satisfying

(i) $\phi(H_i) = V_i$, $i = 1, 2$;

(ii) horizontal (respectively vertical) boundaries of H_i are mapped onto the horizontal (respectively vertical) boundaries of V_i, $i = 1, 2$; and

(iii) if H is a μ-horizontal strip in $H_1 \cup H_2$, then for $i = 1$ or 2
$$\phi^{-1}(H) \cap H_i = \tilde{H}$$
is a non-empty μ-horizontal strip satisfying
$$d(\tilde{H}) \leq \nu\, d(H).$$

Similarly, if V is a μ-vertical strip in $V_1 \cup V_2$, then for $i = 1$ or 2
$$\phi(V) \cap V_i = \tilde{V}$$
is a non-empty μ-vertical strip satisfying
$$d(\tilde{V}) \leq \nu\, d(V).$$

Such a homeomorphism ϕ is called a *shoe*. (One can allow i to range over a general set A; we chose $i = 1, 2$ for the present context.)

The Smale horseshoe is a basic example of a shoe; cf. SMALE [1967].

Theorem B.5. *Let ϕ be a shoe. Let $\Lambda = \bigcap\limits_{-\infty < n < \infty} \phi^n(V_1 \cup V_2)$. Then Λ is a non-empty invariant set for ϕ and $\phi \,|\, \Lambda$ is conjugate to the shift automorphism on two symbols.*

In particular, it follows that ϕ has infinitely many periodic points in Λ and there is a point $p \in \Lambda$ whose orbit is dense in Λ.

Proof of B.5. Let $i_0, i_{-1}, \ldots, i_{-n}$ be a sequence of $n + 1$ 0's or 1's. Define $H_{i_0,\ldots,i_{-n}}$ in terms of sequences of length n inductively by
$$H_{i_0, i_{-1},\ldots,i_{-n}} = H_{i_0} \cap \phi^{-1}(H_{i_{-1},\ldots,i_{-n}}).$$
Thus $H_{i_0, i_{-1},\ldots,i_{-n}}$ is μ-horizontal of width $\leq \nu^n \cdot$ constant. If $i = i_0, i_{-1}, i_{-2}, \ldots$ is a (one sided) sequence of 0's or 1's, then
$$H_{i-} = \bigcap_{n=0}^{\infty} H_{i_0, i_{-1},\ldots,i_{-n}}$$
is the graph of a μ-horizontal map by lemma B.1. Similarly if $i+ = i_1, i_2, \ldots$, define $V_{i_1,\ldots,i_n}$ by
$$V_{i_1,\ldots,i_n} = V_{i_1} \cap \phi(V_{i_2,\ldots,i_n})$$
and obtain the μ-vertical curve
$$V_{i+} = \bigcap_{n=1}^{\infty} V_{i_1,\ldots,i_n}.$$

If $i = \ldots i_{-2}, i_{-1}, i_0, i_1, i_2, \ldots$ is a bi-infinite sequence, then by lemma B.2 V_{i+} and H_{i-} meet in precisely one point, denoted $\tau(i)$. [Lemma B.3 implies τ is continuous from $\{0, 1\}^Z$ to Q.] Thus $\tau: \{0, 1\}^Z \to \Lambda$ and $\tau^{-1} \circ \phi \circ \tau$ is shift automorphism by construction. ∎

Sector Bundles

Assume that ϕ in definition B.4 is C^1. We now give a condition that implies property (iii) in that definition.

Definition B.6. Let $0 < \lambda < 1$ and for $(x, y) \in Q$, let

$$S_\lambda^u(x, y) = \{(\xi, \eta) \in Z \times \mathbb{R} \mid \|\xi\| \leq \lambda |\eta|\}$$

and

$$S_\lambda^s(x, y) = \{(\xi, \eta) \in Z \times \mathbb{R} \mid |\eta| \leq \lambda \|\xi\|\}.$$

If R is a closed subset of Q, we call

$$S_\lambda^u = \bigcup_{(x,y) \in R} S^u(x, y) \text{ and } S_\lambda^s = \bigcup_{(x,y) \in R} S_\lambda^s(x, y)$$

the unstable and stable λ-sector bundles over R.

Consider the following condition on ϕ:

(iii)' There exists a μ, $0 < \mu < \dfrac{1}{2}$ such that

(a) the unstable (respectively stable) μ-sector bundle over $H_1 \cup H_2$ (respectively $V_1 \cup V_2$) is mapped to itself by $d\phi$ (respectively $d\phi^{-1}$) and

(b) if $(x_0, y_0) \in H_i$ for $i = 1$ or 2 and $(\xi_0, \eta_0) \in S_\mu^u(x_0, y_0)$, then $|\eta_1| \geq \mu^{-1} |\eta_0|$ where $(\xi_1, \eta_1) = d\phi(x_0, y_0) \cdot (\xi_0, \eta_0)$, and if $(x_1, y_1) \in V_i$ for $i = 1$ or 2 and $(\xi_1, \eta_1) \in S_\mu^s(x_1, y_1)$, then $\|\xi_0\| \geq \mu^{-1} \|\xi_1\|$.

Condition (b) says that vectors in S_μ^u are vertically expanded by a factor μ^{-1} by $d\phi$ while vectors in S_μ^s are horizontally contracted by a factor μ^{-1}.

Proposition B.7. *Let ϕ satisfy* (i) *and* (ii) *of definition* B.4 *and be C^1. Then* (iii)' *implies* (iii).

Proof. Let γ be the graph of a μ-horizontal map in H_i and let $\gamma' = \gamma \cap V_j$ where i, j are 1 or 2. Consider $\phi^{-1}(\gamma')$; we claim it is the graph of a μ-horizontal map. See Figure 8. Because the boundaries of H_j and V_j correspond under ϕ, $\phi^{-1}(\gamma')$ covers all of B. Also, if $v = (\xi, \eta)$ is tangent to γ', then $|\eta| \leq \mu \|\xi\|$ because γ' is μ-horizontal. Thus by (iii)' (a), the same is true of vectors tangent to $\phi^{-1}(\gamma')$. Thus, by the mean value theorem, if (x_1, y_1) and (x_2, y_2) lie on $\phi^{-1}(\gamma')$, then $\|y_1 - y_2\| \leq \mu \|x_1 - x_2\|$. Thus $\phi^{-1}(\gamma')$ is the graph of a μ-horizontal map. This implies the μ-horizontality of $\tilde{H}$ in B.4(iii).

It remains to check the condition on the diameters. Let p_1 and p_2 be points on different horizontal boundaries of $\tilde{H} = \phi^{-1}(H) \cap H_i = \phi^{-1}(H \cap V_i)$ with the same x-component. The image of the μ-vertical segment joining p_1 and p_2 is a μ vertical curve. Let $Z_1 = \phi(p_1)$ and $Z_2 = \phi(p_2)$. By B.3, $\|Z_1 - Z_2\| \leq (1-\mu)^{-1} d(H)$.

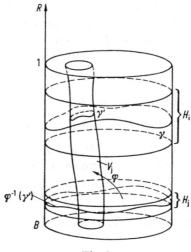

Fig. 8

Let $p(t) = (1 - t) p_1 + p_2$ and $Z(t) = \phi(p(t)) = (x(t), y(t))$. Since $p(t)$ is vertical, (iii)'(b) gives $|\dot{y}| \geq \mu^{-1} |\dot{p}| = \|p_1 - p_2\| > 0$. Thus $\dot{y}$ does not change sign, so

$$\|p_1 - p_2\| \leq \int_0^1 |\dot{y}|\, dt = \mu\, |y(1) - y(0)| \leq \mu\, \|Z_1 - Z_2\| \leq \mu(1 - \mu)^{-1}\, d(H).$$

This verifies $d(\tilde{H}) \leq \nu\, d(H)$ with $\nu = \mu(1 - \mu)^{-1}$; since $0 < \mu < 1/2$ we have $0 < \nu < 1$. The assertions for vertical strips are similar. ∎

We remark that if ϕ is sufficiently close to a volume preserving map (which happens in our example for ε small), then Λ is actually hyperbolic for ϕ.

Homoclinic Points

Now we apply the machinery just developed to prove theorem 2. Introduce local coordinates (x, y) near p, for x in a neighborhood of 0 in a Banach space Z and y in a neighborhood of 0 in $\mathbb{R}$ such that $W^s(p)$ is the codimension 1 submanifold $y = 0$ and $W^u(p)$ is the curve $x = 0$.

Write

$$P(x, y) = (U(x, y), V(x, y)).$$

Then invariance of $W^s(p)$ and $W^u(p)$ implies $U(x, 0) = 0$ and $V(0, y) = 0$. Thus, at $p = (0, 0)$,

$$DP(0, 0) = \begin{pmatrix} A & 0 \\ 0 & B \end{pmatrix}$$

where $A: Z \to Z$ and $B \in \mathbb{R}$. The spectral hypotheses imply that we can define a norm on Z such that $\|A\| < 1$ and can assume $B > 1$. (See, for example, Marsden & McCracken [1976], Sec. 2A.) Let $Q_a = \{(x, y) \mid \|x\| \leq a \text{ and } |y| \leq a\}$.

Near q points can similarly be coordinatized by pairs of points in $W^s(p)$ and $W^u(p)$ which can in turn be assumed to be linear near q. Relative to these coordinates, let R be defined in a manner like Q_a but on one side of $W^s(p)$ (which makes sense as $W^s(p)$ has codimension 1); i.e. relative to coordinates near q,

$$R = \{(x, y) \mid \|x\| < b, \|y\| < b, y \geq 0\}.$$

We can choose b sufficiently small so that for some integers $l, m > 0$

$$A = P^{-m}(R) \subset Q_a \text{ and } B = P^l(R) \subset Q_a;$$

l and m are large enough so that $P^l(q) \in Q_a$ and $P^{-m}(q) \in Q_a$. We shall locate the desired set Λ inside B (or, equivalently, inside A). See Figure 9. The rectangle R is chosen small enough and l is chosen large enough so the sides of A and B are C^r close to the coordinate planes.

One now proceeds in several steps; fix a μ, $0 < \mu < \dfrac{1}{2}$.

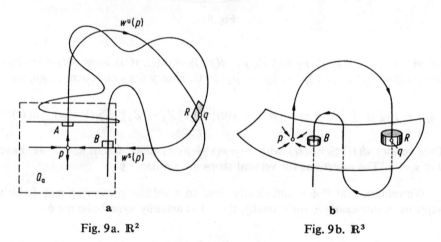

Fig. 9a. $\mathbb{R}^2$ Fig. 9b. $\mathbb{R}^3$

Lemma B.8. *There are disjoint horizontal strips H_1 and H_2 in B and an integer n such that $U_1 = P^n(H_1)$ and $U_2 = P^n(H_2)$ are disjoint vertical strips crossing A from one side to the other.*

The idea is simply to take horizontal strips in B and map them repeatedly forward (keeping track only of things in Q_a). The spectral hypothesis implies that horizontal distances shrink and vertical distances stretch. The details in MOSER [1973, p. 186–188] are readily adapted to the present context.

Lemma B.9. $V_1 = P^{m+l}(U_1 \cap A)$ and $V_2 = P^{m+l}(U_2 \cap A)$ are disjoint vertical strips in B.

Here we follow our vertical strips in A out to R and back to B. The verticality is maintained if R is small enough. Shrink B horizontally so that P^n maps H_i homeomorphically onto V_i, $i = 1, 2$.

Lemma B.10. *Let $\phi = P^N$ where $N > n + m + l$, is sufficiently large, restricted to $H_1 \cup H_2$. Then ϕ satisfies condition* (i) *and* (ii) *of B.4 and* (iii)' *following B.6.*

This is now straightforward using the given spectral conditions on $DP(p)$. Again, see MOSER [1973] for details. Thus we have set up a map ϕ to which Theorem B.5 applies, producing the desired set Λ.

References

R. ABRAHAM & J. MARSDEN [1978] "Foundations of Mechanics," 2nd Edition, Addison-Wesley.

V. ARNOLD [1964] Instability of dynamical systems with several degrees of freedom. Dokl. Akad. Nauk. SSSR **156**, 9–12.

V. ARNOLD [1966] Sur la géometrie différentielle des groupes de Lie de dimension infinie et ses applications à l'hydrodynomique des fluides parfaits. Ann. Inst. Fourier, Grenoble **16**, 319–361.

J. CARR [1980] Application of Center Manifolds, Springer Applied Math. Sciences (to appear).

J. CARR & M. Z. H. MALHARDEEN [1979] Beck's problem, SIAM J. Appl. Math. **37**, 261–262.

P. CHERNOFF & J. MARSDEN [1974] "Properties of Infinite Dimensional Hamiltonian Systems," Springer Lecture Notes in Math., no. 425.

S. N. CHOW, J. HALE & J. MALLET-PARET [1980] An example of bifurcation to homoclinic orbits, J. Diff. Eqn's. **37**, 351–373.

C. CONLEY & J. SMOLLER [1974] On the structure of magneto hydrodynamic shock waves, Comm. Pure Appl. Math. **27**, 367–375, J. Math. of Pures et. Appl. **54** (1975) 429–444.

D. EBIN & J. MARSDEN [1970] Groups of diffeomorphisms and the motion of an incompressible fluid, Ann. of Math. **92**, 102–163.

J. GOLLUB [1980] The onset of turbulence: convection, surface waves, and oscillators, Springer Lecture Notes in Physics.

J. GUCKENHEIMER [1979] On a codimension two bifurcation (preprint).

O. GUREL & O. RÖSSLER (eds) [1979] "Bifurcation theory and applications in scientific disciplines," Ann. of N.Y. Acad. Sciences, vol. **316**.

B. HASSARD [1980] Computation of invariant manifolds, in Proc. "New Approaches to Nonlinear Problems in Dynamics", ed. P. HOLMES, SIAM publications.

E. HILLE & R. PHILLIPS [1957] "Functional Analysis and Semigroups", A.M.S. Colloq. Publ.

M. HIRSCH, C. PUGH & M. SHUB [1977] "Invariant Manifolds", Springer Lecture Notes in Math. no. 583.

P. HOLMES [1979a] Global bifurcations and chaos in the forced oscillations of buckled structures, Proc. 1978 IEEE Conf. on Decision and Control, San Diego, CA, 181–185.

P. HOLMES [1979b] A nonlinear oscillator with a strange attractor, Phil. Trans. Roy. Soc. A **292**, 419–448.

P. HOLMES [1980a] Averaging and chaotic motions in forced oscillations, SIAM J. on Appl. Math. **38**, 65–80.

P. HOLMES [1980b] Space and time-periodic perturbations of the sine-Gordon equation (preprint).

P. HOLMES & J. MARSDEN [1978] Bifurcation to divergence and flutter in flow induced oscillations; an infinite dimensional analysis, Automatica **14**, 367–384.

P. HOLMES & J. MARSDEN [1979] Qualitative techniques for bifurcation analysis of complex systems, in GUREL & RÖSSLER [1979], 608–622.

T. KATO [1977] "Perturbation Theory for Linear Operators" 2nd Ed., Springer.

N. KOPELL & L. N. HOWARD [1976] Bifurcations and trajectories joining critical points, Adv. Math. **18**, 306–358.

M. LEVI, F. C. HOPPENSTADT & W. L. MIRANKER [1978] Dynamics of the Josephson junction, Quart. of Appl. Math. **36**, 167–198.

V. K. MELNIKOV [1963] On the stability of the center for time periodic perturbations, Trans. Moscow Math. Soc. **12**, 1–57.

F. MOON & P. HOLMES [1979] A magneto-elastic strange attractor, J. Sound and Vibrations **65**, 275–296.

J. MOSER [1973] "Stable and Random Motions in Dynamical Systems," Ann. of Math. Studies no. 77, Princeton Univ. Press.

S. NEWHOUSE [1974] Diffeomorphisms with infinitely many sinks, Topology **12**, 9–18.

S. NEWHOUSE [1979] The abundance of wild hyperbolic sets and non-smooth stable sets for diffeomorphisms, Publ. I.H.E.S. **50**, 100–151.

J. RAUCH [1979] Qualitative behavior of dissipative wave equations. Arch. Rational Mech. An. **62**, 77–91.

J. ROBBIN [1971] A structural stability theorem, Ann. of Math. **94**, 447–493.

I. SEGAL [1962] Nonlinear Semigroups, Ann. of Math. **78**, 334–362.

Y. SHIZUTA [1980] On the classical solutions of the Boltzmann equations, Comm. Pure. Appl. Math. (to appear).

S. SMALE [1963] Diffeomorphisms with many periodic points, in "Differential and Combinatorial Topology" (ed. S. S. CAIRNS), Princeton Univ. Press, 63–80.

S. SMALE [1967] Differentiable dynamical systems, Bull. Am. Math. Soc. **73**, 747–817.

W. Y. TSENG & J. DUGUNDJI [1971] Nonlinear vibrations of a buckled beam under harmonic excitation. J. Appl. Mech. **38**, 467–476.

I. VIDAV [1970] Spectra of perturbed semigroups, J. Math. An. Appl. **30**, 264–279.

Theoretical and Applied Mechanics
Cornell University
Ithaca, New York
and
Department of Mathematics
University of California
Berkeley

(Received August 25, 1980)

SECOND ORDER AVERAGING AND BIFURCATIONS TO SUBHARMONICS IN DUFFING'S EQUATION†

C. Holmes

Center for Applied Mathematics

AND

P. Holmes

Department of Theoretical and Applied Mechanics, Cornell University, Ithaca, New York 14853, U.S.A.

(*Received* 3 *October* 1980)

Periodic motions near an equilibrium solution of Duffing's equation with negative linear stiffness can evolve, lose their stability, and undergo period doubling bifurcations as excitation amplitude, frequency and damping are varied. For bifurcations to period two it is shown that these can be either sub- or supercritical, depending upon the excitation frequency. The analysis is carried out by the averaging method, and, to retain important non-linear effects, averaging must be taken to second order. Some remarks on higher order subharmonics are made.

1. INTRODUCTION

Duffing's equation with negative linear stiffness and small periodic excitation and damping is

$$\ddot{x} - x + x^3 = \bar{\gamma} \cos \omega t - \bar{\delta}\dot{x}, \qquad 0 \leq \bar{\gamma}, |\bar{\delta}| \ll 1. \tag{1}$$

This equation arises in models of the forced vibration of buckled beams [1–3] and in electrical circuits [4, 5]. In previous analytical and computer studies [2, 6] it has been shown that the separatrix loop given by the level curve of the Hamiltonian function for the unperturbed system ($\bar{\gamma} = \bar{\delta} = 0$),

$$H(x, \dot{x}) = (\dot{x}^2/2) - (x^2/2) + (x^4/4) = 0, \tag{2}$$

breaks in a complex manner under forcing, leading to chaotic motions, including infinite sets of periodic orbits of arbitrarily long period. However, analogue computer experiments [2] show that, as force ($\bar{\gamma}$) is increased for fixed damping ($\bar{\delta}$), a sequence of period-doubling bifurcations occurs, in which periodic orbits of period $2\pi/\omega$ arising from the equilibrium solutions at $(x, \dot{x}) = (\pm 1, 0)$ lose their stability and are replaced by orbits of period $4\pi/\omega$, $8\pi/\omega$, etc. Such period doublings have been observed in many autonomous and non-autonomous differential systems as well as in maps and they appear to play an important role in the onset of chaotic motions in deterministic dynamical systems.

In this paper the first such bifurcation, to orbits of period $4\pi/\omega$, is studied and it is shown that it is by no means as simple as one might expect. After suitable transformations, carried out in section 1.2, the averaging method is applied. To retain crucial non-linear terms,

† This work was supported by NSF grant ENG 78-02891.

averaging must be carried out to second order, and this theory is reviewed in section 2. In section 3 the averaging process for this example is carried out and then the resulting autonomous system is analyzed in section 4. In section 5 some comments are made on higher order bifurcations and orbits of period $6\pi/\omega$. Finally, in section 6 the results are interpreted in terms of the Poincaré map of equation (1).

1.1. THE VARIATIONAL EQUATION

The aim is to study the behavior of solutions of equation (1) near a periodic orbit with $\bar{\gamma}, \bar{\delta} \neq 0$, and small. In this way one will be able to investigate the stability of, and bifurcations from, the given periodic motion. Suppose one has such a solution, $\hat{x}(t)$. Then one can set $x = \hat{x} + \varepsilon y$, $\varepsilon \ll 1$, and substitute into equation (1), first replacing the parameters $\bar{\gamma}$ and $\bar{\delta}$ by $\varepsilon\gamma$ and $\varepsilon\delta$, respectively, where γ and δ are $O(1)$ quantities. Using the fact that $\hat{x}(t)$ is a solution of equation (1), one obtains

$$\ddot{y} + \{3\hat{x}^2(t) - 1\}y = -\varepsilon\{\delta\dot{y} + 3\hat{x}(t)y^2\} - \varepsilon^2 y^3. \tag{3}$$

To study equation (3), which is a modified, non-linear Mathieu equation, one must approximate the desired periodic solution $\hat{x}(t)$ to $O(\varepsilon^2)$. To do this one can use regular perturbation methods, setting

$$\hat{x}(t) = 1 + \varepsilon\hat{x}_0(t) + \varepsilon^2\hat{x}_1(t) + O(\varepsilon^3), \tag{4}$$

substituting this into equation (1) and equating powers of ε to obtain

$$\ddot{\hat{x}}_0 + 2\hat{x}_0 = \gamma\cos\omega t, \qquad \ddot{\hat{x}}_1 + 2\hat{x}_1 = -\delta\dot{\hat{x}}_0 - 3\hat{x}_0^2. \tag{5a, b}$$

The leading term of equation (4) is chosen to be unity since what is of interest here is the periodic orbit which converges on the fixed point $(\hat{x}, \dot{\hat{x}}) = (1, 0)$ as $\varepsilon \to 0$. A similar expansion and analysis holds for the periodic orbits near $(\hat{x}, \dot{\hat{x}}) = (-1, 0)$, and therefore only one of these two symmetric orbits need be considered.

To solve equation (5a) one must specify ω. For the orbits of period two and three, as is shown below, one requires $\omega \approx 2\sqrt{2}$ and $\omega \approx 3\sqrt{2}$ respectively and thus equation (5a) is non-resonant and the desired particular solution,

$$\hat{x}_0(t) = \{\gamma/(2-\omega^2)\}\cos\omega t = \Gamma\cos\omega t, \tag{6}$$

is obtained immediately. Note that the complementary solution is set equal to zero, since only the orbit of period $2\pi/\omega$ is of interest. Performing a similar analysis for equation (5b), and using equation (6), gives

$$\hat{x}_1(t) = -(3\Gamma^2/4) + \{\Gamma\delta\omega/(2-\omega^2)\}\sin\omega t - \{3\Gamma^2/(4-8\omega^2)\}\cos 2\omega t. \tag{7}$$

Substituting equations (6) and (7) into equation (4) and the whole into equation (3) finally yields an equation governing the evolution of the disturbance y, accurate up to $O(\varepsilon^2)$:

$$\begin{aligned}\ddot{y} + 2y &= \varepsilon\{-\delta\dot{y} - 6\Gamma(\cos\omega t)y - 3y^2\} \\ &\quad + \varepsilon^2\{-(3\Gamma^2\cos^2\omega t + 6[-(3\Gamma^2/4) + \{\Gamma\delta\omega/(2-\omega^2)\}\sin\omega t \\ &\quad - \{3\Gamma^2/(4-8\omega^2)\}\cos 2\omega t])y - 3\Gamma(\cos\omega t)y^2 - y^3\} + O(\varepsilon^3) \\ &\triangleq \varepsilon f_0(y, \dot{y}, t) + \varepsilon^2 f_1(y, \dot{y}, t) + O(\varepsilon^3).\end{aligned} \tag{8}$$

This is the equation to be studied in section 3. First the necessary theory of averaging is reviewed.

2. AVERAGING TO THE SECOND ORDER

Suppose that one has a differential equation

$$\dot{\mathbf{x}} = \varepsilon \mathbf{f}(\mathbf{x}, t) + \varepsilon^2 \mathbf{g}(\mathbf{x}, t), \qquad \mathbf{x} \in \mathcal{R}^n, \qquad 0 < \varepsilon \ll 1, \tag{9}$$

with $\mathbf{f}$ and $\mathbf{g}$ T-periodic in t and sufficiently well behaved (C^2 will do). The well known averaging theorem, or Krylov–Bogoliubov (K–B) method, asserts that equation (9) can be replaced by its averaged counterpart

$$\dot{\mathbf{x}} = \varepsilon \mathbf{f}_0(\mathbf{x}) + O(\varepsilon^2), \tag{10}$$

where $\mathbf{f}_0(\mathbf{x}) = (1/T) \int_0^T f(x, t) \, dt$, and that important local information on periodic orbits of equation (9) can be deduced from the equilibrium solutions of equation (10). However, clearly any information at $O(\varepsilon^2)$ is lost in this process, and in our example certain important information enters at precisely this order. The averaging process must therefore be studied in more detail. Our treatment follows that of Hale [7], and may be applied in more general situations, for example, when $\mathbf{f}$ and $\mathbf{g}$ are quasi-periodic.

The passage from equation (9) to equation (10) relies on a lemma which states that there exists a smooth near-identity transformation,

$$\mathbf{x} = \mathbf{y} + \varepsilon \mathbf{u}(\mathbf{y}, t), \qquad \mathbf{x}, \mathbf{y} \in \mathcal{R}^n, \tag{11}$$

where $\mathbf{u}$ is to be chosen subsequently. In choosing $\mathbf{u}$ we will actually construct the transformation explicitly. Differentiating equation (11) with respect to t, using the chain rule and substituting in equation (9), one obtains the transformed system

$$[\mathbf{I} + \varepsilon \mathbf{Du}]\dot{\mathbf{y}} = \varepsilon \mathbf{f}(\mathbf{y} + \varepsilon \mathbf{u}, t) + \varepsilon^2 \mathbf{g}(\mathbf{y} + \varepsilon \mathbf{u}, t) - \varepsilon \dot{\mathbf{u}}(\mathbf{y}, t). \tag{12}$$

Here $\mathbf{Du}$ denotes the matrix of partial derivatives $[\partial u_i / \partial y_j]$.

Expanding the right-hand side of equation (12) in a Taylor's series about $\mathbf{y}$ gives

$$[\mathbf{I} + \varepsilon \mathbf{Du}]\dot{\mathbf{y}} = \varepsilon [\mathbf{f}(\mathbf{y}, t) - \dot{\mathbf{u}}(\mathbf{y}, t)] + \varepsilon^2 [\mathbf{Df}(\mathbf{y}, t)\mathbf{u}(y, t) + \mathbf{g}(\mathbf{y}, t)] + O(\varepsilon^3). \tag{13}$$

One can now split the function f into its mean $\mathbf{f}_0(\mathbf{y})$ and an oscillating part $\tilde{\mathbf{f}}(\mathbf{y}, t)$ ($\mathbf{f}(\mathbf{y}, t) \equiv \mathbf{f}_0(\mathbf{y}) + \tilde{\mathbf{f}}(\mathbf{y}, t)$) and note that, if $\mathbf{u}(y, t)$ is chosen such that $\dot{\mathbf{u}}(\mathbf{y}, t) = \tilde{\mathbf{f}}(\mathbf{y}, t)$, then the $O(\varepsilon)$ part of the right-hand side of equation (13) becomes simply the average $\varepsilon \mathbf{f}_0(\mathbf{y})$. Finally, multiplying through by the inverse of the near-identity operator $\mathbf{I} + \varepsilon \mathbf{Du}$, one has

$$\dot{\mathbf{y}} = \varepsilon \mathbf{f}_0(\mathbf{y}) + \varepsilon^2 [\mathbf{Df}(\mathbf{y}, t)\mathbf{u}(y, t) - \mathbf{Du}(\mathbf{y}, t)\mathbf{f}_0(\mathbf{y}) + \mathbf{g}(\mathbf{y}, t)] + O(\varepsilon^3). \tag{14}$$

If $\mathbf{f}(\mathbf{y}, t)$ is expanded in a Fourier series in t then $\tilde{\mathbf{f}}(\mathbf{y}, t)$ and hence $\mathbf{u}(\mathbf{y}, t)$ may be easily calculated, as in our example below. Generally $\mathbf{u}$ and hence $\mathbf{Du}$ are chosen to have zero mean, and thus the only terms of non-zero mean at $O(\varepsilon^2)$ in equation (14) are $\mathbf{D}\tilde{\mathbf{f}}(\mathbf{y}, t)\mathbf{u}(y, t) + \mathbf{g}(\mathbf{y}, t)$, and one may now apply averaging to these terms (making use implicitly of a second transformation $\mathbf{y} = \mathbf{z} + \varepsilon^2 \mathbf{v}(\mathbf{z}, t)$) to obtain the second order averaged system

$$\dot{\mathbf{z}} = \varepsilon \mathbf{f}_0(\mathbf{z}) + \varepsilon^2 (\mathbf{g}_0(\mathbf{z}) + [\mathbf{D}\tilde{\mathbf{f}}(\mathbf{z}, t)\mathbf{u}(\mathbf{z}, t)]_0), \tag{15}$$

where the subscript $[\cdot]_0$ denotes the time average and terms of $O(\varepsilon^3)$ have been neglected.

The smoothness of the transformations $\mathbf{u}, \mathbf{v}$ and the smallness of ε then guarantee that a study of equation (15) reveals substantial information about equation (9). In particular, if equation (15) possesses an equilibrium point $\mathbf{z}_0$ for which the linearized system has eigenvalues with non-zero real parts (i.e., a hyperbolic sink, saddle or source), then equation (9) possesses a periodic orbit of the same stability type near $\mathbf{x}_0 = \mathbf{z}_0$. Moreover, if a parameterized family of systems (15) exhibit bifurcations of equilibrium points as the parameter(s) vary, then the corresponding periodic orbits of equation (9) exhibit similar bifurcations at nearby parameter values and the bifurcation curves of the full and averaged systems are related by a smooth near-identity transformation. In fact one has the following theorem (cf. the book by Hale [7]).

AVERAGING THEOREM. Consider the differential equation (9) *and the associated* (*second order*) *averaged system* (15). *Suppose that* (15) *has a hyperbolic fixed point* z_0 *then, for* $\varepsilon > 0$ *sufficiently small,* (9) *has a hyperbolic periodic orbit* $x_\varepsilon(t) = z_0 + O(\varepsilon^2)$ *of corresponding stability type. Moreover, if* $z_0^s(t)$ *is a solution of system* (15) *with* $\lim_{t \to \infty} z_0^s(t) = z_0$ *then there exists a solution* $x_\varepsilon^s(t)$ *of equation* (9) *such that* $x_\varepsilon^s(t) = z_0^s(t) + O(\varepsilon^2)$ *for all* $t \in [t_0, \infty)$ *for some fixed* t_0. *Similar results apply for solutions* $z_0^u(t)$, $x_\varepsilon^u(t)$ *such that* $\lim_{t \to -\infty} z_0^u(t) = z_0$ *with the domain of validity* $t \in (-\infty, t_0]$.

The second part of this result seems not to be well known and so the proof is sketched here.

Existence of hyperbolic periodic solutions $x_\varepsilon(t)$ near z_0 follows from consideration of the full and averaged problems as autonomous systems with phase space $\mathcal{R}^n \times S^1$. Letting the $O(\varepsilon^2)$ term of system (15) be denoted by $h_0(z)$ and the $O(\varepsilon^3)$ time periodic remainder as $p(z, t)$, one can write the full system as

$$z' = f_0(z) + \varepsilon h_0(z) + \varepsilon^2 p(z, \theta/\varepsilon), \qquad (z, \theta) \in \mathcal{R}^n \times S^1, \qquad \theta' = 1, \tag{16}$$

where time has been rescaled with $\tau = \varepsilon t$ and $z' = dz/d\tau$. Associated with this system there is a *Poincaré map* [6, 8, 9], P_ε, defined on the global cross section $\Sigma = \{(z, \theta) | \theta = 0\}$. P_ε is defined as follows. Let $(z(t, z_1), t) \in \mathcal{R}^n \times S^1$ denote the solution of system (16) based at $(z_1, 0) \in \Sigma$; then $P_\varepsilon(z_1) = z(T/\varepsilon, z_1)$. Thus $P_\varepsilon(z_1)$ is the next point at which the solution starting at z_1 intersects Σ, the "time of flight" being εT, the rescaled period.

Let P_0 denote the map obtained when the $\varepsilon^2 p$ term is dropped, as in the averaged equation. Then, by hypothesis, P_0 has a hyperbolic fixed point at z_0 and hence $Id - DP_0(z_0)$ is invertible. Since P_ε is a small perturbation of P_0, the inverse function theorem ensures that P_ε has a similar hyperbolic fixed point near z_0 and hence system (16) has a hyperbolic periodic orbit as stated. (See references [6, 8] for more discussion of Poincaré maps and averaging.)

Existence of orbits asymptotic to $x_\varepsilon(t)$ as $t \to t \pm \infty$ follows from the stable manifold theorem [8, 9], which provides local results near $z_0(x_\varepsilon(t))$ in a fixed neighbourhood of size δ which are valid on infinite time intervals, and from simple perturbation arguments away from z_0, since here the perturbing vector field $\varepsilon^2 p(z, t)$ is small compared to $f_0(z) + \varepsilon h_0(z)$. Sanders [10] has provided a neat proof in which the asymptotics are carefully computed.

3. APPLICATION OF THE AVERAGING METHOD

Prior to application of the averaging theorem, one must first transform equation (8) into the standard form of equation (9). To do this one can employ the invertible van der Pol "rotation" transformation

$$\begin{pmatrix} u \\ v \end{pmatrix} = A \begin{pmatrix} y \\ \dot{y} \end{pmatrix}, \qquad \begin{pmatrix} y \\ \dot{y} \end{pmatrix} = A^{-1} \begin{pmatrix} u \\ v \end{pmatrix}, \qquad A = \begin{bmatrix} \cos k\omega t & -\dfrac{\sin k\omega t}{k\omega} \\ -\sin k\omega t & -\dfrac{\cos k\omega t}{k\omega} \end{bmatrix}, \tag{17}$$

where the rational number $k = m/n$ is to be chosen subsequently and will determine the order of resonance one wishes to study. Using the transformation in (17), one can write equation (8) as

$$\begin{pmatrix} \dot{u} \\ \dot{v} \end{pmatrix} = -\frac{1}{k\omega}\{\varepsilon(\Omega y + f_0(y,\dot{y},t)) + \varepsilon^2 f_1(y,t)\}\begin{pmatrix} \sin k\omega t \\ \cos k\omega t \end{pmatrix} + O(\varepsilon^3), \tag{18}$$

where y and $\dot{y}$ are given as functions of u, v by equation (17), and

$$\Omega = (k^2\omega^2 - 2)/\varepsilon \tag{19}$$

should be at most of $O(1)$, so that the right-hand side of equation (18) contains terms of $O(\varepsilon)$ and $O(\varepsilon^2)$ only, as required.

First one can consider order ε averaging. Expanding equation (18) gives to $O(\varepsilon)$,

$$\begin{pmatrix} \dot{u} \\ \dot{v} \end{pmatrix} = -\frac{\varepsilon}{k\omega}\left\{\Omega\begin{pmatrix} ucs - vs^2 \\ uc^2 - vcs \end{pmatrix} + \delta k\omega\begin{pmatrix} us^2 + vcs \\ ucs + vc^2 \end{pmatrix} - 6\Gamma\bar{c}\begin{pmatrix} ucs - vs^2 \\ uc^2 - vcs \end{pmatrix}\right.$$
$$\left. - 3\begin{pmatrix} u^2c^2s - 2uvs^2c + v^2s^3 \\ u^2c^3 - 2uvc^2s + v^2s^2c \end{pmatrix}\right\} + O(\varepsilon^2), \tag{20}$$

where $c \triangleq \cos k\omega t$, $s \triangleq \sin k\omega t$ and $\bar{c} \triangleq \cos \omega t$. To study subharmonics of order 2 one sets $k = 1/2$, so that the transformation (17) takes a sinusoidal subharmonic of order 2, $y = a \cos(\omega t/2) + b \sin(\omega t/2)$, to a fixed point $(u, v) = (a, -b)$. (In fact such a fixed point is always accompanied by its image under rotation through π: $(-a, b)$, as will be seen.) Similarly, an almost sinusoidal subharmonic will become an almost constant solution, which is transformed by the averaging process to a constant solution.

Setting $k = 1/2$, so that $\cos k\omega t = \cos(\omega t/2)$ and $\sin k\omega t = \sin(\omega t/2)$, and employing various trigonometrical relationships one obtains the mean and oscillating components of $\mathbf{f}(\mathbf{x}, t)$ for this problem:

$$\frac{d}{dt}\begin{pmatrix} u \\ v \end{pmatrix} = \varepsilon \mathbf{f}_0(u,v) + \varepsilon \tilde{\mathbf{f}}(u,v,t) + O(\varepsilon^2), \tag{21a}$$

$$\mathbf{f}_0 = \frac{1}{\omega}\begin{bmatrix} -\delta\omega/2 & 3\Gamma + \Omega \\ 3\Gamma - \Omega & -\delta\omega/2 \end{bmatrix}\begin{pmatrix} u \\ v \end{pmatrix}, \tag{21b}$$

$$\tilde{\mathbf{f}} = \frac{1}{\omega}\left\{\Omega\begin{pmatrix} -us_2 - vc_2 \\ -uc_2 + vs_2 \end{pmatrix} + \frac{\delta\omega}{2}\begin{pmatrix} uc_2 - vs_2 \\ -us_2 - vc_2 \end{pmatrix} + 3\Gamma\begin{pmatrix} us_4 - 2vc_2 + vc_4 \\ 2uc_2 + uc_4 - vs_4 \end{pmatrix}\right.$$
$$\left. + \frac{3}{2}\begin{pmatrix} u^2(s_1 + s_3) - 2uv(c_1 - c_3) + v^2(3s_1 - s_3) \\ u^2(3c_1 + c_3) - 2uv(s_1 + s_3) + v^2(c_1 - c_3) \end{pmatrix}\right\}, \tag{21c}$$

where $s_j \triangleq \sin(j\omega t/2)$ and $c_j \triangleq \cos(j\omega t/2)$. The $O(\varepsilon)$ averaged system is thus simply the linear constant coefficient equation

$$\begin{pmatrix} \dot{u} \\ \dot{v} \end{pmatrix} = \frac{\varepsilon}{\omega}\begin{bmatrix} -\delta\omega/2 & 3\Gamma + \Omega \\ 3\Gamma - \Omega & -\delta\omega/2 \end{bmatrix}\begin{pmatrix} u \\ v \end{pmatrix}. \tag{22}$$

Elementary analysis reveals that the origin $(u, v) = (0, 0)$ is an attracting fixed point (a sink) if $\Delta \stackrel{\text{def}}{=} \delta^2\omega^2/4 - (9\Gamma^2 - \Omega^2) > 0$ and a saddle point if $\Delta < 0$. When $\delta = 0$ the origin is a center for $|3\Gamma| < |\Omega|$ and a saddle for $|3\Gamma| > |\Omega|$. The two dimensional surface $\Delta = 0$ in the (δ, γ, ω) parameter space is thus a bifurcation surface for equation (22). However, since all the $O(\varepsilon)$ non-linear terms vanish in the averaging procedure, one obtains no information on the evolution of post-bifurcation regimes, including the desired period two orbits. One must therefore carry out second order averaging. Note, however, that the bifurcation curve given by $\Delta = \delta^2\omega^2/4 - (9\Gamma^2 - \Omega^2) = 0$ agrees with that obtained by regular perturbation methods for the standard Mathieu equation, if equation (8) is put into that form and truncated at $O(\varepsilon)$ (cf. reference [11], section 5.3.5).

Since one requires the transformation function $\mathbf{u}(\mathbf{y}, t)$ of equation (11) given by $\dot{\mathbf{u}} = \tilde{\mathbf{f}}$, one integrates $\tilde{\mathbf{f}}$ of expression (21c) term by term, noting that $\int s_j \, dt = -(2/j\omega)c_j$ and $\int c_j \, dt = (2/j\omega)s_j$. After tedious but elementary calculations, one obtains the average $[D\tilde{\mathbf{f}}(\mathbf{z})\mathbf{u}(\mathbf{z})]_0$ for this problem as

$$\frac{1}{2\omega^3}\left\{\begin{array}{l} -6\Gamma\delta\omega u + (2\Omega^2 + 12\Gamma\Omega + \frac{\delta^2\omega^2}{2} + 9\Gamma^2)v + 60v(u^2 + v^2) \\ -(2\Omega^2 - 12\Gamma\Omega + \frac{\delta^2\omega^2}{2} + 9\Gamma^2)u + 6\Gamma\delta\omega v - 60u(u^2 + v^2) \end{array}\right\}. \qquad (23)$$

Averaging the $O(\varepsilon^2)$ term of expression (18) directly gives

$$g_0(z) = \frac{1}{\omega}\left\{\begin{array}{l} \frac{3\delta\omega\Gamma}{2-\omega^2}u + 3\Gamma^2 v - \frac{3v}{4}(u^2+v^2) \\ -3\Gamma^2 u - \frac{3\delta\omega\Gamma}{2-\omega^2}v + \frac{3u}{4}(u^2+v^2) \end{array}\right\}, \qquad (24)$$

and addition of expressions (23) and (24) then yields the final averaged system correct to $O(\varepsilon^2)$:

$$\begin{pmatrix}\dot{u}\\\dot{v}\end{pmatrix} = \frac{\varepsilon}{\omega}\begin{bmatrix} -\frac{\delta\omega}{2} & 3\Gamma+\Omega \\ 3\Gamma-\Omega & -\frac{\delta\omega}{2} \end{bmatrix}\begin{pmatrix}u\\v\end{pmatrix} + \frac{\varepsilon^2}{\omega}\left\{\begin{bmatrix} \frac{6\delta\Gamma(\omega^2-1)}{\omega(2-\omega^2)} & \frac{6\Gamma\Omega}{\omega^2} - \left(3\Gamma^2 + \delta^2/4 + \frac{\Omega^2 + 9\Gamma^2/2}{\omega^2}\right) \\ \frac{6\Gamma\Omega}{\omega^2} + \left(3\Gamma^2 + \delta^2/4 + \frac{\Omega^2 + 9\Gamma^2/2}{\omega^2}\right) & -\frac{6\delta\Gamma(\omega^2-1)}{\omega(2-\omega^2)} \end{bmatrix}\begin{pmatrix}u\\v\end{pmatrix} + \left(\frac{30}{\omega^2} - \frac{3}{4}\right)(u^2+v^2)\begin{pmatrix}v\\-u\end{pmatrix}\right\}, \qquad (25)$$

in which $\Gamma = \gamma/(2-\omega^2)$, $\Omega = [(\omega^2/4)-2]/\varepsilon$ ($= O(1)$ at most), $\gamma = \bar{\gamma}/\varepsilon$ and $\delta = \bar{\delta}/\varepsilon$. Note that expression (25) takes the form

$$\dot{u} = (a+b)u + (c+d)v + \alpha(u^2+v^2)v, \qquad \dot{v} = (c-d)u + (a-b)v - \alpha(u^2+v^2)u, \qquad (26)$$

or, in polar co-ordinates ($r = (u^2+v^2)^{1/2}$, $\theta = \arctan(v/u)$),

$$\dot{r} = r(a + b\cos 2\theta + c\sin 2\theta), \qquad \dot{\theta} = c\cos 2\theta - b\sin 2\theta - d - \alpha r^2, \qquad (27)$$

where a, b, c, d and α are found by reference to equation (25) (also see equation (33), below).

Note that the perturbed motion $x(t)$ of the original equation (1) is given approximately by

$$x(t) = \hat{x}(t) + \varepsilon y(t) = \hat{x}(t) + \varepsilon\{u(t)\cos(\omega t/2) - v(t)\sin(\omega t/2)\}, \qquad (28a)$$

and, in terms of the polar co-ordinates, this becomes

$$x(t) = \hat{x}(t) + \varepsilon r(t)\cos\{(\omega t/2) + \theta(t)\}. \qquad 28b)$$

Since (u, v) and (r, θ) evolve in slow time εt, these motions are slowly modulated sinusoids of period $4\pi/\omega$ superimposed on the basic period $2\pi/\omega$ response $\hat{x}(t)$.

4. BIFURCATION TO SUBHARMONICS OF PERIOD TWO

The averaged equations (25) always possess the trivial equilibrium solution $(u, v) = (0, 0)$, corresponding to the original periodic motion $\hat{x}(t)$ whose stability and bifurcations we wish to study. Any additional non-trivial equilibria of equation (25) correspond to subharmonics of period 2 and such equilibria always come in pairs, related by a phase shift of π. To see this, consider equations (25)–(27). Setting the final derivatives equal to zero, dividing by v and equating the resulting expressions for u/v gives

$$\alpha^2 r^4 + 2d\alpha r^2 + (a^2 + d^2) - (c^2 + b^2) = 0, \qquad (29)$$

where $r^2 = u^2 + v^2$. Solving for r gives

$$r = +\sqrt{(1/\alpha)(-d \pm \sqrt{c^2 + b^2 - a^2})}. \qquad (30)$$

For non-trivial solutions to exist, the two conditions

$$c^2 + b^2 - a^2 \geq 0, \qquad (-d \pm \sqrt{c^2 + b^2 - a^2})/\alpha > 0 \qquad (31a, b)$$

must be met. If one such solution for r does exist, then examination of equation (27) shows that there are in fact two distinct solutions, $(r, \theta) = (r_i, \theta_i), (r_i, \theta_i + \pi)$. Including the trivial solution, equation (25) can therefore have one, three or five equilibria. Each additional *pair* of equilibria corresponds to a *single* subharmonic of period two given (approximately) by

$$x(t) = \hat{x}(t) + \varepsilon r_i \cos\{(\omega t/2) + \theta_i\}, \qquad i = 1, 2. \qquad (32)$$

The bifurcation sets upon which these pairs of equilibria are created and annihilated are given by equations (31a, b), with equalities replacing the inequalities. These curves meet at the point $c^2 + b^2 - a^2 = d^2 = 0$.

To study the bifurcations and associated phase portraits of equation (25) one must consider the scaling in more detail. From equation (25), the coefficients a, b, c, d and α are as follows:

$$a = -\frac{\varepsilon\delta}{2}, \qquad b = \frac{\varepsilon^2 6\delta\Gamma(\omega^2-1)}{\omega^2(2-\omega^2)}, \qquad c = \frac{\varepsilon 3\Gamma}{\omega} + \frac{\varepsilon^2 6\Gamma\Omega}{\omega^3},$$

$$d = \frac{\varepsilon\Omega}{\omega} + \varepsilon^2\left(\frac{\Omega^2 + \delta^2\omega^2/4 + 9\Gamma^2/2}{\omega^3} + 3\Gamma^2\right), \qquad \alpha = \frac{\varepsilon^2}{\omega}\left(\frac{30}{\omega^2} - \frac{3}{4}\right). \qquad (33)$$

Thus, using $\varepsilon\Omega = \omega^2/4 - 2$ and returning to the original damping $\bar{\delta} = \varepsilon\delta$ and force amplitude $\bar{\gamma} = \varepsilon\gamma$, and recalling that $\Gamma = \gamma/(2-\omega^2)$, one has

$$a = -\bar{\delta}/2, \qquad b = \frac{6\bar{\delta}\bar{\gamma}(\omega^2-1)}{\omega^2(2-\omega^2)^2}, \qquad c = \frac{3\bar{\gamma}(3\omega^2-8)}{2\omega^3(2-\omega^2)},$$

$$d = \frac{(\omega^2-8)}{16\omega}(5\omega^2-8) + \frac{\bar{\delta}^2}{4\omega} + \frac{3\bar{\gamma}^2(2\omega^2+3)}{2(2-\omega^2)^2\omega^2}. \qquad (34)$$

Also, $\alpha \approx 3\varepsilon^2/2\sqrt{2}$, since $\omega^2 = 8 + O(\varepsilon)$.

In Figure 1 the stability boundaries or bifurcation set in $\bar{\gamma}, \omega$ space for $\bar{\delta} = 0.1$ are shown. In region I only the trivial solution $(u, v) = 0$ exists and it is a sink. In region II, the trivial solution is a saddle point and a pair of non-trivial sinks exists corresponding to a stable subharmonic of order two. In region III the trivial solution is again stable and two subharmonics, one stable and the other of unstable (saddle-type) exist. This pair appears with amplitude $r_i = O(1/\sqrt{\varepsilon})$ in a "non-local" bifurcation as the curve $c^2 + b^2 - a^2 = 0$ is

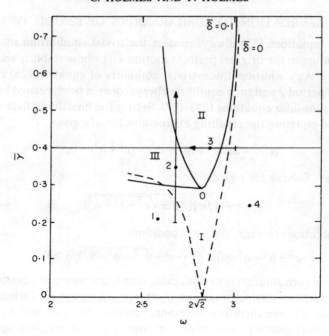

Figure 1. The bifurcation set for subharmonics of order two. $\bar{\delta} = 0.1$ and $\bar{\delta} = 0$. The points 1 to 4 refer to parameter choices for the phase portraits of Figure 3 and the light lines to the paths taken through the parameter space for Figures 2(a) and (b).

crossed from region I to III. Note, however, that the associated subharmonics $x = \hat{x} + \varepsilon r_i \cos\{(\omega t/2) + \theta_i\}$ (equation (28)) have amplitude $|\hat{x}| + O(\sqrt{\varepsilon})$ and thus are near the basic solution $\hat{x}$ as required.

Upon passing from region I to II, the non-trivial stable subharmonic appears in a supercritical bifurcation, and upon passing from region III to II the saddle-type subharmonic disappears in a subcritical bifurcation. These are examples of the two common types of local bifurcation to subharmonics of period two [9, 12, 13]. At the "organizing center", O, where all three bifurcation curves meet, the five equilibria of the averaged equations all coalesce in a doubly degenerate (weakly stable) fixed point.

In Figure 2 bifurcation or branching diagrams are shown illustrating the evolution of the subharmonics as one crosses between regions. In Figure 3 sketches are presented of the phase portraits of the averaged equation in the various regions. Note that the direction of

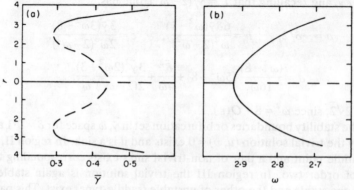

Figure 2. Bifurcation diagrams. (a) Passing through regions I, III, II; (b) passing through regions I, II. ——, Stable (sink) solutions; ---, unstable (saddle) solutions.

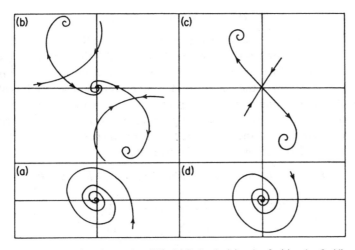

Figure 3. Phase portraits of equation (25). (a) Point 1; (b) point 2; (c) point 3; (d) point 4.

the spiral sink at $(0, 0)$ changes as ω passes through the subharmonic resonance $2\sqrt{2}$ (see Figures 3(a) and (d)). Below $2\sqrt{2}$ the natural rotation speed of the unforced ($\bar{\gamma} = 0$) problem is faster than that of the forcing frequency, and hence motion is anticlockwise in the rotating co-ordinate frame introduced by the transformation (17). For $\omega > 2\sqrt{2}$ the natural motion is slower and hence the averaged system displays clockwise rotation near $(u, v) = (0, 0)$.

In Figure 1 the bifurcation set for zero damping (the Hamiltonian case) is also shown. Note that region I is now confined to the right of the diagram. All the points which were stable sinks for $\bar{\delta} > 0$ are now centers and the associated phase portraits are filled with periodic orbits and contain heteroclinic and homoclinic saddle connections (see Figure 4). This bifurcation picture is similar to that known to occur for two degrees of freedom Hamiltonian systems near resonances of order 2 [13, 14]. Further comments on these degenerate phase portraits are made in section 6.

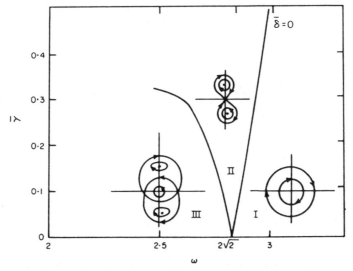

Figure 4. The Hamiltonian case, $k = 1/2$: bifurcation set and phase portraits in the three regions.

5. OTHER SUBHARMONICS

A study of the equations averaged at $O(\varepsilon^2)$ reveals that the only other subharmonics occurring near $x(t) = \hat{x}(t)$ for this problem are those of order three, or period $6\pi/\omega$. These occur for the choices $k = 1/3$ and $1/5$, *but they only occur if the damping $\bar{\delta}$ is of order ε^2*, otherwise the $-\delta/2\binom{u}{v}$ term in the equations dominates and no non-trivial solutions exist. One can expect that averaging to higher order would reveal the presence of yet other, higher subharmonics, provided that the damping were of sufficiently low order.

Supposing then that $\bar{\delta} = \varepsilon^2 \delta$ and repeating the averaging process to $O(\varepsilon^2)$ with $k = 1/3$, one obtains the averaged system in polar co-ordinates,

$$\dot{r} = r(a' + b'r \sin 3\theta), \qquad \dot{\theta} = -(d' + \alpha r^2) + b'r \cos 3\theta, \qquad (35)$$

$$a' = -\frac{3\bar{\delta}}{4}, \qquad b' = \frac{\varepsilon 9\bar{\gamma}(54 - \omega^2)}{8\omega^3(2 - \omega^2)} = -\frac{3\varepsilon\bar{\gamma}}{64\sqrt{2}} + O(\varepsilon^3),$$

$$d' = \frac{(\omega^2 - 18)}{6\omega} - \frac{27\bar{\gamma}^2(5\omega^2 + 36)}{20\omega^3(2 - \omega^2)^2} = \frac{(\omega^2 - 18)}{2\sqrt{2}} - \frac{63\bar{\gamma}^2}{5120\sqrt{2}} + O(\varepsilon^3),$$

$$\alpha' = \frac{9\varepsilon^2}{8\omega^3}(90 + \omega^2) = \frac{9\varepsilon^2}{4\sqrt{2}} + O(\varepsilon^3). \qquad (36)$$

Note that to obtain the correct orders, one requires $\omega = 3\sqrt{2} + O(\varepsilon^2)$ in this case, and then the coefficients a', b', d' and α' are of $O(\varepsilon^2)$, since the $O(\varepsilon)$ terms vanish identically in the averaging procedure. (Taking $k = 1/5$ gives a similar system, with slightly different coefficients.) The non-trivial solutions of equation (35) are

$$r = +\sqrt{(1/2\alpha')(b'^2 - 2\alpha'd') \pm \sqrt{(b'^2 - 2\alpha'd')^2 - 4(d'^2 - a'^2)}}, \qquad (37)$$

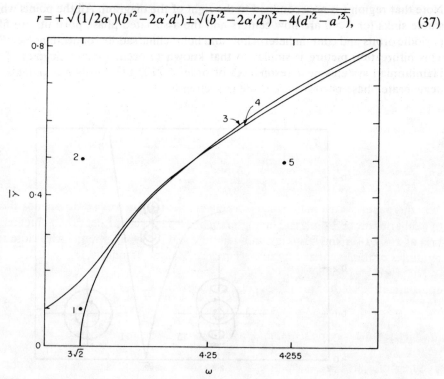

Figure 5. The bifurcation set for subharmonics of order three. The points 1 to 5 refer to parameter choices for the phase portraits of Figure 6.

and occur in sets of three with $\theta = \theta_i$, $\theta_i + 2\pi/3$, $\theta_i + 4\pi/3$. Each set corresponds to a single subharmonic of order three.

Performing an analysis similar to that of section 4 gives the bifurcation set and phase portraits of Figures 5 and 6. Since ε appears explicitly in the expression for b', $\varepsilon = 0 \cdot 01$ has been taken for purposes of presentation. Other values of ε (and δ) will yield qualitatively similar results. Note that unless $\varepsilon^2 \delta = \bar{\delta} = 0$ the trivial solution at the origin remains a sink for all parameter values and does not undergo bifurcation, although orbits asymptotic to it do suffer a change of direction, as in the order two case. In the Hamiltonian case ($\bar{\delta} = 0$), one has the 3:1 resonance (cf. [13–15]), in which, at $d' = 0$, the trivial equilibrium becomes a degenerate saddle point. The Hamiltonian phase portraits are sketched in Figure 7.

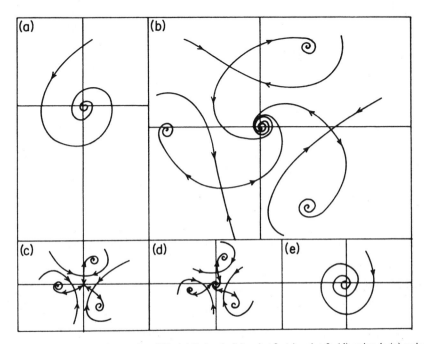

Figure 6. Phase portraits of equation (35). (a) Point 1; (b) point 2; (c) point 3; (d) point 4; (e) point 5.

6. CONCLUSIONS AND COMMENTS ON THE ANALYSIS

In this paper the averaging theorem [7] has been used to study the existence and stability of subharmonic solutions to Duffing's equation (1) occurring near the "unforced" equilibria at $x = \pm 1$ for small forcing and damping and for forcing frequencies close to integer multiples of the linearized natural frequency at these equilibria.

The averaging theorem yields an autonomous planar vector field which provides an approximation to the true time dependent system

$$\dot{\mathbf{z}} = \varepsilon \mathbf{f}_0(\mathbf{z}) + \varepsilon^2 \mathbf{h}_0(\mathbf{z}) + \varepsilon^3 \mathbf{p}(\mathbf{z}, t), \tag{38}$$

in which $\mathbf{p}$ is a T-periodic remainder. Solutions of system (38) induce a Poincaré or time T map, $\mathbf{P}_\varepsilon$, on a global cross section $\Sigma = \{\mathbf{z}, t | t = 0, T, 2T, \ldots\}$ as described in references [6, 8, 9]. Since the vector field is periodic, a fixed point of $\mathbf{P}_\varepsilon$ corresponds to a periodic solution of period T; similarly a periodic point of period k corresponds to a subharmonic of period kT.

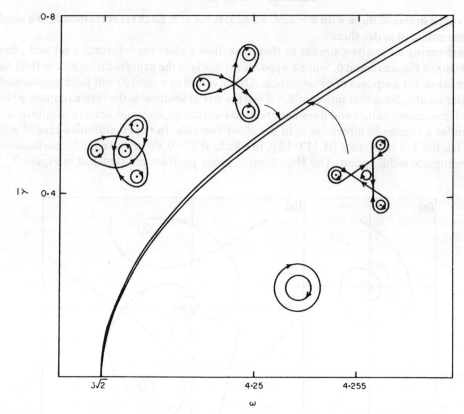

Figure 7. The Hamiltonian case, $k = 1/3$: bifurcation sets and phase portraits in the various regions.

Along with system (38) one has the (truncated) averaged system

$$\dot{z} = \varepsilon \mathbf{f}_0(\mathbf{z}) + \varepsilon^2 \mathbf{h}_0(\mathbf{z}), \tag{39}$$

and one can define an associated Poincaré map $\mathbf{P}_0$, where $\mathbf{P}_0(\mathbf{z}_1) = \mathbf{z}(T, \mathbf{z}_1)$ and $\mathbf{z}(t, \mathbf{z}_1)$ denotes the solution of system (38) based at $\mathbf{z}(0) = \mathbf{z}_1$. Since for $0 < \varepsilon \ll 1$ system (39) is a small perturbation of system (38), $\mathbf{P}_0$ provides a good approximation of $\mathbf{P}_\varepsilon$. These are the ideas behind the proof of the averaging theorem sketched in section 2.

In Figures 3, 4, 6 and 7 phase portraits of the averaged equations have been sketched. These phase portraits, and their solution curves, especially their separatrices, are related to the map $\mathbf{P}_0$ as follows: $\mathbf{P}_0$ is obtained by letting solutions of system (39) flow for time T; thus an invariant (solution) curve of system (39) is an invariant curve for $\mathbf{P}_0$. Hence the saddle separatrices of (39) provide approximations for the saddle separatrices or stable manifolds [8, 9, 12] of $\mathbf{P}_0$, and hence for the domains of attraction. The validity of such approximations has been discussed in reference [6], where it has been argued that the saddle connections or homo- and heteroclinic orbits [8, 12, 13] typically found in averaged Hamiltonian systems are expected to break in a complicated manner, yielding transverse homoclinic points and Smale horseshoes [16], when the unaveraged term $\varepsilon^3 \mathbf{p}(\mathbf{z}, t)$ is restored. However, calculations such as those in reference [6] suggest that, for damping greater than of exponential order ($\bar{\delta} = k e^{-c/\varepsilon}$) the manifolds will in fact not intersect. Certainly for $\bar{\delta} = O(\varepsilon)$ or $O(\varepsilon^2)$, our averaged results can be expected to provide a good approximation, qualitatively correct, of the true Poincaré map and hence of the full system's behaviour.

Finally some comments can be made on the "non-local" bifurcations in which subharmonics appear in pairs at finite amplitude. The phase portrait of the unperturbed system (1) with $\varepsilon = 0$ is easily obtained from the level curves of the Hamiltonian

$$H(x, \dot{x}) = (\dot{x}^2/2) - (x^2/2) + (x^4/4) = \text{const.} \tag{40}$$

The open sets surrounding the centers at $(x, \dot{x}) = (\pm 1, 0)$ are filled with continuous families of periodic orbits whose periods increase monotonically from $2\pi/\sqrt{2}$ to ∞ as one moves outward. Thus, if the forcing period $2\pi/\omega$ is a little greater than some integer multiple $2\pi k/\sqrt{2}$, it will be exactly in resonance at order k with one of the periodic orbits near the center. The general theory of Hamiltonian systems [13–15] then leads one to expect that this "resonant torus" will break up into a resonance band dotted with k pairs of saddle and center type subharmonics of order k. Figures 4 and 7 illustrate this for $k = 2$ and 3. In this paper it has been shown that this behavior persists for the damped system, provided that the damping is not too large, and also it has been shown how the break-up of the resonant torus interacts with the harmonic response $(u, v) = (0, 0)$. The present analysis is restricted to small amplitude solutions (near $\hat{x}(t) = 1 + O(\varepsilon)$). In related work [17] the large amplitude resonant tori are being studied by perturbation of the exact elliptic function solutions of the unperturbed Duffing system. In this work it is found that the behavior described in the present paper is merely the center piece of an infinite sequence of subharmonic bifurcations limiting on the chaotic homoclinic orbits described in reference [2]. In this connection, see also references [18, 19].

REFERENCES

1. W. Y. TSENG and J. DUGUNDJI 1971 *Transactions of the American Society of Mechanical Engineers Journal of Applied Mechanics* **38**, 467–476. Nonlinear vibrations of a buckled beam under harmonic excitation.
2. P. J. HOLMES 1979 *Philosophical Transactions of the Royal Society London* **A292**, 419–448. A nonlinear oscillator with a strange attractor.
3. F. C. MOON and P. J. HOLMES 1979 *Journal of Sound and Vibration* **65**, 275–296. A magnetoelastic strange attractor. (See also 1980 *Journal of Sound and Vibration* **69**, 339.)
4. Y. UEDA 1980 in *New Approaches to Nonlinear Problems in Dynamics* (Editor P. J. Holmes). S.I.A.M. Publications. Steady motions exhibited by Duffing's equations.
5. C. HAYASHI 1964 *Nonlinear Oscillations in Physical Systems*. New York: McGraw-Hill.
6. P. J. HOLMES 1980 *Society of Industrial and Applied Mathematics, Journal of Applied Mathematics* **38**, 65–80. Averaging and chaos in forced oscillations.
7. J. K. HALE 1969 *Ordinary Differential Equations*. New York: Wiley.
8. D. R. J. CHILLINGWORTH 1976 *Differential Topology with a View to Applications*. London: Pitman.
9. F. TAKENS 1973 *Communication 2, Mathematics Institute, Rijksuniversiteit Utrecht*. Introduction to global analysis.
10. J. A. SANDERS 1980 *Vrije Universiteit, Amsterdam, Report*. Note on the validity of Melnikov's method and averaging.
11. A. NAYFEH and D. T. MOOK 1979 *Nonlinear Oscillations*. New York: Wiley.
12. F. TAKENS 1974 *Communication 3, Mathematics Institute Rijksuniversiteit, Utrecht*, 1–59. Forced oscillations and bifurcations.
13. R. ABRAHAM and J. E. MARSDEN 1978 *Foundations of Mechanics*. New York: Addison-Wesley, second edition.
14. J. A. SANDERS 1978 *Ph.D. Thesis, University of Utrecht, The Netherlands*. On the theory of nonlinear resonance.
15. V. I. ARNOLD 1978 *Mathematical Methods of Classical Mechanics*. New York: Springer-Verlag.
16. S. SMALE 1967 *Bulletin of the American Mathematical Society* **73**, 747–817. Differentiable dynamical systems.

17. B. D. GREENSPAN and P. J. HOLMES 1980 (in preparation). Subharmonic and homoclinic bifurcations in forced oscillations.
18. V. K. MELNIKOV 1963 *Transactions of the Moscow Mathematical Society* **12**, 3–57. Stability of the center to time periodic perturbations.
19. S. N. CHOW, J. K. HALE and J. MALLET-PARET 1980 *Journal of Differential Equations* **37**, 351–373. An example of bifurcation to homoclinic orbits.

Physica 16D (1985) 1-13
North-Holland, Amsterdam

SUBHARMONIC AND HOMOCLINIC BIFURCATIONS IN A PARAMETRICALLY FORCED PENDULUM

B.P. KOCH and R.W. LEVEN
Sektion Physik / Elektronik, Ernst-Moritz-Arndt-Universität Greifswald, DDR-2200 Greifswald, German Dem. Rep.

Received 19 March 1984
Revised manuscript received 29 November 1984

Depending on the parameters of a parametrically forced pendulum system the boundaries of subharmonic and homoclinic bifurcations are calculated on the basis of the Melnikov method and of averaging methods. It is shown that, as a parameter is varied, repeated resonances of successively higher periods occur culminating in homoclinic orbits. According to the theorem of Smale homoclinic bifurcation is the source of the unstable chaotic motions observed. For some selected parameter sets the theoretical predictions are tested by numerical calculations. Very good agreement is found between analytical and numerical results.

1. Introduction

It is known that simple nonlinear oscillators with dissipation and additional time-dependent force or periodically varying parameters, which can be described by second order non-autonomous ordinary differential equations, exhibit regular periodic motions as well as complicated non-periodic motions. The most interesting are the chaotic motions, which show a sensitive dependence on initial conditions. Though there is a lack of mathematical proofs a lot of numerical investigations and some experiments on physical systems have yielded evidence that stable chaotic motions really exist for certain parameter intervals. Certain types of forced oscillators have been studied which are related to such fields as mechanics, electronics, nonlinear optics, solid state physics, chemical kinetics, biology, population dynamics.

The assumptions concerning the chaotic motions are supported by approximate values for quantities characterizing the chaos, i.e. entropies, fractal dimensions, Lyapunov exponents. These values can be found by numerical calculations or in physical experiments. The observed transitions in parameter space between periodic and aperiodic regions are further hints at the chaotic nature of the aperiodic regime. The same transition types are found in one-dimensional and higher dimensional maps. Rigorous proofs of the existence of stable chaotic motions exist for some of such systems [1, 2]. Homoclinic and heteroclinic points of intersection of stable and unstable manifolds coming from hyperbolic fixed points of the associated Poincaré map are accompanied by complicated motions. Smale [3] has shown that, if the Poincaré map has transverse homoclinic points, then

a) the set of nonwandering points contains an invariant Cantor set;

b) some iterate of the Poincaré map restricted to this Cantor set is equivalent to the shift automorphism on doubly infinite sequences [4].

In dissipative systems, however, the existence of homoclinic points does not imply stable chaotic behaviour, i.e. the existence of chaotic attractors. The invariant Cantor sets discussed by Smale are unstable in the sense that nearby orbits tend away from them. Therefore, as a rule a so-called transient chaos arises which is possibly the precursor of stable chaotic behaviour.

Important is the occurrence of homoclinic tangencies in dissipative systems. These were studied by Newhouse [5, 6]. His work indicates that we can expect an infinite number of sinks for certain parameter values near the homoclinic tangency.

For some oscillator systems with periodic perturbations it can be shown that, as a parameter is varied, repeated resonances of successively higher periods occur culminating in subharmonics of infinite order and homoclinic orbits. There exists a global method within the perturbation theory, the so-called Melnikov method, which allows to predict the occurrence of subharmonic and homoclinic orbits if parameters are varied. This method developed by Melnikov [7] was generalized by Holmes and Marsden [8, 9]. It can be applied to systems with strong nonlinearities and was successfully employed by several authors [10–22]. A detailed description of the Melnikov method was given by Greenspan and Holmes [10, 11], in the book of Guckenheimer and Holmes [12] and by Salam, Marsden and Varaiya [16]. Similar results were obtained by Chow, Hale and Mallet-Paret [23].

An application of the Melnikov method to homoclinic bifurcations of perturbed pendulum and pendulum-like systems is given in refs. 12, 17–22. In the present paper the Melnikov method is applied to the parametrically excited damped pendulum. Our paper focuses, in addition to homoclinic bifurcations, on subharmonic bifurcations, their relationship to homoclinic bifurcations and on numerical results confirming the theoretical predictions.

The paper is organized as follows: section 2 contains the application of the Melnikov method to the parametrically forced pendulum. Concerning the pertinent definitions and theorems we refer the reader to the papers of Greenspan and Holmes [10, 11] and the fundamental textbook of Guckenheimer and Holmes [12]. In section 3 the stability of the subharmonics is examined using second order averaging. The analytic investigations are compared with numerical calculations in section 4. This section illustrates the usefulness of the Melnikov method.

2. The Melnikov functions of the parametrically forced pendulum

The governing equations for a pendulum which is disturbed by parametrical perturbation and dissipation are

$$\dot{x} = v, \quad \dot{v} = -\varepsilon\beta v - (1 + \varepsilon A \cos \omega t)\sin x \qquad (2.1)$$

$((x, v) \in S^1 \times R, 0 < \varepsilon \ll 1, \beta, A, \omega > 0$ with $\varepsilon\beta, \varepsilon A$ small). Eqs. (2.1) describe a simple mechanical system with very complicated dynamics. There have been published a large number of papers where the system (2.1) is investigated applying analytical as well as numerical methods [24–41]. In several of these papers also the occurrence of subharmonic oscillations was discussed. But as a rule, only a few terms of the expansion of the sine term in (2.1) have been considered [28, 29], i.e. the results hold only for small oscillation amplitudes and nothing can be said about the rotating solution regime. Only in the paper of Chester [35] the sine term is fully considered and averaging methods are used to discuss the oscillating as well as the rotating subharmonical solutions. The Melnikov method applied in our paper has the advantage that the relations between subharmonic and homoclinic bifurcations and thus the connection between periodic motion and the complicated behaviour of the system concerned can be clearly demonstrated. Recently, eqs. (2.1) have been investigated numerically for medium and large perturbations [39–41].

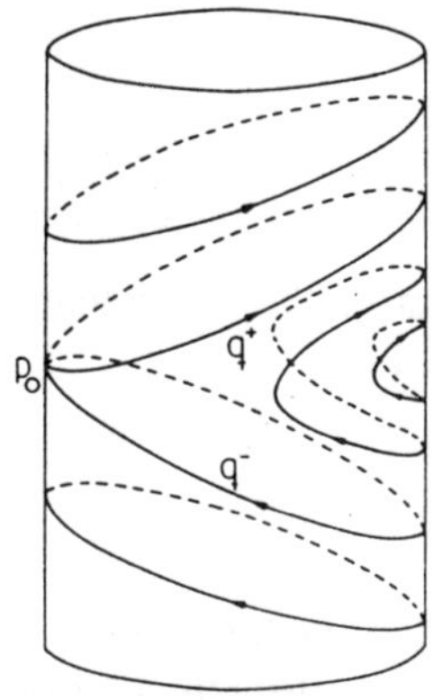

Fig. 1. Phase space trajectories of the unperturbed pendulum system (eqs. (2.1) with $\varepsilon = 0$) on the cylinder $S^1 \times R$.

Depending on the parameters, domains of subharmonical motion as well as regions where the system presumably exhibits chaotic behaviour have been found. Also the transition from periodical to complicated nonperiodical motion has been studied. The unstable chaos reported in this paper has also been observed. In experiments with a mechanical pendulum a periodical vertical displacement of the suspension point [42, 43] also produced the types of motion reported above.

In order to compute the Melnikov functions as defined in ref. 12 one has to know the solutions of the unperturbed problem ($\varepsilon = 0$). With regard to their energy these solutions can be divided into three types: oscillating, homoclinic and rotating motion (see fig. 1).*

For the homoclinic Melnikov function we obtain with the separatrix solution of the unperturbed problem

$$M^{\pm}(t_0) = \int_{-\infty}^{\infty} v(t-t_0)\left[-A\cos\omega t \sin(x(t-t_0)) - \beta v(t-t_0)\right] dt$$

$$= -4A \int_{-\infty}^{\infty} \cos\omega t\, \text{sech}^2(t-t_0)\tanh(t-t_0)\, dt - 4\beta \int_{-\infty}^{\infty} \text{sech}^2(t-t_0)\, dt$$

$$= 2\pi A\omega^2 \text{cosech}\left(\frac{\pi\omega}{2}\right)\sin\omega t_0 - 8\beta. \tag{2.2}$$

The smallest value of A for which $M^{\pm}$ as a function of t_0 has a quadratic zero is given by the relation

$$\pi A\omega^2 \text{cosech}\left(\frac{\pi\omega}{2}\right) = 4\beta. \tag{2.3}$$

This yields

$$\frac{A}{\beta} = R^0(\omega), \tag{2.4}$$

---

*Because in eqs. (2.1) we consider $(x, v) \in S^1 \times R$, we speak of homoclinic motion and homoclinic bifurcation, respectively, instead of heteroclinic motion and heteroclinic bifurcation.

with

$$R^0(\omega) = \frac{4}{\pi\omega^2} \sinh\left(\frac{\pi\omega}{2}\right). \tag{2.5}$$

The above relation characterizes the onset of the homoclinic bifurcation with an accuracy of $\mathcal{O}(1)$ for sufficiently small ε (theorem 4.5.4 of [12]). For

$$\frac{A}{\beta} > R^0(\omega)$$

there exist transverse homoclinic intersection points of the stable and unstable manifolds of the saddle point $(\pi, 0)$ for the Poincaré map of eqs. (2.1) (theorem 4.6.3 of [12]).

The subharmonic Melnikov functions can be computed in the same way with the various periodic solutions of the unperturbed problem:

$$M^{m/n}(t_0) = \int_0^{mT} v(t-t_0,k)\left[-A\cos\omega t \sin(x(t-t_0,k)) - \beta v(t-t_0,k)\right] dt. \tag{2.6}$$

where $T = 2\pi/\omega$ is the period of the external perturbation. The k values follow from the resonance conditions for the oscillating motion

$$4K(k) = \frac{2\pi}{\omega}\frac{m}{n} \tag{2.7}$$

(m, n relatively prime natural numbers) and

$$\frac{2K(1/k)}{k} = \frac{2\pi}{\omega}\frac{m}{n} \tag{2.8}$$

for the rotating solutions. $K(k)$ denotes the complete elliptic integral of the first kind, k is the elliptic modulus.

For the oscillating motion we have the subharmonic Melnikov function

$$M_{\text{osc}}^{m/n}(t_0) = -4Ak^2 \int_0^{mT} \cos\omega t \, \mathrm{dn}(t-t_0,k) \, \mathrm{sn}(t-t_0,k) \, \mathrm{cn}(t-t_0,k) \, dt$$
$$-4\beta k^2 \int_0^{mT} \mathrm{cn}^2(t-t_0,k) \, dt. \tag{2.9}$$

sn, cn and dn are the well-known Jacobi elliptic functions [44]. The two integrals can be computed making use of the method of residues. The first integral vanishes except for $n = 1$ and even m. In this case we obtain

$$M_{\text{osc}}^m(t_0) = 4\pi A\omega^2 \operatorname{cosech}(\omega K') \sin\omega t_0 - 16\beta\left[E(k) - k'^2 K(k)\right], \tag{2.10}$$

with $K' = K(k')$ and $k'^2 = 1 - k^2$. $E(k)$ denotes the complete elliptic integral of the second kind. According to theorem 4.6.3 of [12] the bifurcation condition for the occurrence of subharmonics of period

mT with an accuracy of $\mathcal{O}(1)$ is

$$\frac{A}{\beta} = R_{\text{osc}}^m(\omega), \tag{2.11}$$

with

$$R_{\text{osc}}^m(\omega) = \frac{4}{\pi \omega^2}\left[E(k) - k'^2 K(k)\right] \sinh(\omega K'), \tag{2.12}$$

$m = 2, 4, 6, \ldots$ and k from eq. (2.7). For the rotating motion we obtain

$$M_{\text{rot}}^{m/n}(t_0) = -4Ak \int_0^{mT} \cos \omega t \, \text{dn}\left(k(t-t_0), \frac{1}{k}\right) \text{sn}\left(k(t-t_0), \frac{1}{k}\right) \text{dn}\left(k(t-t_0), \frac{1}{k}\right) dt$$
$$- 4\beta k^2 \int_0^{mT} \text{dn}^2\left(k(t-t_0), \frac{1}{k}\right) dt. \tag{2.13}$$

The first integral vanishes except for $n = 1$. In this case we obtain

$$M_{\text{rot}}^m(t_0) = 2\pi A \omega^2 \operatorname{cosech}\left(\frac{\omega K'(1/k)}{k}\right) \sin \omega t_0 - 8\beta k E\left(\frac{1}{k}\right). \tag{2.14}$$

From this the condition for the occurrence of bifurcations

$$\frac{A}{\beta} = R_{\text{rot}}^m(\omega) \tag{2.15}$$

follows, where

$$R_{\text{rot}}^m = \frac{4kE(1/k)}{\pi \omega^2} \sinh\left(\frac{\omega K'(1/k)}{k}\right), \tag{2.16}$$

$m = 1, 2, 3, \ldots$ and k from eq. (2.8). The same subharmonic bifurcation functions can also be obtained from the results of Chester [35]. But in this paper no explicit formulas like (2.11) and (2.15) are given.

It is of interest to study the limit behaviour of the subharmonic Melnikov functions for $m \to \infty$. In this case we obtain

$$\tfrac{1}{2} \lim_{m \to \infty} M_{\text{osc}}^m(t_0) = \lim_{m \to \infty} M_{\text{rot}}^m(t_0) = M^{\pm}(t_0). \tag{2.17}$$

From this it follows that the homoclinic bifurcation is the limit of a sequence of subharmonic saddle-node bifurcations, the oscillating subharmonics for given β and ω converging from below to the A value of the homoclinic bifurcation and the rotational solutions converging from above. The convergence rates

$$\delta_{\text{osc}} = \lim_{m \to \infty} \frac{R_{\text{osc}}^m - R_{\text{osc}}^{m-2}}{R_{\text{osc}}^{m+2} - R_{\text{osc}}^m} = e^{2\pi/\omega} \tag{2.18}$$

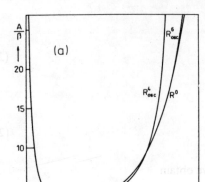 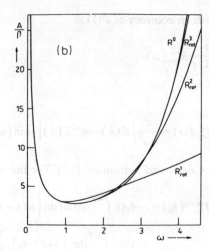

Fig. 2. Subharmonic bifurcation functions for some m values and the homoclinic bifurcation function: a) $R^m_{osc}(\omega)$ and $R^0(\omega)$; b) $R^m_{rot}(\omega)$ and $R^0(\omega)$.

in the oscillating regime and

$$\delta_{rot} = \lim_{m \to \infty} \frac{R^m_{rot} - R^{m-1}_{rot}}{R^{m+1}_{rot} - R^m_{rot}} = e^{2\pi/\omega} \qquad (2.19)$$

in the rotating regime are in accordance with the theory of Gavrilov and Shilnikov [45, 46]. They are simply controlled by the unstable eigenvalue of the saddle of the unperturbed Poincaré map.

Fig. 2 shows some bifurcation functions R^m_{osc} and R^m_{rot}, respectively, and R^0 in dependence upon ω. Only the functions with the lowest m values can be distinguished from R^0 in the scale chosen. It is obvious that the R^2_{osc} curve ends at $\omega = 2$. The reason is that the resonance condition $K(k) = \pi/\omega$ cannot be fulfilled for $\omega > 2$. Of course, also for $\omega > 2$ there exist subharmonics with period $2T$, but in this region they arise via Hopf bifurcation from the origin and therefore they cannot be described with the resonance method applied here.

In the case $\omega > 2$ an expansion of the sine term in (2.1) is justified, and the methods applied in [28, 29] give good results. A similar situation results for the oscillating subharmonics of period $4T, 6T, \ldots$ at $m = 4, 6, \ldots$, respectively.

3. Second order averaging

The Melnikov method yields no assertion on the stability of the subharmonics. Therefore, in order to get this information averaging up to second order is performed. The subharmonic Melnikov functions prove to be essential for the treatment of eqs. (2.1) by means of the averaging theorem in the form given by Hale [47]. With eqs. (2.1) the following transformations are successively performed (details of the method are given in [12]):

1) Transformation to the action–angle variables I, θ of the unperturbed system of (2.1) for the oscillating and rotating solutions, respectively.

2) Transformation $(I, \theta) \to (h, \phi)$ in the neighbourhood of a special resonance $I = I^m$ with the period $mT = 2\pi/\Omega^m$:

$$I = I^m + \sqrt{\varepsilon}\, h, \quad \theta = \Omega^m t + \phi.$$

3) An averaging transformation $(h, \phi) \to (h, \bar{\phi})$ with subsequent averaging up to $\mathcal{O}(\varepsilon)$ yields an autonomous system which according to the averaging theorem, gives results on the existence and the stability of subharmonics of eqs. (2.1) with period mT. Dropping the bars one obtains

$$\dot{h} = \sqrt{\varepsilon}\, \frac{1}{2\pi} M^m\left(\frac{\phi}{\Omega^m}\right) + \varepsilon \overline{F'_m(\phi)} h + \mathcal{O}(\varepsilon^{3/2}), \quad \dot{\phi} = \sqrt{\varepsilon}\, \Omega'(I^m) h + \varepsilon\left(\frac{\Omega''(I^m)}{2} h^2 + \overline{G_m(\phi)}\right) + \mathcal{O}(\varepsilon^{3/2}), \tag{3.1}$$

where $\Omega = \Omega(I)$ is the angular frequency of the closed orbits of the unperturbed system and

$$\Omega'(I^m) = \left.\frac{d\Omega}{dI}\right|_{I = I^m}, \quad \Omega''(I^m) = \left.\frac{d^2\Omega}{dI^2}\right|_{I = I^m}. \tag{3.2}$$

These derivatives are to be taken at the resonance $I = I^m$. Further, we have

$$\overline{F'_m(\phi)} = \frac{1}{mT} \int_0^{mT} F'(\Omega^m t + \phi, I^m, t)\, dt, \quad \overline{G_m(\phi)} = \frac{1}{mT} \int_0^{mT} G(\Omega^m t + \phi, I^m, t)\, dt, \tag{3.3}$$

where

$$F'(\theta, I, t) = \frac{\partial}{\partial I} F(\theta, I, t),$$

$$F(\theta, I, t) = -\frac{1}{\Omega(I)} \left[A \cos \omega t \sin(x(\theta, I)) v(\theta, I) + \beta v^2(\theta, I) \right],$$

$$G(\theta, I, t) = \frac{\partial x(\theta, I)}{\partial I} \left[A \cos \omega t \sin(x(\theta, I)) + \beta v(\theta, I) \right].$$

The subharmonic Melnikov functions M^m are included in the $\mathcal{O}(\sqrt{\varepsilon})$ part of the averaged system (3.1) and are given by M^m_{osc} (see eq. (2.10)) and M^m_{rot} (see eq. (2.14)), respectively. The integration in (3.3) can be carried out in the same way as in section 2 since the integrands are expressed in terms of the solution of the unperturbed problem.

After extensive calculations we obtain

$$\overline{F'_m(\phi)} = AC(m, \omega) \sin m\phi - \beta, \quad \overline{G_m(\phi)} = AC(m, \omega) \frac{1}{m} \cos m\phi, \tag{3.4}$$

with

$$C(m, \omega) = -\frac{\pi^3 m^2}{16 k K(k)} \frac{d}{dk} \left[\frac{1}{K^2(k)} \operatorname{cosech}\left(\frac{m\pi K'(k)}{2K(k)}\right) \right] \tag{3.5}$$

for the oscillating motion ($k < 1$) and

$$C(m, \omega) = -\frac{\pi^3 m^2}{4K(1/k)} \frac{d}{dk}\left[\frac{k^2}{K^2(1/k)} \operatorname{cosech}\left(\frac{m\pi K'(1/k)}{K(1/k)}\right)\right] \quad (3.6)$$

for the rotating regime ($k > 1$). The remaining terms of the eqs. (3.1) are also different for the oscillating and rotating solutions. For the oscillating regime we have

$$\begin{aligned} M^m\left(\frac{\phi}{\Omega^m}\right) &= M_{\text{osc}}^m\left(\frac{\phi}{\Omega^m}\right) = 4\pi A\omega^2 \operatorname{cosech}(\omega K'(k))\sin m\phi - 16\beta\left[E(k) - k'^2 K(k)\right], \\ \Omega'(I^m) &= \left.\frac{d\Omega}{dI}\right|_{I=I^m} = -\frac{\pi^2}{16k^2 k'^2 K^3(k)}\left[E(k) - k'^2 K(k)\right] < 0, \\ \Omega''(I^m) &= \left.\frac{d^2\Omega}{dI^2}\right|_{I=I^m} = -\frac{\pi^3}{128 kK(k)} \frac{d}{dk}\left[\frac{1}{k^2 k'^2 K^3(k)}\left(E(k) - k'^2 K(k)\right)\right], \\ \Omega^m &= \frac{\pi}{2K(k)} = \frac{\omega}{m} \quad (\text{resonance condition}). \end{aligned} \quad (3.7)$$

In the rotating regime we have

$$\begin{aligned} M^m\left(\frac{\phi}{\Omega^m}\right) &= M_{\text{rot}}^m\left(\frac{\phi}{\Omega^m}\right) = 2\pi A\omega^2 \operatorname{cosech}\left(\frac{\omega K'(1/k)}{k}\right)\sin m\phi - 8\beta kE\left(\frac{1}{k}\right), \\ \Omega'(I^m) &= \left.\frac{d\Omega}{dI}\right|_{I=I^m} = -\frac{\pi^2 k^2 E(1/k)}{4(1-k^2)K^3(1/k)} > 0, \\ \Omega''(I^m) &= \left.\frac{d^2\Omega}{dI^2}\right|_{I=I^m} = -\frac{\pi^3}{16K(1/k)} \frac{d}{dk}\left(\frac{k^2 E(1/k)}{(1-k^2)K^3(1/k)}\right), \\ \Omega^m &= \frac{\pi k}{K\left(\frac{1}{k}\right)} = \frac{\omega}{m} \quad (\text{resonance condition}). \end{aligned} \quad (3.8)$$

The values of k can be obtained for given m from the resonance conditions. The remaining derivatives with respect to k can be found easily [44]. In order to keep the expressions as short as possible we partially refrained from performing these mathematical operations. If the terms of $\mathcal{O}(\varepsilon)$ are neglected, the system (3.1) is a conservative one with the Hamiltonian

$$H = \sqrt{\varepsilon}\left(\frac{\Omega'}{2}h^2 + V(\phi)\right), \quad \text{where} \quad V(\phi) = -\frac{1}{2\pi}\int M^m\left(\frac{\phi}{\Omega^m}\right) d\phi. \quad (3.9)$$

The investigation of this system is easy and yields as fixed points m saddle points and m centers as soon as the values A/β are larger than those given by the bifurcation conditions (2.11) and (2.15), respectively. Saddles and centers are alternately placed and their positions are given by

$$h = 0, \quad M^m\left(\frac{\phi}{\Omega^m}\right) = 0. \quad (3.10)$$

Equivalent results were obtained by Greenspan and Holmes [10, 11] and Morosov [13, 14] for the Duffing equation with an external force term.

Since the system (3.9) is structurally unstable, it is important to take into consideration the terms of $\mathcal{O}(\varepsilon)$, i.e. to discuss the complete system (3.1). The trace of the matrix L of the linearized system is negative:

$$\operatorname{Tr} L = \varepsilon \overline{F'_m(\phi)} + \varepsilon \frac{\partial}{\partial \phi} \overline{G_m(\phi)} = -\varepsilon \beta. \tag{3.11}$$

This is true for both the oscillatory and the rotational regime. Linearizing eqs. (2.1), the same trace is obtained. For the determinant of the matrix L in the $\mathcal{O}(\varepsilon)$ approximation we obtain in the case of the oscillating motion

$$\det L = 2\varepsilon m \Omega' A \omega^2 \operatorname{cosech}(\omega K') \cos m\phi. \tag{3.12}$$

ϕ is now a stationary solution of (3.1). For the rotating motion the right-hand side of this equation is slightly modified. Supposing $\det L \neq 0$, only saddle points and sinks exist which are placed alternately. The sinks, in dependence upon the value of det L, can be stable nodes or foci. The position of the fixed points is somewhat changed as compared to the conservative system. Expanding the fixed point solutions in the following form:

$$h = \sqrt{\varepsilon}\, h_1 + \varepsilon h_2 + \cdots, \quad \phi = \phi_0 + \sqrt{\varepsilon}\, \phi_1 + \varepsilon \phi_2 + \cdots \tag{3.13}$$

and inserting them in the fixed point equation gives the first correction

$$\phi_1 = 0, \quad h_1 = -\frac{AC}{m\Omega'} \cos m\phi_0, \tag{3.14}$$

where ϕ_0 is the fixed point solution of the conservative system. The Bendixon criterion shows that no closed curves encircling the fixed points exists. Also from eqs. (3.1) we conclude that there are no invariant closed curves surrounding the cylinder $S^1 \times R$. Thus we obtain the qualitative phase portraits of fig. 3. As $m \to \infty$, the factor Ω' in (3.1) becomes unbounded. Therefore, as m increases, averaging is valid for successively smaller ε regions. Analogous results were obtained for the Duffing equation [10, 11].

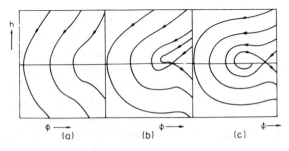

Fig. 3. Qualitative phase portraits of eqs. (3.1) near $(h, \phi) = (0, \pi/2m)$ for $\Omega' < 0$: a) small A/β (no periodic points); b) saddle-node bifurcation; c) large A/β (saddles and sinks). The bifurcation value of A/β is approximately given by eqs. (2.11) and (2.15), respectively.

Table I
Theoretical and numerical bifurcation values of εA for subharmonics in dependence on ω and $\varepsilon\beta$ in the oscillating and rotating regime, respectively

Regime	m	ω	$\varepsilon\beta$	Theoretical value	Numerical value
osc.	2	1.56	0.05	0.11286	0.11283
rot.	1	1.56	0.05	0.17303	0.17301
osc.	2	1.56	0.15	0.3386	0.3377
rot.	1	1.56	0.15	0.5191	0.5187
rot.	1	1.0	0.2	0.6095	0.5977
rot.	1	1.1	0.2	0.6084	0.6008
rot.	1	1.2	0.2	0.6163	0.6114
rot.	1	1.3	0.2	0.6312	0.6280
rot.	1	1.4	0.2	0.6515	0.6494
rot.	1	1.6	0.2	0.7035	0.7027
rot.	1	1.8	0.2	0.7655	0.7653
rot.	1	2.0	0.2	0.8336	0.8336
rot.	1	2.1	0.2	0.8691	0.8693
rot.	1	2.2	0.2	0.9054	0.9056
rot.	1	2.3	0.2	0.9423	0.9426

4. Comparison with numerical calculations and discussion

In order to test the validity of the analytical predictions some comparisons with numerical calculations have been carried out. The first test, presented in table I, is related to the first appearance of certain subharmonics with short periods. As A increases such subharmonics arise from saddle-node bifurcations. In order to determine numerically these bifurcation values, one has to follow the corresponding subharmonic in the direction of decreasing A until it vanishes. For A values slightly above the critical value the change in A has to be carried out in very small steps. To obtain the results given in table I, A was lowered in steps of $\Delta A = 10^{-5}$ ($\Delta A = 10^{-6}$ for the first two examples of table I) using the coordinates of the sink corresponding to a given A as initial values for the following run. The A values, rounded to four (five) decimals, given in table I are the lowest for which the corresponding subharmonics could be detected. The last two columns of table I show a very good agreement, which, surprisingly, is retained if the perturbation parameters in eqs. (2.1) are rather large.

The results of section 2 show that there are regions in the parameter space where many subharmonics exist simultaneously. The asymptotic solutions differ from each other in dependence on the initial conditions chosen. The larger the period of vibration mT, the smaller is the domain of attraction in the phase space [10]. That means subharmonics of long period are difficult to find by numerical methods. In fig. 4 the positions of several periodic points of the Poincaré map representing stable subharmonics of eqs. (2.1) are shown. The parameter values are $\omega = 3.5$, $\varepsilon\beta = 0.005$, $\varepsilon A = 0.6$. Subharmonics with the following periods were numerically detected: $1T, 2T, 3T, 4T$. Other periodic points can be found by the symmetry transformation $(x, \dot{x}) \rightarrow (-x, -\dot{x})$. The origin in this parameter region is stable too.

The detected period 3 solution is characterized by $m/n = 3/2$, i.e. the period is $3T$ and in this period the pendulum performs two rotations. Subharmonics with $n \neq 1$ cannot be described by eqs. (3.1). The oscillating parts of the Melnikov functions and of the second order terms vanish identically for $n \neq 1$. This

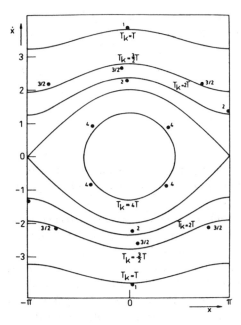

Fig. 4. The positions of several stable periodic points of the Poincaré map of (2.1) for $\omega = 3.5$, $\varepsilon\beta = 0.005$, $\varepsilon A = 0.6$. The periodic points are marked by dots and their m/n value. The corresponding unperturbed resonance level curves are also shown (T_k denotes their period).

at first glance surprising result can be well understood if one takes into account that the Melnikov and the second order averaging method start with solutions of the unperturbed problem which are identical for each rotation, i.e. they have the period mT/n and not mT as have the perturbed solutions. This property of the unperturbed solutions is responsible for the vanishing of the oscillating parts of the Melnikov functions and of the second order terms. To get nonvanishing integrals, one has to start with solutions which are not invariant with respect to translation of time $t' = t + mT/n$. Largely the same is true for the odd solutions

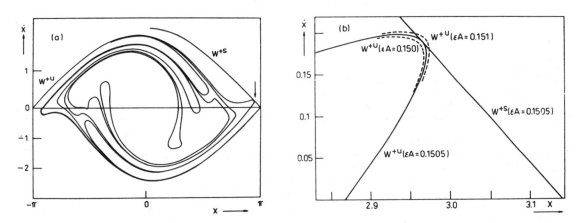

Fig. 5. Parts of the stable and unstable manifolds (W^{+s} and W^{+u}) of $(\pm\pi, 0)$ for $\omega = 1.56$, $\varepsilon\beta = 0.05$, $\varepsilon A = 0.1505$: a) global picture; b) enlarged representation near a homoclinic tangency. Parts of the unstable manifold for $\varepsilon A = 0.150$ and $\varepsilon A = 0.151$ are also included.

Table II
Numerical parameter values for the homoclinic bifurcation.
The theoretical value according to eq. (2.4) is $A/\beta = 3.0102$

$\varepsilon\beta$	A/β
0.15	3.0097 ± 0.0003
0.3	3.0083 ± 0.0003
0.5	3.0081 ± 0.0001

($mT = 1T, 3T, 5T, \ldots$) in the oscillating regime which can be found numerically (see e.g. [41] for $m = 1$). Also in this case the corresponding integrals vanish because of the symmetry property of the unperturbed solutions.

In addition to the examples examined in table I, which are related to periodic solutions, the predictions according to the Melnikov theory concerning homoclinic bifurcations and therefore the beginning of unstable chaotic motions are very interesting. Most important for a concrete system is the answer to the following question: how small must the value of ε be to ensure that intersections of manifolds will result? Numerical calculations can clarify this problem. In fig. 5a parts of the stable and unstable manifolds of $(\pm\pi, 0)$ of the Poincaré map of eqs. (2.1) are shown. The parameter values are $\omega = 1.56$, $\varepsilon\beta = 0.05$, $\varepsilon A = 0.1505$. Eq. (2.4) predicts homoclinic tangencies for $A/\beta = 3.0102$ and transverse homoclinic intersections for $A/\beta > 3.0102$ if ε is sufficiently small. The enlarged representation in fig. 5b clearly shows that the bifurcation value is really close to $\varepsilon A = 0.1505$. In this figure also parts of the unstable manifolds for $\varepsilon A = 0.150$ and $\varepsilon A = 0.151$ are included. The relevant part of the stable manifold does not change noticeably in this parameter region. This shows that the bifurcation value has been predicted with high precision. We have tested the predictions of the Melnikov theory for large perturbations too. The results are given in table II. As in the case of the subharmonics the numerically detected bifurcation values are very close to the theoretical values.

Thus, the numerical examples show that the Melnikov method can quantitatively describe a series of phenomena in perturbed oscillator systems. Good results are also obtained if the perturbation terms cease to be very small.

References

[1] P. Collet and J.P. Eckmann, Iterated Maps on the Interval as Dynamical Systems (Birkhäuser, Basel, 1980).
[2] M. Misiurewicz, Ann. N.Y. Acad. Sci. 317 (1980) 348.
[3] S. Smale, Bull. Amer. Math. Soc. 73 (1967) 747.
[4] J. Moser, Stable and Random Motions in Dynamical Systems, Ann. Math. Studies, No. 77 (Princeton Univ. Press, Princeton, 1973).
[5] S.E. Newhouse, Topology 13 (1974) 9.
[6] S.E. Newhouse, in: Dynamical Systems, Progress in Mathematics, No. 8 (Birkhäuser, Boston, 1980).
[7] V.K. Melnikov, Tr. Moskovsk. Ob-va 12 (1963) 3.
[8] P.J. Holmes and J.E. Marsden, Arch. Rat. Mech. Anal. 76 (1981) 135.
[9] P.J. Holmes and J.E. Marsden, Commun. Math. Phys. 82 (1982) 523.
[10] B.D. Greenspan and P.J. Holmes, in: Nonlinear Dynamics and Turbulence, G. Barenblatt, G. Iooss and D.D. Joseph, eds. (Pitman, London, 1983).
[11] B.D. Greenspan and P.J. Holmes, SIAM J. Math. Anal. in press.
[12] J. Guckenheimer and P.J. Holmes, Nonlinear Oscillations, Dynamical Systems and Bifurcations of Vector Fields, Appl. Math. Sciences 42 (Springer, Berlin, 1983).

[13] A.D. Morosov, Zh. Vychisl. Matem. i Matem. Fiz. 13 (1973) 1134.
[14] A.D. Morosov, Diff. Uravn. 12 (1976) 241.
[15] P.J. Holmes, SIAM J. Appl. Math. 38 (1980) 65.
[16] F.M.A. Salam, J.E. Marsden and P.P. Varaiya, IEEE Trans. Circuits Syst. 30 (1983) 697.
[17] N. Kopell and R. Washburn, IEEEE Trans. Circuits Syst. 29 (1982) 738.
[18] J.A. Sanders, Celestial Mechanics 28 (1982) 171.
[19] Z.D. Genchev, Z.G. Ivanov and B.N. Todorov, IEEE Trans. Circuits Syst. 30 (1983) 633.
[20] F.M.A. Salam and S.S. Sastry, Memorandum No. DUMEM SM 83/02, 10 September 1983, Dep. Mech. Engin. and Mech., Drexel Univ., Philadelphia.
[21] F.M.A. Salam, J.E. Marsden and P.P. Varaiya, IEEE Trans. Circuits Syst. 31 (1984) 673.
[22] V.N. Gubankov, S.L. Ziglin, K.I. Konstantinyan, U.P. Koshelets and G.A. Ovsyannikov, Zh. Eksp. Teor. Fiz. 86 (1984) 343.
[23] S.N. Chow, J.K. Hale and J. Mallet-Paret, J. Diff. Eqns. 37 (1980) 351.
[24] P. Hirsch, ZAMM 10 (1930) 41.
[25] J.J. Stoker, Nonlinear Vibrations (Wiley–Interscience, New York, 1950).
[26] W. Haacke, ZAMM 31 (1951) 161.
[27] P.L. Kapitsa, Zh. Eksp. Teor. Fiz. 21 (1951) 588.
[28] E. Skalak and M.I. Yarymovych, J. Appl. Mech. 27 (1960) 159.
[29] R.A. Struble, J. Appl. Mech. 30 (1963) 301.
[30] F.M. Phelps and J.H. Hunter, Amer. J. Phys. 34 (1966) 533.
[31] J.L. Bogdanoff and S.J. Citron, J. Acoust. Soc. Amer. 38 (1965) 447.
[32] D.J. Ness, Amer. J. Phys. 35 (1967) 964.
[33] J. Dugundji and C.K. Chhatpar, Rev. Roum. Sci. Tech. Mech. Appl. 15 (1970) 741.
[34] B.I. Cheshankov, Priklad. Mat. Mekh. 35 (1971) 343.
[35] W. Chester, J. Inst. Math. Appl. 15 (1975) 289.
[36] H. Troger, ZAMM 55 (1975) T68.
[37] G. Ryland and L. Meirovitch, J. Sound and Vibration 51 (1977) 547.
[38] R. Gradewald and W. Moldenhauer, Abhandl. Akad. Wiss. DDR, Abteilung Math., Naturwiss., Technik, No. 5 (1977) 353.
[39] J.B. McLaughlin, J. Stat. Phys. 24 (1981) 375.
[40] R.W. Leven and B.P. Koch, Phys. Lett. 86A (1981) 71.
[41] A. Arneodo, P. Coullet, C. Tresser, A. Libchaber, J. Maurer and D. D'Humieres, Physica 6D (1983) 385.
[42] B.P. Koch, R.W. Leven, B. Pompe and C. Wilke, Phys. Lett. 96A (1983) 219.
[43] B. Pompe, C. Wilke, B.P. Koch and R.W. Leven, Exp. Technik der Physik 32 (1984) 545.
[44] P.F. Byrd and M.D. Friedman, Handbook of Elliptic Integrals for Engineers and Physicists (Springer, Berlin, 1954).
[45] N.K. Gavrilov and L.P. Shilnikov, Mat. Sbornik 88 (1972) 467.
[46] N.K. Gavrilov and L.P. Shilnikov, Mat. Sbornik 90 (1973) 139.
[47] J.K. Hale, Ordinary Differential Equations (Wiley–Interscience, New York, 1950).

PHYSICAL REVIEW LETTERS

VOLUME 58   27 APRIL 1987   NUMBER 17

Power Spectra of Strange Attractors near Homoclinic Orbits

Victor Brunsden and Philip Holmes

*Departments of Theoretical and Applied Mechanics and Mathematics and Center for Applied Mathematics,
Cornell University, Ithaca, New York 14853*

(Received 3 November 1986)

> Assuming that the chaotic time history of a single variable in a differential equation possessing a strange attractor can be represented as the random superposition of deterministic "structures," we predict the power spectral density. We justify the assumption for perturbations of nonlinear Hamiltonian oscillators and compare our predictions with computations on versions of Duffing's equation.

PACS numbers: 05.45.+b, 02.50.+s, 03.20.+i

Dynamical systems possessing strange attractors have been proposed as models for a number of physical processes which display erratic temporal behavior.[1-3] As well as the abstract theory,[4] there are analytical techniques for the study of global behavior in *specific* systems. In particular, Melnikov's method[2,5] detects transverse homoclinic points in differential equations which are small perturbations of integrable (Hamiltonian) systems. This, with the Smale-Birkhoff homoclinic theorem,[2,4] implies the existence of chaotic motions among the solutions of the equation in question: qualitative information. In contrast, here we propose a method which provides quantitative statistical measures of solutions: We compute power spectra of chaotic motions which are perturbations of homoclinic orbits. Our approach relies on the existence of global homoclinic structures, verifiable by Melnikov theory, and derives from the notion of coherent structures in turbulence theory.[6] It has been proposed before in connection with differential equations,[7] although this earlier work does not provide *a priori* estimates from the unperturbed equations, as does ours. Spectral estimates have also been proposed in connection with the scaling properties of period doubling and halving[8] and with intermittency.[9]

For simplicity, we focus on perturbations of the Duffing equation

$$\dot{x} = y,$$
$$\dot{y} = x - x^3 + \epsilon(\gamma\cos\nu t - \delta y + \beta x^2 y), \quad 0 \leq \epsilon \ll 1, \tag{1}$$

although our ideas are more generally applicable. For $\epsilon = 0$ the unperturbed phase plane of (1) has a pair of homoclinic orbits to the saddle point $(x,y) = (0,0)$: $\Gamma_{-1} \cup \Gamma_{+1}$, solutions in which may be written

$$x_+(t) := s(t) = \sqrt{2}\,\text{sech}\,t, \quad x_-(t) = -s(t) \tag{2}$$

(Fig. 1). The Melnikov method concerns orbits which remain near $\Gamma_{\pm 1}$ when $\epsilon \neq 0$ and involves computation of

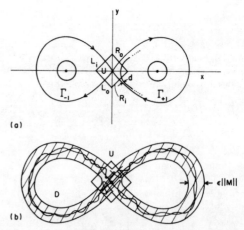

FIG. 1. The Duffing oscillator: (a) Unperturbed phase plane; (b) perturbed Poincaré map showing transverse homoclinic points and trapping region D.

© 1987 The American Physical Society   1699

the function

$$M(\theta) = \int_{-\infty}^{\infty} \dot{s}(t)[\gamma\cos\nu(t+\theta) - \delta\dot{s}(t) + \beta s^2(t)\dot{s}(t)]dt = \sqrt{2}\gamma\pi\nu\,\text{sech}(\tfrac{1}{2}\pi\nu)\sin\nu\theta - 4\delta/3 + 16\beta/15 \qquad (3)$$

[for Eq. (1)]. If M has simple zeros, then, *for sufficiently small ϵ, chaotic motions exist near* $\Gamma_{\pm 1}$.[2,5] If we assign a bi-infinite sequence $\mathbf{a}(z) = \{a_j\}_{-\infty}^{\infty}$ to each initial condition $z = (x(0), y(0))$ which reflects the behavior of the solution $x(t)$ based at z in that $a_k = 0$ or 1 depending on whether the jth maximum of $|x(t)|$ occurs near Γ_{+1} ($x > 0$) or Γ_{-1} ($x < 0$), then *every possible sequence is realized*. Moreover, 0 and 1 have equal probabilities and the process has no memory. Thus, there are solutions which appear indistinguishable from random processes such as coin tossing.[2,4,10]

These chaotic motions do not necessarily constitute a strange attractor[2]: Orbits may escape from the chaotic set and approach stable periodic motions. The homoclinic tangencies predicted by Melnikov analysis *guarantee* that stable periodic orbits exist for residual subsets of parameter values.[10,11] However, these orbits correspond to long periods and are not typically observed. In the following we *assume* that small (numerical) errors destabilize any such sinks. This assumption is essential: There is no proof yet of the existence of a true strange attractor (dense orbit) for a specific differential equation.

If $\delta \approx 4\beta/5 > 0$, almost all orbits starting in some disk are attracted to a neighborhood D of $\Gamma_{\pm 1}$ of width $\epsilon\gamma\nu\,\text{sech}(\tfrac{1}{2}\pi\nu)$; see Fig. 1.[10] This explains our choice of perturbation: We wish to control the solutions as far as possible (but see below). A typical chaotic solution $x(t)$ can thus be approximated by

$$x(t) = \sum_{j=-\infty}^{\infty} (-1)^{a_j} s(t - T_j), \qquad (4)$$

where $a_j \in \{0,1\}$ is the symbol described above, $s(t)$ is the unperturbed homoclinic loop Γ_{+1} [Eq. (2)], and T_j is the time at which the jth maximum in $|x(t)|$ occurs. Moreover, it is reasonable to suppose that $a_j \in \{0,1\}$ and $T_j \in \mathbb{R}$ are random variables.

Multiplying $x(t)$ by a "window" function $g_L(t)$ of compact support [$g_L(t) = 1$, $|t| < L$; $g_L(t) = 0$, $|t| > L$], we rewrite the (integrable) windowed solution $x_L(t)$ as a convolution integral $x_L(t)$:

$$\int_{-\infty}^{\infty} a_L(t'-t)s(t')dt',$$

where

$$a_L(\tau) = g_L(\tau)\sum_j (-1)^{a_j}\delta(\tau + T_j)$$

is a (finite) random sequence of delta functions (shot noise). The Fourier transform of $x_L(t)$ is then the product $\hat{a}_L(f)\hat{s}(f)$ of the transforms[12] and the power spectral density of $x(t)$ is

$$E_x(f) = \lim_{L\to\infty}(1/2L)|\hat{x}_L(f)|^2$$
$$= \lim_{L\to\infty}(1/2L)|\hat{a}_L(f)|^2|\hat{s}(f)|^2. \qquad (5)$$

From the definition we have

$$\hat{a}_L(f) = \int_{-\infty}^{\infty} g_L(\tau)(-1)^{a_j}\delta(\tau+T_j)e^{-i2\pi f\tau}d\tau$$
$$= \sum_{j=-J_1}^{J_2} (-1)^{a_j}e^{i2\pi fT_j}, \qquad (6)$$

where $-J_1$ and J_2 are the largest integers such that $|T_{-J_1}|, |T_{J_2}| < L$; i.e., $J_1 + J_2 = 2L/T$, where $T = \langle T_j - T_{j-1}\rangle$ is the mean gap between events. From (6), we compute

$$|\hat{a}_L(w)|^2 = (J_1 + J_2 + 1) + \sum_{j,k}(\text{cross terms}), \qquad (7)$$

where typical "cross terms" have the form $(-1)^{a_j + a_k}e^{i2\pi f(T_j - T_k)}$, and, since a_j and T_j are independent random variables, by the central limit theorem the sum of these terms is $o(L)$. Thus, substituting into (5), we obtain

$$E_x(f) = (1/T)|\hat{s}(f)|^2. \qquad (8)$$

Goldshtik[7] gives an alternative derivation. For the example in question, from (2) we have

$$E_x(f) = (2\pi^2/T)\text{sech}^2(\pi^2 f)$$
$$\sim (8\pi^2/T)\exp(-2\pi^2 f), \quad f \text{ large}. \qquad (9)$$

We remark that if phase coherence exists (T_j is random but a_j is not) then peaked spectra typical of "noisy periodicity" are predicted.[7]

It remains to estimate T, the mean gap between passages around either Γ_{-1} or Γ_{+1}. Fix a neighborhood U of $(0,0)$ of size μ (Fig. 1). Solutions leaving U through R_0 or L_0 return to U via R_i or L_i after time $\tau_0 \approx 1$ which is independent of ϵ to leading order. Within U, the time spent is controlled by how close solutions are to the stable manifolds of $(0,0)$ on entry. Linear estimates show that this time is $\lambda_+^{-1}\ln(\mu/d)$, where $d < \mu$ is the distance from the stable manifold and $\lambda_+ > 0$ is the expanding eigenvalue of $(0,0)$. d is controlled by the splitting of the manifolds, which the Melnikov calculation shows to be $O(\epsilon M)$ [Eq. (3)].[2,10] Assembling this information, we find that the typical gap between structures is

$$T \approx \text{const} - \lambda_+^{-1}\ln\{\epsilon\max_\theta\|M(\theta)\|\}.$$

In our example, with $\delta = 4\beta/5$, we have $M(\theta) = \sqrt{2}\gamma\pi\nu\,\text{sech}(\tfrac{1}{2}\pi\nu)\sin\nu\theta$ and

$$T \approx \text{const} - \lambda_+^{-1}\ln\{\epsilon\gamma\nu\,\text{sech}(\tfrac{1}{2}\pi\nu)\}, \qquad (10)$$

where $\lambda_+ \approx 1 + \epsilon\delta/2$. Equations (10) and (9) provide our estimates for the power spectrum. The main points are that $E_x(f)$ decays *exponentially* with f, that this functional form is governed by the unperturbed homoclinic orbits $\Gamma_{\pm 1}$, and that the level of $E_x(f) \sim 1/T$ is relatively insensitive to all parameters *except* the (circular) frequency ν of the excitation. We remark that the decay of $E_x(f)$ implies exponential decay of correlation functions.

Numerical experiments were conducted to investigate the predictive value of the theory. Fourth-order Runge-Kutta solutions of (1) were generated with 32–64 integration steps per period $2\pi/\nu$. The resulting time history $x(t)$ was divided into 16 sequential records, each of duration 1613 seconds (4096 data points), which were fast Fourier transformed and averaged to yield power spectra. Double precision arithmetic was used throughout.

For $\gamma = 0$, Melnikov theory predicts that, for parameters β and δ near which $M = 0$ ($\delta = 4\beta/5$), the autonomous perturbed system has an attracting double homoclinic cycle: i.e., $\Gamma_{+1} \cup \Gamma_{-1}$ is an "infinite period" attractor.[2,10] The precise value of $\epsilon\beta$ used was found by numerical search and corresponded to the longest-period motions found. Note how close it is to prediction, despite the size of the parameters. Next, $\epsilon\gamma$ was increased to 0.001, 0.01, and 0.1 in turn and ν varied from 0.5 to 6. Typical durations between homoclinic events were computed to test Eq. (10), which was fitted by a single determination of the unknown constant at $\epsilon = 0.01$, $\nu = 5$, with use of $\lambda_+ \approx 1 + \epsilon\delta/2 = 1.2$. Figure 2 shows that the simple theory behaves reasonably well.

Power spectra were then computed—typical examples are shown, with a time series, in Fig. 3. The general form of the power spectrum is predicted well; in particular the asymptotic slope of $\log_{10}[E_x(f)]$, $2\pi^2 \log_{10} e$ [Eq. (9)], is close to that observed. Spectral levels were obtained from the mean-gap estimate of Eq. (10) after the single fit described above. We observe that the theory consistently over (under) estimates $E_x(f)$ at low (high) frequencies. We suspect that this is due to the fact that the structure $\hat{x}(t)$ is significantly perturbed from that of Eq. (2) for $\epsilon\beta, \epsilon\delta \approx 0.5$. To test this, we integrated the Hamiltonian system $\delta = \beta = 0$, with $\epsilon\gamma = 0.001, 0.01, 0.1$ and obtained spectra whose asymptotic slopes lay within 2% of the prediction.[13] Integrations with lower values of $\epsilon\beta$ and $\epsilon\delta$ show a similar trend. We note that $|E_x(f)|$ covers over ten decades and that, above 1.2 Hz, spectra drop into the numerical noise floor.[13]

We then set $\beta = 0$ and studied the forced, damped Duffing equation.[2,10] Here there is proof of transverse homoclinic orbits and chaotic invariant sets represent-

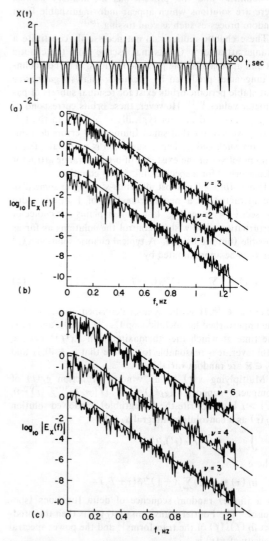

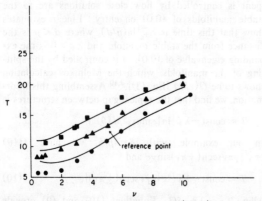

FIG. 2. Mean gaps between maxima of $|x(t)|$ as functions of force level $\epsilon\gamma$ and frequency ν: $\epsilon\gamma = 0.001$ (squares), $\epsilon\gamma = 0.01$ (triangles), $\epsilon\gamma = 0.1$; solid lines, Eq. (10).

FIG. 3. (a) A time history of Eq. (1) for $\epsilon\delta = 0.4$, $\epsilon\beta = 0.498005$, $\epsilon\gamma = 0.001$, $\nu = 1$. (b) Power spectra for $\epsilon\delta = 0.4$, $\epsilon\beta = 0.0498005$, $\epsilon\gamma = 0.01$, and various ν. (c) Power spectra for $\epsilon\gamma = 0.1$ and various ν. Scales are displaced for clarity and theory of Eqs. (9) and (10) is plotted as dashed lines.

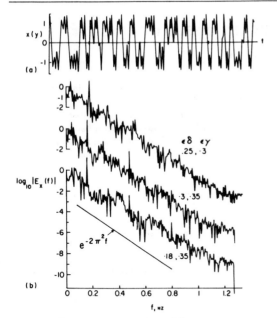

FIG. 4. (a) A time history of Eq. (1) for $\epsilon\delta=0.25$, $\epsilon\beta=0$, $\epsilon\gamma=0.3$, $v=1$. (b) Power spectra for $\epsilon\beta=0$, $v=1$, and various $\epsilon\delta, \epsilon\gamma$.

able as (4) for $\gamma > [4\delta/3\gamma v(2\pi)^{1/2}]\cosh(\frac{1}{2}\pi v)$ and ϵ sufficiently small [cf. Eq. (3)], and for $v \approx 0.8-1.3$ and a range of $\epsilon\gamma$ and $\epsilon\delta$ around 0.2-0.4 chaotic motions are observed. However, solutions can now stray far from the unperturbed homoclinic loops $\Gamma_{\pm 1}$, as the time series of Fig. 4(a) indicates. Nonetheless, our theory is still useful although it is now impossible to estimate T (it is unclear what a "structure" is or how the "mean gap" should be interpreted [Fig. 4(a)]. Consequently, the spectral levels cannot be predicted, but the forms and slopes agree well with Eq. (9); see Fig. 4(b).

To summarize: The assumption of randomly superposed deterministic structures leads to a simple prediction of the power spectral density of a chaotic signal. The functional form of the spectrum is given by the Fourier transform of an individual structure and its level is inversely proportional to the mean gap between the structures appearing in the signal. When the signal is the solution of a perturbed ordinary differential equation possessing homoclinic orbits to a hyperbolic saddle point, both the mean gap and the spectral form can be computed *a priori*. The simple theory is in good agreement with numerical simulations.

A detailed account of this analysis, more examples, and an extension to signals containing multiple structures will be given subsequently.[13]

This work is supported by U.S. Office of Naval Research, Select Research Opportunities IV, Grant No. N00014-85-K-0172, and by the U.S. National Science Foundation Grant No. CME No. 84-02069.

---

[1]*Order in Chaos*, edited by D. Campbell and H. Rose (North-Holland, Amsterdam, 1983).

[2]J. Guckenheimer and P. Holmes, *Nonlinear Oscillations, Dynamical Systems and Bifurcations of Vectorfields* (Springer-Verlag, Berlin/Vienna, 1983).

[3]D. Ruelle and F. Takens, Commun. Math. Phys. **20**, 167 (1971), and **23**, 343 (1971).

[4]S. Smale, Bull. Am. Math. Soc. **73**, 747 (1967).

[5]V. K. Melnikov, Trans. Moscow Math. Soc. **12**, 1 (1963).

[6]B. J. Cantwell, Annu. Rev. Fluid Mech. **13**, 457 (1981).

[7]*Structural Turbulence*, edited by M. A. Goldshtik (Academy of Sciences of the U.S.S.R., Novosibirsk, 1982).

[8]M. J. Feigenbaum, Commun. Math. Phys. **77**, 65 (1980); A. Wolf and J. Swift, Phys. Lett. **83A**, 187 (1981); T. Yoshida and K. Tomita, "Characteristic structures of power spectra in periodic chaos" (unpublished).

[9]Y. Pomeau and P. Manneville, Commun. Math. Phys. **74**, 189 (1980); H. Mori, H. Okamoto, B.-C. So, and S. Kuroki, "Global spectral structures of intermittent chaos" (unpublished).

[10]B. D. Greenspan and P. Holmes, in *Nonlinear Dynamics and Turbulence*, edited by g. Barenblatt, G. Iooss, and D. D. Joseph (Pitman, New York, 1983), Chap. 10, and SIAM J. Math. Anal. **15**, 69 (1984).

[11]S. Newhouse, Publ. Math. Inst. Haut. Etudes Sci. **50**, 101 (1979); C. Robinson, Commun. Math. Phys. **90**, 433 (1983).

[12]D. Robson, *An Introduction to Random Vibration* (Edinburgh Univ. Press, Edinburgh, United Kingdom, 1963).

[13]V. Brunsden, M.Sc. thesis, Cornell University, 1987 (unpublished).

THE MEL'NIKOV TECHNIQUE FOR HIGHLY DISSIPATIVE SYSTEMS*

FATHI M. A. SALAM[†]

Abstract. We present explicit calculations that extend the applicability of the Mel'nikov technique to include a general class of highly dissipative systems. In particular, the dissipation may be in the form of large positive or negative damping. The only required assumption is that each system of this class possesses a homoclinic or a heteroclinic orbit. We also show that sufficiently small time-sinusoidal perturbation of these systems results in (nonempty) transversal intersection of stable and unstable manifolds for all but at most discretely many frequencies. The results are then demonstrated via computer simulation of the highly damped pendulum with constant plus small time-sinusoidal forcing.

Key words. homoclinic or heteroclinic orbits, Mel'nikov technique, transversal intersection, horseshoe chaos

AMS(MOS) subject classifications. 34C15, 34C28, 34C35, 34D30, 34D99, 58F13, 65C99, 35B32, 35G20

0. Introduction. The paper presents explicit calculations that extend the applicability of the Mel'nikov technique to include a general class of highly dissipative systems. In particular the dissipation may be in the form of large positive or negative damping. The basic assumption on the systems of interest is that each possesses a homoclinic or a heteroclinic orbit. However, there need be no small amplitude restriction on the amount of dissipation. The material here (see [17]) generalizes some parts of Salam and Sastry [16], which dealt with the forced and positively damped pendulum in the context of Josephson junction circuits. In contrast to our emphasis, we note that the original work of Mel'nikov [9] had addressed the stability of the center arising in conservative, and nondamped, periodically forced two-dimensional systems. Similarly, all subsequent works, e.g. [4]-[8], [13], have dealt with the same conservative system while allowing the small perturbation to include dissipative terms. Thus the inclusion of damping or dissipation in these systems warrants rederiving the Mel'nikov integral and ensuring its validity.

The key to our results is exploiting the fact that trajectories on the stable or unstable manifolds near a hyperbolic equilibrium are governed by the linearization of the vector field at such an equilibrium. In the interest of brevity, and also to avoid reproducing some of the literature, we do not emphasize the geometric view attained by employing Poincaré maps. We consequently relegate description of such a view to, e.g., Guckenheimer and Holmes [5] and Salam, Marsden and Varaiya [13]. Instead, we emphasize methods of Real Analysis.

Our motivation is practical since many systems in engineering and science are far from being near-Hamiltonian. It is hoped that this work will underline the applicability of the Mel'nikov approach to highly dissipative systems, and consequently would advance the usefulness of the approach to more engineering and physical science problems. For example, in (traditional) feedback control, large amplitude damping is introduced with the intent to stabilize, or improve the performance of, the system. One increases the amplitude of the negative real part of all eigenvalues of the linearization of the vector field about an equilibrium. Then it is presumed that large amplitude

---

* Received by the editors April 22, 1985; accepted for publication (in revised form) June 19, 1986. This research was supported by the National Science Foundation under grant ECS-8596004.

† Department of Electrical Engineering and Systems Science, Michigan State University, East Lansing, Michigan 48824-1226.

Reprinted with permission from the SIAM J. Appl. Math., Vol. 47, No. 2, pp. 232-243
Copyright 1987 by the Society for Industrial and Applied Mathematics. All rights reserved.

damping, which results in a (locally) stable equilibrium, would "damp out" any complicated dynamic behavior such as chaos. The conclusion of this paper implies that such a presumption is false in the class of systems considered, see Salam [12]. Another motivating ground for the author is in the design of, e.g., Josephson junction circuits (see [15] and the references therein) and of (expansions to) interconnected electric power systems (see Salam et al. [14], [15] and the references therein). In design, the conditions leading to the nonempty transversal intersection of stable and unstable manifolds can be used as design constraints.

1. Preliminaries. Consider the system

(1.1) $$\dot{\mathbf{x}} = \mathbf{f}(\mathbf{x}) + \varepsilon \mathbf{g}(\mathbf{x}, t)$$

where $\mathbf{x} = (x_1, x_2)' \in R^2$ and $0 < \varepsilon \ll 1$, i.e. ε is sufficiently small and positive. The functions $\mathbf{f}$ and $\mathbf{g}$ are sufficiently differentiable and bounded on bounded sets; moreover $\mathbf{g}$ is T-periodic in t, i.e. $\mathbf{g}(\mathbf{x}, t) = \mathbf{g}(\mathbf{x}, t+T)$. Call system (1.1) the perturbed system of the (unperturbed) system

(1.2) $$\dot{\bar{\mathbf{x}}} = \mathbf{f}(\bar{\mathbf{x}}); \quad \bar{\mathbf{x}} = (\bar{x}_1, \bar{x}_2)'.$$

We require that system (1.2) possesses a homoclinic orbit, i.e. an orbit that connects a saddle equilibrium point, say $\mathbf{x}_0$, to itself. Denote such an orbit by $\bar{\mathbf{x}}_0(t)$. The homoclinic orbit $\bar{\mathbf{x}}_0(t)$ traces an invariant homoclinic curve as shown dotted in Fig. 1. The curve in fact defines the coincident stable and unstable manifolds of the equilibrium point x_0. For convenience one often suspends system (1.1) to convert it into the autonomous system

(1.1a) $$\dot{\mathbf{x}} = \mathbf{f}(\mathbf{x}) + \varepsilon \mathbf{g}(\mathbf{x}, \phi), \quad \dot{\phi} = 1,$$

where ϕ and t have been identified. Observe that the function $\mathbf{g}$ is T-periodic in t, and hence the variable ϕ belongs to the interval $[0, T)$.

The "perturbed" Poincaré map P_ε is now defined as the mapping of the variable $\mathbf{x}$ over one time-period along the solutions of the differential equation (1.1a), starting at the initial time t_0. The "unperturbed" Poincaré map P_0 is similarly defined when $\varepsilon = 0$ in (1.1a). We remark that the phase portraits of P_0 and (1.2) are identical. In fact the equilibrium point $\mathbf{x}_0$ and its homoclinic invariant curve of (1.2) now become a fixed point $\mathbf{x}_0$ with its homoclinic invariant curve for P_0.

When $\varepsilon \neq 0$, the point $\mathbf{x}_0$, thought of now as the fixed point of the Poincaré map P_0, gets perturbed to a corresponding equilibrium point $\mathbf{x}_\varepsilon$ of the Poincaré map P_ε.

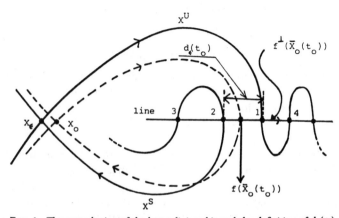

FIG. 1. *The perturbation of the homoclinic orbit and the definition of $d_\varepsilon(t_0)$.*

We denote the periodic trajectory of (1.1) corresponding to $\mathbf{x}_\varepsilon$ by $\mathbf{x}_\varepsilon(t)$. Likewise the unstable and stable manifolds of $\mathbf{x}_0$ get perturbed to the unstable and stable manifolds of $\mathbf{x}_\varepsilon$ of the Poincaré map P_ε. Figure 1 shows the stable and unstable manifolds before and after perturbations. The stable and unstable manifolds for the Poincaré map P_0 are coincident forming the homoclinic manifold which is depicted dotted in Fig. 1. After the perturbation, the stable and unstable manifolds for the Poincaré map P_ε are depicted in Fig. 1 by solid curves. We denote the solutions of (1.1) on the stable, respectively unstable, manifold by $\mathbf{x}^s_\varepsilon(t, t_0)$, respectively $\mathbf{x}^u_\varepsilon(t, t_0)$.

We now present the following three lemmas. The first lemma expresses the periodic solution (1.1) in terms of the constant equilibrium solution of (1.2). The second lemma states the expansion of the solutions of the perturbed unstable, respectively stable, manifold of $\mathbf{x}_\varepsilon(t)$ in terms of the homoclinic orbit and the first variation function $\mathbf{x}^{1u}(t, t_0)$, respectively $\mathbf{x}^{1s}(t, t_0)$. Last, the third lemma presents the linear time-varying differential equation governing the first variations. $D_{\bar{\mathbf{x}}}\mathbf{f}(\bar{\mathbf{x}}_0(t - t_0))$ denotes the Jacobian of the vectorfield $\mathbf{f}$ evaluated along the homoclinic orbit. We defer further details to Mel'nikov [9], Holmes [6], Greenspan and Holmes [4], Salam et al. [13], Salam and Sastry [16] and Guckenheimer and Holmes [5].

LEMMA 1.1. *The periodic solution of* (1.1) *is given as*

$$\mathbf{x}_\varepsilon(t) = \mathbf{x}_0 + O(\varepsilon) \quad \text{for all } t \in (-\infty, \infty).$$

LEMMA 1.2. *There exist functions* $\mathbf{x}^{1u}(t, t_0)$ *and* $\mathbf{x}^{1s}(t, t_0)$ *such that*

(1.3) $\quad \mathbf{x}^u_\varepsilon(t, t_0) = \bar{\mathbf{x}}_0(t - t_0) + \varepsilon \mathbf{x}^{1u}(t, t_0) + O(\varepsilon^2) \quad$ *uniformly in t for $t \in (-\infty, t_0]$,*

(1.4) $\quad \mathbf{x}^s_\varepsilon(t, t_0) = \bar{\mathbf{x}}_0(t - t_0) + \varepsilon \mathbf{x}^{1s}(t, t_0) + O(\varepsilon^2) \quad$ *uniformly in t for $t \in [t_0, \infty)$.*

LEMMA 1.3 (First variation equations). *The functions* $\mathbf{x}^{1u}(t, t_0)$ *and* $\mathbf{x}^{1s}(t, t_0)$ *are governed by*

(1.5) $\quad \dot{\mathbf{x}}^{1u}(t, t_0) = D_{\bar{\mathbf{x}}}f(\bar{\mathbf{x}}_0(t - t_0))\mathbf{x}^{1u}(t, t_0) + \mathbf{g}(\bar{\mathbf{x}}_0(t, t_0), t) \quad$ *for $t \in (-\infty, t_0]$,*

(1.6) $\quad \dot{\mathbf{x}}^{1s}(t, t_0) = D_{\bar{\mathbf{x}}}f(\bar{\mathbf{x}}_0(t - t_0))\mathbf{x}^{1s}(t, t_0) + \mathbf{g}(\bar{\mathbf{x}}_0(t, t_0), t) \quad$ *for $t \in [t_0, \infty)$.*

The separation between $\mathbf{x}^u_\varepsilon(t, t_0)$ and $\mathbf{x}^s_\varepsilon(t, t_0)$ as shown in Fig. 1, is measured by

(1.7) $\qquad d_\varepsilon(t_0) = [1/|\mathbf{f}(\bar{\mathbf{x}}_0(0))|](\mathbf{f}(\bar{\mathbf{x}}_0(0)) \wedge [\mathbf{x}^u_\varepsilon(t_0, t_0) - \mathbf{x}^s_\varepsilon(t_0, t_0)]$

where $\wedge$ represents a wedge product between two (two-dimensional) vectors, i.e., $(z_1, z_2)' \wedge (w_1, w_2)' := z_1 w_2 - z_2 w_1$. The distance $d_\varepsilon(t_0)$ can be explicitly written in terms of its first variation as

(1.8) $\qquad d_\varepsilon(t_0) = [1/|\mathbf{f}(\bar{\mathbf{x}}_0(0))|]\varepsilon M(t_0) + O(\varepsilon^2)$

where

(1.9) $\qquad M(t_0) = \mathbf{f}(\bar{\mathbf{x}}_0(0)) \wedge [\mathbf{x}^{1u}(t_0, t_0) - \mathbf{x}^{1s}(t_0, t_0)].$

The Mel'nikov method evaluates $M(t_0)$, the first variation of the distance $d_\varepsilon(t_0)$ multiplied by the nonzero quantity $|\mathbf{f}(\bar{\mathbf{x}}_0(0))|$. If the function $M(t_0)$ has simple zeros, i.e., if there exists $\tilde{t}_0$ such that

$$M(\tilde{t}_0) = 0 \quad \text{and} \quad dM(\tilde{t}_0)/dt_0 \neq 0,$$

then it follows from the Inverse Function Theorem that the equation $d_\varepsilon(t_0) = 0$ has a unique simple solution in a neighborhood of $\tilde{t}_0$. Consequently, the invariant manifolds, associated with the solutions $\mathbf{x}^u_\varepsilon(t, t_0)$ and $\mathbf{x}^s_\varepsilon(t, t_0)$, intersect nontangentially, i.e. they have nonempty transversal intersection. This transverse intersection becomes explicit and occurs infinitely often when considering the perturbed Poincaré map.

Now we embark on obtaining an expression for $M(t_0)$. Write (1.9) as

(1.10) $$M(t_0) = \Delta^u(t_0, t_0) - \Delta^s(t_0, t_0).$$

Taking note of Lemma 1.3, we define the quantities

(1.11) $$\Delta^u(t, t_0) := \mathbf{f}(\bar{\mathbf{x}}_0(t-t_0)) \wedge \mathbf{x}^{1u}(t, t_0) \quad \text{for } t \in (-\infty, t_0]$$

and

(1.12) $$\Delta^s(t, t_0) := \mathbf{f}(\bar{\mathbf{x}}_0(t-t_0)) \wedge \mathbf{x}^{1s}(t, t_0) \quad \text{for } t \in [t_0, \infty).$$

One can now utilize (1.5) and (1.6) in obtaining an expression of (1.10) that depends on the vector fields $\mathbf{f}$ and $\mathbf{g}$ evaluated along the homoclinic orbit $\bar{\mathbf{x}}_0(t, t_0)$. To that end, one derives the following differential equations by differentiating (1.11) and (1.12) with respect to time. Note that these equations are only defined over their indicated intervals.

(1.13) $$d/dt \, \Delta^u(t, t_0) = a(t-t_0)\Delta^u(t, t_0) + b(t-t_0), \quad t \in (-\infty, t_0],$$

(1.14) $$d/dt \, \Delta^s(t, t_0) = a(t-t_0)\Delta^s(t, t_0) + b(t-t_0), \quad t \in [t_0, \infty)$$

where

$$a(t-t_0) := \text{trace } D_{\bar{x}}\mathbf{f}(\bar{\mathbf{x}}_0(t-t_0)) \quad \text{and} \quad b(t-t_0) := \mathbf{f}(\bar{\mathbf{x}}_0(t-t_0)) \wedge \mathbf{g}(\bar{\mathbf{x}}_0(t-t_0), t).$$

Integrating the linear time-varying differential equations (1.13) and (1.14) over their respective time intervals, one easily obtains

(1.15) $$\Delta^u(t_0, t_0) = \left\{ \exp\left[\int_t^{t_0} a(s-t_0) \, ds \right] \right\} \Delta^u(t, t_0)$$
$$+ \int_t^{t_0} \left\{ \exp\left[\int_{t'}^{t_0} a(s-t_0) \, ds \right] \right\} b(t'-t_0) \, dt', \quad t \in (-\infty, t_0],$$

(1.16) $$\Delta^s(t_0, t_0) = \left\{ \exp\left[-\int_{t_0}^{t} a(s-t_0) \, ds \right] \right\} \Delta^s(t, t_0)$$
$$- \int_{t_0}^{t} \left\{ \exp\left[-\int_{t_0}^{t'} a(s-t_0) \, ds \right] \right\} b(t'-t_0) \, dt', \quad t \in [t_0, \infty).$$

Note that the two equations are identical in form except for the different time interval over which each is valid. To bring attention to the different time intervals, we have deliberately chosen to represent them in the form of (1.15) and (1.16).

It is now clear that the right-hand side of (1.15) and (1.16) depends on the vector fields $\mathbf{f}$ and $\mathbf{g}$ evaluated along the homoclinic trajectory of the "unperturbed" system (1.2).

We note that in the literature it has frequently been assumed that system (1.2) is Hamiltonian, and thus the divergence of $\mathbf{f}$, namely the quantity $a(t-t_0)$ is identically zero. In the non-Hamiltonian case the first term of (1.15) (resp. (1.16)) is assumed to converge to zero as $t \to -\infty$ (resp. $t \to \infty$). The reason often given is that $\Delta^u(t, t_0) \to 0$ as $t \to -\infty$ (resp. $\Delta^s(t, t_0) \to 0$ as $t \to \infty$). This reason is insufficient, however, due to the fact that one of the quantities $\lim_{t \to -\infty} \{\exp[\int_t^{t_0} a(s-t_0) \, ds]\}$ or $\lim_{t \to \infty} \{\exp[-\int_{t_0}^{t} a(s-t_0) \, ds]\}$ equals ∞.

In the next section we will present the remedy to these inadequacies in the form of a lemma and a proposition. In §3, our proofs will be valuable in ensuring the existence of simple zeros of the Mel'nikov function in an application to a class of highly damped systems.

2. Validity of the Mel'nikov integral formula. Consider the linearization of the unperturbed system (1.2) along the homoclinic orbit, namely,

(2.1) $$\dot{\mathbf{y}} = [D_{\bar{x}}\mathbf{f}(\bar{\mathbf{x}}_0(t-t_0))]\mathbf{y}.$$

Now, let $A := \lim_{t \to \pm\infty} D_{\bar{x}}\mathbf{f}(\bar{\mathbf{x}}_0(t-t_0)) = D_{\bar{x}}\mathbf{f}(\mathbf{x}_0)$ and write its characteristic polynomial as (det := determinant):

(2.2) $$\Psi(\lambda) = \lambda^2 - (\text{trace } A)\lambda + \det A = 0.$$

The eigenvalues are then

(2.3) $$\lambda_1 = (\text{trace } A)/2 + (\tfrac{1}{2})\sqrt{(\text{trace } A)^2 - 4 \det A},$$
(2.4) $$\lambda_2 = (\text{trace } A)/2 - (\tfrac{1}{2})\sqrt{(\text{trace } A)^2 - 4 \det A}.$$

Since the linearization is at the saddle point, the eigenvalues are of opposite signs. Consequently, det A must be negative. Trace A however is allowed to be either positive or negative. Specifically, we consider the following two possibilities when det $A < 0$.[1]

Case 1: trace $A > 0$ implies

(2.5) $$\lambda_2 < 0 < \text{trace } A < \lambda_1;$$

Case 2: trace $A < 0$ implies

(2.6) $$\lambda_2 < \text{trace } A < 0 < \lambda_1.$$

The following lemma justifies taking the limit $t \to -\infty$ (resp. $t \to \infty$) of (1.15) (resp. (1.16)).

LEMMA 2.1. $\Delta^u(t, t_0)$, $t \in (-\infty, t_0]$ (*resp.* $\Delta^s(t, t_0)$, $t \in [t_0, \infty)$) *is bounded above and approaches 0 at the exponential rate of* $\exp(\lambda_1 t)$ *as* $t \to -\infty$ (*resp.* $\exp(\lambda_2 t)$ *as* $t \to \infty$).

Proof. We begin with $\Delta^s(t, t_0)$. First, consider $\Delta^s(t, t_0) := \mathbf{f}(\bar{\mathbf{x}}_0(t-t_0)) \wedge \mathbf{x}^{1s}(t, t_0)$ for $t \in [t_0, \infty)$. Note that $\lim_{t \to \infty} \bar{x}_0(t-t_0) = x_0$, the saddle equilibrium point, and hence as $t \to \infty$, $\mathbf{f}(\bar{\mathbf{x}}_0(t-t_0)) \to 0$. It approaches zero at the rate governed by the eigenvalues of the linearization of $\mathbf{f}$ at $\mathbf{x}_0$, namely, $\exp(\lambda_2 t)$. Thus to prove the lemma, it is sufficient to prove that $\mathbf{x}^{1s}(t, t_0)$ is bounded for all $t \in [t_0, \infty)$. Now observe that $\mathbf{x}^{1s}(t, t_0)$ satisfies the (bounded) linear time-varying differential equation (1.6). Thus we only need to show that as $t \to \infty$, $\mathbf{x}^{1s}(t, t_0)$ is bounded.

As $t \to \infty$, $\|\mathbf{x}_\varepsilon(t) - \mathbf{x}_\varepsilon^s(t, t_0)\| \to 0$ and $\bar{\mathbf{x}}_0(t-t_0) \to \mathbf{x}_0$. Hence, from Lemmas 1.1 and 1.2, as $t \to \infty$ one concludes that

(2.7) $$\mathbf{x}^{1s}(t, t_0) \to \text{bounded periodic function}.$$

(The bounded periodic solution satisfies $\dot{\mathbf{x}}^1 = [D_{\bar{x}}\mathbf{f}(\mathbf{x}_0)]\mathbf{x}^1 + \mathbf{g}(\mathbf{x}_0, t)$.) Therefore it follows that $\mathbf{x}^{1s}(t, t_0)$ is bounded for all $t \in [t_0, \infty)$. (A similar argument establishes that $\mathbf{x}^{1u}(t, t_0)$ is bounded for all $t \in (-\infty, t_0]$ and establishes the rate as $t \to -\infty$.) ☐

Now recall that

$$a(t-t_0) := \text{trace } D_{\bar{x}}\mathbf{f}(\bar{\mathbf{x}}_0(t-t_0))$$
$$= \partial f_1(\bar{\mathbf{x}}_0(t-t_0))/\partial x_1 + \partial f_2(\bar{\mathbf{x}}_0(t-t_0))/\partial x_2.$$

To enhance the clarity of the following result, we state a simplifying assumption.

ASSUMPTION A. $a(t-t_0) = \text{trace } A = \text{constant}$, for all t. (This is true if f_1, respectively f_2, is a linear function of x_1, respectively x_2, alone.) Now we state the important *exponential convergence property*.

---
[1] The case det $A > 0$ corresponds to both eigenvalues having the same sign as trace A. This is the case of the linearization at an (associated) node, or a focus when $\{\text{trace } A\}^2 < 4 \det A$.

PROPOSITION 2.1. (a) *As $t \to -\infty$, the expression $\{\exp[\int_t^{t_0} a(s - t_0)\, ds]\}\Delta^u(t, t_0)$ tends exponentially to zero at least at the rate of* $\exp[(\lambda_1 - \text{trace}\, A + \eta_1)t]$, *for arbitrarily small η_1.*

(b) *As $t \to \infty$, the expression $\{\exp[-\int_{t_0}^t a(s - t_0)\, ds]\}\Delta^s(t, t_0)$ tends exponentially to zero at least at the rate of* $\exp[-(\text{trace}\, A - \lambda_2 + \eta_2)t]$, *for arbitrarily small η_2.*

Proof. As $t \to \infty$, $\Delta^s(t, t_0) \to 0$ in the order of $\exp\{\lambda_2 t\}$. By Assumption A we have $\{\exp[-\int_{t_0}^t a(s - t_0)\, ds]\}\Delta^s(t, t_0) = \exp[-(\text{trace}\, A)(t - t_0)]\Delta^s(t, t_0)$. The latter quantity tends to (a constant times) $\exp[-(\text{trace}\, A - \lambda_2)t]$. From (2.5) and (2.6) above, the quantity in parentheses is always positive; hence the exponential goes to zero as $t \to \infty$. Without Assumption A the rate becomes an exponential rate which is slightly larger as in the statement of the proposition.

Similarly, $\{\exp[\int_t^{t_0} a(s - t_0)\, ds]\}\Delta^u(t, t_0)$ tends to $\exp[(\lambda_1 - \text{trace}\, A)t]$. Again, by observing (2.5) and (2.6), the exponential goes to zero as $t \to -\infty$. □

With Lemma 2.1 and Proposition 2.1 established, we are now able to obtain the Mel'nikov formula as

$$
\begin{aligned}
M(t_0) &= \Delta^u(t_0, t_0) - \Delta^s(t_0, t_0) \\
&= \int_{-\infty}^{\infty} \left\{\exp\left[-\int_{t_0}^t a(s - t_0)\, ds\right]\right\} b(t - t_0)\, dt \\
&= \int_{-\infty}^{\infty} \mathbf{f}(\bar{\mathbf{x}}_0(t)) \wedge \mathbf{g}(\bar{\mathbf{x}}_0(t), t + t_0) \left\{\exp\left[-\int_0^t \text{trace}\, D_{\bar{x}}\mathbf{f}(\bar{\mathbf{x}}_0(s))\, ds\right]\right\} dt
\end{aligned}
\tag{2.8}
$$

where in the last step we performed the change of integration variables $t - t_0 \to t$.

Remark 2.1. Proposition 2.1 is valid without Assumption A. The property $a(t - t_0) \to \text{trace}\, A$, as $t \to \pm\infty$, suffices for the proof.

Remark 2.2. Rather than each converging to zero, the two expressions in the proposition need only cancel one another in order to yield (2.8). Such cancellation does take place when the critical element is a normally hyperbolic torus instead of a saddle point, as is the case in the Arnold diffusion phenomenon. A similar point justifying the convergence of integrals of the (vector) Mel'nikov function was mentioned in [14].

3. Application to a class of highly damped systems. We now present an application to a class of highly damped systems that possess a homoclinic orbit. For the purpose of a demonstration, we consider a special structure of the vectorfield. Nevertheless, it should be apparent that the methodology applies to a more general class of dissipative systems. We thus consider the subclass of (1.1) so that

$$
\begin{aligned}
f_1(\mathbf{x}) &= d_1 x_1 + f_1(x_2), & g_1(\mathbf{x}, t) &= 0, \\
f_2(\mathbf{x}) &= f_2(x_1) + d_2 x_2, & g_2(\mathbf{x}, t) &= \sin \omega t.
\end{aligned}
\tag{3.1}
$$

(A cosine function is equally suitable.) The Mel'nikov function, evaluated along the unperturbed homoclinic orbit $\bar{\mathbf{x}}_0(t)$, now reads

$$
M(t_0) = \int_{-\infty}^{\infty} f_1(\bar{\mathbf{x}}_0(t)) \sin \omega(t + t_0) \cdot \{\exp -[d_1 + d_2]t\}\, dt,
\tag{3.2}
$$

or simply

$$
M(t_0) = \{l_1(\omega)\} \cos \omega t_0 + \{l_2(\omega)\} \sin \omega t_0
\tag{3.3}
$$

where the expressions in the brackets are

$$
l_1(\omega) := \int_{-\infty}^{\infty} f_1(\bar{\mathbf{x}}_0(t))\{\exp -[d_1 + d_2]t\} \sin(\omega t)\, dt,
\tag{3.4i}
$$

$$(3.4\text{ii}) \qquad l_2(\omega) := \int_{-\infty}^{\infty} f_1(\bar{\mathbf{x}}_0(t))\{\exp -[d_1+d_2]t\} \cos(\omega t)\, dt.$$

As $t \to \infty$, the bounded vector field component $f_1(\bar{\mathbf{x}}_0(t))$ ($=: \dot{\bar{x}}_{01}(t)$) approaches 0 as $\exp \lambda_2 t$; while as $t \to -\infty$, $f_1(\bar{\mathbf{x}}_0(t))$ approaches 0 as $\exp \lambda_1 t$. Consequently, as can be seen from (2.5) and (2.6), $f_1(\bar{\mathbf{x}}_0(t))\{\exp -[d_1+d_2]t\}$ tends to 0 as $t \to -\infty$. It follows henceforth that the integrals (3.4) are finite, and thus the Mel'nikov function (3.3) is well defined.

Now to show the existence of simple zeros, we need to show that the integrals $l_1(\omega)$ and $l_2(\omega)$ are not both zero. This will be true for all but at most discretely many frequencies. That follows from the analyticity of $l_1(\omega)$ and $l_2(\omega)$ in ω (see Holmes and Marsden [7]). Also, one may interpret the quantity $\{l_2(\omega) - jl_1(\omega)\}$ as the Fourier transform (an analytic function of ω) of the function $\{f_1(\bar{\mathbf{x}}_0(t))\{\exp -[d_1+d_2]t\}\}$, which is bounded, approaches 0 as $t \to \pm\infty$, and is not identically zero (for all t). (See Kopell and Washburn [8], Salam, Marsden and Varaiya [15].)

It is now straightforward to conclude the existence of simple zeros: (3.3) has simple zeros (two in every period) for all but at most a discrete set of ω's.

Remark 3.1. In the case $\mathbf{g}(\mathbf{x}, t) = \mathbf{g}(\mathbf{x}) \cdot \sin(\omega t)$, one obtains (3.3) with

$$(3.5\text{i}) \qquad l_1(\omega) := \int_{-\infty}^{\infty} \mathbf{f}(\bar{\mathbf{x}}_0(t)) \wedge \mathbf{g}(\bar{\mathbf{x}}_0(t))\{\exp -[d_1+d_2]t\} \sin \omega(t)\, dt,$$

$$(3.5\text{ii}) \qquad l_2(\omega) := \int_{-\infty}^{\infty} \mathbf{f}(\bar{\mathbf{x}}_0(t)) \wedge \mathbf{g}(\bar{\mathbf{x}}_0(t))\{\exp -[d_1+d_2]t\} \cos \omega(t)\, dt.$$

$\mathbf{g}(\bar{\mathbf{x}}_0(t))$ is bounded for all t and hence $\mathbf{f}(\bar{\mathbf{x}}_0(t)) \wedge \mathbf{g}(\bar{\mathbf{x}}_0(t))$ approaches 0 in the order of $\exp \lambda_2 t$, as $t \to \infty$, etc. Thus the same argument as above follows. We now note that in [8] a proposition was given that asserts a result analogous to this remark. However, the boundedness and convergence to zero, as $t \to \pm\infty$, of the expression $\mathbf{f}(\bar{\mathbf{x}}_0(t)) \wedge \mathbf{g}(\bar{\mathbf{x}}_0(t))\{\exp -[d_1+d_2]t\}$ was *assumed* to hold.

Remark 3.2. Note that for general bounded functions $\mathbf{f}$ and $\mathbf{g}$, where $\mathbf{g}(\mathbf{x}, t) = \mathbf{g}(\mathbf{x}) \cdot \sin(\omega t)$, one obtains (3.3) with the integrals in the form of (3.5), except now the exponential gets replaced by $\{\exp[-\int_0^t \text{trace } D_{\bar{x}}\mathbf{f}(\bar{\mathbf{x}}_0(s))\, ds]\}$. For t near $\pm\infty$, the exponential is in the order of $\{\exp -(\text{trace } A)t\}$, where $A = D_{\bar{x}}\mathbf{f}(\bar{\mathbf{x}}_0)$ as before and thus this case also follows.

Two difficulties have been overcome in applying the Mel'nikov approach for large damping or dissipation: (1) the validity of the formula for $M(t_0)$ had to be verified by establishing the exponential convergence condition; (2) the convergence of the improper integrals (3.4) had to be verified and then examined for the existence of simple zeros. These questions must be addressed in any application of the Mel'nikov integral. Note that only question two arises in the application to near-Hamiltonian systems (see Mel'nikov [9], Guckenheimer and Holmes [5]).

Therefore the main contribution of this paper is to carefully show that question one is valid for perturbation of non-Hamiltonian dissipative systems. In addition, a more careful treatment of convergence in question two is included than is often given in other publications.

4. An example: the damped pendulum with periodic forcing. Consider the negatively[2] damped pendulum with a constant torque and small sinusoidal forcing

---

[2] For the positively damped pendulum, see Salam and Sastry [16] and Salam [11].

perturbation

(4.1) $$\dot{x}_1 = x_2, \quad \dot{x}_2 = p - \sin(x_1) + d_2 x_2 + \varepsilon \sin \omega t.$$

The unperturbed pendulum is now governed by

(4.2) $$\dot{\bar{x}}_1 = \bar{x}_2, \quad \dot{\bar{x}}_2 = p - \sin(\bar{x}_1) + d_2 \bar{x}_2.$$

Note that along a curve in the (p, d_2)-parameter space, (4.2) possesses an orbit which we denote by $\bar{x}_0(t)$. This orbit is heteroclinic when we consider the plane $(\bar{x}_1, \bar{x}_2)$ and is homoclinic when we consider the cylindrical space $(\bar{x}_1 (\mod 2\pi), \bar{x}_2)$. (See [16] and the references therein.)

The Mel'nikov integral now becomes

(4.3) $$M(t_0) = \int_{-\infty}^{\infty} \bar{x}_{02}(t) \sin \omega(t+t_0) \cdot \{\exp[-d_2]t\} \, dt$$

or

(4.4) $$M(t_0) = \{l_1(\omega)\} \cos \omega t_0 + \{l_2(\omega)\} \sin \omega t_0$$

where the expressions in the brackets are now

(4.5i) $$l_1(\omega) := \int_{-\infty}^{\infty} \bar{x}_{02}(t) \{\exp -d_2 t\} \sin \omega(t) \, dt,$$

(4.5ii) $$l_2(\omega) := \int_{-\infty}^{\infty} \bar{x}_{02}(t) \{\exp -d_2 t\} \cos \omega(t) \, dt.$$

As $t \to \infty$, $\bar{x}_{02}(t)$ becomes of the order of $\exp \lambda_2 t$; while as $t \to -\infty$, it becomes of the order of $\exp \lambda_1 t$. Consequently $\bar{x}_{02}(t)\{\exp -d_2 t\}$ tends to $\{\exp -[d_2 - \lambda_2]t\}$ as $t \to \infty$ and it tends to $\{\exp[\lambda_1 - d_2]t\}$ as $t \to -\infty$. Thus $\bar{x}_{02}(t)\{\exp -d_2 t\}$ tends to zero as $t \to \pm\infty$. Hence, the integrals (4.5) are finite, and the Mel'nikov function (4.4) is well defined. By analyticity the integrals (4.5) are not zero simultaneously for all but at most discretely many frequencies. This suffices to establish the existence of simple zeros.

To verify our conclusions, we have simulated (4.1) using the 4th order Runge-Kutta type formula with integration step maintained at ≤ 0.02 seconds. Figure 2 shows a

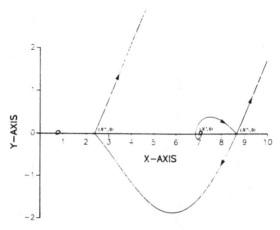

FIG. 2. *A heteroclinic orbit for system* (4.1) *when* $p = 0.7$, $d = 0.6039838$, *and* $\varepsilon = 0$, *using the Runge-Kutta routine.*

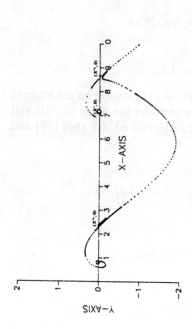

Fig. 3.2. *Poincaré section of 160 initial points along the eigenvector associated with saddle* x^{u_2}.

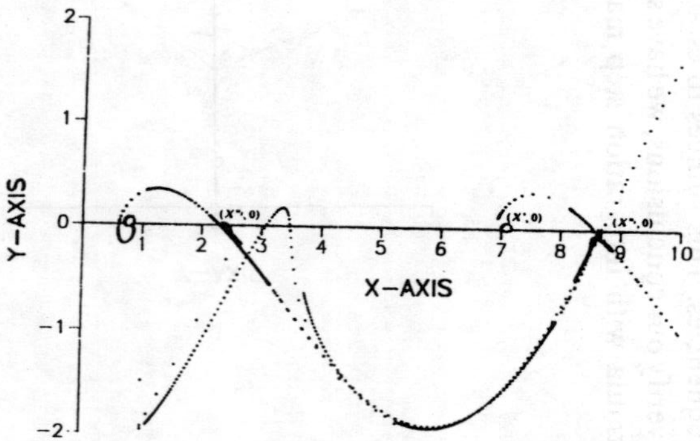

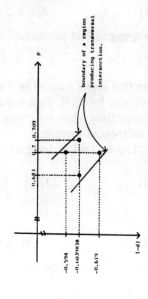

Fig. 4. *A portion of the* (p, d_2)-*parameter space for which* $\bar{\pitchfork}$ *is preserved.*

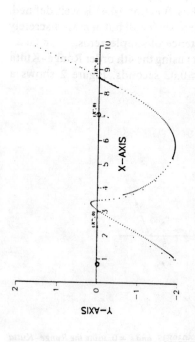

Fig. 3.1. *Poincaré section of 160 initial points along the eigenvector associated with saddle* x^{u_1}.

Fig. 3.3. *Transversal intersection* ($\bar{\pitchfork}$) *of the stable and unstable manifolds for* $\varepsilon = 0.1$ (*Figs. 3.1 and 3.2 superimposed*).

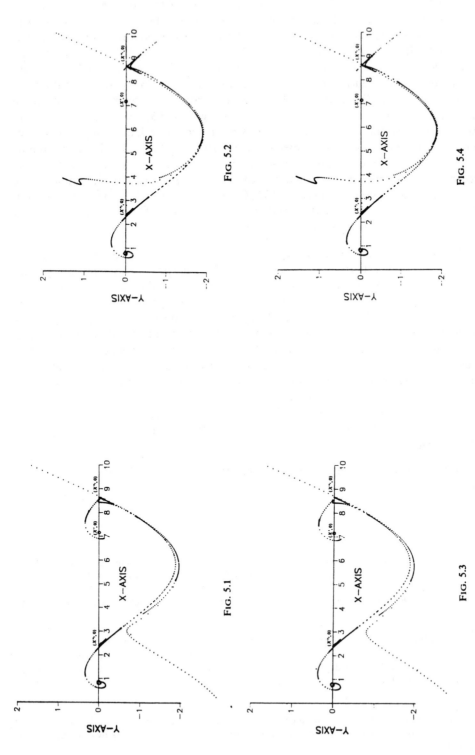

FIGS. 5.1–5.4. *Nontransverse Poincaré sections corresponding to points (p, d_2) on the boundary of the region in Fig. 4.*

heteroclinic orbit[3] of the unperturbed system (4.2) for $p = 0.7$, and $d_2 = 0.6039838$. Figure 2 is a result of the integration of 3 points along the eigenvectors associated with the stable and unstable manifolds of the saddle x^{u2}. For $\varepsilon = 0.1$ in system (4.1) and the same p and d_2 values, we have used 160 initial data points along the eigenvector associated with the stable manifold of the saddle x^{u1}, and 160 points along the eigenvector associated with the unstable manifold of the (replica) saddle x^{u2}. Data points were then stored from each resulting trajectory after every integer multiple of the forcing period. The collection of these data points approximates the Poincaré mapping of system (4.1). The result, shown in Figs. 3.1–3.3, confirms the nonempty transversal intersection.[4] This intersection is preserved for variations in p and d_2 over an open region in the (p, d_2)-parameter space as shown in Fig. 4. The boundary points in the (p, d_2)-space correspond to when the intersection becomes nontransverse according to our numerical simulation. The Poincaré sections corresponding to the near boundary points $(p, d_2) = (0.7, 0.619)$, $(0.7, 0.594)$, $(0.683, 0.6039838)$, $(0.709, 0.6039838)$ are respectively produced in Figs. 5.1–5.4. Note the still maintained transversal intersection of the manifolds near the saddle x^{u2}.

5. Conclusions. Large amplitude damping does not necessarily preclude transversal intersection of stable and unstable manifolds. Some physical systems in engineering and science that possess homoclinic or heteroclinic orbits are highly damped. Consequently, the paper allows for the inclusion of these systems in the class that can be studied via the Mel'nikov approach. Specifically, consider any two-dimensional differential equation that possesses a homoclinic orbit joining a saddle to itself. The paper concludes that for all but at most discrete number of frequencies any small sinusoidal perturbation would give rise to nonempty transversal intersection of stable and unstable manifolds of the perturbed saddle. In essence, any sinusoidal perturbation gives rise to transversal intersection generically.

Our results were derived for perturbations of the form $\mathbf{g}(\mathbf{x}, t) = \mathbf{g}(\mathbf{x}) \sin(\omega t)$. From the viewpoint of applications, it appears valuable to enlarge the class of permissible perturbations.

Acknowledgment. I am grateful to the reviewers for many valuable comments and suggestions.

REFERENCES

[1] R. ABRAHAM AND J. E. MARSDEN, *Foundation of Mechanics*, Benjamin-Cummings, New York, 1978.
[2] V. I. ARNOLD, *Mathematical Methods of Classical Mechanics*, Springer-Verlag, Berlin-Heidelberg-New York, 1978.
[3] ———, *Instability of dynamical systems with several degrees of freedom*, Dokl. Akad. Nauk SSSR, 156 (1964), pp. 9–12.
[4] B. O. GREENSPAN AND P. J. HOLMES, *Homoclinic orbits, subharmonics and global bifurcation in forced oscillations*, in Nonlinear Dynamics and Turbulence, G. Barenhalt ed., Pitman, Boston, MA, 1983.
[5] J. GUCKENHEIMER AND P. J. HOLMES, *Nonlinear Oscillations, Dynamical Systems and Bifurcations of Vector Fields*, Applied Mathematical Sciences, No. 42, Springer-Verlag, Berlin-Heidelberg-New York, 1983.

---

[3] Such an orbit results for all values on a curve in the (p, d_2)-parameter space (see Salam and Sastry [16]). The orbit is homoclinic when considering the cylindrical space on which the two saddles become identified.

[4] The (nonempty) transversal intersection occurs for all $0 < \varepsilon < 0.5$.

[6] P. HOLMES, *Averaging and chaotic motions in forced oscillations*, this Journal, 38 (1980), pp. 68–80; 40 (1980), pp. 167–168.
[7] P. HOLMES AND J. E. MARSDEN, *Horseshoes and Arnold diffusion for Hamiltonian systems on Lie groups*, Indiana Univ. Math. J., 32 (1983), pp. 273–310.
[8] N. KOPELL AND R. B. WASHBURN, *Chaotic motions in the two degree-of-freedom swing equations*, IEEE Trans. Circuits and Systems (1982), pp. 738–746.
[9] V. K. MEL'NIKOV, *On the stability of the center for time periodic perturbations*, Trans. Moscow Math. Soc., 12 (1963), pp. 1–57.
[10] F. M. A. SALAM, *The Mel'nikov technique and Arnold diffusion for a class of perturbed dissipative systems*, Proc. of the American Control Conference (ACC), San Diego, CA, 1984, pp. 258–261.
[11] ———, *The region of chaos in the damped pendulum with torque plus sinusoidal forcing: simulation results*, Phys. Rev. Lett., submitted.
[12] ———, *Feedback stabilization, stability and chaotic dynamics*, 24th IEEE Conference on Decision and Control (CDC), Fort Lauderdale, FL, December 12–14, 1985.
[13] F. M. A. SALAM, J. E. MARSDEN AND P. P. VARAIYA, *Chaos and Arnold diffusion in dynamical systems*, IEEE Trans. Circuits and Systems, CAS-30, 9 (1983), pp. 697–708.
[14] ———, *Arnold diffusion in the swing equations of a power system*, IEEE Trans. Circuits and Systems, CAS-31, 8 (1984), pp. 673–688.
[15] ———, *Arnold diffusion in the dynamics of a 4-machine power system undergoing a large fault*, 22nd IEEE Conference on Decision and Control (CDC), San Antonio, TX, December 1983, pp. 1411–1414.
[16] F. M. A. SALAM AND S. S. SASTRY, *Dynamics of the forced Josephson junction circuit: the regions of chaos*, IEEE Trans. Circuits and Systems, CAS-32, 8 (1985), pp. 784–796.
[17] F. M. A. SALAM, *Explicit inclusion of large damping in the Mel'nikov technique*, Memo. No. DUMEM SM85/02, Dept. of Mechanical Engineering, Drexel Univ., Philadelphia, PA, April 20, 1985.

THE VARIATIONAL APPROACH TO THE THEORY OF SUBHARMONIC BIFURCATIONS

Noboru TAKIMOTO and Hisashi YAMASHIDA

Department of Engineering Science, Tohoku University, Sendai, Japan

Received 9 May 1986
Revised manuscript received 25 November 1986

A multi-time average variational principle is developed for a study of subharmonic bifurcations in a nonlinear ordinary differential equation, and an analytic method is established to calculate the values of the control parameter at the bifurcation points. A specific application is given to the analysis of the Duffing equation $\ddot{x} + 2\lambda\dot{x} + x - 4x^3 = f\cos\omega t$. The connection between period-doubling bifurcations and parametric resonance is clarified. With f the control parameter, the critical values are evaluated analytically at a few bifurcation points with the results in excellent agreement with those of the numerical integration.

1. Introduction

Cascading period-doubling (more generally, subharmonic) bifurcations are one of the routes to chaos in deterministic nonlinear systems. Numerous theoretical and experimental papers have been published on this subject [1–9]. Most theoretical works have been based on purely numerical analysis.

The Duffing equation is one of the systems exhibiting such bifurcations. It takes the form

$$\ddot{x} + 2\lambda\dot{x} + \alpha x + \beta x^3 = f\cos\omega t \quad (\lambda, f > 0), \tag{1.1}$$

and describes forced oscillations in a symmetric nonlinear potential and with friction. The global features of the potential depend on the sign of α and β. For example, the potential has two (locally) stable equilibrium points for $\alpha < 0$ and $\beta > 0$, leading to a static symmetry breaking, whereas it has a single one for $\alpha > 0$ and for both signs of β.

Despite such differences, the numerical integration of (1.1) in most cases yields a chaotic behaviour via subharmonic bifurcations. Holmes [10] studied the case $\alpha < 0$ and $\beta > 0$ with f the control parameter, and found homoclinic orbits playing a crucial role. Huberman and Crutchfield [11] studied the case $\alpha > 0$ and $\beta < 0$ with ω the control parameter, but the importance of a dynamical symmetry breaking was unnoticed. Here "dynamical symmetry breaking" is referred to the fact that the origin $x = 0$ is a stable equilibrium point, with respect to which the potential is symmetric, but it is not the center of oscillations. Parlitz and Lauterborn [12] studied the case $\alpha > 0$ and $\beta > 0$ with ω the control parameter, and noticed the fact that dynamical symmetry breaking is a precursor to period doubling in the absence of static symmetry breaking. This fact was first noticed by D'Humieres et al. [13] in their study of the forced pendulum.

Simplified versions of (1.1) were also studied. Hayashi [14] and Ueda [15] independently carried out the numerical analyses of (1.1) for $\alpha = 0$ and $\beta = \omega = 1$. Their analyses were not systematic enough but the results showed that a subharmonic route to chaos certainly exists.

The question naturally arises whether such bifurcation structures can be derived analytically. Most previous authors are pessimistic in this respect; they claimed that, unless small parameter conditions are met, any conventional analytic method fails to detect important behaviour. However, there are a few papers, in which the possibility of the analytic derivation is discussed. Hsieh [16] applied the direct variational principle to the analysis of (1.1), and derived analytically a criterion for period-tripling bifurcation. However, he did not analyze the stability of the period-tripled state. Itoh [7] and D'Humieres et al. [13] noted that subharmonic oscillations are well known in parametric resonance, and suggested that the bifurcations in question can be described by the Mathieu equation.

The main purpose of this paper is to carry out the analytic integration of (1.1). We shall use the method of the average variational principle of Whitham [18]. This is equivalent in spirit to the classical averaging method due to Krylov and Bogoliubov [19], but the new method is more powerful and convenient in that nonlinear effects such as the frequency renormalization and mode–mode coupling are very easily taken into account.

In section 2 the average variational method will be presented, and then some classical results will be reproduced. V.I. Arnol'd mentioned in ref. 19 in 1977 that the problem of strict justification of the averaging method is still far from being solved. We now feel that it is not very far from being solved, but existence and uniqueness theorems still remain to be established. In section 3, eq. (1.1) will be integrated numerically for $\alpha = 1$ and $\beta = -4$, and with f the control parameter. We shall choose somewhat higher values of ω than in ref. 11 to make the bifurcation structure fairly simple. The analytic integration will be carried out in section 4. We shall find that, for some range of ω and λ, the analytic evaluation of the critical values of f are feasible at the first few bifurcation points on the route to chaos.

2. The average variational principle

2.1. General theory

Consider a one-dimensional non-dissipative, non-autonomous oscillatory system, and denote its Lagrangian by $L(\dot{x}, x, t)$. The least action principle takes the form

$$\int_{t_1}^{t_2} L(\dot{x}, x, t) \, dt = 0, \quad \delta x(t_1) = \delta x(t_2) = 0,$$

and leads to the Euler–Lagrange equation

$$\frac{d}{dt} \frac{\partial L}{\partial \dot{x}} - \frac{\partial L}{\partial x} = 0. \tag{2.1}$$

Oscillations in general contain nonperiodic as well as periodic parts. If only a single frequency ω is involved in the periodic part, one can formally introduce the two time coordinates t and $\theta = \omega t$ to describe each part separately. Namely, x can be regarded as a function of t and θ, such that it is nonperiodic in t and 2π-periodic in θ. To be consistent, any explicit θ-dependence of L must also be 2π-periodic. Then it follows that

$$\dot{x} = x_t + \omega x_\theta \tag{2.2}$$

and

$$\left(\frac{\partial}{\partial t} + \omega \frac{\partial}{\partial \theta} \right) \frac{\partial L}{\partial \dot{x}} - \frac{\partial L}{\partial x} = 0. \tag{2.3}$$

Now regard (2.3) as a partial differential equation for $x(t, \theta)$, depending on the two independent variables t and θ. If $x(t, \theta)$ satisfies (2.3), then $x(t, \omega t)$ evidently satisfies (2.1). Eq. (2.3) is the Euler equation for the variational principle

$$\delta \int_{t_1}^{t_2} dt \frac{1}{2\pi} \int_0^{2\pi} d\theta\, L(x_t + \omega x_\theta, x, \theta, t) = 0, \tag{2.4}$$

provided $\delta x(t, \theta + 2\pi) = \delta x(t, \theta)$ and $\delta x(t_1, \theta) = \delta x(t_2, \theta) = 0$. Indeed, after effecting the variation, we find

$$\frac{\partial}{\partial t} \frac{\partial L}{\partial x_t} + \frac{\partial}{\partial \theta} \frac{\partial L}{\partial x_\theta} - \frac{\partial L}{\partial x} = 0,$$

which, on account of (2.2), is equivalent to (2.3). Eq. (2.4) is rewritten as

$$\delta \int_{t_1}^{t_2} \mathcal{L}(x_t, x, t)\, dt = 0, \tag{2.5}$$

where $\mathcal{L}(x_t, x, t)$ is the average Lagrangian

$$\mathcal{L}(x_t, x, t) = \frac{1}{2\pi} \int_0^{2\pi} d\theta\, L(x_t + \omega x_\theta, x, \theta, t). \tag{2.6}$$

The principle can be easily extended to a multi-periodic case, where $n + 1$ time coordinates t and $\theta_i = \omega_i t$, $i = 1, 2, \ldots, n$ ($\{\omega_i\}$ being incommensurate with each other), are involved. We extend (2.3) to

$$\left(\frac{\partial}{\partial t} + \sum_{i=1}^n \omega_i \frac{\partial}{\partial \theta_i} \right) \frac{\partial L}{\partial \dot{x}} - \frac{\partial L}{\partial x} = 0, \quad \dot{x} = x_t + \sum_{i=1}^n \omega_i x_{\theta_i}, \tag{2.7}$$

and regard it as a partial differential equation for x, depending on $n + 1$ independent variables t and $\{\theta_i\}$. It is evident that if $x(t, \{\theta_i\})$ satisfies (2.7), then $x(t, \{\omega_i t\})$ satisfies (2.1). Eq. (2.7) is the Euler equation for the variational principle

$$\delta \int_{t_1}^{t_2} dt \frac{1}{(2\pi)^n} \int_0^{2\pi} d\theta_1 \cdots \int_0^{2\pi} d\theta_n\, L(\dot{x}, x, \{\theta_i\}, t) = 0, \tag{2.8}$$

provided δx is 2π-periodic in $\{\theta_i\}$ and vanishes at $t = t_1$ and t_2.

Damping effect can also be included easily. Let the original variational principle be of the form

$$\delta \int_{t_1}^{t_2} L(\dot{x}, x, t)\, dt - 2\lambda \int_{t_1}^{t_2} \dot{x}\, \delta x\, dt = 0,$$

where λ is the damping coefficient. The corresponding Euler equation is

$$-\frac{d}{dt} \frac{\partial L}{\partial \dot{x}} + \frac{\partial L}{\partial x} - 2\lambda \dot{x} = 0.$$

In the multi-time formalism, this is written as

$$-\left(\frac{\partial}{\partial t}+\sum_{i=1}^{n}\omega_i\frac{\partial}{\partial \theta_i}\right)\frac{\partial L}{\partial \dot{x}}+\frac{\partial L}{\partial x}-2\lambda\dot{x}=0; \quad \dot{x}=x_t+\sum_{i=1}^{n}\omega_i x_{\theta_i}.$$

This is the Euler equation for the variational principle

$$\int_{t_1}^{t_2} dt\,\mathscr{L}(\dot{x},x,t)-\int_{t_1}^{t_2} dt\,\delta'R(\dot{x},x,t)=0, \tag{2.9}$$

where

$$\mathscr{L}(\dot{x},x,t)=\frac{1}{(2\pi)^n}\int_0^{2\pi} d\theta_1\cdots\int_0^{2\pi} d\theta_n\, L(\dot{x},x,\{\theta_i\},t),$$

$$\delta'R(\dot{x},x,t)=\frac{2\lambda}{(2\pi)^n}\int_0^{2\pi} d\theta_1\cdots\int_0^{2\pi} d\theta_n\,\dot{x}\,\delta x.$$

2.2. Examples

To show how the average variational principle works, a few well-known results will now be reproduced.

Example 1. Forced linear oscillations

The equation to be studied is

$$\ddot{x}+2\lambda\dot{x}+\omega_0^2 x=f\cos\omega t, \tag{2.10}$$

where ω_0 is the natural frequency. Only the underdamping case $\omega_0^2>\lambda^2$ will be treated. This equation is equivalent to

$$\delta\int\left(\frac{\dot{x}^2}{2}-\frac{\omega_0^2}{2}x^2-fx\cos\omega t\right)dt-2\lambda\int\dot{x}\,\delta x\,dt=0. \tag{2.11}$$

The stationary oscillations are represented by

$$x=A\,e^{i\theta}+A^*e^{-i\theta}, \quad \theta=\omega t, \tag{2.12}$$

where A is a constant (the variational parameter). The average variational principle leads to

$$\delta\int\{(\omega^2-\omega_0^2)A^*A-\tfrac{1}{2}f(A^*+A)\}\,dt-2i\omega\lambda\int\{A\delta A^*-\text{c.c.}\}\,dt=0.$$

The Euler equation takes the form

$$(\omega^2-\omega_0^2-2i\omega\lambda)A-f/2=0,$$

whence follows the solution

$$x=\mathrm{Re}\left(\frac{f\,e^{i\omega t}}{\omega^2-\omega_0^2-2i\omega\lambda}\right). \tag{2.13}$$

To describe the transient processes, we must add to the trial function the term

$$x(t,\theta) = a(t)e^{i\theta} + a^*(t)e^{-i\theta}, \quad \theta = \Omega t,$$

where Ω is a constant to be determined by the variational principle itself, and assumed to be incommensurate with ω. The average principle then leads to

$$\delta\int\{(\dot{a}^* - i\Omega a^*)(\dot{a} + i\Omega a) - \omega_0^2 a^* a\}\,dt - 2\lambda\int\{(\dot{a}^* - i\Omega a^*)\delta a + (\dot{a} + i\Omega a)\delta a^*\}\,dt = 0.$$

Effecting the variation with respect to a^*, we find

$$-\ddot{a} - 2(\lambda + i\Omega)\dot{a} + (\Omega^2 - \omega_0^2 - 2i\Omega\lambda)a = 0.$$

We seek the solution in the form $a = (B/2)e^{\mu t}$, where B and μ are constants, and obtain μ from the characteristic equation

$$-\mu^2 - 2(\lambda + i\Omega)\mu + \Omega^2 - \omega_0^2 - 2i\Omega\lambda = 0.$$

We can take μ to be real without loosing generality, and then this is divided into two parts

$$\mu = -\lambda, \quad \Omega^2 = \omega_0^2 - \lambda^2.$$

Then follows the solution

$$x = \operatorname{Re}\left\{A e^{-\lambda t + i\sqrt{(\omega_0^2 - \lambda^2)}\,t}\right\}.$$

Example 2. Van der Pol equation. The van der Pol equation

$$\ddot{x} - 2\lambda(1 - x^2)\dot{x} + \omega_0^2 x = 0$$

is equivalent to

$$\delta\int\left(\frac{\dot{x}^2}{2} - \frac{\omega_0^2}{2}x^2\right)dt + 2\lambda\int(1 - x^2)\dot{x}\,\delta x\,dt = 0.$$

The limit cycle in the lowest approximation is given by

$$x = A e^{i\theta} + A^* e^{-i\theta}, \quad \theta = \Omega t,$$

which yields

$$\delta\int(\Omega^2 - \omega_0^2)A^* A\,dt - 2i\Omega\lambda\int\{(1 - |A|^2)\delta A^* - \text{c.c.}\}\,dt = 0.$$

Effecting the variation, we obtain

$$(\Omega^2 - \omega_0^2)A - 2i\Omega\lambda(1 - |A|^2) = 0,$$

whence follow $\Omega = \omega_0$ and $|A| = 1$. This result is justifiable in the limit of small λ [20].

Example 3. Parametric resonance. The lowest order parametric resonance is described by the Mathieu equation of the form [21],

$$\ddot{x} + 2\lambda \dot{x} + \omega_0^2 \{1 + m \cos(2\omega t + \phi)\} x = 0, \tag{2.14}$$

with m and ϕ being constant. This is equivalent to

$$\delta \int \left(\frac{\dot{x}^2}{2} - \frac{\omega_0^2}{2} x^2 \{1 + m \cos(2\omega t + \phi)\} \right) dt - 2\lambda \int \dot{x} \, \delta x \, dt = 0. \tag{2.15}$$

We take the trial function in the form

$$x = a e^{i\theta} + a^* e^{-i\theta}, \quad \theta = \omega t, \tag{2.16}$$

which will be referred to as the fundamental oscillations. The case where the fundamental frequency differs from ω is treated in appendix A. The exact solution must contain terms with frequencies differing from ω by integral multiples of 2ω, but for small enough m they have no significant effects.

The average variational principle leads to

$$\delta \int \{(\dot{a}^* - i\omega a^*)(\dot{a} + i\omega a) - \omega_0^2 a^* a - \omega_0^2 (h a^{*2} + h^* a^2)/4\} \, dt$$

$$- 2\lambda \int \{(\dot{a}^* - i\omega a^*) \delta a + (\dot{a} + i\omega a) \delta a^*\} \, dt = 0,$$

with $h = m e^{i\phi}$. Effecting the variation with respect to a^*, we find

$$-\ddot{a} - 2(\lambda + i\omega) \dot{a} + (\omega^2 - \omega_0^2 - 2i\omega\lambda) a - (h\omega_0^2/2) a^* = 0. \tag{2.17}$$

We seek the solutions proportional to $e^{\mu t}$ for real μ. The characteristic equation is then written as

$$C(\mu) \equiv (\mu + \lambda)^4 + 2(\omega^2 + \omega_0^2 - \lambda^2)(\mu + \lambda)^2 + (\omega^2 - \omega_0^2 + \lambda^2)^2 - |h\omega^2/2|^2 = 0. \tag{2.18}$$

$C(\mu)$ has the only extreme (a minimum) at $\mu = -\lambda$, and goes to ∞ as $\mu \to \pm \infty$. Accordingly, three cases can be distinguished. (1) If $C(0) < 0$, or

$$|h\omega_0^2/2|^2 > (\omega^2 - \omega_0^2)^2 + (2\omega\lambda)^2, \tag{2.19}$$

then $C(\mu)$ has one real positive and one real negative root, so that the oscillations amplify. $C(\mu)$ has also two complex roots, which do not correspond to any physical reality; they are the solutions of

$$\left\{ (\mu + \lambda)^4 + 2(\omega^2 + \omega_0^2 - \lambda^2)(\mu + \lambda)^2 + (\omega^2 \omega_0^2 + \lambda^2)^2 \right\}^{1/2} = -|h\omega^2/2|,$$

which is evidently self-inconsistent. The appearance of such irrelevant solutions are due to the approximate nature of the trial function (see appendix A for details). (2) If $C(0) > 0$ and $C(-\lambda) < 0$; i.e., if

$$(\omega^2 - \omega_0^2 + \lambda^2)^2 < |h\omega_0^2/2|^2 < (\omega^2 - \omega_0^2)^2 + (2\omega\lambda)^2, \tag{2.20}$$

then $C(\mu)$ has two real negative roots along with two irrelevant ones. (3) If $C(-\lambda) > 0$, or

$$|h\omega_0^2/2|^2 < (\omega^2 - \omega_0^2 + \lambda^2)^2, \tag{2.21}$$

then $C(\mu)$ has no real roots. It is shown in appendix A that the fundamental frequency in this case differs from ω, and no amplification occurs. Thus as the condition for parametric resonance, we obtain the well-known formula (2.19), or

$$|h\omega_0^2/2|^2 > |\omega^2 - \omega_0^2 - 2i\omega\lambda|^2. \tag{2.22}$$

3. The numerical integration of the Duffing equation

Consider a forced nonlinear system described by

$$\ddot{x} + 2\lambda\dot{x} + x - 4x^3 = f\cos\omega t, \tag{3.1}$$

where dimensionless time and length units have been used. Note that the origin is not globally stable, and the system goes to infinity for improperly chosen initial conditions.

To obtain the numerical solutions we first put $\lambda = 0.2$ and $\omega^2 = 2$, and let f be the control parameter. The primary reason for this choice is that the resultant bifurcation structure is simple. For small f, we can choose the initial conditions $\dot{x} = x = 0$ to keep the motion finite. We increase f gradually by a small step $\Delta f = 10^{-3}$ (we let $\Delta f = 10^{-4}$ near the bifurcation points). For large f, the integration diverges for these initial conditions. We have avoided this by using the last computed mechanical states as the initial conditions in the next step of integration.

The computations have been performed on ACOS1000 in Tohoku University. We have used the method of Runge–Kutta and Gill, following the double-precision procedure and choosing $\Delta t = 2^{-4}$ for the time increment. We have discarded all data between $t = 0$ and 1000 as representing the transient processes, and retained those between 1000 and 3200.

The motion for $f < 0.6871 \equiv f_s$ consists of the fundamental oscillations of frequency ω together with its odd harmonics. When f exceeds f_s, symmetry breaking occurs; i.e., the time average $\langle x \rangle$ of x becomes non-vanishing, and at the same time even harmonics appear. We find that the sign of $\langle x \rangle$ depends on the initial conditions. We have confirmed this by observing that, when we change the sign of the initial conditions and also of the external force, then $\langle x \rangle$ changes sign. In other words, a symmetry-broken state has a two-fold degeneracy.

At $f = 0.7509 \equiv f_1$ the first period-doubling bifurcation occurs. This is followed by the subsequent bifurcations at $f_2 = 0.7630$, $f_3 = 0.7668$ and $f_4 = 0.7680$. Fig. 1 shows a state after the 4th period doubling bifurcation. At $f = 0.7690$, a chaotic state starts. It should be noted, however, that it is always a matter of controversy whether such a state really corresponds to a chaos, or it consists of oscillations of very long periods.

When f further increases, the bifurcations proceed in the reversed order. Period-octupled oscillations can be observed at $f = 0.7730$, which turn into period-quadrupled ones at $f = 0.7740$ and then into

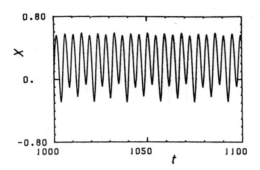
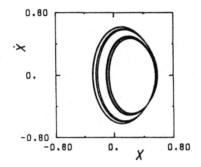
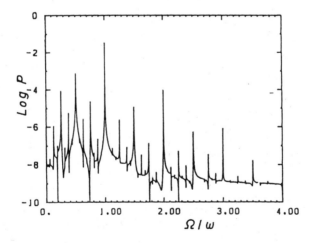

Fig. 1. The state at $f = 0.7680$ after the fourth period-doubling bifurcation.

period-doubled ones at $f = 0.7761$. At $f = 0.7801$ period-quadrupled ones again appear, which turn into a chaotic state at $f = 0.7810$. The system finally goes to infinity for $f > 0.7830$.

In addition to the series of bifurcations just mentioned, we have found the existence of period-tripled oscillations in the range $0.7599 < f < 0.7676$ (fig. 2). The two locally stable states coexist in this range, to which correspond two separate basins in phase space.

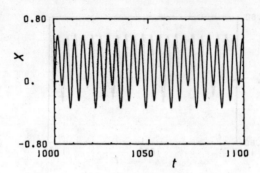

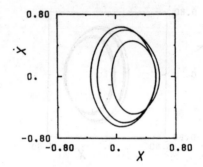

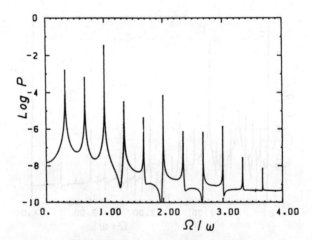

Fig. 2. The period-tripled state at $f = 0.7600$.

We have also found that the bifurcation structure depends strongly on ω. For $\omega^2 > 5.0 = \omega_D^2$ with λ kept at 0.2, the system exhibits no bifurcation at all and directly goes to infinity as f increases. For intermediate values of ω ($4.8 > \omega^2 > 2.2$), the system exhibits only symmetry-breaking bifurcation before it diverges. These features are shown in fig. 3, where the relation between $4\langle x \rangle \equiv y_0$ and f is plotted.

In the following the quantitative features of oscillations as obtained above will be referred to as the experimental results.

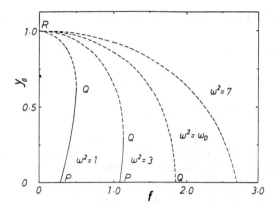

Fig. 3. The relation between $Y_0 = 4\langle x\rangle^2$ and f for $\lambda = 0.2$. Full and dotted lines correspond, respectively, to stable and unstable symmetry-broken states, and dots are the experimental values.

4. The analytic integration of the Duffing equation

4.1. *A symmetric state and its soft-mode stability*

We now apply the variational principle to the analysis of the Duffing eq. (3.1). First consider only the fundamental oscillations

$$x = a_1 e^{i\theta} + a_1^* e^{-i\theta}, \quad \theta = \omega t.$$

Substituting this in the variational principle

$$\delta \int \left(\frac{\dot{x}^2}{2} - \frac{x^2}{2} + x^4 + fx\cos\omega t \right) dt - 2\lambda \int \dot{x}\,\delta x\, dt = 0,$$

and then taking an average with respect to θ, we obtain

$$\delta \int \mathscr{L}\, dt - \int \delta'R = 0,$$

where

$$\mathscr{L} = (\dot{a}_1^* - i\omega a_1^*)(\dot{a}_1 + i\omega a_1) - a_1^* a_1 + 6 a_1^{*2} a_1^2 + f(a_1^* + a_1)/2,$$
$$\delta'R = 2\lambda\{(\dot{a}_1^* - i\omega a_1^*)\delta a_1 + (\dot{a}_1 + i\omega a_1)\delta a_1^*\}.$$

Effecting the variation with respect to a_1^*, we find

$$-\ddot{a}_1 - 2(\lambda + i\omega)\dot{a}_1 + \left(\omega^2 - 1 + 12 a_1^* a_1 - 2i\omega\lambda\right) a_1 + f/2 = 0. \tag{4.1}$$

The steady solution A_1 of (4.1) corresponds to the stationary oscillations and is determined by

$$(\omega^2 - 1 + 12A_1^*A_1 - 2i\omega\lambda)A_1 = -f/2. \tag{4.2}$$

This yields

$$\{(\omega^2 - 1 + Y_1/2)^2 + (2\omega\lambda)^2\}Y_1 = 6f^2, \quad Y_1 \equiv 24A_1^*A_1.$$

To study the local stability of the stationary oscillations, we replace a_1 in (4.1) by $A_1 + a_1$, where a_1 now corresponds to small fluctuations, and then linearize the result with respect to a_1. We then have

$$-\ddot{a}_1 - 2(\lambda + i\omega)\dot{a}_1 + (\omega^2 - 1 + Y_1 - 2i\omega\lambda)a_1 + 12A_1^2 a_1^* = 0. \tag{4.3}$$

This is the same as (2.17), provided ω_0^2 and $h\omega_0^2/2$ are replaced by $1 - Y_1$ and $12A_1^2$, respectively. The term $1 - Y_1$ represents the renormalized fundamental frequency ($-Y_1$ being the frequency correction). The resonance condition (2.22) now implies the instability condition, and yields

$$D \equiv \tfrac{1}{4}Y_1^2 - (\omega^2 - 1 + Y_1)^2 - (2\omega\lambda)^2 > 0. \tag{4.4}$$

There exists a one-to-one correspondence between the sign of D and that of the derivative $Y_1' \equiv dY_1/df$. To show this we differentiate (4.2) with respect to f to obtain

$$\left[\{(\omega^2 - 1 + Y_1/2)^2 + (2\omega\lambda)^2\} + Y_1(\omega^2 - 1 + Y_1/2)\right]Y_1' = 12f.$$

The expression in the square bracket is the same as $-D$, so that

$$Y_1' = -12 f/D. \tag{4.5}$$

Thus it follows the well-known criterion that a stationary state with negative Y_1' corresponds to unstable oscillations.

Next consider higher harmonics, and take

$$x = a_1 e^{i\theta} + a_1^* e^{-i\theta} + a_3 e^{3i\theta} + a_3^* e^{-3i\theta},$$

which yields

$$\mathcal{L} = \sum_{n=1,3} \{(\dot{a}_n^* - in\omega a_n^*)(\dot{a}_n + in\omega a_n) - a_n^* a_n + 6a_n^{*2} a_n^2\} + 4(a_1^3 a_3^* + a_1^{*3} a_3)$$
$$+ 24a_1^* a_1 a_3^* a_3 + f(a_1^* + a_1)/2,$$
$$\delta'R = 2\lambda \sum_{n=1,3} \{(\dot{a}_n^* - in\omega a_n^*)\delta a_n + (\dot{a}_n + in\omega a_n)\delta a_n^*\}.$$

The equations of motion are

$$-\ddot{a}_1 - 2(\lambda + i\omega)\dot{a}_1 + (\omega^2 - 1 + 12a_1^* a_1 + 24a_3^* a_3 - 2i\omega\lambda)a_1 + 12a_1^{*2} a_3 + f/2 = 0,$$
$$-\ddot{a}_3 - 2(\lambda + 3i\omega)\dot{a}_3 + (9\omega^2 - 1 + 24a_1^* a_1 + 12a_3^* a_3 - 6i\omega\lambda)a_3 + 4a_1^3 = 0.$$

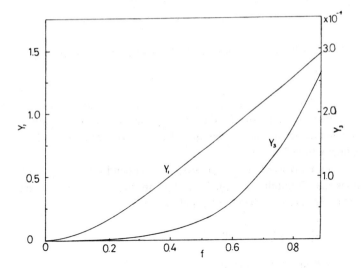

Fig. 4. Symmetric states for $\omega^2 = 2$ and $\lambda = 0.2$.

The steady solutions are determined by

$$\left(\omega^2 - 1 + \tfrac{1}{2}Y_1 + Y_3 - 2i\omega\lambda\right)A_1 + 12A_1^{*2}A_3 + f/2 = 0, \tag{4.6}$$

$$\left(9\omega^2 - 1 + Y_1 + \tfrac{1}{2}Y_3 - 6i\omega\lambda\right)A_3 + 4A_1^3 = 0, \tag{4.7}$$

where $Y_n \equiv 24 A_n^* A_n$. It follows from (4.7) that

$$\frac{Y_3}{Y_1} = \frac{Y_1^2/36}{\left(9\omega^2 - 1 + Y_1 + Y_3/2\right)^2 + (6\omega\lambda)^2}. \tag{4.8}$$

We are concerned with the case, where 3ω is far from the fundamental frequency, so that the ratio (4.8) is very small.

Eqs. (4.6) and (4.7) are solved numerically for Y_1 and Y_3, and the results are plotted in fig. 4. After a straightforward but lengthy analysis, it can be shown that a state with positive Y_1' is stable. Thus all states in fig. 4 are stable.

The approximation improves further by including still higher harmonics $5\omega, 7\omega$ and so on, but their contributions are even smaller.

The stability analysis has so far been limited to the case, where destabilized fluctuations have the same frequency as the fundamental oscillations. This is referred to as the soft-mode instability, whereas the case where they have different frequencies is referred to as the hard-mode instability, and will be discussed in the next subsection.

4.2. Symmetry-breaking bifurcation

A symmetric state coupled with hard-mode fluctuations can be represented approximately by the trial function

$$x = A_1 e^{i\theta} + A_1^* e^{-i\theta} = a e^{i\theta'} + a^* e^{-i\theta'} + a_+ e^{i(\theta+2\theta')} + a_+^* e^{-i(\theta+2\theta')} + a_- e^{i(\theta-2\theta')} + a_-^* e^{-i(\theta-2\theta')}, \tag{4.9}$$

with $\theta = \omega t$, $\theta' = \Omega t$. The first two terms correspond to the stationary oscillations with all harmonics ignored, and the next two terms to the fundamental mode of hard-mode fluctuations with unknown frequency Ω. The remaining terms have been included to take account of the lowest order coupling between hard-mode fluctuations.

A general analysis is quite involved, and is carried out in appendix B. It is found after all that the only hard-mode fluctuations that destabilize a symmetric state are the zero frequency mode ($\Omega = 0$), which we shall denote by the real variable a_0. Then (4.9) simplifies to

$$x = A_1 e^{i\theta} + A_1^* e^{-i\theta} + a_0 + a_2 e^{2i\theta} + a_2^* e^{-2i\theta}. \tag{4.10}$$

Calculations of $\mathscr{L}$ and $\delta' R$ are straightforward:

$$\mathscr{L} = \tfrac{1}{2}\dot{a}_0^2 + \dot{a}_2^* \dot{a}_2 + (4\omega^2 - 1 + Y_1) a_2^* a_2 - \tfrac{1}{2}(1 - Y_1) a_0^2 + 12 a_0 (A_1^2 a_2^* + A_1^{*2} a_2),$$

$$\delta' R = 2\lambda a_0 \delta a_0 + \{(\dot{a}_2^* - 2i\omega a_2^*)\delta a_2 + (\dot{a}_2 + 2i\omega a_2)\delta a_2^*\},$$

where terms more than cubic in fluctuations have been ignored, as they have no concern with the local stability. The equations of motion are

$$-\ddot{a}_0 - 2\lambda \dot{a}_0 + (Y_1 - 1) a_0 + 12(A_1^2 a_2^* + A_1^{*2} a_2) = 0,$$

$$-\ddot{a}_2 - 2(\lambda + 2i\omega)\dot{a}_2 + (4\omega^2 - 1 + Y_1 - 4i\omega\lambda) a_2 + 12 A_1^2 a_0 = 0.$$

Substitution of $a_0, a_2 \propto e^{\mu t}$ (μ being real) gives the characteristic equation

$$C(\mu) \equiv -(\mu + \lambda)^6 + c_2 (\mu + \lambda)^4 + c_1 (\mu + \lambda)^2 + c_0 = 0, \tag{4.11}$$

with

$$c_0 = \Phi_0 \Phi_2^2 - Y_1^2 \Phi_2/2, \quad c_1 = (16\omega^2 - 2\Phi_2)\Phi_0 - \Phi_2^2 + Y_2^2/2,$$

$$c_2 = \Phi_0 + 2\Phi_2 - 16\omega^2,$$

where $\Phi_n = (n\omega)^2 - 1 + Y_1 + \lambda^2$. c_1 and c_2 are negative, if $\omega^2 > 3(-1 + Y_1^2 + \lambda^2)/8$ and $(4\omega^2)^2 + 3(Y_1^2 + \lambda^2 - 1)^2 > Y_1^2/2$. Since Y is at most of order unity, these inequalities are well satisfied for $\omega > 1 > \lambda$. In this case $C(\mu)$ has a maximum at $\mu = -\lambda$, and goes to $-\infty$ as $\mu \to \pm\infty$. Therefore, the inequality $C(0) > 0$ is a necessary and sufficient condition for $C(\mu)$ to have a real positive root. Thus a_0 amplifies if $C(0) > 0$, or

$$F(Y_1) \equiv Y_1 - 1 - \frac{(4\omega^2 - 1 + Y_1) Y_1^2}{2\{(4\omega^2 - 1 + Y_1)^2 + (4\omega\lambda)^2\}} > 0. \tag{4.12}$$

For given ω and λ, the equation $F(Y_1) = 0$ determines the critical value of Y_1, which in turn determines through (4.2) the critical value f_s of f. For $\omega^2 = 2$ and $\lambda = 0.2$, we find $f_s = 0.6906$ in reasonable agreement with the experimental value 0.6871. As will be shown in the next subsection, the discrepancy decreases very rapidly, when the number of modes retained in the trial function increases.

4.3. A symmetry-broken state and its soft-mode stability

Once (4.12) is satisfied, a_0 goes to infinity in the linear approximation, and to study the dynamical properties of a symmetry-broken state we need the full expression for $\mathscr{L}$. We now derive this for the approximate trial function

$$x = a_0 + \sum_{n=1}^{2} \left(a_n e^{in\theta} + a_n^* e^{-in\theta} \right).$$

$\mathscr{L}$ is divided into three parts according to the number of modes involved: $\mathscr{L} = \mathscr{L}_0 + \mathscr{L}_c + \mathscr{L}_c'$. $\mathscr{L}_0$ consists of uncoupled terms

$$\mathscr{L}_0 = \tfrac{1}{2}\dot{a}_0^2 + \sum_{n=1}^{2} \left\{ (\dot{a}_n^* - in\omega a_n^*)(\dot{a}_n + in\omega a_n) - a_n^* a_n \right\} + a_0^4 + 6 \sum_{n=1}^{2} a_n^{*2} a_n^2 - f(a_1^* + a_1)/2.$$

$\mathscr{L}_c$ and $\mathscr{L}_c'$ represent two- and three-mode coupling, respectively

$$\mathscr{L}_c = 12 a_0^2 \sum_{n=1}^{2} a_n^* a_n + 24 a_1^* a_1 a_2^* a_2,$$

$$\mathscr{L}_c' = 12 a_0 \left(a_1^2 a_2^* + a_1^{*2} a_2 \right).$$

The dissipative term takes the obvious form

$$\delta'R = 2\lambda \left[a_0 \delta a_0 + \sum_{n=1}^{2} \left\{ (\dot{a}_n^* - in\omega a_n^*)\delta a_n + (\dot{a}_n + in\omega a_n)\delta a_n^* \right\} \right].$$

The equations of motion are written as

$$-\ddot{a}_0 - 2\lambda \dot{a}_0 + \left(-1 + 4a_0^2 + 24 a_1^* a_1 + 24 a_2^* a_2 \right) a_0 + 12 \left(a_1^2 a_2^* + a_1^{*2} a_2 \right) = 0,$$
$$-\ddot{a}_1 - 2(\lambda + i\omega)\dot{a}_1 + \left(\omega^2 - 1 + 12 a_0^2 + 12 a_1^* a_1 + 24 a_2^* a_2 - 2i\omega\lambda \right) a_1 + 24 a_0 a_1^* a_2 + f/2 = 0, \quad (4.13)$$
$$-\ddot{a}_2 - 2(\lambda + 2i\omega)\dot{a}_2 + \left(4\omega^2 - 1 + 12 a_0^2 + 24 a_1^* a_1 + 12 a_2^* a_2 - 4i\omega\lambda \right) a_2 + 12 a_0 a_1^2 = 0.$$

A steady solution (A_0, A_1, A_2) of (4.13) corresponds to the stationary oscillations with broken symmetry, and is determined by

$$(-1 + Y_0 + Y_1 + Y_2)A_0 + 12\left(A_1^2 A_2^* + A_1^{*2} A_2 \right) = 0,$$
$$\left(\omega^2 - 1 + 3Y_0 + Y_1/2 + Y_2 - 2i\omega\lambda \right) A_1 + 24 A_0 A_1^* A_2 + f/2 = 0, \quad (4.14)$$
$$\left(4\omega^2 - 1 + 3Y_0 + Y_1 + Y_2/2 - 4i\omega\lambda \right) A_2 + 12 A_0 A_1^2 = 0,$$

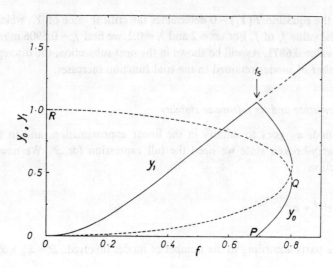

Fig. 5. Symmetry-broken states for $\omega^2 = 2$ and $\lambda = 0.2$.

where $Y_0 = 4A_0^2$ and $Y_n = 24A_n^*A_n$, $n \neq 0$. Evidently, a symmetry-broken state has a two-fold degeneracy ($\pm A_0, A_1, \pm A_2$).

Eqs. (4.14) are solved numerically for the common value $\lambda = 0.2$, and the results are plotted in fig. 5. For $\omega^2 < 4.8$, $\{Y_n\}$ are two-valued functions in a certain range of f; i.e., each curve is divided into two branches limited by Q, where Y_n' diverges. We shall see below that all states on a branch with negative slope Y_0' are soft-mode unstable.

To study the soft-mode stability of the state (A_0, A_1, A_2), we replace a_n in (4.13) by $A_n + a_n$, and linearize the result with respect to a_n. On writing $S_n = (n\omega)^2 - 1 + 3Y_0 + Y_1 + Y_2$, we then have

$$-\ddot{a}_0 - 2\lambda \dot{a}_0 + S_0 a_0 + 24\{(A_1^* A_0 + A_1 A_2^*)a_1 + \text{c.c.}\} + \{(24A_2^* A_0 + 12A_1^{*2})a_2 + \text{c.c.}\} = 0,$$

$$-\ddot{a}_1 - 2(\lambda + i\omega)\dot{a}_1 + (S_1 - i\omega\lambda)a_1 + 24(A_0 A_1 + A_1^* A_2)a_0 + (12A_1^2 + 24A_0 A_2)a_1^*$$
$$+ 24(A_2^* A_1 + A_0 A_1^*)a_2 + 24 A_1 A_2 a_2^* = 0, \quad (4.15)$$

$$-\ddot{a}_2 - 2(\lambda + 2i\omega)\dot{a}_2 + (S_2 - 4i\omega\lambda)a_2 + 12A_2^2 a_2^* + (24A_0 A_2 + 12A_1^2)a_0$$
$$+ 24(A_1^* A_2 + A_0 A_1)a_1 + 24 A_1 A_2 a_1^* = 0.$$

With the use of the notation $a_n^* = a_{-n}$, (4.15) takes the form

$$-\ddot{a}_n - 2(\lambda + in\omega)\dot{a}_n + \sum_{m=-2}^{2} K_{nm} a_m = 0, \quad n = 0, \pm 1, \pm 2 \quad (4.16)$$

with $K_{nm} = K_{mn}^*$. The substitution $a_n \propto e^{\mu t}$ leads to

$$C(\mu) \equiv \sum_{m=0}^{5} c_n (\mu + \lambda)^{2n} = 0, \quad (4.17)$$

with c_n being real (in particular $c_5 = -1$). Note that $C(0) = \text{Det}|K_{nm}|$. Because of the property $c_5 = -1$, we can say even in the absence of any information of the remaining coefficients that the inequality $C(0) > 0$ is a sufficient condition for the soft-mode instability (i.e., $C(\mu)$ has a real positive root). The inequality $C(0) < 0$, on the other hand, provides only a necessary condition for the soft-mode stability.

Again a one-to-one correspondence exists between the signs of $C(0)$ and Y_n'. To see this we differentiate (4.14) with respect to f

$$\sum_{m=-2}^{2} K_{nm} A_m' + \tfrac{1}{2}(\delta_{n,-1} + \delta_{n,1}) = 0.$$

Solving this for A_n', we can derive

$$Y_n' \propto A_n^{*'} A_n + A_n^* A_n' = \Delta_n / C(0),$$

where Δ_n is a polynomial in $\{Y_n\}$, and for given ω and λ depends only on f. If as functions of f, Δ_n and $C(0)$ have no common roots except at P, where both of them vanish, then Y_n' diverges where $C(0)$ vanishes. Thus $C(0)$ vanishes at Q in fig. 5. If $C(0)$ has no multiple roots, then $C(0)$ has different signs on the two branches limited by Q. Namely, one of them corresponds to unstable oscillations.

To determine the sign of Δ_n, we ignore the weakly excited mode A_2 in (4.14). Differentiation then gives

$$S_0 A_0' + 24(A_1' A_1^* + A_1 A_1^{*'}) A_0 = 0,$$
$$(S_1 - 2i\omega\lambda) A_1' + 12 A_1^2 A_1^{*'} + 24 A_0 A_1 A_0' + \tfrac{1}{2} f = 0,$$

which yield

$$A_0' = -6f A_0 / C(0).$$

Therefore, we find

$$Y_0' = 8 A_0' A_0 = -12 f Y_0 / C(0).$$

Note that Δ_0 and $C(0)$ as obtained possess all properties assumed above. Thus a symmetry-broken state with negative slope Y_0' is soft-mode unstable. Therefore according to fig. 3, any symmetry-broken state for $\omega > \omega_D$ is unstable. For $\omega < \omega_D$, on the other hand, all branches have positive slope at P, so that $C(0)$ is negative along PQ, vanishes at Q and becomes positive along QR. Hence the branch QR corresponds to unstable symmetry-broken states. A similar analysis, again ignoring A_2, shows that the inequality $C(0) < 0$ is a necessary and sufficient condition for the soft-mode stability.

The approximation improves by including higher harmonics:

$$x = A_0 + \sum_{n=1}^{N} \left(A_n e^{in\theta} + A_n^* e^{-in\theta} \right).$$

We have evaluated f_s for each N with the results shown in table I. We see that, as N increases, f_s rapidly converges to the experimental value 0.6871.

Table I
Convergence of f_s

N	2	3	4	Exp.
f_s	0.6906	0.6871	0.6871	0.6871

4.4. Period-doubling bifurcations

We are now in a position to study the hard-mode instabilities of a soft-mode stable symmetry-broken state; i.e., a state on the branch PQ. A general analysis is carried out in appendix C, and it is found that the only hard-mode fluctuations that destabilize such a state are the period-doubled mode. Hence the simplest trial function that describes the first period-doubling bifurcation is

$$x = a_0 + a_1 e^{i\theta} + a_1^* e^{-i\theta} + a_{1/2} e^{i\theta/2} + a_{1/2}^* e^{-i\theta/2}. \tag{4.18}$$

The expressions for $\mathscr{L}_0$ and $\mathscr{L}_c$ are exactly similar to those in the previous section, and need not be written down explicitly. The three-mode coupling term is given by

$$\mathscr{L}_c' = 12 a_0 \left(a_1^* a_{1/2}^2 + a_1 a_{1/2}^{*2} \right).$$

The equations of motions are

$$-\ddot{a}_0 - 2\lambda \dot{a}_0 + \left(-1 + 4a_0^2 + 24 a_1^* a_1 + 24 a_{1/2}^* a_{1/2} \right) a_0 + 12 \left(a_1^* a_{1/2}^2 + a_1 a_{1/2}^{*2} \right) = 0,$$

$$-\ddot{a}_1 - 2(\lambda + i\omega) \dot{a}_1 + \left(\omega^2 - 1 + 12 a_0^2 + 12 a_1^* a_1 + 24 a_{1/2}^* a_{1/2} - 2i\omega\lambda \right) a_1$$
$$+ 12 a_0 a_{1/2}^2 + \tfrac{1}{2} f = 0, \tag{4.19}$$

$$-\ddot{a}_{1/2} - 2(\lambda + i\omega/2) \dot{a}_{1/2} + \left(\omega^2/4 - 1 + 12 a_0^2 + 24 a_1^* a_1 + 12 a_{1/2}^* a_{1/2} - i\omega\lambda \right) a_{1/2}$$
$$+ 24 a_0 a_1 a_{1/2}^* = 0.$$

To derive the condition for the period-doubling bifurcation (PDB), we replace a_0 and a_1 in (4.19) by their steady values (A_0, A_1), and then linearize the result with respect to $a_{1/2}$. This yields

$$(-1 + Y_0 + Y_1) A_0 = 0,$$
$$\left(\omega^2 - 1 + 3Y_0 + \tfrac{1}{2} Y_1 - 2i\omega\lambda \right) A_0 + \tfrac{1}{2} f = 0, \tag{4.20}$$
$$-\ddot{a}_{1/2} - 2(\lambda + \tfrac{1}{2} i\omega) \dot{a}_{1/2} + \left(\tfrac{1}{4} \omega^2 - 1 + 3Y_0 + Y_1 - i\omega\lambda \right) a_{1/2} + 24 A_0 A_1 a_{1/2}^* = 0.$$

The last equation describes parametric resonance (see (2.17)), and accordingly the resonance condition

$$6 Y_0 Y_1 > \left(\omega^2/4 - 1 + 3Y_0 + Y_1 \right)^2 + (\omega\lambda)^2 \tag{4.21}$$

is the condition for PDB. Using the relation $Y_0 + Y_1 = 1$, which follows from the first equation of (4.20), we can show that (4.21) never holds if ω is large enough to satisfy

$$\omega^2 > \left[\left\{ (12 + 40\lambda^2)^2 + 216 \right\}^{1/2} - (12 + 40\lambda^2) \right] / 3. \tag{4.22}$$

Table II
Convergence of f_1

N	2	4	6	8	Exp.
f_1	0.7202	0.7551	0.7505	0.7505	0.7505

For $\omega^2 = 2$ and $\lambda = 0.2$, we have evaluated the critical values of Y_0, Y_1 and f_1, and obtained 0.1163, 0.8837 and 0.7207, respectively.

A steady solution $(A_0, A_1, A_{1/2})$ of (4.19) corresponds to the stationary oscillations after the bifurcation, and is determined by

$$\begin{aligned}
&(-1 + Y_0 + Y_1 + Y_{1/2})A_0 + 12(A_1^* A_{1/2}^2 + A_1 A_{1/2}^{*2}) = 0, \\
&(\omega^2 - 1 + 3Y_0 + Y_1/2 + Y_{1/2} - 2i\omega\lambda)A_1 + 12 A_0 A_{1/2}^2 + \tfrac{1}{2}f = 0, \\
&(\omega^2/4 - 1 + 3Y_0 + Y_1 + Y_{1/2}/2 - i\omega\lambda)A_{1/2} + 24 A_0 A_1 A_{1/2}^* = 0.
\end{aligned} \quad (4.23)$$

Again the solution has a two-fold degeneracy $(A_0, A_1, \pm A_{1/2})$. Holmes [10] has already pointed out that subharmonics always occur in multiples.

The analysis of the soft-mode stability of the state $(A_0, A_1, A_{1/2})$ is very involved, but the result is simple; those states with negative slope $Y'_{1/2}$ are soft-mode unstable, while those with positive slope are soft-mode stable.

The approximation improves by including harmonics:

$$x = A_0 + \sum_{n=1}^{N} \left(A_{n/2} e^{in\theta/2} + A_{n/2}^* e^{-in\theta/2} \right). \quad (4.24)$$

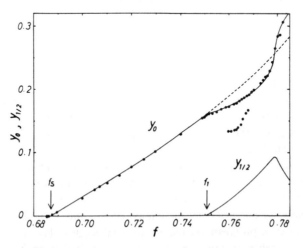

Fig. 6. Period-doubled states for $\omega^2 = 2$ and $\lambda = 0.2$. Full lines are based on the analytic integrations and dots are the experimental results. The experimental results for Y_0 corresponding to the period-tripled states are also shown.

Table III
Convergence of f_2

N	4	6	8	Exp.
f_2	0.7603	0.7630	0.7632	0.7630

We have evaluated $\{Y_{n/2}\}$ and f_1. Table II shows f_1 for each value of N. We see again that f_1 converges rapidly, and the result for $N = 8$ is in excellent agreement with the experimental value 0.7505. Fig. 6 shows Y_0 and $Y_{1/2}$ for $N = 8$.

To study the second PDB, we add to the trial function (4.24) the terms

$$\sum_{n=1}^{N} \left\{ a_{(2n-1)/4} e^{i(2n-1)\theta/4} + a^*_{(2n-1)/4} e^{-i(2n-1)\theta/4} \right\},$$

which correspond to period-quadrupled fluctuation. The procedure from now on is exactly the same as before. Namely, we calculate $\mathscr{L}$ and $\delta'R$, write down the linearized equations of motion for the fluctuations, and derive the characteristic equation $C(\mu) = 0$. Again the inequality $C(0) > 0$ provides a sufficient condition for the bifurcation in question. The critical values of $\{Y_{n/2}\}$ and f are determined by $C(0) = 0$. Table III shows the results for f_2.

We have also found that period-quadrupled oscillations exist in the two ranges $0.7632 < f < 0.7762$ and $0.7801 < f < 0.7964$. In other words no period-quadrupled oscillations exist in the range $0.7762 < f < 0.7801$, a result again in excellent agreement with experiments.

It is now clear how further PDBs occur. The stationary oscillations after the nth bifurcation involves the mode A_m, $m = 2^{-n}$, to which fluctuations $a_{m/2}$ couple resonantly, provided of course that the resonance condition is satisfied, a typical coupling term being $12 A_0 A_m^* (a_{m/2})^2$. As n increases, however, the coupling between harmonics of subharmonics becomes very strong because of small frequency differences between them, and to obtain a reasonably accurate result, many of such modes must be retained in the trial function. The analysis then becomes prohibitively difficult.

5. Conclusions

The numerical integrations of the nonlinear system $\ddot{x} + 2\lambda \dot{x} + x - 4x^3 = f \cos \omega t$ ($\lambda, \omega > 0$) with f the control parameter have shown that the bifurcation structure does not follow a single route to chaos; it simplifies as ω increases. For $\omega > 1 > \lambda$, perturbation expansions are feasible so far as the first few bifurcations are concerned, and the critical values of f obtained analytically are in excellent agreement with those obtained numerically.

The feasibility of the analysis by our method requires a rapid convergence of the Fourier series of the trial functions. This corresponds to the fact that not many sinusoidal modes are in near resonance with the external force and with each other. Most nonlinear systems can possess such properties for properly chosen system parameters. In our example, the choice $\omega^2 \simeq 2$ and $\lambda \simeq 0.2$ corresponds to this. As the bifurcations proceed, however, progressively more sinusoidal modes play an equally important role, and the analysis becomes intractable. In this case, it is crucial for further analysis whether proper periodic functions other than the sinusoidal ones can be found to expand the trial functions.

Now three interesting questions remain. (1) A symmetry-broken state has a two-fold degeneracy, characterized by the signs of $\langle x \rangle$. There must be a deterministic relationship between the initial conditions and the signs of $\langle x \rangle$. Is it possible to determine this analytically? (2) Any post-bifurcation state appears in pair. It is conceivable that the observed chaotic motions consist of a chaotic jump from one state of the pair to another, as was the case in ref. 10. How can this phenomenon be described in our formalism? (3) According to fig. 6, period-tripled bifurcation is an example of inverted bifurcation; i.e., it is characterized by finite amplitude instabilities, and hence cannot be described by a linear stability analysis. The third question is concerned with how our method applies to inverted bifurcations.

Finally it should be mentioned that the period-tripled oscillations as observed in this paper are not identical with those predicted by Hsieh [16] for the following two reasons. (1) If we substitute $r = -4$, $\omega^2 = 2$ and $\alpha = 0.4$ in his condition for the subharmonic oscillations (eq. (V.21) in [16]), we obtain $f < 0.4422$, which is much different from the numerical result $0.7599 < f < 0.7676$ obtained in section 3. (2) The period-tripled oscillations as observed in section 3 occur in the presence of symmetry breaking, whereas those observed in [16] occur in the absence of it.

Appendix A

Higher harmonics in parametric resonance

The exact solution of the Mathieu equation (2.14) is given by

$$x = \sum_n \left\{ b_n e^{i(\theta' + n\theta)} + b_n^* e^{-i(\theta' + n\theta)} \right\}, \quad \theta = \Omega t, \quad \theta' = \omega t, \tag{A.1}$$

where the fundamental frequency Ω is to be determined by the variational principle itself, and b_n corresponds to oscillations with frequency $\Omega + n\omega \equiv \Omega_n$. Provided ω is incommensurate with ω, we can use the multi-time formalism to derive

$$\mathscr{L} = \sum_n \left\{ \left(\dot{b}_n^* - i\Omega_n b_n^* \right)\left(\dot{b}_n + i\Omega_n b_n \right) - \omega_0^2 b_n^* b_n - \left(\omega_0^2/2 \right)\left(h b_n b_{n-1}^* + h^* b_n b_{n-1} \right) \right\},$$

$$\delta' R = 2\lambda \sum_n \left\{ \left(\dot{b}_n^* - i\Omega_n b_n^* \right)\delta b_n + \left(\dot{b}_n + i\Omega_n b_n \right)\delta b_n^* \right\}.$$

The equations of motion are

$$-\ddot{b}_n - 2(\lambda + i\Omega_n)\dot{b}_n + \left(\Omega_n^2 - \omega_0^2 - 2i\Omega_n \lambda \right) b_n - \left(\omega_0^2/2 \right)\left(h b_{n+1} + h^* b_{n-1} \right) = 0, \quad n = 0, \pm 1, \ldots.$$

We seek the solutions proportional to $e^{\mu t}$ for real μ. Substitution yields

$$\left\{ -(\mu + \lambda + i\Omega_n)^2 - \omega_0^2 + \lambda^2 \right\} b_n - \left(\omega_0^2/2 \right)\left(h b_{n+1} + h^* b_{n-1} \right) = 0, \tag{A.2}$$

which leads to the exact characteristic equation of order infinity.

Closed-form analysis of the exact equation is of course infeasible, and we shall solve (A.2) by an expansion in h. We assume that b_0 is dominant, and in the first approximation ignore all other modes. Eq. (A.2) then yields $\mu = -\lambda$ and $\Omega^2 = \omega^2 - \lambda^2$. In the second approximation, we keep the modes $b_{\pm 1}$. They are given approximately by

$$b_1 = \frac{h^*\omega_0^2/2}{\Omega_1^2 - \omega_0^2 + \lambda^2} b_0, \quad b_{-1} = \frac{h\omega_0^2/2}{\Omega_{-1}^2 - \omega_0^2 + \lambda^2} b_0 \tag{A.3}$$

and lead to

$$\mu = -\lambda, \quad \Omega^2 = \omega_0^2 - \lambda^2 + \frac{|h\omega_0^2/2|^2}{2(\omega^2 - \omega_0^2 + \lambda^2)}. \tag{A.4}$$

Further iteration gives the higher order corrections.

The iteration fails when $\omega^2 - \omega_0^2 + \lambda^2$ is of order $|h|^2$, where either b_1 or b_{-1} can be of the same order as b_0. Take $b_1 \simeq b_0$ for definiteness, and ignore all other modes. Eq. (A.2) then gives

$$\begin{aligned}\{-(\mu+\lambda+i\Omega)^2 - \omega_0^2 + \lambda^2\} b_0 - (\omega_0^2/2) h^* b_1 &= 0, \\ \{-(\mu+\lambda+i\Omega_1)^2 - \omega_0^2 + \lambda^2\} b_1 - (\omega_0^2/2) h b_0 &= 0.\end{aligned} \tag{A.5}$$

The characteristic equation takes the form

$$\{-(\mu+\lambda+i\Omega)^2 - \omega_0^2 + \lambda^2\}\{-(\mu+\lambda+i\Omega_1)^2 - \omega_0^2 + \lambda^2\} = |h\omega_0^2/2|^2.$$

Since Ω and μ are real, this is divided into two parts,

$$(\omega + \Omega)(\mu + \lambda)\{(\mu+\lambda)^2 + \omega_0^2 - \lambda^2 - \Omega\Omega_1\} = 0, \tag{A.6}$$

$$\{-(\mu+\lambda)^2 + \Omega^2 - \omega_0^2 + \lambda^2\}\{-(\mu+\lambda)^2 + \Omega_1^2 - \omega_0^2 + \lambda^2\} - 4\Omega\Omega_1(\mu+\lambda)^2 = |h\omega_0^2/2|^2. \tag{A.7}$$

Eq. (A.6) has an obvious solution $\Omega = -\omega$, in which case (A.7) reduces to (2.18). The solutions for $(\mu + \lambda)^2$ are

$$(\mu + \lambda)^2 = -(\omega^2 + \omega_0^2 - \lambda^2) \pm \sqrt{4\omega^2(\omega_0^2 - \lambda^2) + |h\omega_0^2/2|^2}. \tag{A.8}$$

Only the upper sign yields real solutions for μ, provided $(\omega^2 - \omega_0^2 + \lambda^2)2 < |h\omega_0^2/2|^2$. The self-consistency of the result follows from (A.2) and (A.5), which show that $|b_1/b_0| \simeq 1$, and that b_{-1} and b_n for $|n| > 1$ are at most of order h.

The real solutions of (A.6) and (A.7) in the case $(\omega^2 - \omega_0^2 + \lambda^2)^2 > |h\omega_0^2/2|$ are found to be

$$\mu = -\lambda; \quad (\Omega + \omega)^2 = \omega^2 + \omega_0^2 - \lambda^2 \pm \sqrt{4\omega^2(\omega_0^2 - \lambda^2) + |h\omega_0^2/2|^2}. \tag{A.9}$$

The assumption $b_1 \simeq b_0$ requires $\Omega^2 - \omega_0^2 + \lambda^2$ to be of order h, but the expression (A.8) with the upper sign does not comply with this. Thus the relevant solutions are

$$\mu = -\lambda; \quad (\Omega + \omega)^2 = \omega^2 + \omega_0^2 - \lambda^2 - \sqrt{4\omega^2(\omega_0^2 - \lambda^2) + |h\omega_0^2/2|^2}. \tag{A.10}$$

A question naturally arises why irrelevant solutions appear. The answer is as follows. By assumption b_0 has been dominant, but the analysis is exactly similar when a different mode is assumed dominant. Assume for instance that b_1 is dominant, and carry out the transformation $\Omega + 2\omega \to \Omega$ and $b_{n+1} \to b_n$. Then the trial function (A.1) remains unchanged, but the mode b_0 now becomes dominant. Hence the solutions of (A.2) for the transformed variables are exactly the same as before. This means that two solutions, one in which b_0 is dominant and the other in which b_1 is dominant, correspond to one and the same physical reality; they differ only in labelling. The arguments can be generalized to the consequence that, if the data $(\mu, \Omega, \{b_n\})$ satisfy (A.2), then the data $(\mu, \Omega + 2m, \{b_{n+m}\})$ with m an integer also satisfy (A.2). Of course, such properties disappear when the hierarchy of equations (A.2) is truncated. It is now obvious that the irrelevant solutions as observed above are the by-products of the truncation.

Appendix B

The hard-mode instability of a symmetric state

Provided Ω is incommensurate with ω, we can use the mutli-time formalism (2.9) to derive the average Lagrangian for the trial function (4.9). We find that $\mathscr{L}$ is divided into three parts according to the number of modes involved: $\mathscr{L} = \mathscr{L}_0 + \mathscr{L}_c + \mathscr{L}_c'$. $\mathscr{L}_0$ consists of uncoupled terms

$$\mathscr{L}_0 = (\dot{a}^* - i\Omega a^*)(\dot{a} + i\Omega a) + (\dot{a}_+^* - i\Omega_+ a_+^*)(\dot{a}_+ + i\Omega_+ a_+) + (\dot{a}_-^* - i\Omega_- a_-^*)(\dot{a}_- + i\Omega_- a_-)$$
$$- a^*a - a_+^*a_+ - a_-^*a_- + 6(a^{*2}a^2 + a_+^{*2}a_+^2 + a_-^{*2}a_-^2).$$

$\mathscr{L}_c$ and $\mathscr{L}_c'$ represent two and three mode coupling, respectively:

$$\mathscr{L}_c = 24(a^*aa_+^*a_+ + a^*aa_-^*a_- + a_+^*a_+a_-^*a_-) + 24 A_1^* A_1 (a^*a + a_+^*a_+ + a_-^*a_-),$$
$$\mathscr{L}_c' = 12(A_1^2 a^*a_- + A_1^{*2}aa_-^* + A_1^2 aa_+^* + A_1^{*2} a^*a_+).$$

The dissipative part takes the form

$$\delta' R = 2\lambda\{(\dot{a}^* - i\Omega a^*)\delta a + (\dot{a}_+^* - i\Omega_+ a_+^*)\delta a_+ + (\dot{a}_-^* - i\Omega_- a_-^*)\delta a_- + \text{c.c.}\}.$$

The linear stability analysis requires no terms more than cubic in the fluctuating variables. The basic equations are

$$-\ddot{a} - 2(\lambda + i\Omega)\dot{a} + (\Omega^2 - 1 + Y_1 - 2i\Omega\lambda)a + 12(A_1^2 a_-^* + A_1^{*2} a_+) = 0,$$
$$-\ddot{a}_+ - 2(\lambda + i\Omega_+)\dot{a}_+ + (\Omega_+^2 - 1 + Y_1 - 2i\Omega_+\lambda)a_+ + 12 A_1^{*2} a = 0, \quad \text{(B.1)}$$
$$-\ddot{a}_- - 2(\lambda + i\Omega_-)\dot{a}_- + (\Omega_-^2 - 1 + Y_1 - 2i\Omega_-\lambda)a_- + 12 A_1^2 a = 0.$$

We first ignore $a_\pm$. Substitution of $a \propto e^{\mu t}$ leads to the characteristic equation involving the two real unknowns Ω and μ,

$$-\mu^2 - 2(\lambda + i\Omega)\mu + \Omega^2 - 1 + Y_1 - 2i\Omega\lambda = 0.$$

The real and imaginary parts of this equation must vanish separately,

$$\mu^2 + 2\lambda\mu - \Omega^2 + 1 - Y_1 = 0, \quad \Omega(\mu + \lambda) = 0,$$

whence follow the two sets of solutions

(1) $\mu = -\lambda$, $\Omega^2 = 1 - Y_1 - \lambda^2$, provided $Y_1 < 1 - \lambda^2$.

(2) $\mu = -\lambda \pm \sqrt{(\lambda^2 + Y_1^2 - 1)}$, $\Omega = 0$, provided $Y_1 > 1 - \lambda^2$.

The solution (1) corresponds to the damped scillations $x = \text{Re}\{A e^{-(\lambda + i\Omega)t}\}$ which in the limit $Y_1 \to 0$ reduces to (2.13), as it should. The solution (2) corresponds to zero frequency fluctuations, which we shall denote by the real variable a_0. For $\mu > 0$, a_0 amplifies, and the so-called symmetry breaking occurs. The critical value of Y_1 is determined by $\mu = 0$, which yields $Y_1 = 1$.

When the coupling with $a_\pm$ is included, the analysis becomes much more involved, and we consider only the critical case $\mu = 0$, where (B.1) simplifies greatly. Elimination of $a_\pm$ from the simplified equation gives

$$\Omega^2 - 1 + Y_1 - 2i\Omega\lambda - \frac{Y_1^2/4}{\Omega_+^2 - 1 + Y_1 - 2i\Omega_+\lambda} - \frac{Y_1^2/4}{\Omega_-^2 - 1 + Y_1 - 2i\Omega_-\lambda} = 0. \tag{B.2}$$

In writing the trial function (4.9), it has been tacitly assumed that the ratios $|a_\pm/a|$ are small, and hence any solutions of (B.2) not complying with this should be discarded. The smallness of $|a_\pm/a|$ means the smallness of the last two terms of (B.2). We have ignored them in the first approximation to derive the solution $\Omega = 0$ and $Y_1 = 1$. The relevant solutions of (B.2) should differ little from it. It is easy to see that the value $\Omega = 0$ still satisfies the imaginary part of (B.2). Its real part then determines the critical value of Y_1:

$$F(Y_1) \equiv Y_1 - 1 - \frac{(\omega^2 - 1 + Y_1)Y_1^2}{2\{(\omega^2 - 1 + Y_1)^2 + (4\omega\lambda)^2\}} = 0. \tag{B.3}$$

It is evident that μ cannot vanish when $F(Y_1) \neq 0$. In other words, μ cannot change sign inasmuch $F(Y_1) > 0$, or $F(Y_1) < 0$. We have seen that $\mu = -\lambda$ when Y_1 is infinitesimal, where $F(Y_1) < 0$. This leads to the consequence that the hard-mode instability never occurs so far as $F(Y_1) < 0$.

The appearance of the mode a_0 seems in direct contradiction with the assumption that Ω is incommensurate with ω. However, the consistent analysis developed in subsection 4-2 shows that the symmetry breaking really occurs for $F(Y_1) > 0$.

Appendix C

The hard-mode instability of a symmetry-broken state

A symmetry-broken state coupled with hard-mode fluctuations is represented approximately by the trial function

$$x = A_0 + A_1 e^{i\theta} + A_1^* e^{-i\theta} + \sum_{n=-\infty}^{\infty} \left\{ b_n e^{i(\theta' + n\theta)} + b_n^* e^{-i(\theta' + n\theta)} \right\},$$

where $\theta' = \Omega t$ and b_n corresponds to fluctuations with frequency $\Omega + n\omega \equiv \Omega_n$. Assuming Ω to be incommensurate with ω and ignoring terms more than cubic in $\{b_n\}$, we find

$$\mathscr{L} = \sum_n \{(\dot{b}_n^* - i\Omega_n b_n^*)(\dot{b}_n + i\Omega_n b_n) + (-1 + 3Y_0 + Y_1)b_n^* b_n + 24A_0(A_1 b_n b_{n+1}^* + A_1^* b_n^* b_{n+1})$$
$$+ 12(A_1^2 b_n b_{n+2}^* + A_1^{*2} b_n^* b_{n+2})\},$$
$$\delta' R = 2\lambda \sum_n \{(\dot{b}_n^* - i\Omega_n b_n^*)\delta b_n + (\dot{b}_n + i\Omega_n b_n)\delta b_n^*\}.$$

The equations of motion are

$$-\ddot{b}_n - 2(\lambda + i\Omega_n)\dot{b}_n + (\Omega_n^2 - 1 + 3Y_0 + Y_1 - 2i\Omega_n \lambda)b_n$$
$$+ 24A_0(A_1^* b_{n+1} + A_1 b_{n-1}) + 12(A_1^{*2} b_{n+2} + A_1^2 b_{n-2}) = 0, \quad n = 0, 1, \ldots$$

Substituting $b_n \propto e^{\mu t}$, and writing $\psi_0 \equiv -1 + 3Y_0 + Y_1 + \lambda^2$, we have

$$\{-(\mu + \lambda + i\Omega_n)^2 + \psi_0 + \lambda^2\}b_n + 24A_0(A_1^* b_{n+1} + A_1 b_{n-1}) + 12(A_1^{*2} b_{n+2} + A_1^2 b_{n-2}) = 0. \quad (C.1)$$

First assume that a single mode of fluctuations is dominant. Without losing generality, we can assume that a_0 is dominant (see appendix A). Then the hierarchy of equations starts with

$$\{-(\mu + \lambda + i\Omega)^2 + \psi_0\}b_0 + 24A_0(A_1^* b_1 + A_1 b_{-1}) + 12(A_1^{*2} b_2 + A_1^2 b_{-2}) = 0. \quad (C.2)$$

If only three modes b_0 and $b_{\pm 1}$ are retained, (C.1) gives approximately

$$b_1 = \frac{-24A_0 A_1 b_0}{-(\mu + \lambda + i\Omega)^2 + \psi_0}, \quad b_{-1} = \frac{-24A_0 A_1^* b_0}{-(\mu + \lambda + i\Omega_{-1})^2 + \psi_0}. \quad (C.3)$$

To be consistent, the ratios $|b_{\pm 1}/b_0|$ must be small, and this requires

$$6Y_0 Y_1 \ll |(\omega + \lambda + i\Omega_{\pm 1})^2 - \psi_0|^2. \quad (C.4)$$

Substitution of (C.3) in (C.2) gives

$$-(\mu + \lambda + i\Omega)^2 + \psi_0 + \frac{6Y_0 Y_1}{(\mu + \lambda + i\Omega)^2 - \psi_0} + \frac{6Y_0 Y_1}{(\mu + \lambda + i\Omega_{-1})^2 - \psi_0} = 0. \quad (C.5)$$

ψ_0 is of order 0.1 along the branch PQ in fig. 5. The condition (C.4) together with (C.5) then indicates that $|\mu + \lambda + i\Omega|^2$ must also be of order 0.1. Thus only those solutions of (C.5) that comply with

$$(\mu + \lambda)^2 \ll 1, \quad \Omega^2 \ll 1 \quad (C.6)$$

are relevant. The imaginary part of (C.5) takes the form

$$(\mu + \lambda)\left[\Omega + \frac{6Y_0 Y_1 \Omega_1}{\left\{(\mu + \lambda)^2 - \Omega_1^2 - \psi_0^2\right\}^2 + \left\{2\Omega_1(\mu + \lambda)\right\}^2} \right.$$
$$\left. + \frac{6Y_0 Y_1 \Omega_{-1}}{\left\{(\mu + \lambda)^2 - \Omega_{-1}^2 - \psi_0^2\right\}^2 + \left\{2\Omega_{-1}(\mu + \lambda)\right\}^2}\right] = 0. \qquad (C.7)$$

Eq. (C.7) has an obvious solution $\mu = -\lambda$, in which case no instability occurs. The second solution is $\Omega = 0$, and corresponds to soft-mode fluctuations. We have already seen that any state on PQ is soft-mode stable. It is easy to show that (C.7) has no other solutions that comply with (C.6). We can thus conclude that, so far as a single mode of fluctuations is dominant, any state on PQ is hard-mode stable.

Next consider the case, where the two modes b_0 and b_1 couple resonantly with each other. If all other modes are ignored, (C.1) reduces to

$$\left\{-(\mu + \lambda + i\Omega)^2 + \psi_0\right\} b_0 + 24 A_0 A_1^* b_1 = 0,$$
$$24 A_0 A_1 b_0 + \left\{-(\mu + \lambda + i\Omega_1)^2 + \psi_0\right\} b_1 = 1.$$

The characteristic equation is

$$\left\{-(\mu + \lambda + i\Omega)^2 + \psi_0\right\}\left\{-(\mu + \lambda + i\Omega_1)^2 + \psi_0\right\} = 6Y_0 Y_1,$$

which is divided into two parts,

$$(\mu + \lambda)(\Omega + \Omega_1)\left\{-(\mu + \lambda)^2 + \Omega\Omega_1 + \psi_0\right\} = 0, \qquad (C.8)$$
$$\left\{-(\mu + \lambda)^2 + \Omega^2 + \psi_0\right\}\left\{-(\mu + \lambda)^2 + \Omega_1^2 + \psi_0\right\} - 4(\mu + \lambda)^2 \Omega\Omega_1 = 6Y_0 Y_1. \qquad (C.9)$$

Eq. (C.8) has three solutions. The first one $\mu = -\lambda$ corresponds to stable fluctuations. The second one $\Omega + \Omega_1 = 0$, or $\Omega = -\omega/2$ corresponds to period-doubling fluctuations. In this case (C.9) simplifies to

$$C(\mu) \equiv (\mu + \lambda)^4 + 2(\omega^2/4 - \psi_0)(\mu + \lambda)^2 + (\omega^2/4 + \psi_0)^2 - 6Y_0 Y_1 = 0. \qquad (C.10)$$

The quantity $\omega^2/4 - \psi_0$ is positive along PQ, which implies that $C(\mu)$ has a single minimum at $\mu = -\lambda$, and goes to ∞ as $\mu \to \pm\infty$. Hence the inequality $C(0) < 0$ is a necessary and sufficient condition for (C.10) to have one real positive solution for μ. Namely, this is the condition for the so-called period-doubling bifurcation. Note in passing that the ratio $|b_1/b_0|$ is exactly unity at the critical point $C(0) = 0$. The third solution is given by $(\mu + \lambda)^2 = \Omega\Omega_1 + \psi_0$, but when coupled with (C.9), this does not give rise to real solutions for μ and Ω.

We have assumed that Ω is incommensurate with ω, whereas the destabilized mode has frequency $\omega/2$, which is clearly commensurate with Ω. Thus the analysis seems to have apparent internal inconsistencies. However, the consistent analysis carried out in subsection 4-4 shows that the period-doubling bifurcation really occurs when $C(0) < 0$ is satisfied.

References

[1] R.M. May, Nature 261 (1976) 459.
[2] S. Grossmann and S. Thomae, Z. Naturforsch. 32a (1978) 1353.
[3] M.J. Feigenbaum, J. Stat. Phys. 19 (1978) 25.
[4] M.J. Feigenbaum, J. Stat. Phys. 21 (1979) 669.
[5] K. Tomita, Phys. Reports 86 (1982) 113.
[6] R. Shaw, Z. Naturforsch. 369 (1981) 80.
[7] J.P. Eckmann, Rev. Mod. Phys. 53 (1981) 643.
[8] H.L. Swinney, Physica 7D (1983) 3.
[9] M.J. Feigenbaum, Physica 7D (1983) 16.
[10] P. Holmes, Philos. Trans. R. Soc. A292 (1979) 419.
[11] B.A. Huberman and J.P. Crutchfield, Phys. Rev. Lett. 43 (1979) 1743.
[12] U. Parlitz and W. Lauterborn, Phys. Lett. A. 107 (1985) 351.
[13] D. D'Humieres, M.R. Beasley, B.A. Huberman, and A. Libchaber, Phys. Rev. A26 (1982) 3483.
[14] C. Hayashi, Int. J. Non-Linear Mech. 15 (1980) 341.
[15] Y. Ueda, J. Stat. Phys. 20 (1979) 181.
[16] D.Y. Hsieh, J. Math. Phys. 16 (1975) 275.
[17] A. Itoh, Prog. Theor. Phys. 61 (1979) 815.
[18] G.B. Whitham, Linear and Nonlinear Waves (Wiley, New York, 1973).
[19] V. I. Arnol'd, Geometrical Methods in the Theory of Ordinary Differential Equations (Springer, New York, 1977).
[20] N.N. Bogoliubov and Y.A. Mitropolsky, Asymptotic Methods in the Non-Linear Oscillations (Gordon and Breach, New York, 1961).
[21] L.D. Landau and E.M. Lifshitz, Mechanics (Pergamon, Oxford, 1960).

The refined approximate criterion for chaos in a two-state mechanical oscillator

W. Szemplińska-Stupnicka, Warszawa

Summary: In the two-state mechanical oscillator, a mathematical model of a buckled beam, the refined criterion for the system parameter critical values where chaotic motion can be expected is derived. The derivation is based on the assumption of the second approximate solution for the small orbit. It is shown that a simple approximate analysis of the Hill's type variational equation gives the sought stability loss of the resonant solution as the period doubling bifurcation. The stability limits of the resonant and non-resonant solutions are proposed as the boundary of the region where strange phenomena can appear and the refined criterion thus derived is compared to computer simulation results and to other approximate criteria.

Verfeinertes Kriterium für chaotische Bewegung in einem Schwingungssystem mit zwei stabilen Gleichgewichtslagen

Übersicht: Für ein mechanisches Schwingungssystem mit zwei stabilen Gleichgewichtslagen, das ein mathematisches Modell eines Knickstabes darstellt, wird ein verfeinertes Kriterium für die kritischen Systemparameterwerte, bei denen chaotische Bewegung zu erwarten ist, hergeleitet. Die Herleitung geht von der Näherungslösung zweiter Ordnung für kleine Bahnen um das Gleichgewicht aus. Es wird gezeigt, daß eine einfache Näherungsbehandlung der Variationsgleichung vom Hill-Typ den gesuchten Stabilitätsverlust der Resonanzlösung als Verzweigung der Periodenverdopplung liefert. Die Stabilitätsgrenzen der resonanten und nichtresonanten Lösungen werden als Bereichsgrenze, wo chaotische Bewegungen auftreten können, vorgeschlagen. Das derart hergeleitete verfeinerte Kriterium wird mit Computer-Simulationen und anderen Näherungskriterien verglichen.

1 Introduction

Phenomena of chaotic motion in a two-state oscillator governed by an equation in nondimensional form written as

$$\ddot{x} + h\dot{x} - (1 - x^2)\, x/2 = P \cos(\nu t),$$

have been studied extensively in recent literature. In mechanics the equation is interpreted as a mathematical model of a buckled beam and the theoretical and computer analysis is complemented with experiments [1–8].

The system may exhibit two types of periodic motion: oscillations around one of the two stable equilibrium positons (small orbits) and oscillations around all three rest positons $x_{0;1} = 0$, $x_{0;2,3} = \pm 1$, named large orbit (Fig. 1). Chaotic steady-state motion is that as sketched in Fig. 2. Roughly speaking, it manifests itself as irregular hoppings between oscillations around the two stable rest points. Since the time history $x(t)$ does not show any regularity, it is customary to characterize the motion by the aid of a Poincaré map. The Poincaré map, which can be viewed as a stroboscopic picture of the phase portrait plotted at instants of time separated by the period $T = 2\pi/\nu$, shows a complicated structure called strange attractor [2]. Detail observations of routes from the regular periodic motion to the chaotic one reveal that very often a sequence of period doubling bifurcations of the T-periodic small orbit oscillations precedes the chaotic behaviour.

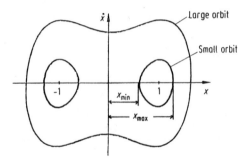

Fig. 1. Two types of periodic oscillations: small and large orbit motions

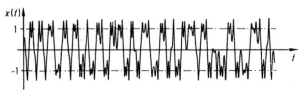

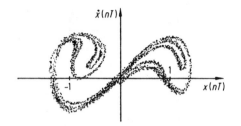

Fig. 2. Chaotic motion: time history and Poincaré map

An estimation of the system parameter critical values $(P, \nu, h)_{cr}$ which give rise to the chaotic phenomena, i.e. a criterion for chaos, is one of numerous problems waiting for further investigations in the field. The criteria considered so far originated from considerations of stability of the T-periodic small orbit oscillations.

The Melnikov criterion should be mentioned first [2, 4, 9]. Here the conclusions on the instability of the small orbit motion are drawn upon considering the structure of Poincaré sections of stable and unstable manifolds of the saddle point. At sufficiently low value of the forcing parameter the unstable manifold W_u is attracted by fixed points $1, 2$ in Fig. 3a. Thus the T-periodic small orbits are stable. As the parameter P/h increases, the stable and unstable manifolds approach each other and eventually intersect themselves transversally (Fig. 3c). The situation when the two manifolds touch each other (see Fig. 3b) is said to be critical — the T-periodic small orbit begins to be unstable. To derive a criterion on separation of the stable and unstable manifolds Mielnikov considered a slightly disturbed Hamiltonian system written as

$$\ddot{x} + f(x) = \varepsilon g(x, \dot{x}, \nu t), \tag{0.1}$$

and proved that the function which provides a good measure of the distance between W_u and W_s exists and can be constructed as power series in the small parameter

$$M_\varepsilon(t_0) = \varepsilon M_1(t_0) + \sum_{n=2,3\ldots} \varepsilon^n M_n(t_0). \tag{0.2}$$

The first term in the series is referred to as Melnikov function $M_1(t_0)$ and is used to estimate critical values of the system parameters, the values providing necessary conditions for chaos to occur.

Another approach to a criterion for chaos was proposed by J. Rudowski and the author in [10]. The starting point was an assumption of harmonic solution for the small orbit motion and a study of its stability by the aid of methods used in the approximate theory of nonlinear oscillations. The criterion thus derived provided simple explicit formulae for the critical parameter values and showed good coincidence with the computer simulation results. A weak point of the theory was the loss of stability of the resonant branch of the considered approximate solution.

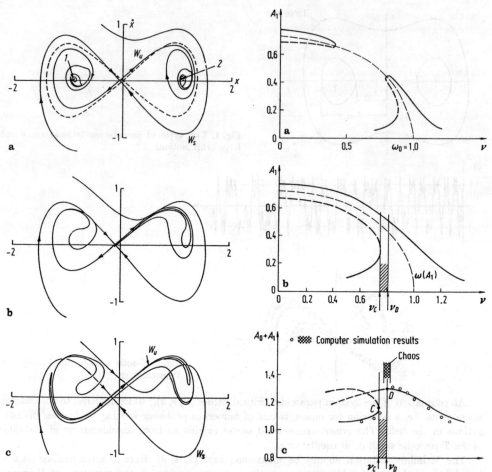

Fig. 3a—c. Poincaré maps showing stable and unstable manifolds of the saddle point near (0, 0), from [4]

Fig. 4a—c. Resonance curves of the first approximate solution; a $A_1(\nu)$ at $P/h < (P/h)_1$, b $A_1(\nu)$ at $P/h > (P/h)_1$, c $A_1 + A_0 \equiv A_{\max}(\nu)$ at $P/h > (P/h)_1$

While the computer simulation indicated clearly that the loss of stability was due to period doubling bifurcation, this type of instability could not be proved for the assumed first approximate solution in the neighbourhood of the considered principal resonance frequency range.

In this paper a refined approximate criterion is developed. This is based on the second approximate solution for the small orbit, the solution which includes the second harmonic component in addition to the fundamental one and the constant term. Now the approximate analysis of Hill's type variational equation gives the period doubling in the region of the principal resonance. The refined approximate criterion for chaos thus derived is compared with computer simulation results and with the two theoretical criteria mentioned above: Melnikov's criterion and the first approximate criterion. The presented numerical examples suggest that the refined theory provides good estimation of the system parameters in the $P - \nu$ plane where chaotic motion appears, while they may give rise to some doubts on the usefulness of the Melnikov criterion.

A brief review of the first approximate criterion is presented first.

2 The first approximate criterion for chaos

The considered two-state mechanical system is governed by an equation written in a non-dimensional form as

$$\ddot{x} + h\dot{x} - (1 - x^2)\, x/2 = P \cos(\nu t - \vartheta), \qquad h > 0 \tag{1}$$

where x denotes the displacement from the unstable rest point $x_{0;1} = 0$ (saddle point). The system has two stable equilibrium positions

$$x_{0;2,3} = \mp 1,$$

and the oscillations around them (small orbit motion) are governed by the equations

$$\ddot{z} + \omega_0^2 z + h\dot{z} \mp \frac{3}{2} z^2 + \frac{1}{2} z^3 = P \cos(\nu t - \vartheta), \qquad \omega_0 = 1, \tag{2}$$

where

$$z = x \mp 1.$$

The first approximate criterion for chaos was constructed upon considering the first approximate periodic solution of (1), (2) near the principal resonance of the small orbit, i.e., at $\nu \approx \omega_0$, in the form

$$\begin{aligned} x(t) &= A_0 + A_1 \cos(\nu t) \quad \text{or} \\ z(t) &= \bar{a}_0 + A_1 \cos(\nu t), \qquad \bar{a}_0 = A_0 \mp 1. \end{aligned} \tag{3}$$

The large orbit motion was described by a single harmonic function as

$$x(t) = A \cos(\nu t).$$

Although it was clear that the legitimacy of the first approximate solution is restricted to small values of the forcing parameter P/h only, the question was put what information can be gained and what conclusions on the possibility of an appearance of strange phenomena can be drawn from the stability analysis of the rough approximation.

First the resonance curves $A_1(\nu)$ and $A_0(\nu)$ were determined by the aid of the harmonic balance procedure. It come out that A_1, A_0, ϑ are given by the formulae

$$A_1 = \frac{P}{\sqrt{\left(1 - \frac{15}{8} A_1^2 - \nu^2\right)^2 + h^2 \nu^2}}, \qquad \operatorname{tg} \vartheta = \frac{-h\nu}{1 - \frac{15}{8} A_1^2 - \nu^2}, \tag{4a}$$

$$A_0^2 = 1 - \frac{3}{2} A_1^2. \tag{4b}$$

By (4a) the behaviour of the softening type of Duffing's oscillator is described with the natural frequency defined as

$$\omega_1(A_1) = 1 - \frac{15}{8} A_1^2, \tag{4c}$$

and hence one expects to obtain typical resonance curves as that sketched in Fig. 4a. Indeed the solutions behave that way provided the forcing parameter P/h does not exceed the critical value

$$\left(\frac{P}{h}\right)_1 = \left(1 - \frac{h^2}{4}\right) \sqrt{\frac{2}{15}}. \tag{5a}$$

At higher values of the parameter P/h the structure of the resonance curves is broken so that no peak resonant amplitude exists (Fig. 4b). The conclusion was drawn that the approximate solution (3) fails to predict the fundamental harmonic amplitude and the attention was focused on another parameter of motion, namely, on the maximal value of the coordinate defined as

$$x_{\max} = A_0 + A_1 \equiv A_{\max}(\nu). \tag{5b}$$

It appeared that true maximal displacements of the small orbit solution were very close to the theoretical values even at high P/h parameter values where chaotic motion was observed (see Fig. 4c).

Consequently the further analysis was focused on this parameter of motion and the zone of frequency between stability limits — point C of the vertical tangent and point D of the horizontal

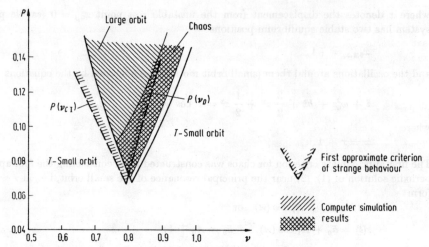

Fig. 5. Region of chaotic motion in the $P - \nu$ plane ($h = 0.0168$)

tangent in Fig. 4c — was suggested as the zone where strange phenomena can be expected. Since the two characteristic frequencies were given as

$$\nu_C^2(h=0) = 1 - \frac{3}{2}\sqrt[3]{\frac{15}{4}P^2}, \qquad \nu_D^2 = \frac{1}{2}\left(1 - h^2 + \sqrt{h^4 - 2h^2 + 15P^2}\right) \qquad (6\text{a})$$

the boundaries of the region with

$$\nu_C < \nu < \nu_D \qquad (6\text{b})$$

were defined by very simple algebraic relations. This is illustrated in Fig. 5 where the true (true meaning determination by digital computer simulation) region of chaotic motion and the theoretical region satisfying the condition (6b) are drawn in the $P - \nu$ plane. The results show surprisingly good coincidence for values of the P parameter close to the lowest critical value P_{cr}, in particular. The true and theoretical chaotic motion zone is also marked in Fig. 4c. Here we see that, indeed, chaos develops from the resonant branch of the small orbit oscillations near the frequency of maximum value of $A_0 + A_1$ (point D) and the chaotic zone can be interpreted as the transition zone between resonant and nonresonant state, replacing the classical jump phenomenon.

While the coincidence of the theoretical and true chaotic region can be regarded as satisfactory, the recognition of point D as the stability limit of the assumed solution is a weak point in presented the theory. The problem is that computer simulation shows evidently that the loss of stability of the T-periodic resonant solution is due to period doubling bifurcation. This cannot be proved for the first approximate solution (3) considered so far.

Further analysis of the computer simulation results indicated that just before the sequence of period doublings the second harmonic component began to grow and contributed significantly to the signal. The observation brought the idea that including the second harmonic component nto the approximate solution might lead to the theoretical explanation of the character of tability loss and to more a accurate criterion for chaos.

3 The second approximate solution

To derive the T-periodic solution involving the second harmonic in addition to the constant term and the fundamental harmonic component a perturbation procedure is applied here. Although the harmonic balance method might have been used as well, it would have required numerical calculations to solve a set of five nonlinear algebraic equations for the unknown three amplitudes and two phases angles. Thus we shift to the perturbation procedure to obtain an explicit form solution. But the procedure is applied in such a way that results are consistent with those obtained in Sect. 2. The condition is satisfied, if appropriate orders of smallness are set down to nonlinear terms in (1). It comes out that the quadratic term is supposed to be much larger than the cubic terms and consequently we rewrite (1) into the form

$$\ddot{z} + \omega_0^2 z + \frac{3}{2} \mu z^2 + \mu^2 \left[\bar{h}\dot{z} + \frac{1}{2} z^3 - \bar{P} \cos(\nu t - \vartheta) \right] = 0 \tag{7}$$

where

$$\mu^2 \bar{h} = h, \quad \mu^2 \bar{P} = P, \quad \omega_0 = 1, \quad 0 < \mu \ll 1.$$

Then the T-periodic solution is sought as power series in the small parameter μ:

$$z = a_1 \cos(\nu t + \varphi) + \mu z_1(t) + \mu^2 z_2(t) + \mu^3 \ldots,$$

$$\frac{da_1}{dt} = \mu D_1(a_1, \vartheta) + \mu^2 D_2(a_1, \vartheta) + \mu^3 \ldots, \tag{8}$$

$$\frac{d\varphi}{dt} = \omega_0 - \nu + \mu E_1(a_1, \vartheta) + \mu^2 E_2(a_1, \vartheta) + \mu^3 \ldots$$

Because the phase angle ϑ is shifted into the forcing term the new phase angle φ for the steady-state solution satisfies the condition

$$\varphi = 0 \quad \text{at} \quad \frac{d\varphi}{dt} = \frac{da_1}{dt} = 0. \tag{9}$$

On inserting (8) and into (7) and equating terms of same power of μ we arrive at the following equations for $z_1(t), z_2(t)$:

$$\ddot{z}_1 + \omega_0^2 z_1 = 2D_1 \omega_0 \sin \Theta + 2E_1 a_1 \omega_0 \cos \Theta - \frac{3}{4} a_1^2 - \frac{3}{4} a_1^2 \cos(2\Theta), \tag{10a}$$

$$\ddot{z}_2 + \omega_0^2 z_2 = (2\omega_0 D_2 + 2D_1 E_1 + \bar{h} a_1 \omega_0) \sin \Theta + (2E_2 \omega_0 + E_1^2)$$

$$\times a_1 \cos \Theta - 3a_1 z_1(t) \cos \Theta - \frac{1}{2} a_1^3 \cos^3 \Theta + \bar{P} \cos(\Theta - \vartheta) \tag{10b}$$

with $\Theta = \nu t = \omega_0 t$. The unknown coefficients D_1, E_1 are determined from the condition of supressing secular terms in (10a) with the results

$$D_1 = E_1 = 0. \tag{11a}$$

It follows that the correction function $z_1(t)$ is

$$z_1(t) = -\frac{3}{4} \frac{a_1^2}{\omega_0^2} + \frac{1}{4} \frac{a_1^2}{\omega_0^2} \cos(2\Theta). \tag{11b}$$

Then by making use of (11) the conditions of vanishing secular terms in the solution for $z_2(t)$ are obtained by equating to zero the coefficients of $\sin \Theta$ and $\cos \Theta$ in the Fourier expansions of the right hand side of (10b):

$$2E_2 \omega_0 a_1 - \frac{3}{8} a_1^3 + \frac{9}{4} \frac{a_1^3}{\omega_0^2} - \frac{3}{8} \frac{a_1^3}{\omega_0^2} + \bar{P} \cos \vartheta = 0, \tag{12a}$$

$$\bar{h} a_1 \omega_0 + \bar{P} \sin \vartheta + 2\omega_0 D_2 = 0$$

where the term $-\dfrac{8}{3} \dfrac{a_1^3}{\omega_0^2}$ accounts for the second harmonic component in the correction function $z_1(t)$. Now it is convenient to make use of the third equation of (8) and the steady-state condition (9) with the result

$$\mu^2 E_2 = \nu - \omega_0, \quad D_2 = 0, \tag{12b}$$

and to introduce the approximation

$$(\omega_0 + \mu^2 \Delta \omega)^2 \approx \omega_0^2 + 2\mu^2 \omega_0 \Delta \omega. \tag{12c}$$

With (12b, c) it is possible to transform (12a) into

$$a_1[\omega_{II}^2(a_1) - \nu^2] = P\cos\vartheta, \qquad -ha_1\nu = P\sin\vartheta, \tag{12d}$$

where

$$\omega_{II}^2(a_1) = \omega_0^2 + \mu^2 a_1^2\left(\frac{3}{8} - \frac{15}{8\omega_0^2}\right).$$

Here $\omega_{II}^2(a_1)$ denotes the second approximate natural frequency of the system. At the value $\omega_0 = 1$ it is reduced to

$$\omega_{II}^2(a_1) = 1 - \frac{3}{2}\mu^2 a_1^2. \tag{13a}$$

With this notation we arrive at formulae analogous to that found in Sect. 2, see (4a):

$$a_1 = \frac{P}{\sqrt{[\omega_{II}^2(a_1) - \nu^2]^2 + h^2\nu^2}}, \qquad \operatorname{tg}\vartheta = \frac{-h\nu}{\omega_{II}^2(a_1) - \nu^2}, \tag{13b}$$

and the second approximate solution for z and x are obtained, respectively, as

$$z_{II}(t) = a_1\cos(\nu t) + \mu z_1(t) = a_0 + a_1\cos(\nu t) + a_2\cos(2\nu t),$$
$$x_{II}(t) = 1 + a_0 + a_1\cos(\nu t) + a_2\cos(2\nu t), \tag{13c}$$
$$a_0 = -\frac{3}{4}\mu a_1^2, \qquad a_2 = \frac{1}{4}\mu a_1^2.$$

To examine the consistency of the solution thus obtained with that found by the harmonic balance procedure two conditions have to be satisfied:

(i) The equation of motion (7) has to be reduced to (2) i.e., we have to put $\mu = 1$.
(ii) The second harmonic component in $z_1(t)$ has to be ignored. Consequently the term $-\frac{3}{8} \times \frac{a_1^2}{\omega_0^2}$ in (12a) is dropped.

Then the resulting two term solution determined by the perturbation method is reduced to

$$z_I(t) = a_0 + a_1\cos(\nu t), \qquad x_I(t) = 1 + a_0 + a_1\cos(\nu t), \qquad a_0 = -\frac{3}{4}a_1^2 \tag{14a}$$

where a_1 is now determined by (13b) with $\omega_{II}^2(a_1)$ replaced by

$$\omega_I^2(a_1) = 1 - \frac{15}{8}a_1^2. \tag{14b}$$

We readily notice that the solution for the fundamental harmonic amplitude is identical with that found by the harmonic balance method, i.e.,

$$a_1 \equiv A_1. \tag{15a}$$

To compare the constant term in the two solutions (3, 4) and (14) one should notice that the perturbation procedure assumes the following approximation to be valid:

$$(1 + a_0)^2 \approx 1 + 2a_0. \tag{15b}$$

If the approximation is accepted, one arrives at the result

$$(1 + a_0)^2 = 1 - \frac{3}{2}a_1^2 = 1 - \frac{3}{2}A_1^2. \tag{15c}$$

It follows that the constant term of (14a) is equal to that determined by (4b):

$$(1 + a_0)^2 = A_0^2. \tag{15d}$$

Now it has been shown that the solution determined by the perturbation scheme is consistent with that obtained by the harmonic balance method considered in Sect. 2, and we turn back to the second approximate solution (13) involving the second harmonic component and examine the structure of the resonance curves first.

From (13a, b) we easily find out that the resonance curves $a_1(\nu)$ have a structure analogous to that of the first approximate solution discussed in Sect. 2. Namely, the peak resonant amplitude exists as long as the forcing parameter does not exceeds a certain critical value. The critical value $(P/h)_1$ is now slightly higher than that given by (5) and is

$$\left(\frac{P}{h}\right)_1 = \left(1 - \frac{h^2}{4}\right)\frac{1}{\sqrt{6}}. \tag{16a}$$

At higher values of P/h where chaotic motion appears, the resonance curves are open as those for undamped systems.

It is also worth noticing that the maximum displacement of the coordinate x, which for the solution (13) is

$$x_{\text{maxII}} = 1 + a_0 + a_1 + a_2 = 1 + a_1 - \frac{1}{2}a_1^2, \tag{16b}$$

does not have a horizontal tangent point when considered as function of frequency ν. Indeed the condition

$$\frac{dx_{\max}}{d\nu} = 0 \quad \text{at} \quad \frac{da_1}{d\nu} \neq 0, \quad \frac{da_1}{d\nu} \neq \infty$$

is satisfied at $a_1 = 1$ and this corresponds to

$$\omega_{\text{II}}^2(a_1) = -\frac{1}{2}.$$

4 Stability of the second approximate solution — period doubling bifurcation

The first order instability regions of solution (13) do not need long considerations: points with vertical tangent on the resonance curves coincide with stability limits of this type. In the considered range of the forcing parameter P/h there is only one such stability limit: this is point C_{II} on the nonresonant branch of the resonance curves. The frequency of the point is determined by formulae analogous to (6a):

$$\nu_{C_{\text{II}}}(h=0) = 1 - \frac{9}{2}a_1^2 \simeq 1 - 2.163 P^{2/3}. \tag{17}$$

The loss of stability of the resonant branch of $a_1(\nu)$ is sought as the period doubling bifurcation, e.g. at the stability limit the T-periodic solution (13) is expected to turn into $2T$-periodic solution.

To examine this form of instability we consider a perturbed solution

$$\tilde{x}(t) = x_{\text{II}}(t) + \delta x \tag{18a}$$

where $x_{\text{II}}(t)$ is given by (13) and the small disturbance $\delta x(t)$ is governed by the linear variational Hill's type equation

$$\delta\ddot{x} + h\,\delta\dot{x} + \delta x\left[\lambda_0 + \sum_{n=1}^{4}\lambda_n \cos(n\nu t)\right] = 0 \tag{18b}$$

where

$$\lambda_0 = 1 - \frac{3}{2}a_1^2 + \frac{57}{64}a_1^4, \quad \lambda_1 = 3a_1 - \frac{15}{8}a_1^3, \quad \lambda_3 = \frac{3}{8}a_1^3, \quad \lambda_2 = \frac{3}{2}a_1^2 - \frac{9}{16}a_1^4,$$

$$\lambda_4 = \frac{3}{64}a_1^4. \tag{18c}$$

Since period of the parametric excitation is $T = 2\pi/\nu$ we may seek a particular solution as [e.g., 11—14]

$$\delta x(t) = e^{\varepsilon t}\varphi(t) \quad (\varepsilon > 0),$$

$$\varphi(t) = \varphi(t + 2T) = \sum_n b_n \cos\left(\frac{n\nu}{2}t + \delta_n\right) \quad (n = 1, 3, 5). \tag{19a}$$

This is the type of instability that may lead to period doubling bifurcation because it brings a build-up of harmonic components of the period $2T$.

In the approximate approach used here one is interested in getting good results upon taking as low number of terms in the Fourier series (19a) as possible. Thus we can start with the solution at the stability limit as

$$\delta x(t)|_{\varepsilon=0} = b_1 \cos\left(\frac{\nu}{2} t + \delta_1\right).$$

It comes out, however, that it gives good approximation close to the region of frequency $\nu \approx 2\omega_0$ and analogous result can be obtained by the averaging method in the second approximation [16].

In this paper we assume the two frequency solution at the stability limit as

$$\delta x(t)|_{\varepsilon=0} = b_{1c} \cos\left(\frac{\nu}{2} t\right) + b_{1s} \sin\left(\frac{\nu}{2} t\right) + b_{3c} \cos\left(\frac{3}{2} \nu t\right) + b_{3s} \sin\left(\frac{3}{2} \nu t\right). \quad (19\text{b})$$

Then we insert (19b) into (18) to find the residual $E(t)$ and apply the harmonic balance principle, i. e. we equate separately to zero coefficients of $\cos(n\nu t/2)$, $\sin(n\nu t/2)$ in the Fourier expansion of $E(t)$. Thus a set of algebraic linear homogeneous equations for b_{nc}, b_{ns} ($n = 1, 3$) is obtained:

$$\begin{aligned}
& b_{1c}\left(\lambda_0 - \frac{\nu^2}{4} + \frac{\lambda_1}{2}\right) + b_{1s} h \frac{\nu}{2} + b_{3c} \frac{\lambda_1 + \lambda_2}{2} = 0, \\
& -b_{1c} h \frac{\nu}{2} + b_{1s}\left(\lambda_0 - \frac{\nu^2}{4} - \frac{\lambda_1}{2}\right) + b_{3s} \frac{\lambda_1 - \lambda_2}{2} = 0, \\
& b_{1c} \frac{\lambda_1 + \lambda_2}{2} + b_{3c}\left(\lambda_0 - \frac{9}{4} \nu^2 + \frac{\lambda_3}{2}\right) + b_{3s} \frac{3}{2} \nu h = 0, \\
& b_{1s} \frac{\lambda_1 - \lambda_2}{2} - b_{3c} h \frac{3}{2} \nu + b_{3s}\left(\lambda_0 - \frac{9}{4} \nu^2 - \frac{\lambda_3}{2}\right) = 0.
\end{aligned} \quad (19\text{c})$$

The zero value of the characteristic determinant of the set provides the relation between ν_1, a_1, h to be satisfied at the stability limit:

$$\Delta = \bar{f}(\nu, h, \lambda_1, \lambda_2, \lambda_3) = f(\nu, h, a_1) = 0. \quad (19\text{d})$$

In the general case $h \neq 0$ this is the fourth order determinant and it leads to a high order polynomial in ν. Here we confine our consideration to $h = 0$ and then examine whether damping can bring significant effects in the considered frequency region. With $h = 0$ the condition (19b) takes the form

$$\begin{vmatrix} \lambda_0 - \dfrac{\nu^2}{4} + \dfrac{\lambda_1}{2} & \dfrac{\lambda_1 + \lambda_2}{2} \\ \dfrac{\lambda_1 + \lambda_2}{2} & \lambda_0 - \dfrac{9}{4} \nu^2 + \dfrac{\lambda_3}{2} \end{vmatrix} = 0. \quad (20\text{a})$$

Therefore the frequency of the bifurcation point is determined as a root of the second order polynomial in ν^2

$$\nu^4 - 2B\nu^2 + C = 0 \quad (20\text{b})$$

where

$$B = \frac{20}{9} \lambda_0 + \lambda_1 + \frac{\lambda_3}{9},$$

$$C = \frac{16}{9}\left[\left(\lambda_0 + \frac{\lambda_1}{2}\right)\left(\lambda_0 + \frac{\lambda_3}{2}\right) - \frac{1}{4}(\lambda_1 + \lambda_2)^2\right]$$

and $\lambda_1, \lambda_2, \lambda_3$ are defined by (18c).

5 Numerical examples and computer simulation results

The resonance curves $a_1 \equiv a_1(\nu)$ for the second approximate solution determined by (13) are drawn in Fig. 6 at two values of the forcing parameter P/h. At $P = 0.04$ the classical structure of the resonance curve is observed. At $P = 0.1$ the condition $P/h > (P/h)_1$, where $(P/h)_1$ is determined by (16a), is satisfied. Hence the resonance curve does not show a peak resonant amplitude and behaves as that of an undamped system. The resonance curve at $h = 0$, $P = 0.1$ was also calculated and, indeed, it can hardly be distinguished from that at $h = 0.1$. Then the boundary of period doubling bifurcation defined by (18c), (20b) is drawn in solid line, the unstable regions denoted as hatched area. We readily see that on decreasing the frequency in the neighbourhood of the principal resonance the T-periodic solution (13) becomes unstable at $\nu = \nu_{PD}$ in the sense determined by (19a, b) e.g., the harmonic components of period $2T$ begin to grow. We conclude that between ν_{PD} and the frequency ν_{CII} corresponding to the vertical tangent point, our T-periodic solution (13) is unstable. This region of frequency is proposed as the refined criterion for system parameters where strange phenomena can be expected — the refined criterion for chaos.

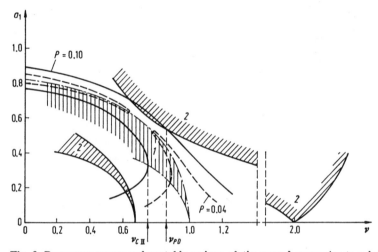

Fig. 6. Resonance curves and unstable regions of the second approximate solution ($h = 0.1$); 1 first unstable region ($h = 0$), 2 the second order instabilities ($h = 0$)

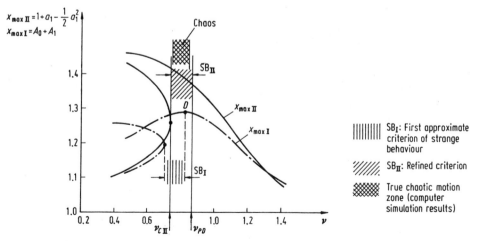

Fig. 7. Resonance curves of the maximal displacement of small orbit ($h = 0.1$; $P = 0.1$)

In Fig. 7 the region of frequency thus determined, the true chaotic motion zone and the first approximate criterion are displayed against the background of the resonance curves of the maximal value x_{max} of the x coordinate, see (16b), (5b). Here we notice that the second approximate solution overestimates x_{max} but it provides a better estimation for the chaotic motion zone. The true chaotic zone falls within the zone calculated by the refined criterion, while it was shifted to the right compared to the first approximate criterion. Then the theoretical and computer simulation results are displayed in the $P - \nu$ plane in Figs. 8 and 9. The theoretical stability limit $P(\nu_{PD})$ drawn here needs some comments. It is seen that it provides a good estimation of the period doubling bifurcation at higher values of ν but it fails to predict it at lower frequency values. To explain and justify the discrepancy it is useful to turn back to Fig. 6 and to interpret the amplitude a_1 as a measure of the parametric modulation λ_n in (18). Thus we readily see that the stability limit ν_{PD} falls into high parametric modulation. Moreover, at low frequency values the parametric modulation is so high that two neighbouring unstable regions appear to be very

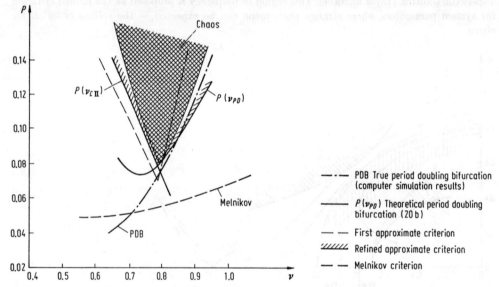

Fig. 8. True chaotic motion zone and the three theoretical criteria in the $P - \nu$ plane ($h = 0.1$)

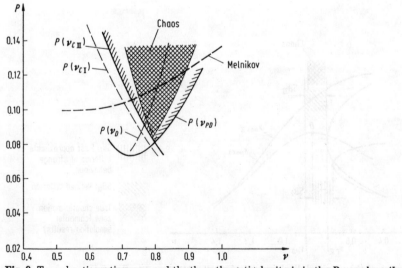

Fig. 9. True chaotic motion zone and the three theoretical criteria in the $P - \nu$ plane ($h = 0.2$)

close to each other. Therefore from the Hill's type equation theory one may conclude that the two term approximation used here is insufficient to estimate the instability correctly. Possible damping effects, neglected in the theory, seem to play no important role because influence of damping is significant at low parametric modulation only.

The fact that the theory presented fails to predict a relation between P and ν for the period doubling bifurcation at $\nu < 0.75$ does not effect, however, the proposed refined criterion for chaos. Indeed, from the point of view of the criterion we are interested at the region where $\nu_{PD} > \nu_{CII}$, e.g., above the point of intersection of theoretical lines for $P(\nu_{PD})$ and $P(\nu_{CII})$ this takes place at $\nu > 0.75$.

Comparison of the two theoretical zones and the true chaotic zone in the $P - \nu$ plane evidently shows that the refined criterion provides a good estimation of the system parameters where chaos can be expected.

It is interesting to compare the results with the Melnikow criterion. The explicit formulae for the critical value of the forcing term were derived in [2] and for the considered system applied in [3]. The criterion says that the necessary condition for chaos is fulfilled, if P exceeds the critical value

$$P_{cr} = h \frac{\sqrt{2}}{\nu 2\pi} \cosh \frac{\pi \nu}{\sqrt{2}}.$$

This critical value is also drawn in Figs. 8 and 9. It is evident that the criterion does not give satisfactory estimation for chaotic region.

6 Concluding remarks

For the two-state mechanical oscillator governed by (1) the search for a simple approximate criterion for the system parameters where chaotic motion can be expected began in [11] and was continued here. In this paper the second approximate T-periodic solution of the small orbit motion was considered and stability analysis provided the refined criterion for chaos.

On accounting for the second harmonic component in the periodic solution the approximate analysis of the variational Hill's type equation made it possible to prove that the stability loss of the resonant branch of the resonance curve is due to period doubling bifurcation. Therefore the approximate theoretical model shows the property of the true system examined by computer simulation in [11]. The theoretical analysis was confined to the evaluation of the first period doubling bifurcation critical point where the $2T$-periodic components began to grow with time but analogous analysis reported in [14, 15] indicated that the same approximate procedure allows to prove further period doublings.

It is worth mentioning that the applied approximate procedure was reduced to very simple algebraic calculations: the period doubling frequency was defined as a root of the second order polynomial.

The boundary of the region on the forcing term-frequency plane where the second approximate T-periodic solution is unstable has been proposed as the criterion for the system parameters where strange phenomena can appear. The theoretical strange behaviour regions thus defined and plotted in the $P - \nu$ plane show very good coincidence with the true chaotic motion zone.

On the contrary the Melnikov criterion does not seem to provide satisfactory estimation of the chaotic zone, in particular at higher values of the damping coefficient (Fig. 9). The results will not be of a great surprise, however, when one realizes that the criterion was derived from the first term in the power series expansion of the complete Melnikov function, see (0.2).

References

1. Holmes, P.: Strange phenomena in dynamical systems and their physical implications. Appl. Math. Modelling 1 (1977) 362—366
2. Holmes, P.: A nonlinear oscillator with a strange attractor. Phil. Trans. R. Soc. London Ser. A 292 (1979) 419—448

3. Moon, F. C.: Experiments on chaotic motion of a forced nonlinear oscillator: Strange attractors. J. Appl. Mech. 47 (1980) 638–644
4. Guckenheimer, J., Holmes, P.: Nonlinear oscillations, dynamical systems and bifurcations of vector fields. Berlin, Heidelberg, New York: Springer 1983
5. Holmes, P.; Moon, F. C.: Strange attractors and chaos in nonlinear mechanics. J. Appl. Mech. 50 (1983) 1021–1032
6. Tseng, W. Y.; Dugundij, J.: Nonlinear vibrations of a buckled beam under harmonic excitation. J. Appl. Mech. 38 (1971) 467–476
7. Moon, F. C.; Li, G.-X.: Fractial basis boundaries and homoclinic orbits for periodic motion in a two-well potential. Phys. Rev. Lett. 55 (1985) 1439–1492
8. Moon, F. C.; Li, G.-X.: The fractial dimension of the two-well potential strange attractor. Physica D 17 (1985) 98–108
9. Melnikov, V. K.: On the stability of the center for time periodic perturbation. Trans. Mosc. Math. Soc. 12 (1963) 1–57
10. Rudowski, J.; Szemplińska-Stupnicka, W.: On an approximate criterion for chaotic motion in a model of a buckled beam. Ing. Arch. 57 (1987) 243–255
11. Hayashi, Ch.: Nonlinear oscillations in physical systems. New York: McGraw-Hill 1964
12. Bolotin, W. W.: Dynamic stability of elastic systems. San Francisco: Holden Day 1964
13. Szemplińska-Stupnicka, W.; Bajkowski, J.: The 1/2 subharmonic resonance and its transition to chaotic motion in a non-linear oscillator. Int. J. Non-Linear Mech. 21 (1986) 401–419
14. Szemplińska-Stupnicka, W.: Secondary resonances and approximate models of transition to chaotic motion in nonlinear oscillators. J. Sound Vib. 183 (1987) 155–172

Received July 20, 1987

Prof. W. Szemplińska-Stupnicka
Polish Academy of Sciences
Institute of Fundamental Technological Research
Swietokrzyska 21
PL-00-049 Warszawa
Poland

ANALYTICAL CONDITION FOR CHAOTIC BEHAVIOUR OF THE DUFFING OSCILLATOR

Tomasz KAPITANIAK [1]

Department of Applied Mathematical Studies and Centre for Nonlinear Studies, University of Leeds, Leeds LS2 9JT, UK

Received 2 November 1989; revised manuscript received 3 January 1990; accepted for publication 4 January 1990
Communicated by A.P. Fordy

A new analytical criterion is developed for the strange chaotic attractor in nonlinear oscillators which show a period-doubling route to chaos. The method is based on approximate analysis and Feigenbaum universal properties of period doubling.

It is well known that the Duffing oscillator

$$\ddot{x} + a\dot{x} + bx + cx^3 = B_0 + B_1 \cos \Omega t \qquad (1)$$

shows chaotic behaviour for certain values of the parameters [1–5]. In many cases it can be shown that the chaotic behaviour is obtained via the period-doubling bifurcation [1,2,5,6]. Recently some attempts to create an analytical criterion which allows us to estimate the domain in the system parameter space has been proposed [6,7]. The criterion by Szemplinska-Stupnicka [7] places the chaotic zone between the vertical tangent of the resonance curve of the second approximate solution and the boundary of stability of the period-two solution. In what follows the limits of stable and unstable period-doubling cascades are proposed as boundaries of the chaotic domain.

First consider the first approximate solution in the form

$$x(t) = C_0 + C_1 \cos(\Omega t + \vartheta), \qquad (2)$$

where C_0, C_1 and ϑ are constants. Substituting eq. (2) into eq. (1) it is possible to determine these constants [6,8,9]. To study the stability of the solution (2) a small variational term $\delta x(t)$ is added to eq. (2) as

$$x(t) = C_0 + C_1 \cos(\Omega t + \vartheta) + \delta x(t). \qquad (3)$$

[1] Permanent address: Institute of Applied Mechanics, Technical University of Lodz, Stefanowskiego 1/15, 90-924 Lodz, Poland.

After some algebraic manipulations, the linearized equation with periodic coefficients for $\delta x(t)$ is obtained,

$$\delta\ddot{x} + a\,\delta\dot{x} + \delta x\,(\lambda_0 + \lambda_1 \cos \Theta + \lambda_2 \cos 2\Theta) = 0, \qquad (4)$$

where

$$\lambda_0 = 3C_0^2 + \tfrac{3}{2}C_1^2, \quad \lambda_1 = 6C_0 C_1,$$
$$\lambda_2 = \tfrac{3}{2}C_1^2, \quad \Theta = \Omega t + \vartheta.$$

In the derivation of eq. (4), for simplicity it was assumed without loss of generality that $b=0$. As we have a parametric term of frequency $\Omega - \lambda_1 \cos \Theta$, the lowest order unstable region is that which occurs close to $\Omega/2 \approx \sqrt{\lambda_0}$ and at its boundary we have the solution

$$\delta x = b_{1/2} \cos(\tfrac{1}{2}\Omega t + \phi). \qquad (5)$$

To determine the boundaries of the unstable region we insert eq. (5) into eq. (4), and the condition of nonzero solution for $b_{1/2}$ leads us to the following criterion to be satisfied at the boundary:

$$(\lambda_0 - \tfrac{1}{4}\Omega^2)^2 + \tfrac{1}{4}a^2\Omega^2 - \tfrac{1}{4}\lambda_1^2 = 0. \qquad (6)$$

From eq. (6) one obtains the interval $(\Omega_1^{(2)}, \Omega_2^{(2)})$, within which period-two solutions exist. Further analysis shows that at Ω_2 we have a stable period-doubling bifurcation for decreasing Ω and at Ω_1 an unstable period-doubling bifurcation for increasing Ω [6]. In this interval we can consider the period-two solution of the form

$$x(t) = A_0 + A_{1/2} \cos(\tfrac{1}{2}\Omega t + \phi) + A_1 \cos \Omega t, \quad (7)$$

where A_0, $A_{1/2}$, A_1 and ϕ are constants to be determined. Again, to study the stability of the period-two solution we have consider a small variational term $\delta x(t)$ added to eq. (7). The linearized equation for $\delta x(t)$ has the following form:

$$\delta\ddot{x} + a\,\delta\dot{x} + \delta x\,[\lambda_0^{(2)} + \lambda_{1/2c} \cos \tfrac{1}{2}\Omega t$$
$$+ \lambda_{1/2s} \sin \tfrac{1}{2}\Omega t + \lambda_{3/2} \cos(\tfrac{3}{2}\Omega t + \phi)$$
$$+ \lambda_{1c}^{(2)} \cos \Omega t + \lambda_{1s}^{(2)} \sin \Omega t + \lambda_2^{(2)} \cos 2\Omega t] = 0, \quad (8)$$

where

$$\lambda_0^{(2)} = 3(A_0^2 + \tfrac{1}{2}A_{1/2}^2 + \tfrac{1}{2}A_1^2),$$
$$\lambda_{1/2c} = 3A_{1/2}(2A_0 + A_1) \cos \phi,$$
$$\lambda_{1/2s} = 3A_{1/2}(A_1 - 2A_0) \sin \phi,$$
$$\lambda_{3/2} = 3A_1 A_{1/2}, \quad \lambda_{1c}^{(2)} = 6A_0 A_1 + \tfrac{3}{2}A_{1/2}^2 \cos 2\phi,$$
$$\lambda_{1s}^{(2)} = -\tfrac{3}{2}A_{1/2}^2 \sin 2\phi, \quad \lambda_2^{(2)} = \tfrac{3}{2}A_1^2.$$

The form of eq. (8) enables us to find the range of existence of a period-four solution, represented by

$$\delta x = b_{1/4} \cos(\tfrac{1}{4}\Omega t + \phi) + b_{3/4} \cos(\tfrac{3}{4}\Omega t + \phi). \quad (9)$$

After inserting eq. (9) into eq. (8) the condition of nonzero solution for $b_{1/4}$ and $b_{3/4}$ gives us the following set of nonlinear algebraic equations for Ω, $\cos \phi$ and $\sin \phi$ to be satisfied for existence:

$$(\lambda_{1/2s} + \lambda_{1s}^{(2)}) - \tfrac{1}{2}(\lambda_{1/2c} - \lambda_{1c}^{(2)})$$
$$\times (-\tfrac{1}{2}a\Omega + \lambda_{1/2s} - \lambda_{3/2} \sin \phi) = 0,$$
$$(\tfrac{9}{8}\Omega^2 + \tfrac{1}{2}\lambda_0^{(2)} + \lambda_{3/2} \cos \phi)$$
$$- \tfrac{1}{2}(\lambda_{1/2c} + \lambda_{1c}^{(2)})(\lambda_{1s}^{(2)} + \lambda_{1/2c}) = 0,$$
$$(-\tfrac{3}{2}a\Omega - \lambda_{3/2} \sin \phi)$$
$$- \tfrac{1}{2}(\lambda_{1/2c} + \lambda_{1c}^{(2)})(\lambda_{1s}^{(2)} + \lambda_{1/2c}) = 0. \quad (10)$$

Solving eq. (10) by a numerical procedure it is possible to obtain $\Omega_1^{(4)}$ and $\Omega_2^{(4)}$, the frequencies of stable and unstable period-four bifurcations. With both stable and unstable period-two and period-four boundaries obtained from eq. (6) and eq. (10), assuming that the Feigenbaum model [10] of period doubling is valid for our system one obtains the following boundaries of the domain where chaotic behaviour may occur:

$$\Omega_1^{(\infty)} = \Omega_1^{(2)} + \frac{\Delta \Omega_1}{1 - 1/\delta},$$
$$\Omega_2^{(\infty)} = \Omega_2^{(2)} - \frac{\Delta \Omega_2}{1 - 1/\delta}, \quad (11)$$

where $\Delta \Omega_1 = \Omega_1^{(4)} - \Omega_1^{(2)}$, $\Delta \Omega_2 = \Omega_2^{(2)} - \Omega_2^{(4)}$, and $\delta = 4.69...$ is the universal Feigenbaum constant. As Feigenbaum's constant is asymptotic and we extrapolate $\Omega_{1,2}^{(\infty)}$ from the period-two stability limits the values of $\Omega_{1,2}^{(\infty)}$ are approximate. The domain where chaotic behaviour can occur is proposed to be between the limits of unstable and stable period-doubling cascades, in the interval $(\Omega_1^{(\infty)}, \Omega_2^{(\infty)})$ and of course to expect chaos one must have

$$\Omega_1^{(\infty)} < \Omega_2^{(\infty)}. \quad (12)$$

In fig. 1 we show the comparison of this analytical estimation of the chaotic domain and the actual (numerically found) chaotic domain obtained by Ueda [1]. Good agreement is seen. In fig. 2 we compare the actual chaotic domain with the domain obtained by the above method. Also the domain obtained by the approximate criterion of Szemplinska-Stupnicka [7] is indicated. Again our approach shows very good agreement with the actual chaotic domain and is better than other analytical estimates. In the domain σ in fig. 2 chaotic behaviour has not been found, as here $\Omega_1^{(\infty)} > \Omega_2^{(\infty)}$, and no chaotic interval estimated by our method exists.

The analytical technique presented in this paper is based on: (a) the approximate period-one, -two and -four solutions and their stability limits computed by harmonic balance method, (b) Feigenbaum's uni-

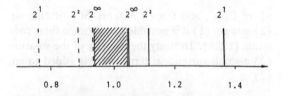

Fig. 1. Stable (solid line) and unstable (broken line) boundaries of period-two and -four bifurcations and chaotic domain for eq. (1). $a=0.77$, $b=0$, $c=1$, $B_0=0.045$, $B_1=0.16$, $\Omega_1^{(2)}=0.77$, $\Omega_2^{(2)}=1.37$, $\Omega_1^{(4)}=0.89$, $\Omega_2^{(4)}=1.12$, $\Omega_1^{(\infty)}=0.93$, $\Omega_2^{(\infty)}=1.05$. Chaotic behaviour has been found for $\Omega \in [0.94, 1.04]$ [1].

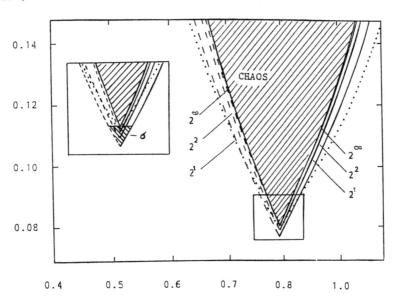

Fig. 2. Chaotic domain of the system (1), $a=0.1$, $b=0.5$, $c=0.5$, $B_0=0$, and analytical criteria, (....) criterion of ref. [7]. In the left-hand side of the figure the enlargement of the box section is shown.

versal constant for the asymptotic ratio of the stability intervals of the 2^n and 2^{n+1} periodic solution.

It can be applied to the class of oscillators for which the harmonic balance method analysis shows the possibility of period-doubling bifurcation ($\lambda_1 \neq 0$ in eq. (4)). Our method can be applied before numerical analysis to estimate the phase space intervals where strange phenomena can take place.

In future works I will try to use the above method to estimate chaotic domains of other nonlinear oscillators.

This work has been supported by the British Council. I am very thankful to J. Brindley, D.G. Knapp and E. Ponce for valuable discussions.

References

[1] Y. Ueda, Ann. NY Acad. Sci. 357 (1980) 422.
[2] Y. Ueda, J. Stat. Phys. 20 (1979) 181.
[3] S. Novak and R.G. Frehlick, Phys. Rev. A 26 (1982) 3660.
[4] J.C. Huang, Y.H. Kao, C.S. Wang and Y.S. Gou, Phys. Lett. A 136 (1989) 131.
[5] S. Sato et al., Phys. Rev. A 28 (1983) 1654.
[6] W. Szemplinska-Stupnicka and J. Bajkowski, Int. J. Nonlin. Mech. 21 (1986) 401.
[7] W. Szemplinska-Stupnicka, Ingenieur-Archiv 58 (1988) 354.
[8] T. Kapitaniak, J. Sound Vibr. 121 (1988) 259.
[9] T. Kapitaniak, Chaos in systems with noise (World Scientific, Singapore, 1988).
[10] M. Feigenbaum, J. Stat. Phys. 19 (1978) 25.

IV. Classical Nonlinear Oscillators: Duffing, Van der Pol and Pendulum

In this section we give examples of chaotic behavior of such classical nonlinear oscillators like: Duffing, Van der Pol and pendulum.

We start with Paper IV.1 which presents computer calculation for the numerical solution of Duffing's and generalized Duffing's equations showing global scaling properties for the bifurcation in parameter space. Scaling properties are discussed in terms of a one-dimensional map.

Paper IV.2 shows resonance curves, bifurcation diagrams, and phase diagrams of the Duffing's equation presenting a periodic recurrence of specific fine structure in the bifurcation set, which is closely connected with the nonlinear resonances of the system.

Paper IV.3 carries out systematic numerical simulation studies which established theoretical chaos threshold values in quantitative agreement with experiments.

Analog computer simulations of Van der Pol's and generalized Van der Pol's (with cubic nonlinearity) equations are presented in Paper IV.4 The chaotic behavior has been found for generalized equation but has not been found for a simple equation with relatively small damping.

Bifurcation diagrams of the driven Van der Pol's equation showing mode-locking and period-doubling cascades are given in Paper IV.5. A generalization of the winding number is used to compute devil's staircase and winding number diagrams of period-doubling cascades.

Chaotic behavior of pendulum is discussed in Paper IV.6. Possible routes from periodic to chaotic oscillations are presented.

The connection between a map of a circle and a forced pendulum is established in Paper IV.7. Some common properties of the circle map and the pendulum are numerically shown.

Periodic solutions of a forced pendulum as well as stability boundaries for symmetric swinging oscillations and their asymmetric descendents are investigated in Paper IV.8. Symmetry breaking bifurcation is a precursor of transition to chaos as has been shown in Paper IV.6.

Universal scaling property in bifurcation structure of Duffing's and of generalized Duffing's equations

Shin-ichi Sato, Masaki Sano, and Yasuji Sawada
Research Institute of Electrical Communication, Tohoku University, Sendai 980, Japan
(Received 2 May 1983)

Computer calculation for the numerical solution of Duffing's and generalized Duffing's equations shows global scaling properties for the bifurcation in parameter space. These scaling properties are discussed in terms of a one-dimensional map. The analysis based on a piecewise linear approximation gave results in good agreement with the experimentally observed scaling behavior.

I. INTRODUCTION

In recent years, many studies have been done on new universal behavior in chaotic dynamical systems. Period doubling, intermittent transition to chaos, and other related phenomena were investigated in detail by many authors from the viewpoint of scaling properties. They are all related to local bifurcation structure which occurs prior to chaos by changing parameters continuously.[1-3] However, a large number of phenomena observed in physical systems, chemical reactions, and biological systems exhibit interesting global bifurcation structure as well as the local bifurcations in parameter space. These global structures correspond to repeating transitions from a phase-locking state to another phase-locking state or to a chaotic state and vice versa.[4] Only a few authors have discussed the global aspects of these bifurcation sets. Among these authors Tomita and Tsuda explained a global bifurcation structure in the Lorentz system by using one-dimensional mapping.[5] They have also succeeded in interpreting and predicting the experimental results for a Belousov-Zhabotinsky reaction in a stirred-flow reactor utilizing similar methodology.[6]

A simplest example presenting a global bifurcation structure is forced nonlinear oscillators which are widely observed in physical,[7-9] chemical, and biological systems. However, scaling properties among these phase-locking regions and chaotic regions have not been noticed until recently.

Recently Kaneko presented similarity and scaling properties of each periodic state in connection with a map of a circle.[10] Almost simultaneously, Sano and Sawada reported a scaling behavior of the bifurcation parameters in a differential system of coupled chemical reaction systems.[11] In this paper we consider Duffing's equation, which represents nonlinear oscillation in an X^4 potential with external periodic force and damping, or generalized Duffing's equation, an oscillation in an X^{2n} potential. As regards Duffing's equation, Ueda has extensively investigated chaotic phenomena mainly concerning the appearance of the homoclinic orbit.[12-14] The system can be a good prototype for studying chaotic phenomena in ordinary differential equations because it presents various types of transitions, e.g., period-doubling bifurcation, sub-critical transition to chaos, and homoclinicity Concerning the structure of global bifurcation sets, Kawakami and Matsuo[15] have pointed out that similarity can be observed in bifurcation sets of Duffing's equation by computer experiment.

The purpose of the present work is to show for the first time the existence and interpretation of scaling properties in global bifurcation sets which can appear in a wide class of forced nonlinear oscillations. We confirmed numerically the similarity and scaling property of global bifurcation sets in Duffing's equation and even in generalized Duffing's equations.

II. EXPERIMENTAL RESULTS

Duffing's equation with cubic nonlinearity is

$$\frac{d^2x}{dt^2}+k\frac{dx}{dt}+x^3=B\cos t, \quad (1)$$

where k is the damping rate and B is the modulation amplitude. Equation (1) can be rewritten as simultaneous equations,

$$\frac{dx}{dt}=y, \quad (2a)$$

$$\frac{dy}{dt}=-ky-x^2+B\cos t. \quad (2b)$$

This system has a symmetry under the transformation S,

$$S: (x,y,t)\rightarrow(-x,-y,t+\pi).$$

Therefore, if ψ is a solution of Eq. (2), then $S\psi$ is also a solution. In order to characterize the bifurcation diagram by topological properties of solution, we introduce the notation P_s, P_a, and X. Periodic and chaotic orbits are denoted by $P_a^{\pm}(n,m)$, $P_s(m,n)$, and X, where by $P(n,m)$ we mean a periodic orbit which cuts the $y=0$ plane from the positive to negative y direction at n different values of x and cuts the $y=0$ plane in the opposite direction at m different values of x. The indices s and a indicate symmetric and antisymmetric solution, and the $\pm$ represents pair solutions connected by the transformation S.

A coarse bifurcation process with increasing B is composed of the alternative appearance of periodic and chaotic states:

$$\cdots \to P \to X \to P \to X \to \cdots .$$

The chaotic region would appear between two periodic states which have topological properties different from each other. When k is small the bifurcation process can be described, in more detail, as

$$\cdots \to P_s(n,n) \to \begin{Bmatrix} P_a^+(n,n) \\ P_a^-(n,n) \end{Bmatrix} \underset{I}{\to} X \underset{D}{\to} \begin{Bmatrix} P_a(n,n+1) \\ P_a(n+1,n) \end{Bmatrix} \underset{I}{\to} X \underset{D}{\to} P_s(n+1,n+1) \to \cdots$$

(double bifurcation process), where the curly brackets mean coexistence of symmetric pair solutions. The route to chaos is period doubling (indicated by I) or subcritical transition (indicated by D). On the other hand, when k is relatively large the bifurcation process is

$$\cdots \to \begin{Bmatrix} P_a(n,n+1) \\ P_a(n+1,n) \end{Bmatrix} \underset{I}{\to} X \underset{D}{\to} P_s(n+1,n+1) \underset{D}{\to} \begin{Bmatrix} P_a(n+1,n+2) \\ p_a(n+2,n+1) \end{Bmatrix} \to \cdots$$

(single bifurcation process). Figure 1 shows some examples of the periodic orbits and Fig. 2 shows an example of transition from the periodic state to the chaotic state. In the chaotic region the fine structure of windows of phase-locking states is inevitably observed in addition to the homoclinic orbits observed by Ueda.[14]

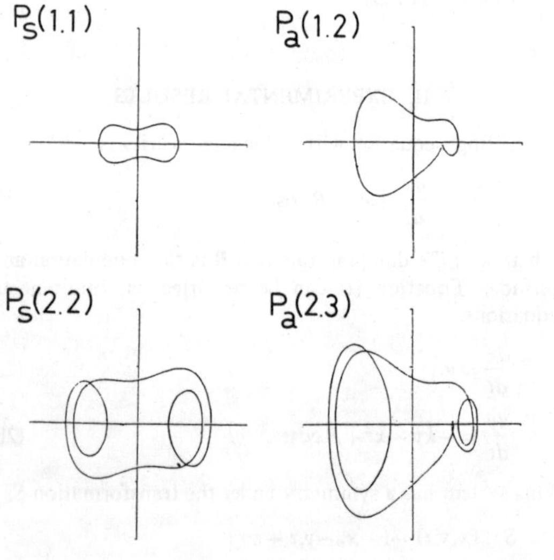

FIG. 1. Periodic orbits of Duffing's equation in phase-locking regions at $k=0.3$. $P(1,1)$ for $B=2.0$, $P(1,2)$ for $B=5.0$, $P(2,2)$ for $B=15.0$, $P(2,3)$ for $B=31.0$, $P(3,3)$ for $B=70.0$, $P(3,4)$ for $B=92.0$.

Present computer experiments have elucidated a series of similar bifurcation structures in k-B parameter space as shown in Fig. 3. A part of this structure in the small-B region was studied before[15] but the region with a large value of B has not been investigated. To examine quantitatively the scaling properties for the bifurcation process, we measured B_m as shown by Fig. 3. B_m represents the initiation value of B for a period-doubling bifurcation from a periodic state $P_a(n,n+1)$ where $m=2n+1$. The logarithmic plot of B_m vs m shown in Fig. 4 exhibits a relation

$$B_m \propto m^\alpha , \qquad (3)$$

where $\alpha = 3.18$ for the ordinary Duffing's equation.

Similar experiments were carried out for the generalized Duffing's system

$$\frac{d^2x}{dt^2} + k\frac{dx}{dt} + x^{2\nu+1} = B\cos t , \qquad (4)$$

where $\nu = 2, 3, 4, \ldots$. The numerical results for these bifurcation sets are shown in Fig. 4. We have obtained $\alpha = 2.57$ for $\nu = 2$, $\alpha = 2.42$ for $\nu = 3$.

In order to understand the scaling behavior of the bifurcation sets we take here a Poincaré mapping obtained from "the maximum absolute point" of x. Let $p=(t,x,y)$ be a solution of Eq. (2). The maximum absolute point is $p(t,x,0)$ in $x > 0$ (but $d^2x/dt^2 < 0$) and in $x < 0$ (but $d^2x/dt^2 > 0$). If we let $p_n=(t_n,x_n,0)$, $p_{n+1}=(t_{n+1},x_{n+1},0)$ be two successive maximum absolute points, then one obtains a discrete dynamical system,

$$T: \; p_n \to p_{n+1} . \qquad (5)$$

Near the bifurcation point from a periodic to a chaotic state the point sequence given by Eq. (5) was found by computer experiments to move on invariant sets which can be approximately regarded as one-dimensional attractors embedded in the x-t plane. Therefore, one can reduce the map given by Eq. (5) to the following one-dimensional map which is useful in order to discuss the topological structure of Duffing's system:

$$f: \; \theta_n \to \theta_{n+1} , \qquad (6)$$

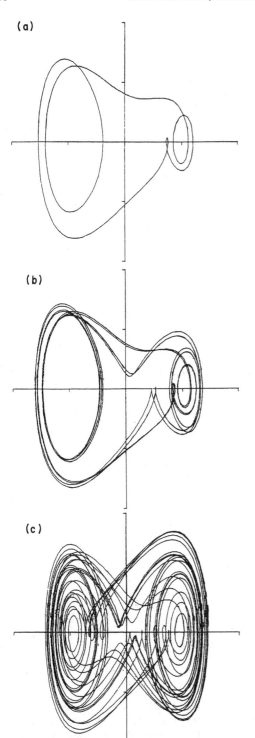

FIG. 2. Transition from a periodic state to chaotic state at $k=0.3$ for (a) $B=31.0$, (b) $B=36.0$, (c) $B=40.0$. System in a phase-locking state displays period-doubling bifurcations and leads to chaos.

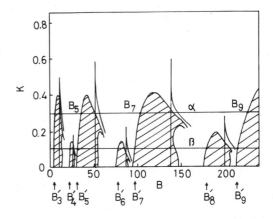

FIG. 3. Phase diagram in k-B parameter space. Parts of the hatched line are the regions where the period of the limit cycle is not 2π, i.e., period-doubling bifurcations, chaotic, and window regions. There exist hysteresis phenomena on a cusp line. Similar bifurcation sets have a scaling law vs the parameter B. Lines α and β are examples of "single bifurcation process" and "double bifurcation process," respectively. B_m and B'_m are the critical values of period-doubling bifurcation.

where $\theta_n = t_n/2\pi$. Figures 5(a) and 5(b) demonstrate the map f for two different values of B. The obtained map can be regarded as

$$f(\theta_n) = \theta_n + \Omega(k,B) + N(\theta_n, K, B) , \qquad (7)$$

where Ω is a constant independent of θ_n and the magni-

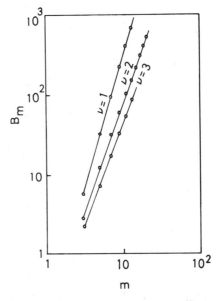

FIG. 4. Scaling law in the generalized Duffing's system. Scaling law obtained by Eq. (4) where $\alpha = 3.18$ for $\nu=1$, $\alpha=2.57$ for $\nu=2$, $\alpha=2.24$ for $\nu=3$.

tude was experimentally found to be proportional to $B^{-1/3}$. Ω has a very weak dependence of k. $N(\theta_n,k,B)$ is a nonlinear contribution in the map which is responsible for the occurrence of chaos.

III. DISCUSSION

In this section we discuss the reason for the scaling law obtained in the preceding section. The map f represented by Eq. (7) reminds us of the map which has been studied by Shenker,[16] but f has a global structure shown by Eq. (3) which is different from the map by Shenker.

At first let us consider the two-dimensional map T for the generalized Duffing's equation and introduce the piecewise linearized version at the small interval $t_n \leq t < t_{n+1}$ where $P_n = (t_n, x_n, 0)$, the maximum absolute point. Thus

$$\frac{d^2x}{dt^2} + k\frac{dx}{dt} + a(x_n)x = B\cos t, \qquad (8)$$

where $a(x_n) = (2\nu+1)x^{2\nu}$. In the real Duffing's system the motion of a forced pendulum is very complicated but Eq. (8) is valid when the interval $[t_n, t_{n+1}]$ is sufficiently small (i.e., B is sufficiently large) and x_n is large. When $a(x_n) > k^2/4$, the solution of the piecewise linear equation is

$$x(t) = A\exp\left[-\frac{k}{2}t\right]\cos(\omega_n t + \delta_1)$$
$$+ \frac{B}{[(a-1)^2 + k^2]^{1/2}}\cos(t + \delta_2), \qquad (9)$$

where A, δ_1 are determined by initial conditions, $\delta_2 = \tan^{-1}[k/(a-1)]$, $\omega_n = [a(x_n) - k^2/4]^{1/2}$. From the definition of map T, initial conditions are $x(t_n) = x_n$, $dx(t_n)/dt = 0$ and

$$\theta_{n+1} \approx \theta_n + \frac{1}{\omega_n}, \qquad (10a)$$

$$x_{n+1} \approx x\left[2\pi\theta_n + \frac{2\pi}{\omega_n}\right], \qquad (10b)$$

where $\theta_n = t_n/2\pi$ (mod 1). θ_n is the phase of the periodic perturbation. When $k \ll 1$, $B \gg 1$, and $x_n \gg 1$, then Eq. (10) is

$$\theta_{n+1} \approx \theta_n + \frac{1}{(2\nu+1)^{1/2}|x_n|^\nu}, \qquad (11a)$$

$$x_{n+1} \approx x_n - \frac{2\pi B}{(2\nu+1)^{3/2}|x_n|^{3\nu}}\sin 2\pi\theta_n. \qquad (11b)$$

We approximate the recursion relation Eq. (11) by a differential equation on which the step space is $d\tau$,

$$\frac{d\theta}{d\tau} \approx \frac{1}{(2\nu+1)^{1/2}|x|^\nu}, \qquad (12a)$$

$$\frac{dx}{d\tau} \approx \frac{2\pi B}{(2\nu+1)^{3/2}|x|^{3\nu}}\sin 2\pi\theta. \qquad (12b)$$

These equations are readily integrated;

$$x \approx (B\cos 2\pi\theta)^{1/(2\nu+1)}. \qquad (13)$$

This result was found to be in good agreement with computer simulations near bifurcation points. The map f defined by Eq. (6) is

$$\theta_{n+1} = f(\theta_n)$$
$$\approx \theta_n + (2\nu+1)^{-1/2}(|B\cos 2\pi\theta_n|)^{-\nu/(2\nu+1)}. \qquad (14)$$

We discuss the global structure of the generalized Duffing's system by using this one-dimensional map.

If there is an m-periodic orbit of f then the dependence of m on B is obtained by applying a theory of intermittency. Approximating the discrete dynamic equation (14) by a differential equation

$$\frac{d\theta}{d\tau} \approx (2\nu+1)^{-1/2}(1/|B\cos 2\pi\theta|)^{\nu/(2\nu+1)} \qquad (15)$$

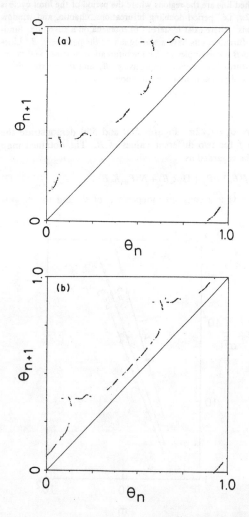

FIG. 5. One-dimensional map obtained by computer simulations. (a) $k=0.3$, $B=130.0$; (b) $k=0.3$, $B=500.0$. Map is expressed as $\theta_{n+1} = \theta_n + \Omega + N(\theta_n)$ where $\Omega \simeq 0.15$ and $N(\theta_n)$ is the nonlinear contribution.

and by integrating Eq. (15) one obtains, as the period,

$$T \equiv \int_0^1 d\tau$$
$$= B^{\nu/(2\nu+1)}(2\nu+1)^{1/2} \int_0^1 |\cos 2\pi\theta|^{\nu/(2\nu+1)} d\theta$$
$$= C(\nu)B^{\nu/(2\nu+1)}, \quad (16)$$

where $C(\nu)$ represents a constant independent of B. Since $T \propto m$ holds roughly in the m-periodic state, the scaling law is

$$B_m \propto m^{(2\nu+1)/\nu} \quad (17)$$

which explains excellently the experimentally observed results. In fact, the values of experimental results and theoretical results $\alpha_t \equiv (2\nu+1)/\nu$ are

$\alpha = 3.18, \quad \alpha_t = 3.0$,
$\alpha = 2.57, \quad \alpha_t = 2.5$,
$\alpha = 2.42, \quad \alpha_t = 2.33\ldots$,

for $\nu = 1, 2$, and 3, respectively.

Usually in the intermittency problem one considers a recursion relation

$$x_{n+1} = x_n + ax_n^z + \epsilon,$$

where ϵ is a small parameter, for which it is well known that the scaling behavior for the m-periodic orbit in the limit of the small parameter approaching the critical value is $m \propto \epsilon^{-1+1/Z}$.[17] In this paper, however, we have discussed an example of abnormal intermittency which can appear in a map of a circle onto itself or a forced nonlinear oscillator in the limit of the rotation number approaching zero. The map in this case can be expressed

$$\theta_{n+1} = f(\theta_n) = \theta_n + \epsilon g(\theta_n, \epsilon), \quad (18)$$

where ϵ is a control parameter and g is an arbitrary function. Considering a small deviation and expanding Eq. (18) by Taylor series around the fixed point of $f(\theta)$ and $\epsilon = 0$, one obtains a map f constituting the class of abnormal intermittency as $\epsilon \to 0$.[10] Those maps belonging to the abnormal intermittency class have the scaling law for m-periodic orbit ($m = 2, 3, 4, \ldots$) where $m \propto \epsilon^{-1}$ as $\epsilon \to 0$ and its scaling law is independent of $g(\theta, \epsilon)$.

---

[1] M. J. Feigenbaum, J. Stat. Phys. 19, 25 (1978); 21, 669 (1979).

[2] Y. Pomeau and P. Manneville, Commun. Math. Phys. 74, 189 (1980).

[3] D. Rand, S. Ostlund, J. Sethna, and E. D. Siggia, Phys. Rev. Lett. 49, 132 (1982).

[4] L. Glass and R. Perez, Phys. Rev. Lett. 48, 1772 (1982).

[5] K. Tomita and I. Tsuda, Prog. Theor. Phys. 69, 185 (1980).

[6] K. Tomita and I. Tsuda, Prog. Theor. Phys. 64, 1183 (1980).

[7] D. D'Humieres, M. R. Beasley, B. A. Huberman, and A. Libchaber, Phys. Rev. A 26, 3483 (1982).

[8] N. F. Pedersen and A. Davidson, Appl. Phys. Lett. 39, 830 (1981).

[9] J. E. Flaherty and F. C. Hoppenstead, Stud. Appl. Math. 58, 5 (1978).

[10] K. Kaneko, Prog. Theor. Phys. 63, 669 (1982).

[11] M. Sano and Y. Sawada, Phys. Lett. A (in press).

[12] Y. Ueda, N. Akamatsu, and C. Hayashi, Trans. I.E.C.E. Jpn. 56A, 218 (1979).

[13] Y. Ueda, in *Proceedings of the Engineering Foundation Conference on New Approaches to Nonlinear Problems in Dynamics*, edited by P. J. Holmes (SIAM, Philadelphia, 1979), p. 311.

[14] Y. Ueda, Trans. I.E.C.E. Jpn. 98A, 167 (1979).

[15] H. Kawakami and J. Matsuo, Trans. I.E.C.E. Jpn. 64A, 1018 (1981).

[16] S. J. Shenker, Physica D 5, 405 (1982).

[17] For example, J. E. Hirsch, B. A. Huberman, and D. J. Scalapino, Phys. Rev. A 25, 519 (1982).

SUPERSTRUCTURE IN THE BIFURCATION SET OF THE DUFFING EQUATION
$\ddot{x} + d\dot{x} + x + x^3 = f\cos(\omega t)$

Ulrich PARLITZ and Werner LAUTERBORN
Drittes Physikalisches Institut, Universität Göttingen, D-3400 Göttingen, Fed. Rep. Germany

Received 7 November 1984
Revised manuscript received 10 December 1984

Resonance curves, bifurcation diagrams, and phase diagrams of the Duffing equation $\ddot{x} + d\dot{x} + x + x^3 = f\cos(\omega t)$ are presented. They show a periodic recurrence of a specific fine structure in the bifurcation set, which is closely connected with the nonlinear resonances of the system.

The numerical investigation of nonlinear dynamical systems in recent years has confirmed many involved ideas on their dynamical behaviour. The most important features found are period-doubling bifurcations, coexistence of attractors with complicated basin structures, and strange attractors with the fundamental property of sensitive dependence on initial conditions. A very simple looking continuous dynamical system, which nevertheless exhibits all these phenomena, is the Duffing equation

$$\ddot{x} + d\dot{x} + x + x^3 = f\cos(\omega t), \quad (1)$$

or

$$\dot{x}_1 = x_2,$$
$$\dot{x}_2 = -dx_2 - x_1 - x_1^3 + f\cos(2\pi x_3),$$
$$\dot{x}_3 = \omega/2\pi, \quad (2)$$

where x_1, x_2 and x_3 are the coordinates in the phase space $R^2 \times S^1$. Equations of this kind have been investigated by many authors [1–10] yielding quite a number of examples for the above-mentioned features of nonlinear dynamics. But despite considerable efforts only particular results have been obtained so far; a global description of the dynamics of Duffing's equations could not yet been given. This paper is intended to improve our understanding of the Duffing equation (1) by emphasizing the important role of the nonlinear resonances.

The first significant result on this type of equations is due to Duffing [1]. He showed, that the eigenfrequency Ω of the system depends on the excitation amplitude f and that two asymptotically orbitally stable solutions may coexist for a certain range of the excitation frequency ω.

These properties lead to the well-known leaning-over of the amplitude resonance curve and to hysteresis jumps. Little is known up to now about the shape of the amplitude resonance curves for medium and high values of the excitation amplitude f. To fill this gap we have numerically investigated the Duffing equation (1) taking $d = 0.2$ and especially considering the question, how the amplitude resonance curves change with increasing excitation amplitude f.

Another method to show the dependence of the system on a control parameter are bifurcation diagrams, which for continuous dynamical systems are based on the associated (global) Poincaré map [11]. Every subharmonic periodic solution of period $T = n \cdot 2\pi/\omega, n \in N$ corresponds to n cyclic points in the Poincaré cross section, which here is given by

$$\Sigma = \{(x_1, x_2, x_3) \in R^2 \times S^1 : x_3 = \text{const}\}.$$

The coordinates in the Poincaré cross section Σ are called X1P and X2P. The bifurcation diagrams show the projection of these cyclic points onto one of the axes of the Poincaré cross section plotted against the control parameter ω.

Reprinted with permission from Phys. Lett., Vol. 107A, pp. 351–355
Copyright (1985), Elsevier Science Publishers B.V.

For the numerical computations of the bifurcation diagrams and the resonance curves the control parameter ω is increased and decreased in small steps ($\Delta\omega = 0.0001-0.001$) and the last computed cyclic point is always used as new initial value. The Poincaré cross section is located at $t = 0 = x_3$.

The Duffing equations belong to the physically interesting class of nonlinear oscillators showing secondary resonances (see ref. [12] for examples of such resonances). Resonance is a physical concept developed in the context of driven oscillators and not yet well defined in the theory of general dissipative dynamical systems. But it seems that this concept should be paid more attention to, as resonance curves are related to the global bifurcation set of nonlinear oscillators. Fig. 1 gives an amplitude resonance curve for a relatively small excitation amplitude ($f = 2$). The maximum amplitude of the oscillation X1M is plotted versus the driving frequency ω. Besides the main resonance R_1 several secondary resonances (denoted by R_3, R_5, R_7) are present. It is a peculiar finding that in contrast to the main resonance R_1, the secondary resonances are steeper at the lower driving frequency side of the resonances and will lean over to the left for increasing excitation amplitude f. The appearance of secondary resonances below the main resonance may be explained through a coincidence of a higher harmonic of the nonlinear oscillation with the main resonance frequency of the system. It can easily be shown that for every periodic solution $x(t)$ of (1) all Fourier coefficients x_k in the Fourier series

$$x(t) = \sum_x x_k \, e^{ik\omega t},$$

do not vanish for odd $k \in Z$. The coefficients x_k with even k are all equal to zero for sufficiently small excitations. For excitation frequencies ω below the main resonance we therefore have a secondary resonance, whenever ωk ($k \in Z$, odd) equals the (amplitude dependent) eigenfrequency Ω of the system. This k is taken to label the secondary resonances by R_k in giving the sequence R_1, R_3, R_5 ... of fig. 1.

If the excitation amplitude f is high enough the closed trajectory may lose its geometrical symmetry [3] and two asymmetric period-1 partner orbits appear (period in units of $2\pi/\omega$). This may be viewed as a kind of *dynamic symmetry breaking* as we have only a one-valley potential and thus no symmetry breaking in the static case.

Fig. 2 shows a bifurcation diagram (X2P versus ω), where for $\omega = 0.88$ (S_2) and $\omega = 0.97$ (S'_2) symmetry bifurcations take place. For higher excitation amplitudes f these first symmetry bifurcations between the

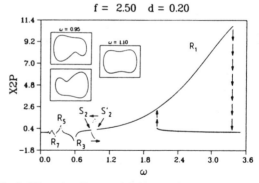

Fig. 2. Bifurcation diagram with the second coordinate X2P of the cyclic points in the Poincaré cross section plotted versus the control parameter ω for $f = 2.5$ and $d = 0.02$. Secondary resonances (denoted by R_1, R_3, R_5, ...) and symmetry bifurcations (denoted by S_2 and S'_2) occur. The two branches of symmetry broken solutions the system follows for increasing and decreasing ω are denoted by arrows. The behaviour of the system at the symmetry bifurcation points depends on the step size of the parameter variation, the basin structure of the two partner orbits created and the errors of the numerical integration. In this bifurcation diagram the system follows the lower symmetry bifurcation branch for increasing ω and the upper branch for decreasing ω, but different behaviour is possible, too (see the branches between S_6 and S'_6 in fig. 3). Inserted are projections of the trajectory in the x_1-x_2 plane showing the geometrical consequences of the symmetry breaking.

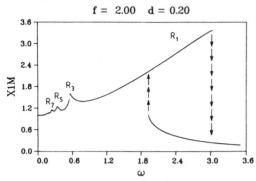

Fig. 1. Amplitude resonance curve showing the maximum X1M of the coordinate X1 of the Duffing equation (1) in dependence on the excitation frequency ω for $f = 2$ and $d = 0.2$. The arrows denote hysteresis jumps of the main resonance R_1 for increasing and decreasing ω.

main resonance R_1 and the first odd secondary resonance R_3 are followed by successive breaking between the odd secondary resonances of increasing order. For the two period-1 attractors in the parameter intervals where the symmetry is broken all even Fourier coefficients are no longer equal to zero. The symmetry bifurcations may therefore be viewed as resonances $\omega = \Omega/k$ for even $k \in Z$. This interpretation is the reason for the even indices of the letters "S" and "S'" denoting the symmetry bifurcation points in fig. 2 and the following diagrams.

Fig. 3 shows the bifurcation diagram for $f = 22$. To obtain the diagram the excitation frequency has been increased and decreased again as in the figure before. For some ω-values strong bifurcations, i.e. creation or annihilation of attractors, take place. The arrows denote the attractors the system follows during variation of the control parameter ω. As typical for 2-parameter systems [13] finite or infinite cascades of period-doubling bifurcations take place in the parameter region, where the orbits have lost their symmetry. Two of these cascades can be seen between the symmetry bifurcation points S_2, S'_2 and S_4, S'_4. They form closed loops by bifurcating back and forth two times at B^1, B'^1 and B^2, B'^2 so that in the innermost loop a period-4 oscillation takes place (only the bifurcations between S_2 and S'_2 are labeled for clarity). It can be shown for Duffing equations with symmetric potentials, that even subharmonics, i.e. subharmonics whose period equals an even multiple of the excitation period, are always asymmetrical. Symmetry breaking is therefore a prerequisite for period-doubling bifurcations.

For higher excitation amplitudes f symmetry and period-doubling bifurcations occur between all odd secondary resonances. The "depth" of the bifurcation cascade, i.e. the number of bifurcations, depends on the value of the excitation parameter f. Furthermore in a small ω-interval two pairs of period-1 partner orbits coexist and hysteresis jumps occur in both symmetry bifurcation branches.

Further increase of the excitation amplitude f leads to "deeper" bifurcation cascades near the even secondary resonances until the period-doubling bifurcations accumulate and a sequence of chaotic attractors appears. At even higher excitation amplitudes f jumps onto coexisting attractors take place (see fig. 4).

The bifurcation scenario sketched above appears first between the main resonance and the first odd secondary resonance. With increasing excitation the same sequence of bifurcations and jumps is observed between all odd secondary resonances. Fig. 4 shows sections of the bifurcation diagram between the main resonance and the first four odd secondary resonances. The excitation amplitudes in fig. 4a–d are chosen in such a way that the bifurcation cascades are at similar stages. A high degree of similarity can be detected between them. Not only are the period-doubling cascades into and out of chaos similar in fig. 4a–d, but also the gaps in between filled with period-3 cascades.

To investigate this similarity between the bifurcation trees, phase diagrams have been calculated by tracing back in computer dialogue sessions the bifurcation points of the preceding diagrams in some region of the $\omega-f$ parameter space. The bifurcation curves of the original period-$1 \cdot 2^n$ attractors, $n \in N$, and the bifurcation curves belonging to the coexisting period-$3 \cdot 2^n$ attractors, $n \in N$, are given separately in figs. 5 and 6. As can be seen in the figures, all bifurcation curves recur periodically and may be collected to what we call resonance horns. To explain this structure of the bifurcation set and give a classification of the resonance horns, we have to consider the general resonance condition $k\omega = n\Omega$. In the preceding discussion we have only considered the resonances of the higher harmonics in the Fourier series of the original

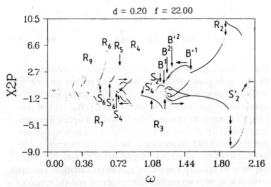

Fig. 3. Bifurcation diagram X2P(ω) for $f = 22.0$ (cf. fig. 2) showing the ω-interval below the main resonance. Resonances, additional symmetry bifurcations (at S_4, S'_4, S_6, S'_6, ...) and period-doubling bifurcations (B^1, B'^1 and B^2, B'^2) occur. In the symmetry bifurcation branches additional hysteresis jumps appear due to the leaning-over of the even secondary resonance R_2 (see also fig. 4 and text).

353

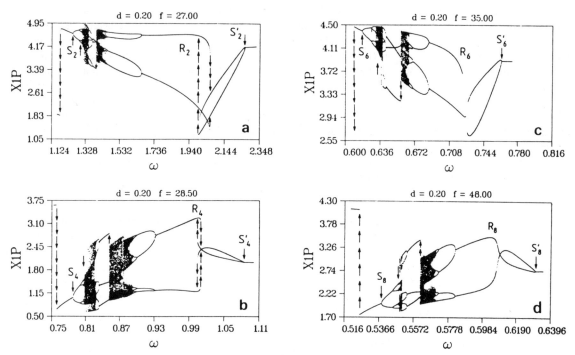

Fig. 4. Sections of the bifurcation diagram X1P(ω) for the frequency region between the first odd resonances R_1, R_3, R_5, R_7 and R_9 (see also fig. 3 and text). (a) The first cascade between R_3 and R_1; $f = 27$. (b) The second cascade between R_5 and R_3; $f = 28.5$. (c) The third cascade between R_7 and R_5; $f = 35$. (d) The fourth cascade between R_9 and R_7; $f = 48$.

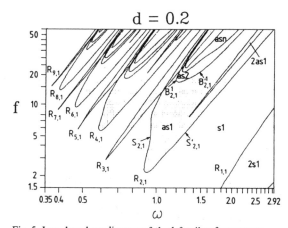

Fig. 5. Log–log phase diagram of the 1-family of resonance horns (see text). In the differently dotted areas there are symmetrical period-1 attractors (s1), two coexisting symmetrical period-1 attractors (2s1), a pair of asymmetric period-1 partner orbits (as1), two pairs of asymmetric period-1 partner orbits (2as1), a pair of period-2 partner orbits (as2) and pairs of period-n partner orbits (asn) with $n = 4, 8, 16, \dots$.

period-1 solutions. In the general case of a period-n oscillation there is a (ultraharmonic) resonance [12] whenever a harmonic of the Fourier series

$$\sum_k x_k \exp[\mathrm{i}k(\omega/n)t]$$

coincides with the eigenfrequency Ω of the oscillator. Each resonance horn in fig. 6 is composed of one or more ultraharmonic resonances of the type $k\omega \approx 3\Omega$. The k-values are connected with the winding numbers of the projected trajectories. The winding number is defined here as the number of maxima or minima of the periodic solution in one period. As in some horns oscillations with different winding number are present the k-values occurring at the lower left-hand side of the resonance horns are taken as labels. The occurrence of different winding numbers, e.g. 5 to 9 in the horn $R_{9,3}$, indicate that these horns should additionally be divided according to the winding number. This suggests the concept of a winding number bifurcation.

In view of these extended results the labels R_1, R_2, .. in the diagrams 1 to 4 used there for simplicity should be replaced by $R_{1,1}$, $R_{2,1}$, ... the first subscript being the winding number and the second the (lowest) period. This has been done in fig. 5. We call the resonance horns in fig. 5 the 1-family of resonance horns and the horns given in fig. 6 the 3-family. In the considered parameter region both families consist of two kinds of resonance horns, one with period doubling to chaos and one without period-doubling sequences. This is a consequence of the potential symmetry. In general, n-families of resonance horns exist, which will be discussed elsewhere.

An interesting feature of the bifurcation curves is their asymptotic behaviour for high excitation amplitudes. In the log–log phase diagrams in figs. 5 and 6 they follow rectilinear lines. Those lines, which belong to comparable bifurcations, have the same slope.

Figs. 5 and 6 indicate a periodicity in the global bifurcation set. Furthermore, the bifurcation diagrams in fig. 4 and other numerical studies show that the shape of the bifurcation trees belonging to a specific family recurs every fourth resonance horn.

The investigation of the Duffing equation (1) shows that even this simple nonlinear dynamical system possesses an unexpectedly rich structure in its dynamical properties. On the other hand we see how extended computations may put some order in the confusing variety of phenomena. While the local structure of the bifurcation set is given by several kinds of (may be incomplete) Feigenbaum–Grossmann scenarios, the global order is governed by the repeated resonances which generate a superstructure in the bifurcation set.

Evidence for a superstructure related to nonlinear resonances has also been found in a model of acoustic turbulence [15], in the Toda oscillator [14], and in a nonlinear bubble oscillator [12]. First indications of a superstructure can be found in experiments on acoustic turbulence [16], and in a nonlinear electronic oscillator [17,18]. We therefore conjecture that this superstructure in the global bifurcation set and its relation to the resonant properties of the system is universal for a large class of forced nonlinear oscillators.

We thank the members of the nonlinear dynamics group at the University of Göttingen, especially A. Kramer, T. Kurz and W. Meyer-Ilse, for many valuable discussions. All computations have been done on the Sperry 1100/82 and the VAX-11/780 of the Gesellschaft für wissenschaftliche Datenverarbeitung, Göttingen.

[1] G. Duffing, Erzwungene Schwingungen bei veränderlicher Eigenfrequenz und ihre technische Bedeutung (Vieweg, Braunschweig, 1918).
[2] P.J. Holmes and D.A. Rand, J. Sound Vibr. 44 (1976) 237.
[3] P. Holmes, Philos. Trans. R. Soc. A292 (1979) 419.
[4] B.A. Hubermann and J.P. Crutchfield, Phys. Rev. Lett. 43 (1979) 1743.
[5] Y. Ueda, J. Stat. Phys. 20 (1979) 181.
[6] Y. Ueda, in: Nonlinear dynamics, eds. R.H.G. Helleman (Annals New York Acad. Sci., New York, 1981) p. 422.
[7] C. Hayashi, J. Nonlinear Mech. 15 (1980) 341.
[8] C. Holmes and P. Holmes, J. Sound Vibr. 78 (1981) 161.
[9] S. Sato, M. Sano and Y. Sawada, Phys. Rev. A28 (1983) 1654.
[10] P. Holmes and D. Whitley, Physica 7D (1983) 111.
[11] J. Guckenheimer and P. Holmes, Nonlinear oscillations, dynamical systems and bifurcations of vector fields, Appl. Math. Sci. Vol. 42 (Springer, Berlin, 1983).
[12] W. Lauterborn, J. Acoust. Soc. Am. 59 (1976) 290.
[13] G.L. Oppo and A. Politi, Phys. Rev. A30 (1984) 435.
[14] W. Meyer-Ilse, Zur Resonanzstruktur nichtlinearer Oszillatoren am Beispiel des Toda Oszillators, PhD thesis Göttingen (1984), in German.
[15] W. Lauterborn and E. Suchla, Phys. Rev. Lett., to be published.
[16] W. Lauterborn and E. Cramer, Phys. Rev. Lett. 47 (1981) 1445.
[17] Th. Klinker, W. Meyer-Ilse and W. Lauterborn, Phys. Lett. 101A (1984) 371.
[18] S.D. Brorson, D. Dewey and P.S. Linsay, Phys. Rev. A28 (1983) 1201.

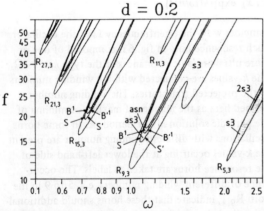

Fig. 6. Log–log phase diagram of the 3-family of resonance horns for $d = 0.2$. The meaning of the differently dotted areas is similar to those in fig. 5.

E. H. Dowell
Dean of Engineering.
Duke University,
Durham, NC 27706

C. Pezeshki
Graduate Student.
Department of Mechanical Engineering
and Materials Science,
Duke University,
Durham, NC 27706

On the Understanding of Chaos in Duffings Equation Including a Comparison With Experiment

The dynamics of a buckled beam are studied for both the initial value problem and forced external excitation. The principal focus is on chaotic oscillations due to forced excitation. In particular, a discussion of their relationship to the initial value problem and a comparison of results from a theoretical model with those from a physical experiment are presented.

Introduction, Background and Motivation

In the present paper the following equation is studied

$$\ddot{A} + \gamma \dot{A} - \frac{A}{2}(1 - A^2) = F(\tau) \qquad (1)$$

This is the particular form of Duffings equation (with a negative linear stiffness) studied by Moon [1]. It is known this equation has solutions with chaotic oscillations under certain conditions. Here we extend the earlier work on Duffings equation and provide an improved understanding of why the chaotic oscillations occur by first considering the initial value problem when $F \equiv 0$. These results are of substantial interest in their own right as well as leading to additional understanding of why chaotic oscillations occur. The present theoretical results are also compared to the physical experiments of Moon. The opportunity to compare the present theoretical results with the experimental data of Moon is also an important motivation for this work.

A physical model is helpful in the interpretation of Duffings equation. Following previous authors [1-4], we interpret equation (1) as describing a one mode oscillation of a buckled beam under the action of a prescribed lateral external force, $F(\tau)$. Other physical systems may also be described by this equation, but they will not be discussed here. As may be seen from equation (1), when $F \equiv 0$ there are three static equilibrium solutions: $A = 0$, $+1$ and -1. It is easily shown that the first of these (an unbuckled beam) is dynamically unstable and the latter two are stable with respect to *infinitesimal* disturbances. It is of great help in understanding the onset of chaos to consider next the stability of the static equilibria, $A = +1$ or -1, with respect to *finite* disturbances. This is done in the following section of the paper and then

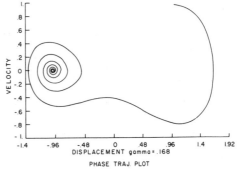

Fig. 1 Phase plane trajectory

chaos due to a harmonic force, $F(\tau)$, is considered directly in subsequent sections.

The Initial Value Problem for the Homogeneous Duffings Equation ($F \equiv 0$)

It is helpful to think first in physical terms. Consider the buckled beam at rest in one of its stable (with respect to *infintesimal* disturbances) static equilibria, say $A = +1$. With prescribed initial conditions,

$$A(\tau \equiv 0) = A_0$$
$$\dot{A}(\tau \equiv 0) = \dot{A}_0$$

consider the transient solution and the final steady state solution as $\tau \to \infty$. Obviously $A(\tau \to \infty) \to +1$ or -1. The question is which of these two solutions is the correct one for given $A_0, \dot{A}_0$. As shall be seen, the answer is in a certain sense unknowable (or to use a more technical term, *uncertain*). Once the reason for this is understood, the occurrence of chaos for certain $F \neq 0$ becomes more understandable, perhaps even expected.

It is possible to construct a diagram (which is called a *shell plot* because of its appearance) that summarizes compactly

---

Contributed by the Applied Mechanics Division and presented at the Winter Annual Meeting, Miami, Fla., November 17-21, 1985 of THE AMERICAN SOCIETY OF MECHANICAL ENGINEERS.

Discussion on this paper should be addressed to the Editorial Department, ASME, United Engineering Center, 345 East 47th Street, New York, N.Y. 10017, and will be accepted until two months after final publication of the paper itself in the JOURNAL OF APPLIED MECHANICS. Manuscript received by ASME Applied Mechanics Division, September 11, 1984; final revision, April 30, 1985. Paper No. 85-WA/APM-27.

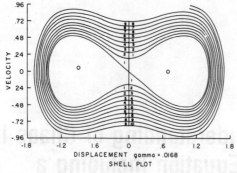

Fig. 2 Shell plot

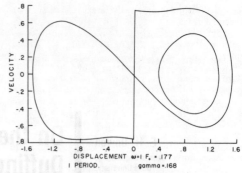

Fig. 3(a) Phase plane trajectory (1 period motion)

which of the two static equilibria solutions will be reached for a given set of initial conditions, A_0, $\dot{A}_0$.

To anticipate the form of the shell plot, consider a specific example of initial conditions and the subsequent solution trajectory. This is shown as a phase plane trajectory in Fig. 1 for $A_0 = 1$ and $\dot{A}_0 = 1$. Because $\dot{A}_0 > 0$, A increases for small time but then decreases for larger time because of the nonlinear restoring stiffness. Indeed, A subsequently becomes negative (the beam moves from one buckled configuration, $A = +1$, to the other, $A = -1$, and beyond). The damping term, $\gamma \dot{A}$, leads to dissipation of energy; thus the beam does not continue to oscillate between and about the two static buckled equilibria, but instead spirals into one of them as $\tau \to \infty$.

From such phase plane trajectories, one can construct the shell plot, which shows the final state of the system as $\tau \to \infty$, $A = +1$ or -1, for given initial conditions, A_0 and $\dot{A}_0$. This is shown (partially) in Fig. 2. Here $\dot{A}_0$ is plotted versus A_0 and various regions are identified with integral values, 0, 1, 2, 3, 4, ... Note there are two disjoint regions associated with each integer value. Consider first the integer zero (0) regions. For definiteness consider the region where $A_0 \geq 0$. If the system starts with A_0, $\dot{A}_0$ within the zero region, the solution spirals into $A = +1$ as $\tau \to 0$ and crosses the $A = 0$ axis zero times. Consider now the 1 region. A solution begun there moves clockwise and crosses the $A = 0$ axis one time and enters the zero region for $A < 0$. Once there, it spirals into $A = -1$ as $\tau \to \infty$. To firmly establish the pattern, finally consider the 2 region. For initial conditions in the 2 region, the phase plane trajectory moves clockwise, crosses the $A = 0$ axis the first time and moves into the 1 region for $A < 0$. It then continues to move clockwise and crosses the $A = 0$ axis a second (and final time) and moves into the 0 region for $A > 0$. Once there it spirals into $A = +1$ as $\tau \to \infty$.

The pattern is now clear. For $A_0 > 0$, initial conditions in an even integer region reach a final state of $A = +1$. Those in an odd integer region reach a final state of $A = -1$. The integer number corresponds to the number of crossings of the $A = 0$ axis during the completion of the motion (phase plane trajectory). For $A_0 < 0$, a similar sequence of events occurs. For initial conditions which lie precisely on a shell boundary, the final configuration would be $A, \dot{A} \to 0$ as $\tau \to \infty$. In practice this will never occur, of course, because the shell boundary curves are of vanishing thickness.

It is interesting to note that a shell plot of any finite extent can be constructed from a single artfully chosen phase plane trajectory, once the zero region is known. The latter region is readily determined by direct calculation.

Now comes the central point. If there is *sufficient* uncertainty in the values of the initial conditions, A_0, $\dot{A}_0$, it is clear from an examination of the shell plot that the final system state, $A = +1$ or -1, is unpredictable, unknowable or

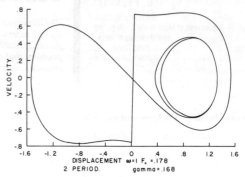

Fig. 3(b) Phase plane trajectory (2 period motion)

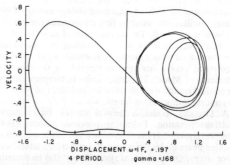

Fig. 3(c) Phase plane trajectory (4 period motion)

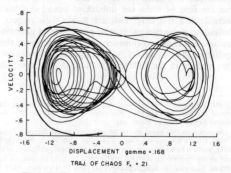

Fig. 3(d) Phase plane trajectory (chaotic motion)

uncertain. This point is made all the more powerful by noting that as the damping becomes even smaller the width of each region of the shell plot (excluding the 0 region) becomes even smaller and vanishes as $\gamma \rightarrow 0$. Hence for any (finite) uncertainty in A_0 or $\dot{A}_0$ the final system state is unpredictable as $\gamma \rightarrow 0$.

Two additional points are worthy of note in concluding this discussion. First, the boundary contours of the shell plot are curves of essentially constant total (kinetic plus potential) energy. Secondly as $\gamma \rightarrow 0$, the boundary curves for the two zero regions correspond to the separatrix of the undamped system.

Although not concerned with chaos, the reader will find Ref. 5 on the initial value problem of interest.

The Continuous Oscillation Problem for the Inhomogenous Duffings Equation ($F \neq 0$)

Here a simple harmonic external force is considered,

$$F = F_0 \sin \omega t \qquad (2)$$

where F_0 is the force amplitude and ω its frequency of excitation. This is not the only force-time history which might be studied. It is, perhaps, the simplest periodic force.

As the reader may note, the initial value problem previously studied can be also thought of as an external force problem. For example, an initial velocity, $\dot{A}_0$, corresponds to an impulsive force,

$$F(\tau) = \dot{A}_0 \delta(\tau) \qquad (3)$$

This suggests that a study of continual impulses, periodically or randomly spaced in time, would be of interest. Nevertheless, only a simple harmonic force will be considered here.

The response of the system will be considered first for fixed frequency, $\omega = 1$, and increasing force amplitude, F_0. The frequency is normalized by the small amplitude natural frequency about the beam buckled equilibrium. For F_0 sufficiently small, it is expected that the response of the system will be a simple harmonic oscillation about one or the other of the two static equilibria, $A = +1$, or -1. For definiteness the initial conditions, $A_0 = 1$, $\dot{A}_0 = 0$ are chosen so that for small F_0 the harmonic response oscillation is about $A = +1$. It is anticipated that, as F_0 increases and the response phase plane trajectory approaches the zero region boundary of the shell plot, interesting response behavior will occur.

Note that for small F_0 the phase plane trajectory is an ellipse indicating a simple harmonic response oscillation. As F_0 increases additional harmonics beyond the fundamental are detected and the phase plane trajectory is distorted from a simple ellipse. See Fig. 3(a) for the result for $F_0 = 0.177$. Also shown for reference is the boundary for the zero region from the previously discussed shell plot.

For $0 \leq F_0 \leq .177$ the response is termed 1 period motion. By that is meant, as the force oscillates through one period, the response also oscillates through one period. For $F_0 = .178$, however, as the force oscillates through one period the response oscillates through only one half a period. For the response to go through one period, the force must oscillate through two periods. Thus this is called 2 period motion. See Fig. 3(b). This change from 1 to 2 period response as $F_0 = 0.177 \rightarrow 0.178$ appears to be a bifurcation.

At a higher F_0, 4 period motion occurs. See Fig. 3(c) for example, and at yet higher F_0, 8 and 16 period motion occurs. Holmes [4] has suggested that 32, etc. period motions occur as F_0 increases further. This may well happen but this behavior has not been observed by the present authors. Possibly this is because the range of F_0 over which the higher period motions occur is very small. This period doubling behavior has been previously described and discussed by Feigenbaum [6] in a more general context.

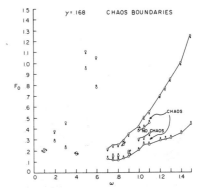

Fig. 4(a) Force amplitude versus frequency of excitation

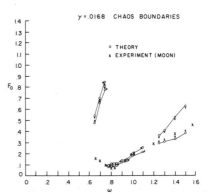

Fig. 4(b) Force amplitude versus frequency of excitation

For $F_0 \geq 0.205$ chaos is observed, i.e., no periodicity is apparent. See Fig. 3(d) which gives results for $F_0 = 0.21$. As Holmes has indicated, for yet higher F_0 the chaos no longer appears and periodicity returns.

It is clear that for F_0 just below the value where chaos first appears the periodic response phase plane trajectory approaches and slightly penetrates the boundary of the zero region shell plot. See Fig. 3(c). Moreover it is clear that for this frequency, $\omega = 1$, chaos occurs when the motion is no longer about only one of the static equilibria, say $A = +1$, but instead encircles both, $A = +1$ and -1. This is called snap-buckling. These observations suggest that the onset of chaos can be associated with periodic motions which penetrate the zero region boundary and thus lead subsequently to motion about both static equilibria points. Moon in an earlier paper [1] suggested a more restricted notion of this sort when he took as an empirical criterion for the onset of chaos that the periodic response maximum velocity (in his calculation he assumed one period motion) must exceed the maximum velocity of the system separatrix. Recall it has been shown here that the zero region boundary of the shell plot corresponds to the system separatrix as $\gamma \rightarrow 0$.

It is speculated, though it remains to be shown, that as $\gamma \rightarrow 0$ *any* penetration of the zero region boundary by the phase plane trajectory leads to chaos. For small, but finite, γ the phase plane trajectory may (slightly) penetrate the zero region boundary before chaos occurs. Hence the penetration of the zero region boundary by a phase plane trajectory at a certain force level may provide a lower bound criterion for the onset of chaos, at least for ω near 1.

For excitation frequencies well away from resonance, in particular for $\omega < < 1$, chaos was found to occur even without

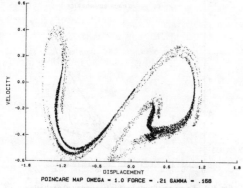

Fig. 5(a) Poincare map; $\omega = 1$, $\gamma = .168$, $F_0 = .21$

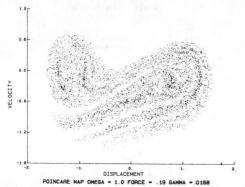

Fig. 5(b) Poincare map; $\omega = 1$, $\gamma = .0168$, $F_0 = .09$

snap-buckling. The minimum F_0 for the onset of chaos occurs for $\omega \cong 0.85$. This is the characteristic frequency for free vibration about both static equilibrium points.

Comparison of Theory and Experiment

Calculations similar to those described in the previous section have been carried out for several excitation frequencies, ω. From these a summary plot may be constructed of the force required to cause snap buckling, period doubling, or chaos versus the excitation frequency. Such a plot for the onset of chaos is shown in Fig. 4(a) and 4(b) for $\gamma = 0.168$ and 0.0168, respectively. The uncertainty in the data is less than a diameter of a circle.

Time integrations using the Runge-Kutta method were performed for frequency values ranging from 0.1 to 1.5 for varying force levels at damping levels of 0.168 and 0.0168. Principal lower and upper force chaos boundaries were found for a discrete series of frequencies by incrementing the first from zero until chaos was observed. See Fig. 4. Force increments of 0.01 and, where necessary, 0.001, were used. All results shown are for simulations started from the initial condition values of one for the displacement and zero for the velocity. Other initial conditions were tried, but no observable effect of initial conditions on chaos boundaries was detected. Of course, the time history details do depend upon initial conditions, particularly in the chaotic regime.

The types of chaos found in the simulations varied from frequency to frequency. However, the form of the Poincare map for a given set of frequencies tended to be of the same type when the lower force steady state periodic phase plane portraits were shaped the same and possessed the same periodicity and when the corresponding upper force portraits were also the same across the frequency band. The lower and upper plane portraits were not identical. Such identification or association of results at several frequencies has led us to connect some data points in Fig. 4(a) and 4(b) by straight lines. Such results suggest that chaos lies in fragmented pockets in the force-frequency plane. These pockets can also have smaller pockets of force-frequency combinations within them that can lead to periodic phase plane orbits.

It is apparent from the situations simulated that chaos can assume many forms. Some of the entries into the chaotic region in the force-frequency plane for certain frequency values were precipitated by beam snap-through; others were not. At certain frequencies, the system went into chaos even before the beam snapped through. Period doubling was observed at some frequencies, e.g., $\omega = 0.9$ and 1.0, but not at others. Chaos appeared at all frequencies simulated, though this does not preclude the possibility of finding frequencies that are chaos-free. A simple boundary cannot be drawn in the force-frequency plane above which there is guaranteed chaos; in fact, the simulations point to the opposite.

The simulations run at high damping levels gave the same qualitative answer as the ones run at low damping. The major difference is that the width of the chaotic band in the force-frequency plane for the low damping case is much narrower than its counterpart for higher damping. The limit of zero damping may be pathological. Another difference is that the higher damping case allows a much richer selection of equilibrium periodic phase plane orbits. As the damping is decreased, the Poincare maps also lose their ordered structure.

The correlation between data obtained from simulation and the data obtained by Moon from his physical experiment also appears to be generally good. See Fig. 4(b). The principal difference is that at $\omega \cong 0.65$, the simulation predicts chaos at much higher force levels, $F_0 \neq 0.45\text{-}0.55$, than those observed by Moon in his physical experiment, $F_0 \cong .17$. It is worthy of note that the simulation predicts that snap-through of the beam occurs at $F_0 \cong .12$ and it is possible that this snap-through was identified as chaos in the physical experiment. At higher frequencies, snap-through and chaos occur at force levels which are much closer together. Of course other factors may enter in including the effects of higher beam modes.

For brevity, we have not shown the large number of phase-plane portraits and Poincare maps that have been calculated. The authors would be pleased to make these available to other investigators who may wish to extend the present study. In Fig. 5(a) and 5(b), two representative Poincare maps are shown for $\omega = 1$ and the two damping values used in this study.

Conclusions and Future Work

Among the conclusions reached based upon the present work are the following:

(1) The initial value problem for a second order homogeneous system is a key to the understanding of higher order systems, including the inhomogeneous second order system.

(2) Chaos is not difficult to find by numerical simulation, however a Feigenbaum (period doubling) sequence may be difficult to find for some parameter conditions.

(3) A comparison between theoretical results for Duffing's equation and (physical) experiments for a buckled beam shows generally good agreement.

(4) Future theoretical studies should consider

- investigating the limit as damping approaches zero; setting the damping identically zero may lead to pathological results
- multimode convergence studies (based upon the results from panel flutter calculations [7, 8], it is expected good convergence will occur)

(5) Future experimental work should attempt to study
- various damping levels
- determination of period doubling conditions
- identification of entire pockets of chaos

Acknowledgment

This work was supported, in part, by NSF Grant MEA-8315193 and AFOSR Grant 83-0346. Drs. Elbert Marsh and Anthony Amos, respectively, are the technical monitors. The authors would like to thank Mr. Michael D'Antonio for his help with the computations. They would also like to thank Professor Francis C. Moon for helpful discussions.

References

1 Moon, F. C., "Experiments on Chaotic Motions of a Forced Nonlinear Oscillator: Strange Attractors," ASME JOURNAL OF APPLIED MECHANICS, Vol. 47, 1980, pp. 638-644.

2 Holmes, P. J., and Moon, F. C., "Strange Attractors and Chaos in Nonlinear Mechanics," ASME JOURNAL OF APPLIED MECHANICS, Vol. 108, 1983, pp. 1021-1032.

3 Tseng, W.-Y., and Dugundji, J., "Nonlinear Vibrations of a Buckled Beam Under Harmonic Excitation," ASME JOURNAL OF APPLIED MECHANICS, Vol. 38, 1971, pp. 467-476.

4 Holmes, P. J., "A Nonlinear Oscillator with a Strange Attractor," *Phil. Trans. of Royal Society*, London, Vol. 292, 1979, pp. 419-448.

5 Reiss, E. L., and Matkowsky, B. J., "Nonlinear Dynamic Buckling of a Compressed Elastic Column," *Quart. Appl. Math.*, Vol. 29, 1971, pp. 245-260.

6 Feigenbaum, M. J., "Quantitative Universality for a Class of Nonlinear Transformations," *J. Stat. Physics*, Vol. 19, 1978, pp. 25-52.

7 Dowell, E. H., "Flutter of a Buckled Plate as an Example of Chaotic Motion of a Deterministic Autonomous System," *J. Sound Vibration*, Vol. 85, 1982, pp. 333-344.

8 Dowell, E. H., "Observation and Evolution of Chaos in an Autonomous System," ASME JOURNAL OF APPLIED MECHANICS, Vol. 51, 1984, pp. 333-344.

Chaotically Transitional Phenomena in the Forced Negative-Resistance Oscillator

YOSHISUKE UEDA AND NORIO AKAMATSU

Abstract—This paper deals with chaotically transitional phenomena which occur in the forced negative-resistance oscillator. Experimental studies using analog and digital computers have been carried out. The difference between the almost periodic oscillations and the chaotically transitional processes is clarified. Various strange attractors representing chaotically transitional processes and their average power spectra are given. They are discussed in detail and compared with the results obtained in the forced oscillatory systems.

I. INTRODUCTION

WHEN A PERIODIC excitation is injected into the self-oscillatory system, either synchronized periodic oscillation or asynchronized nonperiodic oscillation comes out. Among the latter type of oscillations, chaotic oscillation frequently takes place which is different from almost periodic oscillations and is not reproducible in every experiment. However, owing to the perfectly deterministic nature of the equations describing the system, an appearance of such chaotic oscillation has been known very little for a long time. The chaotic oscillations are attributed to both the global structure of the solutions of the deterministic differential equations of the system and the small uncertain factors in the physical system. We have long been studying on this kind of oscillations and have called the phenomenon the chaotically transitional process [1]–[5].

This paper deals with chaotically transitional processes which occur in the computer-simulated systems of the forced negative-resistance oscillator. Regions of various types of frequency entrainment are first shown. Subsequently, the difference between almost periodic oscillations and chaotically transitional processes is explained in detail adducing respective examples. Then, various types of strange attractors and their average power spectra have been given. At the end of this paper they are discussed and are also compared with those obtained in the system exhibited by Duffing's equation.

II. PRELIMINARIES

In this chapter we derive the differential equations from the negative-resistance oscillator with periodic excitation.

Manuscript received April 28, 1980; revised September 3, 1980. This work has been carried out in part under the Collaborating Research Program at the Institute of Plasma Physics, Nagoya University, Japan. The authors wish to express their sincere thanks to the staff of the Institute. This work was presented at the Special Session on Computation and Simulation of Nonlinear Circuits and Systems, Houston, TX.
Y. Ueda is with the Department of Electrical Engineering, Kyoto University, Kyoto 606, Japan.
N. Akamatsu is with the Department of Information Science, Tokushima University, Tokushima 770, Japan.

©1981 IEEE. Reprinted, with permission, from IEEE Trans. Circuits Systems, Vol. CAS-28, No. 3 pp. 217-224; March/1981

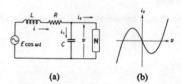

Fig. 1 (a) Negative-resistance oscillator with periodic excitation. (b) Voltage–current characteristic of negative resistance.

Then we present a brief summary of the previous investigations relating to the chaotically transitional processes [1]–[5].

A. Equations Describing the Forced Negative-Resistance Oscillator

In this paper, the negative-resistance oscillator of Fig. 1 is studied. A periodic excitation $E\cos\omega t$ can be injected into the circuit. The differential equations for the system in terms of the variables as shown in the figure are readily derived; that is,

$$L\frac{di}{dt} + Ri + v = E\cos\omega t$$
$$i_1 = C\frac{dv}{dt}, \qquad i = i_1 + i_2. \qquad (1)$$

We now make the assumption that the voltage–current characteristic is given by

$$i_2 = f(v) = -Sv\left(1 - \frac{v^2}{V_s^2}\right) \qquad (2)$$

with $S = 1/R$ and V_s constants, where R is the resistance of the inductor L. Let us introduce new quantities:

$$x = \sqrt{\gamma}\,\frac{v}{V_s}, \qquad \tau = \frac{t}{\sqrt{\gamma LC}}$$
$$B = \gamma\sqrt{\gamma}\,\frac{E}{V_s}, \qquad \nu = \sqrt{\gamma LC}\,\omega$$
$$\mu = \sqrt{\frac{3S}{C}(LS - RC)}, \qquad \gamma = \frac{3LS}{LS - RC} \qquad (3)$$

in terms of which (1) become

$$\frac{dx}{d\tau} = y$$
$$\frac{dy}{d\tau} = \mu(1-x^2)y - x^3 + B\cos\nu\tau. \qquad (4)$$

It should be noted that, though small uncertain factors exist in the real system, there is no room for the introduction of random parameters in the course of the above formulation. In fact, we can see that regular or deterministic oscillations in the system of Fig. 1 are well explained by (4). Moreover, through the series of our investigations on chaotic oscillations, it can be said that the deterministic differential equations can describe the essence of this kind of phenomenon. Accordingly, the deterministic equations (4) can be regarded as the proper mathematical model for the system, as shown in Fig. 1.

B. Brief Summary Relating to the Chaotically Transitional Processes [1]–[5]

Regular or deterministic phenomenon is represented by the single solution of the equations, which is asymptotically stable having a wide basin as compared with random noise in the real system. Whereas chaotic or random one corresponds to a bundle of solutions. Namely, the representative point of the state of the circuit, e.g., voltages and/or currents, keeps wandering in the solution bundle under the influence of small noise in the physical system. Accordingly, the physical state is not determined uniquely within the extent of the bundle of solutions. Thus stochastic properties are introduced into the actual phenomenon. However, it seems that the time evolution characteristic of the resulting chaotic phenomenon depends only on the structure of the bundle of solutions regardless of the nature of uncertain factors of the real system.

In the following we briefly explain the terminology relating to the chaotically transitional processes.

(a) Bundle of Solutions: Let us consider the case in which chaotic phenomenon takes place in the circuit exhibited by (4). These chaotically transitional processes $\{X(\tau)\}$ and $\{Y(\tau)\}$ correspond to a bundle of solutions in the τxy space which is asymptotically orbitally stable and contains infinitely many unstable periodic solutions. However, the complex structure of the solution bundle is not fully understood.

(b) Strange Attractor: A strange attractor is given by the set of points on the xy plane consisting of the cross section of the solution bundle at $\tau = 2n\pi/\nu$ ($n \in Z$). The strange attractor is identical with a closure of unstable manifolds of a saddle point of a diffeomorphism f_λ: $\mathbb{R}^2 \to \mathbb{R}^2$, which is defined by using the solutions of (4). The unstable manifolds of these cases are infinitely long but are confined within the bounded region and present homoclinic structure intersecting with stable manifolds. Homoclinic structure causes the existence of infinitely many periodic points. However, the decomposition of strange attractor is still open.

(c) Discrete Dynamical System: A diffeomorphism on the xy plane into itself is introduced by using the solutions of (4). To see this, let $x = x(\tau, x_0, y_0)$, $y = y(\tau, x_0, y_0)$ be a solution of (4) starting from a point at $p_0 = (x_0, y_0)$ at $\tau = 0$. Let $p_1 = (x_1, y_1)$ be the location of this solution at $\tau = 2\pi/\nu$, i.e., $x_1 = x(2\pi/\nu, x_0, y_0)$, $y_1 = y(2\pi/\nu, x_0, y_0)$; then a C^∞-diffeomorphism

$$f_\lambda: \mathbb{R}^2 \to \mathbb{R}^2$$
$$p_0 \mapsto p_1$$
$$\lambda = (\mu, B, \nu) \quad (5)$$

of the xy plane into itself is defined.

The diffeomorphism f_λ is not a contracting type like one derived from Duffing's equation. Because the system under consideration is of self-oscillatory, while the system exhibited by Duffing's equation is quiescent unless an external periodic force is applied.

(d) Average Power Spectrum: By regarding the process $\{X(\tau)\}$ as the periodic random process $\{X_T(\tau)\}$ with sufficiently long period T, which is a multiple of $2\pi/\nu$, a realization $x_T(\tau)$ is expanded into Fourier series as

$$x_T(\tau) = \frac{1}{2} a_0 + \sum_{m=1}^{\infty} (a_m \cos m\omega_0 \nu \tau + b_m \sin m\omega_0 \nu \tau) \quad (6)$$

where

$$a_m = \frac{2}{T} \int_{-T/2}^{T/2} x_T(\tau) \cos m\omega_0 \nu \tau \, d\tau, \quad m = 0, 1, 2, \cdots$$

$$b_m = \frac{2}{T} \int_{-T/2}^{T/2} x_T(\tau) \sin m\omega_0 \nu \tau \, d\tau, \quad m = 1, 2, 3 \cdots$$

$$T = 2\pi/\omega_0 \nu. \quad (7)$$

From these coefficients, the average power spectrum $\Phi_X(\omega)$ of the random process $\{X(\tau)\}$ can be estimated as follows:

$$\Phi_X(\omega) = \lim_{T \to \infty} \left\langle \frac{1}{T} \left| \int_{-T/2}^{T/2} x_T(\tau) e^{-i\omega\tau} d\tau \right|^2 \right\rangle$$
$$\doteq \frac{2\pi}{\omega_0 \nu} \left\langle \frac{1}{4}(a_m^2 + b_m^2) \right\rangle, \quad \omega = m\omega_0 \nu. \quad (8)$$

It should be emphasized repeatedly that, the average power spectra of the chaotically transitional processes depend practically not on the nature of uncertain factors but on the structure of the bundle of solutions.

III. EXPERIMENTAL RESULTS

By making use of analog and digital computers, we give experimental results on the chaotically transitional phenomena that occur in the computer-simulated systems exhibited by (4). The regions of frequency entrainment are first shown. Subsequently the differences are explained in detail between the almost periodic oscillations and the chaotically transitional processes that occur when the frequency entrainment is not realized. Then the interesting examples of strange attractors are shown together with corresponding average power spectra. At the end of this chapter these results are discussed. Special attention is also directed towards the comparison of the results with those obtained in the system exhibited by Duffing's equation.

A. Regions of Frequency Entrainment

In the forced self-oscillatory system exhibited by (4), the entrainment occurs at various frequencies depending

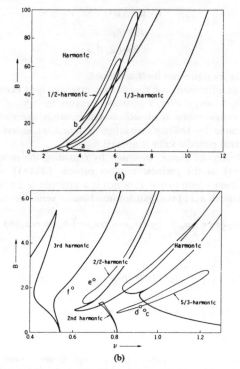
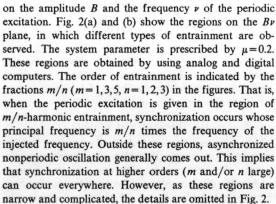

Fig. 2. Regions of frequency entrainment. (a) Harmonic and subharmonic entrainments. (b) Harmonic, higher harmonic, and ultrasubharmonic entrainments.

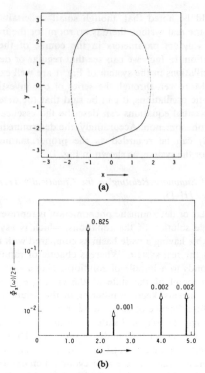

Fig. 3. Almost periodic oscillation which occurs when the periodic excitation is prescribed by $B=1.0$ and $\nu=4.0$. (a) Invariant closed curve. (b) Average power spectrum.

on the amplitude B and the frequency ν of the periodic excitation. Fig. 2(a) and (b) show the regions on the $B\nu$ plane, in which different types of entrainment are observed. The system parameter is prescribed by $\mu=0.2$. These regions are obtained by using analog and digital computers. The order of entrainment is indicated by the fractions m/n ($m=1,3,5, n=1,2,3$) in the figures. That is, when the periodic excitation is given in the region of m/n-harmonic entrainment, synchronization occurs whose principal frequency is m/n times the frequency of the injected frequency. Outside these regions, asynchronized nonperiodic oscillation generally comes out. This implies that synchronization at higher orders (m and/or n large) can occur everywhere. However, as these regions are narrow and complicated, the details are omitted in Fig. 2.

When the system is not entrained, either almost periodic oscillation or chaotically transitional process comes out. Which one occurs depends on the values of B and ν. As it is difficult to draw lines of demarcation, they are also omitted in the figure. However, we add that those regions seem to be separated by the regions of higher order entrainment and that almost periodic oscillations are likely to occur at small values of B for all ν.

B. Difference Between Almost Periodic Oscillations and Chaotically Transitional Processes

In order to explain the difference between almost periodic oscillations and chaotically transitional processes, let us first give an example of almost periodic oscillations. Fig. 3(a) shows an attractor represented by the invariant closed curve C of the diffeomorphism f_λ. From this result, the curve C can be regarded as a smooth simple closed curve of finite length. The amplitude B and the frequency ν of the periodic excitation are given by

$$B=1.0 \text{ and } \nu=4.0$$

which are located just below the region of 1/3 harmonic entrainment and are marked by a in Fig. 2(a).

All the solutions of (4) starting from C at $\tau=0$ form a surface in the τxy space. As the curve C is invariant, the surface between $0 \leqslant \tau < 2\pi/\nu$ can be mapped on the closed torus. Thus the movement of a representative point of (4) restricted on C can be considered in connection with a differential equation on a torus and it is characterized by a rotation number ρ, which gives average advance of an image under the diffeomorphism f_λ of the curve C into itself. The value of ρ for the invariant closed curve C of Fig. 3(a) is approximately estimated by $\rho=0.403,\cdots$. Here let us suppose that $\rho=0.403,\cdots$, is irrational, then x and y of (4) can be represented by almost periodic functions of τ [6], [7].

Though we can not see strictly that ρ is rational or irrational, it seems no difficulty to regard ρ as irrational from a practical point of view. For if ρ is rational and of the form m/n, then the diffeomorphism (5) has n-periodic points on the closed curve C. In the example of Fig. 3(a),

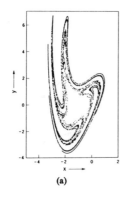

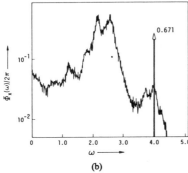

Fig. 4. Chaotically transitional process which occurs when the periodic excitation is prescribed by $B=17.0$ and $\nu=4.0$. (a) Strange attractor. (b) Average power spectrum.

however, the domains of attraction for these periodic points are considered to be narrow due to large n, so they are covered with small noise in the real system. Moreover, in relation to the structural stability, the rotation number ρ can not remain constant but fluctuate slightly according to the small perturbations of the system parameters.

Fig. 3(b) shows the average power spectrum of the almost periodic oscillation represented by the invariant closed curve of Fig. 3(a). We can see no continuous finite (random) components in the figure. There are only four line spectra at $\omega = \nu_0, 4\nu_0 - \nu, \nu,$ and $3\nu_0$, where $\nu_0 = 1.614$ is the angular frequency of the self-excited component and $\nu = 4.0$ is that of the periodic excitation.[1] Numerical values represent the power focussed on those frequencies. From these results, this almost periodic oscillation can be regarded as a slightly modified self-excited oscillation accompanying with the injected periodic component.

Next let us give an example of chaotic oscillations. Fig. 4(a) shows a strange attractor representing chaotically transitional process. This is obtained by plotting the representative points of the computer-simulated system at the instant $\tau = 2n\pi/\nu$ ($n \in Z$) after the transient has vanished. The amplitude B and the frequency ν of the periodic excitation are given by

$$B = 17.0 \quad \text{and} \quad \nu = 4.0$$

which are located just outside the region of harmonic entrainment and are marked by b in Fig. 2(a). This point b is also in the encroached portion of the region of the 1/2-harmonic entrainment. Different from the attractor of Fig. 3(a), this strange attractor is not a simple closed curve but is identical with a closure of unstable manifolds of saddles of the diffeomorphism f_λ.[2] As mentioned before, the representative point of the physical state wanders chaotically among the bundle of solutions starting from the strange attractor at $\tau = 0$. In this manner, stochastic properties are brought into the physical phenomena resulting from small uncertain factors in the real system. Fig. 4(b) shows the average power spectrum for this chaotically transitional process. We can see in the figure the continuous finite components of the spectrum representing the stochastic properties.

C. Strange Attractors and Average Power Spectra

In this section we give some interesting strange attractors which occur in the computer-simulated system exhibited by (4). The average power spectra corresponding to them are also estimated. They are shown in Figs. 5–8. The amplitude B and the frequency ν of the periodic excitation are as follows:

Fig. 5: $B = 1.0$ and $\nu = 0.94$ (Point c in Fig. 2(b))
Fig. 6: $B = 1.2$ and $\nu = 0.92$ (Point d in Fig. 2(b))
Fig. 7: $B = 2.4$ and $\nu = 0.7$ (Point e in Fig. 2(b))
Fig. 8: $B = 2.0$ and $\nu = 0.6$ (Point f in Fig. 2(b)).

In the same manner as in the preceding section, the location of these parameters are also shown in Fig. 2(b).

D. Discussion

In this section we discuss the experimental results obtained in the preceding sections.

1) The intrinsic difference between the almost periodic oscillations and the chaotically transitional processes can be summarized as follows.

The almost periodic oscillation is represented by a smooth invariant closed curve of finite length, which consists of only one minimal set representing almost periodic functions (ρ: irrational) or contains some periodic groups of finite order representing ultra-subharmonic functions (ρ: rational).[3] In any case, small uncertain factors of the real system bring no remarkable stochastic properties into the corresponding physical phenomenon.

On the other hand, chaotically transitional process is represented by a strange attractor, which contains infinitely many unstable minimal sets.[4] This global structure

---

[1] Actually, line spectra exists at $\omega = \nu - 2\nu_0, \cdots$, but they are omitted as their powers are small. It should be added that, when $B = 0$, the angular frequency of the self-excited oscillation is given by $\nu_0 = 1.617$ and the amplitude (fundamental) by $r_1 = 1.824$.

[2] The global phase plane structure of the diffeomorphism f_λ of this case was shown in fig. 7 of [1]. This picture is reproduced in the Appendix (Fig. 9).
[3] In this case, periodic solutions corresponding to these periodic groups are composed of the same frequency components.
[4] Among them, the existence of periodic groups of various orders has been guaranteed by the homoclinic structure. These periodic solutions have different frequency components one another.
There may exist minimal sets different from the periodic groups, but the details are open.

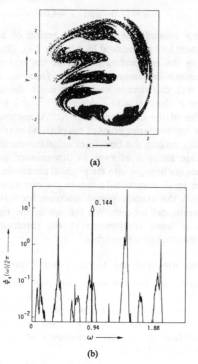

(a)

(b)

Fig. 5. Chaotically transitional process which occurs when the periodic excitation is prescribed by $B=1.0$ and $\nu=0.94$. (a) Strange attractor. (b) Average power spectrum.

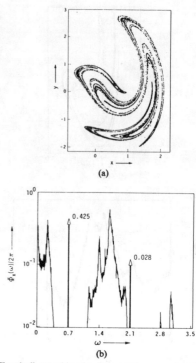

(a)

(b)

Fig. 7. Chaotically transitional process which occurs when the periodic excitation is prescribed by $B=2.4$ and $\nu=0.70$. (a) Strange attractor. (b) Average power spectrum.

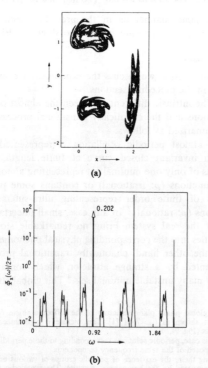

(a)

(b)

Fig. 6. Chaotically transitional process which occurs when the periodic excitation is prescribed by $B=1.2$ and $\nu=0.92$. (a) Strange attractor. (b) Average power spectrum.

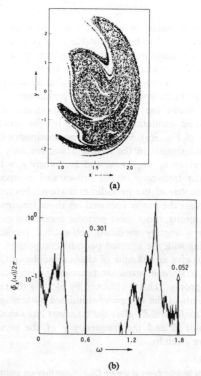

(a)

(b)

Fig. 8. Chaotically transitional process which occurs when the periodic excitation is prescribed by $B=2.0$ and $\nu=0.60$. (a) Strange attractor. (b) Average power spectrum.

TABLE I

Point in Fig. 2	Strange Attractor	Periodic Excitation B	Periodic Excitation ν	Power Total Power	Power Periodic Component	Power Chaotic Component	Remarks
		0	1.617	1.67	1.67	0	Self-Excited Oscillation
a	Fig. 3a	1.0	4.0	1.66	1.66	0	Almost Periodic Oscillation
b	Fig. 4a	17.0	4.0	2.74	1.34	1.40	Chaotically Transitional Process
c	Fig. 5a	1.0	0.94	1.01	0.29	0.72	Chaotically Transitional Process
d	Fig. 6a	1.2	0.92	1.02	0.40	0.62	Chaotically Transitional Process
e	Fig. 7a	2.4	0.70	1.24	0.90	0.34	Chaotically Transitional Process
f	Fig. 8a	2.0	0.60	1.04	0.71	0.33	Chaotically Transitional Process

of the solutions itself brings stochastic properties into the actual phenomenon and the small uncertain factors of the real system do not paly very important role.

Corresponding to these facts, continuous finite (random) components of the average power spectrum do not appear in the almost periodic oscillations, but they do appear in the chaotically transitional processes.

2) We can see seven blocks in the strange attractor of Fig. 5(a), which are connected with one another. In the corresponding average power spectrum of Fig. 5(b), continuous finite components exist in some limited parts on the ω-axis and are distributed nearly symmetrically with respect to the injected frequency $\omega = \nu = 0.94$. The highest peak of them appears at $\omega = \nu_0' = 1.48$. The self-excited oscillation ($\nu_0 = 1.617$) is interpreted to be transformed into these random components. Taking the relation $\nu_0'/\nu \doteq 11/7$ into account, this chaotically transitional process has developed from the 11/7-harmonic entrainment.

Except that the strange attractor is separated into three parts, the nearly same characters with $\nu_0'/\nu \doteq 5/3$ are recognized in the result of Fig. 6. This chaotically transitional process has developed from the 5/3-harmonic entrainment.

Though the phenomena are chaotic, these two examples preserve vestiges of almost periodic oscillations.

3) The strange attractors in Figs. 4, 5, and 6 encircle sources of the diffeomorphism f_λ, while the strange attractor in Fig. 7(a) does not. This implies that the self-oscillatory component in the case of Fig. 7 has been suppressed by the injected signal. And this strange attractor resembles that of Fig. 3 in [2] exhibited by Duffing's equation

$$\frac{d^2x}{dt^2} + k\frac{dx}{dt} + x^3 = B\cos t. \quad (9)$$

The average power spectrum of Fig. 7(b) has continuous finite components which are distributed to the range $\omega = 0 \sim 0.4$, $1.2 \sim 2.0$ and have the peaks at $\omega = 0.24$, 1.65.

The strange attractor of Fig. 8(a) is so complicated that the existence of sources can not readily be seen in the figure. However, the average power spectrum looks like that of Fig. 7(b). That is, continuous finite components extended over $\omega = 0 \sim 0.4$, $1.2 \sim 1.8$ and the peaks at $\omega = 0.30$, 1.51.

In the above two examples, the normalized ranges of continuous finite components are nearly equal to $\omega/\nu = 0 \sim 1.0$, $2.0 \sim 3.0$ and their peaks appear in the neighborhood of $\omega/\nu = 0.3 \sim 0.5$, $2.3 \sim 2.5$.

These facts coincide with the results of the average power spectra exhibited by Duffing's equation (9) reported in [2], [3]. This nature is probably due to the identical cubic restoring terms contained in (4) and (9). In the example of Fig. 4, however, both damping and restoring terms interact each other and generate distinctive chaotic character which cannot be found in the system exhibited by Duffing's equation.

4) The powers of the chaotically transitional processes $\{X(\tau)\}$ together with their periodic and chaotic components are listed in Table I. For reference the powers of the self-excited oscillation and the almost periodic oscillation of Fig. 3 are also given. From these results, we can see that the ratios of chaotic component to total power are comparatively large in the cases of Figs. 5 and 6. As mentioned above, these two cases retain traces of almost periodic oscillations.

5) We have confirmed that there appear almost periodic oscillations but no chaotic phenomena in the computer-simulated system exhibited by van der Pol's equation with periodic excitation, i.e.,

$$\frac{d^2x}{dt^2} - \mu(1-x^2)\frac{dx}{dt} + x = B\cos\nu t, \quad 0 < \mu < 1. \quad (10)$$

Considering the abovementioned matters collectively, we conclude that the nonlinear restoring term brings

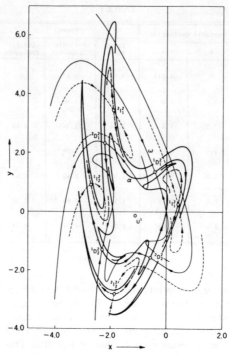

Fig. 9. Global phase plane structure of the diffeomorphism f_λ, the system parameters being $\mu=0.2$, $B=17.0$, and $\nu=4.0$.

chaotic properties in the second-order forced oscillatory systems.

IV. Conclusion

Chaotically transitional phenomena which occur in the forced negative-resistance oscillator have been studied by using analog and digital computers. The differential equation describing the system is of second order with periodic excitation in which both damping and restoring terms are nonlinear.

Difference between almost periodic oscillations and chaotically transitional processes is explained in detail adducing the respective examples.

Various types of strange attractors and their average power spectra are given. They are investigated in detail by comparing with the results obtained in the system exhibited by Duffing's equation.

Considering these results collectively, the role of nonlinear damping term and nonlinear restoring term are also clarified in the second-order forced oscillatory systems.

Appendix

Fig. 9 shows the global phase plane structure of the diffeomorphism f_λ, the system parameters being $\mu=0.2$, $B=17.0$, and $\nu=4.0$. In the figure, the symbol ${}^i D_j^n$ indicates the ith directly unstable n-periodic point and the subscript $j=1,2,\cdots,n$ represents the order of the successive movement of the images under the diffeomorphism f_λ in the group. Similar symbols are applied to the inversely unstable (I) and completely unstable (U) periodic points. The points D and I are called the saddles and U is the source of the diffeomorphism f_λ.

Acknowledgment

The authors wish to express their thanks to Prof. R. W. Liu of Notre Dame University and Prof. K. S. Chao of Texas Tech University for their thoughtful considerations. Thanks are also due to Prof. S. Utku of Duke University, Prof. P. Holmes and Prof. F. Moon, both of Cornell University, for their helpful and valuable discussions.

References

[1] Y. Ueda et al., "Computer simulation of nonlinear ordinary differential equations and nonperiodic oscillations," Trans IECE Japan, vol. 56-A, no. 4, pp. 218–225, Apr. 1973 (English translation) Scripta, pp. 27–34.
[2] Y. Ueda, "Random phenomena resulting from nonlinearity—In the system described by Duffing's equation," Trans. IEE Japan, vol. 98-A, no. 3, pp. 167–173, Mar. 1978.
[3] ——, "Randomly transitional phenomena in the system governed by Duffing's equation," J. Statistical Physics, vol. 20, no. 2, pp. 181–196, 1979.
[4] ——, "Steady motions exhibited by Duffing's equation—A picture book of regular and chaotic motions—," to appear in Proc. Engineering Foundation Conf. on New Approaches to Nonlinear Problems in Dynamics, Monterey, CA, Dec. 9–14, 1979.
[5] ——, "Explosion of strange attractors exhibited by Duffing's equation," to appear in the Annals of the New York Academy of Sciences, Proc. Int. Conf. on Nonlinear Dynamics, Dec. 17–21, 1979.
[6] N. Levinson, "Transformation theory of nonlinear differential equations of the second order," Ann. Math., vol. 45, pp. 723–737, 1944, vol. 49, p. 738, 1949.
[7] E. A. Coddington and N. Levinson, Theory of Ordinary Differential Equations. New York: McGraw-Hill, 1955, ch. 17.

Yoshisuke Ueda was born in Kobe, Japan, on December 23, 1936. He received the B.E. degree in 1959, the M.E. degree in 1961 and the D.E. degree in 1965, all in electrical engineering from Kyoto University, Kyoto, Japan.

He was a Research Associate from 1964 to 1967, a Lecturer from 1967 to 1971 at Kyoto University. Since 1971 he is an Associate Professor of Kyoto University. His major interests have been on nonlinear phenomena in electrical and electronic circuits. He is currently interested in the dynamic behavior of magnetic lines of force in synchronous generators in electric power systems.

Dr. Ueda is a member of the Institute of Electrical Engineers of Japan, the Institute of Electronics and Communication Engineers of Japan, the Mathematical Society of Japan, and the Japan Association of Automatic Control Engineers.

Norio Akamatsu was born in Tokushima, Japan, on September 9, 1943. He received the B.S. degree in electrical engineering from Tokushima University, Tokushima, Japan, in 1966, and the M.S. and Ph.D. degrees in electrical engineering from Kyoto University, Kyoto, Japan, in 1968 and 1974, respectively.

From 1971 to 1974 he was an Associate Lecturer of Electrical Engineering at Tokushima University, Tokushima, Japan. Since 1975 he has been an Associate Professor at the Department of Information Science and Systems Engineering of Tokushima University. His research interests are in the field of nonlinear system analyses by electronic instruments and their applications to medical science.

Dr. Akamatsu is a member of the Institute of Electronics and Communication Engineers of Japan.

Period-doubling cascades and devil's staircases of the driven van der Pol oscillator

Ulrich Parlitz and Werner Lauterborn

Drittes Physikalisches Institut, Universität Göttingen D-3400 Göttingen, Federal Republic of Germany
(Received 12 February 1987)

Bifurcation diagrams of the driven van der Pol oscillator are given showing mode-locking and period-doubling cascades. At low driving amplitudes locking regions occur following Farey sequences. At high driving amplitudes this relationship is destroyed due to the appearance of period-doubling cascades and coexisting attractors. A generalization of the winding number is used to compute devil's staircases and winding-number diagrams of period-doubling cascades. The winding numbers at the period-doubling bifurcation points constitute an alternating sequence that converges at the accumulation point of the cascade.

I. INTRODUCTION

The van der Pol oscillator

$$\ddot{x} + d(x^2-1)\dot{x} + x = a\cos(\omega t)$$

or equivalently written

$$\begin{aligned}\dot{x}_1 &= x_2 \\ \dot{x}_2 &= -d(x_1^2-1)x_2 - x_1 + a\cos(2\pi x_3), \\ \dot{x}_3 &= \frac{\omega}{2\pi}\end{aligned} \quad (1)$$

is one of the most intensely studied systems in nonlinear dynamics.[1-14] It serves as a basic model of self-excited oscillations in physics, electronics, biology, neurology, and many other disciplines. Many efforts have been made to approximate the solutions of (1) (Refs. 2–9) or to construct simple maps that qualitatively describe important features of the dynamics.[10,11] Although some of these maps possess strange attractors no investigation of the original equation (1) concerning chaotic solutions is known to the authors. In this paper we therefore want to give examples of bifurcation diagrams of the driven van der Pol oscillator (1) showing, besides other features, complete period-doubling cascades. A new quantity called (generalized) winding number that has been introduced recently in connection with nonlinear resonances of driven dissipative oscillators[15] is used to describe the topological changes of the local flow around a period-doubling orbit. Furthermore we present a devil's staircase based on this (generalized) winding number and discuss its (fractal) dimension.

II. BIFURCATION DIAGRAMS

The following bifurcation diagrams show the strobed amplitude of the oscillations (i.e., projections of the attractors in the Poincaré cross section onto the coordinate X_{1p} of the cross section) versus the excitation frequency ω. The damping parameter is held constant at $d=5$. Figure 1 shows a bifurcation diagram for $a=1$ and $0 < \omega < 1.5$. Quasiperiodic and periodic oscillations (mode-locked states) occur. All trajectories lie on an invariant torus within the three-dimensional phase space.

Figure 2(a) shows the Poincaré cross section of a typical quasiperiodic attractor. The corresponding orbit $\{(x_{1p}^n, x_{2p}^n), n \in \mathbb{Z}\}$ of the Poincaré map is restricted to an invariant circle and may therefore be described by a one-dimensional circle map, called *attractor map*. The angles Θ_n of the orbit points $(x_{1p}^n, x_{2p}^n)(n=1,2,3,\dots)$ with respect to the origin within the invariant circle are used to parametrize the attractor map $\Theta_n \mapsto \Theta_{n+1}$. The graph of the map obtained in this way is shown in Fig. 2(b).

When the driving amplitude a is increased the periodic windows become larger. Figure 3 shows a bifurcation diagram for $a=2.5$. Between the ω intervals where mode locking with (small) odd periods $1,3,5,7,\dots$ takes place periodic orbits with large periods and quasiperiodic orbits occur. A section of Fig. 3 showing details of the parameter interval between the period-3 and the period-5 oscillations is given in Fig. 4. Numerical investigations of the other intervals (e.g., period-5 to period-7, period-7 to period-9, etc.) have shown that all intervals evolve in the

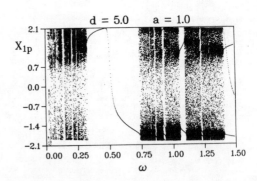

FIG. 1. Bifurcation diagram for $a=1$ showing the first coordinate X_{1p} of the attractor in the Poincaré cross section versus the excitation frequency ω that has been increased in small steps. After each step the last solution has been used as new initial value. All oscillations with even periods occur by pairs, where only one of the coexisting attractors is plotted here and in the following diagrams.

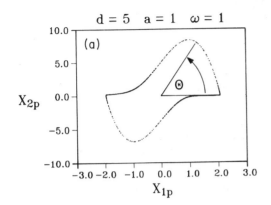

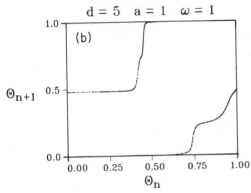

FIG. 2. (a) Poincaré cross section of an attractor lying on an invariant torus in phase space. (b) Attractor map of the attractor shown in Fig. 2(a).

same way, when the excitation amplitude a is varied. This phenomenon is similar to the superstructure observed in the bifurcation sets of the Duffing equation,[15,16] the Toda oscillator,[17] and the driven pendulum.[18] The superstructure of the bifurcation set of a driven nonlinear oscillator arranges a specific fine structure of bifurcation curves (surfaces) in the parameter space in a repetitive or-

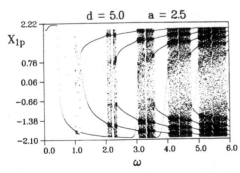

FIG. 3. Bifurcation diagram for $a=2.5$ (compare Fig. 1). Between the extended locking regions of period $1,3,5,7,\ldots$ parameter intervals with large-period oscillations occur each of them undergoing the same bifurcation scenario when the excitation amplitude a is increased.

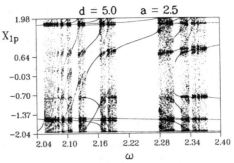

FIG. 4. Bifurcation diagram for $a=2.5$, showing an enlarged section of the bifurcation diagram given in Fig. 3. The parameter interval between the period-3 and the period-5 oscillation given here may be viewed as a prototype of all the large-period intervals in Fig. 3.

der that is closely connected with the nonlinear resonances of the system. The ω interval shown in Fig. 4 may be viewed as a prototype of all the other intervals between the entrainment regions with (small) odd periods occurring in Fig. 3.

As the van der Pol oscillator is a symmetric system oscillations with even periods must occur as pairs of two asymmetric coexisting solutions.[19] In all bifurcation diagrams given here only one of these two partner orbits is plotted. The coexistence of asymmetric attractors is a feature of the van der Pol oscillator that differs from the scenario of the ordinary sine circle map.[20] Figure 5 shows the largest Lyapunov exponent of the Poincaré map versus the driving frequency ω. It is nonpositive in the whole ω interval, i.e., no chaotic states occur in this parameter range. A similar investigation of the larger ω interval shown in Fig. 3 led to the same result.

III. GENERALIZED WINDING NUMBERS

Besides bifurcation diagrams plots showing the winding number in dependence on the excitation frequency are very useful to analyze the complicated parameter dependence of mode-locked oscillations. In Ref. 5 we introduced a definition of a winding number based on the torsion of the local flow around a given orbit. In contrast to

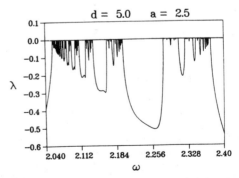

FIG. 5. Largest Lyapunov exponent λ_{max} versus driving frequency ω (compare Fig. 4 and Fig. 7).

the ordinary definition of the winding number[13,14] this local concept does not need an invariant torus in phase space. In the following we briefly want to motivate the definition of this new quantity. Let γ be an orbit of the van der Pol oscillator associated with the solution $x(t)=(x_1(t),x_2(t),x_3(t))$ of equation (1) and let γ' be a neighboring orbit of γ given by the solution $z(t)=x(t)+y(t)$ of (1). The perturbation $y(t)$ is assumed to be (infinitesimally) small. Then the local torsion of the flow is described by the rotations of the difference vector $y(t)$ about γ (compare Fig. 6). The time evolution of $y(t)$ is given by the variational equation (2) of (1)

$$\dot{y}_1 = y_2 ,$$
$$\dot{y}_2 = -(2dx_1x_2+1)y_1 - d(x_1^2-1)y_2 , \quad (2)$$
$$\dot{y}_3 = 0 .$$

To describe the torsion of the local (i.e., linearized) flow only the first two equations of (2) are of importance (because y_3 = const). Using polar coordinates $(y_1,y_2) = (r\cos\alpha, r\sin\alpha)$ we obtain the (reduced) variational equations in polar coordinates (3)

$$\dot{r} = r\{[1-(2dx_1x_2+1)]\sin\alpha\cos\alpha$$
$$\quad -d(x_1^2-1)\sin^2\alpha\} ,$$
$$\dot{\alpha} = -(2dx_1x_2+1)\cos^2\alpha - d(x_1^2-1)\sin\alpha\cos\alpha \quad (3)$$
$$\quad -\sin^2\alpha .$$

The mean angular velocity Ω of the difference vector $y(t)$ is given by

$$\Omega(\gamma) = \lim_{t\to\infty}\frac{1}{t}\int_0^t \dot{\alpha}\,dt' = \lim_{t\to\infty}\frac{\alpha(t)-\alpha(0)}{t} . \quad (4)$$

We call $\Omega(\gamma)$ the *torsion frequency* of the orbit γ. From (2) it is easy to see that $\Omega(\gamma)$ is always negative. We use the torsion frequency Ω_h of the driving harmonic oscillator

$$\Omega_h = -\omega \quad (5)$$

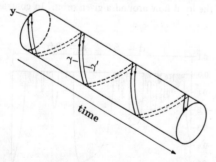

FIG. 6. A trajectory γ and its neighboring orbit γ'. The torsion frequency measures the mean rotation frequency of the difference vector y with respect to γ. If the attractor lies on an invariant torus the generalized winding number (6) equals the ordinary winding number because the number of "windings" of the trajectory γ equals the number of rotations of the difference vector y.

to define the *winding number* $w = w(\gamma)$ as

$$w(\gamma) = \frac{\Omega(\gamma)}{\Omega_h} . \quad (6)$$

The quantity w is called winding number because it equals the ordinary winding number as long as an invariant torus exists (compare Fig. 6). Especially in those cases where the (Poincaré) cross section of the torus differs strongly from the shape of a circle w is much easier to compute than the ordinary winding number. Therefore definition (6) may be useful for the investigation of conservative systems, too. An alternative derivation of (4) using the QR decomposition of the linearized flow map and further details can be found in Ref. 15. In contrast to already existing concepts of winding or rotation numbers our winding number w is also well defined for systems that do not possess an invariant torus in phase space or where the torus is broken. It enables the description of physical systems with coexisting attractors and the definition of winding numbers of strange attractors. We conjecture that in the case of chaos the limit (4) exists in the same sense and under the same conditions as that of the Lyapunov exponents. For periodic oscillations with period (number) m we call

$$n = mw \quad (7)$$

the *torsion number* of the closed orbit γ. The torsion number is a suitable quantity to classify saddle node and period-doubling bifurcation curves (surfaces) in the parameter space of one-dimensional driven dissipative oscillators. Furthermore it may be used to give an exact definition of resonance that does not depend on the existence of an invariant torus.[15] We call a periodic oscillation *resonant* when it possesses an integer torsion number.[21]

IV. A DEVIL'S STAIRCASE

Figure 7 shows a winding-number diagram corresponding to the bifurcation diagram in Fig. 4. Every "step" on the "devil's staircase" is associated with a rational value of w. The numerator of this rational number w is the torsion number n of the orbit and the denominator its period m. A diagram showing the inverse period $1/m$ versus the excitation frequency is given in Fig. 8 to elucidate the

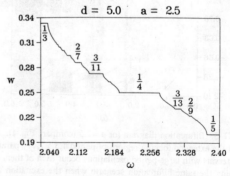

FIG. 7. Winding-number diagram for $a=2.5$ corresponding to Fig. 4 showing a devil's staircase.

self-similarity of the staircase. As can be seen from the diagrams in Fig. 7 and Fig. 8 the winding number in the locking region follows Farey sequences up to high order. Bak, Bohr, and Jensen[20] showed numerically that in the case of the one-dimensional sine circle map the complement of the locked states may become a fractal set with dimension $D = 0.868\ldots$. In this case the staircase is called a complete devil's staircase. Meanwhile the value $D = 0.868\ldots$ for the dimension of the complete devil's staircase has also been found in a hydrodynamical experiment[22] and other systems. A renormalization approach has confirmed the conjecture that this value of D is universal for a certain class of one-dimensional maps.[23] Unfortunately it is not easy to compare these results of the circle map theory with the corresponding scaling behavior of general systems such as the van der Pol oscillator. In the case of the circle map the critical curve in parameter space where the staircase becomes complete is a well-known straight line. In general, however, almost nothing is known about this curve. Even theorems concerning its smoothness features or algorithms to trace it do not seem to exist. Only some methods to locate it approximately in the parameter space are mentioned in the literature (e.g., Refs. 20 and 22). When we apply the technique described in Ref. 23 to the devil's staircase in Fig. 7 we obtain approximants D^n of the fractal dimension that range between 0.7 and 0.9. These results may be interpreted in a way that is compatible with the conjecture that $0.868\ldots$ is a universal dimension of complete devil's staircases of continuous systems, too. Details of this investigation will be given elsewhere.

V. PERIOD-DOUBLING CASCADES

When the excitation amplitude a is increased further the locking intervals in the diagram become wider and intervals with smaller periods compress or remove those with larger periods. Then the invariant torus is destroyed and symmetry breaking and (first finite) period-doubling cascades occur. Figure 9 shows a sequence of trajectories in the projection onto the (x_1, x_2) plane of the $\mathbb{R}^2 \times S^1$ phase space. In Fig. 9(a) a symmetry-broken trajectory of (basic) period 4 is given. It successively period doubles to period 4×2 [Fig. 9(b)], period 4×2^2 [Fig. 9(c)], and to a chaotic orbit [Fig., 9(d)]. The Poincaré cross section of this chaotic attractor consists of four very thin islands. Figure 10 shows the Poincaré cross section of the chaotic attractor at $\omega = 2.466$ (compare Figs. 11 and 12). A part of the attractor is blown up to emphasize its very thin structure. The Lyapunov dimension has been determined to $D_L = 1.014$. Figure 11 shows a bifurcation diagram of this period-doubling cascade into chaos. In Fig. 12 an enlargement of the period-doubling cascade and diagrams of the corresponding Lyapunov exponents and winding numbers are given. In the period-doubling cascade the winding number is constant near bifurcation points.[15]

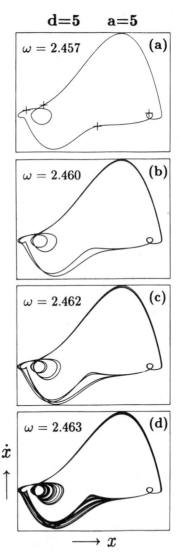

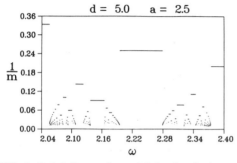

FIG. 8. Period diagram for $a = 2.5$ showing the inverse $1/m$ of the period m of the oscillation vs the excitation frequency ω. (Compare Figs. 4, 5, and 7.)

FIG. 9. Period-doubling sequence of a (basic) period-4 attractor. The damping parameter d and the driving amplitude a equal 5. (Compare Figs. 11 and 12). (a) Period-4 attractor. The orbit repeats after 4 periods of the driving as indicated by the crosses. (b) Period-4×2^1 attractor. (c) Period-4×2^2 attractor. (d) Period-4 chaos.

The height of these steps in the winding-number diagram is given by a simple formula similar to the result for the torsion number in a period doubling cascade.[15,24] At the kth period-doubling bifurcation point of the cascade shown in Fig. 11 the winding number w takes the value

$$w_k = w_\infty + \frac{(-1)^k}{3m_0 2^k}, \quad (8)$$

where w_∞ is the winding number at the accumulation point of the period-doubling cascade. w_∞ is given by the (basic) winding number w_0 and the (basic) period m_0 of the locking region (Arnold tongue) where the period-doubling cascade takes place,

$$w_\infty = w_0 - \frac{1}{3m_0}. \quad (9)$$

The period doublings shown in Figs. 9 and 12 occur in the locking region with $w_0 = \frac{1}{4}$ and $m_0 = 4$. Therefore w_∞ equals $\frac{1}{6}$ and the winding numbers are $w_1 = \frac{1}{8}$, $w_2 = \frac{3}{16}$, $w_3 = \frac{5}{32}, \ldots$ (see Fig. 12). This parameter dependence of the winding number in a period-doubling

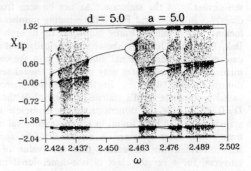

FIG. 11. Bifurcation diagram of the period-3 to period-5 interval for $a=5$ showing complete period-doubling cascades into chaos. Owing to the symmetry of the system for each period-doubling cascade a counterpart exists (which is reached from other intial conditions) that is not plotted here.

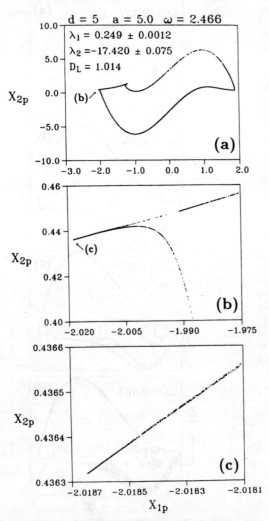

FIG. 10. Cross section of the strange attractor at $x_3 = 0$ for $d=5$, $a=5$, and $\omega = 2.466$ (compare Figs. 11 and 12). Inserted in the plot of the whole attractor (a) are the Lyapunov exponents λ_1 and λ_2 of the Poincaré map and the Lyapunov dimension D_L which equals almost one. This very small value of D_L is consistent with the very thin structure of the attractor shown in the enlargements (b) and (c).

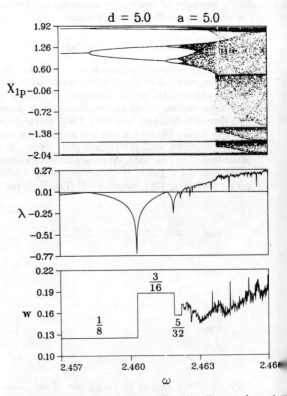

FIG. 12. Enlargement of the bifurcation diagram shown in Fig. 8 and the corresponding evolution of the largest Lyapunov exponent λ and the winding number w.

cascade describes the evolution of the invariant manifolds of the attractor.[17,24] Some period-doubling cascades of the van der Pol oscillator (e.g., the period-13×2^n cascade between $\omega = 2.4711$ and 2.4765 in Fig. 11 with $w_0 = \frac{2}{13}, w_1 = \frac{5}{26}, w_2 = \frac{9}{52}, \ldots$) do not obey the law (8) but instead the very similar formula,

$$w_k = w_\infty - \frac{(-1)^k}{3m_0 2^k},$$

$$w_\infty = w_0 + \frac{1}{3m_0}. \qquad (10)$$

The two (empirical) recursion schemes (8) and (10) apply to the period-doubling cascades of many other nonlinear oscillators, too.[17,25] That two kinds of winding-number sequences occur may be understood by looking at the logistic map. There the winding number is simply given as the relative number of R's in the $R-L$ string of the symbolic description of the dynamics, and the formulas (8) and (10) are immediate consequences of the construction law for the symbolic strings upon period doubling. Beyond the accumulation point of the period-doubling cascade the winding number describes the geometry of the strange attractor. Details will be given in a forthcoming paper.[25]

At even higher excitation amplitudes the bifurcation diagram becomes very complicated due to a multitude of coexisting attractors and period doubling cascades. As an example, Fig. 13 shows a bifurcation diagram for $a = 40$. At $\omega = 5.06 \ldots$ the period-1 attractor undergoes a Hopf bifurcation and an invariant torus in phase space is created. In those parts of the diagram where period-doubling cascades occur the torus is destroyed again or the period-doubling attractors coexist with the torus. A detailed investigation of this part of the parameter space will probably yield further interesting results.

VI. CONCLUSION

Mode-locking phenomena and period-doubling cascades of the driven van der Pol oscillator have been investigated. In both cases a new quantity called (generalized) winding

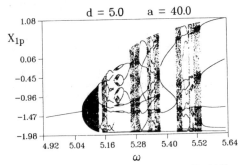

FIG. 13. Bifurcation diagram for $a=40$. A Hopf bifurcation and period-doubling cascades occur.

number was used to describe the dynamical behavior of the system and its parameter dependence. As long as all attractors lie on an invariant torus the winding-number diagrams show the well-known devil's staircase scenario. For large driving amplitudes, however, the invariant torus may be destroyed and period-doubling cascades into chaos occur. The winding number w_k at the period-doubling points constitute an alternating sequence converging at the accumulation point of the period-doubling cascade. This sequence describes the folding and unfolding process of the invariant manifolds.[17,24] For large driving amplitudes the Farey ordering is destroyed and many periodic, quasiperiodic, and chaotic attractors coexist. Details of this part of the parameter space will be published elsewhere.

ACKNOWLEDGMENTS

This work was supported by the Stiftung Volkwagenwerk. We thank the members of the Nonlinear Dynamics Group at the Third Physical Institute of the University of Göttingen, especially T. Kurz, K. Geist, and M. Wiesenfeldt, for many valuable discussions. All computations have been carried out on a SPERRY 1100 and a VAX-11/780 of the Gesellschaft für wissenschaftliche Datenverarbeitung, Göttingen.

---

[1]B. van der Pol, Philos. Mag. **43**, 700 (1927).

[2]M. L. Cartwright and J. E. Littlewood J. London Math. Soc. **20**, 180 (1945).

[3]E. M. El-Abbasy, Proc. R. Soc. Edinburgh Sect. A **100**, 103 (1985).

[4]N. Levinson, Ann. Math. **50**, 127 (1949).

[5]J. P. Gollub, T. O. Brunner, and B. G. Danly, Science **200**, 48 (1978).

[6]P. J. Holmes and D. A. Rand, Q. Appl. Math. **35**, 495 (1978).

[7]J. Grasman, E. J. M. Veling, and G. M. Willems, SIAM (Soc. Ind. Appl. Math.) J. Appl. Math. **31**, 667 (1976).

[8]J. Grasman, M. J. W. Jansen, and E. J. M. Veling, North Holland Math. Studies **31**, 93 (1978).

[9]J. Grasman, Q. Appl. Math. **38**, 9 (1980).

[10]J. Guckenheimer, Physica 1D, 227 (1980).

[11]J. Grasman, N. Nijmeijer, and E. J. M. Veling, Physica **13D**, 195 (1984).

[12]J. E. Flaherty and F. C. Hoppensteadt, Stud. Appl. Math. **58**, 5 (1978).

[13]J. Guckenheimer and P. J. Holmes, *Nonlinear Oscillations, Dynamical System, and Bifurcations of Vector Fields* (Springer, Berlin, 1983).

[14]H. G. Schuster, *Deterministic Chaos* (Physik Verlag, Weinheim, 1984).

[15]U. Parlitz and W. Lauterborn, Z. Naturforsch. **41A**, 605 (1986).

[16]U. Parlitz and W. Lauterborn, Phys. Lett. **107A**, 351 (1985).

[17]T. Kurz and W. Lauterborn (unpublished).

[18]K. Schmidt and H. G. Schuster (unpublished).

[19]J. W. Swift and K. Wiesenfeld, Phys. Rev. Lett. **52**, 705 (1984).

[20]M. H. Jensen, P. Bak, and T. Bohr, Phys. Rev. A **30**, 1960

(1984).

[21]In Ref. 15 we proposed as resonance criterion the existence of a rational winding number (5). That definition also included all cases where the torsion number is a noninteger rational number. The inclusion of this possibility was not intended.

[22]J. Stavans, F. Heslot, and A. Libchaber, Phys. Rev. Lett. **55**, 596 (1985).

[23]P. Cvitanovic, M. H. Jensen, L. P. Kadanoff, and I. Proccacia, Phys. Rev. Lett. **55**, 343 (1985).

[24]P. Beiersdorfer, Phys. Lett. **100A**, 379 (1984).

[25]U. Parlitz and W. Lauterborn (unpublished).

Chaotic states and routes to chaos in the forced pendulum

D. D'Humieres, M. R. Beasley,[*] B. A. Huberman,[†] and A. Libchaber

*Groupe de Physique des Solides, Ecole Normale Superieure, 24 rue Lhomond,
75231 Paris, Cedex 05, France*
(Received 17 June 1982)

An experimental study of the chaotic states and the routes to chaos in the driven pendulum as simulated by a phase-locked-loop electronic circuit is presented. For a particular value of the quality factor ($Q=4$), for which the chaotic behavior is found to be rich in structure, the state diagram (phase locked or unlocked) is established as a function of driving frequency and amplitude, and the nature of the chaos in these states is investigated and discussed in light of recent models of chaos in dynamical systems. The driven pendulum is found to exhibit symmetry breaking as a precursor to the period-doubling route for chaos. Although period doubling is found to be fairly common in the phase-locked states of the pendulum, it does not always manifest itself in complete bifurcation cascades. Intermittent behavior between two unstable phase-locked states is also commonly observed.

I. INTRODUCTION

Recently a large number of theoretical calculations, simulations, and experiments have been carried out on various nonlinear systems in an effort to understand chaos and the routes to chaos in such systems. In this paper we present the results of an investigation of the forced pendulum, for which the control parameters are the driving amplitude and frequency. As is well known, the forced pendulum is isomorphic to many other familiar nonlinear systems, such as Josephson junctions and the phase-locked-loop configuration of a voltage-controlled oscillator, or VCO. In fact, the experimental work reported here was carried out using a phase-locked loop. Our study was stimulated by the earlier work of Huberman, Crutchfield, and Packard[1] and of Kautz[2] who have also studied this problem in the context of Josephson junctions.

It is essential to point out right at the outset that the seemingly simple situation of a forced pendulum is quite complex due to the fact that the space of variables is large. Besides the driving amplitude (which can have a dc component or bias in addition to the ac drive) and the driving frequency, one has the resonant frequency Ω_0 and the quality factor Q of the pendulum as important parameters. Indeed, it is the interplay between the driving force and the natural modes of the pendulum that results in chaos. Moreover, because of the periodic nature of the restoring potential, both running and oscillatory motions are possible. In order to deal as simply as possible with such complexity in this paper, we first establish the "state" diagram of the forced pendulum as a function of the driving amplitude and frequency for a particular value of Q where the chaotic behavior is found to have rich structure. We then establish the nature of these states (e.g., phase locked or not, periodic or chaotic) and investigate the nature of transitions between them.

In broad outline our results are in agreement with the earlier work of Huberman et al.,[1] and Kautz,[2] but much more complete in the zero-bias case. In particular, we have found some new results of interest. They are as follows:

(a) Prior to going chaotic the pendulum is found to break its spatial symmetry and oscillate with a larger amplitude to one side than the other. This symmetry-breaking phenomenon appears to be an inherent part of the period-doubling cascade in this particular system.

(b) The period-doubling cascade is found to be a generic phenomena for the forced pendulum. It occurs in the oscillating states of the pendulum but also in the rotating regime where the pendulum rotation frequency is phase-locked to the driving frequency. In these cases the chaotic behavior is characterized by a power spectrum $S(\omega)$ peaked at each subharmonic of the drive frequency but $S(\omega) \to 0$ as $\omega \to 0$, implying that phase locking is maintained at dc.

(c) Another chaotic state commonly observed appears to be related to random transitions between two phase-locked states that have become unstable. The two states may be oscillating or rotating. In this case the chaotic behavior is characterized by a mean white-noise spectrum and hence is associated with the loss of phase locking.

(d) A third kind of chaotic behavior arises when the driving frequency is much smaller than the low-amplitude resonance frequency of the pendulum ($\Omega \ll \Omega_0$). If the amplitude of the driving force becomes larger than the critical value at which the pendulum begins to rotate, the motion is a combination of positive and negative rotations in between which the pendulum undergoes damped oscillations. The ensuing sensitivity on initial conditions for every rotation leads to a chaotic state with a white-noise spectrum.

The main body of this paper is organized as follows. In Sec. II we review the equations of motion of the forced pendulum, establishing notation and providing a correspondence between several useful isomorphisms. Next, we establish the state diagram (Sec. III) followed by a discussion of the various transitions between these states and the routes to chaos we observed (Secs. IV and V). The details of the actual experimental equipment and procedures are contained in the appendixes, along with some of the mathematical details of the theoretical interpretations.

II. EQUATIONS OF MOTION FOR THE FORCED PENDULUM

The equation of motion for the forced pendulum is of the form

$$a\ddot{\theta} + b\dot{\theta} + c\sin\theta = \Gamma(t) , \quad (1)$$

where a is the moment of inertia, b the damping constant, $c\sin\theta$ the restoring torque, and $\Gamma(t)$ the driving torque. The corresponding variables for pendula, Josephson junctions, and phase-locked loops are given in Table I for convenience, as it is frequently helpful from the physical point of view to use the "language" of these various systems in discussing our results. Specifically, for a phase-locked loop, such as the one used in our experimental work, the accessible variables are the summing voltage V and the feedback current I_{FB}, which are proportional to $\dot{\theta}$ and $\sin\theta$, respectively.

As usual, it is convenient to transform this equation into dimensionless form. The two characteristic times are $\tau_0 = b/c$ and $\Omega_0^{-1} = (c/a)^{-1/2}$, where τ_0 is the exponential damping time for the highly viscous pendulum (i.e., when the $\ddot{\theta}$ term can be neglected) and Ω_0 is the low-amplitude natural oscillatory frequency of the undamped pendulum. This leads to two possible dimensionless equations:

$$\beta\ddot{\theta} + \dot{\theta} + \sin\theta = \gamma(\tau) \quad (2)$$

or

$$\ddot{\theta} + \frac{1}{Q}\dot{\theta} + \sin\theta = \gamma(\tau) , \quad (3)$$

where $\beta = Q^2 = ac/b^2$, $\gamma(\tau) = \Gamma(\tau)/c$, and the time τ has been normalized by τ_0 and Ω_0^{-1} in Eqs. (1) and (2), respectively. In general, the forcing term (torque) is given by

$$\gamma(t) = \gamma_0 + \gamma_1\cos\omega\tau + \gamma_N(\tau) , \quad (4)$$

where ω is the normalized frequency, γ_0 the dc forcing term, γ_1 the amplitude of the ac driving force, and γ_N is a noise term. In this work Q is fin-

TABLE I. Corresponding variables for pendula, Josephson junctions, and phase-locked loops.

	Pendulum	Josephson junction	Phase-locked loop
θ	Angular position	Quantum phase difference	Phase difference between the oscillators
$\dot{\theta}$	Angular velocity	$\dfrac{2eV}{\hbar}$	kV
a	Inertia momentum	$\hbar C/2e$	C/k
b	Viscous damping	$\hbar/2eR$	$1/kR$
$c\sin\theta$	Restoring torque	Josephson current I_J	Feedback current I_{FB}
c		Critical current I_c	V_1/R_S
$\Gamma(t)$	Applied torque	Applied current $I(t)$	Applied current $V_E(t)/R_E$
$\Omega_0 = \sqrt{c/a}$	Natural frequency	$(2eI_c/\hbar C)^{1/2}$	$(kV_1/R_SC)^{1/2}$
$\tau_0 = b/c$	Damping time	$\hbar/2eRI_c$	R_S/kRV_1
$Q = (ac/b^2)^{1/2}$	Quality factor	$(2eI_cR^2C/\hbar)^{1/2}$	$(kV_1CR^2/R_S)^{1/2}$

ite and we shall use Eq. (3). However, in the limit of large damping $\beta = Q^2 \to 0$, Eq. (2) is more appropriate.

III. STATE DIAGRAM OF THE FORCED PENDULUM

In order to provide a simple framework in which to understand our results, we present here the state diagram of the forced pendulum for the particular conditions we studied most carefully. The diagram was obtained directly from experiment by taking traces of the summing point voltage of our phase-locked loop (see Appendix A) as a function of the drive frequency for various amplitudes. A typical sequence of traces is shown in Fig. 1.

As can be readily seen, the pendulum exhibits two basic types of states: phase-locked states in which there are no fluctuations in $\dot{\theta} \propto V$ in the dc limit and unlocked states for which fluctuations are evident near dc. For example, consider the trace for $\gamma_1 = 1.5$, i.e., 1.5 times the critical torque. As the driving frequency is reduced, the system, initially in an oscillatory state phase locked to the driving frequency, breaks its symmetry and jumps to a new state in which it is still phase locked but now with a nonzero $\langle \dot{\theta} \rangle$. (As we shall see later, the symmetry is actually broken before the state change takes place.) The motion in this new state corresponds to a periodic running solution in which the pendulum rotates 2π successively each period of the driving force. The other phase-locked states are similar, involving only different numbers of net rotations in each driving period. These states are analogs of the zero-bias ac Josephson steps seen in Josephson junctions and follow the relation $\dot{\theta} \propto V = k^{-1} n \omega$ indicated by the straight lines in the figure, where n is an integer and k is the voltage-to-frequency conversion factor of the VCO. They are stable in the presence of small dc bias. Some higher-order steps are also seen at lower frequencies, as are occasional steps of the form $\dot{\theta} \propto V = k^{-1} (p/q) \omega$, where p and q are integers. The hysteresis present between decreasing and increasing frequency is shown by the arrows.

The unlocked states are clearly chaotic as shown by the finite noise power spectral density at dc evident in the figure. We should point out that they are not unrelated to the phase-locked $(V = k^{-1} n \omega)$ states, however. For $\omega \lesssim 1$ they serve as the transitions between phase-locked states of different n, and at low frequencies the steps themselves ($n = 0$ and 1 for $\gamma_1 = 1.5$) appear chaotic. The undulatory behavior seen at low frequencies in Fig. 1 persists to very low drive frequencies as is illustrated in Fig. 2, which shows the behavior in this region on an expanded scale. The overall state diagram that emerges from a complete family of traces such as these is shown in Fig. 3 for the particular case $Q = 4$.

It is important to note that, whereas this state diagram is closely related to the bifurcation diagram

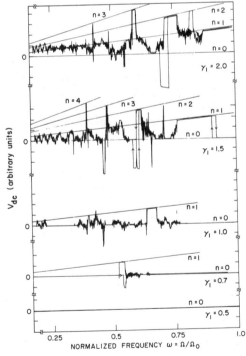

FIG. 1. Voltage fluctuations of the simulator circuit near dc as a function of the driving frequency for various driving amplitudes. Note pattern of phase-locked and -unlocked states. Straight lines indicate phase-locked steps satisfying the relationship $V = n k^{-1} \omega$.

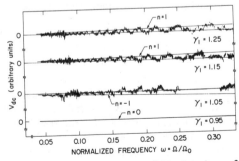

FIG. 2. Extension of the data of Fig. 1 to lower frequencies.

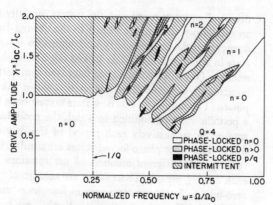

FIG. 3. State diagram for the driven pendulum with $Q=4$ and $\gamma_0=0$.

of Huberman et al.,[1] it is not exactly the same. Period-doubling cascades (pitchfork bifurcations) and their associated chaos are not seen in Figs. 1—3 because they have a noise power spectral density such that $S(\omega)\to 0$ as $\omega\to 0$. They do exist, however, and arise in the phase-locked states shown in Fig. 3. We shall return to this point later. The structure seen in Fig. 3 does clarify the content of the generally chaotic region noted by Huberman et al.,[1] although, as seen here, that region extends to much lower frequencies than found by those authors. Recent numerical simulations by Pedersen and Davidson[3] yield a similar qualitative picture.

IV. PHASE-LOCKED STATES

As seen in Fig. 3, phase-locked states occur at low ac drive and high frequencies ($n=0$) and in the form of stripes ($n \geq 0; n=p/q$) throughout the generally chaotic region at high drives and low frequencies. In this section we discuss the nature of the chaos observed in these phase-locked regions and of the transitions observed between the phase-locked states and the intermittent regions in Fig. 3.

A. Phase-locked oscillating states ($n=0$)

Consider first the large portions of $n=0$ region at low γ_1 and large ω where the motion is entirely contained in the first potential minimum. As discussed by Huberman et al.,[1] the basic response of a pendulum in this regime is that of a highly nonlinear oscillator for which, in the presence of sufficiently strong ac drive, the system exhibits chaos preceded by the period-doubling cascade analogous

to that present in a single-well anharmonic oscillator.[4] This behavior is best understood by considering the response of a pendulum as a function of frequency of various ac drive amplitudes. The filtered (fundamental) amplitude response of our simulator as a function of ω for various γ_1 is shown in Fig. 4. The evolution of the phase-space portraits θ vs $\sin\theta$ (i.e., the signals I_{FB} and V from the phase-locked loop) for increasing γ_1 at fixed $\omega=0.67$ are illustrated in Figs. 5 and 6 along with the noise power spectra for those cases ($\gamma_1=0.54$ and 0.68) where the response is chaotic.

In order to interpret these results it is helpful to compare them with those expected from classical perturbation theory. In a linear analysis the solution to Eq. (3) can be written in the form

$$\theta = -\alpha \sin(\omega t - \phi), \qquad (5)$$

where

$$\alpha e^{-i\phi} = \frac{\gamma_1}{i(1-\omega^2) - \omega/Q} \qquad (6)$$

leading to a resonance for $Q > 1/\sqrt{2}$ at a frequency $\omega = (1 - 1/2Q^2)^{1/2}$ with a maximum amplitude

$$\alpha = \frac{\gamma_1 Q}{(1 - 1/4Q^2)^{1/2}}. \qquad (7)$$

When the nonlinearity is included the problem becomes less trivial, but for small γ_1 it can still be described by a perturbation expansion. We consider solutions of the type

$$\theta = -\sum_{n=0}^{\infty} \alpha_{2n+1} \sin[(2n+1)\omega\tau - \phi_n] \qquad (8)$$

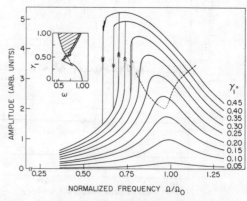

FIG. 4. Observed filtered (fundamental) amplitude response of the simulator circuit for various driving amplitudes. Dashed line shows domain of broken symmetry in the pendulum motion. Insert shows the observed bifurcation diagram similar to that reported in Ref. 1.

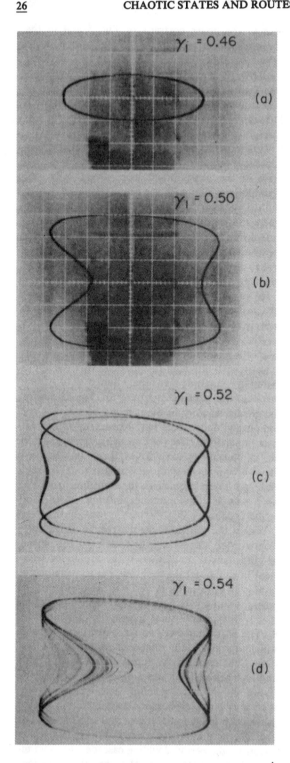

FIG. 5. (a) Evolution of the phase-space portraits ($\dot{\theta}$ vs $\sin\theta$) for increasing driving amplitude γ_1 at fixed frequency $\omega=0.67$. (b) Note the symmetry breaking, (c) followed by period doubling, (d) followed by chaos.

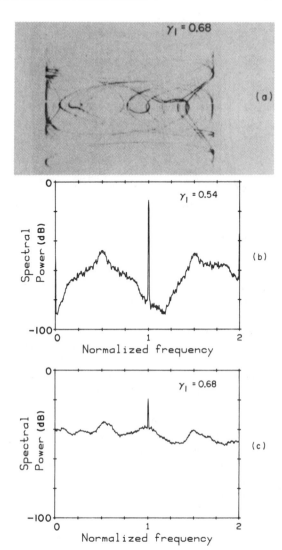

FIG. 6. (a) Phase-space portrait corresponding to the intermittent domain of Fig. 3 at $\omega=0.67$. (b) and (c) show the spectral power density corresponding to Figs. 4(d) and 5(a). Note different behavior as $\omega\to 0$. Note also that in this figure and all other power spectra the frequency axis is normalized to the drive frequency.

as it is easy to show that for small amplitudes the even terms do not appear. Limiting ourselves to the first-order terms we obtain

$$-\omega^2\alpha+2J_1(\alpha)=\gamma_1\sin\phi , \quad (9)$$

$$-\frac{\alpha\omega}{Q}=\gamma_1\cos\phi , \quad (10)$$

where the J_n are Bessel functions of the first kind. In Fig. 7 we show the numerical solution of Eqs. (9)

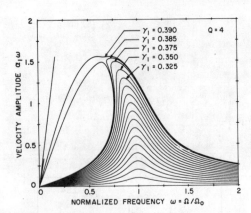

FIG. 7. Calculated fundamental response using perturbation theory as described in text.

and (10) for the amplitude of the velocity term $\dot{\theta}$ [i.e., $\alpha_1\omega$ as $\dot{\theta} = -\alpha_1\omega\cos(\omega\tau - \phi_1)$] as a function of ω for various drive amplitudes.

Comparing Figs. 4 and 7 we see that for $\gamma_1 < 0.385$ there is good agreement between the observed behavior and the results obtained with perturbation theory. We observe the usual nonlinear resonance shape, with a shift of the resonance peak to lower frequencies. Above a critical value $\gamma_1 \simeq 0.3$ the resonance becomes S shaped with three solutions, only two of which are stable, and hysteresis develops in the experimental curve.

Clearly, however, for $\gamma_1 \geq 0.39$ the simple analytical solution is no longer valid, since even including higher-order odd terms, it is not possible to explain the following two important phenomena which appear in the experiment:

(1) For $\omega \lesssim 1$ at the higher driving amplitudes the resonance curve shows a small distortion which is related to the generation of second harmonics. In fact, that is just the signature of a bifurcation of the pendulum at which an asymmetry appears in the angular motion. This asymmetry is clearly evident in the phase-space portrait of Fig. 5(b) and exists above the dashed line shown in Fig. 4. This symmetry breaking of the pendulum oscillations appears to be an unavoidable precursor of the period-doubling cascade for this system. This phenomena has been overlooked or ignored in previous discussions of chaos in symmetric anharmonic potentials. It is possible to analytically derive the onset of the symmetry breaking as shown in Appendix B.

(2) The period-doubling cascade also leaves its signature. In Fig. 4 it corresponds to the small region before the instability point of the S-shaped curve. The period-doubling cascade itself is shown both in the phase portraits of Fig. 5 and in the power spectrum of Fig. 6(b). For the sequence shown in these figures (increasing γ_1 at fixed $\omega = 0.67$), the transition at the end of the chaotic region associated with the period-doubling cascade is into an intermittent state [e.g., Figs. 6(a) and 6(c)]. The nature of this intermittent state will be discussed in greater detail in Sec. V.

We should note that in Figs. 5 and 6 the period-doubling cascade is incomplete, presumably due to noise in the phase-locked loop. Specifically, the asymptotic evolution of the cascade is not observed and the system goes directly into the chaotic region where the inverse cascade takes place. The signature of this inverse cascade can be seen in the power spectrum shown in Fig. 6(b), i.e., a noise spectrum peaked at each subharmonic of the drive frequency which goes to zero as $\omega \to 0$, as expected.

In general, we found that for low γ_1 and $0.6 \lesssim \omega \lesssim 1$, where the motion of the pendulum is entirely in one well, the $n = 0$ state became unstable via the period-doubling route as found by Huberman et al.[1] (see inset of Fig. 4). The results of Fig. 5 are typical, although it was found that the application of a small dc bias often allowed us to go further into the cascade (e.g., to $\omega/8$). The reason for the greater stability in the presence of a bias is not completely understood but appears to reflect (at least in part) the effect of external symmetry breaking in preventing noise-driven transitions between the two equivalent symmetry-broken states at zero bias. At lower frequencies the transition from the $n = 0$ state was discontinuous and not preceded by period doubling.

Within the generally chaotic regime of the state diagram, the striped regions with $n = 0$ correspond to phase-locked oscillations with amplitudes sufficient to go beyond the first well, but in a symmetric fashion so as to result in $\langle \dot{\theta} \rangle = 0$. Two phase-space portraits and power spectra ($\omega = 0.67$; $\gamma_1 = 0.76$ and $\gamma_1 = 1.94$) typical of these regions are presented Fig. 8. The particular cases correspond to situations in which the pendulum rotates one time 2π and one time -2π, respectively, during three [8(a) and 8(b)] and one period [8(c) and 8(d)] of the ac drive, as shown by the corresponding noise spectra. Note that, as in Rayleigh-Bernard experiment,[5] or for the parametric pendulum,[6] the period tripling showed Figs. 8(a) and 8(b) occurs not in the chaotic region following the period-doubling cascade, but as an independent route to chaos. However, the transitions away from this period three state were observed to follow a period-doubling pattern (i.e., 3×2^n).

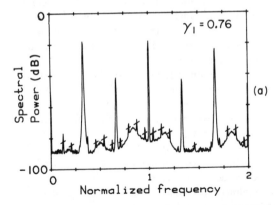

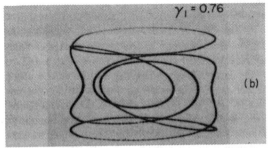

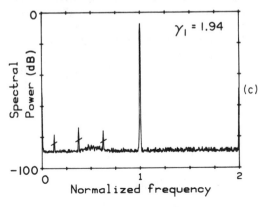

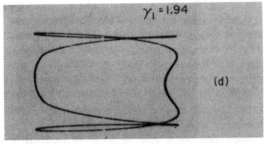

FIG. 8. Phase-space portraits and associated spectral power for two symmetric ($n=0$) large-amplitude rotating states. (a) and (b) correspond to a period-tripled state. (c) and (d) correspond to a simple period-one state. Note that the spectral lines with slash marks are due to 50-Hz pick-up.

B. Phase-locked rotating states ($n \geq 1$)

There are many phase-locked rotating (i.e., periodic running) states evident in Fig. 3. In general, we find that the behavior of the pendulum as one crosses such states (either in driving frequency or amplitude) is not universal in that no single pattern is observed. Period doubling is common place but not necessarily in the form of fully developed cascades. Consider, for example, the phase-space orbits shown in Fig. 9, which illustrate the evolution of the system as γ_1 is increased across the $n=2$ state at $\omega=0.67$. The basic structure of these orbits is very typical, other states differing essentially only in the number of loops on the top (or bottom) of the orbit. Each loop corresponds to a 2π rotation of the pendulum. Note, however, that as γ_1 increases monotonically, the orbits exhibit the sequence $\omega/2 \rightarrow \omega \rightarrow \omega/2$. This type of behavior is not understood but may simply reflect a nonmonotonic relationship between the physical control parameter γ_1 and that governing the pitchfork bifurcations in the Feigenbaum[7] theory based on the logistic equation.

In any event, it is clear from the phase-space orbits that the dynamics associated with period doubling is essentially the same for the $n=0$ and $n \neq 0$ states (compare Figs. 5 and 9.) The orbits differ only in the existence of loops on the orbit for $n > 1$. In particular, without exception period doubling is seen to arise when the velocity of the pendulum ($V \propto \dot{\theta}$) goes through zero just as the pendulum approaches the "up" position ($I \propto \sin\theta \rightarrow 0$).

In order to put the above ideas on a more mathematical basis, consider a perturbation-expansion solution of Eq. (3) for a phase-locked rotating state. Because of the phase locking, the solutions have the form

$$\dot{\theta} = n\omega - \omega \sum_{p=1}^{\infty} p\alpha_p \cos(p\omega\tau - \phi_p) \quad (\langle \dot{\theta} \rangle = n\omega) \tag{11}$$

and thus

$$\theta = \theta_0 + n\omega\tau - \sum_{p=1}^{\infty} \alpha_p \sin(p\omega\tau - \phi_p) . \tag{12}$$

To first order, the solution has been found by Pedersen et al.[8]:

$$\theta = \theta_0 + n\omega\tau - \alpha \sin(\omega\tau - \phi) . \tag{13}$$

Writing $\theta_n = \theta_0 + n\phi$ one obtains

$$\gamma_0 = J_n(\alpha)\sin\theta_n + n\omega/Q , \tag{14}$$

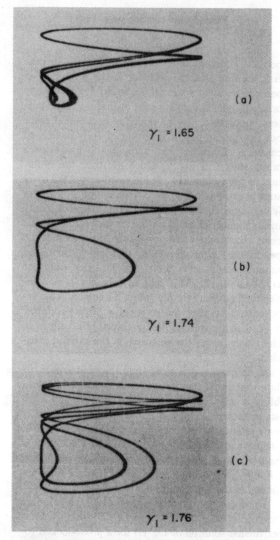

FIG. 9. Phase-space portraits of phase-locked periodic running states ($n \geq 1$) illustrating nonmonotonic period-doubling pitch fork bifurcations.

the driving frequency but with an angular velocity which is sinusoidally modulated by the term $\alpha \sin(\omega\tau + \phi)$. It is this term that is associated with the complex behavior of the pendulum, i.e., the period-doubling cascade.

One can develop this picture further and relate it to the period-doubling cascade of the phase-locked oscillating pendulum for $n=0$ and $\gamma_0 = 0$ [$\theta_0 = 0$ and Eq. (15) reduce to Eqs. (9) and (10)] described previously. Indeed, in Eq. (15) there are resonance terms for the imaginary and real parts of $\gamma_1 e^{i\phi}$. These nonlinear resonance terms presumably lead to the period-doubling cascade for reasons similar to the nonlinear resonance effects of the oscillating pendulum.

Going one step further we can analyze the stability of the solution (13). As shown in Appendix B such an analysis leads to Mathieu's equation. The essential point here is that both parameters a and q of the Mathieu equation depend in a complex manner on γ_1, the amplitude of the driving force. This is why one usually observes period-doubling phenomena which evolves in a nonmonotonic manner such as illustrated in Fig. 9. The forced pendulum is in this respect strikingly different from the parametric pendulum,[6] where the period-doubling cascade evolves fully. In this case, in the Mathieu equation for the stability of the solution, only the q term depends on the forcing amplitude and varies linearly with it. Thus, except for the $n=0$ state, the forced pendulum is not, in general, a simple system for displaying the period-doubling cascade.

V. NONPHASE-LOCKED STATES

A. Low-frequency regime

As seen in Fig. 3, the generally chaotic region of the state diagram extends down to low frequencies such that $\omega \ll 1$. This is in contrast to the original result of Huberman *et al.*,[1] in which there was found to be a cutoff at $\omega \simeq Q^{-1}$. Although the exact reason for this discrepancy is not known, it appears likely to be due to the simulator circuit used in Ref. 1, which was restricted in the number of 2π rotations it could simulate. We should point out, however, that some role of noise in inducing a chaotic response in this region can not be completely ruled out. In a similar vein, it is possible that transients due to the frequency sweep also play a role in the behavior observed. Finally, it should be noted that we did not determine exactly how low in

$$\gamma_1 e^{i\phi} = -\alpha \left[\frac{\omega}{Q} \left(1 + \frac{2n^2}{\alpha^2} \right) - \frac{2n}{\alpha^2} \gamma_0 \right.$$
$$\left. + i \left(\omega^2 + \frac{2J_n'(\alpha)}{\alpha} \cos\theta_n \right) \right]. \quad (15)$$

Thus, we arrive at a set of three implicit equations [(14) and (15)] that cannot be solved analytically to obtain the unknowns α, ϕ, and θ_n. However, considering Eq. (13), we see that it implies that under the oscillating constraint the pendulum rotates with an angular frequency which is n times

frequency this region extends.

On the basis of the traces in Fig. 2, we surmise that any structure in the state diagram in this region must be on a very fine scale and associated with the undulations between the $n=0$ and 1 states evident in the figure. This extreme sensitivity to the control parameters is also shown in Fig. 10, which shows the real-time response of our circuit and the associated noise spectra for two nearly identical values of γ_1. Note the striking 30-dB noise rise in going from 10(c) to 10(d) in a power spectrum that is otherwise the same. Note also the slight differences in the ringing down transients from cycle to cycle in Fig. 10(b) but not in Fig. 10(a).

Better insight into the nature of this behavior can be obtained by considering first the response of the pendulum to an applied dc bias. Since we are considering low ac drive frequencies, the motion of interest should be closely related to that at dc. This problem has been analyzed by McCumber[9] in the context of the resistively shunted junction (RSJ) model of Josephson junctions. The dc I-V (i.e., $\langle \dot{\theta} \rangle - \gamma_0$) curves obtained on our simulator corresponding to the situation studied by McCumber are shown in Fig. 11. The main features are as follows.

The asymptotic behavior for large γ_0 is such that $\langle \dot{\theta} \rangle = \gamma_0 Q$. In effect, for $\gamma_0 \gg 1$ the pendulum rotates at a high angular velocity, thus the nonlinear $\sin\theta$ term can be neglected in Eq. (1), and it is easy to show that the pendulum reaches a limiting angular velocity $\dot{\theta} = \gamma_0 Q$ in a time of order $1/Q$. For $Q \gtrsim 1$ the $\langle \dot{\theta} \rangle$ vs γ_0 curve exhibits hysteresis, which can be simply understood. When the torque is increased from zero to the critical value $\gamma_0 = 1$ because of the inertia term, the pendulum starts abruptly to rotate at an angular average velocity $\langle \dot{\theta} \rangle \simeq Q$. Upon decreasing the torque, the inertia of the pendulum causes it to keep rotating even for $\gamma_0 < 1$ down to a critical value γ_{0C}. At this critical value of γ_0 the pendulum reaches the unstable position with zero velocity; from there it relaxes toward a stable equilibrium position through a damped limit cycle with a characteristic time $\tau = Q/\Omega_0$. The critical torque value γ_{0C} has been calculated numerically by McCumber as a function of $\beta = Q^2$. The asymptotic behavior of γ_{0C} for $Q \to \infty$ is $\gamma_{0C} \sim \sqrt{2}/Q$.

Let us return now to the driven pendulum at low driving frequencies $\omega \ll Q^{-1}$. In this limit the pendulum responds adiabatically and experiences the motions described above. Thus, for $\gamma_1 \gtrsim 1$ the pendulum successively rotates and oscillates as shown in Fig. 10(a) or 10(b) (evidently with positive and negative rotations). Since the driving period is much longer than the damping time of the oscillations, as the drive amplitude passes through zero the pendulum almost reaches its equilibrium position before the next rotation. Paradoxically, in this low-frequency region where the pendulum follows the excitation in a more or less adiabatical way, a chaotic state exists.

As already intimated above, the crucial point in

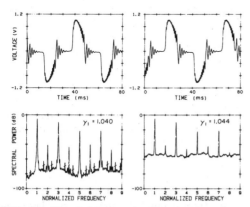

FIG. 10. Observed temporal voltage traces and associated spectral power of the chaotic and nonchaotic states observed at low driving frequencies. Note that in this figure the frequency is normalized to the driving frequency ($\omega = 0.034$), not the natural frequency of the circuit and that the spectral lines with slash marks are due to 50-Hz pick-up.

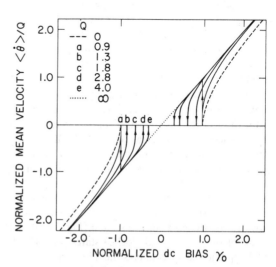

FIG. 11. Traces of the dc response of the simulator circuit as a function of finite dc bias γ_0 for various values of Q.

the dynamics of the pendulum associated with the observed chaos at these low frequencies is the transition from rotation to ringing down. More specifically, it is the occasional "extra" rotations [see transition on the far right of Fig. 10(b)] that lead to the observed noise. Apparently, for particular values of γ_1, the ringing-down portion of the cycle becomes very sensitive to the exact manner (i.e., the "initial conditions") in which the ringing down begins. Moreover, as for the period doubling seen on the phase-locked steps, the "dangerous" part of the orbit is when the pendulum approaches the unstable position with small angular velocity. Unfortunately, none of the general modes of chaos with which we are familiar seem able to describe this type of chaotic state.

B. Intermediate frequencies—intermittency between unstable phase-locked states

As shown in Fig. 3, in between the phase-locked states there are chaotic regions where, as seen in Fig. 1, the noise spectrum is finite even at dc. The chaotic stripes appear to be associated with intermittency between unstable phase-locked states. Similar intermittent behavior has been noted by Kautz[2] and by Ben-Jacob and co-workers[10,11] in the present of a dc bias.

A typical noise spectrum of one of these regions was shown in Fig. 6(c). The spectrum has two important characteristics. First of all, it extends down to dc, as already noted. Second, it has an amplitude which is much larger than the noise associated with the period-doubling cascade. [Compare Figs. 6(b) and 6(c).] Finally, it exhibits broad "resonances" centered around the unstable periodic phase-locked orbits, the width of which are presumably related to the lifetime of the periodic states.

In Fig. 12 we show the phase-space portrait of one of these chaotic states [Fig. 12(b)] compared with those of the phase-locked states between which it is observed [Figs. 12(a) and 12(c)]. It is clear from the figure that in this particular type of chaotic state, the motion involved can be viewed as an intermittency between two unstable phase-locked states as asserted above. Figure 13 shows another example of such a chaotic state [Fig. 13(a)], along with its associated Poincaré section [Fig. 13(b)] and those of the closest phase-locked states [Fig. 13(c)], with γ_1 adjusted so as to bias them in a chaotic condition. In our experiment, the Poincaré section of the intermittent states was always found to have the structure illustrated in this figure, namely, two

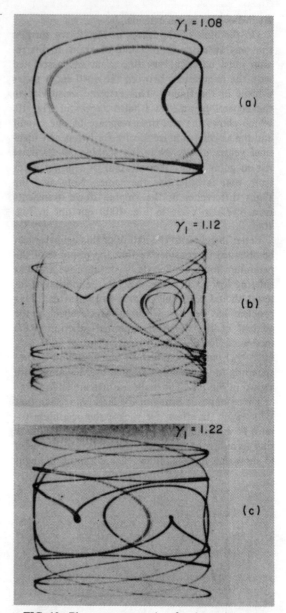

FIG. 12. Phase-space portraits of an intermittent state (b) along with the two nearby phase-locked states [(a) and (c)] to which it is related.

strange attractors connected by an important transient, even if the intermittent state was entered directly from a purely periodic state. Hence, Fig. 13 illustrates the important role phase noise plays in this intermittent state. Specifically, the presence of the strange attractors of phase-locked states could be the signature of an important phase randomization after each change of state. Note, however, that careful examination of Fig. 13(b) shows that the

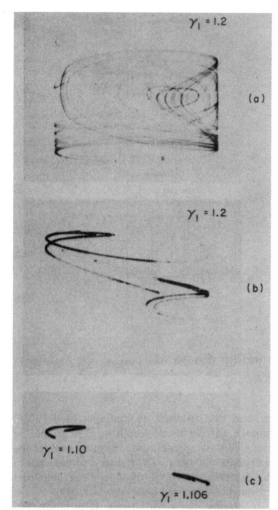

FIG. 13. Another example of the (a) phase-space portrait of an intermittent state, (b) along with its corresponding Poincaré section, (c) and the Poincaré sections of the two phase-locked states to which it is related. Note also the twofold structure in the Poincaré section of (b). These Poincaré sections were obtained directly from oscilloscope traces by means of a z-axis modulation synchronized to the drive signal.

Poincaré section of the full intermittent state has a double-line structure like that found in strange attractors derived from two-dimensional mappings of the Henon-type.[12] This double-line structure is not present in the related Poincaré sections of Fig. 13(c).

To date we have not found a model or mapping that properly describes this intermittent state. The model of Ben-Jacob, Goldhirsh, and Imry[11] has many of the features we observe but does not include the random-phase noise at the start of each laminarlike segment that is seen experimentally. We feel this phase noise is an essential feature of the observed behavior. This intermittency may be related to tangent bifurcations such as discussed by Yorke and Yorke[13] and Manneville and Pomeau[14] but it does not agree with the particular case analyzed in detail by Hirsh, Huberman, and Scalapino,[15] which deals with the familiar one-dimensional map with a quadratic maximum. In this latter case, as the control parameter is increased, the intermittency is followed by an odd period-doubling cascade—a behavior not seen in our experiments. The connection between the observed behavior and the sinusoidal map (which does exhibit periodic running solutions and multiwell diffusion) discussed recently by various authors,[16-18] is also not yet clear. Finally, in a somewhat unrelated vein, we note that nowhere in our experiment have we observed the quasiperiodic route to chaos proposed by Ruelle and Takens.[19]

VI. EFFECT OF dc BIAS

The presence of a dc bias γ_0 strongly influences the response of the pendulum to a periodic driving force, but to the extent that we have been able to establish, it does not introduce any fundamentally different states or new types of chaos. That is, there are both phase-locked and -unlocked states of the type found in zero-bias as shown in Fig. 3. The phase-locked states correspond to ac Josephson steps and the unlocked states can be intermittent. Smooth transitions between phase-locked states are also observed under some conditions as the dc bias is increased, but they correspond to the uninteresting situation where there is no interaction (i.e., locking or tendency to lock) between the drive frequency and the average rotation frequency of the pendulum. The power spectrum under these conditions consists of lines at the drive frequency, the rotation frequency, their harmonics and mixing products, with no subharmonics or broadband noise.

The effect of a small dc bias (i.e., biasing within a given potential well) is simple and easy to understand. It has four principal effects. External symmetry breaking is introduced, the resonant frequency is lowered, the quality factor is decreased, and the critical ac torque required to produce rotation is reduced. This behavior can be understood trivially in terms of the linear ac response expected for a biased pendulum. The equilibrium position is given by $\theta_0 = \sin^{-1}\gamma_0$, from which it follows that $\omega_0 \to (1-\gamma_0^2)^{1/4}$ and $Q \to Q(1-\gamma_0^2)^{1/4}$. The net effect of all this is, roughly speaking, simply to scale the state diagram of Fig. 3 along the γ_1 and ω axes. Note, however, that as $\gamma_0 \to 1$, $Q \to 0$ and one expects the structure in the state diagram to be strongly modified and eventually all chaotic behavior to cease. This limit was not carefully explored in our study.

ACKNOWLEDGMENTS

Two of us (M.R.B. and B.A.H.) would like to thank all of the members of the Groupe de Physique des Solides de l'Ecole Normale Superieure for having provided both the facilities and the warm hospitality that made this collaboration possible. All of us would like to express our thanks to A. Launay for his invaluable technical help with the development of the circuits used in these experiments. Finally, one of us (M.R.B.) would like to acknowledge the support of the U. S. National Science Foundation during the final stages of the completion of this work.

APPENDIX A: SIMULATOR CIRCUIT

To simulate a driven pendulum we have used an analog circuit of the phase-locked-loop variety described by Bak[20] in the context of a simulator for the resistively shunted junction model of a Josephson junction. The basic functional diagram of the circuit is shown in Fig. 14 and the detailed circuit diagram in Fig. 15. In the circuit the output of a voltage-controlled oscillator (VCO) is mixed with a reference oscillator, passed through a low-pass fil-

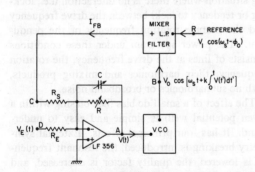

FIG. 14. Schematic diagram of the phase-locked loop used in our simulations. Typical circuit values are $R=11$ kΩ, $C=100$ nF, $k=4\pi \times 10^3$ s^{-1}/V which lead to a natural frequency $\Omega_0/2\pi=600$ Hz and a $Q=4$. We also used $R_E=R_S=10$ kΩ and a reference frequency $\omega_s/2\pi=100$ kHz.

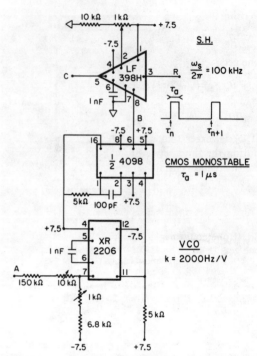

FIG. 15. Detailed circuit diagram for phase-locked loop.

ter, and then fed back to the input of the VCO through an operational amplifier. The ac and dc bias inputs are applied at the summing point of the operational amplifier. The resistor and capacitor in the feedback circuit of the operational amplifier provide the inertia and damping of the loop.

In practice, the actual circuit we used was patterned after that developed by Henry and Prober,[21] in which the mixing and low-pass filtering functions are accomplished using a sample-and-hold (SH) circuit. The merits and demerits of this approach have been discussed by those authors.

If the reference signal is $V_1 \sin(\omega_s t + \phi_0)$, the output of the sample-and-hold circuit sampled at the time τ_n is $V_1 \sin(\omega_s \tau_n + \phi_0)$. During the hold time between τ_n and the next sampling time τ_{n+1}, the input $V(t)$ of the VCO is given by

$$\frac{1}{C}\frac{dV}{dt} + \frac{1}{R}V + \frac{V_1}{R_S}\sin(\omega_s \tau_n + \phi_0) = -\frac{V_E(t)}{R_E} .$$
(A1)

The sampling time τ_{n+1} is related to τ_n and to the input of the VCO by

$$\omega_s(\tau_{n+1} - \tau_n) + k \int_{\tau_n}^{\tau_{n+1}} V(t)dt = 2\pi .$$
(A2)

Writing

$$\theta_n = (2n+1)\pi - \omega_s \tau_n - \phi_0 \tag{A3}$$

one gets

$$\begin{aligned}\theta_{n+1} &= \theta_n + 2\pi - \omega_s(\tau_{n+1} - \tau_n) \\ &= \theta_n + k \int_{\tau_n}^{\tau_{n+1}} V(t)dt \ .\end{aligned} \tag{A4}$$

If ω_s is larger than the highest frequency of $V(t)$ during the time between τ_n and τ_{n+1}, $V(t)$ can be considered as constant and written

$$V(t) = \lim_{\omega_s \to \infty} \left[\frac{1}{k} \left[\frac{\theta_{n+1} - \theta_n}{\tau_{n+1} - \tau_n} \right] \right] = \frac{\dot{\theta}}{k} \ . \tag{A5}$$

So, if one could neglect the time between sampling, one gets

$$\dot{\theta} = kV(t) \ , \tag{A6}$$

$$C \frac{dV}{dt} + \frac{1}{R} V + \frac{V_1}{R_S} \sin\theta = -\frac{V_E(t)}{R_E} \ , \tag{A7}$$

or

$$\frac{C}{k} \ddot{\theta} + \frac{1}{kR} \dot{\theta} + \frac{V_1}{R_S} \sin\theta = -\frac{V_E(t)}{R_E} \ , \tag{A8}$$

where V_1 is the peak amplitude of the reference. With the notations of Sec. II one gets

$$a = C/k, \ b = 1/kR, \ c = V_1/R_S \ ,$$

$$\Gamma(t) = -V_E(t)/R_E \ .$$

It is also interesting to note, referring back to Eq. (A4), that for a periodic $V(t)$ the integral of $V(t)$ can be calculated explicitly and then Eq. (A4) becomes a discrete mapping reflecting the actual behavior of an ideal VCO fed back through an ideal sample-and-hold circuit.

From the simulation point of view, this circuit has two drawbacks associated with the detailed operation of the sample-and-hold circuit. Specifically, because of the finite aperture time and an overshoot in the response of the SH, the sampling transient leads to an error in I_{FB} that depends on the phase between the VCO and the reference signal. An important consequence of this problem is the introduction of two discontinuities in the slope of I_{FB} near its extrema and of an asymmetry in the amplitude of I_{FB} between its positive and negative half-cycles. Note, however, that while such asymmetries may dictate which way the symmetry breaking of the simulated pendulum goes, by carefully studying the effects of an applied dc bias, we conclude that the observed symmetry breaking is intrinsic to the forced pendulum and not an idiosyncrasy of our simulator circuit.

APPENDIX B: STABILITY OF PHASE-LOCKED STEPS

In Sec. IV we have investigated solutions of Eq. (3) of the form

$$\theta^* = \theta_n + n(\omega\tau - \phi) - \alpha \sin(\omega\tau - \phi) \tag{B1}$$

with θ_n, α, and ϕ related to γ_0, γ_1, and n by Eqs. (13)–(15). Such solutions will be observed only if they are stable, i.e., only if any small perturbation $\delta\theta$ of θ^* is damped. Writing $\theta = \theta^* + \delta\theta$, for $\delta\theta \ll 1$ one gets

$$\delta\ddot{\theta} + \frac{1}{Q}\delta\dot{\theta} + \cos\theta^* \delta\theta = 0 \ . \tag{B2}$$

Limiting the $\cos\theta^*$ expansion to the first two terms, one obtains the following Mathieu equation:

$$\frac{d^2 y}{dz^2} + (a - 2q\cos 2z)y = 0 \ , \tag{B3}$$

where

$$a = \frac{4}{\omega^2}\left[J_n(\alpha)\cos\theta_n - \left[\frac{1}{2Q}\right]^2 \right] \ , \tag{B4}$$

$$q = \frac{4}{\omega^2}\left[\left[\frac{n}{\alpha} - J_n(\alpha)\cos\theta_n \right]^2 + [J'_n(\alpha)\sin\theta_n]^2 \right]^{1/2} \ , \tag{B5}$$

$$z = \tfrac{1}{2}(\omega\tau - \phi + \nu_n) \ , \tag{B6}$$

$$y = \delta\theta \, e^{-\tau/2Q} \ , \tag{B7}$$

$$\tan\nu_n = \alpha J'_n(\alpha)\tan\theta_n / nJ_n(\alpha) \ . \tag{B8}$$

However, if $n=0$ and $\gamma_0=0$, then $\theta_n=0$, Eq. (B5) leads to $q=0$, and the expansion of $\cos\theta^*$ must be taken one order higher in which case (B4)–(B6) are replaced by

$$a = \frac{1}{\omega^2}\left[J_0(\alpha) - \frac{1}{(2Q)^2} \right] \ , \tag{B9}$$

$$q = \frac{1}{\omega^2} J_2(\alpha) \ , \tag{B10}$$

$$z = \omega\tau - \phi \ . \tag{B11}$$

Hence, we find that the stability of θ^* will have the same kind of behavior[22] as the solutions of the corresponding Mathieu equation. If a and q belong in the stability domain of the Mathieu equation,[23] θ^* is stable. If, on the other hand, the solutions of Eq. (B3) are unstable, only two kinds of bifurcations can occur, either with period π or 2π for z. Now the interpretation of these bifurcations depends on the symmetry of θ^*. If $n=0$ and $\gamma_0=0$, z and $\omega\tau$ are related by Eq. (B11) and the bifurcations occurs with periods ω or 2ω. The first case corresponds to a jump toward another stable state still given by Eq. (B1) (if one exists). A vivid illustration of this bifurcation is given in Figs. 4 and 7, where the negative slope part of the S-shaped resonance is unstable, leading to the hysteretic behavior in Fig. 4. When the other bifurcation occurs, the solution of (B2), which grows, has a frequency twice the drive frequency and, as a consequence, a dc component.[23] Hence, in this case, the system goes to a state where $\langle \theta \rangle$ is nonzero and this bifurcation corresponds to a symmetry breaking of the pendulum motion for $\gamma_0=0$. If $n\neq 0$ or $\gamma_0\neq 0$, then the motion of the pendulum is already asymmetric, z and $\omega\tau$ are related by Eq. (B6), and the bifurcations occur with periods ω or $\omega/2$. The first one was described above. The second corresponds to a growth of solution with frequency half the drive frequency, i.e., to period doubling.

From this analysis we conclude that, for the forced pendulum, before any period-doubling bifurcation the symmetry of the motion must be broken either with an external torque γ_0 or by a previous bifurcation.

---

*Permanent address: Department of Applied Physics, Stanford University, Stanford, California 94305.

†Permanent address: Xerox Palo Alto Research Center, Palo Alto, California 94304.

[1] B. A. Huberman, J. P. Crutchfield, and N. Packard, Appl. Phys. Lett. 37, 750 (1980).

[2] R. C. Kautz, J. Appl. Phys. 52, 624 (1981).

[3] N. F. Pedersen and A. Davidson, Appl. Phys. Lett. 39, 830 (1981).

[4] B. A. Huberman and J. P. Crutchfield, Phys. Rev. Lett. 43, 1743 (1979).

[5] A. Libchaber and J. Maurer, J. Phys. Coll. 41, 51 (1980).

[6] A. Arneodo, P. Coullet, C. Tresser, A. Libchaber, J. Maurer, and D. d'Humieres (unpublished).

[7] M. J. Feigenbaum, Phys. Lett. 74A, 375 (1979).

[8] N. F. Pedersen, O. H. Soerensen, B. Dueholm, and J. Mygend, J. Low. Temp. Phys. 38, 1 (1980).

[9] D. E. McCumber, J. Appl. Phys. 39, 3113 (1968).

[10] E. Ben Jacob, Y. Braiman, and R. Shainsky, Appl. Phys. Lett. 38, 822 (1981).

[11] E. Ben Jacob, I. Goldhirsh, and Y. Imry (unpublished).

[12] E. Henon, Commun. Math. Phys. 50, 69 (1976).

[13] J. A. Yorke and E. D. Yorke, J. Stat. Phys. 21, 263 (1979).

[14] P. Manneville and Y. Pomeau, Phys. Lett. 75A, 1 (1979).

[15] J. E. Hirsh, B. A. Huberman, and D. J. Scalapino, Phys. Rev. A 25, 519 (1982).

[16] T. Geisel and J. Nierwetberg, Phys. Rev. Lett. 48, 7 (1982).

[17] S. Grossman (unpublished).

[18] M. Schell, S. Fraser, and R. Kapral (unpublished).

[19] D. Ruelle and F. Takens, Commun. Math. Phys. 20, 167 (1970).

[20] C. K. Bak and N. F. Pedersen, Appl. Phys. Lett. 22, 149 (1973).

[21] R. W. Henry and D. E. Prober, Rev. Sci. Instrum. 52, 912 (1981).

[22] We must point out that Eq. (B3) is only an approximation of Eq. (B2). As a consequence, the choice between sets of Eqs. (B4)–(B8) or Eqs. (B9)–(B11) is not exactly given by conditions $n=0$, $\gamma_0=0$ but by comparison between the value of q given by Eqs. (B5) and (B10) with a domain where the two values are close and in which Eq. (B3) cannot describe the stability of θ^*.

[23] *Handbook of Mathematical Functions*, 7th ed., edited by M. Abromowitz and I. Stegun (Dover, New York, 1970), p. 721.

JOSEPHSON JUNCTIONS AND CIRCLE MAPS

Per Bak*

Physics Department, Brookhaven National Laboratory, Upton, NY 11973, U.S.A.
NORDITA, Blegdamsvej 17, Copenhagen, Denmark

and

Tomas Bohr, M. Høgh Jensen and P. Voetmann Christiansen

H.C. Ørsted Institute, Universitetsparken 5, Copenhagen, Denmark

(*Received* 28 *March* 1984 *by* A. Zawadowski)

The return map of a differential equation for the current driven Josephson junction, or the damped driven pendulum, is shown numerically to be a circle map. Phase locking, noise and hysteresis, can thus be understood in a simple and coherent way. The transition to chaos is related to the development of a cubic inflection point. Recent theoretical results on universal behavior at the transition to chaos can readily be checked experimentally by studing I–V characteristics.

RECENTLY, THERE HAS BEEN an enormous theoretical activity in studying the transition to chaos by utilizing discrete maps [1–5]. In particular it has been suggested that the transition to chaos in systems with two competing frequencies can be described by a simple map of the circle onto itself [2–6], and universal scaling behavior was found at the point where chaos sets in.

Unfortunately there has not been parallel experimental activity, mostly because of the difficulties in firmly establishing the connection between extremely complicated dynamical systems and the very simple discrete maps. The resistively shunted Josephson junction is an example of a system which can be driven to a chaotic state by perturbing it with an rf microwave signal [7, 8], and indeed numerical simulations have indicated that the noise arises as a solution to a deterministic differential equation [9–11]. It has been suggested that the phase locking (subharmonic steps [12]) and the noise might be described by the circle map [5], but no real connection has yet been demonstrated.

Here we present numerical results providing the missing link by showing that the *return maps for a differential equation of the resistively shunted Josephson junction is a circle map* for a wide range of parameters including the transition to chaos. The transition to chaos and many other phenomena such as the subharmonic steps and hysteresis can thus be directly understood by taking over theoretical results for the circle map and, vice versa, *recent theoretical results on universality in the transition to chaos can be checked experimentally* on this device. In fact, because of the extreme precision with which voltages can be measured, the Josephson junction might well become a model system for studying these phenomena. The transition to chaos is brought about by the development of a cubic inflection point in the underlying circle map.

Consider the resistively and capacitively shunted Josephson junction driven by a d.c. current I and an r.f. microwave field with amplitude A. In appropriate units, the phase difference θ across the junction obeys the equation [9–11]

$$\ddot{\theta} + G\dot{\theta} + \sin\theta = I + A\sin\omega_{ext}t, \quad (1)$$

where $G = 1/R\sqrt{h/2eCI_c}$, R is the resistance, C the capacitance, I_c the critical current and ω_{ext} the frequency of the signal. The time unit is $1/\omega_0 = \sqrt{hC/2eI_c}$. The voltage across the junction is given by the Josephson equation [13]

$$V = RI_c G\dot{\theta}. \quad (2)$$

Actually, the equation (2) also describes the simple damped driven pendulum perturbed by an external periodic force so the experiment could in principle (but with much less accuracy) be performed on such a simple mechanical set-up. For I large enought the "pendulum" will rotate with a characteristic frequency or winding number W.

Now, imagine that after n cycles of the perturbing r.f. field we measure the values θ_n and $\dot{\theta}_n$ of the phase and its derivative. Since θ_n and $\dot{\theta}_n$ contain all information on the system, the values of θ and

---
* Present address: Cornell University, Ithaca, New York.

$\dot{\theta}$ after the $n + 1$st cycle are functions of θ_n and $\dot{\theta}_n$

$$(\theta_{n+1}, \dot{\theta}_{n+1}) = F(\theta_n, \dot{\theta}_n), \quad (3)$$

where F is the generally two-dimensional, return map. Because of the dissipative nature of the problem one might hope that after a transient period the return map collapses to a *one*-dimensional mapping which must generally have the form

$$\theta_{n+1} = \theta_n + \Omega + g(\theta_n) \, (\text{mod } 2\pi), \quad (4)$$

where $g(\theta_n) = g(\theta_n + 2\pi)$ because the original equation is invariant under $\theta \to \theta + 2\pi$ [14]. We have separated a constant driving term Ω from an oscillating term g. A map of the form (4) is called a circle map since it describes the motion of the phase θ on the circle $0 < \theta < 2\pi$. In its simplest form, the periodic function is a sine function

$$g(\theta_n) = -K \sin \theta_n, \quad (5)$$

but the general features are expected to be universal and will not depend on the specific function g [2–5].

Thus, starting from some initial values of θ and $\dot{\theta}$ we compute the phase θ_n after n cycles of the r.f. signal. We are thus watching the pendulum with "stroboscobic light" using the external frequency as a clock [15]. We can then plot the return map (θ_n, θ_{n+1}). The winding number W of the trajectory is defined by

$$W = \lim_{n \to \infty} \frac{\theta_n - \theta_0}{2\pi n}.$$

The numerical integration was performed using an Adams predictor-corrector method of order up to twelve.

Figure 1(a) shows the return map obtained for $G = 1.576, I = 1.4, \omega_{\text{ext}} = 1.76$ and $A = 1$. *Note first that the return map is one-dimensional.* The curve seems to be filled up ergodically indicating that the winding number is indeed irrational (~ 0.38). The periodic nature of the curve shows that *the return map is a circle map.* By varying the parameters G and I one can thus tune the winding number to any value, and the return map varies in a smooth way (even, as we shall see, when the system enters the chaotic regime). For rational W only a finite number of points are generated, which are confined to the smoothly varying curve.

All the essential information of the differential equation is contained in the circle map, and the evolution can be studied by iterating the map rather than continuing the cumbersome integration. Taking over previous results for the circle map, we see that as long as the function $g(\theta) + \theta$ is monotonically increasing the iteration leads to regular behavior. The winding number in this regime is either rational, indicating a frequency locking (or subharmonic step) with $\omega = P/O \, \omega_{\text{ext}}$, or, with finite probability, irrational [16].

Figure 1(b) shows the return map calculated for values of I and G which have been reduced in a way that keeps the winding number almost constant. The mapping now assumes zero slope at one point, i.e. its inverse has developed a *cubic singularity indicating a critical point for the transition to chaos*. The critical line in (I, G) space can be constructed by performing calculations for several winding numbers. Drawing upon knowledge from the circle map [5] we predict that along this line the winding number exhibits a complete "devil's staircase", i.e. the frequency is essentially always locked. This feature was clearly borne out in our numerical integration: near the transition line the points θ_n would always enter into a limit cycle sequence which makes it difficult to trace up the return map. Because of the Josephson equation $\langle V \rangle \sim \langle \dot{\theta} \rangle \sim W$, the frequency locking can be studied experimentally by directly measuring the voltage jumps along the transition line.

For smaller values of G and I [Figs. 1(c) and 1(d)] the return map develops a local maximum signalling the onset of *chaos*. The return map is no longer invertible and the iteration is not necessarily monotonic. The winding number in this regime for the circle map can depend on the initial values of θ which explains the *hysteresis* observed in the junction [8] and further the existence of a local maximum leads to infinite series of bifurcations [1]. As seen on Fig. 1(c) the curve develops "wiggles" in the chaotic regime. When proceeding futher into the chaotic regime these wiggles become more pronounced [Fig. 1(d)]. Figure 1(e) shows the return map as computed for a set of parameters giving a different winding number $W \simeq 0.48$. Again the horizontal tangent indicates a transition to chaos. The major change from Fig. 1(b) is simply a shift in the y direction indicating an increased winding number.

Figure 1 clearly establishes the connection between the resistively shunted Josephson junction and the circle map. The complicated behavior of the junction can be inferred from the map: The phase locking in the circle map reflects the subharmonic steps in the junction; the chaos in the circle map represents the noise, and the dependence of the starting point in the chaotic regime describes the hysteresis. And, most importantly, in the light of the many studies of noise in Josephson junctions [9–11]: this approach tells us clearly when to expect chaos, namely when the return map becomes non-invertible.

Extensive theoretical work has been performed on the transition to chaos in the circle map. Shenker [2] has studied the mapping near the point where the

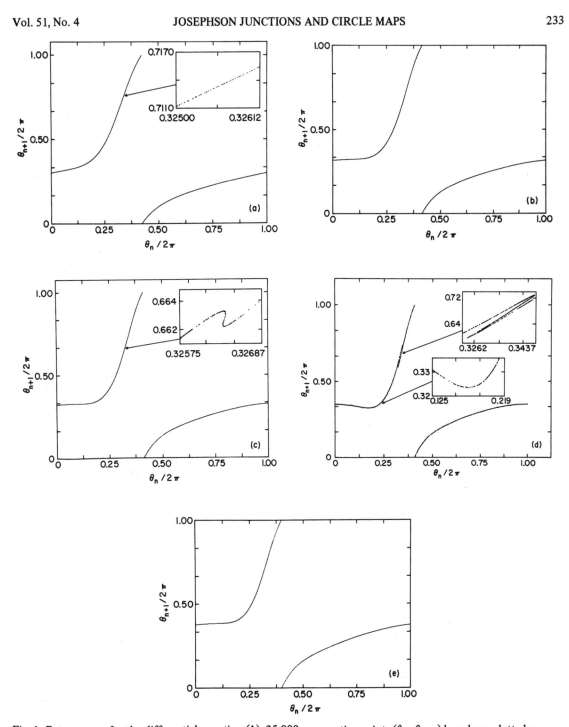

Fig. 1. Return maps for the differential equation (1). 25,000 consecutive points (θ_n, θ_{n+1}) have been plotted on each curve; the first 1000 points have been omitted. $A = 1$. (a)–(d): $\omega_{ext} = 1.76$ and $W \sim 0.38$; (e): $\omega_{ext} = 1.772$ and $W \sim 0.48$. (a) $G = 1.576$ and $I = 1.4$. The function $\theta_{n+1}(\theta_n)$ is monotonically increasing indicating regular behavior. The inset is a magnification emphasizing the 1d character. (b) $G = 1.293$ and $I = 1.225$. The function develops a cubic inflection point indicating the transition to chaos. The inset shows an enlargement of the curve around the inflection point. (c) $G = 1.253$ and $I = 1.2$ and (d) $G = 1.081$ and $I = 1.094$: The map develops a local minimum and "wiggles" (insets) indicating chaotic behavior. (e) Return map for $G = 1.2$ and $I = 1.25$. The map is critical as indicated by the cubic inflection point.

winding number approaches the Golden mean and found scaling behavior characterized by universal indices near the critical point where chaos sets in. The transition has been elegantly treated by means of a renormalization group technique by Feigenbaum et al. [3] and by Rand et al. [4]. Jensen et al. [5] have found universal scaling behavior for the phase locking structure at the critical line. Through the connection with equation (1) established above, these results give clear predictions about the nature of the I–V characteristics of the resistively shunted Josephson junction. Previous measurements clearly show that it is possible to find a multitude of phase locked intervals [8] and thus experimentally to check our predictions. The transition to chaos is caused by overlap of an infinity of resonances, so the transition line can be identified as the line where hysteresis and chaos near the *small* intervals (or at the edges of large intervals) sets in. We urge that such experiments, or analogue numerical simulations, be performed.

In this work we have concentrated on the behavior at and below criticality, where our numerical work indicates that the study of simple circle maps gives quantitative information about the differential equation (1). The behavior above critically, as well as a detailed survey of the transition at different winding numbers, will be presented in a forthcoming publication [17]. Preliminary numerical investigation show that the return maps above criticality become fractal in nature, with fractal dimension less than unity.

In conclusion, we believe that we have provided the link between a complicated dynamical system, and simple model calculations on discrete mappings, yielding better understanding of the former and greater credibility of the latter.

Acknowledgements — We are indebted to A. Bishop, M. Feigenbaum, Y. Imry and M. Levinsen for informative discussions. The work was supported in part by the Danish Natural Science Research Council and The Division of Materials Sciences U.S. Department of Energy under contract DE-AC02-76H00016 and NSF Grant DMR 8020929.

REFERENCES

1. M.J. Feigenbaum, *J. Stat. Phys.* **19**, 25 (1978); **21**, 669 (1979).
2. S.J. Shenker, *Physica* **5D**, 405 (1982).
3. M.J. Feigenbaum, L.P. Kadanoff & S.J. Shenker, *Physica* **5D**, 370 (1982); M.J. Feigenbaum & B. Hasslacher, *Phys. Rev. Lett.* **49**, 605 (1982).
4. D. Rand, S. Ostlund, J. Sethna & E. Siggia, *Phys. Rev. Lett.* **49**, 132 (1982); S. Ostlund, D. Rand, J. Sethna & E. Siggia, *Physica* **8D**, 303 (1983).
5. M.H. Jensen, P. Bak & T. Bohr, *Phys. Rev. Lett.* **50**, 1637 (1983).
6. G.M. Zaslavsky, *Phys. Lett.* **69A**, 145 (1978); L. Glass & R. Perez, *Phys. Rev. Lett.* **48**, 1772 (1982).
7. R.Y. Chiao, M.J. Feldman, D.W. Peterson, B.A. Tucker & M.T. Levinsen, *Future Trends in Superconducting Electronics,* AIP Conf. Proc. **44**, 259 (1978).
8. V.N. Belykh, N.F. Pedersen & O.H. Soerensen, *Phys. Rev.* **B16**, 4860 (1977).
9. Y. Braiman, E. Ben-Jacob & Y. Imry, *SQUID 80* (Edited by H.D. Hahlbohm & H. Lubbig) Walter de Gruyter, Berlin (1980); E. Ben-Jacob, Y. Braiman, R. Shainsky & Y. Imry, *Appl. Phys. Lett.* **38**, 822 (1981); E. Ben-Jacob, I. Goldhirsch, Y. Imry & S. Fishman, *Phys. Rev. Lett.* **49**, 1599 (1982); R.L. Kautz, *J. Appl. Phys.* **52**, 6241 (1981).
10. B.A. Huberman, J.P. Crutchfield & N.H. Packard, *Appl. Phys. Lett.* **37**, 751 (1980); D. D'Humieres, M.R. Beasley, B.A. Huberman & A. Libchaber, *Phys. Rev.* **A26**, 3483 (1982); N.F. Pedersen & A. Davidson, *Appl. Phys. Lett.* **39**, 830 (1981); M. Cirillo & N.F. Pedersen, *Phys. Lett.* **90A**, 150 (1982); W.J. Yeh & Y.H. Kao, *Phys. Rev. Lett.* **49**, 1888 (1982); A.H. MacDonald & M. Plischke, *Phys. Rev.* **B27**, 201 (1983); W.J. Yeh & Y.H. Kao, *Appl. Phys. Lett.* **42**, 299 (1983). These papers deal with the case $I = 0$ which is less relevant here.
11. M.T. Levinsen, *J. Appl. Phys.* **53**, 4294 (1982).
12. P.E. Gregers-Hansen, E. Hendricks, M.T. Levinsen & G. Fog Pedersen, *Proc. Appl. S-C. Conf. Annapolis*, p. 597 (1972).
13. B.D. Josephson, *Phys. Lett.* **1**, 251 (1962).
14. In the overdamped case where the coefficient to the θ term is zero this relation is trivial. J.R. Waldram & P.H. Wu, *J. Low Temp. Phys.* **47**, 363 (1982) found that the function g is zero in this case, so there is no mode locking and chaos.
15. This is fundamentally different from a conventional time series analysis which fails to reveal the 1d map for the system discussed here [K. Fesser, A.R. Bishop & P. Kumar, *Appl. Phys. Lett.* **43**, 123 (1983)].
16. M.R. Herman, *Lecture Notes in Mathematics* **597**, 271 (1977).
17. P. Bak, T. Bohr & M.H. Jensen (to be published).

Physica D 31 (1988) 252–268
North-Holland, Amsterdam

RESONANCE AND SYMMETRY BREAKING FOR THE PENDULUM

John MILES*

Institute of Geophysics and Planetary Physics, University of California, San Diego, La Jolla, CA 92093, USA

Received 22 September 1987
Revised manuscript received 26 February 1988
Communicated by J.E. Marsden

Periodic solutions of the differential equation for periodic forcing of a lightly damped pendulum are obtained on the alternative hypotheses that (i) the contributions of second and higher harmonics to the average Lagrangian are negligible or (ii) the solution is close to that for free oscillations. The resonance curves (amplitude or root-mean energy vs. driving frequency) and stability boundaries for symmetric swinging oscillations and their asymmetric descendents (following symmetry breaking) are determined for $\delta \ll 1$ and $\varepsilon = \mathcal{O}(\delta)$, where δ is the ratio of actual to critical damping, and ε is the ratio of the maximum external moment to the maximum gravitational moment. Resonance, as defined by synchronism between the external moment and the damping moment, is found to be impossible, and the conventional resonance curve separates into two branches, if $\varepsilon > \varepsilon_* = 3.28\delta + \mathcal{O}(\delta^3)$, which condition is necessary for normal symmetry breaking. A numerical, Fourier-series determination of the resonance curve and bifurcation points for $\delta = \frac{1}{8}$ and $\varepsilon = \frac{1}{2}$ is presented in an appendix by P.J. Bryant.

1. Introduction

I consider here periodic solutions of the pendulum equation

$$\ddot{\theta} + 2\delta\dot{\theta} + \sin\theta = \varepsilon \sin\omega t, \tag{1.1}$$

where θ is the angular displacement from the (downward) vertical, $\dot{\theta} \equiv d\theta/dt$, t is the dimensionless time, for which the scale is the inverse natural frequency, ω is the ratio of the driving frequency to the natural frequency, δ is the damping ratio, and ε is the ratio of the applied moment to the maximum gravitational moment.

This fundamental problem, the analytical study of which goes back to Huygens [1], is isomorphic to such modern nonlinear oscillators as the Josephson junction and is arguably the simplest mechanical system for which the forced periodic motion may become chaotic**. The transition of this motion to chaos has been studied by D'Humieres et al. [4], who give references to earlier studies. They report numerical integrations for both swinging ($\langle\dot{\theta}\rangle = 0$) and running ($\langle\dot{\theta}\rangle \neq 0$) oscillations for $\delta = \frac{1}{8}$ and $0 < \varepsilon < 2$ ($\delta = \frac{1}{2}Q^{-1}$ and $\varepsilon = \gamma_1$ in their notation), remark that the transition from swinging oscillations to chaos is preceded by symmetry breaking, and, following Pedersen et al. [5], establish implicit stability boundaries through reference to Mathieu's equation. [They also allow for a non-zero mean (d.c.) component of the

---

*Supported in part by the Physical Oceanography, Applied Mathematics and Fluid Dynamics/Hydraulics programs of the National Science Foundation, NSF Grant OCE81-17539, by the Office of Naval Research under Contract N00014-84-K-0137, 4322318 (430), and by the DARPA Univ. Res. Init. under Appl. and Comp. Math. Program Contract N00014-86-K-0758 administered by the Office of Naval Research.
**The Duffing oscillator [2] also provides such a model, but it is not as readily realized as the pendulum. See also [3], where the stability of resonant motion of a spherical pendulum is analyzed.

excitation, which is relevant for the Josephson junction, but of lesser interest for the pendulum.] Experimental observations of asymmetric and chaotic motion of the pendulum are reported by Blackburn et al. [6].

My primary analytical goals are the determination of the resonance curves – i.e. either the amplitude or the root-mean energy of a periodic oscillation versus ω – and the bifurcation points on these curves, in particular the symmetry-breaking locus in the δ, ε-plane at which symmetric swinging oscillations bifurcate to asymmetric swinging oscillations. By "symmetric" I imply $\theta(t + \tfrac{1}{2}T) = -\theta(t)$, where $T \equiv 2\pi/\omega$ is the fundamental period.

I assume $0 < \delta \ll 1$, $\varepsilon = \mathcal{O}(\delta)$ and $0 < \omega < 1$ [note that weak nonlinearity would require $\theta = \mathcal{O}(\delta^{1/2})$, $\varepsilon = \mathcal{O}(\delta^{3/2})$ and $\omega = 1 + \mathcal{O}(\delta)$] and expand θ in the form

$$\theta = \sum_n q_n \theta_n (t - \omega^{-1}\phi). \tag{1.2}$$

The q_n and $q_\phi \equiv \phi$ are generalized coordinates that may be determined either through Galerkin's method or by invoking Hamilton's principle in the form

$$-\partial \langle L \rangle / \partial q_n = \langle Q_n \rangle, \tag{1.3}$$

where

$$\langle L \rangle = \langle \tfrac{1}{2}\dot{\theta}^2 - 1 + \cos\theta \rangle \tag{1.4}$$

is the average Lagrangian ($\langle\ \rangle$ signifies an average over the period $2\pi/\omega$), and

$$\langle Q_n \rangle = \langle (\varepsilon \sin\omega t - 2\delta\dot{\theta})(\partial\theta/\partial q_n) \rangle \tag{1.5}$$

is the average generalized force corresponding to q_n. Letting $q_n = \phi$ and involving $\partial\langle L \rangle/\partial\phi = 0$, we obtain

$$\varepsilon \langle \dot{\theta} \sin\omega t \rangle = 2\delta \langle \dot{\theta}^2 \rangle, \tag{1.6}$$

which describes the balance between the work done by the external moment and the dissipation and is a counterpart of the classical symmetry that the invariance of the Lagrangian under a translation of time implies conservation of energy.

The simplest expansion functions are $\sin n\omega t$ and $\cos n\omega t$, for which (1.2) is a Fourier series. I consider the truncation $\theta = \theta_0 + \alpha \sin(\omega t - \phi)$ in section 2. The resulting equation for θ_0 may be satisfied by choosing $\theta_0 = 0$, $\alpha = \alpha_0 = 2.40$ (the smallest zero of the Bessel function J_0) with $0 < |\theta_0| < \pi$, or $\theta_0 = \pi$, which correspond, respectively, to symmetric swinging, asymmetric swinging, or inverted oscillations [7].

I consider symmetric swinging oscillations of the form $\theta = \alpha \sin(\omega t - \phi)$ and calculate the corresponding family of resonance curves, $\alpha = \alpha(\omega; \delta, \varepsilon)$, in section 3; see figs. 1(a, b). It follows from (1.6) that resonance, as defined by $\phi = \tfrac{1}{2}\pi$ (so that the external moment is in phase with the damping moment and $\omega = \omega_f$, the frequency of free oscillations), is possible if and only if $\varepsilon < \varepsilon_*$, where $\varepsilon_*/\delta \approx 3.16$ in the approximation of section 3 and 3.28 in the exact (for $\delta \downarrow 0$) calculation of section 5. The maximum of the resonance curve coincides with $\phi = \tfrac{1}{2}\pi$ only in the limit $\delta \downarrow 0$.

If $\varepsilon < \varepsilon_\times = \varepsilon_* + \mathcal{O}(\delta^2)$ the resonance curve has lower and upper ($\alpha \lessgtr \alpha_\times$ in fig. 1b) branches. The lower branch, which I describe as *normal*, is similar to that shown in textbooks for a soft-spring oscillator; if δ

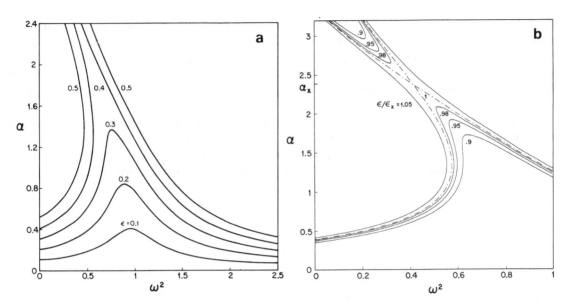

Fig. 1. (a). The resonance curves (3.2) for $\delta = \frac{1}{8}$ and $\varepsilon = 0.1(0.1)0.5$. (b). The resonance curves (3.2) for $\delta = \frac{1}{8}$ and $\varepsilon/\varepsilon_\ast = 0.9, 0.95, 0.98, 1(-\cdot-), 1.05$. All points above $\alpha = \alpha_s = 2.40 > \alpha_\ast$, between the turning points for $\varepsilon < \varepsilon_\ast$, or above the turning point on the left branch for $\varepsilon/\varepsilon_\ast > 1$ correspond to unstable states.

and $1 - (\varepsilon/\varepsilon_\ast)$ are sufficiently small (e.g. $\varepsilon/\varepsilon_\ast = 0.95$ in fig. 1b) this branch is triple-valued between two turning points (at which $d\omega/d\alpha = 0$), and the segment that connects these turning points comprises only unstable states. The upper ($\alpha > \alpha_\ast$) branch, which I describe as *anomalous*, has a single turning point that divides this upper branch into upper and lower segments, and the latter comprises only unstable points; it also has a single symmetry-breaking bifurcation (which I also describe as anomalous) just above the turning point, and all states above this bifurcation are unstable. In brief, the anomalous branch of the resonance curve for $\varepsilon < \varepsilon_\ast$ comprises only a very small interval of stable states, which prove to be difficult to realize (the lower segment of the normal resonance curve for prescribed δ, ε and ω has a much larger basin of attraction).

If $\varepsilon > \varepsilon_\ast$ the resonance curve (see fig. 1b, $\varepsilon/\varepsilon_\ast = 1.05$) consists of two separate branches, the lower of which comprises stable/unstable states below/above the remaining turning point and the upper of which comprises stable/unstable states below/above a symmetry-breaking bifurcation at $\alpha = \alpha_s$. I describe this bifurcation as *normal*, in contrast to the anomalous symmetry-breaking bifurcation for $\varepsilon < \varepsilon_\ast$.

The symmetric oscillations bifurcate at $\alpha = \alpha_s$ and $\omega = \omega_s$ to asymmetric oscillations (for which $\alpha > \alpha_s$ and $\omega < \omega_s$). I consider these solutions in section 4 and their stability in appendix A. They are stable only in a rather small parametric domain and become unstable through a period-doubling bifurcation, which leads to a period-doubling cascade.

The analysis in sections 2–4 is limited, like those of [4] and [5], by the aforementioned truncation of the Fourier series that describe the exact periodic solutions of (1.1). It is possible to proceed analytically by including higher harmonics, but the results are too complicated to offer any significant advantage over a numerical procedure that retains a large number of terms in the Fourier series. Such a solution is reported by Bryant in appendix B for $\delta = \frac{1}{8}$ and $\varepsilon = \frac{1}{2}$.

An alternative to a Fourier-series expansion is to base (1.2) on a perturbation of the solution of (1.1) for free oscillations, for which $\delta = \varepsilon = 0$ and $\theta \equiv \theta_f(t)$ is an elliptic function. Such a procedure may be expected to be useful for those states that are close to the *spine* of the resonance curve, $\omega = \omega_f(\alpha)$, which is possible only if $\delta \ll 1$, $\varepsilon = \mathcal{O}(\delta)$ and $\omega < 1$ and excludes the skirts of the resonance curve (on which the results of section 3 are accurate). I develop the $\mathcal{O}(\delta)$ component of this perturbation in section 5 and use the results to show that the errors in the approximations of section 3 to ε_*/δ, α_* and ω_* in the limit $\delta \downarrow 0$ with $\varepsilon = \mathcal{O}(\delta)$ are, respectively, 4, 3 and 7% and that the normal symmetry-breaking bifurcation is close to that observed by D'Humieres et al. [4] for $\delta = \frac{1}{8}$ and $\varepsilon = 0.5$ and to that calculated by Bryant (appendix B).

It appears from a comparison of the results of sections 2–5 and Bryant's numerical analysis that the approximation $\theta = \alpha \sin(\omega t + \phi)$ in section 3 provides a reliable description of the forced periodic motion of the pendulum for $\varepsilon < \varepsilon_*$ and $\alpha < \alpha_*$ and that the perturbation of the free oscillation of section 5 provides a reliable description of that motion in the parametric neighborhood of symmetry breaking.

2. Truncated Fourier expansion

Following D'Humieres et al. [4] and Pederson et al. [5], we first approximate the steady-state (following the decay of transients) solution of (1.1) by

$$\theta = \theta_0 + \alpha \sin(\omega t - \phi) \quad (-\pi < \theta_0 \leq \pi, \alpha > 0, 0 < \phi < \pi). \tag{2.1}$$

Invoking (1.3)–(1.5), we obtain

$$J_0(\alpha)\sin\theta_0 = 0, \quad 2\delta\alpha\omega = \varepsilon\sin\phi, \quad 2J_1(\alpha)\cos\theta_0 - \alpha\omega^2 = \varepsilon\cos\phi, \tag{2.2a, b, c}$$

where, here and subsequently, $J_n \equiv J_n(\alpha)$ is a Bessel function of the first kind.

It follows from (2.2a) that there are three distinct solutions of the form (2.1): $\theta_0 = 0$, which corresponds to symmetric oscillations; $\alpha = \alpha_s = \alpha_0$ (the subscript s signifies *symmetry breaking*) and $0 < |\theta_0| < \pi$, which corresponds to asymmetric oscillations; $\theta_0 = \pi$, which corresponds to symmetric oscillations of the inverted pendulum [7].

A preliminary estimate of the error in the approximation (2.1) may be obtained by comparing the frequency of the free oscillations ($\omega \equiv \omega_f$) implied by (2.2c) for $\delta = \varepsilon = \theta_0 = 0$,

$$\omega_f^2 = 2\alpha^{-1}J_1(\alpha) \equiv \Omega(\alpha) = 1 - \tfrac{1}{8}\alpha^2 + \alpha^4/192 + \cdots, \tag{2.3}$$

with the exact result (in which α is the amplitude of the oscillation)

$$\omega_f^2 = [(2/\pi)K(\sin\tfrac{1}{2}\alpha)]^{-2} = 1 - \tfrac{1}{8}\alpha^2 + \tfrac{7}{8}\alpha^4/192 + \cdots, \tag{2.4}$$

Table I
Comparison of the approximation (2.3) with the exact result (2.4) for the frequency of free oscillations.

α	ω_f^2 (2.3)	ω_f^2 (2.4)
0	1	1
$\pi/4$	0.925	0.925
$\pi/2$	0.721	0.718
$3\pi/4$	0.449	0.428
2.405	0.432	0.411
π	0.181	0

where K is an elliptic integral of the first kind. It appears from table I that the error in (2.3) is less than 5% for $\alpha \leq \alpha_0 = 2.40$ (the smallest zero of J_0), to which range we restrict most of the subsequent development. But note that the error in (2.3) is of second order in the amplitude of the third harmonic in (2.1) by virtue of symmetry and Rayleigh's principle.

3. Symmetric swinging oscillations

Setting $\theta_0 = 0$ and eliminating the phase angle ϕ between (2.2b, c), we obtain the resonance curve(s)

$$\alpha^2\left[(\beta - \Omega)^2 + 4\delta^2\beta\right] = \varepsilon^2 \quad (\beta \equiv \omega^2), \tag{3.1}$$

where Ω is defined by (2.3). Solving (3.1) for β, we obtain (see fig. 1a)

$$\beta = \Omega - 2\delta^2 \pm (\varepsilon^2\alpha^{-2} - 4\delta^2\Omega + 4\delta^4)^{1/2}, \tag{3.2}_\pm$$

where the alternative signs are vertically ordered. Resonance, as defined by $\phi = \tfrac{1}{2}\pi$ (so that the external moment is in phase with the damping moment), occurs at $\beta = \Omega$, for which $\varepsilon = 2\delta\alpha\Omega^{1/2}$. It follows that resonance (as defined here) is impossible if $\varepsilon > \varepsilon_*$, where (cf. section 5)

$$\varepsilon_* = 2\delta(2\alpha J_1)^{1/2}_{\max} = 3.16\delta, \quad \alpha_* = \alpha_0 = 2.40, \quad \omega_* = \Omega_0^{1/2} = 0.657. \tag{3.3a, b, c}$$

The maximum (if any) of the resonance curve, $\alpha = \alpha_m$, occurs on the spine if and only if $\delta^2 \ll 1$, in which case $(3.2)_\pm$ may be approximated by

$$\beta = \Omega \pm (\varepsilon^2\alpha^{-2} - 4\delta^2\Omega)^{1/2} + \mathcal{O}(\delta^2). \tag{3.4}_\pm$$

The two branches $(3.2)_\pm$ join, and the resonance curve $\alpha = \alpha(\beta)$ has a maximum, at the smaller (or smallest) root of

$$2\delta\alpha(\Omega - \delta^2)^{1/2} = \varepsilon, \quad \beta = \Omega - 2\delta^2 \quad (\alpha = \alpha_m) \tag{3.5a, b}$$

if and only if $\varepsilon < \varepsilon_\times(\delta)$, where (see fig. 2) $\varepsilon_\times$ is the smaller of the critical values determined by the requirement that the solution of (3.5) be real, which implies the parametric relations

$$\varepsilon_\times = 2\delta\alpha_\times[J_2(\alpha_\times)]^{1/2}, \quad J_0(\alpha_\times) = \delta^2, \tag{3.6a, b}$$

or the requirement that $\beta > 0$, which implies

$$\varepsilon_\times = 2J_1(\alpha_\times), \quad \Omega(\alpha_\times) = 2\delta^2. \tag{3.7a, b}$$

These two conditions coincide, and $\varepsilon_\times(\delta)$ has a maximum of $\varepsilon_\times = \varepsilon_1 = 1.164$, at $\delta = \delta_1 = 0.562$ ($\alpha_\times = \alpha'_1 = 1.841$); the condition (3.6)/(3.7) is the more critical for $\delta \lessgtr \delta_1$. The parameter $\alpha_\times$ decreases monotonically along $\varepsilon = \varepsilon_\times(\delta)$ from α_0 at $\delta = 0$ through α'_1 at $\delta = \delta_1$ to 0 at $\delta = 1/\sqrt{2}$. We assume $\delta < \delta_1$, so that $\alpha_\times$ and $\varepsilon_\times$ are determined by (3.6), throughout the subsequent development. Expanding (3.6) about $\delta = 0$ and

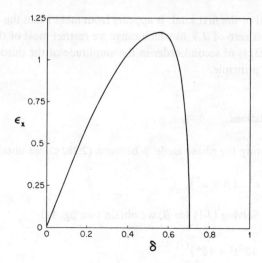

Fig. 2. The domin of resonant maxima, $\varepsilon < \varepsilon_\times(\delta)$, as determined by (3.6) for $0 < \delta \le 0.562$ or by (3.7) for $0.562 \le \delta < 0.707$.

neglecting $\mathcal{O}(\delta^4)$, we obtain

$$\alpha_\times = \alpha_*(1 - 0.80\delta^2), \quad \omega_\times = \omega_*(1 - 1.5\delta^2), \quad \varepsilon_\times = \varepsilon_*(1 - 1.16\delta^2). \qquad (3.8a, b, c)$$

It can be shown, by expanding (3.2) about $\alpha = \alpha_\times$ and $\beta = \beta_\times$, that the resonance curve for $\varepsilon = \varepsilon_\times$ crosses itself at $\alpha = \alpha_\times$ and $\beta = \beta_\times$ and that the resonance curves in the neighborhood of this crossing ($|\varepsilon - \varepsilon_\times|, |\alpha - \alpha_\times|, |\beta - \beta_\times| \ll 1$) approximate hyperbolae; see fig. 1b. The curves above $\alpha = \alpha_\times$ for $\varepsilon < \varepsilon_\times$, which (see section 1) I describe as anomalous, have minima at the smallest root of (3.5a) that exceeds $\alpha_\times$; however, only a small subinterval, bounded above by a symmetry-breaking bifurcation and below by a turning-point bifurcation (see appendix A), of each of these anomalous curves comprises stable states. Moreover, these stable states are difficult to attain in consequence of the much stronger attraction of the stable states on the lower segments ($\alpha < \alpha_+$ – see below) of the normal resonance curves. The anomalous resonance curves are implicitly excluded in the subsequent development (although some parts thereof may be valid for $\varepsilon < \varepsilon_\times$ and $\alpha > \alpha_\times$).

If $\varepsilon < \varepsilon_\times$ and δ is sufficiently small the resonance curves for $\varepsilon < \varepsilon_\times$ and $\alpha < \alpha_\times$ have two turning points, $(\beta_\pm, \alpha_\pm)$, $\alpha_+ < \alpha_-$ and $\beta_+ > \beta_-$, at which $d\beta/d\alpha = 0$. This condition, in conjunction with (3.2)$_-$ and the Bessel-function identities $J_0 + J_2 = \Omega$ and $J_0 - J_2 = 2J_1'$, yields

$$\varepsilon^2 = 2\alpha^2 J_2 \{ J_2 + 2\delta^2 \pm [(J_2 + 2\delta^2)^2 - 4\delta^2 \Omega]^{1/2} \} \quad (\alpha = \alpha_\pm), \qquad (3.9)_\pm$$

for the determination of $\alpha_\pm$. The corresponding results for $\beta_\pm$ are plotted in fig. 3; see also (4.4). The intermediate branch, $\beta_- < \beta < \beta_+$ and $\alpha_+ < \alpha < \alpha_-$, corresponds to unstable, and therefore unobservable, motions (see appendix A). The two turning points coincide for $\varepsilon = \varepsilon_c(\delta)$ (the subscript c signifies *critical*), which is determined by the parametric equations

$$\delta = \tfrac{1}{2}[(J_0 + J_2)^{1/2} - (J_0 - J_2)^{1/2}], \quad \varepsilon_c = 2\alpha J_2^{1/2} \Omega^{1/4} \delta^{1/2} \quad (\alpha = \alpha_+ = \alpha_-) \qquad (3.10a, b)$$

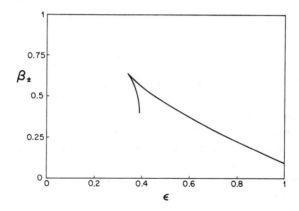

Fig. 3. The turning points $\beta_\pm$, $\beta_+ > \beta_-$, as determined from (3.2)$_-$ and (3.9) for $\alpha = \alpha_\star$ and $\delta = \frac{1}{8}$. Note that β_- terminates at $\varepsilon = \varepsilon_\star = 0.40$ and $\beta_- = \beta_\star = 0.41$ due to the restriction $\alpha < \alpha_\star$; however, the solution of (3.2)$_-$ and (3.9)$_-$ actually admits a continuation into $\varepsilon < \varepsilon_\star$ and $\beta_- < \beta_\star$, corresponding to the turning point on the anomalous branch of the resonance curve.

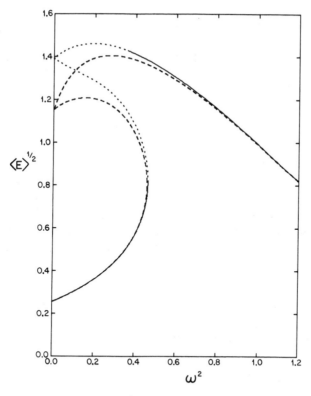

Fig. 4. Resonance curves for symmetric oscillations for $\delta = \frac{1}{8}$ and $\varepsilon = 0.5$, as computed in appendix B (—/ $\cdots$ for stable/unstable states) and (3.2) (– – –).

and increases monotonically from $\varepsilon_c \sim 2^{7/2}\delta^{3/2}$ as $\delta \downarrow 0$ to $\varepsilon_c = 1.164$ at $\delta = 0.398$. (The maximum value of ε_c is equal to that of ε_*, 1.164, but it occurs at $\delta = 0.398$ in contrast to $\delta = 0.562 = 0.398\sqrt{2}$.) The resonance curve is single valued if $\varepsilon < \varepsilon_c$ or if $\delta > 0.398$.

If $\varepsilon > \varepsilon_*$ the two branches $(3.2)_\pm$ remain separate. The right branch $(3.2)_+$ is monotonic $(d\alpha/d\omega < 0)$, while the left branch $(3.2)_-$ has a single turning point and intersects $\omega = 0$ at the two points determined by $2J_1(\alpha) = \varepsilon$ if $\varepsilon < \varepsilon_1 = 1.164$ or disappears if $\varepsilon > \varepsilon_1$. The upper (above the turning point) segment of the left branch corresponds to unstable solutions (appendix A).

It appears from the comparison of (2.3) with the exact result (2.4) and of (3.3) with the exact results in section 5 for $\delta \downarrow 0$ that the present results for symmetric oscillations are in error by at most a few percent for $\varepsilon < \varepsilon_*$ and $\alpha < \alpha_*$. But the contributions of the higher harmonics, neglected in the truncation (2.1), increase rapidly as ε increases above ε_*, and it then is preferable to plot $\langle E \rangle^{1/2}$, where $\langle E \rangle^{1/2}$, the root-mean energy of the oscillation, is given by

$$\langle E \rangle^{1/2} = \langle \tfrac{1}{2}\dot\theta^2 + 1 - \cos\theta \rangle^{1/2} = \langle \tfrac{1}{4}\alpha^2\beta + 1 - J_0(\alpha) \rangle^{1/2} \qquad (3.11)$$

in the present approximation and tends to $\alpha/\sqrt{2}$ in the limit $\alpha \downarrow 0$. This result is graphically indistinguishable from the results of Bryant's numerical analysis (appendix B) for $\delta = \tfrac{1}{8}$ and $\varepsilon \leq 0.3$ and only marginally distinguishable for $\varepsilon \leq 0.4$, but differs significantly therefrom for $\varepsilon > 0.5$; see fig. 4.

4. Symmetry breaking and asymmetric oscillations

It follows from (2.2a) that symmetry is broken at $\alpha = \alpha_s = \alpha_0$, for which (3.2) yields

$$\beta = \Omega_0 - 2\delta^2 \pm \alpha_0^{-1}(\varepsilon^2 - \varepsilon_s^2)^{1/2} \equiv \beta_s, \qquad (4.1a)_\pm$$

where

$$\varepsilon_s^2 = 4\delta^2\alpha_0^2(\Omega_0 - \delta^2) = \varepsilon_*^2 + \mathcal{O}(\delta^8) \qquad (4.1b)$$

and $\Omega_0 \equiv \Omega(\alpha_0) = 0.432$. Neglecting the $\mathcal{O}(\delta^8)$ difference between ε_s^2 and ε_*^2, we infer from (4.1a) that the symmetry-breaking bifurcations occur on the separated branches (since $\alpha_s > \alpha_*$) of the resonance curve. The left bifurcation point lies on the upper (unstable) segment of the left branch $(3.2)_-$ and therefore is of no further interest. The remaining bifurcation lies on the right branch $(3.2)_+$, which is unstable for $\alpha > \alpha_s$ (appendix A). The asymmetric oscillations follow

$$\beta = \Omega_0 c_0 - 2\delta^2 + \left[(\varepsilon/\alpha_0)^2 - 4\delta^2\Omega_0 c_0 + 4\delta^4\right]^{1/2}, \qquad (4.2)$$

where $c_0 \equiv \cos\theta_0$. Solving (4.2) for c_0, we obtain

$$c_0 = \left[\beta - (\varepsilon^2\alpha_0^{-2} - 4\delta^2\beta)^{1/2}\right]/\Omega_0, \qquad (4.3)$$

from which it follows that $d|\theta_0|/d\beta < 0$.

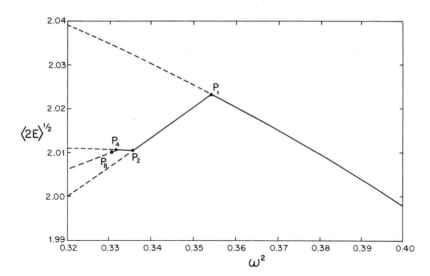

Fig. 5. Bifurcating resonance curves for symmetric, asymmetric, period 2, and period 4 oscillations (—/--- for stable/unstable states).

Letting $\delta = \frac{1}{8}$ and $\varepsilon = \frac{1}{2}$ in (4.1), we obtain $\varepsilon_s = 0.39$ and $\omega_s = 0.73$. This compares with $\omega_s = 0.595$ obtained by Bryant (see Appendix B) through numerical integration and with $\omega_s = 0.613$ obtained in subsection 5.1 below. (Fig. 5b of [4], for which $\delta = \frac{1}{8}$ and $\varepsilon = 0.50$, suggests $\omega_s \simeq 0.67$, but Bryant's results appear to be the more accurate.) The principal reason for the relatively poor accuracy of (4.1) appears to be its neglect of the second harmonic.

The stable range of θ_0 appears to be quite small (e.g. $|\theta_0| < 15°$ for $\delta = \frac{1}{8}$), and the asymmetric oscillations become unstable through a period-doubling bifurcation (see appendix A), which is followed, for increasing ε or decreasing ω, by a period-doubling cascade. Bryant (appendix B) has calculated the first three bifurcations (period doubling, quadrupling and octupling) in this cascade for $\delta = \frac{1}{8}$ and $\varepsilon = \frac{1}{2}$; see fig. 5, in which the locus of stable asymmetric oscillations appears as $P_1 P_2$.

If $\delta \ll 1$ the bifurcation point β_s may lie to the left of the bifurcation point β_+, in which case the solution of (1.1) may be asymptotic (as $\omega t \uparrow \infty$) to either a symmetric oscillation on the lower branch of the resonance curve, $(3.2)_-$, or an asymmetric oscillation, and which of these two states (symmetric or asymmetric) is realized for $\beta_s < \beta < \beta_+$ then depends on the initial conditions. Letting $\alpha = \alpha_+ \downarrow 0$ in $(3.2)_-$ and (3.9) on the hypothesis that $\varepsilon = \mathcal{O}(\delta)$ and eliminating α_+, we obtain

$$\beta_+ = 1 - \tfrac{3}{8}(4\varepsilon)^{2/3} + \left[\tfrac{1}{192} + \tfrac{1}{2}\left(\frac{\delta}{\varepsilon}\right)^2\right](4\varepsilon)^{4/3} + \mathcal{O}(\delta^2). \tag{4.4}$$

The corresponding approximation to β_s is, from (4.1a),

$$\beta_s = \Omega_0 + \alpha_0^{-1}(\varepsilon^2 - \varepsilon_s^2)^{1/2} + \mathcal{O}(\delta^2). \tag{4.5}$$

Letting $\varepsilon = \varepsilon_s$ (4.1b), we find that $\beta_s < \beta_+$ if $\delta < 0.20$.

5. The neighborhood of resonance

The solution of (1.1) for $\delta = \varepsilon = 0$ (free oscillations) is given by

$$\theta = 2\sin^{-1}(k\,\text{sn}\,t) \equiv \theta_f(t), \quad \sin\tfrac{1}{2}\alpha = k, \quad \omega_f = \tfrac{1}{2}\pi/K, \tag{5.1a, b, c}$$

where sn is a Jacobi elliptic sine of modulus k, α is the amplitude of the oscillation, and $K = K(k)$ is a complete elliptic integral of the first kind (the notation is that of [8]). We pose the corresponding solution for $\delta \ll 1$ and $\varepsilon = \mathcal{O}(\delta)$ in the form

$$\theta = \theta_f(t_\phi) + \sum_{1}^{N}\theta_n(t_\phi) + \mathcal{O}(\delta^{N+1}), \quad t_\phi = t - \omega^{-1}\phi, \quad \omega = \omega_f(k), \tag{5.2a, b, c}*$$

where $\theta_n = \mathcal{O}(\delta^n)$. Substituting (5.2) into (1.1), equating $\mathcal{O}(\delta^n)$ terms, invoking $\varepsilon = \mathcal{O}(\delta)$, and shifting t by ϕ/ω ($t_\phi \to t$), we obtain the inhomogeneous Lamé equation

$$\mathscr{L}\theta_n \equiv \ddot{\theta}_n(t) + (1 - 2k^2\sin^2 t)\theta_n(t) = f_n(t), \tag{5.3}$$

where

$$f_1 = \varepsilon\sin(\omega t + \phi) - 4\delta k\,\text{cn}\,t, \quad f_2 = k\theta_1^2\,\text{sn}\,t\,\text{dn}\,t - 2\delta\dot{\theta}_1, \ldots . \tag{5.4a, b}$$

The solution of (5.3), subject to the symmetry condition $\theta_n(t + 2K) = -\theta_n(t)$, admits the expansion

$$\theta_n(t) = \sum_{m=0}^{\infty}\left[A_n^{2m+1}Ec_1^{2m+1}(t) + B_n^{2m+1}Es_1^{2m+1}(t)\right], \tag{5.5}$$

where the Lamé functions Ec_1^{2m+1} and Es_1^{2m+1} are even and odd (in t) eigenfunctions of the Sturm-Liouville operator $\mathscr{L}$ [9]. In particular, $Ec_1^1 = \text{cn}\,t$ and $Es_1^1 = \text{sn}\,t$ satisfy

$$\mathscr{L}\,\text{cn}\,t = 0, \quad \mathscr{L}\,\text{sn}\,t = -k^2\,\text{sn}\,t. \tag{5.6a, b}$$

It follows from (5.3) and (5.6a) that f_n must be orthogonal to cn t:

$$\langle f_n\,\text{cn}\,t\rangle = 0. \tag{5.7}$$

[It also follows from (5.6a) that the solution of (5.3) may be reduced to quadrature by letting $\theta_n = \psi_n(t)\,\text{cn}\,t$, but this does not appear to be advantageous in the present development.] If $n \geq 2$ (5.7) determines A_{n-1}^1; for $n = 1$ it implies [cf. (1.6)]

$$\varepsilon\sin\phi = 2\delta R(\omega), \tag{5.8a}$$

where

---

*The solution (5.2) may be altered to describe subharmonic resonance by replacing (5.2c) by $\omega = m\omega_f$ ($m = 1, 2, \ldots$). The principal change in the subsequent development is the replacement of q by q^m in (5.7b) et seq.

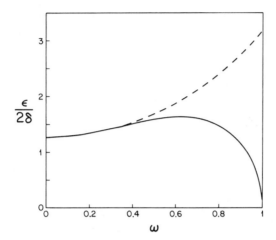

Fig. 6. The ratio $\varepsilon/2\delta$ at resonance ($\phi = \tfrac{1}{2}\pi$) in the limit $\delta \downarrow 0$ (5.8) (—) compared with Melnikov's criterion (5.23) (– – –) for bifurcation to transverse homoclinic orbits.

$$R(\omega) = \frac{2k\langle \operatorname{cn}^2 t\rangle}{\langle \operatorname{cn} t \cos \omega t\rangle} = \left(\frac{E - k'^2 K}{\tfrac{1}{2}\pi}\right)(q^{-1/2} + q^{1/2}), \qquad (5.8\mathrm{b})$$

$k' \equiv (1 - k^2)^{1/2}$, $E = E(k)$ is a complete elliptic integral of the second kind (E stands for energy in the subsequent development only when fenced by $\langle\ \rangle$), and $q = q(k)$ is the Jacobi nome.

The function $R(\omega)$, which is plotted in fig. 6, exhibits the limiting behaviours,

$$R \sim 2k \sim \alpha \quad (k \downarrow 0), \quad R \sim (4/\pi)\cosh\left(\tfrac{1}{2}\pi\omega\right) \equiv R_{\mathrm{M}}(\omega) \quad (k \uparrow 1), \qquad (5.9\mathrm{a,b})$$

where R_{M} is the Melkinov function (see below). The counterpart of R for the trial function (2.1), [$\theta_0 = 0$ and $\phi = \tfrac{1}{2}\pi$ in (2.2)]

$$R_1 = \alpha \Omega^{1/2} = (2\alpha J_1)^{1/2} \qquad (5.10)$$

is compared with R, qua function of α, in fig. 7. R has a single maximum of 1.642, which implies $\varepsilon_*/\delta = 3.284$, at $\alpha = \alpha_* = 2.479$ ($\omega_* = 0.616$), which compare with the approximations $\varepsilon_*/\delta = 3.16$ at $\alpha_* = 2.40$ ($\omega_* = 0.66$) implied by (5.10). The former approximations are exact for $\delta \downarrow 0$; the latter approximations to $\varepsilon_*/\alpha_*/\omega_*$ are in error by 3.7/3.2/7.1%.

The formal calculation of the expansion coefficients in (5.5) is straightforward by virtue of the orthogonality of the Lamé functions, but only the truncation

$$\theta_n(t) = A_n \operatorname{cn} t + B_n \operatorname{sn} t \qquad (5.11)$$

leads to tractable results. (A more tractable procedure for higher approximations, following Ince's [9] construction of the Lamé functions, is to expand θ_n in $\cos[(2m+1)\operatorname{am} t]$ and $\sin[(2m+1)\operatorname{am} t]$.) For $n = 1$ we obtain

$$A_1 = \frac{\delta \langle \operatorname{cn}^2 t \operatorname{dn} t\rangle}{k \langle \operatorname{sn}^2 t \operatorname{cn}^2 t \operatorname{dn} t\rangle} = \frac{4\delta}{k}, \quad B_1 = -\varepsilon \cos\phi B(\omega), \qquad (5.12\mathrm{a,b})$$

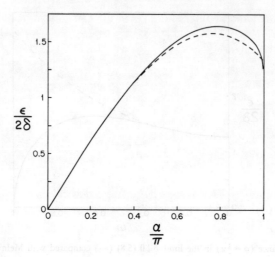

Fig. 7. The ratio $\varepsilon/2\delta$ for resonance in the limit $\delta \downarrow 0$, as given by the exact result (5.8b, 5.1b) (—) and the approximate result (5.10) (- - -).

where

$$B(\omega) = \frac{\langle \sin \omega t \, \mathrm{sn}\, t \rangle}{k^2 \langle \mathrm{sn}^2 t \rangle} = \left(\frac{\pi k^{-1}}{K-E} \right) \left(\frac{q^{1/2}}{1-q} \right). \tag{5.12c}$$

Combining (5.2), (5.8), (5.11) and (5.12), we obtain the approximation

$$\theta = 2\sin^{-1}(k\,\mathrm{sn}\,\hat{t}) \pm B(\varepsilon^2 - 4\delta^2 R^2)^{1/2} \mathrm{sn}\,\hat{t} + \mathcal{O}(\delta^2), \quad \hat{t} = t - \omega^{-1}\phi + 2\delta k^{-2}, \tag{5.13}_{\pm}$$

wherein the upper/lower sign corresponds to $\phi \gtrless \frac{1}{2}\pi$. The corresponding approximation to the mean energy is

$$\langle E \rangle \equiv \langle \tfrac{1}{2}\dot{\theta}^2 + 1 - \cos\theta \rangle = 2k^2 \pm 2kB(\varepsilon^2 - 4\delta^2 R^2)^{1/2}(E/K) + \mathcal{O}(\delta^2). \tag{5.14}_{\pm}$$

Letting $\delta = \tfrac{1}{8}$ and $\varepsilon = \tfrac{1}{2}$, we find that $(5.14)_+$ is within 1–2% of Bryant's (appendix B) upper branch of the resonance curve (fig. 5) for $0 \le \omega \le 0.75$; e.g. $(5.14)_+$/Bryant gives $\langle E \rangle^{1/2} = 1.429/1.430$ for $\omega = 0.595$. The maximum amplitude given by $(5.13)_+$ may exceed π, but only on the unstable segment of the resonance curve.

It is evident from the shift $2\delta k^{-2}$ in $\hat{t}$ that the approximation (5.13) is not uniformly valid as $k \downarrow 0$ ($\omega \uparrow 1$), in which limit the true resonance curve ceases to approximate its spine. This deficiency is muted in the approximations (5.14) and (5.21) below, but the approximations in this section are expected to hold only in the neighborhood of the spine.

5.1. Symmetry breaking

Now suppose that

$$\theta = \theta_s(t - \omega^{-1}\phi;\,\omega) + \theta_a(t - \omega^{-1}\phi;\,\omega), \tag{5.15}$$

where θ_s is the symmetric, periodic solution of (1.1), and θ_a is an antisymmetric perturbation that satisfies

$$|\theta_a| \ll 1, \quad \theta_a(t+2K) = \theta_a(t). \tag{5.16a, b}$$

Substituting (5.15) into (1.1), linearizing in θ_s shifting t by ϕ/ω, letting

$$\theta_a(t) = \psi(t) \operatorname{dn} t, \tag{5.17}$$

and approximating θ_s by $\theta_f + \theta_1$, we obtain

$$(\operatorname{dn}^2 t \psi')' + p \operatorname{dn}^2 t \psi = 0, \quad p = k'^2 - 2k\theta_1(t) \operatorname{sn} t \operatorname{dn} t + \mathcal{O}(\delta^2). \tag{5.18a, b}$$

The corresponding perturbation of the average Lagrangian yields

$$\langle L_a \rangle = \tfrac{1}{2} \langle \operatorname{dn}^2 t (\psi'^2 - p\psi^2) \rangle. \tag{5.19}$$

The simplest trial function for (5.19) is $\psi = C$, for which $\partial \langle L_a \rangle / \partial C = 0$ implies

$$\langle p \operatorname{dn}^2 t \rangle = 0. \tag{5.20}$$

Invoking (5.18b) and the truncation (5.11) for θ_1, we obtain the symmetry-breaking condition

$$\varepsilon_s \cos \phi = -\tfrac{1}{2} \frac{k'^2 \langle \operatorname{dn}^2 t \rangle}{kB \langle \operatorname{sn}^2 t \operatorname{dn}^3 t \rangle} = -\frac{2}{\pi^2} \left(\frac{1-k^2}{1-\tfrac{3}{4}k^2} \right) E(K-E)(q^{-1/2} - q^{1/2}). \tag{5.21}$$

We remark that, by virtue of its variational derivation, the error in (5.21) associated with the approximation $\psi = C$ is $\mathcal{O}(p^2)$, which implies the restriction $k'^2 \ll 1$ and suggests that the dominant error in (5.21) is associated with the truncation implicit in (5.11)*.

Let $\delta = \tfrac{1}{8}$ and $\varepsilon = \tfrac{1}{2}$ in (5.8) and (5.21), we obtain $\omega_s = 0.638$, which compares with Bryant's (private communication) $\omega_s = 0.595$ from numerical integration. An *ad hoc* correction for $\mathcal{O}(\delta^2)$, based on (4.1a), yields $\omega_s = [(0.638)^2 - 2\delta^2]^{1/2} = 0.613$. D'Humieres et al. [1, fig. 5b] observed symmetry breaking in their numerical integrations for $\delta = \tfrac{1}{8}$, $\omega = 0.67$ and $\varepsilon_s = 0.50$ (the bifurcation occurred for $0.46 < \varepsilon < 0.50$, but apparently much closer to 0.50 than to 0.46) but Bryant's results appear to be more accurate.

5.2. Melnikov's criterion

Melnikov's criterion for the transition to chaos through bifurcation from the homoclinic orbit [$k = 1$ in (5.1a)]

$$\sin \tfrac{1}{2}\theta = \tanh t, \quad \dot\theta = 2 \operatorname{sech} t \tag{5.22a, b}$$

is given by [11, section 4.5]

$$\varepsilon_M / 2\delta = \int_0^\infty \dot\theta^2 \, dt \bigg/ \int_0^\infty \dot\theta \cos \omega t \, dt = (4/\pi) \cosh(\tfrac{1}{2}\pi\omega) \equiv R_M(\omega), \tag{5.23}$$

---

*It appears that the results in this section also provide a valid approximation to the anomalous symmetry-breaking bifurcation.

which exceeds R everywhere in $0 < \omega < 1$ but is quite close thereto for $\omega < 0.4$ (see fig. 6). Letting $\delta = \tfrac{1}{8}$ and $\varepsilon = \tfrac{1}{2}$ in (5.23), we obtain $\omega = 0.651$, which compares with Bryant's (Appendix B) $\omega = 0.575$ for the transition to chaos. It should be emphasized, that Melnikov's criterion effectively assumes that the transition to chaos occurs for $\phi = \tfrac{1}{2}\pi$ (on the spine of the resonance curve), which is not possible for $\varepsilon > \varepsilon_*$ except in the limit $\omega \downarrow 0$. $[\mathcal{O}(k'^2)$ is neglected, but $\omega \sim \tfrac{1}{2}\pi/\ln(4/k')$ is not assumed to be small, in the derivation of (5.23).] Moreover, it is associated with the subharmonic-bifurcation cascade $T, 3T, 5T, \ldots$, which contrasts with the period-doubling cascade $T, 2T, 4T, \ldots$ that follows symmetry breaking.

Acknowledgement

The numerical work for figs. 1–3 and 6–8 was carried out by Ms. Diane Henderson. I am indebted to Dr. Peter Bryant for several helpful comments.

Appendix A: Stability of truncated-Fourier approximations

We explore the stability of the forced solutions of (1.1) by considering

$$\theta = \Theta(\tau) + \mathcal{R}\{e^{\lambda t}P(\tau)\} \quad (|P| \ll 1,) \quad \tau = \omega t - \phi, \qquad (A.1a,b)$$

where Θ is the solution to be tested, $\mathcal{R}$ signifies *the real part of*, λ is a small parameter, the real part of which must be non-positive for stability, and P is periodic in τ. Substituting (A.1) into (1.1), linearizing in P, approximating Θ by (2.1), neglecting the third and higher harmonics [as is consistent with (2.1)] in the Fourier expansion of $\cos(\theta_0 + \alpha \sin \tau)$, and introducing

$$\mu \equiv \beta^{-1/2}(\lambda + \delta), \quad \kappa \equiv \beta^{-1}(\lambda^2 + 2\delta\lambda + J_0 c_0), \quad q_1 \equiv -\beta^{-1} J_1 s_0, \quad q_2 \equiv \beta^{-1} J_2 c_0, \qquad (A.2a,b,c,d)$$

where $c_0 \equiv \cos \theta_0$ and $s_0 \equiv \sin \theta_0$, we obtain

$$P'' + 2\mu P' + (\kappa + 2q_1 \sin \tau + 2q_2 \cos 2\tau) P = 0. \qquad (A.3)$$

Remarking that (A.3) reduces to Mathieu's equation if μ and either q_1 or $q_2 = 0$, and guided by the perturbation solution of that equation near one of its stability boundaries in a κ, q-plane [11], we posit the expansions*

$$P = P_0 + P_1 + P_2 + \cdots, \quad \kappa = \kappa_0 + \kappa_1 + \kappa_2 + \cdots \qquad (A.4a,b)$$

where

$$P_m, \kappa_m = \mathcal{O}(q^m), \quad q_1, q_2, \mu = \mathcal{O}(q), \quad q = \max(q_1, q_2). \qquad (A.5a,b,c)$$

---

*The radius of convergence of the q-expansion of the dominant ($n = 0$) eigenvalue for Mathieu's equation, for which $\mu = q_1 = 0$, is $q = |q_2| = 1.47$ [12]. The maximum value of $|q_2|$ in the present context is realized at the symmetry-breaking bifurcation, where $J_0 = 0$, $c_0 = 1$ and $\beta = J_2 + \mathcal{O}(\delta^2)$, which imply $q_2 = 1$.

Substituting (A.4) into (A.3) and equating terms of like order in q, we obtain

$$P_0'' + \kappa_0 P_0 = 0, \tag{A.6a}$$

$$P_1'' + \kappa_0 P_1 = -2\mu P_0' - (\kappa_1 + 2q_1 \sin \tau + 2q_2 \cos 2\tau) P_0, \tag{A.6b}$$

$$P_2'' + \kappa_0 P_2 = -2\mu P_1' - (\kappa_1 + 2q_1 \sin \tau + 2q_2 \cos 2\tau) P_1 - \kappa_2 P_0, \ldots. \tag{A.6c}$$

It follows from Floquet theory that P may be of period 2π or 4π (periods of $2n\pi$, $n > 2$, are possible but irrelevant in the present context) and hence that

$$\kappa_0 = n^2, \quad P_0 = A_n \cos n\tau + B_n \sin n\tau, \tag{A.7a,b}$$

where n is an integer (period 2π) or half-integer (period 4π). Substituting (A.7) into (A.6b), we obtain

$$P_1'' + n^2 P_1 = -2\mu n(-A_n s_n + B_n c_n) - \kappa_1 (A_n c_n + B_n s_n)$$
$$- q_1 [A_n(s_{n+1} - s_{n-1}) + B_n(c_{n-1} - c_{n+1})] - q_2 [A_n(c_{n-2} + c_{n+2}) + B_n(s_{n+2} + s_{n-2})], \tag{A.8}$$

wherein $c_n \equiv \cos n\tau$ and $s_n \equiv \sin n\tau$. The requirement that P_1 be periodic implies that the coefficients of c_n and s_n in (A.8) must vanish (to prevent the secular growth of P_1); requiring the determinant of the resulting homogeneous equations in A_n and B_n to vanish, we obtain

$$\kappa_1^2 = \delta_{n\frac{1}{2}} q_1^2 + \delta_{n1} q_2^2 - (2\mu n)^2, \tag{A.9}$$

where δ_{mn} is the Kronecker delta.

Combining (A.2b), (A.7a) and (A.9) in (A.4b) and neglecting $\mathcal{O}(q^2)$, we obtain

$$\lambda^2 + 2\delta\lambda + S_n = 0, \tag{A.10a}$$

where

$$S_0 = J_0 c_0, \quad 4n^2 \beta S_n = (J_0 c_0 - n^2 \beta)^2 + 4n^2 \beta \delta^2 - \delta_{n\frac{1}{2}} (J_1 c_0)^2 - \delta_{n2} (J_2 c_0)^2. \tag{A.10b,c}$$

The necessary and sufficient conditions for stability then are $S_n \geq 0$ for $n = 0, 1, \frac{1}{2}$. The stability boundaries for $n = 2, \ldots$ and $\frac{3}{2}, \ldots$ are inaccessible, and the corresponding perturbations are stable, in the present parametric domain.

Stability of symmetric oscillations

Setting $\theta_0 = 0$ and invoking $J_0 + J_2 \equiv \Omega$ in (A.10), we find that the necessary and sufficient conditions for the stability of the symmetrical oscillation $\theta = \alpha \sin \tau$ are

$$J_0 \geq 0, \quad (\beta - \Omega)^2 + 2J_2(\beta - \Omega) + 4\delta^2 \beta \geq 0. \tag{A.11a,b}$$

The condition $J_0 > 0$ ($\alpha < \alpha_0$) is satisfied everywhere on those resonance curves with maxima ($\alpha < \alpha_\times$, $\varepsilon < \varepsilon_\times$) but is violated on all resonance curves above $\alpha = \alpha_0$. The symmetry-breaking (pitchfork) bifurcation at $J_0 = 0$ ($\alpha = \alpha_0$) leads to instability through the growth of $A_0 \exp(\lambda \omega_0 t)$ in (A.1), which implies a transition from the symmetric oscillation $\theta = \alpha \sin \tau$ to the asymmetric oscillation $\theta = \theta_0 + \alpha_0 \sin \tau$.

It follows from (3.2) that (A.11b) is satisfied in $\alpha < \alpha_\times$ except on those segments of the resonance curve(s) to the left of the spine on which $d\beta/d\alpha < 0$ – in particular on the intermediate branch between the turning points for $\varepsilon < \varepsilon_\times$. The turning-point (saddle-node) bifurcations lead to instability through the growth of $\exp(\lambda\omega_0 t)(A_1\cos\tau + B_1\sin\tau)$, which implies a transition from an unstable symmetric oscillation at the lower/upper turning point of the resonance curve to a stable symmetric oscillation on the upper/lower branch at $\beta = \beta_\pm$ (the tuning-hysteresis phenomenon that is typical for any nonlinear, lightly damped oscillator).

It also follows from (3.2) that (A.11b) is satisfied for $\varepsilon < \varepsilon_\times$ and $\alpha > \alpha_\times$ (above the crossing in fig. 1(b)) only on those segments of the right branches of the resonance curves that are above the turning points but below $\alpha = \alpha_0$. The narrowness of the latter domain suggests that the stable states therein would be difficult to attain, which conjecture is supported by the failure of numerical integrations of (1.1) to produce stable oscillations near $\varepsilon = \varepsilon_\times$ and $\beta = \beta_\times$.

Stability of asymmetric oscillations

Considering next the stability of the asymmetric oscillation $\theta = \theta_0 + \alpha_0 \sin\tau$, we set $\alpha = \alpha_0$, $J_0 = 0$, $J_1 = \frac{1}{2}\alpha_0\Omega_0$ and $J_2 = \Omega_0$ in (A.10) to obtain the necessary and sufficient conditions

$$\beta^2 - \Omega_0^2 c_0^2 + 4\delta^2\beta \geq 0, \quad \left(\tfrac{1}{4}\beta\right)^2 - \left(\tfrac{1}{2}\alpha_0\Omega_0 s_0\right)^2 + \delta^2\beta \geq 0. \tag{A.12a, b}$$

Invoking (4.2) and neglecting $\mathcal{O}(\delta^4)$, as is consistent with (A.10), we find that (A.12b) is more critical than (A.12a) and may be approximated by

$$|s_0| \leq (2\alpha_0\Omega_0)^{-1}(\beta + 8\delta^2) < 0.208 + 3.85\delta^2, \tag{A.13}$$

where the second inequality follows from the upper bound $\beta < \Omega_0$. Letting $\delta = \frac{1}{8}$, we obtain $0.96 < c_0 < 1$. The transition to instability is through a period-doubling bifurcation.

The approximations (A.12) and (A.13), like those in section 4, appear to be only qualitatively valid for ε in the symmetry-breaking range. E.g., (A.3) implies a minimum of $c_0 = 0.96$ for stability at $\delta = \frac{1}{8}$, whereas Bryant (appendix B) obtains $c_0 = 0.87$ for symmetry breaking at $\delta = \frac{1}{8}$ and $\varepsilon = \frac{1}{2}$.

Appendix B: Extended Fourier-series calculation

By Peter J. BRYANT,
University of Canterbury, Christchurch 1, New Zealand

Symmetric oscillations about a zero mean with a period equal to the forcing period are described by the Fourier series

$$\theta = \sum_{k=1}^{\infty} \left[a_{2k-1}\cos(2k-1)\omega t + b_{2k-1}\sin(2k-1)\omega t \right], \tag{B.1}$$

which generalizes to

$$\theta = \theta_0 + \sum_{k=1}^{\infty} \left[a_k\cos(k\omega t/m) + b_k\sin(k\omega t/m) \right] \tag{B.2}$$

for asymmetric oscillations about a non-zero mean with a period m times the forcing period. Truncated forms of the Fourier series (B.1) or (B.2) are substituted into (1.1) with trial values for θ_0 and the Fourier amplitudes a_k, b_k (all k), and the trial values are improved by Newton's method to yield fully nonlinear, periodic solutions of (1.1). These are solutions in the sense that they satisfy (1.1) to an arbitrary precision over an array of collocation points spanning the periodic time interval. [We plan to report subsequently in more detail about the method and the wide range of solutions of (1.1) found by it.]

The values of $\langle E \rangle^{1/2}$ obtained from (B.1) through (i) the numerical method and (ii) the elimination of α between (3.2) and (3.11) are compared in fig. 4 for $\delta = \frac{1}{8}$ and $\varepsilon = 0.5$. The solid/dotted lines are the exact (within the accuracy of the plot) stable/unstable resonance curves for symmetric oscillations. The dashed lines are the resonance curves derived from (3.2) and (3.11). The left branch of the exact resonance curve becomes unstable at the turning point $\beta = 0.464$, $\langle E \rangle^{1/2} = 0.806$, in good agreement with the transition at $\beta = \beta_-$ predicted in section 3. The right branch of the exact resonance curve becomes unstable at $\beta = 0.354$, $\langle E \rangle^{1/2} = 1.430$, where bifurcation to asymmetric oscillations occurs, in good agreement with the transition predicted in section 5.

The exact resonance curves near this bifurcation for the different forms of oscillation are drawn in fig. 5. Starting from the right, the solid line is the stable resonance curve shown in fig. 4, continuing on as the unstable dotted section beyond the first bifurcation point P_1. The solid line $P_1 P_2$ is the exact resonance curve for stable asymmetric oscillations, with an unstable section beyond P_2. The point P_2, at $\beta = 0.336$, $\langle E \rangle^{1/2} = 1.422$, is the first period-doubling point. The section $P_2 P_4$ is the exact resonance curve for stable oscillations of twice the forcing period, continuing on in the dotted section beyond P_4 as unstable oscillations. The point P_4, at $\beta = 0.332$, $\langle E \rangle^{1/2} = 1.422$, is the period-quadrupling point, from which $P_4 P_8$ is the exact resonance curve for stable oscillations of four times the forcing period. The point P_8, at $\beta = 0.331$, is the period-octupling point and is followed by a smaller section (not drawn) of stable oscillations of eight times the forcing period. This sequence from symmetric oscillations through symmetry-breaking to period-doubling, illustrated here by the bifurcating resonance curves for the different forms of oscillation, was observed by D'Humieres et al. [4, fig. 5] as a sequence of phase-space portraits.

References

[1] Christiaan Huygens, Horologium Oscillatorium (1673), vol. 18 of OEuvres complètes (Nijhoff, The Hague, 1934); Christiaan Huygens' The Pendulum Clock or Geometrical Demonstrations Concerning the Motion of Pendula as Applied to Clocks, translated by R.J. Blackwell (Iowa State University Press, 1986).
[2] B.A. Huberman and J.P. Crutchfield, Phys. Rev. Lett. 43 (1979) 1743.
[3] J. Miles, Physica D 11 (1984) 309.
[4] D. D'Humieres, M.R. Beasley, B.A. Huberman and A. Libchaber, Phys. Rev. A 26 (1982) 3483.
[5] N.F. Pederson, O.H. Sorenson, B. Dueholm and J. Mygind, J. Low Temp. Phys. 38 (1980) 1.
[6] J.A. Blackbum, Yang Z-j, S. Vik, H.J.T. Smith and M.A.H. Nerenberg, Physica D 26 (1987) 385.
[7] J. Miles, Phys. Lett. A (sub judice).
[8] P.F. Byrd and M.D. Friedman, Handbook of Elliptic Integrals for Engineers and Physicists (Springer, Berlin, 1954).
[9] E.L. Ince, Proc. Roy. Soc. Edinb. 60 (1940) 47, 83.
[10] J. Guckenheimer and P. Holmes, Nonlinear Oscillations, Dynamical Systems, and Bifurcations of Vector Fields (Springer, New York, 1983), sections 4.5, 6.
[11] A.H. Nayfeh, Introduction to Perturbation Techniques (Wiley–Interscience, New York, 1981), section 11.4.
[12] M. Abramowitz and I.A. Stegun, Handbook of Mathematical Functions (National Bureau of Standards, Washington, 1964), p. 725.

V. Other Oscillatory Systems

This section describes other chaotic oscillatory systems. Most of the examples in this section have important applications in engineering.

We start with Paper V.1 which describes experiments on chaotic behavior of a buckled beam. The first experimentally obtained Poincaré map is presented.

Next two papers: V.2 and V.3 investigate chaotic behavior of a driven impact oscillator. The period-doubling and intermittent transitions to chaos have been identified.

Harmonic, subharmonic and chaotic motions of a piecewise linear oscillator are found in Paper V.4. Bifurcations leading to chaotic states are analyzed.

Complex dynamics of advanced compliant off-shore structures with subharmonic resonances and chaotic motion is investigated in Paper V.5. The resonances of bilinear and impact oscillators are used to illustrate this behavior.

The possibility of chaotic oscillations during the cutting process is reported in Paper V.6. The simple model of the cutting process is investigated numerically.

Paper V.7 considers the stability of modes at rest in nonlinear mechanical system which has one active mode undergoing chaotic oscillations. Analytical results for the almost sure stability of the inactive mode are obtained numerically and compared with simulation results.

A numerical study presented in Paper V.8 shows that chaotic motion is possible from periodic forcing of an elastic-plastic beam. The results suggest that geometrical and material nonlinearities in mechanical problems may lead to extreme sensitivity to initial conditions.

Two last articles in this section: V.9 and V.10 deal with nonlinear electronic circuits. In both of them chaotic behavior was found in experimental investigation of a physical system as well as in the numerical analysis of an appropriate mathematical model.

Reprinted from JOURNAL OF APPLIED MECHANICS, Vol. 47, No. 3, September 1980

F. C. Moon

Associate Professor,
Department of Theoretical and Applied Mechanics,
Cornell University,
Ithaca, N. Y. 14853. Mem. ASME

Experiments on Chaotic Motions of a Forced Nonlinear Oscillator: Strange Attractors[1]

The forced vibrations of a buckled beam show nonperiodic, chaotic behavior for forced deterministic excitations. Using magnetic forces to buckle the beam, two and three stable equilibrium positions for the postbuckling state of the beam are found. The deflection of the beam under nonlinear magnetic forces behaves statically as a butterfly catastrophe and dynamically as a strange attractor. The forced nonperiodic vibrations about these multiple equilibrium positions are studied experimentally using Poincare plots in the phase plane. The apparent chaotic motions are shown to possess an intricate but well-defined structure in the Poincare plane for moderate damping. The structure of the strange attractor is unravelled experimentally by looking at different Poincare projections around the toroidal product space of the phase plane and phase angle of the forcing function. An experimental criterion on the forcing amplitude and frequency for strange attractor motions is obtained and compared with the Holmes-Melnikov criterion and a heuristic formula developed by the author.

Introduction

There has been growing interest in nonperiodic, steady-state solutions of nonlinear differential equations in applications to atmospheric dynamics [1], electrical circuits [2], and elastic structures [3, 4]. The equations governing these systems are deterministic while for certain control parameters chaotic motions appear. The importance of these motions is twofold. First, conventional methods for finding steady-state solutions to nonlinear differential equations such as perturbation schemes, and averaging techniques must be abandoned or modified since they assume periodic solutions. The second point is that in many physical systems observation of chaotic behavior is often ascribed to some randomness in the problem parameters. The existence of strange attractor motions of deterministic systems may obviate the need for the existence of random "demons" in certain dynamical problems.

Mathematicians have used the name "strange attractor" to denote bounded, chaotic, nonperiodic solutions of deterministic, nonlinear differential equations in contrast to more predictable motions such as those near equilibrium points and limit cycles. Strange attractor oscillations have been found for third-order autonomous differential equations by Lorenz [1] in developing an atmospheric dynamics model. Strange attractor solutions have been found in analog computer simulations of Duffing's equation by Ueda [2, 5] and Holmes [3, 4]. Ueda, [2], has also observed chaotic behavior in analog computer solutions of a forced Van der Pol oscillator.

Mechanical examples of continuous, nonperiodic, bounded motions in deterministic systems can be found in the vibrations of buckled or curved plates and beams. These motions occur when the vibration amplitude becomes large enough to cause the beam or plate to "snap-through." Tseng and Dugundji [6] have studied the nonlinear vibrations of a buckled beam with fixed ends and observed both periodic and nonperiodic motions. They refer to the latter as continuous, "intermittent" snap-through under harmonic excitations. In a recent paper, the author and Holmes [7] examined the nonlinear forced vibrations of a cantilevered beam which is buckled by magnetic forces. The harmonic excitation of this model exhibited chaotic snap-through behavior similar to strange attractor motions found in analog computer studies. In [6, 7], the Galerkin approximation was used to reduce the nonlinear beam equations to Duffing's equation with harmonic excitation.

The work of Tseng and Dugundji [6] was an extension of previous work by Cummings [8], and Eisley [9] on large amplitude vibrations of buckled and curved plates. Cummings treated the snap-through

---

[1] Research supported in part by a grant from the National Science Foundation, Engineering Mechanics Division, Grant No. 76-23627.

Contributed by the Applied Mechanics Division for presentation at the Winter Annual Meeting, Chicago, Ill., November 16–21, 1980, of The AMERICAN SOCIETY OF MECHANICAL ENGINEERS.

Discussion on this paper should be addressed to the Editorial Department, ASME, United Engineering Center, 345 East 47th Street, New York, N. Y. 10017, and will be accepted until December 1, 1980. Readers who need more time to prepare a discussion should request an extension from the Editorial Department. Manuscript received by ASME Applied Mechanics Division, May, 1979; final revision, January, 1980. Paper No. 80-WA/APM-2.

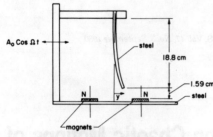

Fig. 1 Sketch of experimental apparatus

problem under a pulsed load but did not examine the intermittent snap-through discussed in [6].

The present paper is an extension of the work of Moon and Holmes [7] on the chaotic vibrations of a cantilevered beam buckled by magnetic forces. Although there are potential applications for the study of magnetically buckled structures, such as in fusion reactors (see, e.g., [10]), its main interest in this paper is as one of the simplest mechanical examples of strange attractor motions which can be easily studied experimentally. In the previous work analysis of the magnetic and elastic forces led to a Duffing-type equation for the first mode approximation. Briefly, the beam was assumed to have a magnetization $\mathbf{M}$ induced by the magnetic field of external magnets $\mathbf{B}^0$. The magnetic field acting on the beam creates distributed magnetic forces and couples

$$\mathbf{F} = \mathbf{M} \cdot \nabla \mathbf{B}^0$$
$$\mathbf{C} = \mathbf{M} \times \mathbf{B}^0 \qquad (1)$$

When a one mode Galerkin approximation is used, a magnetic energy potential can be found in terms of $\mathbf{M}, \mathbf{B}^0$

$$\mathcal{W} = -\frac{1}{2} \int \mathbf{M} \cdot \mathbf{B}^0 dv \qquad (2)$$

This potential is nonlinear in the modal amplitude "a" and is expanded in a Taylor series in "a;"

$$\mathcal{W} = \frac{1}{2} \gamma a^2 + \frac{1}{4} \beta a^4 + \frac{1}{6} \eta a^6. \qquad (3)$$

The resulting nonlinear modal magnetic forces when added to linear modal elastic forces lead to a nonlinear differential equation of the form,

$$\ddot{a} + \delta \dot{a} + \alpha a + \beta a^3 + \eta a^5 = \Omega^2 A_0 \cos \Omega t \qquad (4)$$

where a is the modal amplitude of the first bending mode of the cantilevered beam, A_0 is the vibration amplitude of the forced support motion, and Ω is the frequency of the support motion. Structural damping is represented by δ. The control parameters in this problem are the spacing of the magnets, damping, forcing amplitude, and frequency. As a static problem only the spacing of the magnets is relevant. The elastic bending stiffness, magnet field strength, and positions of the magnets relative to the beam form a four parameter system which admit one to five equilibrium positions of the tip. For a given beam stiffness, the locus of points in the plane of magnet spacing parameters for which the number of equilibrium points changes is known as a butterfly catastrophe, [11], and is shown in Fig. 2. For the three equilibrium state case only two are stable and the governing nondimensionalized differential equation takes the form

$$\ddot{A} + \gamma \dot{A} - \frac{1}{2}(1 - A^2)A = f \cos \omega t \qquad (5)$$

where the following nondimensional groups are noted:

$$\gamma = \delta/\omega_0, \quad A = a/x_0$$
$$f = \frac{\Omega^2 A_0}{\omega_0^2 x_0}$$
$$\omega = \Omega/\omega_0$$

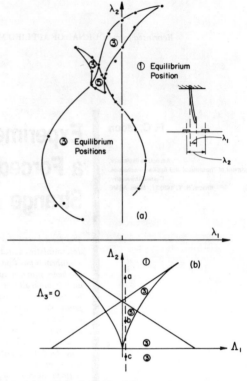

Fig. 2 (a) Equilibrium state regions in the plane of magnet spacing parameters; experimental data. (b) Ideal "butterfly" catastrophe set for a potential $\mathcal{V} = a^6 + \Lambda_4 a^4 + \Lambda_3 a^3 + \Lambda_2 a^2 + \Lambda_1 a$ with $\Lambda_3 = 0, \Lambda_4 < 0$

The amplitude is normalized by x_0 and the time by $2\pi/\omega_0$ where x_0 is the static position of the beam tip and ω_0 is the frequency for small vibrations about the buckled position.

For fixed damping and frequency the motion for small forcing amplitudes is periodic but for larger amplitudes becomes chaotic with the beam tip jumping from one equilibrium position to the other as shown in Fig. 3. A phase plane picture is shown in Fig. 4 and it is clear that a continuous history of the motion has very little structure.

In this paper experiments are described which attempt to characterize the behavior of this chaotic motion and to determine the critical parameters for which one might expect chaotic behavior from a second-order, single-degree-of-freedom system. Theoretical attempts have been made recently to determine the nature of the attracting set. In [4] Holmes has developed a necessary criterion for strange attractor motions which determines the minimum forcing amplitude as a function of forcing frequency. This criterion is compared with experimentally determined parameters as well as another theoretical criterion posited by the author. Experimental Poincare plots in the phase plane are used to partially unravel the strange attractor.

Description of Experimental Apparatus

The apparatus consisted of a steel (ferromagnetic) cantilevered beam suspended vertically. The clamped end was attached to a vibration shaker, (Fig. 1) while permanent magnets, 2.54 cm (1 in.) in diameter, were placed below the free end of the beam. The dimensions of the beam were 18.8 cm (7.4 in.) long, 9.5 mm (3/8 in.) wide, and 0.23 mm (0.009 in.) thick. The magnets had a 0.18 Tesla magnetic field normal to the magnet face and rested on a steel base.

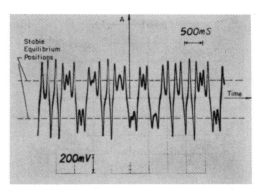

Fig. 3 Bending strain versus time for forced chaotic vibrations of the buckled beam

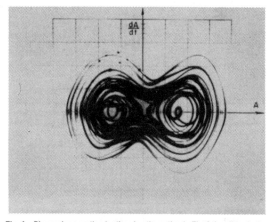

Fig. 4 Phase plane motion for the chaotic motion in Fig. 3; bending strain, horizontal axis and time rate of strain, vertical axis

Strain gages were attached to the beam near the clamped end while a linear variable differential transformer was attached to the shaker platform to measure the forced vibration amplitude of the beam base. Data were recorded on a storage oscilloscope.

To display motion in the phase plane and to perform a Poincaré map, a differentiator was used. In order to avoid spurious differentiation from high frequency noise, a low pass active filter was built (Bessel filter) with a 3 db drop in amplitude at 40 hz and less than 1 percent error in phase shift in the operating region. The experiments were performed at driving frequencies below 15 hz. With no magnets the beam had natural frequencies of 4.6, 26.6, and 73.6 hz. The amplitude and phase shift of the differentiator was checked carefully over the range of driving frequencies.

To perform a Poincaré map, a storage oscilloscope was used. The scope trace intensity was modulated by a pulse triggered by the vibration shaker amplitude. The bending strain was displayed on the horizontal axis of the scope while the time rate of strain controlled the vertical displacement of the scope trace. By modulating the trace intensity in synchronization with a particular phase of the vibrator motion, a dot would appear with every cycle and the set of dots over time would provide a Poincaré map or section of the motion.

Results

A number of different experimental methods were used to characterize the nonperiodic motion of the beam including time histories, Fourier analysis, zero crossing times distribution, Poincaré maps, and determination of chaotic motion threshold for driving amplitude and frequency.

Static Bifurcations. Static buckling experiments were done as reported in [7] to determine the critical values of magnet spacing λ_2 and magnetic offset λ_1 at which the number of equilibrium positions changed (Fig. 2(a)). The locus of points in the λ_1, λ_2 plane where the *number* of equilibrium solutions changes is known as a *catastrophe set*, [11]. In classic symmetric buckling problems this set is simply a point, namely, the buckling load. However when one allows other parameters to vary, such as geometric imperfections, the set becomes a curve, surface, or hypersurface in the parameter space. The dimension of the hyperspace depends on the potential energy function.

For this problem the magnetic potential (3) implies that four parameters will be sufficient to describe all the possible bifurcations. The theoretical set is called a butterfly catastrophe, [11], and a two-dimensional section is shown in Fig. 2(b), for the potential,

$$\mathcal{V} = a^6 + \Lambda_4 a^4 + \Lambda_3 a^3 + \Lambda_2 a^2 + \Lambda_1 a \tag{6}$$

The projection shown in Fig. 2(b) is for $\Lambda_3 = 0$, $\Lambda_4 < 0$. The number of equilibrium positions in each region is shown by the circled numbers.

Comparison of the experimental and theoretical catastrophe sets

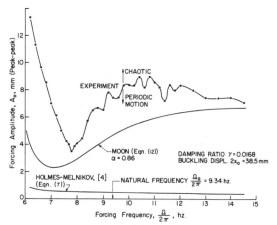

Fig. 5 Experimental and theoretical thresholds for spontaneous chaotic motion-moderate damping, $\gamma = 0.0168$

shows good qualitative agreement. One can imagine a change of parameters $\Lambda_1(\lambda_1, \lambda_2)$ and $\Lambda_2(\lambda_1, \lambda_2)$ which will transform the experimental "rabbit" catastrophe Fig. 2(a), into the topologically identical butterfly set, Fig. 2(b).

Most of the dynamic experiments were run in the three equilibria regimes in the λ_1, λ_2 plane with two stable and one unstable equilibria. However a few tests were performed for the five point case. In the latter case, three are stable equilibria while two are saddle-type points and are unstable.

Experimental Criteria for Chaotic Motions. Next the range of vibration base amplitudes and frequencies for chaotic motions was determined. These data were obtained by fixing the frequency and varying the shaker amplitude. For small motions periodic orbits of period one would occur. For larger amplitudes, period one, two, three, four, or more times the driving period might occur. At a sufficiently high amplitude, chaotic motions would occur. Such motions might not persist. Thus, if a periodic motion were disturbed by deflection of the beam, a chaotic motion like that in Fig. 4 might appear and decay to the periodic orbit. However a threshold would occur where the beam would spontaneously jump out of periodic motion into nonperiodic or chaotic motions.

This threshold amplitude of shaker motion is shown in Fig. 5 for different shaker frequencies and damping. The lowest amplitude for

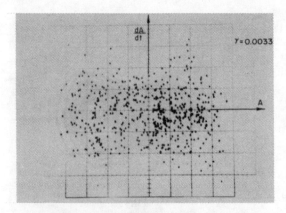

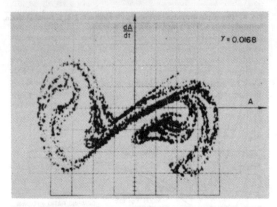

Fig. 6 Experimental Poincaré map of chaotic motion for low damping; experimental Poincaré map of chaotic motion for moderate damping

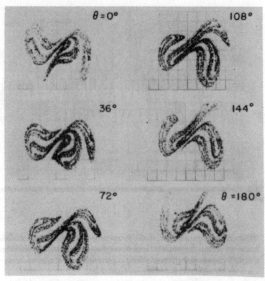

Fig. 7 Poincaré maps of a strange attractor for different phase synchronization with forcing function

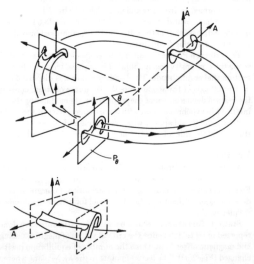

Fig. 8 Sketch of strange attractor surfaces in the product space of Poincaré plane and forcing amplitude phase

chaos occurs at a forcing frequency below the natural frequency of the linearized buckled beam. A comparison of the experimental criteria with theoretical predictions is discussed in a later section of this paper.

There is some belief that an upper criterion exists where the forced motion changes from nonperiodic to periodic, [4], but this was not observed within the range of shaker amplitudes available to the author.

Poincaré Plots. A time history of the bending strain for nonperiodic motions is shown in Fig. 3. This oscilloscope trace shows vibration about the two stable equilibrium positions, the transition between them, and oscillations about all three equilibrium positions. A phase plane portrait of this motion is shown in Fig. 4. Motion about the left, right, and all three equilibrium points can be seen but it is clear that the plane will become dense with these traces, making any characterization of the motion difficult to interpret. Instead of looking at the motion for all times, one can choose to observe the position in the phase plane at certain multiples of periods of the forcing motion. A sequence of points or dots on the oscilloscope will appear called a Poincaré map.

A period one Poincaré map is shown in Fig. 6(a) for low damping. One can see that there appears to be little global structure, though locally small clusters of straight lines of dots can be seen. Thus the forced nonperiodic motions of the near Hamiltonian system do not reveal much structure or order in the Poincaré map.

To increase the damping, a 0.05 mm (0.002 in.) thick stainless steel strip was glued to the beam which increased the damping from $\gamma = 0.0033$ to $\gamma = 0.0168$.

The Poincaré map for the moderate damping case is shown in Fig. 6(b). Here one can see that the period one Poincaré map shows a great deal of order resembling a line wrapped back and forth on itself. In analog computer studies [3], Holmes has shown that this parallel line structure continues to exist when local regions of the phase plane are magnified, suggesting the property of a Cantor set.

The Poincaré map in Fig. 6(b) depends on the phase angle of the driving motion, θ. For the symmetric problem examined here, the map should invert itself when $\theta \rightarrow \theta + \pi$. The change of shape of the strange attractor for different θ where $0 \leq \theta \leq \pi$, is shown in Fig. 7. One can see that although the figure at $\theta = \pi$ appears to be rotated, the evolution of this change shows that the "arms" of the attracting set deform in such a way as to invert the shape.

It should be noted that the motion of the beam must pierce all of these maps so that the lines in the Poincaré plane become sheets in

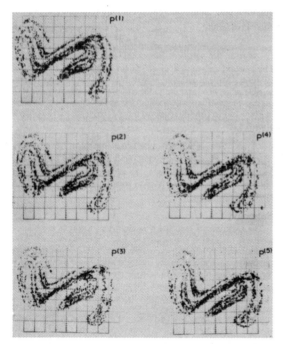

Fig. 9 Experimental multiperiod Poincare maps of the same chaotic motion

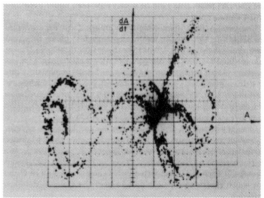

Fig. 10 Experimental strange attractor in the Poincare plane for the five equilibria case

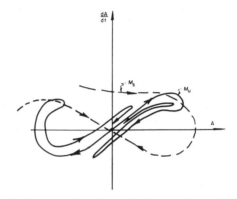

Fig. 11 Intersection of the stable manifold M_s and unstable manifold M_u of the Poincare map in the phase plane (see [4])

the toroidal product space of the Poincare plane and forcing phase θ, Fig. 8.

It is remarkable that a Poincare map synchronized with the forcing motion appears to organize what appears to be chaotic behavior. It should be clear that if the map is slightly unsynchronized, points from different P_θ maps of the synchronized maps will project onto the unsynchronized map. The resulting attracting set will appear as a blur and the structure will be lost.

One of course can obtain multiperiod Poincare maps $P^{(n)}$ defined by

$$\left\{ x, \dot{x} \mid \theta = \frac{2n\pi}{\omega}, \frac{4n\pi}{\omega} \ldots \right\}.$$

We note that the maps $P^{(n)}$, $n \geq 2$ are contained in the P^1 map but the question arises as to whether the structure of the strange attractor as seen in the higher period maps will look like that of the P^1 map. Experiments were carried out for $n = 2, 3, 4, 5$ and indeed the structure of the attractor looks identical to the P^1 map as shown in Fig. 9. For example, the only difference between $P^{(5)}$ and P^1 is that $P^{(5)}$ took 20 min to obtain, while P^1 took around four minutes of data. This illustrates again that although the motion appears to be chaotic in continuous observations, stroboscopic, or P^n maps reveal highly structured features of this motion.

"Butterfly" Strange Attractor. While most of the experiments were carried out for the single saddle and double sink or three equilibrium point case, a few experiments were performed for the double saddle or five equilibria case. As had been discussed by Holmes [4] and Ueda [2], the strange attracting set in the Poincare plane seems to be organized about the unstable manifold of the saddle point. When two saddles are present in the phase plane of the unforced motion, one would expect two organizing centers to appear in the Poincare map for the chaotic motions. Experimentally this has been observed as can be seen in Fig. 10.

Fourier Analysis. Frequency analyses of these chaotic motions were carried out by digitizing the data and using a fast Fourier transform [7]. The chaotic motions exhibit a continuous spectrum of frequencies below the driving frequencies, including subharmonics. Similar results were reported earlier by Ueda [5] and Holmes [4] for a forced Duffing's equation.

Threshold Criteria for Chaotic Motions

Various qualitative analyses have demonstrated the existence and characteristics of chaotic motions in deterministic nonlinear systems. However there is at present no theory to predict for what range of parameters these chaotic motions will occur. The engineer would like a chaotic "Reynolds number" or an equivalent parameter below which periodic motions would be insured and above which chaotic, nonperiodic motions would occur in the forced nonlinear oscillator.

In [4] Holmes has presented a necessary criteria for the strange attractor based on the work of Melnikov [12]. For chaotic motions to occur, the forcing amplitude, driving frequency, and damping in (5) must satisfy the relation

$$f_1 = \frac{\gamma \sqrt{2}}{3\pi\omega} \cosh\left(\frac{\pi\omega}{\sqrt{2}}\right). \qquad (7)$$

In [4] Holmes showed that the Poincare map itself has a saddle point and that as the forcing amplitude is increased the stable and unstable manifolds of the saddle of the Poincare map intersect, giving rise to infinitely many intersections or homoclinic points as shown in Fig. 11. It has been shown in [4] that two arbitrarily close points in the Poincare phase plane can be widely separated under iterations of the Poincare map.

Comparison of the Holmes-Melnikov criterion with experimental thresholds in the driving amplitude-frequency plane for fixed damping is shown in Fig. 5. While the theoretical threshold (7) gives a lower bound it does not compare well with measurements for low damping. Also it predicts that the minimum forcing amplitude occurs at a driving frequency greater than the natural frequency while experiments indicate that for low damping the minimum occurs below the natural frequency. There is evidence however that this criterion may give good results for high damping ($\gamma > 0.1$), [4].

It should be pointed out that the experimental thresholds were for "spontaneous" departure from forced periodic motion about a buckled state to chaotic motions. Experiments have shown that under certain initial conditions chaotic motion can occur below the spontaneous threshold. Thus it is possible that initial conditions could be found for which chaotic motion could occur near the Holmes-Melnikov criterion.

As just mentioned, it has been predicted and observed experimentally that multiperiod subharmonic forced oscillations are often precursors to the strange attractor behavior. Thus a better criterion may be found by studying the stability of subharmonic motions as has been reported by Hayashi [13]. Unfortunately the stability criterion as reported in [13] does not use f and ω as parameters and it is difficult to determine if a subharmonic stability criterion will compare well with the threshold in Fig. 5. But this seems to be a worthwhile direction to go in.

Finally, we propose a hueristic criterion based in part on a perturbation solution for forced periodic motion, and experimental observations. First we observe that the criterion sought governs the transition from forced periodic to nonperiodic motion. Thus, before chaotic motions occur, the response amplitude and velocity are known functions of forcing amplitude and frequency, i.e., $\langle x^2 \rangle = g(f, \omega)$, where $\langle \ \rangle$ indicates time averaged. If a critical amplitude of $\langle x^2 \rangle$ or $\langle \dot{x}^2 \rangle$ can be found, then f and ω can be related when chaotic motion is incipient.

To find the response function we write (5) about the buckled position $A = 1$ or -1. If we denote the motion about $A = 1$ by

$$A - 1 = X/x_0$$

where x_0 is the static deflected position of the tip of the beam, then the equation of motion takes the form

$$\ddot{X} + \gamma \dot{X} + X\left(1 + \frac{3}{2}\mu X + \frac{1}{2}\mu^2 X^2\right) = x_0 f \cos(\omega t + \phi_0) \quad (8)$$

where $\mu = 1/x_0$. The parameter μ will act as a perturbation parameter, while the phase angle ϕ_0 will be adjusted so that the first-order motion is proportional to $\cos \omega t$. Then using either Duffing's method, or Linstedt's perturbation method [13], [14] one assumes a solution of the form

$$X = C_0 \cos \omega t + \mu(C_1 + C_2 \cos \omega t) + \mu^2 X_1 \quad (9)$$

The resulting force-response relation is found to be

$$\left(\frac{C_0}{x_0}\right)^2 \left\{\left[(1 - \omega^2) - \frac{3}{2}\left(\frac{C_0}{x_0}\right)^2\right]^2 + \gamma^2 \omega^2\right\} = f^2 \quad (10)$$

Finally, it has been observed that over a limited range of frequencies close to the natural frequency in the post buckled state, the periodic motion seems to change to chaotic at a *critical velocity*. This velocity was not measured. However we hazard a guess that the critical velocity is near the maximum velocity on the separatrix for the phase plane motion of the undamped, unforced oscillator. This is certainly a guess, but this velocity is a characteristic of the beam and independent of the force. In nondimensional units $(dA/dt)_{max} = \frac{1}{2}$. Thus we assume that near the chaotic threshold

$$\frac{\omega C_0}{x_0} = \frac{\alpha}{2} \quad (11)$$

where α is near but less than unity. These assumptions lead to a criterion of the form

$$f_1 = \frac{\alpha}{2\omega}\left\{\left[(1 - \omega^2) - \frac{3}{8}\frac{\alpha^2}{\omega^2}\right]^2 + \gamma^2 \omega^2\right\}^{1/2} \quad (12)$$

The constant α gives us a "cheat" factor with which to fit the data. However what is remarkable, at least to the author, is that with reasonable values of α (near unity) the criterion compares very well with the experimental data both qualitatively and quantitatively as shown in Fig. 5. In Fig. 5 the dimensional forcing frequency and amplitude are given by

$$\Omega = \omega_0 \omega, \quad A_0 = x_0 f \omega_0^2/\Omega^2.$$

It is left to theoreticians to determine whether the assumptions implicit in (12) are at all related to the subharmonic stability criterion of Hayashi or the Holmes-Melnikov equation (7), or whether (12) is simply a fortuitous guess.

Summary

The experiments reported here show that a simple mechanical structure can exhibit nonperiodic or chaotic motions even when the forcing inputs are highly deterministic. The results are consistent with qualitative analysis and analog computer solutions of a deterministic forced Duffing's equation. In both the mechanical structure and the analog studies reported earlier, the nonperiodic motion exhibits a remarkable mathematical structure in the Poincare plane, resembling a sheet folded infinitely many times about the saddle point with properties of a Cantor set. This structure is preserved in higher-order maps of period two or higher. However the maps must be exactly synchronized with the phase of the driving motion or else this organized structure will be blurred or washed out. Experimentally the structure of the strange attracting set is more readily observed in moderate to highly damped systems.

The attempt by Holmes [4] to develop a dynamical "catastrophe" set of parameters for which strange attractor motions will occur seems to give a lower bound for the driving force. An ad-hoc criterion developed in this paper gives a set of driving amplitudes and frequencies closer to the experimental set. However the effects of other parameters such as initial conditions have yet to be explored.

The extension of this work to other nonlinear and multistate mechanical systems should reveal similar phenomena. In experiments on the dynamics of a magnetically levitated model on a rotating guidance track the author has observed similar chaotic behavior [15]. Lateral "rattling" motions of trains may also fall into this class of problems.

Whatever the specific example however, it is clear from these experiments and those of others that what appears to be "random" or chaotic motion in many mechanical systems may be governed by deterministic mathematical models and controlled by nonrandom parameters.

Strange attractor dynamics in other engineering systems such as chemical reactors and aerospace applications have been reported in [16], including a two-dimensional mechanical oscillator with chaotic behavior [17].

Acknowledgment

The author wishes to thank Prof. P. Holmes and Prof. R. Rand of Cornell University and Prof. Ueda of Kyoto University for helpful discussions. Thanks are also due to Stephen King, research engineer, Cornell University for the design of some of the electronic equipment.

References

1 Lorenz, E. N., "Deterministic Nonperiodic Flow," *Journal of the Atmospheric Sciences*, Vol. 20, 1963, pp. 130–141.

2 Ueda, Y., Hayashi, C., and Akamatsu, N., "Computer Simulation of Nonlinear Ordinary Differential Equations and Nonperiodic Oscillations," *Electronics and Communications in Japan*, Vol. 56-A, No. 4, 1973.

3 Holmes, P. J., "Strange Phenomena in Dynamical Systems and Their Physical Implications," *Applied Mathematical Modelling*, Vol. 1 (7), 1977, pp. 362–366.

4 Holmes, P. J., "A Nonlinear Oscillator With a Strange Attractor," *Philosophical Trans. of Royal Society*, London, Vol. 292, No. 1394, Oct. 1979, pp. 419-448.

5 Ueda, Y., "Randomly Transitional Phenomena in the System Governed by Duffing's Equation," *Journal of Statistical Physics*, Vol. 20, 1979, pp. 181-196.

6 Tseng, W.-Y., and Dugundji, "Nonlinear Vibrations of a Buckled Beam Under Harmonic Excitation, ASME JOURNAL OF APPLIED MECHANICS, Vol. 38(2), 1971, pp. 467-476.

7 Moon, F. C., and Holmes, P. J., "A Magnetoelastic Strange Attractor," *Journal of Sound and Vibration*, Vol. 65(2), 1979, pp. 276-296.

8 Cummings, B. E., "Large Amplitude Vibration and Response of Curved Panels," *AIAA Journal*, Vol. 2(4) Apr. 1964, pp. 709-716.

9 Eisley, J. G., "Large Amplitude Vibration of Buckled Beams and Rectangular Plates," *AIAA Journal*, Vol. 2(12), Dec. 1964, pp. 2207-2209.

10 Moon, F. C., "Problems in Magnetosolid Mechanics," *Mechanics Today*, Vol. 4, Chapter 5, 1978.

11 Poston, T., and Stewart, I., *Catastrophe Theory and Applications*, Pitman, London, 1978.

12 Mel'nikov, V. K., "On the Stability of the Center for Time Periodic Perturbations," *Trans. Moscow Math. Soc.*, Vol. 12(1), 1963, pp. 1-57.

13 Hayashi, C., *Nonlinear Oscillations in Physical Systems*, McGraw-Hill, New York, 1964.

14 Stoker, J. J., *Nonlinear Vibrations in Mechanical and Electrical Systems*, Interscience Publ., N.Y., 1950.

15 Moon, F. C., "Static and Dynamic Instabilities in Mag-Lev Model Experiments," *Proceedings of the Conference on Noncontacting Suspension and Propulsion Systems for Advanced Ground Transportation*, M.I.T., Sept. 1977, Wormley, D., ed., printed by the U.S. Dept. of Transportation.

16 Holmes, P., ed., *New Approaches to Nonlinear Problems*, SIAM Publishing, to appear.

17 Moon, F. C., "Strange Attractor Vibrations in One and Two-Degree-of-Freedom Elastic Systems", to appear in reference [16].

CHAOS AFTER PERIOD-DOUBLING BIFURCATIONS IN THE RESONANCE OF AN IMPACT OSCILLATOR

J.M.T. THOMPSON and R. GHAFFARI
Department of Civil Engineering, University College London, London, UK

Received 7 May 1982
Revised manuscript received 18 June 1982

A linear, viscously damped oscillator that rebounds elastically whenever the displacement drops to zero is a well-defined quasi-linear deterministic system arising throughout the physical sciences. It exhibits a family of sub-harmonic resonant peaks between which we delineate cascades of period-doubling bifurcations leading to chaotic regimes typical of a strange attractor.

Much current interest in mathematics and physics has been generated by the strange chaotic response of simple deterministic dynamical systems. We might quote for example the significant recent contributions in Physics Letters of Eilbeck et al. [1] working on the driven sine-Gordon equation, Hao and Zhang [2] examining the forced Brusselator tri-molecular chemical reaction, and Li et al. [3] studying a simple mapping. Earlier pioneering work is due to Lorenz [4], Ruelle and Takens [5], Hénon [6], Rossler [7], Holmes [8], Haken [9] and others, as reported in [10].

We present here a new example of the chaotic motions of a strange attractor that has arisen in marine engineering where the repeated slackening of a mooring line introduces a discontinuity in the stiffness of, for example, an articulated oil-loading tower driven by ocean waves [11,12]. Systems such as this can sometimes be modelled as an impacting oscillator that rebounds elastically whenever the displacement X drops to zero: when forced sinusoidally we show that chaotic regimes can be generated by a cascade of period-doubling bifurcations. This result is of immediate interest to physicists and engineers because approximations to such an impacting system are often encountered in the real world. It may also be of considerable interest to mathematicians, because, being quasi-linear, the system might be tractable to topological analysis, as with the Smale horseshoe.

We can write the oscillator equation [11,12] as

$$\ddot{X} + 2(\zeta/\eta)\dot{X} + \tfrac{1}{4}\eta^{-2}X = \eta^{-2}\sin\tau, \qquad (1)$$

where the non-dimensional displacement X cannot take negative values. Here a dot denotes differentiation with respect to the scaled time τ, ζ is the damping ratio defined with respect to the effective natural frequency of the unforced and undamped rebounding system, and η is the ratio of the forcing frequency to this effective natural frequency. When X drops to zero during any motion, the velocity $\dot{X}$ is assumed to be instantaneously reversed in a perfectly elastic rebound. Notice that the magnitude of the sinusoidal forcing has been incorporated into the definition of the non-dimensional displacement X: the system thus has the useful quasi-linear property that the strength of the forcing only influences the behaviour through the scaling of the displacement X.

Defining y as half the maximum value of X during a steady periodic oscillation, the resonance response curve of y against η has been determined as shown in fig. 1. Here and elsewhere in this letter the damping is held constant at the quite high value of $\zeta = 0.1$. Well-defined and repeatable resonant peaks occur as shown, corresponding to a fundamental response ($n = 1$) at $\eta \doteq 1$, and sub-harmonic resonances of order n (=2, 3, 4, ...) at $\eta \doteq n$.

These peaks, and our subsequent investigation of chaotic motions, were determined using a very precise

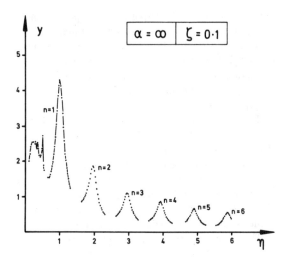

Fig. 1. The resonance response curve of the impact oscillator, showing the fundamental $n = 1$ resonance, and the sub-harmonic peaks for $n = 2$ to 6.

digital computer programme [11] that determines after each impact the new arbitrary constants of the complementary function of the known analytical solution. This solution is then evaluated at fine time intervals to accurately locate the next impact when X is again zero.

The plotted peaks of the figure were determined automatically by the computer using a sub-routine that inspected the Poincaré mapping points in the space of $(X, \dot{X})$ at $\tau = 0, 2\pi, 3\pi, \ldots i\pi, j\pi, \ldots$. When the sub-routine observed that the mapping points were repeating, within a specified close tolerance, so that for example

$$(X_i, \dot{X}_i) \doteq (X_{i+n}, \dot{X}_{i+n}), \qquad (2)$$

it reported, after suitable further checks, that it had found a stable steady state periodic sub-harmonic of order n (the special case of $n = 1$ being a fundamental response at the forcing frequency).

It is worth observing here that the sub-harmonics of order n in the vicinity of the peaks at $\eta \doteq n$ all involve only one impact per response cycle. This is not the case away from the peaks.

The automatic procedure failed repeatedly in the regions between the peaks, and more careful interactive procedures had to be employed. In particular the gap between the $n = 4$ and $n = 5$ resonant peaks has

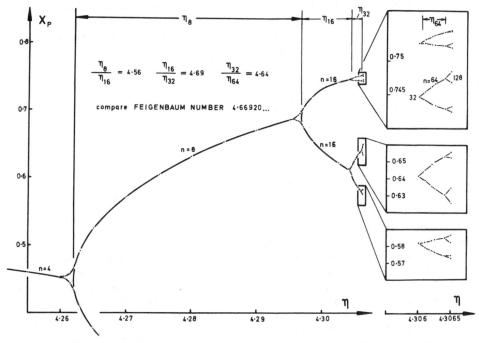

Fig. 2. Period-doubling bifurcations leading to chaos, shown on a plot of the X value of an arbitrarily chosen Poincaré point, X_P, versus the frequency ratio.

been explored with great precision, and repeated branching of the stable steady-state periodic solutions was observed as shown in fig. 2.

This figure follows the X coordinate, X_P, of an arbitrarily chosen $n = 4$ mapping point, as η increases. We see period-doubling bifurcations from the original $n = 4$ to $n = 128$ as the control parameter increases towards 5. A similar reversed cascade of period-doubling bifurcations starting with $n = 5$ at $\eta = 4.554$ is observed as the control parameter is *decreased*, yielding subharmonics of order 5, 10, ...

These two bifurcating cascades are in a sense incompatible, since one has the values $n = 2, 4, 8, 16,...$ while the other has the values of $n = 5, 10, 20, 40,...$, and it is difficult to see how they will eventually interact.

This issue seems to be resolved by the appearance of a *chaotic regime* in which we have been unable to detect any periodic behaviour at all. The Poincaré map for $\eta = 4.500$, for example, has been examined carefully, and the number of distinct points seems to increase without limit. We have here, for example, run the computer for 1000 forcing cycles without plotting any points to get to a "steady state" followed by point-plotting for a further 480 forcing cycles, without observing periodicity.

The hand-like region of development in the phase space is reminiscent of the strange attractor mapping of Hénon [6] and that found by Holmes [8] in his comprehensive study of the chaotic resonance of a buckled beam. We conclude provisionally that we have a chaotic dynamical regime governed by a strange attractor. This conclusion is strengthened by the observation that the η range of the $n = 8$ sub-harmonic in fig. 2 is approximately 4.6 times the range of the $n = 16$ sub-harmonic. This ratio of 4.6 is close to a number characteristic of period-doubling bifurcations leading to a strange attractor, as indicated.

In conclusion, we have observed for our simple quasi-linear deterministic impact oscillator, well defined resonant peaks for sub-harmonics of order 1 to 6. Between all of these peaks our automatic sub-harmonic detection routine has failed. We have explored carefully the region of the frequency ratio between the $n = 4$ and $n = 5$ peaks, and observed a cascade of period-doubling bifurcations as we move towards the centre from either end of the region. Separating the seemingly incompatible cascades, we find a region of chaotic response with apparently no steady state periodic solutions, governed presumably by a strange attractor. In intermediate regions around the centre, the strange attractor co-exists with stable sub-harmonic solutions, the solution obtained depending only on the starting conditions represented by $(X_0, \dot{X}_0)$ at $\tau = 0$.

This generation of chaos by as cascade of period-doubling bifurcations is similar to that observed by Hao and Zhang [2] using a modified stroboscopic sampling technique, and similar to that apparently controlling the onset of certain types of hydrodynamic turbulence. We feel this result will be of immediate interest to physicists, engineers and mathematicians in the rapidly growing field of chaotic dynamics.

The present system arose as a limiting case of $\alpha = \infty$ in the study of a driven *bilinear* oscillator with two different linear stiffnesses of ratio α, details of which are given in refs. [11,12]. The largest finite α value that we have studied in detail so far is $\alpha = 10$, and with the damping given by $\zeta = 0.1$ the bilinear system has been shown to exhibit multiple steady states and period-doubling bifurcations but apparently *no* chaotic behaviour. Limited semi-analytic solutions using Fourier series for the simple periodic solutions are given for the *bilinear* oscillator by Maezawa [13], and for the *impact* oscillator by Lean [14]. A similar oscillator (with the

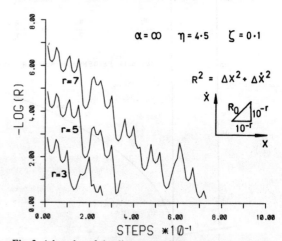

Fig. 3. A log plot of the divergence of the motions resulting from adjacent starts. Here the measure of divergence, R, is the distance of separation in the phase space. Three runs are shown, corresponding to three different values of the starting gap, R_0. All runs start nominally on the attractor, located by a previous long computation.

additional complications of a biasing force and a coefficient of restitution) is discussed by Senator [15] who writes the analytical matching conditions for a simple closed periodic solution: unlike our more general time integration, which embraces transient behaviour plus stable steady states, Senator's scheme can pick up both stable and unstable periodic solutions, and some appropriate stability analyses were presented.

A divergence study of motions from closely spaced starts of our impact oscillator is finally shown in fig. 3, which shows noisy *exponential* divergence leading to a rapid lack of correlation.

This research into the sub-harmonic resonances of moored marine systems is supported by the Science and Engineering Research Council of Great Britain under its marine technology programme: we acknowledge the contributions of Dr. A.R. Bokaian and Dr. J.S.N. Elvey to this continuing work.

References

[1] J.C. Eilbeck, P.S. Lomdahl and A.C. Newell, Phys. Lett. 87A (1981) 1.
[2] B. Hao and S. Zhang, Phys. Lett. 87A (1982) 267.
[3] T. Li, M. Misiurewicz and J.A. Yorke, Phys. Lett. 87A (1982) 271.
[4] E.N. Lorenz, J. Atmos. Sci. 20 (1963) 130.
[5] D. Ruelle and F. Takens, Commun. Math. Phys. 20 (1971) 167.
[6] M. Hénon, Commun. Math. Phys. 50 (1976) 69.
[7] O.E. Rossler, Ann. NY Acad. Sci. 316 (1979) 376.
[8] P.J. Holmes, Philos. Trans. R. Soc. London 292A (1979) 419.
[9] H. Haken, ed., Chaos and order in nature (Springer, Berlin, 1981).
[10] J.M.T. Thompson, Instabilities and catastrophes in science and engineering (Wiley, Chichester, 1982).
[11] J.M.T. Thompson, A.R. Bokaian and R. Ghaffari, Sub-harmonic resonance of a bilinear oscillator with applications to moored marine systems (1982), to be published.
[12] J.M.T. Thompson, A strange attractor in the resonance of an impact oscillator (1982), to be published.
[13] S. Maezawa, Bull. Japan. Soc. Mech. Eng. 4 (1961) 201.
[14] G.H. Lean, Trans. R. Inst. Naval Arch. 113 (1971) 387.
[15] M. Senator, J. Acoust. Soc. Am. 47 (1970) 1390.

DEVIL'S ATTRACTORS AND CHAOS OF A DRIVEN IMPACT OSCILLATOR

H.M. ISOMÄKI[1], J. VON BOEHM and R. RÄTY
Department of General Sciences, Helsinki University of Technology, SF-02150 Espoo 15, Finland

Received 30 October 1984

The motion of a driven elastic impact oscillator: $\ddot{x} + 0.4\dot{x} + x = \cos(\omega t)$, $x > 0$ and $\dot{x}(t^+) = -\dot{x}(t^-)$ at $x = 0$, is studied for $\omega \approx 2-4$. The oscillator exhibits Feigenbaum's bifurcations (computed $\delta \approx 4.70$), the Feigenbaum and intermittent transitions to chaos, crises in chaos and a strong hysteresis region for $\omega \approx 3.18-3.20$ where the impact/period ratios of a group of attractors show the Devil's staircase behaviour with locking values between 3/5 and 3/4.

Since the recent observations that simple nonlinear dissipative driven oscillators exhibit chaotic motion [1–9] having universal properties of one-dimensional non-invertible maps [10–12] considerable efforts have been made to find simple physical systems that can be solved with an accuracy comparable with that used when iterating one-dimensional maps. Such systems are the nonlinear oscillator driven by a periodic impulsive force [13–17] and the sinusoidally driven impact oscillator [18–26]. The studies on the impact oscillators reported thus far have concentrated mainly in the neighbourhood of the resonances where the motion is of the one-impact period-p type. However, less effort has been paid to study the regions between the resonances where the motion is of the general i-impact period-p or chaotic type. The purpose of this letter is to report and discuss the detailed results of a study on a damped fully elastic sinusoidally driven impact oscillator between the fundamental resonance and the first subharmonic one. The good general accuracy of our calculation allows us to study the chaotic motion in very fine details. The impact oscillator studied exhibits several of the properties of one-dimensional non-invertible maps like the Feigenbaum and intermittent routes to chaos [11,12,27] as well as sudden changes in the chaotic attractor called crises by Grebogi et al. [28–30]. Moreover, a complex behaviour with strong hysteresis and a peculiar group of attractors

coexisting with other attractors is found. We call this group of attractors the Devil's attractors because the impact/period ratios

$$q = i/p \qquad (1)$$

of the attractors show the interesting Devil's staircase behaviour, i.e. q locks at all rational numbers between 3/5 and 3/4. [In eq. (1) i is the number of impacts during the period p given in units $2\pi/\omega$ where ω is the angular frequency of the driving force.] The Devil's staircases [31] have been observed earlier in the Frenkel–Kontorova model [32–34], in the ANNNI model of magnetic moments [35–38], in the one-dimensional Ising model [39], in the circle map [40] and in the Josephson junction [41].

The dimensionless equation of motion of the impact oscillator studied in this letter reads between the impacts as

$$\ddot{x} + 0.4\dot{x} + x = \cos(\omega t), \qquad (2)$$

where $x > 0$. Always when the particle hits the wall at $x = 0$ Newton's rule of an elastic impact

$$\dot{x}(t_i^+) = -\dot{x}(t_i^-) \qquad (3)$$

is applied. This rule is responsible for the non-linearity. The nice thing with the impact oscillator is that the solution $x(t)$ to eq. (2) is known exactly between the impacts. The impact time t_i is iterated from the transcendental equation

[1] Supported by the Academy of Finland.

$x(t_i) = 0$ (4)

using the Newton–Raphson method with a double precision accuracy ($\sim 10^{-17}$). The procedure is repeated until the stationary solution, i.e. the attractor, is reached. The above procedure produces the map

$T_{n+1} = f(T_n)$, (5)

where T is the time between two successive impacts. If (for a large n) $T_{n+1} + T_{n+2} + ... + T_{n+i}$ equals $p \cdot 2\pi/\omega$ we have the i-impact period-p steady state motion denoted by (i, p) in the following.

When increasing ω from $\omega = 2$ the (1, 1) limit cycle is found first. The Feigenbaum bifurcations start at $\omega = 2.64$ leading to chaos at $\omega = 2.74$. The limit cycles in the Feigenbaum sequence are of the form $(2^n, 2^n)$, $n = 1, 2, 3, ...$. These findings are in agreement with refs. [20,22]. We find for the bifurcation ratio $\Delta\omega(p = 2^6)/\Delta\omega(p = 2^7)$ the value 4.70 that agrees closely with Feigenbaum's universal $\delta = 4.669$ [11,12]. This indicates that the present bifurcation sequence follows closely the universal features of the non-invertible one-dimensional maps.

Between $\omega = 2.74$ and 3.13 the chaotic region is interrupted by a sudden jump at $\omega \approx 2.75$ (crisis) and periodic windows. When ω is increased beyond 3.15 quite a rich behaviour is found. Fig. 1 shows the amplitude response diagram in this region (the amplitude X is defined as the maximum value of $x(t)$ of the attractor). The diagram consists of a sudden jump A → B in the chaotic attractor (crisis), two opposite hysteresis loops (C → D → E → F → C and G → H → I → J → K → G) and the peculiar Devil's attractors L → M (denoted by dots in fig. 1) whose q's vary with ω as the Devil's staircase. We find Feigenbaum's bifurcation sequences leading to chaos from C to B and from the neighbourhood of E towards F. From H to I there is an incomplete Feigenbaum's sequence (7, 10), (14, 20), (28, 40) and a return to (14, 20) before the jump to J. From J via K to G there is only one bifurcation (1, 2) → (2, 4) in agreement with refs. [20,22].

The Devil's staircase for the q's of the attractors L → M is presented in fig. 2. q seems to lock at all rational numbers between 3/4 and 3/5. This was checked up to rational numbers $n/14$ (n integer), i.e. there is a plateau for the reduced values of $q = 2/3$, 3/4, 3/5, 5/7, 5/8, 7/10, 7/11, 8/11, 8/13, 9/13 and

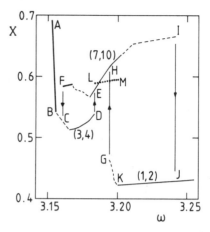

Fig. 1. Amplitude response diagram. The amplitude X is defined as the maximum value of $x(t)$. ω is the angular frequency of the driving force. The thick solid, thin solid and broken lines denote the chaotic attractor, the periodic attractor and the Feigenbaum sequence, respectively. The thin vertical lines with arrows denote hysteresis jumps. The chain of dots denotes the Devil's attractors. The numbers in parentheses (i, p) denote i impacts during the period p.

9/14 [1]. We find that the general form of an arbitrary Devil's attractor (i, p) can be presented as

$(i, p) \triangleq (\frac{5}{3}i - p)*(3, 4) + (p - \frac{4}{3}i)*(3, 5)$. (6)

---

[1] The plateau of $q = 8/13$ is so short that a fully stable attractor was not observed yet.

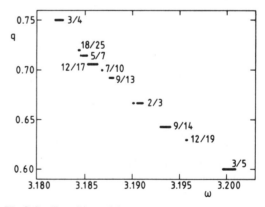

Fig. 2. Devil's staircase of the attractors L → M. q denotes the reduced impact/period ratio.

344

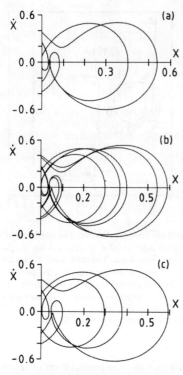

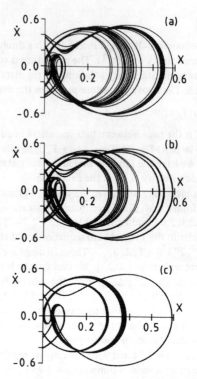

Fig. 3. Limit cycles of the Devil's attractors. (a) The basic limit cycle (3, 4) at $\omega = 3.1828$. (b) The limit cycle (6, 9) at $\omega = 3.19$. (c) The basic limit cycle (3, 5) at $\omega = 3.2$.

Fig. 4. Chaotic attractors close to the limit cycles of fig. 3. (a) $\omega = 3.18425$, (b) $\omega = 3.1899$ and (c) $\omega = 3.19957395$.

As an example the Devil's attractors (3, 4), (6, 9) $\triangleq 1*(3, 4) + 1*(3, 5)$ and (3, 5) are shown in figs. 3a, 3b and 3c, respectively. It can be shown that when the coefficients $\frac{5}{3}i - p$ and $p - \frac{4}{3}i$ in eq. (6) attain all (positive) integer values then $q = i/p$ covers all rational numbers between 3/4 and 3/5. In this sense the Devil's staircase would be complete. For real completeness the plateaus should also fill the whole range from 3.18 to 3.201 [31,33]. However, the regions $\omega = 3.1828$ ... 3.1834, 3.188 ... 3.190, 3.194 ... 3.1955 and 3.196 ... 3.1996 are relatively sparse. This leads us to suspect that the filling is incomplete.

Some of the plateaus of the Devil's staircase in fig. 2 are broken. This is due to the period-changes (like period-doublings) within such a plateau, for example (for decreasing ω) (12, 18) → (18, 27) in the $q = 2/3$ plateau and (15, 21) → (30, 42) in the $q = 5/7$ plateau. Some of the Devil's attractors L → M in fig. 1 end in chaos. As an example the chaotic attractors close to the periodic attractors of fig. 3 are presented in fig. 4. By using the map of eq. (5) and its higher order maps it is straightforward to find the scenarios of these attractors. The attractors of figs. 4a and 4b have developed via Feigenbaum's bifurcations and the attractor of fig. 4c has developed via intermittency.

In conclusion, we have solved in detail the motion of the damped sinusoidally driven elastic impact oscillator in the whole ω-range between the two adjacent resonances (1, 1) and (1, 2). The motion has turned out to be extremely rich displaying many properties of great current interest in nonlinear dynamics, like the intermittent and Feigenbaum routes to chaos, crises in chaos and the Devil's attractors (the impact/period ratios of these form the Devil's staircase). Moreover, some of the plateaus of the staircase are broken due to the changes in the periods and may end in chaos.

345

The authors would like to thank Professor M.A. Ranta for his interest and support and Mrs. T. Aalto for typing the manuscript.

References

[1] P. Holmes, Philos. Trans. R. Soc. A292 (1979) 419.
[2] B.A. Huberman and J.P. Crutchfield, Phys. Rev. Lett. 43 (1979) 1743.
[3] Y. Ueda, J. Stat. Phys. 20 (1979) 181.
[4] K. Tomita and T. Kai, J. Stat. Phys. 21 (1979) 65.
[5] P.S. Linsay, Phys. Rev. Lett. 47 (1981) 1349.
[6] J. Testa, J. Pérez and C. Jeffries, Phys. Rev. Lett. 48 (1982) 714.
[7] F.T. Arecchi and F. Lisi, Phys. Rev. Lett. 49 (1982) 94.
[8] D. D'Humieres, M.R. Beasley, B.A. Huberman and A. Libchaber, Phys. Rev. A26 (1982) 3483.
[9] R. Räty, J. von Boehm and H.M. Isomäki, Phys. Lett. 103A (1984) 289.
[10] N. Metropolis, M.L. Stein and P.R. Stein, J. Comb. Theory (A) 15 (1973) 25.
[11] M.J. Feigenbaum, J. Stat. Phys. 19 (1978) 25.
[12] M.J. Feigenbaum, J. Stat. Phys. 21 (1979) 669.
[13] L. Glass and R. Pérez, Phys. Rev. Lett. 48 (1982) 1772.
[14] R. Pérez and L. Glass, Phys. Lett. 90A (1982) 441.
[15] L. Glass, M.R. Guevara, A. Shrier and R. Pérez, Physica 7D (1983) 89.
[16] D.L. Gonzalez and O. Piro, Phys. Rev. Lett. 50 (1983) 870.
[17] D.L. Gonzalez and O. Piro, Phys. Lett. 101A (1984) 455.
[18] J.M.T. Thompson and R. Ghaffari, Phys. Lett. 91A (1982) 5.
[19] J.M.T. Thompson and R. Ghaffari, Phys. Rev. A27 (1983) 1741.
[20] J.M.T. Thompson, A.R. Bokaian and R. Ghaffari, IMA J. Appl. Math. 31 (1983) 207.
[21] S.W. Shaw and P. Holmes, Phys. Rev. Lett. 51 (1983) 623.
[22] S.W. Shaw and P.J. Holmes, J. Sound Vibrat. 90 (1983) 129.
[23] F.C. Moon and S.W. Shaw, Int. J. Non-Linear Mech. 18 (1983) 465.
[24] H. Frosch and H. Büttner, Two coupled impact oscillators (1984), unpublished.
[25] M.B. Hindmarsh and D.J. Jefferies, J. Phys. A17 (1984) 1791.
[26] H.M. Isomäki, J. von Boehm and R. Räty in: Proc. Tenth Intern. Conf. on Nonlinear oscillations (Varna, September 1984) pp. 12–17.
[27] P. Manneville and Y. Pomeau, Physica 1D (1980) 219.
[28] C. Grebogi, E. Ott and J.A. Yorke, Phys. Rev. Lett. 48 (1982) 1507.
[29] C. Grebogi, E. Ott and J.A. Yorke, Phys. Rev. Lett. 50 (1983) 935.
[30] C. Grebogi, E. Ott and J.A. Yorke, Physica 7D (1983) 181.
[31] B.B. Mandelbrot, Fractals: form, change and dimension (Freeman, San Francisco, 1977).
[32] S. Aubry, in: Solitons and condensed matter physics, eds. A. R. Bishop and T. Schneider (Springer, Berlin, 1978) p. 264.
[33] S. Aubry, J. Phys. C16 (1983) 2497.
[34] S. Aubry, Physica 7D (1983) 240.
[35] J. von Boehm and P. Bak, Phys. Rev. Lett. 42 (1979) 122.
[36] P. Bak and J. von Boehm, Phys. Rev. B21 (1980) 5297.
[37] P. Bak, Phys. Rev. Lett. 46 (1981) 791.
[38] M.H. Jensen and P. Bak, Phys. Rev. B27 (1983) 6853.
[39] P. Bak and R. Bruinsma, Phys. Rev. Lett. 49 (1982) 249.
[40] M.H. Jensen, P. Bak and T. Bohr, Phys. Rev. Lett. 50 (1983) 1637.
[41] W.J. Yeh, D.-R. He and Y.H. Kao, Phys. Rev. Lett. 52 (1984) 480.

Journal of Sound and Vibration (1983) **90**(1), 129–155

A PERIODICALLY FORCED PIECEWISE LINEAR OSCILLATOR†

S. W. Shaw and P. J. Holmes

Department of Theoretical and Applied Mechanics, Cornell University, Ithaca, New York 14853, U.S.A.

(*Received* 15 *December* 1982)

A single-degree of freedom non-linear oscillator is considered. The non-linearity is in the restoring force and is piecewise linear with a single change in slope. Such oscillators provide models for mechanical systems in which components make intermittent contact. A limiting case in which one slope approaches infinity, an impact oscillator, is also considered. Harmonic, subharmonic, and chaotic motions are found to exist and the bifurcations leading to them are analyzed.

1. INTRODUCTION

Mechanical systems in which moving components make intermittent contact with each other often give rise to equations of motion containing piecewise linear terms. Such systems have been the subject of several investigations, because of the apparently simple nature of the non-linearity. Standard texts on non-linear oscillations such as those of Minorsky [1] and Andronov *et al.* [2] contain analyses of these systems, although only free oscillations are considered. In contrast, periodically forced systems with non-linear restoring forces are studied in the present paper. Single degree of freedom oscillators of this type were studied by Maezawa and Furukawa [3] and Dragoni and Repaci [4], who considered symmetric restoring forces. The non-symmetric case with linear damping is to be considered in the present paper. Maezawa [5] and Maezawa *et al.* [6] previously studied the harmonic and superharmonic response of such a system using a Fourier series method. Klotter [7] obtained harmonic motions of several piecewise linear systems using a Galerkin method, but the stability of these harmonic solutions was not discussed. Thompson [8] has studied harmonic and subharmonic motions using numerical methods. An experimental investigation of both symmetric and non-symmetric restoring forces has been done by Robinson [9], who has observed many of the phenomena we analyze in this paper. Moreover, such piecewise linear systems have not been used to model the motion of beams with non-linear boundary conditions [10–12].

When one stiffness approaches infinity the system becomes an *impact oscillator*. Senator [13] studied such a system with a constant restoring force and energy loss upon impact. He discussed single impact periodic motions and their stability. Holmes [14] considered a similar system (a mass bouncing on a vibrating table) and found not only harmonic and subharmonic motions, but also bounded non-periodic "chaotic" ones. The impact oscillator considered here was studied in detail using numerical methods by Thompson [15, 16]. He found cascades of period doubling bifurcations from harmonic and subharmonic solutions, apparently leading to non-periodic motions. These period doubling bifurcations seem similar to those found in one dimensional maps [17, 18].

† This work was supported by NSF Grant MEA-80-17570.

2. DESCRIPTION OF THE SYSTEM

2.1. THE GENERAL SYSTEM

Consider the simple system shown in Figure 1. A mass m is attached to a linear spring of stiffness k_1 and a linear dashpot with damping factor c. When the displacement, x,

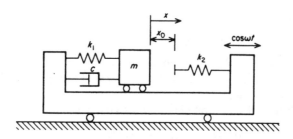

Figure 1. The physical system.

exceeds a certain value, x_0, a second linear spring, k_2, contacts m. Without loss of generality one can assume $x_0 \geq 0$. The two springs give rise to an overall restoring force which is piecewise linear. When the system is externally excited by a harmonic force, the non-dimensionalized equation of motion may be written as

$$\ddot{x} + 2\alpha\dot{x} + H(x) = \beta \cos(\omega t), \qquad (1)$$

where

$$H(x) = \begin{cases} x, & x < x_0 \\ \tilde{\omega}^2 x + (1 - \tilde{\omega}^2) x_0, & x \geq x_0 \end{cases},$$

$\tilde{\omega}^2$ is the stiffness ratio $(k_1 + k_2)/k_1 = \omega_+^2/\omega_-^2$, α is the damping coefficient, which we shall generally assume to be less than 1 (subcritical), β is the forcing amplitude, and ω is the forcing frequency. An overdot indicates differentiation with respect to the non-dimensional time t. Figure 2(a) shows the function $H(x)$.

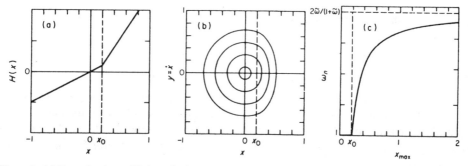

Figure 2. (a) Restoring force $H(x)$ vs. displacement x; (b) undamped ($\alpha = 0$), unforced ($\beta = 0$) phase portrait, velocity (y) vs. displacement; (c) frequency ω_n vs. amplitude x_{MAX}.

Local solutions of this equation are known explicitly on each side of $x = x_0$. Such solutions can be repeatedly matched at $x = x_0$ to obtain a global solution of equation (1). Piecing together these known solutions is not directly possible however, since, as will be seen, the times of flight in each region cannot be found in closed form.

The two equations are

$$\ddot{x} + 2\alpha\dot{x} + x = \beta \cos(\omega t) \quad \text{for } x \leq x_0, \tag{2}$$

$$\ddot{x} + 2\alpha\dot{x} + \tilde{\omega}^2 x + (1-\tilde{\omega}^2)x_0 = \beta \cos(\omega t) \quad \text{for } x \geq x_0. \tag{3}$$

The solution to equation (3) based at $x(t_0) = x_0$ and $\dot{x}(t_0) = y_0$, for $\alpha < 1$, is

$$x_+(t; t_0, y_0) = e^{-\alpha(t-t_0)}[A_+ \cos(\Omega_+(t-t_0)) + B_+ \sin(\Omega_+(t-t_0))]$$
$$+ \gamma_+ \cos(\omega t) + \delta_+ \sin(\omega t) - x_0(1-\tilde{\omega}^2)/\tilde{\omega}^2, \tag{4}$$

where $\gamma_+ = (\tilde{\omega}^2 - \omega^2)\beta/\Delta_+$, $\delta_+ = (2\alpha\omega)\beta/\Delta_+$, $\Delta_+ = (\tilde{\omega}^2 - \omega^2)^2 + (2\alpha\omega)^2$, $\Omega_+^2 = \tilde{\omega}^2 - \alpha^2$, $A_+ = -\gamma_+ c_0 - \delta_+ s_0 + x_0/\tilde{\omega}^2$, $B_+ = (1/\Omega_+)[y_0 + \alpha x_0/\tilde{\omega}^2 + s_0(\gamma_+\omega - \delta_+\alpha) - c_0(\gamma_+\alpha + \delta_+\omega)]$, $c_0 = \cos(\omega t_0)$, and $s_0 = \sin(\omega t_0)$.

Similarly the solution of equation (2) based at $x(t_1) = x_0$ and $\dot{x}(t_1) = y_1$ is

$$x_-(t; t_1, y_1) = e^{-\alpha(t-t_1)}[A_- \cos(\Omega_-(t-t_1)) + B_- \sin(\Omega_-(t-t_1))] + \gamma_- \cos(\omega t) + \delta_- \sin(\omega t), \tag{5}$$

where $\gamma_- = (1-\omega^2)\beta/\Delta_-$, $\delta_- = (2\alpha\omega)\beta/\Delta_-$, $\Delta_- = (1-\omega^2)^2 + (2\alpha\omega)^2$, $\Omega_-^2 = 1 - \alpha^2$, $A_- = x_0 - \gamma_- c_1 - \delta_- s_1$, $B_- = (1/\Omega_-)[y_1 + \alpha x_0 + s_1(\gamma_-\omega - \delta_-\alpha) - c_1(\gamma_-\alpha + \delta_-\omega)]$, $c_1 = \cos(\omega t_1)$, and $s_1 = \sin(\omega t_1)$.

The difficulty in joining solutions (4) and (5) together to obtain the global solution is that the crossing times (when $x(t_i) = x_0$) are not known explicitly. These times are roots of the equations

$$x_+(t; t_0, y_0) = x_0 \quad \text{and} \quad x_-(t; t_1, y_1) = x_0. \tag{6, 7}$$

When $\beta = 0$ and $\alpha = 0$, the unforced, undamped system has a phase portrait like that shown in Figure 2(b). The phase plane is filled with a continuous family of closed orbits. For $x_{\max} < x_0$ the orbits are simple harmonic motions of period 2π. For $x_{\max} > x_0$ the orbits consist of two pieces of ellipses jointed at $x = x_0$. The natural frequency of oscillation as a function of amplitude is shown in Figure 2(c). It is 1 up to $x_{\max} = x_0$ and then begins to increase. As the amplitude becomes large, the frequency approaches an asymptotic value, since the gap (between $x = 0$ and x_0) becomes negligible. This asymptotic value is obtained from the case $x_0 = 0$, considered below, since for large amplitude motions $|x_{\max}| \gg |x_0|$.

The reader should note that orbits are once differentiable along $x = x_0$. This result is obtained directly from equation (1), and follows from the Lipschitz-continuity of the function $H(x)$. Henceforth we shall adopt the notation that $y = \dot{x}$.

2.2. THE $x_0 = 0$ CASE

In the $x_0 = 0$ case the forcing amplitude may be scaled out by letting $x \to \beta x$. Also in this case the unforced system has a natural frequency which is independent of the amplitude of oscillation. This is an important difference between this and most other non-linear systems (including the $x_0 \neq 0$ case). The damped natural frequency in this case becomes

$$\Omega_n = 2\Omega_+\Omega_-/(\Omega_+ + \Omega_-), \tag{8}$$

where $\Omega_-^2 = 1 - \alpha^2$ and $\Omega_+^2 = \tilde{\omega}^2 - \alpha^2$. The undamped natural frequency is obtained by setting $\alpha = 0$ in equation (8) and is

$$\omega_n = 2\tilde{\omega}/(1 + \tilde{\omega}). \tag{9}$$

The undamped, unforced phase plane for this case is shown in Figure 3.

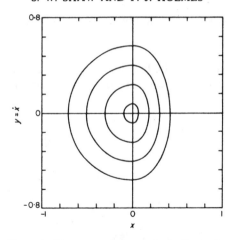

Figure 3. Phase portrait for $\alpha = 0$, $\beta = 0$, $x_0 = 0$.

2.3. THE IMPACT LIMIT

When $k_2 \to \infty$ the resulting system will be referred to as an *impact oscillator* [15, 16]. Letting $\varepsilon = 1/\bar{\omega}$ and rescaling equation (1) yields, for $x > x_0$,

$$\ddot{x} + 2\alpha\varepsilon\dot{x} + x = \varepsilon^2\beta \cos(\bar{\omega}\tau) + x_0(1-\varepsilon^2), \qquad (10)$$

where $\bar{\omega} = \omega/\omega_+$ and $\tau = \omega_+ t$. The solution in this rescaled time variable is (cf. equation (4))

$$x(\tau) = x_0 + y_0(1 - \alpha\varepsilon(\tau - \tau_0))\sin(\tau - \tau_0) + O(\varepsilon^2). \qquad (11)$$

The total time spent in $x > x_0$ may be computed by solving for the first root $\tau_1 > \tau_0$ of $x(\tau_1) = x_0$. Letting $\tau_1 = T_{10} + \varepsilon T_{11} + O(\varepsilon^2)$ and solving, one obtains $\tau_1 = \tau_0 + \pi + O(\varepsilon^2)$. Hence, from equation (11), if the velocity at impact is y_0, immediately after the impact it is

$$y_1 = \dot{x}(\tau_1) = -(1 - \alpha\varepsilon\pi)y_0 + O(\varepsilon^2). \qquad (12)$$

The factor $(1 - \alpha\varepsilon\pi)$ is the leading part of the Taylor series expansion of the exponential damping decay $e^{-\alpha\varepsilon\pi}$ and represents a loss of energy during impact. In what follows we shall denote this coefficient of restitution as r and write

$$y_1 = -ry_0, \qquad (13)$$

thus obtaining the standard impact rule (cf. reference [19]). Although as derived here $r = 1 - O(\varepsilon)$, we shall generalize and allow r to range from 0 to 1. This allows us to account for other losses during the impact.

3. THE POINCARÉ SECTION AND RETURN MAP

To study equations of this type we shall employ the method of a Poincaré section. For a one degree of freedom periodically forced oscillator one has a three dimensional extended phase space with co-ordinates (x, y, t). The vector field defined by equation (1) is easily seen to be $2\pi/\omega$ periodic in t. A natural place to slice this space is at the points of discontinuity in stiffness: i.e., at $x = x_0$. We define the Poincaré section as

$$\Sigma = \{(x, y, t) | x = x_0, y > 0\}. \qquad (14)$$

Orbits in the phase space will be studied by considering the mapping

$$\mathbf{P}:\Sigma \to \Sigma \qquad (15)$$

induced by solutions of equation (1). Figure 4 shows the phase space and the section Σ. This section is used since the solutions of the linear equations are known explicitly on

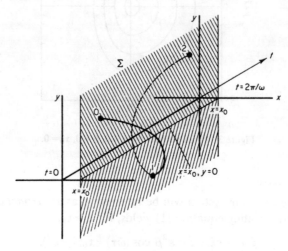

Figure 4. Phase space (x, y, t) showing the Poincaré section Σ. Note the $2\pi/\omega$ periodicity in t.

each side of Σ. It should be noted that the flow of the differential equation is everywhere transverse to Σ. This is easily seen by considering the vector field in (x, y, t) space on Σ given by

$$\dot{x} = y, \qquad \dot{y} = -2\alpha y + x_0 + \beta \cos(\omega t), \qquad \dot{t} = 1. \qquad (16)$$

for $y > 0$, $\dot{x} \neq 0$ and thus the flow is transverse to Σ. However, $\mathbf{P}$ is not always well defined and is not generally "onto", since some points in Σ are mapped onto the line $x = y = 0$. This leads to discontinuities in the mapping, as will be seen.

The section Σ considered here should be distinguished from the more common section [20]

$$\Sigma^{t_0} = \{(x, y, t) | t = t_0, \mod 2\pi/\omega\}. \qquad (17)$$

On Σ^{t_0} a point which is mapped on to itself after k iterates of the return map $\mathbf{P}^{t_0}: \Sigma^{t_0} \to \Sigma^{t_0}$ is known to correspond to a (subharmonic) orbit with period $k(2\pi/\omega)$ for the differential equation. In contrast, points on Σ are never mapped back to themselves, since the t (time) co-ordinate continues to increase linearly. Periodic (subharmonic) orbits correspond to points $(\bar{y}, \bar{t})$ with images $\mathbf{P}(\bar{y}, \bar{t}) = (\bar{y}, \bar{t} + 2\pi k/\omega)$; if $k = 1$ one has a harmonic (period $2\pi/\omega$) response. Note that orbits of the same period $k(2\pi/\omega)$ might contain different numbers of impacts in the same time $2\pi k/\omega$. In what follows we are mainly concerned with single impact, period k motions. However, in both cases, the stability of periodic points of the return map is the same as the stability of the corresponding orbits, and bifurcations of these orbits may be studied by considering bifurcations of the periodic points of the return map.

The mapping $\mathbf{P}$ cannot be written down explicitly. Consider an orbit starting at point $0 = (t_0, y_0)$ as shown in Figure 4. From point 0 to point $1 = (t_1, y_1)$ the motion is governed by equation (4) (note that point 1 is not in Σ but is in $\{(x, y, t) | x = x_0, y < 0\}$). Here one must solve equation (6) for the x_0 crossing time t_1,

$$x_+(t_1; t_0, y_0) = x_0, \tag{18}$$

where t_1 is the first root larger than t_0. Once t_1 has been determined then one can immediately compute

$$y_1 = \dot{x}_+(t_1; t_0, y_0) \tag{19}$$

and point 1 has been determined. A similar procedure but with the solution $x_-(t; t_1, y_1)$ of equation (5) takes the orbit from point 1 to point 2. The mapping **P** can then be written schematically as a difference equation in the form

$$t_{n+2} = f(t_n, y_n), \qquad y_{n+2} = g(t_n, y_n). \tag{20}$$

Here the functions f and g are necessarily $2\pi/\omega$ periodic in t_n, but since their nature depends on the roots of transcendental equations, they cannot be written down explicitly. This implies that, in general, periodic points of the map cannot be analytically determined. It will be seen, however, that in the impact limit certain periodic orbits can be found analytically, and even in the general case one can study the behavior of some of the periodic points without explicitly finding them. This is done by examining their stability and the bifurcations they undergo.

4. PERIODIC ORBITS AND LOCAL BIFURCATIONS

4.1. GENERAL CASE

The stability of a periodic point is determined by the eigenvalues of the first derivative of the map evaluated at that point. Bifurcations occur when the linearized map is degenerate: i.e., at least one eigenvalue has unit modulus [21]. We denote a period one point by $(\bar{t}, \bar{y})$ where

$$\bar{t} = f(\bar{t}, \bar{y}) - 2\pi/\omega, \qquad \bar{y} = g(\bar{t}, \bar{y}). \tag{21}$$

The first derivative of the map **DP** is given by

$$\mathbf{DP} = \begin{bmatrix} \dfrac{\partial f}{\partial t} & \dfrac{\partial f}{\partial y} \\ \dfrac{\partial g}{\partial t} & \dfrac{\partial g}{\partial y} \end{bmatrix} \overset{\text{def}}{=} \left[\dfrac{\partial (f, g)}{\partial (t, y)} \right]. \tag{22}$$

DP can be computed directly by using implicit differentiation. We outline that calculation here.

Differentiating equation (18) with respect to t_0 and y_0, using equation (4), gives $\partial t_1/\partial(t_0, y_0)$, as follows:

$$\dfrac{\partial t_1}{\partial t_0} = \dfrac{e^{-\alpha(t_1 - t_0)}}{\Omega_+ y_1}[\Omega_+ y_0 c_+ + N_0 s_+], \qquad \dfrac{\partial t_1}{\partial y_0} = \dfrac{-e^{-\alpha(t_1 - t_0)}}{\Omega_+ y_1} s_+, \tag{23, 24}$$

where

$$c_+ = \cos(\Omega_+(t_1 - t_0)), \qquad s_+ = \sin(\Omega_+(t_1 - t_0)), \quad \text{and} \quad N_0 = \beta c_0 - \alpha y_0 - x_0.$$

Next, taking derivatives of equation (19) using the first time derivative of equation (4) and equations (16), (23) and (24) gives

$$\dfrac{\partial y_1}{\partial t_0} = \dfrac{e^{-\alpha(t_1 - t_0)}}{\Omega_+ y_1}[(N_1 N_0 + \Omega_+^2 y_0 y_1) s_+ + c_+ \Omega_+ (y_0 N_1 - y_1 N_0)], \tag{25}$$

$$\frac{\partial y_1}{\partial y_0} = \frac{e^{-\alpha(t_1-t_0)}}{\Omega_+ y_1}[\Omega_+ y_1 c_+ - N_1 s_+], \tag{26}$$

where $N_1 = \beta c_1 - \alpha y_1 - x_0$, and $c_1 = \cos(\omega t_1)$. Equations (23)–(26) give the four components of $[\partial(t_1, y_1)/\partial(t_0, y_0)]$.

A similar calculation carried out between points 1 and 2 yields $[\partial(t_2, y_2)/\partial(t_1, y_1)]$. One simply replaces the subscripts 0, 1, and + by 1, 2, and −, respectively, in equations (23)–(26).

From the chain rule one has

$$\left[\frac{\partial(f, g)}{\partial(t_0, y_0)}\right] = \left[\frac{\partial(t_2, y_2)}{\partial(t_0, y_0)}\right] = \left[\frac{\partial(t_2, y_2)}{\partial(t_1, y_1)}\right]\left[\frac{\partial(t_1, y_1)}{\partial(t_0, y_0)}\right], \tag{27}$$

and equation (27) then gives the desired components of **DP**:

$$\partial f/\partial t_0 = \partial t_2/\partial t_0$$
$$= [e^{-\alpha(t_2-t_0)}/\Omega_+\Omega_- y_2][-s_+s_-(\Omega_+^2 y_0) + s_-c_+(\Omega_+ N_0) + s_+c_-(\Omega_- N_0) + c_+c_-(\Omega_+\Omega_- y_0)], \tag{28}$$

$$\partial f/\partial y_0 = \partial t_2/\partial y_0 = [e^{-\alpha(t_2-t_0)}/\Omega_+\Omega_- y_2][-s_-c_+\Omega_+ - s_+c_-\Omega_-], \tag{29}$$

$$\partial g/\partial t_0 = \partial y_2/\partial t_0 = [e^{-\alpha(t_2-t_0)}/\Omega_+\Omega_- y_2][s_+s_-(\Omega_-^2 y_2 N_0 - \Omega_+^2 y_0 N_2)$$
$$+ \Omega_+ s_-c_+(N_0 N_2 + \Omega_-^2 y_0 y_2) + \Omega_- s_+c_-(N_0 N_2 + \Omega_+^2 y_0 y_2)$$
$$+ \Omega_+\Omega_- c_+c_-(y_0 N_2 - y_2 N_0)], \tag{30}$$

$$\partial g/\partial y_0 = \partial y_2/\partial y_0 = [e^{-\alpha(t_2-t_0)}/\Omega_+\Omega_- y_2][-s_+s_-(\Omega_-^2 y_2) - s_-c_+(\Omega_+ N_2) - s_+c_-(\Omega_- N_2)$$
$$+ c_+c_-(\Omega_+\Omega_- y_2)]; \tag{31}$$

where $c_- = \cos(\Omega_-(t_2-t_1))$, and $s_- = \sin(\Omega_-(t_2-t_1))$. This matrix has the determinant

$$D = (y_0/y_2) e^{-2\alpha(t_2-t_0)}, \tag{32}$$

and trace

$$T = [e^{-\alpha(t_2-t_0)}/\Omega_+\Omega_- y_2][-s_+s_-(y_0\Omega_+^2 + y_2\Omega_-^2) + s_-c_+\Omega_+(N_0 - N_2)$$
$$+ s_+c_-\Omega_-(N_0 - N_2) + c_+c_-\Omega_+\Omega_-(y_0 + y_2)]. \tag{33}$$

One can now evaluate D and T on a period one orbit: i.e., one sets $y_2 = y_0$ and $t_2 - t_0 = 2\pi/\omega$ to obtain

$$\bar{D} = e^{-4\pi\alpha/\omega}, \quad \bar{T} = e^{-2\pi\alpha/\omega}\left[-s_+s_-\left\{\frac{(\Omega_+^2 + \Omega_-^2)}{\Omega_+\Omega_-}\right\} + 2c_+c_-\right]. \tag{34, 35}$$

$\bar{D}$ and $\bar{T}$ determine the eigenvalues λ_i of **DP** evaluated on the period one point via the expression

$$\lambda_{1,2} = \tfrac{1}{2}(\bar{T} \pm \sqrt{\bar{T}^2 - 4\bar{D}}). \tag{36}$$

These eigenvalues determine the stability of the period one point [21, 22]. If λ_1 and λ_2 lie inside the unit circle C, then the fixed point is stable, while if either one lies outside C the fixed point is unstable. As one varies the system parameters, eigenvalues may pass through the unit circle, at which point a bifurcation occurs. Note that since $\lambda_1\lambda_2 = \bar{D}$ and $\bar{D} < 1$, the only possible way for an eigenvalue to pass through C is either through $+1$ or through -1: i.e., no Hopf bifurcations to doubly periodic motions can occur [21].

The condition that $\lambda = \pm 1$, from equation (36), is

$$\bar{D} \mp \bar{T} + 1 = 0. \tag{37}$$

Using equations (34) and (35) in (37) one obtains

$$e^{-4\pi\alpha/\omega} \mp 2 e^{-2\pi\alpha/\omega}[-\hat{\Omega}s_+s_- + c_+c_-] + 1 = 0, \tag{38}$$

where $\hat{\Omega} = \frac{1}{2}\{(\Omega_+^2 + \Omega_-^2)/\Omega_+\Omega_-\}$. This equation is quadratic in $e^{-2\pi\alpha/\omega}$ and, in order for α and ω to be real and positive, it must have real roots between 0 and 1. This gives a necessary condition for a ($\lambda = \pm 1$) bifurcation to occur:

$$\pm(c_+c_- - \hat{\Omega}s_+s_-) > 1. \tag{39}$$

Up to this point no assumptions or approximations have been made.

Equation (38) for the bifurcation condition can now be examined. Note that the equation depends on the system parameters α, ω, and $\tilde{\omega}$ while it is apparently independent of x_0 and β. However, it is also important to note that equation (38) contains the terms $s_\pm$ and $c_\pm$ which depend on the times of flight $(t_1 - t_0)$ and $(t_2 - t_1)$, and hence implicitly upon x_0 and β as well as the other parameters. This presents a difficulty, since the times of flight are not known (although $(t_2 - t_0) = 2\pi/\omega$ is known). The values of t_0, t_1, and t_2 are determined by the system parameters, but through the unknown functions f and g (equation (20)).

We now make an important assumption regarding these times of flight based on observations of numerical simulations of the system. This assumption appears valid only *for x_0 small compared with the maximum amplitude of the period one orbit*. It was observed that the times of flight for a period one orbit are distributed approximately as in the unforced problem. (Thus, the orbit spends less time on the side of x_0 with the greater stiffness than it does on the side with smaller stiffness.) We *assume*, based on these observations, that for x_0 small,

$$(t_1 - t_0) = \pi\Omega_n/\omega\Omega_+ \quad \text{and} \quad (t_2 - t_1) = \pi\Omega_n/\omega\Omega_-. \tag{40}$$

Using equations (40) one obtains

$$\Omega_+(t_1 - t_0) = \Omega_-(t_2 - t_1) = \pi\Omega_n/\omega \quad \text{and}$$

$$s_+ = s_- = \sin(\pi\Omega_n/\omega) \stackrel{\text{def}}{=} s, \qquad c_+ = c_- = \cos(\pi\Omega_n/\omega) \stackrel{\text{def}}{=} c. \tag{41}$$

Using equations (41) in equation (38) then gives

$$e^{-4\pi\alpha/\omega} \mp 2 e^{-2\pi\alpha/\omega}[-\hat{\Omega}s^2 + c^2] + 1 = 0, \tag{42}$$

an expression involving only α, ω, and $\tilde{\omega}$. Also, condition (39) becomes

$$\pm(c^2 - \hat{\Omega}s^2) > 1. \tag{43}$$

One therefore sees immediately that, since $0 \leq c^2 \leq 1$ and $\hat{\Omega}s^2 \geq 0$, only the $\lambda = -1$ bifurcation is possible (under our assumption regarding the times of flight). This is a *flip bifurcation* [21], in which a period two orbit branches out from the bifurcation point, generally in one of two ways. Figure 5(a) shows a *supercritical* flip bifurcation in which a stable period two orbit appears and the period one orbit becomes unstable. Figure 5(b) shows a *subcritical* flip bifurcation in which an unstable period two orbit merges with the stable period one orbit and an unstable period one orbit remains. Higher order terms must be computed to determine which type of bifurcation occurs.

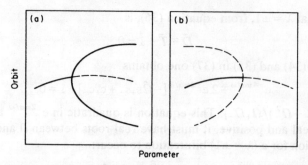

Figure 5. A (a) supercritical and (b) subcritical flip bifurcation.

The bifurcation condition (42) for the $\lambda = -1$ case can be solved for $e^{-2\pi\alpha/\omega}$ and then for α/ω to obtain

$$\alpha/\omega = -(1/2\pi) \ln \left[(\hat{\Omega} s^2 - c^2) - \sqrt{(\hat{\Omega} s^2 - c^2)^2 - 1} \right]. \quad (44)$$

We have taken the "minus" root of the quadratic since it is the one that gives $\alpha, \omega > 0$. Note that α and ω appear in both sides of this equation. It may be solved numerically by choosing $\tilde{\omega}$ and ω and solving for the corresponding bifurcation value of α by a simple root finding method. This was done for several values of $\tilde{\omega}$ over a range of ω and the results are shown in Figure 6 as curves in $(\alpha, \omega/\omega_n)$ space for several values of

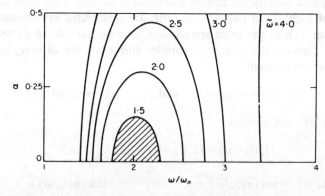

Figure 6. $\lambda = -1$ bifurcation curves for finite stiffness ratio, shown in $(\alpha, \omega/\omega_n)$ space for values of $\tilde{\omega}$.

$\tilde{\omega}$. For $\tilde{\omega} = 1 \cdot 5$ the region in which the period one orbit is unstable is shaded. Note that as $\tilde{\omega} \to 1$, these curves collapse around the point $\alpha = 0$, $\omega/\omega_n = 2$. This agrees with the known result that $\lambda = -1$ for the linear oscillator only at that point. In order to determine the type of flip bifurcation which occurs as a bifurcation curve is crossed, we now consider a special case.

Making the assumption that the forcing frequency ω is almost twice the value of the natural frequency Ω_n, i.e., $\omega = 2\Omega_n + \varepsilon$, $|\varepsilon| \ll 1$, one obtains

$$s = \sin(\pi \Omega_n/\omega) = 1 + O(\varepsilon^2) \quad \text{and} \quad c = \cos(\pi \Omega_n/\omega) = O(\varepsilon). \quad (45)$$

This, along with the small damping assumption $\alpha \ll 1$, reduces equation (44) to the simple expression

$$\alpha/\omega = |(1/2\pi) \ln(\tilde{\omega})| + O(\alpha^2, \varepsilon^2). \quad (46)$$

A computation using center manifold methods, outlined in the Appendix, shows that this bifurcation is supercritical. Therefore a stable period two orbit exists just below the bifurcation curves shown in Figure 6.

In carrying out this analysis, we have assumed that a period one orbit exists. This follows from a simple continuation argument. When $\tilde{\omega}=1$, one has a linear oscillator and a unique (harmonic) $2\pi/\omega$ periodic solution exists for $\omega \neq 1$. Moreover, if $\alpha > 0$, this orbit is stable and will therefore continue to exist for $\tilde{\omega}$ close to one, since the map **P** has a stable period one point for $\tilde{\omega}=1$, at which $(\mathbf{Id}-\mathbf{DP})$ is invertible. Use of the implicit function theorem [23] then implies that **P** continues to have a period one point for $\tilde{\omega}$ near one. The only way in which this orbit can cease to exist, as $\tilde{\omega}$ increases, is by coalescence with another (unstable) period one motion [21] in which case an eigenvalue of **DP** evalued at the orbit would reach +1. But we have shown that (under our time of flight assumption) this cannot occur. Therefore a period one orbit exists for all $\tilde{\omega}$. The numerical results shown below, and analysis of the $\tilde{\omega} \to \infty$ limit, bear this out.

The times of flight assumption (40) is approximately true only for x_0 small compared to the maximum amplitude of the period one orbit. For $x_0 \neq 0$, as ω is increased, this orbit will shrink and eventually become simply the steady state solution of equation (2). The value of ω at which this occurs can be determined by setting x_0 equal to the maximum amplitude of that steady solution and solving the resulting equation for ω.

For large damping, $\alpha > \tilde{\omega} > 1$, one must use the overdamped solutions of equations (2) and (3). A calculation of **DP** in that case proceeds just as in the undamped ($\alpha < 1 < \tilde{\omega}$) case. When the damping becomes very large, $\alpha \gg \tilde{\omega}$, and x_0 is very small ($x_0 = O(1/\alpha^2)$), it can be shown that

$$\bar{T} = 1 - (\pi/2\alpha\omega)(1+\tilde{\omega}^2) + O(1/\alpha^2), \qquad (47)$$

while $\bar{D}$ is exponentially small. Thus the eigenvalues are, from equation (36),

$$\lambda_1 = 0 \quad \text{and} \quad \lambda_2 = \bar{T}, \qquad (48)$$

up to exponentially small terms. One concludes that *no bifurcations occur* as α is increased to large values. Here we have assumed that the times of flight are equal and are both π/ω. This is so since one expects the solution leaving x_0 to be very close to the steady state solution of the appropriate equation (2) or (3). These steady solutions are of size $O(1/\alpha)$ and both are approximately 90° out of phase with the force. Thus their phases match well at $x = x_0$. Their velocities also match well (to $O(1/\alpha^2)$), due to the large damping. Since one expects the period one orbit to lie near these solutions on either side of $x = x_0$, the modified time of flight assumption seems reasonable.

4.2. GENERAL CASE, DIGITAL SIMULATION

The matching of solutions (4) and (5) described previously can be done easily on a computer. Initial conditions (t_0, y_0) are set in Σ and the equation $x_+(t; t_0, y_0)$ then determines the motion until $x(t)$ reaches x_0 again. The computer then solves for the x_0 crossing time, t_1, using a simple Newton–Raphson method [24] on equation (18); t_1 is used to compute the velocity from equation (19). The new time and velocity are used as initial conditions in $x_-(t; t_1, y_1)$, which gives the motion exactly until $x = x_0$ once more. The procedure is repeated at length to obtain a solution of equation (1). From this global time solution one easily obtains the iterates of the mapping **P** by recording the values of (t, y) at each x_0 crossing for which $y > 0$. Note that this solution is considerably more accurate than the usual numerical solutions of ordinary differential equations [24], the only approximations being made at the x_0 crossing points, which can be easily computed to high precision. The analytical results from the previous section are verified by using this simulation.

First one can verify equation (46), the simplified bifurcation condition at $\omega \simeq 2\Omega_n$ and $\alpha \ll 1$. Taking values of $\tilde{\omega}$ of 1·2, 1·4, 1·6 and 1·8 one sets $\omega = 4\omega/(1+\tilde{\omega}) = 2\omega_n = 2\Omega_n + O(\alpha^2)$. The damping is then varied in small increments and the steady state solution observed. In this way one finds the actual bifurcation value to moderate precision. These values are then checked with those computed from equation (46). Table 1 and Figure 7 show good agreement between the predicted and the actual bifurcation points. The times of flight are all found to be within 5% of values assumed on the basis of equation (40) with $\alpha \ll 1$.

TABLE 1

Bifurcation values

$\tilde{\omega}$	$\omega = 2\omega_n$	α_{cr} from equation (46)	α_{cr} from digital simulation	$\bar{t}, \bar{y}$ (all are ±0·003)	a
1·2	2·182	0·063	0·066 ± 0·002	0·738, 0·605	0·07
1·4	2·333	0·125	0·128 ± 0·002	0·708, 0·554	0·16
1·6	2·462	0·184	0·190 ± 0·002	0·680, 0·512	0·26
1·8	2·571	0·241	0·250 ± 0·002	0·660, 0·475	0·37

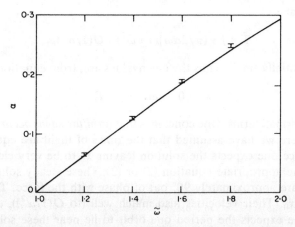

Figure 7. Bifurcation values of α vs. $\tilde{\omega}$ for $x_0 = 0$, $\beta = 1$ and $\omega = 2\omega_n$. Solid curve is theory (equation (46)). Error bars ⊺ from digital simulation at $\tilde{\omega} = 1\cdot 2$, 1·4, 1·6 and 1·8.

As α is lowered below the period doubling value, for parameter values $\tilde{\omega} < 1\cdot 5$, $x_0 = 0$, $\omega \simeq 2\Omega_n$, one finds no further bifurcations. This is in contrast with the period doubling cascades found in many non-linear systems. When $\tilde{\omega}$ becomes large, however, other bifurcations can occur. The limit $\tilde{\omega} \to \infty$ will subsequently be studied in detail.

We now present an example using the more general bifurcation condition (44). Setting $\tilde{\omega} = 4$ (a stiffness ratio of 16) and $\alpha = 0\cdot 125$, we solved equation (44) numerically to determine $\omega_{\text{bif}} \cong 2\cdot 26$. Figures 8(a) and (b) show that the bifurcation actually occurs between $\omega = 2\cdot 40$ and $\omega = 2\cdot 42$. The error is due to the fact that the actual times of flight for the period one orbit near the bifurcation value differ from the assumed values by 6%.

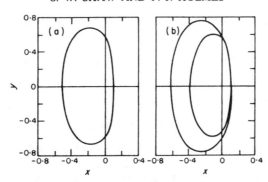

Figure 8. (a) Stable period one orbit at $\omega = 2\cdot40$, $\alpha = 0\cdot125$, $x_0 = 0$, $\beta = 1$, and $\tilde{\omega} = 4$, from digital simulation; (b) stable period two orbit at $\omega = 2\cdot42$, $\alpha = 0\cdot125$, $\beta = 1$, $x_0 = 0$, and $\tilde{\omega} = 4$, from digital simulation. Both in projected phase plane (x, y).

By using digital simulations, other subharmonic orbits are also found to exist. Figures 9(a) and (b) show the coexistence of single impact stable period one and stable period three orbits for $x_0 = 0$, $\alpha = 0\cdot026$, $\tilde{\omega} = \sqrt{2}$ and $\omega = 3\cdot5$. We conjecture that the period three orbit appeared in a saddle-node bifurcation [21] and that an unstable period three orbit also exists at these parameter values. Analysis of this bifurcation is much more difficult since no reasonable assumption regarding the times of flight can be made and higher iterates of the mapping (multiple impacts) appear to be involved.

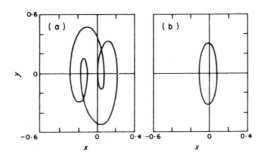

Figure 9. Coexistence of stable period three and stable period one at $\tilde{\omega} = \sqrt{2}$, $\alpha = 0\cdot026$, $\beta = 1$, $x_0 = 0$ and $\omega = 3\cdot5$. (a) Period 3; (b) period 1. Both in projected phase plane (x, y).

4.3. THE IMPACT LIMIT, ANALYSIS

In the impact limit an important simplification occurs, since the time of flight during the impact is taken to be zero. This allows more analysis to be done on periodic orbits. We also rescale: $x \to \beta x$, and take unit forcing amplitude.

The return map for the impact oscillator is very similar to the one for the general system. From points 0 to 1 in Figure 4 one uses the impact limit, i.e.,

$$t_1 = t_0, \quad \text{and} \quad y_1 = -ry_0. \tag{49}$$

From points 1 to 2 one uses the same mapping as described in the general case. Thus the mapping **P** still cannot be written down explicitly. As before, however, one can compute **DP** analytically. Moreover, in this limiting case one can compute periodic points corresponding to single impact, period n orbits directly. Such orbits correspond to those motions which strike the wall (the very stiff spring k_2) and then remain in $x < x_0$ for exactly $2\pi n/\omega$ in time and then strike the wall again with the same velocity as the previous impact. The conditions for the existence of such an orbit are

$$t_2 - t_0 = 2\pi n/\omega, \qquad y_2 = y_0 = -y_1/r, \quad \text{with} \quad y_0 > 0. \qquad (50, 51)$$

These two conditions allow one to compute the period n point $(\bar{t}, \bar{y})$. First one writes equation (7), using equations (5), (50) and (51), to obtain

$$0 = -x_0\Gamma_- + \Lambda y_1 + s_1(\delta\Gamma_- + \Lambda\gamma\omega) + c_1(\gamma\Gamma_- - \Lambda\delta\omega), \qquad (52)$$

where we have dropped the minus subscripts on γ, δ, and Ω and where $\Gamma_\pm = 1 - Ec \pm \alpha\Lambda$, $\Lambda = Es/\Omega$, $E = e^{-2\pi n\alpha/\omega}$, $s = \sin(2\pi n\Omega/\omega)$, and $c = \cos(2\pi n\Omega/\omega)$. Next one writes equation (51) using equation (50) and the time derivative of equation (5) to obtain

$$0 = y_1(-1 - r + r\Gamma_+) + \Lambda r x_0 + s_1 r(\gamma\omega\Gamma_+ - \Lambda\delta) - c_1 r(\delta\omega\Gamma_+ + \Lambda\gamma). \qquad (53)$$

Since y_1 appears in a linear manner in both equations (52) and (53), it may be eliminated to obtain a single equation involving only t_1 as an unknown (in the terms $c_1 = \cos(\omega t_1)$ and $s_1 = \sin(\omega t_1)$):

$$0 = (x_0/\Lambda)[r\Lambda^2 - \Gamma_-\psi] + s_1[r\gamma\omega\Gamma_+ - \Lambda\delta r + (\psi/\Lambda)(\delta\Gamma_- + \Lambda\gamma\omega)]$$
$$+ c_1[-r\delta\omega\Gamma_+ - \Lambda\gamma r + (\psi/\Lambda)(\gamma\Gamma_- - \Lambda\delta\omega)], \qquad (54)$$

where $\psi = 1 + r - r\Gamma_+$. Straightforward association of terms allows this equation to be written as

$$0 = X + s_1 Y + c_1 Z, \qquad (55)$$

which has a solution

$$\bar{t}_1 = (1/\omega)[\arctan(Y/Z) + \arccos(-X/W)], \qquad (56)$$

where $W = \sqrt{Y^2 + Z^2}$. This expression gives the time (i.e., forcing phase) at impact on the period n orbit. The velocity just after impact $\bar{y}_1$ is then easily computed by using either equation (52) or equation (53).

It is important to note that a solution obtained as described above only satisfies $x(t_1) = x_0$ and $\dot{x}(t_1) = -rx(t_2)$. If the value of $\bar{y}_1 = \dot{x}(t_1)$ is positive, then the solution corresponds to a non-physical, or "penetrating" orbit [13]. The orbits for $\bar{y}_1$ negative must also be checked since nowhere has one been assured that the desired x_0 crossing is the first on the orbit. In fact, the above conditions can be satisfied after several x_0 crossings, for some parameter values. Care must be taken to determine which of these orbits are physically possible.

Knowing the periodic point, one can now compute its stability. As before, one breaks the calculation of **DP** into two parts, from point 0 to 1 and from point 1 to point 2. Here

$$\left[\frac{\partial(t_1, y_1)}{\partial(t_0, y_0)}\right] = \begin{bmatrix} 1 & 0 \\ 0 & -r \end{bmatrix}, \quad \text{and} \quad \mathbf{DP} = \left[\frac{\partial(t_2, y_2)}{\partial(t_1, y_1)}\right]\begin{bmatrix} 1 & 0 \\ 0 & -r \end{bmatrix}, \qquad (57, 58)$$

where $[\partial(t_2, y_2)/\partial(t_1, y_1)]$ was derived above in the analysis of the finite stiffness case. From this calculation one finds that **DP** has determinant

$$D = (-ry_1/y_2) e^{-2\alpha(t_2 - t_0)}, \qquad (59)$$

and trace

$$T = [e^{-\alpha(t_2 - t_0)}/\Omega y_2][(N_1 + rN_2)\sin(\Omega(t_2 - t_0)) + \Omega(y_1 - ry_2)\cos(\Omega(t_2 - t_0))]. \qquad (60)$$

Evaluating on a period n orbit gives

$$\bar{D} = r^2 E^2 \quad \text{and} \quad \bar{T} = (E/\Omega\bar{y}_0)[(1+r)(\bar{c}_0 + x_0)s - 2r\bar{y}_0\Omega c], \qquad (61, 62)$$

where $\bar{y}_0 = -\bar{y}_1/r$ and $\bar{c}_0 = \cos(\omega \bar{t}_0) = \cos(\omega \bar{t}_1)$. Using equation (36) one can determine the eigenvalues and thus the stability of the period n point. The above analysis is quite general.

In order to study this impact system in more detail we shall consider a special case in which the equations simplify significantly: we shall take $\alpha = 0$ and $x_0 = 0$ in what follows. This case corresponds to the wall being at the origin and there being no damping in the $x < 0$ motion. Note that the coefficient of restitution continues to provide an energy loss mechanism.

Proceeding exactly as in the general case, one obtains a simplified version of equation (56):

$$\bar{t}_1 = (1/\omega) \arctan[+(1-r)(1-c)/\{-(1+r)\omega s\}]. \tag{63}$$

Solving for $\bar{y}_1$ yields

$$\bar{y}_1 = 2\omega r(1-c)/(1-\omega^2)\{[(1-c)(1-r)]^2 + [\omega s(1+r)]^2\}^{1/2}. \tag{64}$$

Note that requiring $\bar{y}_1 < 0$ implies that $\omega > 1$; thus these period n single impact orbits do not exist for forcing frequencies below $\omega = 1$. The determinant and trace on this orbit simplify to

$$\bar{D} = r^2 \quad \text{and} \quad \bar{T} = [s^2(1+r)^2(1-\omega^2) - 4rc(1-c)]/2(1-c), \tag{65}$$

respectively. It should be noted that in this case $\Omega = 1$, $s = \sin(2\pi n/\omega)$, and $c = \cos(2\pi n/\omega)$.

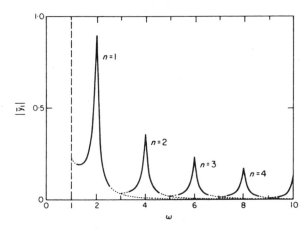

Figure 10. Resonance curve $|\bar{y}_1|$ vs. ω for simple impact period n orbits. Solid portions correspond to stable orbits, dotted portions correspond to unstable orbits. Parameter values are $\alpha = 0$, $x_0 = 0$, $r = 0.8$.

Figure 10 shows a typical resonance curve for the impact oscillator with $r = 0.8$. The graph shows the magnitude of the velocity just after impact on period n orbits versus forcing frequency ω. The solid part of the curves represent stable orbits while the dashed parts represent unstable ones. Penetrating orbits were eliminated by checking enough points on the orbit to be sure that $t = t_1 + 2\pi n/\omega$ was the first $x = 0$ crossing time. These curves, produced analytically, show much similarity to those produced numerically by Thompson [15, 16]. However, he also found multiple impact periodic orbits.

As before, $\bar{D} < 1$, so that only $\lambda = \pm 1$ bifurcations occur. Upon using equations (37) and (65) the $\lambda = \pm 1$ bifurcation condition, written as a quadratic in r, becomes

$$r^2 \mp 2r\left(\frac{s^2(1-\omega^2) - 2c(1-c)}{s^2(1-\omega^2) + 2(1-c)}\right) + 1 = 0. \tag{66}$$

The requirement that this equation have a root between 0 and 1 shows that only the $\lambda = -1$ case can occur. (This equation should be compared with equation (42) in the finite stiffness case.) Here r does not appear in the coefficients of the quadratic and the equation may be solved exactly for the bifurcation curves. The parameters in equation (66) are the forcing frequency ω, the period of the orbit n, and, of course, the coefficient of restitution r. For each n one can vary ω and record those roots of the quadratic which fall between 0 and 1. In this way one constructs, for each n, a bifurcation curve in r–ω space. These curves are shown in Figure 11. A calculation, which is outlined in the Appendix, shows that the bifurcations are supercritical. Therefore one expects a stable, two impact, orbit of period $2n$ to appear just outside both edges of each stable period n region in Figure 11.

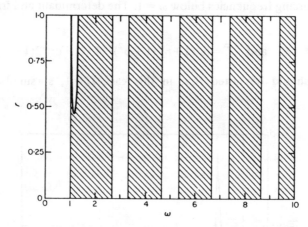

Figure 11. $\lambda = -1$ bifurcation curves of single impact orbits for $x_0 = 0$, $\beta = 1$, and $\alpha = 0$ in r–ω space. Shaded regions are stable.

The $n = 2, 3, 4, \ldots$ regions of stable impact orbits appear to be nearly identical. They are approximately vertical strips centered at $\omega = 2n$. However, although the right side of the $n = 1$ region is similar to those for $n > 1$, the left side is much different. There exists a small region above $r \approx 0.45$ and just to the right of $\omega = 1$ in which the period one single impact orbit is unstable. At certain parameter values in this region there coexist stable *two* impact period two orbits (resulting from the flip bifurcation) and stable *three* impact period two orbits, both found by using the digital simulation explained below. Note that as $\omega = 1$ is approached from the right, the period one orbit is stable for all values of $r < 1$ (the unstable regions meet at $\omega = 1$, $r = 1$ and for r near 1 the stable region becomes very thin). As ω passes through 1 into $\omega < 1$, period one orbits become no longer physically possible.

Other multiple impact periodic orbits also exist in this system [15]. However, the analysis of their existence, stability, and the bifurcations from them requires numerical or approximate techniques, since the times of flight between each impact are not known explicitly. We now describe such a digital simulation for this system.

4.4. THE IMPACT LIMIT, DIGITAL SIMULATION

The digital computer simulation of the impact oscillator is very similar to that for the finite stiffness case. One composes the linear solution (5) for $x < x_0$ with the impact relationship (49) in a straightforward manner. In this case each iterate of **P** requires the solution of one (instead of the previous two) transcendental equations, for the time of impact with the wall. This digital simulation is the means by which we obtained the following results for the impact oscillator, including those in the following section, in which chaotic motions were observed.

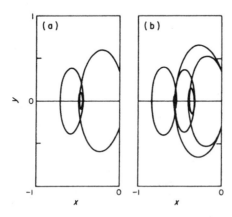

Figure 12. Period doubling of impact oscillator; $r = 0.8$, $x_0 = 0$, $\beta = 1$, $\alpha = 0$ and $n = 3$. (a) Stable period three orbit at $\omega = 5.37$; (b) stable period six orbit at $\omega = 5.35$. Both in projected phase plane (x, y).

Here we present two typical periodic motions of this system. Figure 12(a) shows a stable period 3 orbit of single impact type found by the above analysis and generated by using the digital simulation. The parameter values are $r = 0.8$ and $\omega = 6.63$ ($x_0 = 0$ along with $\alpha = 0$). These values near the bifurcation point $r = 0.8$, $\omega \cong 6.635$ determined from equation (66). Figure 12(b) shows a two impact period six orbit at $r = 0.8$, $\omega = 6.65$. This orbit appeared at the flip bifurcation. The period three orbit is of course still present, but is now unstable.

We next turn our attention to further bifurcations which cannot be studied analytically but require the use of the digital simulation.

5. FURTHER PERIOD DOUBLINGS, HORSESHOES AND STRANGE ATTRACTORS

5.1. CASCADES OF PERIOD DOUBLINGS AND OTHER TRANSITIONS

So far we have been able to find analytically single impact, period n motions and the bifurcations from them to period $2n$ double impact motions. The regions in Figure 11 which have no stable single impact orbits will now be examined in greater detail. The following results should be compared with those of Thompson [15, 16].

In Figure 10 (with $r = 0.8$) the region between the stable $n = 1$ orbit and the stable $n = 2$ orbit lies in the range $\omega \cong 2.6533$ to $\omega \cong 3.3535$. Figure 13 shows the results of digital simulations in that range. The maximum displacement between each impact is plotted against ω. One sees that the period one orbit undergoes further period doublings as ω is increased. This cascade of period doublings is typical of many non-linear systems. Approximate bifurcation values of ω were recorded and checked against the general

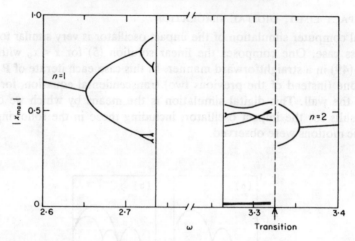

Figure 13. Bifurcation diagram between the stable $n = 1$ and $n = 2$ regions. Note period doubling sequence on the left and the transition and period doublings on the right. $\alpha = 0$, $\beta = 1$, $x_0 = 0$, $r = 0.8$, $\omega = 2.65$ to 3.36. Plot is of maximum x values in between impacts, x_{MAX} vs. ω.

asymptotic result due to Feigenbaum [17, 18],

$$F_\infty = \lim_{j \to \infty} \frac{\omega_{j+2} - \omega_{j+1}}{\omega_{j+1} - \omega_j} \stackrel{\text{def}}{=} \lim_{j \to \infty} F_j = 4.669\ldots, \qquad (67)$$

which one expects to hold for period doubling cascades in contracting ($\bar{D} < 1$) maps of this type [17]. Here ω_j is the bifurcation value from a period 2^j orbit to a period 2^{j+1} orbit.

Although the first such number produced, $F_1 = 4.76 \pm 0.05$, is remarkably close to the expected limit, the succeeding two numbers appear to diverge and further numbers cannot be computed with sufficient accuracy (see Table 2). Nonetheless, there is clear digital evidence for an accumulation value, and orbits up to period 32 have been observed.

TABLE 2

Bifurcation values of ω for period $2^j \to 2^{j+1}$, and resulting Feigenbaum numbers F_j [17, 18]

j	ω_j	F_j
0	2·65335†	
1	2·7158 ± 0·0001	
2	2·72890 ± 0·00003	4·76 ± 0·05
3	2·73131 ± 0·00003	5·4 ± 0·2
4	2·73175 ± 0·00003	5·0 ± 1·0

† Known from equation (63).

Figure 14 shows some of the orbits in the period doubling sequence. Figure 14(a) shows a simple period one orbit in the projected (x, y) phase space at $\omega = 2.64$. Figure 14(b) shows the same orbit as a fixed point in Σ (note that we have taken the variable t modulo $2\pi/\omega$). Figure 14(c) shows the period 2 orbit at $\omega = 2.69$, and Figure 14(d) shows the orbit in Σ. The period doublings occur very rapidly with increase of ω as shown, by the period four orbit of Figures 14(e) and (f) at $\omega = 2.725$ and the period eight orbits shown in Figures 14(g) and (h) at $\omega = 2.731$.

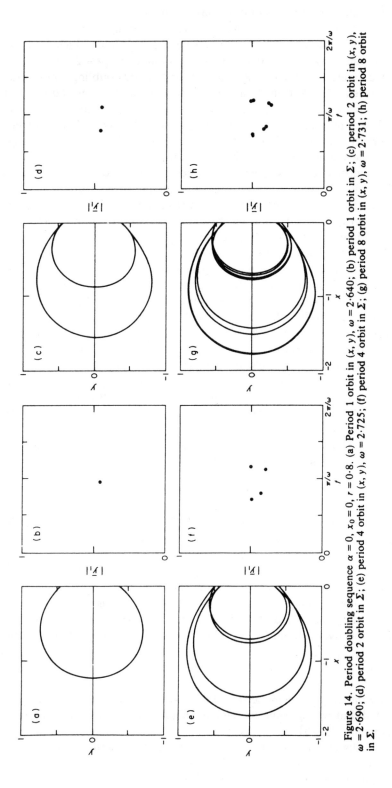

Figure 14. Period doubling sequence $\alpha = 0$, $x_0 = 0$, $r = 0.8$. (a) Period 1 orbit in (x, y), $\omega = 2.640$; (b) period 1 orbit in Σ; (c) period 2 orbit in (x, y), $\omega = 2.690$; (d) period 2 orbit in Σ; (e) period 4 orbit in (x, y), $\omega = 2.725$; (f) period 4 orbit in Σ; (g) period 8 orbit in (x, y), $\omega = 2.731$; (h) period 8 orbit in Σ.

Equation (67) implies that the bifurcation values accumulate in a geometric series and hence at some finite value, ω_{ACC}, there is an orbit of "period 2^∞": i.e., a non-periodic orbit [17]. Based on the measured period doubling values of ω one can predict an approximate value for the accumulation point of $\omega_{ACC} \cong 2 \cdot 732$.

However, the $n = 2$ solution, after its first period doubling for *decreasing* ω, can undergo a different type of transition. As ω is lowered and the period four solution develops, a point on the orbit becomes tangent at the origin and a degenerate impact occurs corresponding to a singular point for the mapping $\mathbf{P}$. This will be discussed below. Thereafter, further bifurcations occur as shown in Figure 13 (including period doublings) until the solution appears to become non-periodic.

Thompson [16] has shown similar results for the case $r = 1$, $\alpha \neq 0$, $x_0 = 0$. He studied the region between $n = 4$ and the $n = 5$ stable orbits using a similar digital simulation and found the same qualitative structure as shown in Figure 13. The "left" branch ($n = 1$ in the present case, $n = 4$ in Thompson's [16]) period doubles in a straightforward manner. The "right" branch ($n = 2$ here and Thompson's $n = 5$ [16]) doubles once to a double impact orbit. The orbit then passes through a singularity in the map and becomes a three impact orbit of the same period (period four ($T = 8\pi/\omega$) in this study); see Figure 13. Thereafter it appears to continue period doubling until non-periodic motions occur.

The transition from the period four double impact orbit to a period four triple impact orbit is *not* a bifurcation in the usual sense, as encountered in smooth maps [21]. As pointed out earlier, our map is not "onto" and consequently has discontinuities associated with orbits which leave Σ and return at $x = y = 0$. The transition referred to above occurs precisely when the double impact orbit passes through $x = y = 0$. The nature of such transitions is determined by the particular map and its discontinuities and there is no general theory as in the case of "smooth" bifurcations [21]. However, clearly such transitions play an important role in the dynamics of piecewise linear systems and deserve further study.

The behavior in the unstable bands between $n = 2$ and $n = 3$, $n = 3, 4$, etc., seems essentially the same as that in the first such band.

5.2. THE EXISTENCE OF HORSESHOES

In this section we take $r = 1$ (the dissipationless case). Orbits are considered as iterates of the mapping $\mathbf{P}$ and are shown as sequences of points in Σ. As in Figure 14, we take $t \bmod 2\pi/\omega$, so that an n impact periodic orbit appears as a fixed point of the nth iterate of $\mathbf{P}$, or as a cycle of n periodic points of $\mathbf{P}$.

The location and stability of the period one point $(\bar{t}, \bar{y})$ is known exactly: within the range $\omega \simeq 2 \cdot 6533 - 3 \cdot 3535$ it is an unstable fixed point of saddle type for the mapping $\mathbf{P}$. We now examine the stable and unstable manifolds of $(\bar{t}, \bar{y})$.

The stable (W^s) and unstable (W^u) manifolds are defined as those sets of points which are respectively forward and backward asymptotic to $(\bar{t}, \bar{y})$ under iterates of $\mathbf{P}$:

$$W^s = \{\mathbf{x} \in \Sigma | \mathbf{P}^n(\mathbf{x}) \to (\bar{t}, \bar{y}) \text{ as } n \to +\infty\}, \quad W^u = \{\mathbf{x} \in \Sigma | \mathbf{P}^n(\mathbf{x}) \to (\bar{t}, \bar{y}) \text{ as } n \to -\infty\}. \tag{68}$$

Since the map is piecewise smooth, but discontinuous, these sets are themselves disconnected, but we continue to refer to them as manifolds, as in the usual case of smooth mappings [23, 25].

The digital simulations indicate that the stable and unstable manifolds intersect, in the dissipationless case, for all values of ω for which $(\bar{t}, \bar{y})$ is unstable. Therefore a complicated invariant set, a Smale horseshoe, exists immediately after the bifurcation, but perhaps in some high iterate of $\mathbf{P}$ [25, 26].

A simple introduction to the horseshoe and its implications for chaos in iterated maps such as the present one can be found in reference [14] or [23]. More complete mathematical treatments have been given by Moser [26] and Guckenheimer and Holmes [25]. The main conclusion one can draw from the presence of horseshoes is that the map possesses a complicated invariant set Λ which contains (a) a countable infinity of unstable periodic orbits, including orbits of arbitrarily long periods; (b) an uncountable infinity of bounded, non-periodic orbits, and (c) a dense orbit. The horseshoe acts as a "chaotic saddle point", since while most orbits approaching it eventually leave its neighborhood, they do so in a manner which is extremely difficult to predict. We shall return to the effects and physical implications of such invariant sets in the next section.

For certain values of ω above the accumulation value ω_{ACC}, we can clearly demonstrate a horseshoe for the second iterate of $\mathbf{P}$, $\mathbf{P}^2$. For example, Figure 15 shows a plot, generated by the digital simulation, of W^s and W^u for $\omega = 2 \cdot 8$. In this case $(r = 1)$ $\bar{t} = \pi/\omega$ and the reflectional symmetry about the line $\bar{t} = \pi/\omega$ should be noted. To see the horseshoe more clearly the area of interest is shown in Figure 16. The "primed" points

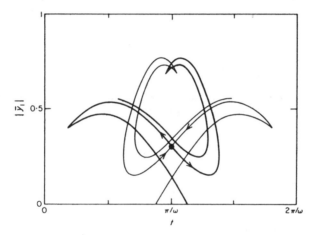

Figure 15. W^u (heavy line) and W^s (light line) in the dissipationless case $r = 1$, $\alpha = 0$, $x_0 = 0$, and $\omega = 2 \cdot 8$. Note transversal intersections.

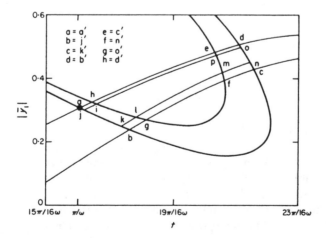

Figure 16. W^u (heavy line) and W^s (light line), as in Figure 15, showing region of interest and the horseshoe. Primed points are images of unprimed points under $\mathbf{P}^2$.

are the images of the unprimed points. The segments $\overline{oj}$ and $\overline{kn}$ are approximate preimages of segments $\overline{bg}$ and $\overline{fc}$. These segments were generated by approximating $\overline{bg}$ and $\overline{fc}$ by straight lines and iterating them twice under $\mathbf{P}^{-1}$: i.e., once by $\mathbf{P}^{-2}$. ($\mathbf{P}^{-1}$ is obtained in the digital simulation by simply running time backwards.) We define "horizontal" strips $H_1 = adoj$ and $H_2 = kncb$ along with "vertical" strips $V_1 = ahgb$ and $V_2 = edcf$. Then by considering the images of H_1 and H_2 under $\mathbf{P}^2$, we see that

$$\mathbf{P}^2(H_i) = V_i, \qquad i = 1, 2, \tag{69}$$

and thus we have a topological horseshoe [25].

The invariant set Λ of points which never leave the two strips H_i under forward or backward iteration of $\mathbf{P}^2$ is obtained by intersection of all the images

$$\Lambda = \bigcap_{n=-\infty}^{\infty} (\mathbf{P}^2)^n (H_1 \cup H_2). \tag{70}$$

It can be shown to be a Cantor set with the properties outlined earlier (cf. reference [14]). The hyperbolicity and persistence of the horseshoe will not be dealt with in this paper [25, 26]; cf. references [14] and [25].

5.3. CHAOTIC MOTIONS AND STRANGE ATTRACTING SETS

In this final section we present a digital simulation for the parameter values $x_0 = 0$, $\alpha = 0$, $r = 0.8$, and $\omega = 2.8$, for which an apparently persistent chaotic motion was observed. Similar motions appear to occur for large sets of parameter values in ω ranges for which the period n single impact orbits are unstable. In Figure 17(a) we show a

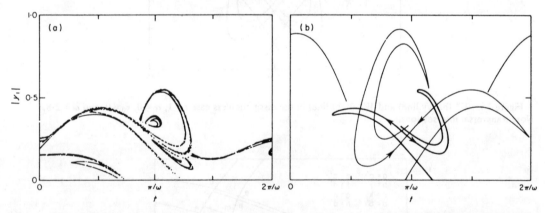

Figure 17. (a) Strange attracting set at $x_0 = 0.0$, $\alpha = 0$, $r = 0.8$ and $\omega = 2.8$; (b) W^u (heavy line) and W^s (light line) at the same parameter values.

segment of a typical orbit of $\mathbf{P}$ containing 4000 points. Note that the orbit appears to lie on a well defined (set of ?) curves. Plotting the stable and unstable manifold of the saddle point (the $n = 1$ orbit) in Figure 17(b) we see that this set of curves is indistinguishable from the unstable manifold. Similar observations have been made in many earlier papers (see references [14, 20, 25], for example). The conjecture that there is an attracting set equal to the closure of this unstable manifold is irresistible, but we have not been able to prove it in this case (it can be proved for the Poincaré map of the Duffing equation [27]). However, even if it is true, Newhouse's work [28] shows that the attracting set may contain large, even infinite, sets of stable periodic orbits, so that it is not truly chaotic, for in such a situation almost all orbits will be asymptotically periodic, albeit with very long periods.

It is easy to establish the existence of an attracting set for the map **P**. We take the region $R = \{t, y | 0 \leq y \leq L\}$ in the cylindrical ($2\pi/\omega$ periodic in t) phase space. If L is taken sufficiently large, then we can show that $\mathbf{P}(R) \subset R$, since a solution striking the wall $x = 0$ with velocity $y_0 > 0$ at time t_0 immediately thereafter leaves with velocity $-ry_0$ and next strikes the wall at the velocity

$$y_2 = \frac{1}{1-\omega^2} \cos \omega t_0 \sin (t_2 - t_0) + \left(\frac{\omega}{1-\omega^2} \sin \omega t_0 - ry_0\right) \cos (t_2 - t_0) - \frac{\omega}{1-\omega^2} \sin \omega t_2, \tag{71}$$

obtained from equation (5), with $\alpha = 0$, $\beta = 1$ and $t_1 = t_0$. Here $t_2 - t_0$ is the (unknown) time of flight. From equation (71),

$$y_2 \leq \frac{1+2\omega}{|1-\omega^2|} + ry_0 \tag{72}$$

and thus, provided $r < 1$, one has

$$y_2 < y_0 \tag{73}$$

if y_0 is sufficiently large. We define the (closed) attracting set Ω as the intersection of all forward images of R:

$$\Omega = \bigcap_{n=0}^{\infty} \mathbf{P}^n(R). \tag{74}$$

(Note that points on the lower boundary $y = 0$ of R may not be mapped into the interior, so Ω might contain points on this boundary.)

Now Ω might simply be a fixed point or a periodic orbit, but as shown in the previous section, it can also contain horseshoes. In this situation we refer to it as a *strange attracting set*. The orbit shown in Figure 17(a) is asymptotic to such a set, but as remarked above there is no guarantee that such an orbit might not be eventually periodic: the attracting set might contain stable high period orbits (it certainly does, for some parameter values, if the period doubling sequences accumulate as described).

In the case of one dimensional maps there is a fairly complete general theory and it has been shown that genuinely chaotic attracting sets, containing no stable periodic orbits, exist for large sets of such maps (cf. reference [17]). Unfortunately, very few results are available in the two dimensional case, and in general we have only digital evidence for the existence of strange attractors. The present example provides a little more evidence of this type.

We end with the remark that, if we let energy losses in the impact become large, then the coefficient of restitution r tends to zero and the map becomes one dimensional (since the velocity y returns to zero after every impact). Such a map is more amenable to analysis than the full two dimensional map, and we plan to study it in a subsequent paper.

6. CONCLUSIONS AND DISCUSSION OF RESULTS

In this paper we have studied a periodically forced single degree of freedom non-linear oscillator. The non-linearity is piecewise linear and thus explicit solutions are known on each side of the point of discontinuity in slope, x_0. This allows one to consider the motion as iterates of a map **P**, each iterate corresponding to the state crossing x_0 with velocity

$y \geq 0$. Although **P** cannot be written out explicitly, and in fact is not "onto", one can compute its derivative **DP**, almost everywhere, i.e., except at points of discontinuity.

In the case of finite stiffness ratio we have used **DP**, with a simple assumption (equation (40)) regarding the times of flight on each side of x_0, to show that stable, single impact, period one orbits can undergo supercritical flip bifurcations [21]. This results in the appearance of a stable, two impact, period two orbit. Digital simulations of the system verify the analysis and also demonstrate that other bifurcations occur. However, these do not seem amenable to direct analysis. Investigations using numerical techniques could be employed for further study of these bifurcations.

When the stiffness ratio becomes large, the system is referred to as an impact oscillator [15, 16]. Using a simple impact rule, we have analytically found single impact orbits of period n and investigated their stability. Again, these single impact orbits undergo supercritical flip bifurcations and stable, period two, double impact orbits appear. We then used a digital simulation to study further bifurcations of this system. Further period doublings were found to occur in an accumulating sequence [17, 18]. In the dissipationless case the existence of a complicated "chaotic" invariant set, a horseshoe, was demonstrated for the second iterate of the map $\mathbf{P}^2$ [25]. At a set of parameter values near those for which the horseshoe appears, a strange attracting set was observed. Orbits asymptotic to such a set are either non-periodic or of extremely long period.

The results obtained for the impact oscillator show qualitative agreement with digital simulations done by Thompson [15, 16]. Also, recent experiments by Robinson [9], using a circuit to model equations of this type, show many similar features. Especially interesting are the transitions Robinson found which seem to correspond to the singularities of our mapping **P**. Preliminary experimental results of our own, using a cantilever beam with a non-linear boundary condition (cf. reference [12]), verify that a physical system with finite stiffness ratio does in fact exhibit period doubling as well as subharmonics of order three. Equation (1) provides a model for such behavior.

This singular nature of the mapping **P** induced by solutions of equation (1) is of interest. Discontinuous maps arise in these piecewise linear systems as well as in other non-linear problems such as relaxation oscillations [1, 2]. Little is known of what occurs globally in such maps and further analysis should be done in this area.

REFERENCES

1. N. MINORSKY 1962 *Nonlinear Oscillations*. Huntington, New York: Krieger Publishing.
2. A. A. ANDRONOV, A. A. VITT and S. E. KHAIKEN 1966 *Theory of Oscillators*. Reading, Massachusetts: Addison Wesley.
3. S. MAEZAWA and S. FURUKAWA 1973 *Bulletin of the Japanese Society of Mechanical Engineering* **16**, 931–941. Superharmonic resonance in piecewise linear systems.
4. R. DRAGONI and A. REPACI 1979 *Mechanics Research Communications* **6**, 283–238. Influence of viscous–coulomb damping on a system with steps.
5. S. MAEZAWA 1961 *Bulletin of the Japanese Society of Mechanical Engineering* **4**, 201–229. Steady forced vibrations of unsymmetrical piecewise linear systems.
6. S. MAEZAWA, H. KUMANO and Y. MINAKUCHI 1980 *Bulletin of the Japanese Society of Mechanical Engineering* **23**, 68–75. Forced vibrations in an unsymmetric linear system excited by general periodic forcing functions.
7. K. KLOTTER 1953 in *Symposium on Nonlinear Circuit Analysis, Polytechnic Institute of Brooklyn*, April 1953, 234–257. Steady state vibrations in systems having arbitrary restoring and damping forces.
8. J. M. T. THOMPSON, A. R. BOKAIAN and R. GHAFFARI 1982 *Preprint, Department of Civil Engineering, University College, London*. Subharmonic resonance of a bilinear oscillator with applications to moored marine systems.

9. F. N. H. ROBINSON 1982 *Preprint, Clarendon Laboratory, Oxford, U.K.* An experimental study of the catastrophic behavior of a nonlinear differential equation.
10. T. WATANABE 1978 *Journal of Mechanical Design* **100**, 487–491. Forced vibrations of continuous system with nonlinear boundary condition.
11. H. G. DAVIES 1980 *Journal of Sound and Vibration* **68**, 479–487. Random vibration of a beam impacting stops.
12. F. C. MOON and S. W. SHAW 1983 (to appear) *International Journal of Nonlinear Mechanics.* Chaotic vibrations of a beam with nonlinear boundary condition.
13. M. SENATOR 1970 *Journal of the Acoustical Society of America* **47**, 1390–1397. Existence and stability of periodic motions of a harmonically forced impacting system.
14. P. HOLMES 1982 *Journal of Sound and Vibration* **84**, 173–189. The dynamics of repeated impact with a sinusoidally vibrating table.
15. J. M. T. THOMPSON 1982 *Preprint, Department of Civil Engineering, University College, London.* A strange attractor in the resonance of an impact oscillator.
16. J. M. T. THOMPSON and R. GHAFFARI 1982 *Physics Letters* **91A**, 5–8. Chaos after period doubling bifurcations in the resonance of an impact oscillator.
17. P. COLLET and J. P. ECKMANN 1980 *Iterated Maps on the Interval as Dynamical Systems, Progress in Physics, No. 1.* Boston, Massachusetts: Birkhauser.
18. M. FEIGENBAUM 1979 *Journal of Statistical Physics* **19**, 25–52. Quantitative universality for a class of nonlinear transformations.
19. J. P. MERIAM 1975 *Dynamics* New York: John Wiley and Sons, Inc.
20. P. J. HOLMES 1979 *Philosophical Transactions of the Royal Society* **A292**, 419–448. A nonlinear oscillator with a strange attractor.
21. J. GUCKENHEIMER 1980 in *Dynamical Systems, CIME Lectures Bressanone, Italy, June 1978, Progress in Mathematics No. 8,* 115–231. Boston, Massachusetts: Birkhauser.
22. C. S. HSU 1977 *Advances in Applied Mechanics* **17**, 245–301. On nonlinear parametric excitation problems.
23. D. R. J. CHILLINGWORTH 1976 *Differential Topology with a View to Applications.* London: Pitman Press.
24. G. DAHLQUIST and Å. BJORK 1974 *Numerical Methods* (translated by N. Anderson). London: Prentice–Hall.
25. J. GUCKENHEIMER and P. HOLMES 1983 *Nonlinear Oscillations, Dynamical Systems, and Bifurcations of Vector Fields.* New York: Springer Verlag.
26. J. MOSER 1973 *Stable and Random Motions in Dynamical Systems.* Princeton University Press.
27. P. J. HOLMES and D. C. WHITLEY 1983 in *Proceedings Year of Concentration in Partial Differential Equations and Dynamical Systems, University of Houston, Texas.* On the attracting set of Duffing's equation; I: analytical methods for small force and damping.
28. S. E. NEWHOUSE 1980 in *Dynamical Systems, CIME Lecture Bressanone, Italy, June 1978, Progress in Mathematics No. 8,* 1–114. Boston, Massachusetts: Birkhauser.
29. J. CARR 1981 *Applications of Centre Manifold Theory, Applied Mathematical Sciences No. 35.* New York: Springer-Verlag.

APPENDIX: NORMAL FORM AND CENTER MANIFOLD CALCULATIONS

We outline here the calculations which determine the types of flip bifurcation (sub- or super-critical) which occur in the finite stiffness and the impact problems.

Defining local variables,

$$\boldsymbol{\rho} = \begin{pmatrix} \xi \\ \eta \end{pmatrix} = \begin{pmatrix} t - \bar{t} \\ y - \bar{y} \end{pmatrix}, \tag{A1}$$

one can expand the mapping (20) near the period one point in a Taylor series,

$$\boldsymbol{\rho} = \begin{pmatrix} \xi \\ \eta \end{pmatrix} \to \begin{pmatrix} f_t & f_y \\ g_t & g_y \end{pmatrix} \begin{pmatrix} \xi \\ \eta \end{pmatrix} + \begin{pmatrix} \tfrac{1}{2} f_{tt} \xi^2 + f_{ty} \xi \eta + \tfrac{1}{2} f_{yy} \eta^2 + \cdots \\ \tfrac{1}{2} g_{tt} \xi^2 + g_{ty} \xi \eta + \tfrac{1}{2} g_{yy} \eta^2 + \cdots \end{pmatrix}, \tag{A2}$$

or

$$\rho \to \mathbf{DP}\rho + \begin{pmatrix} R(\rho) \\ S(\rho) \end{pmatrix},$$

where the derivatives are evaluated at the periodic point. A similarity transformation

$$\mathbf{z} = \begin{pmatrix} u \\ v \end{pmatrix} = \mathbf{B}^{-1}\rho \tag{A3}$$

puts equation (A2) into diagonalized (or normal) form,

$$\mathbf{z} \to \begin{pmatrix} \lambda_1 & 0 \\ 0 & \lambda_2 \end{pmatrix} \mathbf{z} + \begin{pmatrix} P(\mathbf{z}) \\ Q(\mathbf{z}) \end{pmatrix}, \tag{A4}$$

where $\lambda_{1,2}$ are the eigenvalues of $\mathbf{DP}$, $\mathbf{B}$ is the corresponding matrix of eigenvectors, and

$$\begin{pmatrix} P(\mathbf{z}) \\ Q(\mathbf{z}) \end{pmatrix} = \mathbf{B}^{-1} \begin{pmatrix} R(\mathbf{Bz}) \\ S(\mathbf{Bz}) \end{pmatrix}. \tag{A5}$$

At the bifurcation point $\lambda_1 = -\bar{D}$ and $\lambda_2 = -1$ (only flip bifurcations are considered here) so that

$$u \to -\bar{D}u + P(u,v) = -\bar{D}u + \tfrac{1}{2}P_{uu}u^2 + P_{uv}uv + \tfrac{1}{2}P_{vv}v^2 + \cdots, \tag{A6}$$

$$v \to -v + Q(u,v) = -v + \tfrac{1}{2}Q_{uu}u^2 + Q_{uv}vu + \tfrac{1}{2}Q_{vv}v^2 + \cdots. \tag{A7}$$

Since $|\bar{D}| < 1$ there is linear contraction (locally) in the u direction, but in the v direction behavior depends on non-linear terms. A local center manifold [29] given by

$$u = h(v) \tag{A8}$$

exists on which the local behaviour can be reduced, by substituting equation (A8) into equation (A7), to a one dimensional mapping, where w is the co-ordinate on $h(v)$:

$$w \to -w + Q(h(w), w). \tag{A9}$$

Equations (A6)–(A8) combine to give a functional equation for $h(v)$:

$$h(-v + Q(h(v), v)) + \bar{D}h(v) - P(h(v), v) = 0. \tag{A10}$$

Expanding $h(v)$, $P(u,v)$, and $Q(u,v)$ in Taylor series at the origin, one obtains, from equation (A10),

$$h(v) = \left(\frac{\tfrac{1}{2}P_{vv}}{(1+\bar{D})}\right) v^2 + O(|v|^3). \tag{A11}$$

Using equations (A9) and (A11) one then obtains

$$w \to -w + \tfrac{1}{2}Q_{vv}w^2 + [\tfrac{1}{6}Q_{vvv} + \tfrac{1}{2}Q_{uv}P_{vv}/(1+\bar{D})]w^3 + O(|w|^4), \tag{A12}$$

which governs the local behavior of the two dimensional map at the bifurcation point. The second iterate of equation (A12) is

$$w \to w - [\tfrac{1}{2}Q_{vv}^2 + \tfrac{1}{3}Q_{vvv} + Q_{uv}P_{vv}/(1+\bar{D})]w^3 + O(|w|^4), \tag{A13}$$

from which one sees that the stability of $w = 0$ at the bifurcation point (which corresponds to the periodic point $(\bar{l}, \bar{y})$) depends on the non-linear coefficient

$$a = \tfrac{1}{2}Q_{vv}^2 + \tfrac{1}{3}Q_{vvv} + Q_{uv}P_{vv}/(1+\bar{D}). \tag{A14}$$

For $a < 0$ $w = 0$ is weakly unstable and one has a subcritical flip bifurcation. For $a > 0$ $w = 0$ is weakly stable and one has a supercritical flip bifurcation [21]. Note from equation (A14) that it is sufficient to show that

$$\tfrac{1}{3}Q_{vvv} + Q_{uv}P_{vv}/(1+\bar{D}) > 0 \tag{A15}$$

for supercriticality.

The coefficient a depends on the terms Q_{vv}, Q_{uv}, P_{vv} and Q_{vvv}. The first three are linear combinations of the second derivatives of f and g and Q_{vvv} is a linear combination of their third derivatives. The expressions for them are determined from equation (A5) and are quite lengthy.

In order that these bifurcations be non-degenerate (cf. reference [25]) the eigenvalue passing through -1 must do so with non-zero speed as the parameters are varied. If the bifurcation curves for the finite stiffness case (Figure 6) and the impact oscillator (Figure 11) are crossed transversally as the parameters change, then this non-degeneracy condition is met.

We now outline the normal form and center manifold calculations and show the results for these two cases.

GENERAL SYSTEM (FINITE STIFFNESS)

Here we consider the case $x_0 = 0$, $\alpha \ll 1$, $\omega = 2\omega_n = 2\Omega_n + O(\alpha^2)$ for simplicity. We also use the time of flight assumption given by equation (40). The second and third derivatives of f and g can be computed analytically and evaluated to $O(\alpha^2)$. They are complicated expressions and are not given here. The transformation (A5) is greatly simplified here by using equation (45) to obtain, from equations (28)–(31) (at the bifurcation point and on the period one orbit),

$$\partial f/\partial t_0 = \partial t_2/\partial t_0 = f_t = -(\Omega_+/\Omega_-)\,\mathrm{e}^{-2\pi\alpha/\omega} + O(\alpha^2) = -1 + O(\alpha^2), \tag{A16}$$

$$\partial f/\partial y_0 = \partial t_2/\partial y_0 = f_y = O(\alpha^2), \tag{A17}$$

$$\partial g/\partial t_0 = \partial y_2/\partial t_0 = g_t = [(\Omega_-^2 - \Omega_+^2)/\Omega_+\Omega_-]\,\mathrm{e}^{-2\pi\alpha/\omega}\bar{N}_0 + O(\alpha^2), \tag{A18}$$

$$\partial g/\partial y_0 = \partial y_2/\partial y_0 = g_y = -(\Omega_-/\Omega_+)\,\mathrm{e}^{-2\pi\alpha/\omega} + O(\alpha^2) = -\bar{D} + O(\alpha^2) \tag{A19}$$

where $\bar{D} = \mathrm{e}^{-4\pi\alpha/\omega}$. One sees that t is an eigendirection (to be associated with v) and the similarity transformation $\mathbf{B}$ is given by

$$\mathbf{B} = \begin{bmatrix} 0 & 1 \\ 1-\bar{D} & (\bar{D}-1)/g_t \end{bmatrix}, \tag{A20}$$

to order α^2. The required terms in this case are

$$\tfrac{1}{2}Q_{vv} = \tfrac{1}{2}f_{tt} + nf_{ty} + (n^2/2)f_{yy}, \qquad Q_{uv} = mf_{ty} + mnf_{yy},$$

$$\tfrac{1}{6}Q_{vvv} = \tfrac{1}{6}f_{ttt} + (n/2)f_{tty} + (n^2/2)f_{tyy} + (n^3/6)f_{yyy},$$

$$\tfrac{1}{2}P_{vv} = -(n/2m)f_{tt} - (n^2/m)f_{ty} - (n^3/3m)f_{yy} + (1/2m)g_{tt} + (n/m)g_{ty} + (n^2/2m)g_{yy}, \tag{A21}$$

where $m = 1 - \bar{D}$ and $n = g_t/(1 + \bar{D})$ and all derivatives are evaluated on the periodic orbit. The coefficient a, given by equation (A14), was computed for a few values of $\tilde{\omega}$ for $\omega = 2\omega_n$ and α given by equation (46). It should be made clear that the calculation of a numerical value for a requires knowledge of the fixed point value $(\bar{t}, \bar{y})$. This can be found only by use of the digital simulation. Thus the analytical work described above must be combined with results from the digital simulation in order to compute a. Table

1 shows values of a to $O(\alpha^2)$ for some sample values of $\tilde{\omega}$. In all cases $a > 0$, and the bifurcations are consequently supercritical flip bifurcations as shown in Figure 5(a). This analysis verifies only a few discrete points on the curves of Figure 6 and in no way guarantees that all of the flip bifurcations in this problem are supercritical. However, no examples of subcritical bifurcations were observed in the digital simulations.

IMPACT OSCILLATOR

For the impact oscillator we restrict our attention to the case $x_0 = 0$, $\alpha = 0$. Here the coefficient a can be computed with no approximation and without use of the digital simulation. This is so because (1) one can find the bifurcation point exactly by using equation (66), (2) the fixed point is known exactly from equations (63) and (64), and (3) the times of flight are known exactly. The required terms here involve all of the second and third derivatives of f and g; no simplifications occur as in the finite stiffness case. The matrix **B** is

$$\mathbf{B} = \frac{1}{r^2 - 1} \begin{bmatrix} -1 - f_t & -f_y \\ f_t - r^2 & f_y \end{bmatrix} \tag{A22}$$

and the required terms are

$$\tfrac{1}{2} Q_{vv} = k[-(p/2)f_{tt} - pqf_{yt} - (pq^2/2)f_{yy} + \tfrac{1}{2}g_{tt} + qg_{yt} + (q^2/2)g_{yy}],$$

$$Q_{uv} = k[-pf_{tt} - p(p+q)f_{yt} - p^2qf_{yy} + g_{tt} + (p+q)g_{yt} + pqg_{yy}],$$

$$\tfrac{1}{6} Q_{vvv} = k[-(p/6)f_{ttt} - (pq/2)f_{tty} - (pq^2/2)f_{tyy} - (pq^3/6)f_{yyy}$$

$$+ \tfrac{1}{6}g_{ttt} + (q/2)g_{tty} + (q^2/2)g_{tyy} + (q^3/6)g_{yyy}],$$

$$\tfrac{1}{2} P_{vv} = k[(q/2)f_{tt} + q^2 f_{ty} + (q^3/2)f_{yy} - \tfrac{1}{2}g_{tt} - qg_{ty} - (q^2/2)g_{yy}], \tag{A23}$$

where $k = 1/f_y(r^2 - 1)$, $p = -(r^2 - f_t)/f_y$, $q = (-1 - f_t)/f_y$, and all derivatives are again evaluated on the periodic orbit. Sample calculations were performed and the results are shown in Figure A1. As in the finite stiffness case, the bifurcations are supercritical flips.

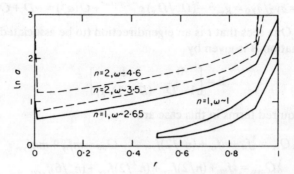

Figure A1. Plots of $\ln a$ vs. r, showing supercriticality for $n = 1$ and 2.

Proc. R. Soc. Lond. A **387**, 407–427 (1983)
Printed in Great Britain

Complex dynamics of compliant off-shore structures

By J. M. T. Thompson

*Department of Civil Engineering, University College London,
Gower Street, London WC1E 6BT*

(*Communicated by Sir Henry Chilver, F.R.S. – Received* 14 *December* 1982)

The dynamics of advanced compliant off-shore structures can be extremely complex owing to the inherent nonlinearities.

Subharmonic resonances can coexist with stable small-amplitude solutions, the response observed depending solely on the starting conditions of the motion. So care must be taken in digital, analogue and model studies to explore a comprehensive set of initial conditions. 'Efficient' automated digital computations could miss an entire subharmonic peak by locking onto a coexisting small-deflexion fundamental solution.

Chaotic, non-periodic motions of strange attractors can also arise in well defined deterministic resonance problems. The waveforms of these look like the result of a stochastic process, and because of their extreme sensitivity to initial conditions a statistical description must be sought. Genuinely chaotic solutions can be identified by looking for nearby period-doubling bifurcations, and for the exponential divergence of adjacent starts leading to a loss of correlation. The period-doubling cascades give rise to subharmonics of arbitrarily high order close to the chaotic régimes, so the duration of digital, analogue and laboratory experiments must be long and chosen with care.

The resonance of simple bilinear and impact oscillators is used as a vehicle to illustrate these general ideas.

1. Introduction

In the design of advanced off-shore production facilities, the oil industry has been forced to turn to *compliant* systems, which, like reeds in the wind, avoid unacceptably high stresses by deforming in a flexible manner under wave and current loading. For such structures, large working deflexions are then the rule, rather than the exception, and the dynamics are inherently nonlinear in nature. New and unexpected dynamic instability phenomena can thus arise, as with the potential Mathieu instabilities of tethered, buoyant (tension-leg) production platforms, predicted so spectacularly by Rainey (1978).

Two particular sources of high nonlinearity are heavy catenary cables, and light mooring lines. The former are used to stabilize certain deep-water production facilities. They have a highly nonlinear, but continuous, load–extension characteristic, which must be included in the static and dynamic modelling of these systems. Light mooring lines, on the other hand, can become slack during dynamic motions, giving rise to a discontinuity in stiffness that, as we shall see, must be viewed as a very severe form of nonlinearity.

An example of the latter, which motivated the present research, gave rise to unexpected subharmonic resonances in laboratory model tests on a common type of articulated mooring tower. Such a tower is basically an inverted pendulum, pinned to the sea bed, standing vertically under its own buoyancy. When a massive oil tanker is moored to the tower, the tanker can be regarded as a fixed object during wave-induced tower oscillations. The repeated slackening of the mooring line during large motions now gives a discontinuity in stiffness to the tower system. For positive deflexions towards the tanker, the restoring moment is due to buoyancy alone, while for negative deflexions away from the tanker, the restoring forces on the tower are due to the buoyancy plus the elasticity of the mooring line. This system can therefore be very adequately modelled as a *bilinear* oscillator that has simply a different linear stiffness for positive and negative deflexions.

An analogous bilinear problem, in which the stiffness ratio is very high, concerns the motion of vessels moored against stiff fenders in a harbour. Wave-induced motions towards the fenders are here resisted by their own high elastic stiffness, while motions away from the fenders can involve a loss of contact, leaving just the weak elastic restoring forces of the mooring lines. A valid assumption here (Lean 1971) is that the fenders are infinitely stiff compared with the mooring lines, giving us a limiting form of the bilinear oscillator that we shall call an *impact* oscillator. This oscillator just experiences an elastic rebound whenever the displacement tries to become negative. It is significant to our later demonstration of chaotic motions in the forced vibrations of the impact oscillator that very irregular motions of moored vessels have been reported, even in quite regular ocean waves (Kilner 1960).

We present here an overview of the results of a detailed study of the resonances of the bilinear and impact oscillators when they are driven by simple deterministic sinusoidal forcing. We find that very complex subharmonic resonances are generated, with a dangerous complicating feature that is never encountered in *linear* vibration studies, and is indeed quite uncommon even in nonlinear dynamics.

Specifically we find that beneath the subharmonic resonant peaks there is often a continuous coexisting stable fundamental solution of low amplitude, which can be picked up by appropriate starting conditions. A given system, with a given set of parameters, has then two (and often many more) coexisting stable periodic dynamic solutions, the one observed in a given time integration being dependent solely on the starting conditions of the motion. This is potentially dangerous for designers using a small number of one-off, *ad hoc* simulations because the chosen starting conditions employed might unluckily be those that generate the low amplitude fundamental solution: a dangerous subharmonic resonance could then be missed. This conclusion applies equally to *digital* time integrations, *analogue* simulations and *model* tests.

An entire subharmonic peak could also be quite easily missed by systematic or automated digital computations that use a found solution at a particular frequency ratio η to guide the search for a solution at an adjacent ratio $\eta + \Delta\eta$. Such 'efficient' computations often use the determined phase between the response and the forcing at η to guide the choice of starting conditions in the search for a stable steady solution at $\eta + \Delta\eta$. This type of efficient computing has an in-built tendency to follow an already determined sequence of solutions, and the danger we foresee is that an automated process of this type could step along one of the continuous small-amplitude fundamental *curves* that seem to lie beneath most of the resonant peaks.

These dangers of missing a subharmonic resonance can only be avoided by making simulations from a variety of starting conditions chosen in a systematic manner for the problem in hand. In this connexion, it should not be supposed that a large (or small) amplitude start will inevitably lead to a large (or small) response. The catchment areas of the competing steady-state stable solutions in the space of the starting conditions are extremely complex, and are as much concerned with phase as with amplitude.

The second important feature that we feel has general relevance in the analysis of compliant structures, arises in the resonance of the impact oscillator. For certain régimes of the frequency ratio η there are no steady-state stable periodic solutions at all, and a *chaotic* output is observed even for the deterministic sinusoidal forcing under consideration. *Chaos* of this type, associated with what topologists call a *strange attractor* in the phase space, is the most complex form of dynamical behaviour so far observed in deterministic dynamical systems. It is a topic of intense current research in mathematics and physics, and its detailed statistical properties are as yet only partially understood (Eckmann 1981; Ott 1981).

Arising as it does from a deterministic system, the response from a given start is of course *strictly* determinate, but it is not periodic, and after the decay of transients it looks like the result of a steady state stochastic process. The degree of determinism is moreover very small because the chaotic motions are extremely sensitive to initial conditions. Dynamic responses from two adjacent starts diverge exponentially with time, and very rapidly become uncorrelated. Because of this, only a statistical description of the motion is at all meaningful.

The onset of the chaotic η régime is typified by the appearance, under slowly varying η, of an infinite cascade of period-doubling bifurcations that lead in a finite η interval to an accumulation point exhibiting an infinity of unstable periodic solutions (Feigenbaum 1978; Thompson & Ghaffari 1982). There thus exist stable subharmonics of arbitrarily high order in the vicinity of the chaotic régime, so that the *duration* of simulations, be they digital, analogue or experimental, must be sufficiently long and chosen with care.

Much of the current research activity in chaotic dynamics has been reported over the last few years in *Physics Letters* (see for example Thompson & Ghaffari

1982), and it is becoming clear that practically all nonlinear oscillators exhibit at least some parameter régimes of chaos. So, as Rossler observed in 1979, chaos is really a quite *typical* form of behaviour for dynamical systems in a phase space of three or more dimensions. Notice here that a non-autonomous forced mechanical oscillator such as ours has a three-dimensional phase space spanned by displacement, velocity and time: typical trajectories in this space never cross, as they most certainly do in the phase projection spanned by just the displacement and its time derivative.

So in studying the forced motions of compliant nonlinear off-shore structures, designers and analysts should be aware that a chaotic response does not necessarily imply a bad piece of digital or analogue programming, or a failed experiment. They should learn to look for, and identify, the chaotic solutions of a strange attractor, and prepare to examine its statistical properties. The two most common tests that can be used to identify a genuine chaotic régime involve a search for period-doubling bifurcations, or an examination of the divergence from adjacent starts, and both of these will be illustrated later in this paper.

It is because they seem to raise important *general issues* such as these, that we shall now look in some detail at the behaviour of bilinear and impact oscillators, as being typical examples of driven compliant marine systems (Thompson & Bokaian 1983).

2. Dimensionless forms of the oscillators

A convenient dimensionless form for our two oscillators can be written as (Thompson *et al.* 1983)

$$\eta^2 \ddot{X} + 2\eta \zeta \dot{X} + \beta X = \sin \tau, \tag{1}$$

where
$$\beta = (1+\alpha^{\frac{1}{2}})^2/4\alpha \quad \text{for} \quad X > 0 \tag{2}$$

and
$$\beta = (1+\alpha^{\frac{1}{2}})^2/4 \quad \text{for} \quad X < 0. \tag{3}$$

Here X is the displacement, non-dimensionalized with respect to a convenient static deflexion; ζ is the damping ratio, defined in the conventional way with respect to an undamped *bilinear* frequency obtained by averaging the periodic times; and η is the ratio of the forcing frequency to this bilinear frequency. The ratio of the two linear stiffnesses is written as α, and finally τ is the time, scaled to make the apparent forcing frequency equal to unity. A dot denotes differentiation with respect to τ.

As the stiffness ratio α becomes infinite, we have in the limit the *impact* oscillator governed by equation (1) with $\beta = \frac{1}{4}$ for $X > 0$. For this oscillator the displacement X can never be negative, and the system is assumed to have a perfect elastic rebound, in which the velocity $\dot{X}$ is instantaneously reversed whenever X drops to zero.

The bilinear and impact oscillators are both quasilinear, and the magnitude of the sinusoidal forcing plays essentially no role, having already been incorporated as a scaling factor into the definition of X. However, we shall see that in all other respects the behaviour has a decidedly nonlinear character.

3. NUMERICAL COMPUTATIONS

Throughout this work we have used a digital computer that performs time integrations by using the known analytical solutions, complementary functions plus particular integrals, for each half of the $(X, \dot{X})$ phase projection. This involves evaluating the current known solution at fine time intervals until X is on the point of changing sign for the first time. The switching values of $\dot{X}$ and τ for $X = 0$ are then estimated to a specified precision, these being then the starting conditions for the next half-space. For the limiting case of the impact oscillator, a simple rebound instruction is substituted.

A given *system* is thus specified by the damping ratio ζ, the stiffness ratio α, and the frequency ratio η, and for given starting conditions specified by $(X_0, \dot{X}_0)$ at $\tau = 0$, we can run the time integration and watch for the arrival at a steady state.

The settling to a steady state of stable periodic motion is best detected by studying the values of $(X, \dot{X})$ whenever τ is a multiple of 2π, the so-called Poincaré mapping. For example, once we find that

$$X[2\pi j] = X[2\pi(j+n)] \quad (4)$$

and
$$\dot{X}[2\pi j] = \dot{X}[2\pi(j+n)], \quad (5)$$

we know that we have found a steady subharmonic of order n, the case of $n = 1$ being simply a fundamental response. The computer is programmed to make this, and some further checks automatically, and once a stable steady periodic solution is identified in this way, the amplitude parameter y, defined as half the positive-peak to negative-peak range of X, is recorded.

We have used this procedure to plot the resonance response surface of $y(\eta, \alpha)$ shown in figures 1 and 2 for the fixed damping ratio corresponding to $\zeta = 0.1$. Here we see the main fundamental resonance, $n = 1$, at $\eta \approx 1$, followed by the regular subharmonic resonances of order $n = 2, 3, 4$ at $\eta \approx 2, 3$ and 4 respectively. The 'irregular' resonances for $\eta < 1$ are all fundamental responses ($n = 1$) with, however, an increasing number of switching points per cycle as η decreases. Despite their ragged appearance at this scale, the responses are perfectly periodic and repeatable, and are in no way chaotic.

The simple situation depicted in these figures is complicated by the fact that the final steady state is dependent on the starting conditions, and the equally-likely low-amplitude coexisting solutions beneath the resonant peaks are shown for $\alpha = 10$ in figure 3. Here we see in the top $y(\eta)$ response diagram that beneath the $n = 4$ resonant peak there is a continuous, coexisting *stable* fundamental solution with $n = 1$. As a variant on this, we have detected stable $n = 2$ and $n = 4$ subharmonics of low amplitude under the $n = 3$ resonance. It is these continuous low-amplitude curves that could lead to erroneous results from one-off or 'efficient' automated computations.

The lower graphs of figure 3 show the movements of the steady-state Poincaré points $(X_P, \dot{X}_P)$, there being of course n such points for a subharmonic of order n. The bifurcation from a stable $n = 1$ solution to a stable $n = 2$ solution at $\eta \approx 1.5$ is nicely seen in these two diagrams.

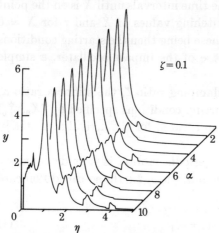

FIGURE 1. The resonance response surface of the bilinear oscillator, showing the fundamental and regular subharmonic resonances.

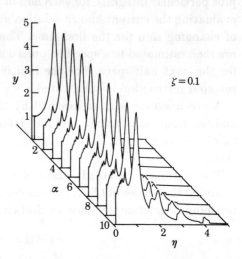

FIGURE 2. The resonance response surface of the bilinear oscillator, showing the fundamental and irregular resonances.

4. ANALYTICAL SOLUTIONS

Our digital time integrations, and some back-up analogue studies, simply follow the physical transient behaviour of the system, and finally attain a *stable* steady state; but for a complete picture it would be helpful to also locate any *unstable* periodic solutions. We could then, for example, test the rather obvious theory that an unstable $n = 1$ solution persists beyond the critical value of $\eta \approx 1.5$ in figure 3, being subsequently restabilized to join the stable trace under the $n = 4$ peak.

Now a completely unstable solution (corresponding to a source) could be located by running our time integrations backwards, but most of the unstable periodic solutions are likely to correspond to saddles, so this technique is of limited value. An alternative is to seek periodic solutions analytically. We could, for example, choose a general starting point $(X_0, \dot{X}_0)$ at $\tau = 0$ and write down the analytical conditions that at $\tau = 2\pi n$, after $2m$ switches in the sign of X, we should be back

Complex dynamics of compliant off-shore structures

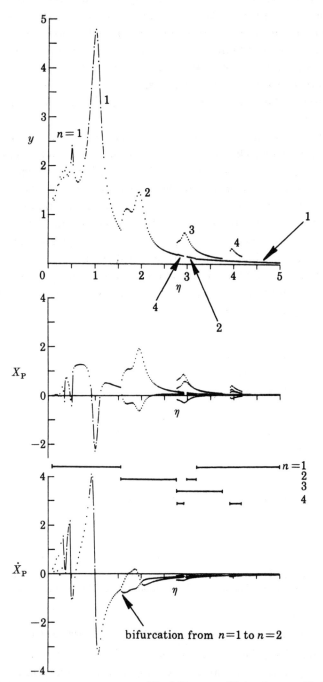

FIGURE 3. The resonance response curve of the bilinear oscillator for $\alpha = 10$, $\zeta = 0.1$ showing the coexisting stable steady states. The two lower diagrams show the movement of the Poincaré points.

at the starting values of X and $\dot{X}$. So by specifying in advance that we are looking for a subharmonic of order n with $2m$ switching points, we would arrive at a set of simultaneous nonlinear transcendental equations to solve on a computer for X_0 and $\dot{X}_0$. Even a rather limited survey using this technique would be difficult and very time-consuming, especially as we would want to explore high values of n and m; and this procedure does of course supply no answers at all about the length of *transients*, the *stability* of the located periodic solutions, or the existence of *chaotic* régimes.

A similar, but in many ways more ingenious scheme due to Tee (1983) has however been used to check some of our stable solutions and to extend some of our paths into their unstable régimes. In this way we have confirmed the existence of periodic $n = 1$ solutions beneath the $n = 2$ and $n = 3$ resonant peaks for $\alpha = 10$ and $\zeta = 0.1$. Since we have been unable to pick up these solutions in our time integrations, even by an efficient skilful choice of appropriate starting conditions by using adjacent Poincaré points, they are presumably unstable. A similar analytical technique is used by Shaw & Holmes (1982) in their recent study of our impact oscillator modified to have a coefficient of restitution rather than viscous damping as an energy sink: their work is also note-worthy for their analytical stability investigations.

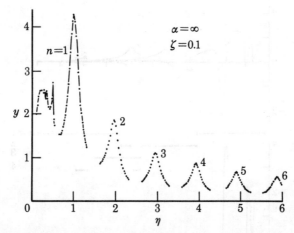

FIGURE 4. The resonance response curve for the impact oscillator, showing the fundamental, irregular and subharmonic peaks.

5. Numerical results for the impact oscillator

We have determined the resonance response diagram of figure 4 for the limiting case of infinite α by using our automated digital time integrations. On this figure we can see the main fundamental peak, and the regular subharmonic resonances up to $n = 6$. These, and the irregular resonances for $\eta < 1$, look like straightforward modifications of our $\alpha = 10$ curve. We have, however, observed no co-

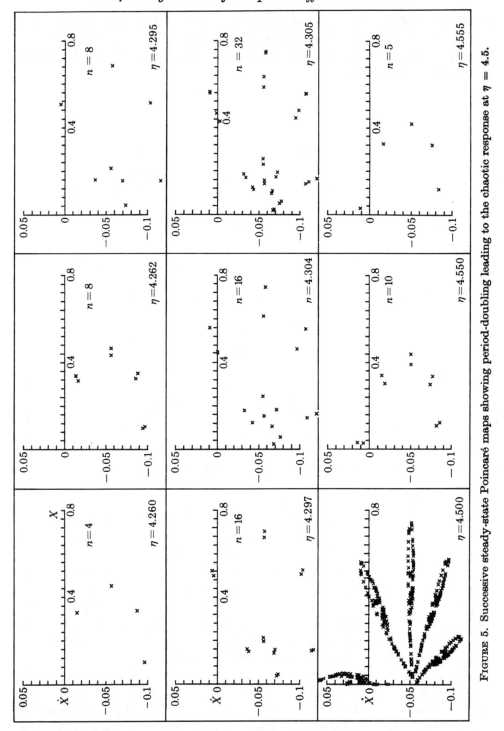

FIGURE 5. Successive steady-state Poincaré maps showing period-doubling leading to the chaotic response at $\eta = 4.5$.

existing solutions *beneath* the resonant peaks. *Between* the subharmonic peaks, our automatic procedure failed repeatedly to supply satisfactory results, and we were forced to use more interactive procedures and choose steps and precision appropriate to each run.

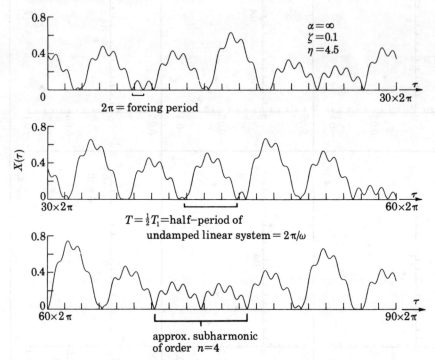

FIGURE 6. The waveform of the steady-state chaotic response of the impact oscillator, after 1480 cycles.

We have looked carefully, in this manner, at the gap between the $n = 4$ and $n = 5$ resonant peaks, and figure 5 shows the sequence of steady-state Poincaré maps. The first shows an $n = 4$ subharmonic that bifurcates under increasing η to an $n = 8$ solution. Repeated period-doubling bifurcations are then observed leading through $n = 16, 32$, etc. to the chaotic picture at $\eta = 4.5$. Under decreasing η we have detected only one period-doubling bifurcation from $n = 5$ to $n = 10$, leading us towards the chaotic régime from above. Further study of this feature is necessary as the approach from above seems to be rather different from the period-doubling cascade from below (Shaw & Holmes 1982).

The chaotic Poincaré mapping at $\eta = 4.5$ was determined by running the computer for 1000 forcing cycles before plotting any points, in the hope of getting to a state of 'steady chaos'. We have then plotted the $(X, \dot{X})$ points at intervals of $\Delta\tau = 2\pi$ for 480 forcing cycles. There is apparently no stable periodic state, and the ensuing chaotic waveform is shown in figure 6. This does indeed look like

Complex dynamics of compliant off-shore structures

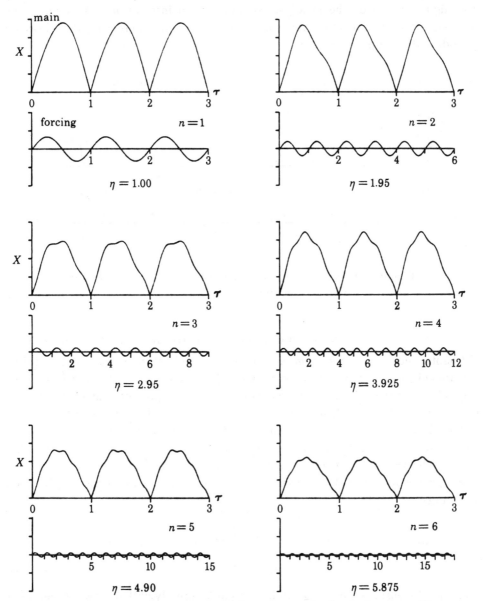

FIGURE 7. The waveforms at the actual resonant peaks for the impact oscillator; $\alpha = \infty$, $\zeta = 0.1$.

a steady state of chaos, for which a statistical description would have to be sought: it probably involves a wandering between many unstable periodic solutions, and an approximate unstable subharmonic of order $n = 4$ is indicated.

We can compare this chaotic waveform with the single-impact resonant waves of figure 7, and the double-impact solutions of figure 8.

The cascade of period-doubling bifurcations is shown on a plot of X_P against

η in figure 9, *one* of the steady-state Poincaré points for $n = 4$ having been selected arbitrarily to start the sequence. Notice that here we have detected period-doubling up to $n = 128$.

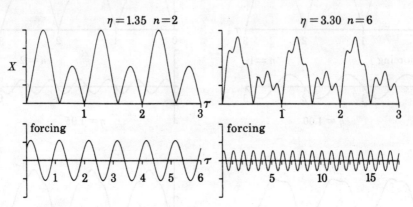

FIGURE 8. Double-impact waveforms of the impact oscillator away from the resonant peaks; $\alpha = \infty$, $\zeta = 0.1$.

6. IDENTIFICATION OF CHAOS BY A PERIOD-DOUBLING SCENARIO

Chaotic motions of a strange attractor in the phase space of a deterministic dynamical system are currently under active study by mathematicians and physicists, but their theory is by no means complete, at present. Indeed, the precise definitions of *chaos* and *strange attractor* vary somewhat from writer to writer.

Identification of these phenomena therefore relies quite heavily on a number of scenarios, sketched for example by May (1976); Eckmann (1981) and Ott (1981). The scenario that our impact oscillator seems to fit is that of Feigenbaum (1978), which draws on the behaviour of the very simple one-dimensional iterative mapping

$$X_{i+1} = \Lambda X_i (1 - X_i), \tag{6}$$

an equivalent form of which is given by Grebogi *et al.* (1982) as

$$x_{i+1} = C - x_i^2. \tag{7}$$

This one-dimensional mapping can be compared directly with our two-dimensional Poincaré mapping of the impact oscillator, with the control parameter Λ or C corresponding to our η.

The behaviour of the mapping (6) is summarized schematically in figure 10. By increasing the value of the control parameter Λ, the steady state of the mapping exhibits an *infinite cascade* of period-doubling bifurcations, leading, in a finite Λ interval, to an accumulation point involving an infinity of periodic steady states. Feigenbaum (1978) shows this to be a general property of more complex maps, and with use of arguments of scale invariance, he shows that as

Complex dynamics of compliant off-shore structures

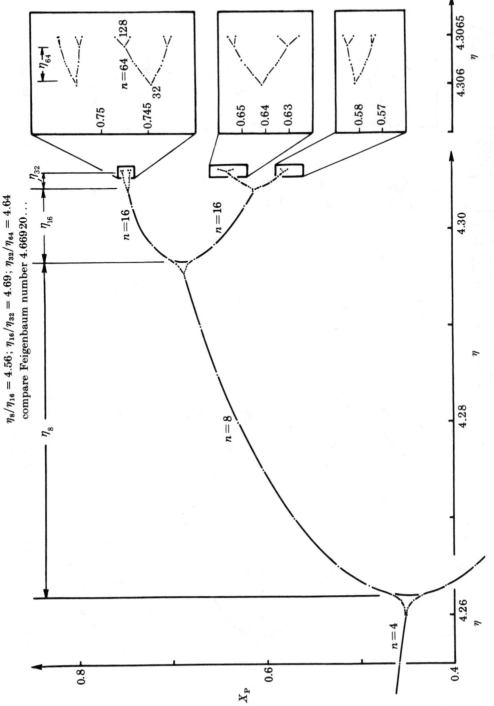

FIGURE 9. Bifurcation diagram for the approach to chaos via period-doubling bifurcations in the impact oscillator.

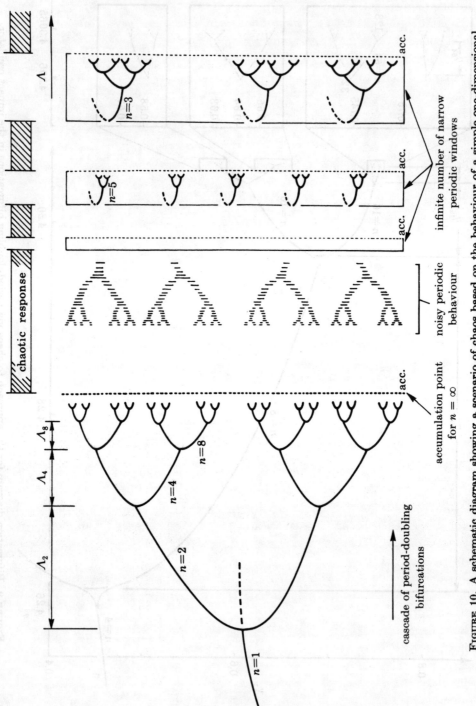

FIGURE 10. A schematic diagram showing a scenario of chaos based on the behaviour of a simple one-dimensional mapping. Note: $\Lambda_k/\Lambda_{2k} \to 4.66920\ldots$ as $k \to \infty$.

the accumulation point is approached the ratio of successive Λ intervals tends to a universal number, 4.66920 ....

It is most significant that our η intervals seem to be in *good agreement* with this, even at some distance from the accumulation point, as shown in figure 9.

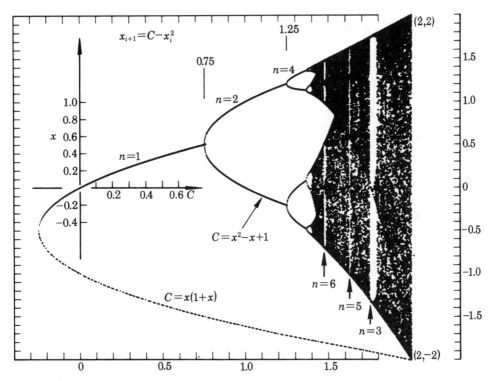

FIGURE 11. A computed diagram showing the chaotic régime of the simple one-dimensional iterative mapping. Note: 500 iterations before plotting; 200 plotted at each C in chaotic régimes; $\Delta C = 0.005$.

Beyond the first accumulation point, the response of (6) looks like noisy periodic behaviour governed by an inverse cascade of (noisy) bifurcations. Careful study, however, shows the existence of an infinite number of very narrow periodic windows in which strictly periodic behaviour is observed. Each window starts at a tangent bifurcation (akin to a fold or limit point in elastic stability) and contains an infinite cascade of period-doubling bifurcations leading to an accumulation point as shown.

The real, computed, bifurcation diagram based on the equivalent equation (7) is shown in figure 11 with C as the single control parameter. Here, in analogy with figure 5, we have evaluated 500 iterations at each fixed C value before plotting, to eliminate transient effects. We have then plotted the next 200 points for each C value in the chaotic régimes. The parameter C is incremented by an amount $\Delta C = 0.005$ between calculations.

The $n = 1$ and $n = 2$ solution curves are easily written down analytically as indicated on the figure, and we note that the broken region of the $n = 1$ curve is unstable, and is therefore not picked up numerically. Three narrow periodic windows corresponding to cascades based on $n = 3, 5, 6$ are seen in the general chaotic régime.

It is our observed period-doubling cascade with the correct Feigenbaum number (figure 9) that gives us our first good reason to suppose that our impact oscillator does indeed exhibit a truly chaotic régime (Thompson & Ghaffari 1983a, b).

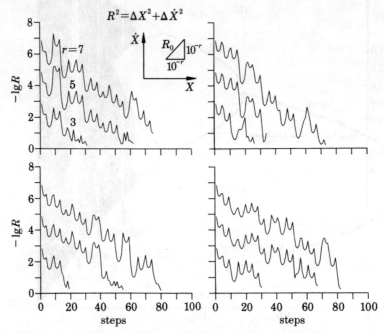

FIGURE 12. Four divergence studies in the steady-state chaos of the impact oscillator; $\alpha = \infty$, $\eta = 4.5$, $\zeta = 0.1$.

7. IDENTIFICATION OF CHAOS BY EXPONENTIAL DIVERGENCE

A second standard test for chaos is to look at the divergence of adjacent starts in one of the located steady-state chaotic time integrations.

The results of such a test for our impact oscillator (Thompson & Ghaffari 1982, 1983a, b) are shown in figure 12. Here, starting at $\tau = 0$ at a point $(X_0, \dot{X}_0)$, and then at a point $(X_0 + 10^{-r}, \dot{X}_0 + 10^{-r})$ we have observed the distance R between the two subsequent motions for four different choices of $(X_0, \dot{X}_0)$ on the located steady-state attractor. For each of these four choices we have taken $r = 3, 5, 7$ and plotted $-\lg R$ against the steps of $\Delta \tau = 2\pi$.

The noisy straight lines on these logarithmic plots confirm that the adjacent solutions diverge exponentially (but noisily) before becoming completely uncorrelated.

A similar result for the Hénon (1976) strange attractor is shown by Thompson (1982), following Feit (1978). The Liapunov numbers and characteristic exponents for three presumed strange attractors, including the present impact oscillator, are summarized in figure 13.

This demonstration of noisy exponential divergence, with a characteristic exponent similar in magnitude to those of other strange attractors, is the second reason for our presumption of chaos in the impact oscillator.

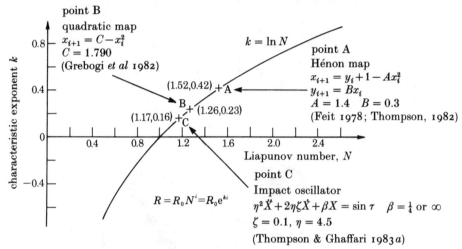

FIGURE 13. Characteristic exponents and Liapunov numbers of three presumed strange attractors.

8. GENERAL DISCUSSION

We have aimed to show, through the particular study of our bilinear and impact oscillators, that the dynamics of compliant off-shore marine structures can be very complex. Unexpected subharmonic resonances can coexist with stable small-amplitude periodic solutions, so in digital, analogue and experimental studies care must be taken to vary the starting conditions in a systematic manner, so that all of the coexisting competing solutions are observed.

By thinking of the starting conditions in the space of $(X_0, \dot{X}_0)$ at time zero, the catchment regions of the various competing steady states are likely to be spirals, as in figure 14, which relates to Duffing's equation. This figure shows the flow of the variational equation in the Van der Pol plane in the presence of two competing sinks. By thinking of this as our present $(X, \dot{X})$ space, we see that as $A^2 = X^2 + \dot{X}^2$ becomes large, the catchment regions of the two stable steady-state sinks are intermeshing spirals.

The second feature we would draw attention to is the existence of chaotic solutions to deterministic nonlinear differential equations. This is illustrated by our study of the impact oscillator, and we have demonstrated two tests that should be used to establish the existence of a truly chaotic régime. These involve the demonstration of period-doubling bifurcations, and the determination of the exponential divergence of adjacent starts. A third test would involve the study of the power spectrum of the response, along the lines of the definitive treatment of Holmes (1979) who established a strange attractor in the resonance of a buckled structure.

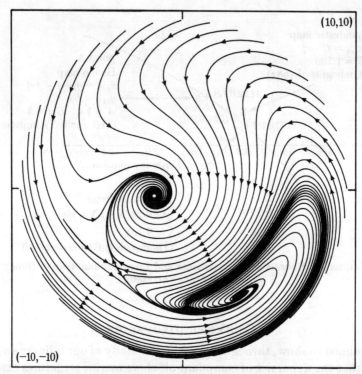

FIGURE 14. Phase portrait of the variational equation (Holmes & Rand 1976) of Duffing's equation in the Van der Pol plane, showing the catchment regions of the two competing stable sinks.

We would finally emphasize that the two phenomena that we have outlined, namely the coexistence of multiple steady states and the appearance of chaos, can arise at one and the same time. Thus a chaotic solution governed by a strange attractor in the phase space can coexist with a stable periodic solution, as shown for our impact oscillator in figure 15.

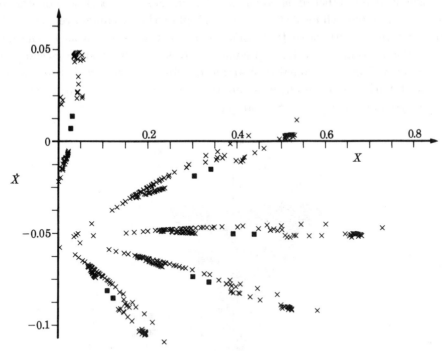

FIGURE 15. A strange attractor, ×, coexisting with a stable, $n = 10$, subharmonic ■, in the response of the impact oscillator; $\eta = 4.55$.

9. CONCLUSIONS AND IMPLICATIONS FOR OFF-SHORE DESIGNERS

The dynamical behaviour of modern and proposed compliant marine systems is intrinsically more complex than that of conventional rigid structures. For this reason, designers and analysts should be alerted to a number of dangers that can arise in the interpretation of digital, analogue and model investigations.

The first point we wish to emphasize is that unexpected subharmonic resonances often *coexist* with stable small-amplitude fundamental motions, the outcome observed depending only on the *starting conditions* of the simulation. So a limited number of one-off studies could easily miss an important resonance: this should be avoided by always systematically varying the starting conditions to ensure that all possible coexisting motions are revealed. This warning is perhaps even more relevant to users of 'efficient' systematic computer programs that use a found solution at a certain forcing frequency as the 'first guess' for a solution at a slightly different frequency. Such programs have an in-built tendency to lock-on to a particular solution curve. The curve found might then easily be the stable small-amplitude fundamental solution that lies under many of the resonant peaks, so that one or more *entire peaks* could be missed by this seemingly efficient procedure. Designers should thus always supplement any such automated 'efficient' computations by a number of one-off *ad hoc* simulations using a variety of starting conditions. More research needs to be done to explore the catchment regions of the competing stable solutions so that some guidance can be given as to the best choice of the alternative starts. A preliminary result in this respect is shown in figure 16. This shows the domains of attraction for coexisting $n = 1$ and $n = 4$ solutions at fairly small amplitudes: no Duffing-like spirals are seen here, but these may possibly emerge at larger amplitudes.

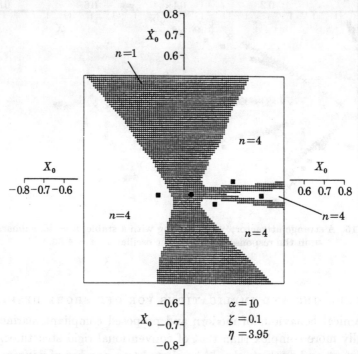

FIGURE 16. Catchment regions (domains of attraction) for the two competing solutions at the $n = 4$ resonant peak. Starting conditions leading to the small-amplitude $n = 1$ solution are indicated by the hatched region, which is composed of small crosses showing each trial made. Starting conditions leading to the large-amplitude resonant $n = 4$ solution are indicated by the blank region, trials in this region having been made on the same fine mesh as the crosses. The steady-state Poincaré mapping points are indicated.

The second warning concerns the existence of *chaotic solutions* to even well-posed deterministic dynamical systems. Here an observed waveform may look like the output of a stochastic process, and when arising in a digital, analogue or laboratory study it might easily be brushed aside as erroneous. Off-shore analysts should therefore be aware that such a chaotic, non-periodic result can indeed represent a correct solution, and they should be familiar with three tests that can positively identify a genuine chaotic output.

The first test, illustrated in the present paper, uses the Feigenbaum (1978) scenario that suggests that a parameter régime of chaos will usually be surrounded by period-doubling bifurcations that generate subharmonics of arbitrarily high order. The second test, also fully illustrated in this paper, relies on the fact that a genuine chaotic solution will have an extreme sensitivity to initial starting conditions with adjacent starts diverging according to a noisy exponential law and rapidly becoming uncorrelated. A third test, not specifically used in our present work, involves the study of the power spectrum of the waveform; a genuine chaotic solution will usually have a broad-band white-noise spectrum.

The existence of a true chaotic régime having been established, the extreme sensitivity to the starting conditions means that a statistical description of the 'steady-state' chaotic output must be sought, and further study is needed to see precisely what form this should take in a marine technology context. The existence of high-order subharmonics, and the lengthy transients known to exist both within and close to a chaotic régime, mean that the duration of simulations must be carefully considered and will usually need to be longer than those of more conventional problems.

References

Eckmann, J. P. 1981 *Rev. mod. Phys.* **53**, 643–654.
Feigenbaum, M. J. 1978 *J. statist. Phys.* **19**, 25–52.
Feit, S. D. 1978 *Communs math. Phys.* **61**, 249–260.
Grebogi, C., Ott, E. & Yorke, J. A. 1982 *Phys. Rev. Lett.* **48**, 1507–1510.
Hénon, M. 1976 *Communs math. Phys.* **50**, 69–77.
Holmes, P. J. 1979 *Phil. Trans. R. Soc. Lond.* **292** A, 419–448.
Holmes, P. J. & Rand, D. A. 1976 *J. Sound Vib.* **44**, 237–253.
Kilner, F. A. 1960 In *Coastal Engineering, 7th Conf.* Hague, 723–745.
Lean, G. H. 1971 *Trans. R. Instn. nav. Archit.* **113**, 387–399.
May, R. M. 1976 *Nature, Lond.* **261**, 459–467.
Ott, E. 1981 *Rev. mod. Phys.*, **53**, 655–671.
Rainey, R. C. T. 1978 *Trans. R. Instn nav. Archit.* **120**, 59–80.
Rossler, O. E. 1979 *Ann. N.Y. Acad. Sci.* **316**, 376–392.
Shaw, S. W. & Holmes, P. J. 1982 *A peroidically forced piecewise linear oscillator*, Report, Dept. of Theoretical and Applied Mechanics, Cornell University.
Tee, G. J. 1983 Periodic oscillations of a bilinear oscillator, with reference to moored marine systems. Engineering Structures, submitted.
Thompson, J. M. T. 1982 *Instabilities and catastrophes in science and engineering*. Chichester: Wiley.
Thompson, J. M. T. & Bokaian, A. R. 1983 Dangers, complexities and chaos in the dynamics of compliant marine structures. To be presented at 11th I.F.I.P. Conference on System Modelling and Optimization, Copenhagen, July 25, 1983.
Thompson, J. M. T., Bokaian, A. R. & Ghaffari, R. 1983 Subharmonic resonance of a bilinear oscillator with applications to moored marine systems. (To be published.)
Thompson, J. M. T. & Ghaffari, R. 1982 *Physics Lett.* A **91**, 5–8.
Thompson, J. M. T. & Ghaffari, R. 1983*a* *Phys. Rev.* A, **27**, 1741–1743.
Thompson, J. M. T. & Ghaffari, R. 1983*b* Complex dynamics of bilinear systems: bifurcational instabilities leading to chaos. In *Collapse: the buckling of structures in theory and practice* (ed. J. M. T. Thompson and G. W. Hunt). (Proceedings of the IUTAM Symposium, University College London, August 1982). Cambridge University Press.

CHAOS GENERATED BY THE CUTTING PROCESS

Igor GRABEC

Faculty of Mechanical Engineering, pp. 394, Ljubljana, Yugoslavia

Received 15 April 1986; revised manuscript received 22 May 1986; accepted for publication 30 May 1986

> Due to nonlinearity and coupling of cutting forces chaotic oscillations take place in an elastic manufacturing machine. A transition from quasiperiodic via chaotic to synchronised anharmonic oscillations with increasing cutting intensity is demonstrated by spectral distributions and the correlation exponent.

Manufacturing processes are well-known generators of irregular machine vibrations, therefore it is rather surprising that so little attention is paid to them in current studies of chaos [1,2]. The aim of this article is to show that the development of chaotic vibrations is quite generally to be expected when inertia, elasticity and deformation or friction play dominant roles in a manufacturing machine. For this purpose we present the most simple model of an orthogonal cutting machine and explain its properties by chaotic dynamics [3].

A scheme of the cutting zone is shown in fig. 1. A workpiece is pushed with input velocity v_i against the tool, where the chip is formed. The force generated due to the plastic flow of material is the resultant of cutting and friction components F_x, F_y, which are normal and parallel to the tool surface. This force is transmitted over the tool, machine frame and the remaining part of the specimen which can be treated altogether as an elastic two-dimensional structure. Its dynamics can be approximately described by

$$m\ddot{x} + k_x x = F_x, \qquad (1)$$
$$m\ddot{y} + k_y y = F_y, \qquad (2)$$

in which m and k represent the inertia and elasticity of the structure. F_y is related to F_x by the friction coefficient K:

$$F_y = KF_x. \qquad (3)$$

Due to shear in the cutting zone the velocity of the material flow is reduced, so that in the direction of friction it can be represented by

$$v_f = v/R, \qquad (4)$$

where the reduction factor R is equal to cot ϕ, with ϕ the typical shear angle.

The properties of the cutting force, the friction coefficient and the shear angle have been extensively studied [4] and for a wide class of materials can be approximately described by the empirical relations [3]

$$F = F_{x^0}(h/h_0)\left[c_1(v/v_0 - 1)^2 + 1\right], \qquad (5)$$
$$K = K_0\left[c_2(v_f R/v_0 - 1)^2 + 1\right]\left[c_3(h/h_0 - 1)^2 + 1\right], \qquad (6)$$

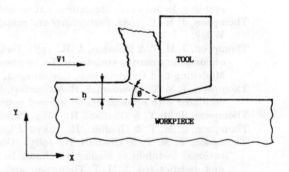

Fig. 1. Scheme of the cutting zone.

$$R = R_0 \left[c_4 (v/v_0 - 1)^2 + 1 \right]. \quad (7)$$

The following set of specific parameters approximately corresponds to a material like low-carbon-content steel:

$$c_1 = 0.3, \quad c_2 = 0.7, \quad c_3 = 1.5, \quad c_4 = 1.2,$$
$$h_0 = 0.25 \text{ mm}, \quad v_0 = 6.6 \text{ m s}^{-1},$$
$$K_0 = 0.35, \quad R_0 = 2.2. \quad (8)$$

The parameter F_{x^0} is proportional to the chip width which can be arbitrarily chosen and represents the intensity of cutting.

When the tool is moving, the cut depth, the flow and friction velocity must be expressed by

$$h(t) = h_i - y(t), \quad v(t) = v_i - \dot{x}(t),$$
$$v_f(t) = v(t)/R(t) - \dot{y}(t), \quad (9)$$

and the empirical relations must be supplemented by the conditions

$$F_x = 0, \quad \text{for } h < 0 \text{ or } v < 0,$$
$$K(-v_f) = -K(v_f). \quad (10)$$

In order to complete the problem we assume that at the start of the cutting the tool is in rest:

$$x_0 = y_0 = 0, \quad \dot{x}_0 = \dot{y}_0 = 0. \quad (11)$$

The dependence of the cutting force on velocity and cut depth is nonlinear. Besides these both force components are interrelated. Eqs. (1), (2) therefore represent coupled nonlinear oscillators in two dimensions. Due to the negative slope of F_x for $0 \leq v \leq v_0$ self-excited nonlinear oscillations can develop in the cutting process. The corresponding phase space is four-dimensional, therefore transition to chaos is possible. In order to demonstrate this we have first introduced normalised variables:

$$X = x/h_0, \quad Y = y/h_0,$$
$$T = tv_0/h_0, \quad V_x = dX/dT, \quad (12)$$

and the parameters

$$A = k_x h_0^2 / mv_0^2, \quad B = k_y h_0^2 / mv_0^2,$$
$$F_0 = F_{x^0} h_0 / mv_0^2. \quad (13)$$

The dynamic equations (1), (2) were then solved

Fig. 2. The records of tool displacement components calculated for $F_0 = 0.75$.

numerically for $A = 1$, $B = 0.25$, $v_i = 0.5v_0$, $h_i = 0.5h_0$ and various values of F_0 [3]. At a low intensity of cutting, $F_0 \approx 0.25$, quasiperiodic oscillations develop from the initial conditions. At $F_0 \approx 0.5$ a transition to chaotic vibrations is established. With a further increase of the cutting intensity $F \to 1$ chaotic oscillations are changed into intensive, synchronised anharmonic oscillations [3]. Fig. 2 shows the records of the tool displacement components for a typical example of well-developed chaos at $F_0 = 0.75$. The corresponding phase portraits representing projections of a strange attractor to the (X, V_x) and (Y, V_y) planes are shown in fig. 3 The transition from quasiperiodic to chaotic and synchronised anharmonic oscillations is accompanied by characteristic changes of spectral distributions which are shown in fig. 4. For a well-developed chaotic state a broad band background is characteristic, while the sharp peaks correspond to quasiperiodic oscillations. The same property was also observed experimentally [3].

The dimensionality of the strange attractor was characterized by the correlation exponent ν [5]. Its dependence on the cutting intensity is shown in

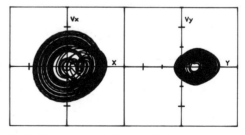

Fig. 3. Phase portraits of chaotic oscillations for $F_0 = 0.75$.

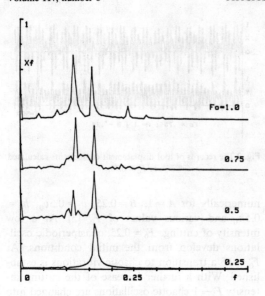 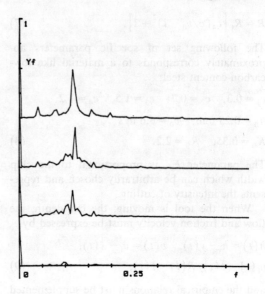

Fig. 4. The distribution of the spectral amplitude determined by FFT, calculated for various values of F_0.

table 1. The value of a correlation exponent close to 2 corresponds to quasiperiodic or synchronised anharmonic oscillations; it then jumps to approximately 3 with the transition to chaos.

The properties of forces, dissipated power and return map will be published in a more comprehensive article [3].

A cutting manufacturing machine represents a typical example of a dissipative, energy-transmitting structure in which instability and chaos can develop. The properties of the cutting force and that produced by dry friction of solids are very similar and probably of the same origin. It is therefore quite probable that the stick–slip effect caused by friction in one dimension can lead to chaotic behavior in multicomponent mechanical systems. Besides the inhomogeneity of input materials and the irregularity of their surfaces chaotic dynamics should therefore be considered as one of the main reasons for stochasticity observed in manufacturing systems.

Table 1

F_0	ν
0.25	2.22
0.5	2.70
0.75	2.98
1	2.15

References

[1] S.A. Tobias, Schwingungen und Werkzeugmaschinen (Carl Hauser, Munich, 1961).
[2] H. Haken, Comp. Meth. Appl. Mech. Eng. 52 (1985) 635.
[3] I. Grabec, Mach. Tool Des. Res., submitted for publication.
[4] W.F. Hastings, P.L.B. Oxley and M.G. Stevenson, Proc. 12th Int. Machine tool design and research Conf., Manchester, 1971 (MacMillan) p. 507.
[5] P. Grassberger and I. Procaccia, Phys. Rev. Lett. 50 (1983) 346; Phys. Rev. 28 (1983) 2591.

Journal of Sound and Vibration (1990) **138**(3), 421-431

THE STABILITY OF MODES AT REST IN A CHAOTIC SYSTEM

S. R. Hsieh AND S. W. Shaw

Department of Mechanical Engineering, Michigan State University, East Lansing, Michigan 48824, U.S.A.

(*Received* 12 *December* 1988, *and in final form* 25 *July* 1989)

The stability of modes at rest in non-linear mechanical systems which have one active mode undergoing chaotic oscillations is considered. A general formulation is developed for these systems. Results from the theory of almost sure stability (Infante [1]) for stochastic systems are briefly reviewed in order to determine the minimal statistical measures of the chaos which are required to determine the stability of the inactive modes. A specific non-linear two-degree-of-freedom system is then considered for which these measures can be explicitly estimated by using a method developed by Brunsden et al. [2, 3]. Analytical results for the almost sure stability of the inactive mode are obtained and compared with simulation results.

1. INTRODUCTION

The local stability of modes at rest in non-linear systems has been considered in several contexts. The original works were focused on situations where the active modes were undergoing periodic oscillations; this leads to Mathieu-Hill equations for the amplitudes of the inactive modes [4]. The case of stochastic behavior of the active modes, typically arising from stochastic inputs, is closer to the case considered in this paper. Such work has emphasized the use of *almost-sure-stability* and requires some quantitative information about the statistics of the dynamic response of the active mode [1, 5-11]. Based on various specific assumptions, Caughey and Gray [5], Infante [1], Kozin [7-9], Ariaratnam [10, 11] and others have developed results on almost-sure-stability. Among these results, the one developed by Infante is employed, which provides conservative results but requires only the mean and mean-square values of the random coefficients. Other results exist which provide less conservative stability estimates, but these require more statistical information.

Of interest here is the stability of inactive modes which are parametrically driven by *chaotic motions* in the active mode. These situations arise quite naturally in parametrically excited systems which are undergoing chaotic oscillations. Specific physical systems of this type which are known to exhibit chaos are parametrically driven surface waves in fluids [12, 13] (although, typically, at least two modes are active in the chaotic motion), rotating beams [14], and buckled beams and plates which are excited by end loads.

The main contribution in the present work is that it links together results from the theory of almost-sure-stability with chaotic motions. The missing piece for the analysis of these cases is that, typically, the statistical properties of chaos can be determined either in only very simple cases, or computed directly via simulations [15, 16]. However, recent works by Brunsden et al. [2, 3] have provided a means by which one can estimate some statistical measures for one of the benchmark chaotic systems: Duffing's equation with negative linear stiffness. They employed a variation of Melnikov's method [17], chapter 4 to estimate the power spectral density function for certain solutions of Duffing's equation.

421

© 1990 Academic Press Limited

In this paper these results are used to determine conditions for the local almost-sure-stability of inactive modes in multi-degree-of-freedom systems for which Duffing's equation describes the active mode dynamics.

In the course of the analysis it will be necessary to make certain assumptions regarding the statistical nature of some chaotic motions. These (or similar) assumptions are required if one is to employ results from the theory of stochastic processes for chaotic systems since no rigorous results regarding the stationarity, etc., of chaos are known.

The paper is arranged as follows: in section 2 a general formulation is provided for the equations which govern small-amplitude dynamics of modes at rest in a multi-degree-of freedom system (see the paper by Ariaratnam [11]); in section 3 a brief review of the simplest theory for almost-sure-stability (from Infante [1]) is provided; section 4 is a review of the derivation of the power spectral density function for the active mode which is modeled by Duffing's equation (from Brunsden et al. [2, 3]); in section 5 a two-degree-of-freedom example is presented, explicit criteria for the almost-sure-stability of the inactive mode are developed, and results from a simulation study are reported; and in section 6 the paper is closed with a discussion.

2. FORMULATION

Consider a system the dynamics of which are governed by a set of non-linear ordinary differential equations of the form

$$\ddot{x}_1 + 2\xi_1\dot{x}_1 + \omega_1^2 x_1 + f_1(x_1, x_2, \ldots, x_n, t) = Q(t),$$
$$\ddot{x}_i + 2\xi_i\dot{x}_i + \omega_i^2 x_i + f_i(x_1, x_2, \ldots, x_n, t) = 0, \qquad i = 2, 3, \ldots, n, \qquad (2.1)$$

which can be used to describe the motion of a class of n-degree-of-freedom non-linear, oscillatory systems. If the non-linear functions $f_2, \ldots, f_n$ have the property

$$f_i(x_1, 0, 0, \ldots, 0, t) = 0, \qquad i = 2, 3, \ldots, n, \qquad (2.2)$$

then equations (2.1) admit a solution of the form

$$x_1(t) = \bar{x}_1(t), \qquad x_i(t) = 0, \qquad i = 2, 3, \ldots, n, \qquad (2.3)$$

which corresponds to a single-mode motion with the unexcited modes remaining at rest. The differential equation

$$\ddot{\bar{x}}_1 + 2\xi_1\dot{\bar{x}}_1 + \omega_1^2 \bar{x}_1 + f_1(\bar{x}_1, 0, 0, \ldots, 0, t) = Q(t) \qquad (2.4)$$

describes the motion of the active mode.

In order to investigate the local stability of solution (2.3), the dynamics of a small perturbation of it are considered. This perturbation is expressed as

$$x_1(t) = \bar{x}_1(t) + \mu q_1(t), \qquad x_i(t) = \mu q_i(t), \qquad i = 2, 3, \ldots, n, \qquad (2.5)$$

where $|\mu| \ll 1$. Substituting equation (2.5) into equation (2.1) and employing equation (2.4) yields the following equations which describe the linearized dynamics of the perturbations:

$$\ddot{q}_i + 2\xi_i\dot{q}_i + \omega_i^2 q_i + \sum_{j=1}^{n} \frac{\partial f_i}{\partial x_j}(\bar{x}_1, 0, 0, \ldots, 0)q_j + O(\mu) = 0, \qquad i = 1, 2, 3, \ldots, n. \qquad (2.6)$$

Upon assuming that the non-linear functions $f_1, f_2, \ldots, f_n$ also have the property

$$(\partial f_i/\partial x_j)(\bar{x}_1, 0, 0, \ldots, 0) = \delta_{ij}\phi_j(t), \qquad i, j = 1, 2, \ldots, n, \qquad (2.7)$$

in which δ_{ij} is the Kronecker delta, equations (2.6) take the form

$$\ddot{q}_i + 2\xi_i\dot{q}_i + (\omega_i^2 + \phi_i(t))q_i = 0, \qquad i = 1, 2, \ldots, n, \qquad (2.8)$$

which are linear uncoupled differential equations with time-varying coefficients. The stability of the trivial solution of equations (2.8) corresponds to the local stability of solution (2.3). If condition (2.7) does not hold, one must consider a coupled set of linear, time-varying differential equations. In structural applications this uncoupling can be achieved by using modal co-ordinates for the modes at rest.

Note that if $q_i \to 0$ as $t \to \infty$ for $i = 2, 3, \ldots, n$, then the full solution will be asymptotic to the single-mode solution for sufficiently small initial disturbances. The solution for q_1 indicates the local stability of $\bar{x}_1(t)$ relative to small disturbances restricted to the active mode.

3. ALMOST-SURE-STABILITY OF LINEAR STOCHASTIC DIFFERENTIAL EQUATIONS

In this section a criterion is dealt with that pertains to the almost-sure-stability (that is, stability with probability one) of a linear system described by the equation

$$\ddot{q} + 2\xi\dot{q} + [\omega^2 + \phi(t)]q = 0, \tag{3.1}$$

in which $\phi(t)$ is a random process. The following development is an adaptation of those of Infante [1] and Ariaratnam [11]. It is assumed that the function $\phi(t)$ (1) is continuous on the interval $0 < t < \infty$ with probability 1, (2) is weakly stationary (that is, the first and second moments are stationary), and (3) satisfies an ergodic property which guarantees the equality of time averages and ensemble averages with probability 1.

The substitution of $y = q \exp(\xi t)$ into equation (3.1) results in the governing equation

$$\ddot{y} + [\omega^2 - \xi^2 + \phi(t)]y = 0. \tag{3.2}$$

With the notation $y_1 = y$ and $y_2 = \dot{y}$, consider the positive-definite function

$$v = \alpha^2 y_1^2 + y_2^2, \tag{3.3}$$

which is the square of the norm of the vector $\{\alpha y_1, y_2\}$. The parameter α is to be determined. The time derivative of $v(t)$ along the solution trajectories of equation (3.2) is given by

$$\dot{v} = -2[\alpha^2 + \xi^2 - \omega^2 - \phi(t)]y_1 y_2,$$

and can be bounded as follows:

$$\dot{v} \leq [|\alpha^2 + \xi^2 - \omega^2 - \phi(t)|(1/\alpha)]v(t). \tag{3.4}$$

Integration along the time axis results in the following bound for $v(t)$:

$$v(t) \leq v(0) \exp\left\{\int_0^t \left[|\alpha^2 + \xi^2 - \omega^2 - \phi(t)|\frac{1}{\alpha}\right] dt\right\}. \tag{3.5}$$

Also, by the given assumptions,

$$\lim_{t \to \infty} \frac{1}{t} \int_0^t [|\alpha^2 + \xi^2 - \omega^2 - \phi(t)|/\alpha] dt = E[|\alpha^2 + \xi^2 - \omega^2 - \phi(t)|/\alpha], \tag{3.6}$$

with probability 1. Therefore, equation (3.5) takes the form

$$v(t) \leq v(0) \exp(tE[|\alpha^2 + \xi^2 - \omega^2 - \phi(t)|/\alpha]\}. \tag{3.7}$$

The inverse transformation from $y(t)$ back to $q(t)$ yields

$$E[|\delta - \phi(t)|] \leq 2\xi(\delta + \omega^2 - \xi^2)^{1/2} \tag{3.8}$$

for the almost-sure asymptotic stability of the trivial solution $q = 0$ in equation (3.1); the parameter $\delta = \alpha^2 + \xi^2 - \omega^2$ has been introduced for convenience for the following analysis. The Schwarz inequality is utilized in equation (3.8), resulting in the following condition:

$$E[\phi^2(t)] < 4\xi^2(\delta + \omega^2 - \xi^2) - \delta^2 + 2\delta E[\phi(t)]. \tag{3.9}$$

Since $\delta = \alpha^2 + \xi^2 - \omega^2$ is yet to be determined by choice of α, the right side of equation (3.9) can be optimized with respect to δ to obtain the largest region of stability. This procedure gives the sufficient stability condition

$$E[\phi^2(t)] - (E[\phi(t)])^2 < 4\xi^2\{\omega^2 + E[\phi(t)]\}, \tag{3.10}$$

as determined by Infante [1]. Equation (3.10) provides a sufficient criterion for almost-sure-stability of equation (3.1) and can be computed explicitly if the mean and mean-square value of $\phi(t)$ are known. Moreover, when $E[\phi(t)] = 0$, then $E[\phi^2(t)]$ is the mean-square value of the function $\phi(t)$ and can be calculated by

$$E[\phi^2(t)] = \int_0^\infty S(f)\,df, \tag{3.11}$$

where $S(f)$ is the power spectral density function of $\phi(t)$ [18, 19]. This result will be useful for an example.

It is interesting to note that condition (3.10) is related to Lyapunov exponents in the following way: it insures that the Lyapunov exponents associated with the modes at rest all have negative real parts.

4. THE POWER SPECTRAL DENSITY FUNCTION FOR DUFFING'S EQUATION

In this section a specific equation is to be examined, which can undergo chaotic motions, and for which the power spectral density can be estimated. These results are from the work of Brunsden et al. [2, 3]. The specific equation to be considered is a version of Duffing's equation with negative linear stiffness; it will be used to model the active mode of the multi-mode system.

The equation to be considered is

$$\ddot{u} - u + \alpha u^3 = \varepsilon(-\zeta\dot{u} - \delta u^2 \dot{u} + \gamma \cos \omega t). \tag{4.1}$$

For $\alpha > 0$ (this is not the same α as from section 3) this represents a buckled-beam type Duffing oscillator with a van der Pol type damping term. Written in first order form with $z = (u, \dot{u}) = (u, v)$ it becomes

$$\dot{z} = f(z) + \varepsilon g(z, t), \qquad z \in \mathbf{R}^2, \tag{4.2}$$

$$f(z) = \begin{pmatrix} f_1(u, v) \\ f_2(u, v) \end{pmatrix} = \begin{pmatrix} v \\ u - \alpha u^3 \end{pmatrix}, \qquad g(z, t) = \begin{pmatrix} g_1(u, v, t) \\ g_2(u, v, t) \end{pmatrix} = \begin{pmatrix} 0 \\ \gamma \cos(\omega t) - \zeta v - \delta u^2 v \end{pmatrix}.$$

The parameter ε will be considered to be small: $0 \leq \varepsilon \ll 1$.

For $\varepsilon = 0$ the unperturbed phase plane of equation (4.2) has a pair of homoclinic orbits $\Gamma_{\pm 1}$ to the saddle point $P_0 = (0, 0)$. Solutions on the unperturbed homoclinic orbits based at $(\pm\sqrt{2\alpha}, 0)$ at $t = 0$ are given by

$$q_+^0(t) = (u^0(t), v^0(t)) = (\sqrt{2\alpha}\,\mathrm{sech}\,(t), -\sqrt{2\alpha}\,\mathrm{sech}\,(t)\tanh\,(t)), \qquad q_-^0(t) = -q_+^0(t). \tag{4.3a, b}$$

Melnikov's method describes the fate of orbits which remain near $\Gamma_{\pm 1}$ when $\varepsilon \neq 0$. It involves the computation of the Melnikov function, $M(t_0)$; if $M(t_0)$ has simple zeros then the stable and unstable manifolds of the $\varepsilon \neq 0$ continuation of P_0, $P_\varepsilon(t)$, intersect transversely and chaotic motions exist near $\Gamma_{\pm 1}$ [17]. (These chaotic motions do not constitute a strange attractor since almost all orbits generally escape from the neighborhood of $\Gamma_{+1} \cup \Gamma_{-1}$.) The Melnikov function is given by [17]

$$M(t_0) = \int_{-\infty}^{\infty} f(q_+^0(t)) \wedge g(q_+^0(t), t+t_0)\, dt, \qquad (4.4)$$

where the wedge product is defined as $a \wedge b = a_1 b_2 - a_2 b_1$. For the equation being considered,

$$M(t_0) = \int_{-\infty}^{\infty} v^0 (\gamma \cos(\omega(t+t_0)) - \xi v^0 + \delta u^{0^2} v^0)\, dt$$

$$= \gamma\sqrt{2\alpha}\, \pi\omega \sin(\omega t_0) \operatorname{sech}\left(\frac{\pi\omega}{2}\right) - \tfrac{4}{3}\xi\alpha + \tfrac{16}{15}\alpha^2 \delta. \qquad (4.5)$$

In order to keep motions trapped in a neighborhood of $\Gamma_{+1} \cup \Gamma_{-1}$, the perturbation is designed so that the last two terms in equation (4.5) cancel. This implies that almost all orbits starting near $\Gamma_{+1} \cup \Gamma_{-1}$ are trapped in a neighborhood D of $\Gamma_{+1} \cup \Gamma_{-1}$ of width $\varepsilon\gamma\sqrt{2\alpha}\, \pi\omega \operatorname{sech}(\pi\omega/2)$ [2, 3]. Under these circumstances a bi-infinite sequence $\{a_n\}_{n=-\infty}^{\infty}$, with $a_n = 0$ or 1, is assigned to each initial condition $z(0)$ which reflects the behavior of the resulting solution $z(t)$. A typical solution $z(t)$ can thus be approximated by

$$z(t) = \sum_{j=-\infty}^{\infty} (-1)^{a_j(t)} q_+^0(t - T_j). \qquad (4.6)$$

This approximation assumes that the chaotic motion can be modeled as a sequence of near-homoclinic motions, randomly switching between Γ_+ and Γ_-. The term $a_j(t)$ represents a discrete random variable taking the value of 0 or 1 with equal probability (i.e., $\Pr(a_j = 1) = \Pr(a_j = 0) = \tfrac{1}{2}$), depending on whether the jth extreme value of $u(t)$ occurs near $+\sqrt{2\alpha}\ (a_j = 0)$ or near $-\sqrt{2\alpha}\ (a_j = 1)$. T_j is a real random variable which marks the time at which the jth maximum in $|u(t)|$ occurs and $q_+^0(t)$ is the unperturbed homoclinic solution on Γ_+. Moreover, the random processes $a_j(t)$ and $T_{j+1} - T_j$ are assumed to be uncorrelated and stationary [2, 3]. These assumptions are consistent with those in section 3 and allow one to apply results from almost-sure-stability to motions modeled by equation (4.1). They cannot be rigorously proved, but are reasonable and have proved useful in computing power spectra of chaotic attractors; see the papers of Brunsden et al. [2, 3] for more details and numerical results.

Since we wish to apply Fourier transforms, $z(t)$ is multiplied by a window function $\{g_L(t) = 0, |t| > L,\ g_L(t) = 1, |t| < L\}$ and the windowed solution $Z_L(t)$ is written as a convolution integral $Z_L(t) = \int_{-\infty}^{\infty} a_L(\eta - t) z(\eta)\, d\eta$, where $a_L(\eta) = g_L(\eta) \sum_j (-1)^{a_j} \delta(\eta + T_j)$ is a finite random sequence of delta functions (i.e., shot noise).

Taking the Fourier transform of equation (4.6) gives the power spectral density function of $z(t)$ as

$$S_z(f) = \lim_{k \to \infty} \frac{1}{2k} |\bar{a}_k(f)|^2 \cdot |\bar{q}_+^0(f)|^2, \qquad (4.7)$$

where $\bar{a}_k(f)$ and $\bar{q}_+^0(f)$ denote the Fourier transforms of $a_k(t)$ and $q_+^0(t)$, respectively. Using the definition of $a_k(t)$ and the central limit theorem gives the power spectral density function in the form

$$S_z(f) = (1/T) 2\pi^2 \alpha \operatorname{sech}^2(\pi^2 f), \qquad (4.8)$$

where T is the mean time-gap between passages (approximately) around Γ_{+1} or Γ_{-1}. To compute the value of T the unperturbed system is linearized near the saddle point P_0. Then for a neighborhood D near the saddle point the mean gap between such "events" is the summation of the time spent inside and outside this neighborhood during an average event. The time spent inside the region D is controlled by how close the solutions are to the stable manifold of $P_\varepsilon(t)$ on entry. By linear estimates, this time is given by $\ln(\mu/d)/\lambda_+$, where d represents the distance from the stable manifold on entry, μ is the size of region D, and λ_+ is the outgoing eigenvalue of P_0. For small ε, the time spent by the solution outside the region D is $O(1)$ and is independent of ε to leading order. By assembling this information and recalling that d is controlled by the splitting of the manifolds, which the Melnikov calculation shows to be $O(\varepsilon M(t_0))$, one finds that the typical gap between structures is

$$T \simeq K - \ln(\varepsilon \sup_{t_0} |M(t_0)|)/\lambda_+, \tag{4.9}$$

where K is a constant which can be determined by a single experiment (K is the sum of the time spent outside of the region D and $\ln(\mu)/\lambda_+$). With expressions (4.9) and (4.8), the power spectral density is known approximately in terms of the system parameters. By use of expression (3.11) the mean square can be determined, and hence almost-sure-stability results can be obtained for systems which have dynamics with an active mode governed by a Duffing equation of the form (4.2), with parameter conditions as provided above.

5. EXAMPLE

A mechanical system which consists of a lumped mass of mass M with an attached pendulum of mass m is shown in Figure 1. The equations of motion describing the

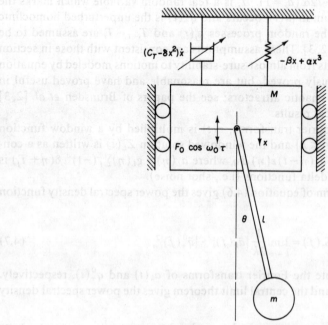

Figure 1. Model of the mechanical system.

dynamics of the vertical displacement of M, $x(t)$, and the angular displacement of M, $\theta(t)$, are

$$(M+m)\ddot{x} - (ml\sin\theta)\ddot{\theta} - ml\dot{\theta}^2\cos\theta - \bar{\beta}x + \bar{\alpha}x^3 + \bar{C}_T\dot{x} - \bar{\delta}x^2\dot{x} = F_0\cos(\omega_0\tau), \quad (5.1a)$$

$$(-ml\sin\theta)\ddot{x} + ml^2\ddot{\theta} + \bar{C}_R\dot{\theta} + mgl\sin\theta = 0. \quad (5.1b)$$

Upon introducing the rescaled variables and parameters

$$x = lx_1, \quad \rho = m/(M+m), \quad \alpha = \bar{\alpha}l^2/\bar{\beta}, \quad \varepsilon\xi = [\bar{C}_T/(M+m)][1/\omega_1],$$
$$\varepsilon\delta = [(\bar{\delta}l^2)/(M+m)][1/\omega_1], \quad \varepsilon\gamma = F_0/(\bar{\beta}l),$$
$$\omega_1\tau = t, \quad \omega_1^2 = \bar{\beta}/(M+m), \quad \omega = \omega_0/\omega_1, \quad k = (g/l)(1/\omega_1)^2,$$
$$C_R = (\bar{C}_R/ml^2)(1/\omega_1),$$

equations (5.1) take on the following non-dimensional form, with $x_2 = \theta$:

$$\ddot{x}_1 - (\rho\sin x_2)\ddot{x}_2 - \rho\dot{x}_2^2\cos(x_2) - x_1 + \alpha x_1^3 + \varepsilon\xi\dot{x}_1 - \varepsilon\delta x_1^2\dot{x}_1 = \varepsilon\gamma\cos(\omega t), \quad (5.2a)$$

$$-\sin(x_2)\ddot{x}_1 + \ddot{x}_2 + C_R\dot{x}_2 + k\sin(x_2) = 0. \quad (5.2b)$$

Here ε represents a small parameter: i.e., it is assumed that the damping and forcing associated with mass M are small. Equations (5.2) admit a stationary solution of the form

$$x_1(t) = \bar{x}_1(t), \quad x_2(t) = 0. \quad (5.3)$$

To investigate the stability of this solution one applies a small perturbation to equations (5.3):

$$x_1(t) = \bar{x}_1(t) + \mu q_1(t), \quad x_2(t) = \mu q_2(t), \quad (5.4a, b)$$

where $0 < |\mu| \ll |\varepsilon| \ll 1$. Substituting equations (5.4) into equations (5.2), expanding in powers of μ, and recalling that $\bar{x}_1(t)$ satisfies the equation

$$\ddot{\bar{x}}_1 + \varepsilon\xi\dot{\bar{x}}_1 - \varepsilon\delta\bar{x}_1 + \alpha\bar{x}_1^3 = \varepsilon\gamma\cos\omega t, \quad (5.5)$$

one finds the following first order variational equations which govern q_1 and q_2:

$$\ddot{q}_2 + C_R\dot{q}_2 + (k - \ddot{\bar{x}}_1)q_2 = 0, \quad \ddot{q}_1 + q_1 + 3\alpha\bar{x}_1^2 q_1 + \varepsilon\xi\dot{q}_1 - 2\varepsilon\delta q_1\bar{x}_1\dot{\bar{x}}_1 - \varepsilon\delta\dot{q}_1\bar{x}_1^2 = 0. \quad (5.6a, b)$$

Local stability of the trivial solution of equation (5.6a) implies local stability of solutions (5.3). Note that if $x_2 \to 0$ as $t \to \infty$ then the full equation of motion (5.2) reduces to equation (5.5). The stability of $\bar{x}_1(t)$, determined by examining equation (5.6b), cannot be determined by using the present methods and it is not even clear what the "stability" or "instability" of $\bar{x}_1(t)$ would imply since it is chaotic in nature and necessarily has unstable and stable components: that is, it represents motion on a strange attractor which is itself attracting, but solutions in the attractor are mutually divergent. In fact, attempts to prove that q_1 remains bounded fail, as they must, since $\bar{x}_1$ is chaotic. Thus only equation (5.6a) is of interest: i.e., only the Lyapanov exponents associated with θ need to be considered.

Based on the criterion obtained in section 3, the solutions (5.3) are almost-surely stable if

$$E[\phi^2(t)] - (E[\phi(t)])^2 < C_R^2(k + E[\phi(t)]), \quad (5.7)$$

where $\phi(t) = \ddot{\bar{x}}_1(t)$. The mean value of $\ddot{\bar{x}}_1(t)$, the acceleration of the lumped mass m is zero, that is $E[\ddot{\bar{x}}_1(t)] = 0$, since it has been assumed that there is no bias in the a_j's. Equation (5.7) then simplifies to

$$E[\ddot{\bar{x}}_1^2(t)] < C_R^2 k. \quad (5.8)$$

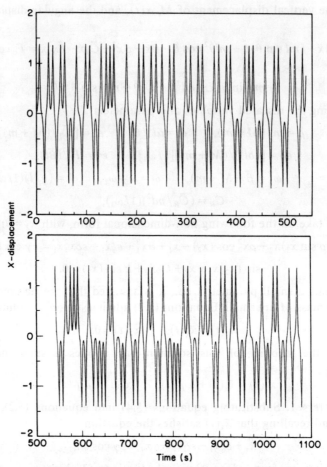

Figure 2. Chaotic response of the mechanical system (x-component with $\theta = 0$). $\varepsilon = 0 \cdot 01$, $F_0 = 1 \cdot 00$, $\omega_0 = 1 \cdot 00$, $\xi = 40 \cdot 0$, $\delta = 49 \cdot 80$.

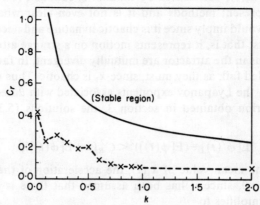

Figure 3. Simulation (∗ - - - ∗) and analytical (———) results for the almost-sure-stability of solution (5.3). $\varepsilon = 0 \cdot 01$, $\omega = 1 \cdot 00$, $\gamma = 1 \cdot 00$, $\xi = 40 \cdot 0$, $\delta = 49 \cdot 8$, $\mu = 0 \cdot 10$, $\alpha = 1 \cdot 00$.

The left side of equation (5.8) is the mean square value of $\ddot{\bar{x}}_1(t)$ which can be calculated via the relation

$$E[\ddot{\bar{x}}_1^2(t)] = \int_0^\infty (2\pi f)^4 S_z(f) \, df, \tag{5.9}$$

where $S(f)$ is the spectral density function of $\bar{x}_1(t)$. By using the results obtained in section 4, one obtains

$$E[\ddot{\bar{x}}_1^2(t)] = \int_0^\infty (2\pi f)^4 \frac{2\pi^2}{T} \alpha \operatorname{sech}^2(\pi^2 f) \, df = \frac{\alpha}{T} \frac{32}{\pi^4} \int_0^\infty \xi^4 \operatorname{sech}^2 \xi \, d\xi = \left(\frac{14}{15}\right)\left(\frac{\alpha}{T}\right). \tag{5.10}$$

Equations (5.8) and (5.10) reveal that the stationary solution (5.3) is stable with probability one if

$$C_R^2 k > (14/15)(\alpha/T), \tag{5.11}$$

where T is the mean gap between "events", α is the non-dimensionalized non-linearity parameter, k is the non-dimensionalized spring constant, and C_R is the non-dimensionalized damping coefficient for the θ-component.

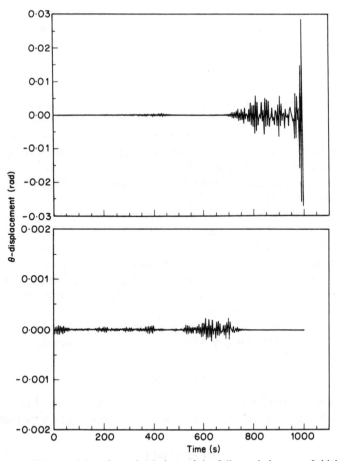

Figure 4. Response of the pendulum from simulations of the full coupled system. Initial data: $\theta = 0.0001$, $\dot\theta = 0.0000$. (a) Unstable response: $(\alpha/\beta) = 1.00$, $F_0 = 1.00$, $\omega_0 = 1.00$, $k = 0.9806$, $\xi = 40.0$, $\delta = 49.80$, $C_R = 0.07$; (b) asymptotically stable response: $(\alpha/\beta) = 1.00$, $F_0 = 1.00$, $\omega_0 = 1.00$, $k = 0.9806$, $\xi = 40.0$, $\delta = 49.80$, $C_R = 0.08$.

Results from numerical experiments conducted on equation (5.2) are presented in order to investigate the predictive value of the theory as given by equation (5.11). To this end, a chaotic solution of equation (5.5) must first be observed and used to compute T. By letting $\alpha = 1.00$, $\xi = 40.0$, $\delta = 49.8$, $\varepsilon = 0.01$, $\gamma = 1.00$ and $\omega = 1.00$ in equation (5.5) and using initial conditions near the origin, a chaotic motion is observed; it is shown in Figure 2. With these parameter values and $\rho = 0.10$ in equation (5.2) several cases were investigated by varying the values of C_R and k and determining the region of asymptotic stability of solution (5.3) for th full coupled equations (5.2). The numerical results and the analytic predictions are plotted in Figure 3 for comparison.

The results shown in Figure 3 were obtained by providing the system with small initial conditions. For the $(x, \dot{x})$ co-ordinates this corresponds to starting near the chaotic solution of (5.6b) (since such motions frequently return to a small neighborhood of the origin). For the $(\theta, \dot{\theta})$ co-ordinates small initial conditions were used since the local stability of $\theta \equiv 0$ was of interest. The stability bounds determined from simulations were determined by searching for inconclusive results above which (in (C_R, k) space) the motion was stable and below which it was unstable. This parameter range was quite narrow and the data points shown represent quite precise values for the stability bounds. In Figure 4 are shown θ vs. t for both an unstable and a stable case (note the difference in the θ scales used).

The results from these simulations are consistent with those from the analysis, since above the theoretical curve the pendulum damping is more than sufficient to result in the stability of the $\theta = 0$ solution.

6. DISCUSSION

The procedure considered in this paper is an outline illustration of one method for determining the stability of modes at rest in a chaotic system. In other cases one will be able to carry out similar studies, possibly by obtaining the required statistical information via direct simulations of the active mode or modes. Such a procedure could provide better stability estimates since more detailed statistics can be obtained. The example presented here was chosen since the stability bound was explicitly computable by combining the results of Brunsden *et al.* [2, 3] and Infante [1].

The non-linear damping term in equation (4.2) was required for the analysis in order to bound the motions for $0 < \varepsilon \ll 1$ near the $\varepsilon = 0$ homoclinic orbit. This term represents a van der Pol type of damping and does not arise naturally in structural oscillations (since it can add energy to the system). However, recent work by Brunsden *et al.* [2] indicates that the spectral density function obtained for the present example ($\delta \neq 0$) is a good measure for that obtained for the usual Duffing oscillator ($\delta = 0$); this has even been confirmed by using simulations and physical experiments [2, 3].

As methods such as those in section 4 are developed more fully, the use of results from the theory of the almost-sure-stability will be more widely applicable to chaotic systems.

ACKNOWLEDGMENT

This work was supported by grants from NSF (MSM 8613294) and DARPA. Also, the authors are grateful to N. Sri Namachchivaya for several helpful comments.

REFERENCES

1. E. F. INFANTE 1968 *American Society of Mechanical Engineers Journal of Applied Mechanics* **35**, 7–12. On the stability of some linear nonautonomous random systems.

2. V. BRUNSDEN, J. COTTREL and P. HOLMES 1989 *Journal of Sound and Vibration* **130**, 1-25. Power spectra of chaotic vibrations of a buckled beam.
3. V. BRUNSDEN and P. HOLMES 1987 *Physical Review Letters* **58**, 1699-1702. Power spectra of strange attractors near homoclinic orbits.
4. R. F. HENRY and S. A. TOBIAS 1961 *Journal of Mechanical Engineering Science* **3**, 163-173. Modes at rest and their stability in coupled nonlinear systems.
5. T. K. CAUGHEY and A. H. GRAY, JR. 1965 *American Society of Mechanical Engineers, Journal of Applied Mechanics*, 365-372. On the almost sure stability of linear dynamic systems with stochastic coefficients.
6. R. A. IBRAHIM 1985 *Parametric Random Vibration*. New York: John Wiley.
7. F. KOZIN 1989 *Probabilistic Engineering Mechanics* (to appear). Some results on stability of stochastic dynamical systems.
8. F. KOZIN 1969 *Automatica* **5**, 95-112. A survey of stability of stochastic systems.
9. F. KOZIN 1963 *Journal of Mathematical Physics* **42**, 59-67. On almost-sure-stability of linear systems with random coefficients.
10. S. T. ARIARATNAM and B. L. LY 1989 *American Society of Mechanical engineers, Journal of Applied Mechanics* **56**, 175-178 Almost-sure-stability of some linear stochastic systems.
11. V. T. ARIARATNAM 1987 *Nonlinear Stochastic Dynamic Engineering Systems, IUTAM*. Stochastic stability of modes at rest in coupled nonlinear systems.
12. P. J. HOLMES, 1986 *Journal of Fluid Mechanics*, **162**, 365-388. Chaotic motions in a weakly nonlinear model for surface waves.
13. S. CILIBERTO and J. GOLLUB 1985 *Journal of Fluid Mechanics* **158**, 381-398. Chaotic mode competition in parametrically forced surface waves.
14. S. W. SHAW 1988 *Journal of Sound and Vibration* **124**, 329-343. Chaotic dynamics of a slender beam rotating about its longitudinal axis.
15. T. KAPITANIAK 1987 *Journal of Sound and Vibration* **114**, 588-592. Quantifying chaos with amplitude probability density function.
16. A. J. LICHTENBERG and M. A. LIEBERMAN 1983 *Regular and Stochaotic Motion*. New York: Springer-Verlag.
17. J. GUCKENHEIMER and P. HOLMES 1983 *Nonlinear oscillations, Dynamical Systems and Bifurcations of Vector Fields*. New York: Springer-Verlag.
18. N. C. NIGAM 1983 *Introduction to Random Vibrations*. Cambridge, Massachusetts: MIT Press.
19. J. D. ROBSON 1964 *An Introduction to Random Vibration*. Edinburgh University Press.

B. Poddar*

F. C. Moon
Professor,
Mem. ASME

S. Mukherjee
Professor,
Mem. ASME

Department of Theoretical and Applied
Mechanics,
Cornell University,
Ithaca, NY 14853

Chaotic Motion of an Elastic-Plastic Beam

A numerical study is presented here which suggests that chaotic motion is possible from periodic excitation of an elastic-plastic beam. Poincaré maps of the motion reveal a fractal-like structure of the attractor. The results suggest that geometric and material nonlinearities in solid mechanics problems may lead to extreme sensitivity to small changes in parameters and resulting unpredictability. These results may explain the total disagreement of nine finite element codes in the analysis of the transient response of an elastic-plastic beam, that has been reported recently by Symonds and his coworkers.

Introduction

Recent research in dynamical systems has shown that the time histories of deterministic nonlinear systems can be very sensitive to initial conditions and may exhibit chaotic or random-like behavior (see, e.g., Holmes and Moon, 1983). The dynamic behavior of elastic-plastic structures is a highly nonlinear system and may, in some problems, have very sensitive solutions which make the precise time history unpredictable. In two recent papers (Symonds and Yu, 1985; Symonds et al. 1985), the authors have presented numerical results for the transient response of an elastic-plastic beam using nine different finite element codes and have obtained different time histories from all the codes. Considering all the codes, Symonds and Yu (1985) report that "12 solutions may be considered strictly comparable." For a certain narrow range of initial displacements, the subsequent free motion appears impossible to predict and the different computer programs could not even agree as to the final equilibrium configuration of the beam. Clearly, these results suggest extreme sensitivity of the system to initial conditions, such that the slightest uncertainty in initial conditions causes extreme variation in the response.

In Symonds' problem, a beam has fixed pin supports at both ends and is subjected to a transient force. In the two papers quoted above, Symonds et al. (1985) have studied this problem by both finite element analysis and also by employing a simplified Shanley (1947) model. This paper describes further analysis of this problem, again using a Shanley model, for the case of free and forced vibrations of the beam. Evidence is presented here that the "counter-intuitive" problem of Symonds represents an example of an elastic-plastic chaotic vibration problem. In such problems both the geometric and constitutive stress-strain nonlinearities play an important role. While a highly idealized model for the plasticity in the beam is used here (the Shanley model), it is expected that similar unpredictable phenomena may be a generic problem involving the dynamics of inelastic structures with geometric nonlinearities.

Governing Differential Equations

The Shanley (1947) model, as used here, approximates a pin-ended beam of length 2ℓ, with a uniform rectangular cross section of area $A = bh'$, by two rigid links, each of length ℓ (Fig. 1). These links are pinned at the ends and are joined together at the center by an elastic-plastic element. This elastic-plastic element has two short flanges placed at distances $h/2$ above and below the beam axis, respectively. Each flange is elastic-perfectly plastic with a yield stress, both in tension or compression, of σ_y. Following Symonds et al. (1985),

$$\dot{a} = \ell\dot{\phi}, \quad \dot{e} = \ell\phi\dot{\phi} \tag{1}$$

where ϕ is the angular deflection of a link of the beam, a is the

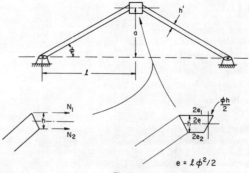

Fig. 1

---

*Currently Assistant Professor, Department of Mechanical Engineering, University of Miami, Coral Gables, FL 33124. Assoc. Mem. ASME

Contributed by the Applied Mechanics Division for publication in the JOURNAL OF APPLIED MECHANICS.

Discussion on this paper should be addressed to the Editorial Department, ASME, United Engineering Center, 345 East 47th Street, New York, N.Y. 10017, and will be accepted until two months after final publication of the paper itself in the JOURNAL OF APPLIED MECHANICS. Manuscript received by ASME Applied Mechanics Division, May 16, 1986; final revision July 15, 1987.

transverse central deflection, and $2e$ is the extension of the center line of the beam. The semiextension rates of the flanges, therefore, become

$$\dot{e}_1 = (a + h/2)\dot{a}/\ell, \; \dot{e}_2 = (a - h/2)\dot{a}/\ell \quad (2)$$

the subscripts 1 and 2 denoting the upper and lower flanges, respectively. The force rates $\dot{N}_1$ and $\dot{N}_2$ in the flanges are

$$\dot{N}_1 = \begin{cases} \dfrac{E'}{\ell^2}(a+h/2)\dot{a} & \text{elastic} \\ \\ 0 & \text{plastic} \end{cases} \quad (3)$$

and similarly for $\dot{N}_2$ with $(a + h/2)$ replaced by $(a - h/2)$ in the above. The assumption here is that the strain rates, concentrated in the flanges of the plastic element, are obtained by dividing $\dot{e}_1$ and $\dot{e}_2$ by the semilength ℓ of the beam. Also, in the above, $E' = EA/2$ in terms of the effective Shanley model Young's modulus E and cross section area A of the beam.

The governing differential equation of motion of the beam is written for the case where a is moderately large so that the membrane forces N_1 and N_2 must be included in the analysis. The resulting equation is

$$\ddot{a} + 2\zeta\omega_R \dot{a} + \frac{3}{\rho\ell^2} Na - \left(\frac{3}{\rho\ell^2}\right) M = f(t) \quad (4)$$

where $N = N_1 + N_2$ and $M = (N_2 - N_1)h/2$ are the resultant axial force and moment, respectively, on half the beam, ρ is the mass per unit length of the beam, and ζ is the damping factor. Also, $f(t) = 1.5F(t)/\rho$ in terms of $F(t)$, the transverse force per unit length applied on the beam, and

$$\omega_R = \frac{h}{\ell^2}\sqrt{\frac{1.5E'}{\rho}},$$

the natural frequency of small elastic vibrations of the beam about its initial equilibrium position.

It is convenient to write equation (4) as a system of four first order differential equations

$$\dot{y}_1 = y_2$$
$$\dot{y}_2 = -2\zeta\omega_R y_2 - \frac{3}{\rho\ell^2} y_1 y_3 + \frac{3}{\rho\ell^2} y_4 + f(t)$$
$$\dot{y}_3 = \dot{N}_1 + \dot{N}_2$$
$$\dot{y}_4 = (\dot{N}_2 - \dot{N}_1)h/2 \quad (5)$$

where $a \equiv y_1$, $\dot{a} \equiv y_2$, $N \equiv y_3$, $M \equiv y_4$, and $\dot{N}_1$ and $\dot{N}_2$ are given by equation (3). It is easy to show that in this case the yield surface in force space is diamond shaped and is bounded by the pairs of lines

$$|N/N_y - M/M_y| = 1, \; |N/N_y + M/M_y| = 1$$

where $N_y = A\sigma_y$ and $M_y = Ah\sigma_y/2$.

It is very important to note here that the governing equation (4) has the quantities N and M which depend on the history of motion. Hence, it is convenient to rewrite equation (4) as a system of four first order differential equations.

Initially y_1 and y_2 are prescribed and y_3 and y_4 are determined from y_1 assuming monotonic deflection up to the initial value of y_1.

Equilibrium Points, Stability and Frequency Analysis

As mentioned before, the natural frequency of small elastic vibrations is ω_R, which is given below equation (4). A very interesting situation arises if elastic oscillations follow touching of the yield surface, as is often the case for free vibrations, especially with damping. Such a situation in force space is depicted in Fig. 2. If purely elastic oscillations follow the last

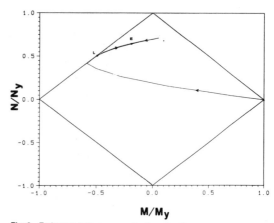

Fig. 2 Trajectory in force space for free vibrations: $a_o = 0.9$cm, $\dot{a}_o = 0$, $\zeta = 0.2$

touching of the yield surface at L, the system remembers the values M_L, N_L, and a_L at this point. Now the governing equation is (4) with the integrated values

$$N = (E'/\ell^2)(a^2 - a_L^2) + N_L,$$
$$M = -(h^2E'/2\ell^2)(a - a_L) + M_L \quad (6)$$

Substituting the above into equation (4) and collecting terms, one obtains the equation

$$\ddot{a} + 2\zeta\omega_R \dot{a} + \alpha a + \beta a^3 - \delta = 0 \quad (7)$$

where $f(t)$ has been set to zero. The coefficients α, β, and δ in the above are

$$\alpha = (3/(\rho\ell^2))[N_L - E'a_L^2/\ell^2 + h^2E'/(2\ell^2)]$$
$$\beta = (3/(\rho\ell^2))(E'/\ell^2)$$

and

$$\delta = (3/(\rho\ell^2))[M_L + h^2E'a_L/(2\ell^2)] \quad (8)$$

It is clear that α and δ depend upon the situation at L, and hence on the history of the motion.

Equation (7) is a form of Duffing's equation with an extra term δ. The behavior of solutions of the standard Duffing's equation, with respect to fractal basins and chaos, have been studied by many researchers. Recent papers on the subject, of particular importance to this work, are those by Moon and Li (1985) and by Dowell and Pezeshki (1986). It is important to mention again that the coefficients α and δ, in equation (7), are history dependent. Consequently, the equilibrium points of equation (7) are history dependent.

Equation (7) has three equilibrium points. A standard analysis shows that if $\alpha^3 > -(27/4)\delta^2\beta$, there is only one real equilibrium point. Otherwise, there are three. Thus, for example, if $\alpha > 0$, only one equilibrium point exists (note that β is always positive, while α and δ can have either sign). The closed form expressions for these equilibrium points are quite complex. With $\zeta = 0$, however, it is easy to show that the stability of an equilibrium point a_E is determined by γ where

$$\gamma = \pm\sqrt{-(\alpha + 3a_E^2\beta)}$$

With $\alpha + 3a_E^2\beta > 0$, the two values of γ are imaginary and the equilibrium point has periodic vibrations around it. In this, case, $\zeta > 0$ leads to the equilibrium point being stable. If $\alpha + 3a_E^2\beta < 0$, the equilibrium point is unstable.

All the transient simulations considered later in this paper admit three real equilibrium points $a_E^{(1)} > a_E^{(2)} > a_E^{(3)}$. Of these, $a_E^{(1)} > 0$, $a_E^{(3)} < 0$ and both are stable, while the intermediate

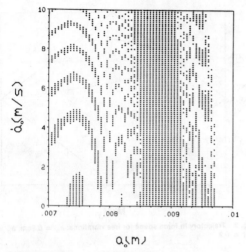

Fig. 3 Dependence of numerical simulation of the transient solution on initial conditions

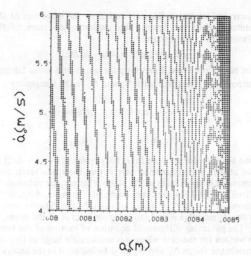

Fig. 4 Enlargement of a small region of Fig. 3

a_E is unstable. Depending on the initial conditions, the solution in the presence of damping settles down to one of the two stable equilibrium points. The phase plane trajectory corresponding to Fig. 2, for example, has the features of Fig. 1 in Dowell and Pezeshki (1986). In this case the equilibrium value $a_E < 0$. Of course, a_E can be greater than zero for other initial conditions.

The frequency of small amplitude elastic oscillations, following last touching of the yield surface at L, can be calculated from equation (7). Writing $a = a_E + \eta$ substituting the above into equation (7) and linearizing with respect to η, the damped natural frequency of small amplitude oscillations about this equilibrium point is obtained as

$$\omega_d = \omega_n \sqrt{1-(\zeta')^2} \text{ with } \omega_n = \sqrt{(3a_E^2\beta + \alpha)}$$

and

$$\zeta' = \zeta \omega_R/\omega_n \qquad (9)$$

It is clear from equations (8) and (9) that this natural frequency is history dependent.

Numerical Simulations

Parameter Values. The parameter values used here are the same as those from Symonds et al. (1985). These are

$\ell = 10$cm, $b = 2$cm, $h' = 0.4$cm, $h = 0.68h'$

$\rho = 0.216$ kg/m, $E = 120 \times 10^9$ N/m^2

$\sigma_y = 0.3 \times 10^9$ N/m^2

It should be noted that the Young's modulus of the beam material is 80×10^9 N/m^2 and the number above is the effective Young's modulus for the Shanley model as suggested by Symonds et al. (1985).

Natural Frequencies. The equations (5) are integrated numerically by using the Runge-Kutta method with time-step control. The first check on the computer program for numerical simulations has been to calculate a_E and ω_d for two cases—(a) corresponding to Fig. 2 with $a_o = 0.009$m, $\dot{a}_o = 0$ where the solution settles down to $a_E < 0$, and (b) with $a_o = 0.011$m, $\dot{a}_o = 0$ which gives $a_E > 0$. A damping factor $\zeta = 0.2$ was included in the numerical simulations in order to get small amplitude oscillations. The numerical simulation frequencies below correspond to those obtained from the time period of a small amplitude cycle before the beam settles down at its equilibrium configuration. The results are summarized below.

	a_E (cm)		ω_d (rad/s)	
	Numerical Simulation	From equation (7) with $\ddot{a} = \dot{a} = 0$	Numerical Simulation	From equation (9)
(a)	-0.4087	-0.4112	4187.6	4238
(b)	0.8209	0.8204	9622.7	9615

It should be mentioned here that the three calculated equilibrium points for case (a) above are $+0.5376$, -0.1288, and -0.4087, of which the intermediate one is unstable and the other two are stable. Similarly, for case (b) the a_E values are 0.8209, -0.557, and -0.765, of which the first and last are stable.

Free Vibrations. Moon and Li (1985) have discussed the phenomenon of extreme sensitivity of solutions of Duffing's equation to the initial values of a and $\dot{a}$. In predictable problems, the boundary between two basins of attraction in initial condition space is smooth. In chaotic problems, however, the boundary becomes fractal and subsequent motion is extremely sensitive to initial conditions (see, e.g., Grebogi et al., 1983). Symonds et al. (1985) report extreme sensitivity of the solution of this beam problem to a_o (they took ζ and $\dot{a}_o$ equal to zero) in the approximate range $0.85 < a_o < 0.98$ cm. In recent work, Genna and Symonds (1987) include damping in their simulations and report anamolous behavior over a wider range of a_o (approximately for $0.70 < a_o < 0.98$ cm).

This question of extreme dependence of the transient solution to initial conditions has been studied in this work in the following way. The initial condition space $0.7 < a_o < 1$ cm, $0.0 < \dot{a}_o < 10$ m/s has been studied. In all cases the beam starts from an initial configuration above the horizontal and any nonzero initial velocity is upwards as well. For a given simulation, the following steps have been carried out.

(a) Choose a_o, $\dot{a}_o$. Set $\zeta = 0$.
(b) Compute the solution for a certain number of cycles and find a_{max} and a_{min}. If $a_{max} < 0$ then the behavior is clearly anamolous since the beam vibrates entirely below its original straight configuration.

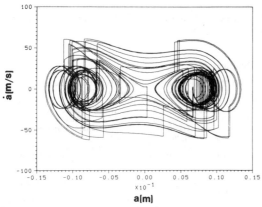

Fig. 5 Chaotic forced vibrations—first 35 forcing cycles in phase space

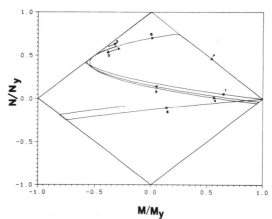

Fig. 6 Chaotic forced vibrations—first 1.5 forcing cycles in force space: $a_o = 0.009$, $\dot{a}_o = 0.0$, $\Delta\dot{a} = 38.0$ m/s, $\omega_f = 2.5\omega_R = 3926.0$ rad/s

(c) If $a_{max} > 0$, introduce $\zeta = 0.10$ and let the solution converge to a_E. Anamolous behavior is obtained if $a_E < -0.1 (a_{max} - a_{min})$.

Condition (c) is chosen to ascertain that the beam does indeed end up below its original horizontal configuration in order for the simulation to merit the anamolous classification. The results are shown in Figs. 3 and 4 where a dot denotes anamolous and a blank regular (expected) behavior. Figure 4 is an enlargement of a small region of Fig. 3. The extreme sensitivity of the simulations, to initial conditions, is evident. With $\dot{a}_o = 0$, the range of a_o, for anamolous behavior, is approximately $0.73 < a_o < 0.97$, which is in good agreement with Genna and Symonds (1987). It should be emphasized that, as expected, the details of Figs. 3 and 4 would be affected by the choice, among others, of ζ and the specific criteria used to define counterintuitive behavior. The observation of extreme sensitivity of the solution to initial conditions, however, is a salient feature of this behavior, and this feature would be retained in different types of simulations. It is clear that in this range of initial conditions, the slightest change in initial conditions can cause a drastic change in the response, and attempts at obtaining detailed numerical solutions to the problem are meaningless.

Another observation is that for all the undamped free vibration problems simulated in this work, the final limit cycle motion involves touching of the yield surface in force space without part of the trajectory moving on it. It is felt that such is the case in general. The conjecture, is, as yet, unproven but is considered highly likely.

Forced Vibrations. Two of the hallmarks of chaotic behavior are a positive Lyapunov exponent and a fractal dimension of the strange attractor on which the solution moves. One way to observe the fractal properties of an attractor is to obtain a Poincaré section of the motion. In order to explore the strange attractor one must allow the trajectory to explore many regions of the attracting set. This requires a long time simulation. For this reason, a periodically excited problem has been considered here in which the beam receives alternating positive and negative impulses. The forcing function here is a periodic series of these impulses. Each impulse is assumed to cause sudden velocity changes of $\Delta\dot{a}$ in its direction. As has been described often in the chaos literature (e.g., Dowell and Pezeshki, 1986), a suitable combination of $\Delta\dot{a}$ and forcing frequency ω_f can lead to the response being unable to decide between two equilibrium positions. Such a case is depicted in the phase plane (Fig. 5, first 35 forcing cycles) and in the force space (Fig. 6, first 1.5 forcing cycles). Physically,

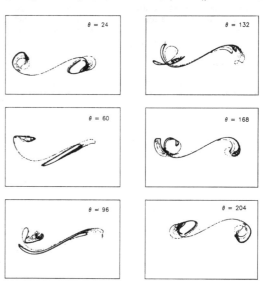

Fig. 7 Chaotic forced vibrations—Poincaré maps for 5000 forcing cycles. Same simulation as in Fig. 6.

once the beam becomes plastic, its length increases and the beam acts like an arch with new equilibrium positions above and below the original undeformed straight shape. The shallow arch dynamics is similar to a particle in a double-well potential which is known to exhibit chaotic vibrations under periodic excitation (see Tseng and Dugundji, 1971; Moon, 1980).

Poincaré sections were obtained by stroboscopically looking at the motion synchronous with the periodic impulses and projecting the resulting set of points into the amplitude-velocity plane $(a, \dot{a})$. The Poincaré sections for this problem use 5000 points and show a characteristic fractal structure (Fig. 7; see also Moon, 1980). These maps correspond to initial phase intervals of $\pi/5$ and, as expected, the picture for $\theta = 204$ deg is the reversed version of that for $\theta = 24$ deg. An enlargement of one of the Poincaré sections from Fig. 7 is shown in Fig. 8. The structure of the section is more clearly apparent in this larger picture. It should be observed here that the trajectory in force space continues to touch or move on the yield surface during this entire simulation of 5000 cycles, showing that the motion, in this case, remains elasto-plastic throughout.

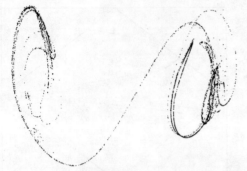

Fig. 8 Chaotic forced vibrations—Poincaré map for $\theta = 24$ deg from Fig. 7

It should be reemphasized, in conclusion, that the chaotic motion demonstrated in this paper from the simplified Shanley model is expected to be of a generic nature. Similar unpredictable phenomena are expected in the dynamics of structures in the presence of material and/or geometric nonlinearities. Research in this general direction is currently in progress at Cornell University.

Acknowledgments

This work was supported in part by a grant from the Air Force Office of Scientific Research, Aerospace Sciences Division, to Cornell University. The computing of Poincaré maps has been carried out on the NSF supported Cornell National Supercomputer Facility.

References

Dowell, E. H., and Pezeshki, C., 1986, "On the Understanding of Chaos in Duffing's Equation Including a Comparison with Experiment," ASME JOURNAL OF APPLIED MECHANICS, Vol. 53, pp. 5–9.

Genna, F., and Symonds, P. S., 1987, Private Communications.

Grebogi, C., Ott, E., and Yorke, J. A., 1983, "Fractal Basin Boundaries, Long-Lived Chaotic Transients, and Unstable-Unstable Pair Bifurcations," *Physical Review Letters*, Vol. 50, pp. 935–938.

Holmes, P. J., and Moon, F. C., 1983, "Strange Attractors and Chaos in Nonlinear Mechanics," ASME JOURNAL OF APPLIED MECHANICS, Vol. 50, pp. 1021–1032.

Moon, F. C., 1980, "Experiments on Chaotic Motions of a Forced Nonlinear Oscillator: Strange Attractors," ASME JOURNAL OF APPLIED MECHANICS, Vol. 47, pp. 638–644.

Moon, F. C., and Li, G.-X., 1985, "Fractal Basin Boundaries and Homoclinic Orbits for Periodic Motion in a Two-Well Potential," *Physical Review Letters*, Vol. 55, pp. 1439–1442.

Shanley, F. R., 1947, "Inelastic Column Theory," *Journal of the Aeronautical Sciences*, Vol. 14, pp. 261–267.

Symonds, P. S., and Yu, T. X., 1985, "Counterintuitive Behavior in a Problem of Elastic-Plastic Beam Dynamics," ASME JOURNAL OF APPLIED MECHANICS, Vol. 52, pp. 517–522.

Symonds, P. S., McNamara, J. F., and Genna, F., 1985, "Vibrations of a Pin-Ended Beam Deformed Plastically by Short Pulse Excitation," *Material Nonlinearity in Vibration Problems*, Sathyamoorthy, M., ed., AMD-Vol. 71, ASME, NY, pp. 69–78.

Tseng, W.-Y., and Dugundji, 1971, "Nonlinear Vibrations of a Buckled Beam Under Harmonic Excitation," ASME JOURNAL OF APPLIED MECHANICS, Vol. 38, pp. 467–476.

Chua's Circuit Family

SHUXIAN WU

*Chaos has been widely reported and studied using Chua's circuit. This **tutorial** paper presents a generalization of this well-known chaotic circuit into a hugh 6-parameter family of potentially chaotic circuits, henceforth referred to a **Chua's Circuit Family**. The highlight of this tutorial consists of a collection of **three-dimensional phase portraits** associated with two representative members of this family. These phase portraits show how the associated vector fields evolve from **order** into **chaos**.*

I. INTRODUCTION

Chua's circuit (Fig. 1(a)) is widely regarded as a textbook example of *chaos* for the following reasons:

a) It is the *simplest autonomous circuit which can become chaotic*. Indeed, it contains only three energy storage elements[1] (the minimum number needed for a dynamic system to be chaotic) and only one nonlinear element of the simplest type; namely a two-terminal piecewise-linear resistor.[2]

b) The chaotic behavior of Chua's circuit has been observed by computer simulation [1] and verified by laboratory measurements [2].

c) Chua's circuit has been the subject of several in-depth mathematical analyses and its *chaotic* nature has been rigorously proved [3], [4]. Indeed, Chua's circuit is the only known example of a physical system which has been shown to be chaotic using three different approaches: *computer simulation, laboratory experiments*, and *mathematical analysis*.[3] Chaos is such a complicated phenomenon that very

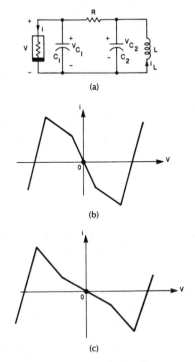

Fig. 1. Chua's circuit. (a) Circuit made of one linear resistor R, two linear capacitors C_1 and C_2, one linear inductor L, and a two-terminal resistor. (b) v–i characteristic for Chua's Circuit 1. (c) v–i characteristic for Chua's Circuit 2.

Manuscript received December 28, 1986; revised January 12, 1987.
The author is with the Department of Radio Electronics, Beijing Normal University, Beijing, The Peoples Republic of China.
IEEE Log Number 8714770.

[1] A circuit is said to be *autonomous* if it contains only time-invariant elements and dc sources. It can be proved that the most complicated behavior that can occur in an autonomous circuit having less than 3 energy storage elements is a *limit cycle*; i.e., a periodic oscillation.

[2] Many other autonomous circuits have been reported in the recent literature to be chaotic. However, they are more complicated than Chua's circuit because some contain a *nonlinear* capacitor (or inductor), or *multi-terminal* devices (e.g., transistors, unijunction transistors, op-amps, etc.), or more than one nonlinear element, or linear elements with *negative* parameters ($-R$, $-L$, or $-C$).

[3] While there are now literally thousands of papers on chaos, almost all of them have yielded to only one or two of these three complementary methods of attack. Even the classic Lorenz equation has not been proved to be chaotic in a completely rigorous manner (only a *discrete* version of Lorenz equation has been proved to be chaotic [5]).

few mathematical tools are presently available for its analysis. Consequently, most published papers on chaos are restricted to either computer simulation or experiments. Moreover, except for electrical circuits where excellent *lumped* circuit models are available [6]–[8], most chaotic physical systems reported to date are so extremely complex (e.g., turbulence in fluids, atmosphere, etc.) that just taking measurements is in itself a major undertaking, let alone modeling the system accurately. This situation explains why most of the published literature on Chaos to date are restricted to computer simulations of *hypothetical* nonlinear differential, or discrete, equations. The few equations which purported to have been derived from some real phys-

© 1987 IEEE. Reprinted, with permission, from (Proceedings of the IEEE, Vol. 75, No. 8, pp. 1022–1032; August/1987).

ical systems represent at best a gross idealization and drastic simplification of the system.

d) Chua's circuit exhibits an immense variety of nonlinear dynamical phenomena, including many typical "bifurcations" and "routes to chaos" observed in other systems. Consequently, it can be regarded as a *prototype* model of chaos which provides a quick and broad introduction to the subject of chaos for the nonspecialists.

Although much has already been written about the bifurcation phenomena in Chua's circuit [9]-[12], all published trajectories and strange attractors associated with this circuit are two-dimensional projections. For the nonspecialist on chaos, it would be more illuminating to display a family of accurately drawn trajectories emanating from a uniform grid of initial points (V_{C_1}, V_{C_2}, I_L) in the *three-dimensional* state space, where V_{C_1}, V_{C_2}, and I_L denote the voltage across capacitor C_1, the voltage across capacitor C_2, and the current in inductor L in Fig. 1(a), respectively. Such a picture is called a *phase portrait* [6]. Our main objective of this *tutorial* paper is to present a selection of some of the more revealing of these phase portraits.

II. PORTRAITS—FROM ORDER TO CHAOS

Depending on the choice of the *v-i* characteristic for the nonlinear resistor in Fig. 1(a), Chua's circuit can exhibit many distinct forms of strange attractors. Most of the papers concerning this circuit that have been published to date are based on the *v-i* characteristic shown in Fig. 1(b), which can be realized in the laboratory by using 2 op-amps [2], 1 op-amp, and 2 diodes [13], or 2 transistors and 2 diodes [14], in addition to linear resistors.

Chaos has also been observed with the *v-i* characteristic shown in Fig. 1(c) [15].[4] To avoid confusion, we will henceforth refer to the circuit in Fig. 1(a) as *Chua's Circuit 1* if the *v-i* characteristic in Fig. 1(b) is chosen, or *Chua's Circuit 2* if the *v-i* characteristic in Fig. 1(c) is chosen. Both circuits are described by the following state equations:

$$\dot{V}_{C_1} = \frac{1}{RC_1}(V_{C_2} - V_{C_1}) - \frac{1}{C_1}g(V_{C_1})$$
$$\dot{V}_{C_2} = \frac{1}{RC_2}(V_{C_1} - V_{C_2}) + I_L \quad (1)$$
$$\dot{I}_L = -\frac{1}{L}V_{C_2}$$

where $g(V_{C_1})$ is given by Fig. 1(b) for Chua's Circuit 1 and by Fig. 1(c) for Chua's Circuit 2.

Due to space limitation, we will present only a few carefully selected phase portraits associated with (1). Each phase portrait corresponds to a particular choice of values of R, C_1, C_2, L, and a particular choice of the nonlinear function $g(V_{C_1})$. Space limitation precludes a more extended discussion of the bifurcation phenomena which evolved into each phase portrait, as well as the various pertinent routes to chaos associated with Chua's circuit. The readers are therefore urged to read [9]-[11] and [15] for additional details.

[4]We have built this circuit in the laboratory and confirmed the chaotic phenomena reported in [15].

A collection of 11 phase portraits associated with Chua's Circuit 1 is shown in Figs. 2-12. The phase portrait in Fig. 2 corresponds to the case when the circuit parameters are chosen to be a point in the *blue* region in the multi-color $\alpha-\beta$ bifurcation diagram in [3, p. 1072]. In this case, the circuit is *completely stable* [7] and all trajectories approach either the equilibrium point on top or the bottom (P^+ or P^- in the *inset* labeled "1" in [3, p. 1072] of the state space. Here, the resistance R is chosen large enough so that its associated load line intersects the left-most and right-most segments of the *v-i* characteristic in Fig. 1(b). Since the small-signal resistances at these operating points are positive, it is obvious that the corresponding equilibrium points of (1) are *locally asymptotically stable*, i.e., all three eigenvalues associated with the Jacobian matrix of (1) at P^+ or P^- are in the open left-half plane. Note that the phase portrait is *odd symmetric*. It can be proved that no oscillation exists in this case.

The phase portrait in Fig. 3 corresponds to the case when the circuit parameters are chosen to be a point in the *dark-green* region in the above cited $\alpha-\beta$ bifurcation diagram. In this case, a small stable limit cycle exists in a neighborhood of the upper equilibrium point P^+, and another, its odd-symmetric image, appears in a neighborhood of P^-. This case corresponds to the *inset* labeled "2" in [3, p. 1072]. In addition, note that there exist two larger limit cycles: an *unstable* (saddle-type) limit cycle depicted by the "hollow" double-line loop and a larger *stable* limit cycle depicted by the bold solid-line loop. The unstable limit cycle cannot be observed experimentally, or even by computer simulation in reverse time because the instability is of *saddle* type [5]. Its existence, however, has been proved in [3] and can be found numerically by the "shooting method" [16]. Its dynamic range is found to lie within the range of the three inner segments of the *v-i* characteristic. In contrast, the dynamic range associated with the outermost limit cycle covers all five segments, including the two outer segments. Consequently, readers interested in looking at the time waveforms associated with this limit cycle must use the five-segment *v-i* characteristic in Fig. 1(b), and not the three-segment characteristic chosen in [3], [4], [13].

The phase portrait in Fig. 4 corresponds to the case where the circuit parameters are chosen to be a point in the *light-green* region in the $\alpha-\beta$ bifurcation diagram (see *inset* "3"). In this case, the pair of odd-symmetric "1-revolution" limit cycles in Fig. 3 have evolved and bifurcated into a pair of "2-revolution" limit cycles. This phenomenon is referred to in the literature on chaos as a "period-doubling" phenomenon because the "period" of the oscillation just after the bifurcation occurs is (approximately) equal to *twice* the period of the oscillation immediately prior to the bifurcation. This phenomenon can be dramatically demonstrated by looking at the evolution of the "frequency spectrum" of any one of the three signal waveforms: at the point of bifurcation, a new frequency line at *half* the earlier frequency suddenly appears in the spectrum analyzer. Just as in Fig. 3, an unstable limit cycle and a stable limit cycle are also present in the phase portrait in Fig. 4. In fact, these two limit cycles will appear in *all* subsequent phase portraits displayed in this paper and we will not identify or comment on them any further to avoid redundancy.

The phase portrait in Fig. 5 corresponds to the case where the circuit parameters are chosen to be a point in the *light-*

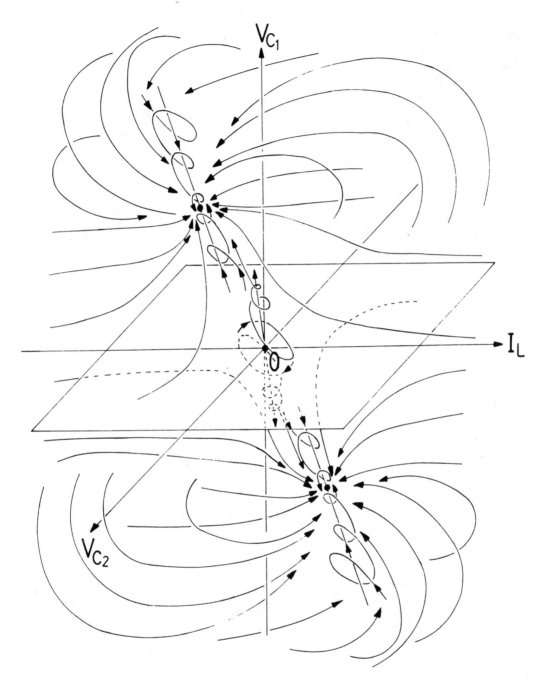

Fig. 2. Phase portrait for Chua's Circuit 1 when the circuit parameters are chosen so that all trajectories tend to one of two stable equilibrium points.

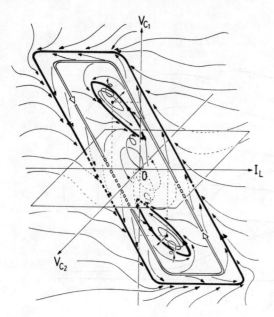

Fig. 3. Phase portrait for Chua's Circuit 1 when the circuit parameters are chosen so that a *period-1* oscillation exists in a neighborhood of the upper equilibrium point P^+ and the lower equilibrium point P^-.

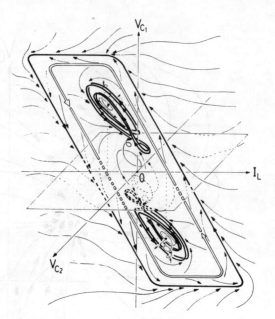

Fig. 5. Phase portrait for Chua's Circuit 1 when the circuit parameters are chosen so that a *period-4* oscillation exists in a neighborhood of the upper equilibrium point P^+ and the lower equilibrium point P^-.

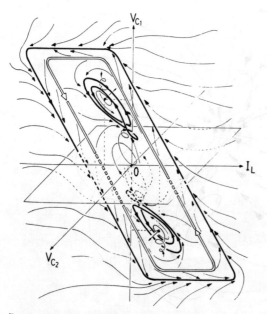

Fig. 4. Phase portrait for Chua's Circuit 1 when the circuit parameters are chosen so that a *period-2* oscillation exists in a neighborhood of the upper equilibrium point P^+ and the lower equilibrium point P^-.

green region in the α–β bifurcation diagram. In this case, the pair of odd-symmetric "2-revolution" limit cycles in Fig. 4 have evolved and bifurcated into a pair of "4-revolution" limit cycles. The period just after this bifurcation occurs is again (approximately) equal to *twice* the period of the oscillation immediately prior to the bifurcation, or approximately *4 times* the period of the "single-revolution" limit cycle (immediately before it bifurcates into a period-2 oscillation) in Fig. 3.

The above period-doubling phenomenon repeats itself in quick succession (i.e., the *increment in the parameter value before the next bifurcation* occurs decreases rapidly and in fact converges to a limit point) so that the *period* of these odd-symmetric pair of limit cycles becomes 2, 4, 8, 16, $\cdots$, 2^n, etc., and becomes *infinite* at some critical parameter values. This limiting waveform is no longer periodic and its frequency spectrum becomes *continuous*, rather than discrete in the periodic case. The phase portrait depicting the situation shortly after this limiting bifurcation occurs is shown in Fig. 6. The circuit parameters in this case correspond to a point in the *dark-yellow* region (see *inset* "4") in the α–β bifurcation diagram. The resulting waveforms of V_{C_1}, V_{C_2}, and I_L are no longer periodic and are said to be *chaotic*. However, in spite of its *nonperiodic* nature, all nearby trajectories are seen to converge towards some well-defined three-dimensional geometrical structure in the state space.

Such a limiting structure is called a *strange attractor* because strictly speaking, the trajectory does *not* really go through all points enclosed within the strange attractor, but rather comes *arbitrarily close* to it in the sense that given *any* point P on the strange attractor and a ball of arbitrarily small (but positive) radius centered about P, any trajectory starting from some nearby initial point necessarily intersects the ball at some future time. Hence a strange attractor is not completely "solid" but rather resembles a strange geometrical "pumice-like" set of points. The mathemati-

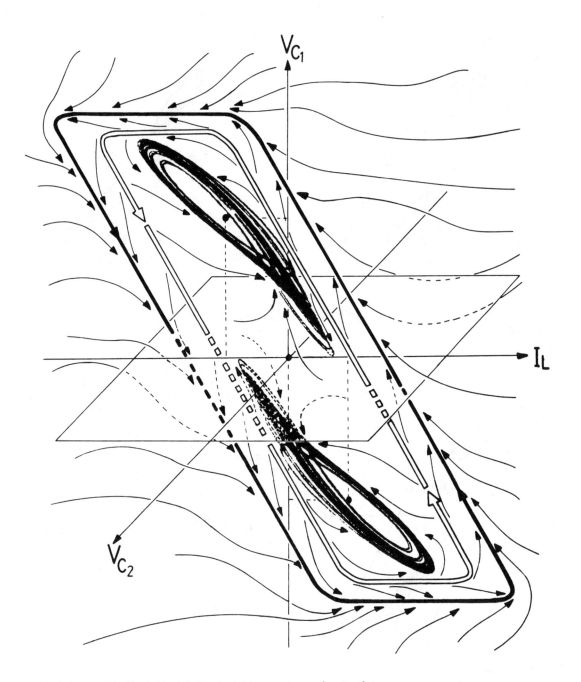

Fig. 6. Phase portrait for Chua's Circuit 1 when the circuit parameters are chosen so that a "strange attractor" resembling a Rössler spiral-type attractor appears.

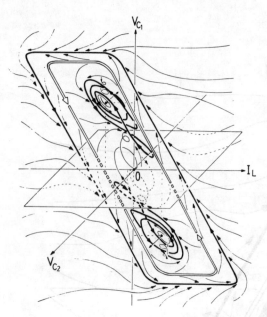

Fig. 7. Phase portrait for Chua's Circuit 1 when the circuit parameters are chosen so that a *period-3* "periodic window" orbit appears.

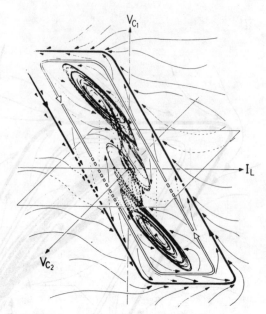

Fig. 9. Phase portrait for Chua's Circuit 1 when the circuit parameters are chosen so that the two Rössler screw-type attractors have grown in size to the point where they almost touch each other.

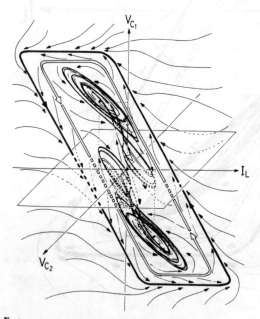

Fig. 8. Phase portrait for Chua's Circuit 1 when the circuit parameters are chosen so that a strange attractor resembling a Rössler screw-type attractor appears.

cian Mandelbrot has christened such a set as a *fractal* and even shows how one could calculate its "fractal" *dimension*, which in this case turns out to be an intuitively reasonable *fraction* between 2 and 3. The strange attractor in Fig. 6, as well as the one in Figs. 10 and 11, does not lie on a plane so its dimension has to be greater than 2. Since it is not "completely solid," its dimension must be less than 3. The two odd-symmetric strange attractors shown in Fig. 6 are similar topologically to a "spiral-type" attractor first discovered by Rössler using a completely different system of nonlinear differential equations [18].

The preceding sequence of phase portraits can be visualized as a sequence of "snapshots" taken as we *decrease* the value of the parameter R while keeping all other parameters fixed [9], [10]. The above phase portrait (Fig. 6) represents the first event where chaos occurs. As we decrease the value of R further, the Rössler spiral-like attractor evolves by changing its shape in a continuous way until some critical value where it suddenly bifurcates back into some *periodic orbit*. Such an orbit is called a *periodic window* because a further *small* change in the tuning parameter would cause the trajectory to become chaotic once more. The phase portrait shown in Fig. 7 shows one such periodic orbit, and its odd-symmetric clone. Since it has 3 revolutions, it is called a *period-3 orbit*. The circuit parameters in this case correspond to a point in the *light-yellow* region (see *inset* "5") in the α–β bifurcation diagram.

The phase portrait shown in Fig. 8 corresponds to the case where the circuit parameters are chosen just beyond the range of the above periodic window. The resulting odd-symmetric pair of strange attractors are similar to the screw-type attractor first observed by Rössler [18] using a different equation. As we tune the value of R further, the phase portrait in Fig. 8 is seen to evolve with both Rössler "screw-type" attractors increasing their size in a continuous way. In all cases, they remain bounded within the "hollow" *unstable* periodic orbit, which in turn is surrounded by the larger "solid" stable periodic orbit. The phase portrait in Fig. 9 corresponds to the parameter values where these two

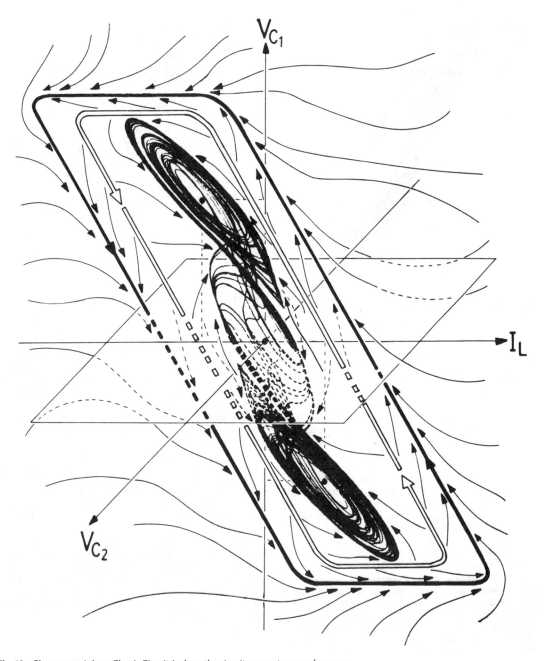

Fig. 10. Phase portrait from Chua's Circuit 1 where the circuit parameters are chosen so that the Rössler screw-type attractors collided with each other and merged into a new strange attractor called the *double scroll*.

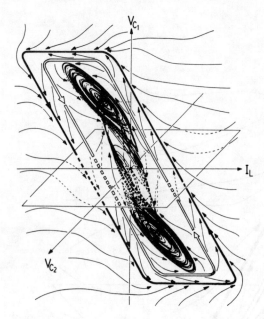

Fig. 11. Phase portrait from Chua's Circuit 1 where the circuit parameters are chosen so that the "holes" in the double scroll attractor are completely filled with trajectories.

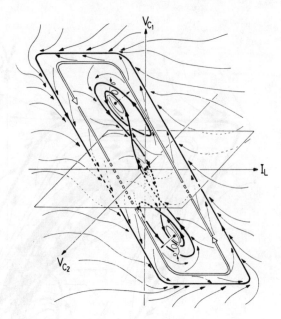

Fig. 12. Phase portrait from Chua's Circuit 1 where the circuit parameters are chosen within the range of a periodic window corresponding to a *single* contiguous periodic orbit.

attractors have grown so big that they would *collide* with the slightest further reduction in the value of R.

The phase portrait shown in Fig. 10 corresponds to the case where the circuit parameters are chosen to lie in the *dark-pink* region (see *inset* "6") of the α-β bifurcation diagram. Here, the two Rössler screw-type attractors in Fig. 9 have since collided with each other and a single contiguous strange attractor has emerged. From computer-generated cross sections of this attractor [13] and from a rigorous mathematical analysis of the "Poincaré map" in [3], we found the cross sections consist of two thin layers of points, each layer resembling a *spiral*. Consequently, the strange attractor in Fig. 10 is called the *double scroll* [13].

The phase portrait shown in Fig. 11 corresponds to the case where the circuit parameters are carefully chosen to coincide with the *inset* "7" in the α-β bifurcation diagram. Note that the two *holes* in the double scroll attractors of Fig. 10 have disappeared. Such a "hole-filling" orbit is rare because it occurs only for some precisely chosen set of circuit parameters—in contrast to the earlier phase portraits, which remain qualitatively unchanged over some *range* of circuit parameters. This "hole-filling" orbit is an example of a *heteroclinic orbit* [3], [4]. It can be proved mathematically that the existence of a heteroclinic orbit implies the associated system of differential equations ((1) in this case) is *chaotic* in a rigorous mathematical sense.

Let us examine the more robust double-scroll attractor in Fig. 10 again. As we decrease the value of R further, the double scroll evolves and eventually bifurcates into a periodic window. The resulting periodic orbit, however, is quite different from those of Fig. 7 in that there is only one contiguous orbit. One such periodic orbit is shown in the phase portrait in Fig. 12. The circuit parameters in this case corresponds to the *inset* "8" in the α-β bifurcation diagram in [3, p. 1072]. A further reduction in the value of R will lead to an almost endless variety of qualitatively distinct periodic windows sandwiched between a sequence of qualitatively similar double scrolls. This infinite collection of distinct periodic orbits and the interposed double scrolls are vividly depicted in the multi-color α-β diagrams in [3, figs. 17 and 18]. A display of the associated phase portraits though instructive is clearly impractical here.

In fact, due to space limitation, we have room for only one more phase portrait. To give the reader some feeling of the typical trajectories in Chua's Circuit 2 (Fig. 1(c)), the last phase portrait in Fig. 13 corresponds to a choice of circuit parameters where the circuit has three limit cycles. Note that while the two larger limit cycles are similar qualitatively to the preceding phase portraits, the *innermost* limit cycle is different. As we tune one of the circuit parameters, this period-1 orbit would go through a sequence of period-doubling bifurcations and eventually become chaotic. The associated strange attractor, however, differs from the Rössler's type attractors, or the double-scroll attractor, in a significant way. In fact, it is topologically similar to the strange attractor reported in [19].

III. A Prototype Family of Chaotic Circuits

It has been shown in [3] that it is possible to systematically generate an *infinite* number of chaotic circuits of different topologies but which are *equivalent* to Chua's Circuit 1 (respectively, Chua's Circuit 2) in the sense that they have *identical qualitative behaviors*. More precisely, two circuits are said to be equivalent if and only if the phase portrait (or vector field) of one circuit when drawn on a jelly-like cube can be deformed *smoothly* (i.e., without cutting the cube) into the phase portrait (or vector field) of the other circuit.

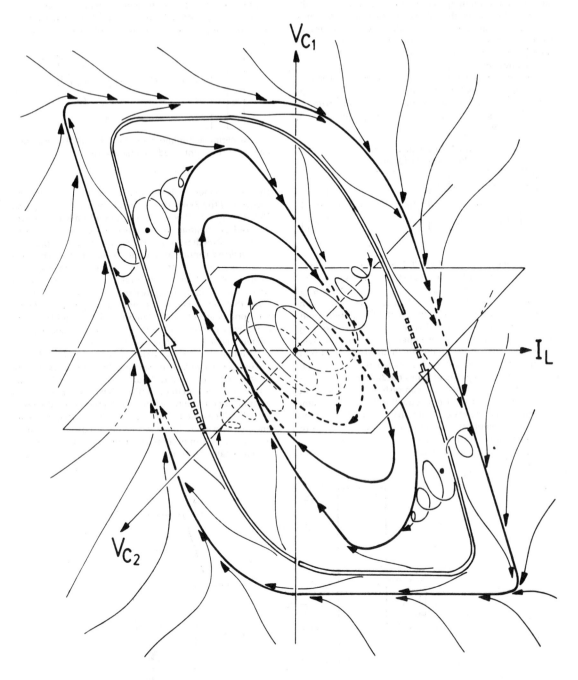

Fig. 13. Phase portrait from Chua's Circuit 2 where the circuit parameters are chosen to give a *period-1* periodic orbit.

Although these infinitely many equivalent circuits have different state equations, they all have three equilibrium points which are symmetrically placed with respect to the origin. Moreover, the Jacobian matrix evaluated at corresponding equilibrium points of these equivalent circuits have *identical* eigenvalues.

As a matter of fact, the mathematical theory in [3] shows that both of Chua's Circuits 1 and 2 and their equivalent family members can be systematically *embedded within a much larger family of circuits* described by the following state equation:

Normal Form Equation for
Chua's Circuit Family

$$\begin{bmatrix} \dot{x} \\ \dot{y} \\ \dot{z} \end{bmatrix} = \begin{bmatrix} a_{11} & a_{12} & a_{13} \\ a_{21} & a_{22} & a_{23} \\ a_{31} & a_{32} & a_{33} \end{bmatrix} \begin{bmatrix} x \\ y \\ z \end{bmatrix} - [|z+1| - |z-1|] \begin{bmatrix} b_1 \\ b_2 \\ b_3 \end{bmatrix} \quad (2)$$

The 12 coefficients a_{jk}, $j, k = 1, 2, 3$, and b_j, $j = 1, 2, 3$ in (2) are specified by *explicit* formulas in [3] in terms of only six eigenvalue parameters $\{\alpha_0, \beta_0, \gamma_0, \alpha_1, \beta_1, \gamma_1\}$, where $\alpha_0 \pm j\beta_0$ and γ_0 denote the complex-conjugate and real eigenvalues, respectively, associated with the equilibrium point at the origin, and $\alpha_1 \pm j\beta_1$ and γ_1 denote the corresponding eigenvalues associated with either one of the two symmetrically (with respect to the origin) located equilibrium points. The state equation (2) is called the *normal form equation* of this hugh circuit family because for *each* choice of eigenvalue parameters, we can systematically generate an infinite number of *distinct* equations having the same eigenvalue parameters and such that all equations in this family have identical qualitative behavior as the corresponding normal form equation.

The normal form equation (2) can be synthesized by the circuit shown in Fig. 14(a), henceforth referred to as *Chua's Circuit Family*, where N denotes a *linear* time-invariant resistive four-port described by the following *hybrid* representation:

$$\begin{bmatrix} i_1 \\ i_2 \\ v_3 \\ i_4 \end{bmatrix} = \begin{bmatrix} -C_1 a_{11} & -C_1 a_{12} & -C_1 a_{13} & C_1 b_1 \\ -C_2 a_{21} & -C_2 a_{22} & -C_2 a_{23} & C_2 b_2 \\ -L a_{31} & -L a_{32} & -L a_{33} & L b_3 \\ 0 & 0 & -1 & 0 \end{bmatrix} \begin{bmatrix} v_1 \\ v_2 \\ i_3 \\ v_4 \end{bmatrix}. \quad (3)$$

The two-terminal nonlinear resistor in Fig. 14(a) is described by the current-controlled v-i characteristic

$$v = |i + 1| - |i - 1|. \quad (4)$$

This v-i characteristic is shown in Fig. 14(b) and can be synthesized by the Zener-diode circuit shown in Fig. 14(c) [6]. The *linear* four-port N in Chua's Circuit Family can be synthesized by linear active circuits using either op-amps or transistors. Needless to say, the circuit in Fig. 14 represents a *normalized* canonical realization. Additional renormalization is needed to transform this into a practical realization. Clearly, there exists infinitely many other circuits which are *qualitatively equivalent* to that of Fig. 14 in the sense that their *eigenvalues* about corresponding equilibrium points are identical. Such circuits should therefore logically be classified as members of an equivalent class of circuits, namely, the *Chua's circuit family*.

Note that the Chua's Circuit Family in Fig. 14(a) defines a *six-parameter family of piecewise-linear circuits*, two of them being equivalent to Chua's Circuits 1 and 2. The significance of Chua's circuit family is that the members of this gigantic circuit family can be interpreted as *prototypes* of virtually all chaotic phenomena so far reported in the literature, as well as those yet to be discovered. Such an immense family of potentially chaotic circuits are yet to be systematically studied.

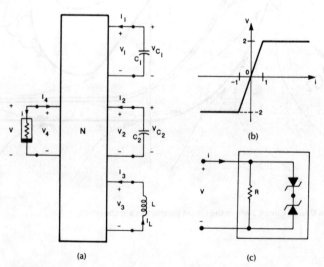

Fig. 14. (a) Chua's Circuit Family. (b) v-i characteristic for the nonlinear resistor. (c) Circuit which realizes the v-i characteristic in (b) when $R = 2\,\Omega$.

References

[1] T. Matsumoto, "A chaotic attractor from Chua's circuit," *IEEE Trans. Circuits Syst.*, vol. CAS-31, pp. 1055–1058, Dec. 1984.

[2] G.-Q. Zhong and F. Ayrom, "Experimental confirmation of chaos from Chua's circuit," *Int. J. Circuit Theory Appl.*, vol. 13, pp. 93–98, Jan. 1985.

[3] L. O. Chua, M. Komuro, and T. Matsumoto, "The double scroll family," *IEEE Trans. Circuits Syst.*, vol. CAS-33, pp. 1072–1118, Nov. 1986.

[4] I. Mees and P. B. Chapman, "Homoclinic and heteroclinic orbits in the double scroll attractor," to appear in *IEEE Trans. Circuits Syst.*, vol. CAS-34, no. 9, Sept. 1987.

[5] J. Guckenheimer and P. Holmes, *Nonlinear Oscillations, Dynamical Systems, and Bifurcations of Vector Fields*. New York, NY: Springer-Verlag, 1983.

[6] L. O. Chua, C. A. Desoer, and E. S. Kuh, *Linear and Nonlinear Circuits*. New York, NY: McGraw-Hill, 1987.

[7] L. O. Chua, "Nonlinear circuits," *IEEE Trans. Circuits Syst.* (Centennial Special Issue), vol. CAS-31, pp. 69–87, Jan. 1984.

[8] ——, "Device modelling via basic nonlinear circuit elements," *IEEE Trans. Circuits Syst.*, vol. CAS-27, pp. 1014–1044, Nov. 1980.

[9] G.-Q. Zhong and F. Ayrom, "Periodicity and chaos in Chua's circuits," *IEEE Trans. Circuits Syst.*, vol. CAS-32, pp. 501–503, May 1985.

[10] T. Matsumoto, L. O. Chua, and M. Komuro, "The double scroll bifurcations," *Int. J. Circuit Theory Appl.*, vol. 14, pp. 117–146, Apr. 1986.

[11] M. E. Broucke, "One-parameter bifurcation diagram for Chua's circuit," *IEEE Trans. Circuits Syst.*, vol. CAS-34, pp. 208–209, Feb. 1987.

[12] L. Yang and Y. Liao, "Self-similar bifurcation structures from Chua's circuit," *Int. J. Circuit Theory Appl.*, 1987.

[13] T. Matsumoto, L. O. Chua, and M. Komuro, "The double scroll," *IEEE Trans. Circuits Syst.*, vol. CAS-32, pp. 797–818, Aug. 1985.

[14] T. Matsumoto, L. O. Chua, and K. Tokumasu, "Double scroll via a two-transistor circuit," *IEEE Trans. Circuits Syst.*, vol. CAS-33, pp. 828–835, Aug. 1986.

[15] T. S. Parker and L. O. Chua, "The dual double scroll equation," to appear in *IEEE Trans. Circuits Syst.*, vol. CAS-34, no. 9, Sept. 1987.

[16] L. O. Chua and P. M. Lin, *Computer-Aided Analysis of Electronic Circuits: Algorithms and Computational Techniques*. Englewood Cliff, NJ: Prentice-Hall, 1975.

[17] B. Mandelbrot, *The Fractal Geometry of Nature*. New York, NY: Freeman, 1983.

[18] O. E. Rössler, "Continuous chaos-four prototype equations," *Ann. N.Y. Acad. Sci.*, vol. 31, pp. 376–392, 1979.

[19] C. T. Sparrow, "Chaoss in a three-dimensional single-loop feedback system with a piecewise-linear feedback function," *J. Math. Anal. Appl.*, vol. 83, pp. 275–291, 1981.

Shuxian Wu was born in Beijing, China, on October 2, 1957. She received the B.S. degree in electrical engineering from the Beijing Normal University in 1982 and the M.S. degree from the same university in 1985.

She has served as Assistant Professor at the Beijing Normal University since 1985. Her research interests include nonlinear dynamics and chaotic systems on which she has published several papers.

Chaos Via Torus Breakdown

TAKASHI MATSUMOTO, FELLOW, IEEE, LEON O. CHUA, FELLOW, IEEE, AND RYUJI TOKUNAGA

Abstract —Chaos has been observed from a four-element *autonomous* circuit whose only nonlinear element is a two-terminal resistor characterized by a three-segment piecewise-linear v–i characteristic. Both laboratory measurements and computer simulations have confirmed the chaotic behavior to have resulted from the *breakdown* of a "quasi-periodic" attractor (torus) into a "folded torus."

A two-parameter bifurcation diagram is carefully constructed to predict and explain various observed bifurcation phenomena, such as rotation number, devil's staircase, and Arnold tongue.

I. INTRODUCTION

TORUS BREAKDOWN has been observed for an extremely simple *real physical* system and the observation has been *confirmed* by computer simulation with much greater accuracy. The physical system is a *third-order autonomous* circuit which contains *only one nonlinear element*: a three-segment piecewise-linear resistor. Previous works on torus breakdown have been, to the best of our knowledge, *either laboratory measurement only* [1], *or simulation only* [2]–[4]. In this paper, we will report *both* of them for our circuit. This has been possible because of the simplicity of our circuit, which enables us to build it easily and to derive a very accurate model for simulation. It should also be noted that previous systems for which torus breakdown has been observed are either nonautonomous [1], [2] or higher order [3], [4]. Since a two-torus cannot be embedded into $\mathbb{R}^2$, the circuit reported in this paper is of the *minimal* dimension.

In Section II, we will report two basic mechanisms of torus breakdown observed from our circuit, by showing a series of one-parameter bifurcations. The two mechanisms are as follows.

i) *Phase-locking* (periodic state) and *torus* (quasi-periodic state) initially appear and disappear alternately many times. Suddenly, however, a phase-locking state bifurcates into a *chaotic attractor*, rather than the usual torus.

ii) After a repeated bifurcation between a phase-locking (periodic) and *a toroidal* (quasi-periodic) state, a periodic state bifurcates into a period-doubling cascade and culminating in a *chaotic attractor*.

In Section III, we will analyze the bifurcations reported in Section II. We will also describe the relative positions of the equilibria and their associated eigenspaces and explain how a solution trajectory can traverse on the surface of a torus. Piecewise-linearity of the dynamics simplifies our analysis in a significant manner.

Finally, in Section IV, we will give a detailed *two-parameter bifurcation diagram* where both rotation number [5] and bifurcation phenomena are indicated. In the diagram, one can also clearly identify various *Arnold tongues* [6]. Furthermore, the one-parameter bifurcations from the previous sections can be identified as a one-dimensional linear subspace of this two-parameter bifurcation space, thereby rendering the many phenomena from Section III somewhat more obvious.

II. OBSERVATIONS OF TORUS BREAKDOWN

A. The Circuit

Consider the circuit in Fig. 1(a) where the nonlinear resistor is characterized by Fig. 1(b) and where the capacitance on the right-hand side has a negative value $-C_1$. It is easily realized by the circuit of Fig. 2.[1] The dynamics of the circuit in Fig. 1(a) is governed by the state equation

$$\begin{aligned} C_1 \frac{dv_{C_1}}{dt} &= -g(v_{C_2} - v_{C_1}) \\ C_2 \frac{dv_{C_2}}{dt} &= -g(v_{C_2} - v_{C_1}) - i_L \\ L \frac{di_L}{dt} &= v_{C_2} \end{aligned} \quad (1)$$

where v_{C_1}, v_{C_2}, and i_L denote, respectively, the voltage across C_1, the voltage across C_2, and the current through L. The function $g(\cdot)$ denotes the v–i characteristic of the nonlinear resistor and is described analytically by

$$g(v) = -m_0 v + 0.5(m_0 + m_1)[|v + E_1| - |v - E_1|]. \quad (2)$$

To simplify our analysis, let us transform (1) into the following dimensionless form:

$$\begin{aligned} \frac{dx}{dt} &= -\alpha f(y - x) \\ \frac{dy}{dt} &= -f(y - x) - z \\ \frac{dz}{dt} &= \beta y \end{aligned} \quad (3)$$

Manuscript received June 5, 1986.
T. Matsumoto and R. Tokunaga are with the Department of Electrical Engineering, Waseda University, Tokyo 160, Japan.
L. O. Chua is with the Department of Electrical Engineering and Computer Sciences, University of California, Berkeley, CA 94720.
IEEE Log Number 8612447.

[1] Note that the resistive subcircuit N (subcircuit enclosed by the box) is designed in such a way that if one connects a (positive) capacitance to the right-hand port, it acts as a negative capacitance in the overall dynamics.

©1987 IEEE. Reprinted, with permission, from IEEE Trans. Circuits Systems, Vol. CAS-34, No. 3, pp. 240–253; March/1987

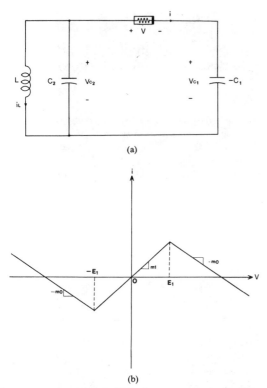

Fig. 1. An extremely simple third-order autonomous circuit which exhibits the torus breakdown phenomenon. (a) Circuitry. (b) Nonlinear resistor $v-i$ characteristic.

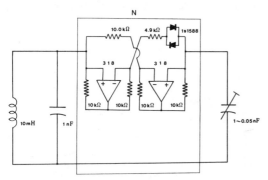

Fig. 2. Physical realization of the circuit in Fig. 1.

where

$$x \triangleq v_{C_1/E_1}, y \triangleq v_{C_2/E_1}, z \triangleq i_L/C_2 E_1, \alpha \triangleq C_2/C_1,$$
$$\beta \triangleq 1/LC_2,$$
$$a \triangleq m_0/C_2, b \triangleq m_1/C_2 \quad (4)$$
$$f(x) = -ax + 0.5(a+b)(|x+1| - |x-1|). \quad (5)$$

The dynamics associated with (3) depends on four parameters: a, b, α, and β. In this section, we will *choose α as our bifurcation parameter* by fixing the other parameters as follows:

$$a = 0.07, \quad b = 0.1, \quad \beta = 1. \quad (6)$$

B. *Bifurcations*

Fig. 3 shows the bifurcation phenomena observed from the circuit of Fig. 2 in the laboratory. They show 13 qualitatively distinct trajectories projected onto the (v_{C_2}, v_{C_1})-plane in the order of decreasing C_1 or increasing α. Here, v_{C_1} is the vertical axis and v_{C_2} is the horizontal axis. Fig. 4 shows 13 *cross sections* of the corresponding trajectories at $i_L = 0$, where the cross section in the region $v_{C_2} < 0$ is shown. Figs. 5 and 6 give the digital computer confirmation of Figs. 3 and 4, respectively, obtained by solving the state equation (3). The remarkable agreement between experiment and simulation shows that the *ideal* circuit of Fig. 1(a) is an excellent model of the physical circuit in Fig. 2.

In order to explain these bifurcation phenomena, first note that (3) has three equilibria all located (symmetrically) on the v_{C_1}-axis, namely,

$$O \triangleq (0,0,0), P^+ \triangleq (1+b/a, 0, 0), P^- \triangleq (-1-b/a, 0, 0). \quad (7)$$

Within the range of the parameter values that we are looking at, the eigenvalues at O (resp. $P^{\pm}$) consist of one real γ_0 (resp. γ_1) and a complex-conjugate pair $\sigma_0 \pm j\omega_0$ (resp. $\sigma_1 \pm j\omega_1$). The following observations provide a qualitative explanation of the various bifurcation phenomena shown in Figs. 3–6:

i) It can be shown (see Section III) that $P^{\pm}$ are *always unstable*. Equilibrium point O is stable for $\alpha < 0$, but loses its stability at $\alpha = 0$, whereupon a periodic attractor is born (see Figs. 3–6(a)). This, however, is *not* a Hopf bifurcation. The *real* eigenvalue γ_0 *becomes positive* for $\alpha > 0$, while σ_0 remains negative.

ii) A further increase in the value of α eventually gives rise to a *two-torus* (see Figs. 3–6(b)). To further confirm that this trajectory is indeed a torus, we calculated the following three Lyapunov exponents [7] associated with this trajectory[2]:

$$\mu_1 = 0, \quad \mu_2 = 0, \quad \mu_3 = -0.00675. \quad (8)$$

Since no Lyapunov exponent in (8) is positive, the system is not chaotic. However, only one Lyapunov exponent is negative, and the solution is not a periodic attractor either. The presence of two zero Lyapunov exponents therefore provides a further confirmation that the trajectory in Fig. 3(b) is indeed a two-torus, namely, *quasi-periodic*, solution. It appears that the *Poincaré map* of the periodic attractor born at $\alpha = 0$, has undergone a *Hopf bifurcation* [5], as depicted in Fig. 7, thereby giving birth to a two-torus. The above torus is observed in the half space $x > 0$. The symmetry of (3), however, suggests the existence of another

---

[2]An explicit formula is given in [7] for calculating the Lyapunov exponents. One can also use another method given in [9]. The results, of course, coincide with each other.

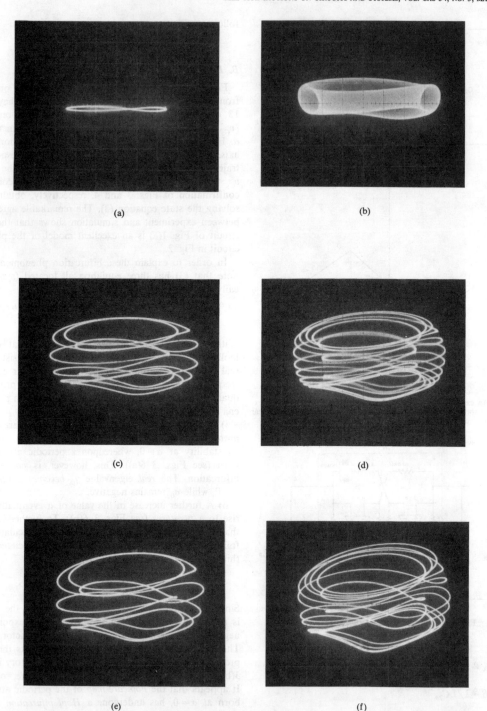

Fig. 3. Attractors observed from the circuit of Fig. 2 projected onto the (v_{C_2}, v_{C_1})-plane. Here v_{C_2} is the vertical axis. See Fig. 5 for value of α used. Horizontal scale (except (m)): 0.5 V/division. Vertical scale (except (m)): 0.5 V/division. In (a)–(l), only one of two attractors is shown. (a) Period-1 (0:1) orbit. (b) Two-Torus after Hopf bifurcation of the Poincaré map. (c) Period-8 (1:8) orbit. (d) Period-15 (2:15) orbit. (e) Period-7 (1:7) orbit. (f) Period-13 (2:13) orbit. (g) Period-6 (1:6) orbit. (h) Chaotic attractor resulting from intermittency. (i) Period-5 (1:5) orbit. (j) Period-10 (2:10) orbit after period doubling. (k) Chaotic attractor resulting from period doubling cascades. (l) Folded torus chaotic attractor. (m) Double scroll attractor. Horizontal scale: 1.0 V/division. Vertical scale: 1.0 V/division.

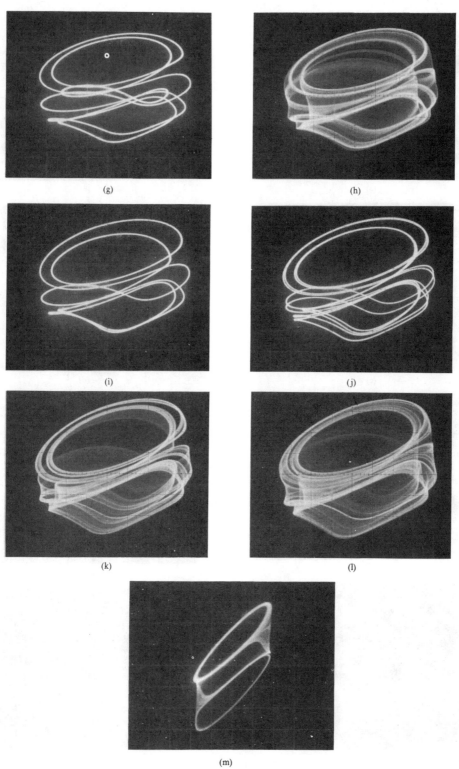

Fig. 3. Continued

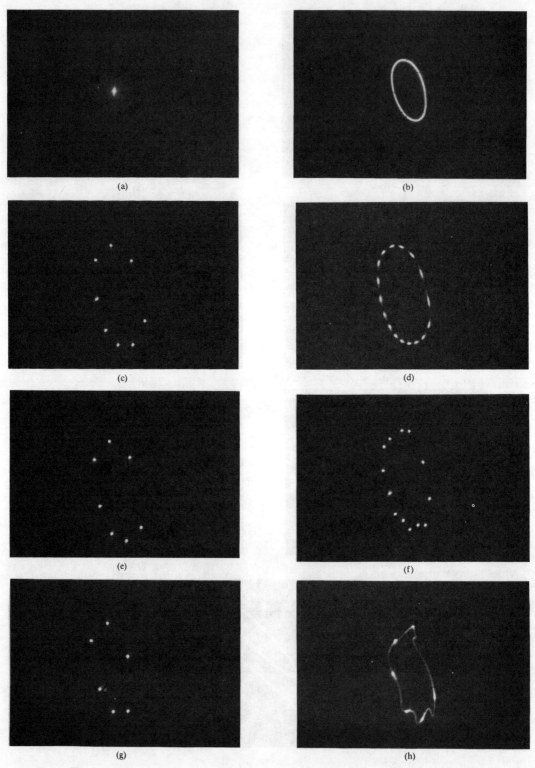

Fig. 4. The cross sections at $i_L = 0, v_{C_2} < 0$, of the corresponding trajectories from Fig. 3. Vertical axis is v_{C_1} and the horizontal axis is v_{C_2}.

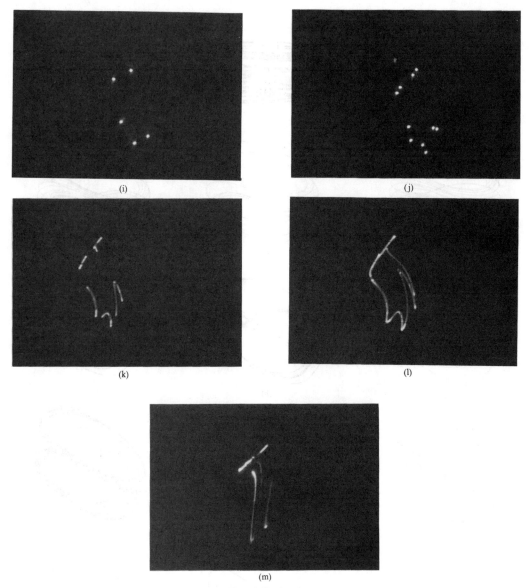

Fig. 4. Continued

torus in the half space $x < 0$. This observation has been confirmed, both by experiment and by simulation.

iii) As we increase α further, we observed that the two-torus and the periodic attractor (phase-locking) alternatively appear and disappear many times. Figs. 3–6(c)–(f) give a sample of some of the periodic attractors in this bifurcation sequence.

iv) For the parameter range $11.0 < \alpha < 13.4$, we observed a period-6 phase-locking (Figs. 3–6(g)). After this, the *two-torus fails to reappear*. The following is one of the scenarios following the disappearance (or breakdown) of a torus.

The First Scenario

At $\alpha \doteq 13.4$, the period-6 attractor suddenly bifurcates into a chaotic attractor.

In Section IV, we will present some evidence which indicates that this chaotic phenomenon is typical of that found in intermittency chaos [8] (Figs. 3–6(h)).

v) At $\alpha \doteq 13.4$, the situation becomes a little more complicated; namely, another *periodic attractor* of period-5 is born out of a *saddle-node bifurcation* [5] (Figs. 3–6(i)), and

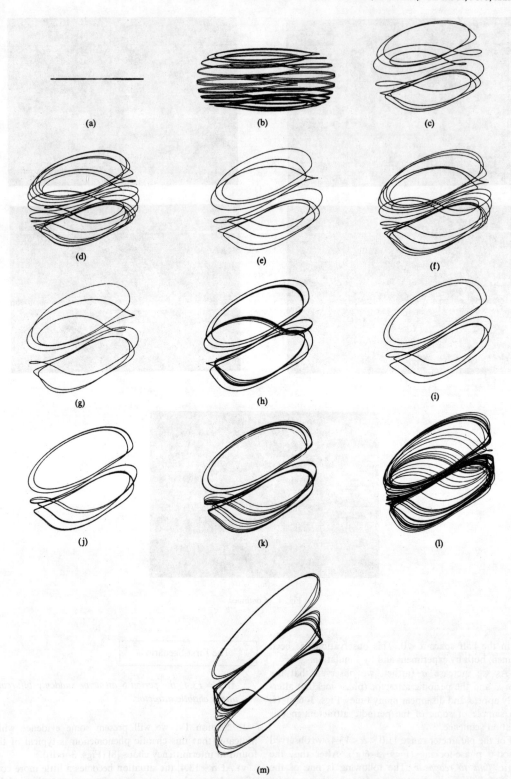

Fig. 5. The digital computer confirmation of Fig. 3. (a) $\alpha = 0.5$, (b) $\alpha = 2.0$, (c) $\alpha = 8.0$, (d) $\alpha = 8.8$, (e) $\alpha = 9.6$, (f) $\alpha = 10.8$, (g) $\alpha = 13.0$, (h) $\alpha = 13.4$, (i) $\alpha = 13.4$, (j) $\alpha = 13.45$, (k) $\alpha = 13.52$, (l) $\alpha = 15.0$, (m) $\alpha = 33.0$.

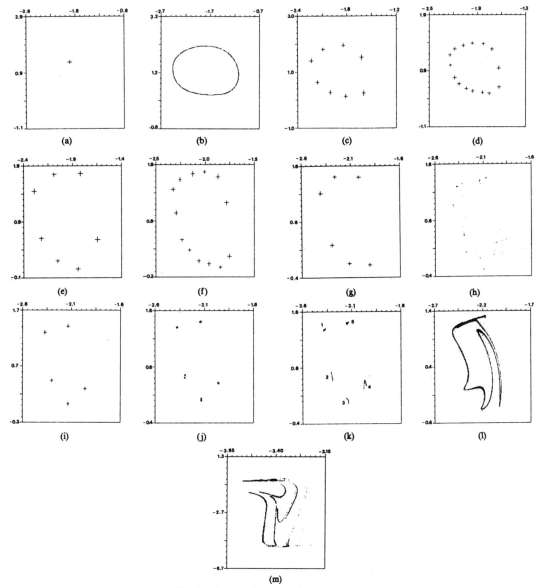

Fig. 6. The digital computer confirmation of Fig. 4.

this period-5 attractor *coexists* with the chaotic attractor described above.

vi) Since a saddle-node bifurcation occurs at $\alpha \doteq 13.4$, another period-5 closed orbit of the saddle-type is born simultaneously. Since this orbit is unstable, it is not observable experimentally.

vii) At $\alpha \doteq 13.5$, the saddle-type closed orbit *collides* with the chaotic attractor and the chaotic attractor disappears. Clearly, this is a boundary crisis [16].

viii) The period-5 attractor survives after collision (i.e., for $\alpha > 13.4$), and the two-torus is no longer observed. The following observations provide the second scenario following the torus break down.

The Second Scenario

At $\alpha \doteq 13.44$, a period doubling cascade is initiated (Figs. 3–6(j)) and the solution eventually bifurcates into a chaotic attractor consisting of five "islets" (Figs. 3–6(k)).

In Fig. 6(k), the islets are numbered according to the order of visitation of the trajectory. Note that islet 4

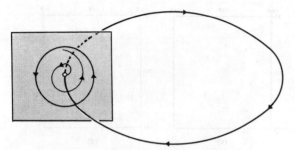

Fig. 7. Hopf bifurcation for the Poincaré map.

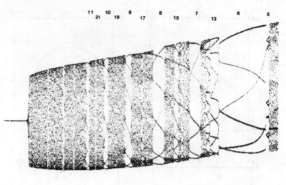

Fig. 8. Bifurcation diagram where the horizontal axis is α, and the vertical axis is v_{C_1} on the cross section at $i_L = 0$, $v_{C_2} < 0$.

appears to have resulted from a "stretching" and "folding" transformation.

ix) As we increase α beyond $\alpha \doteq 13.44$, we found the islets *merged* into one chaotic attractor[3] (Figs. 3–6(l)), which appears to be a *"folded torus"* [8]. Its Lyapunov exponents are

$$\mu_1 = 0.027, \mu_2 = 0, \mu_3 = -0.1134. \qquad (9)$$

Note that the first exponent is *positive* as expected. Note also that the symmetry of (3) implies that there exists another folded torus in the half space $v_{C_1} < 0$.

x) It is rather interesting to observe that a further increase of α resulted in the two folded tori *merging together* and giving rise to *a double scroll attractor*[4] [9], [10], [17] (Figs. 3–6(m)).

xi) By increasing the value of α even further, we observed that the double scroll *suddenly disappeared*. This sudden death signifies a *boundary crisis* [9]; namely, there is yet another saddle-type periodic orbit (surrounding the double scroll) which collides with the double scroll. Beyond this value of α, no further attractor was observed.

C. Period-Adding Sequence

The phase-locked periodic states shown in Figs. 3–6 are related to each other in accordance with some definite law which can be derived (empirically) from the one-dimensional bifurcation diagram given in Fig. 8. Here, the horizontal axis is α, and the vertical axis represents v_{C_1} with $i_L = 0$ and $v_{C_2} < 0$, i.e., the points in Fig. 8 corresponding to each α represent a cross section of Figs. 6 and 7 at $i_L = 0$, where $v_{C_2} < 0$. Fig. 8 reveals the following interesting properties.

i) There is a sequence of phase-locked states whose period *decreases by exactly one*: 11, 10, $\cdots$, 6.

ii) In between period n and period $n-1$, there is a phase-locked state of period $2n-1$.

These *period-adding* properties have also been observed from several simple *nonautonomous* circuits [12]–[14]. They will be explained in Section IV in terms of the property of *rotation numbers*.

---

[3] This appears to be an interior crisis [16].
[4] In [10] and [17], the double scroll is born by merging a pair of Rössler screw-type attractors.

III. ANALYSIS

In this section, we will examine some of the phenomena reported in the previous section from a more analytical perspective. Let us partition the $x - y - z$ state space into three parallel regions R_1, R_0, and R_{-1} separated by boundaries B_1 and B_{-1}, respectively, where

$$R_1 \triangleq \{(x, y, z)|y - x < -1\}$$
$$R_0 \triangleq \{(x, y, z)||y - x| < 1\}$$
$$R_{-1} \triangleq \{(x, y, z)|y - x > 1\}$$
$$B_1 \triangleq \{(x, y, z)|y - x = -1\}$$
$$B_{-1} \triangleq \{(x, y, z)|y - x = 1\}. \qquad (10)$$

The Jacobian matrix associated with (3) in R_0 (resp. R_1, R_{-1}) is given by

$$A_0 \triangleq \begin{bmatrix} \alpha b & -\alpha b & 0 \\ b & -b & -1 \\ 0 & \beta & 0 \end{bmatrix}$$

$$\text{resp. } A_1 \triangleq \begin{bmatrix} -\alpha a & \alpha a & 0 \\ -a & a & -1 \\ 0 & \beta & 0 \end{bmatrix}. \qquad (11)$$

One can easily show that O (resp. $P^{\pm}$) is stable if and only if

$$\alpha > 0, b(1-\alpha) > 0, b\alpha\beta < 0, (1-\alpha)\beta < 0$$
$$(\text{resp. } \alpha > 0, a(1-\alpha) > 0, a\alpha\beta < 0, (1-\alpha)\beta < 0). \quad (12)$$

A. Divergence Zero Boundary

It follows from (12) that equilibrium O loses its stability at $\alpha = 0$. This is not a Hopf bifurcation, however, as mentioned in Section II-B, namely, the real eigenvalue γ_0 becomes positive, while the real part σ_0 of the complex-conjugate pair remains negative. There is, however, more to it. Consider the divergence div f of the vector field f associated with (3)

$$\text{div } f = \begin{cases} \text{div}_0 f = -b(1-\alpha), (x, y, z) \in R_0 \\ \text{div}_1 f = a(1-\alpha), (x, y, z) \in R_1 \text{ or } R_{-1} \end{cases} \quad (13)$$

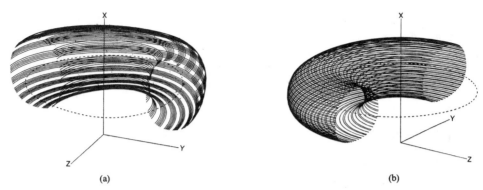

Fig. 9. (a) Repelling torus in the half space $z < 0$ and 0:1-periodic attractor at $\alpha = 0.5$. (b) Attracting torus in the half space $z < 0$ and 0:1-periodic repeller at $\alpha = 2.0$.

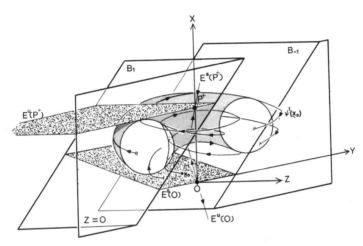

Fig. 10. The relative positions of the eigenspaces and the two-torus.

Note that the $\text{div}_0 f$ and $\text{div}_1 f$ exchange their signs at $\alpha = 1$, namely,

for $\alpha < 1, \text{div}_0 f < 0, \text{div}_1 f > 0,$
for $\alpha = 1, \text{div}_0 f = 0, \text{div}_1 f = 0,$
for $\alpha > 1, \text{div}_0 f > 0, \text{div}_1 f < 0.$ (14)

We have observed (by simulation) that, for $\alpha < 1$, a *periodic attractor coexists* with a *repelling torus* (Fig. 9(a)), while, for $\alpha > 1$, a *periodic repeller coexists* with an *attracting torus* (Fig. 9(b)). At $\alpha = 1$, neither attractor nor repeller is observed as expected since the circuit in this case has no dissipative mechanism. Therefore, a *Hopf bifurcation*[5] for the *Poincaré map* occurs at $\alpha = 1$ and it gives birth to an attracting two-torus.

B. Trajectory on the Torus

Recall that the eigenvalues at O (resp. $P^{\pm}$) consist of one real γ_0 (resp. γ_1) and a complex-conjugate pair $\sigma_0 \pm j\omega_0$

---

[5] If α is increased, a *supercritical* Hopf bifurcation occurs at $\alpha = 1$, i.e., an attracting periodic orbit bifurcates into an attracting torus. On the other hand, if α is decreased, a *subcritical* Hopf bifurcation occurs at $\alpha = 1$, i.e., a repelling periodic orbit bifurcates into a repelling torus.

(resp. $\sigma_1 \pm j\omega_1$). In particular, at $\alpha = 2, \beta = 1$,

$$\gamma_0 = 0.14786, \sigma_0 = -0.048886, \omega_0 = 1.0060$$
$$\gamma_1 = -0.10425, \sigma_1 = 0.034426, \omega_1 = 1.0030. \quad (15)$$

Let $E^s(O)$ (resp. $E^u(O)$) denote the eigenspace corresponding to γ_0 (resp. $\sigma_0 \pm j\omega_0$). Similarly, let $E^u(P^{\pm})$ (resp. $E^s(P^{\pm})$) denote the eigenspace corresponding to $\sigma_1 \pm j\omega_1$ (resp. γ_1). Fig. 10 describes relative positions of these sets. While the relative positions of the eigenvalue in (14) are identical to those of the *double scroll family* [15], there are two subtle differences.

i) The magnitude of $|\gamma_1|$ is not as large as in [9], and, hence, "the flattening" of the attractor onto $E^u(P^{\pm})$ is relatively weak.

ii) $E^s(O)$ and $E^u(P^{\pm})$ are almost parallel to each other.

Let ψ^t be the flow generated by (3) and pick an initial condition x_0 near O above $E^s(O)$ but not on $E^u(O)$. Since $\gamma_0 > 0, \psi^t(x_0)$ starts moving up (with respect to the x-axis) while rotating clockwise around $E^u(O)$ (Fig. 10). Since (3) is linear in $R_0, \psi^t(x_0)$ eventually hits B_1 and enters R_1. Because of the relative position of $E^s(P^+)$,

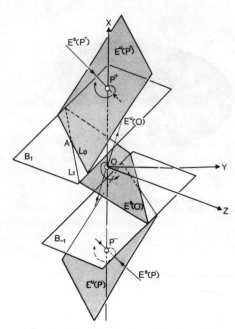

Fig. 11. The relative positions of the eigenspaces.

$\psi^t(x_0)$ moves up further while rotating around, this time, $E^s(P^+)$. Since $\sigma_1 > 0$, the solution $\psi^t(x_0)$ *increases* its magnitude of oscillation and eventually enters R_0. Then, because of the relative positions of R_0 and R_1, $\psi^t(x_0)$ starts moving downward (with rotation), it eventually hits B_{-1} and then flattens itself against $E^s(O)$ while rotating around $E^u(O)$. Since $\sigma_0 < 0$, the solution *decreases* its magnitude of oscillation and gets into the original neighborhood of O. This process then repeats itself *ad infinitum* but never returns to the original point and, hence, the associated loci eventually covers the surface of a two-torus.

Next, let us explain the situation where the double scroll is observed. Typical parameter values are $\alpha = 30, \beta = 1$. Various sets are shown in Fig. 11 where

$$L_0 \triangleq E^u(P^+) \cap B_1$$
$$L_1 \triangleq E^s(O) \cap B_1$$
$$A \triangleq L_0 \cap L_1$$
$$L_{-0} \triangleq E^u(P^-) \cap B_{-1}$$
$$L_{-1} \triangleq E^s(O) \cap B_{-1}$$
$$A_- \triangleq L_{-0} \cap L_{-1}. \qquad (16)$$

Observe that the angle between $E^s(O)$ and $E^u(P^\pm)$ is much larger than in Fig. 10. The behavior of trajectories in the double scroll is described in [9]. Recall that point A of Fig. 11 plays an important role in determining the "fate" of a trajectory in the double scroll. In Fig. 10, however, point A is located far from the torus and play no significant role.

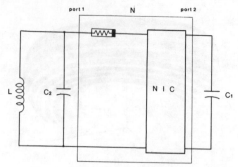

Fig. 12. Circuit equivalent to that of Fig. 1(a).

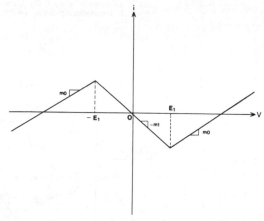

Fig. 13. The $v-i$ characteristic observed across port 2 of the circuit of Fig. 12 where the inductor is short circuited.

C. Folded Torus and Double Scroll

It should be noted that for the present parameter values, the two points A and A_-, which played an important role in [9], [10], and [17], are located far beyond the region where the torus is observed. However, these points keep getting closer to the attractors as one increases α so that eventually A and A_- touch the folded torus attractors. When this situation occurs, the two folded tori cannot stay away from each other and must therefore merge into one attractor, namely, the double scroll. This is because if a trajectory (in the attractor) hits A or a point to the left of A in Fig. 11, then it has to descend and eventually hits B_{-1}.

D. Circuit Theoretic Interpretation

We can give an intuitive circuit theoretic interpretation of the dynamics when a two-torus is observed. Fig. 12 shows a circuit which is equivalent to the circuit of Fig. 1(a), where the NIC denotes a *negative impedance converter* and N denotes the resistive two-port terminated by the dynamic elements. Note that at dc equilibrium, the $v-i$ characteristic across port 2 (L is a short circuit at an equilibrium) is given by Fig. 13 in view of the property of the NIC. We can view the circuit in Fig. 12 as made up of three components: i) The L-C_2 tank circuit, ii) C_1, and iii)

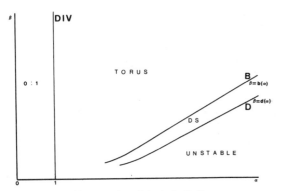

Fig. 14. Several important boundaries in the (α, β) parameter space.

N, which couples the first two. When this circuit operates in the *active* region of the nonlinear resistor, i.e., $|v| > |E_1|$, the subcircuit N supplies energy to the L-C_2 tank circuit while dissipating *the energy* stored in the capacitor C_1. In this case the magnitude of the oscillation of the L-C_2 tank circuit *increases* while that across C_1 *decreases* (relative to P^+). On the other hand, when this circuit operates in the passive region of the nonlinear resistor, i.e., $|v| < |E_1|$, the subcircuit N *dissipates the energy* stored in the tank circuit while *supplying energy* to the capacitor C_1. In this case, the magnitude of oscillation of the tank circuit *decreases* while that across C_1 increases (relative to O).

IV. TWO-PARAMETER BIFURCATION DIAGRAM

In this section, *we vary β as well as α*. We will pay particular attention to how the rotation number varies. The *rotation number* ρ is defined for a homeomorphism h on a circle [5], namely,

$$h: S^1 \to S^1,$$

$$\rho \triangleq \lim_{n \to \infty} \frac{h^n(x) - x}{n} \quad \text{for any } x \in S^1. \quad (17)$$

If ρ is *rational*, i.e., $\rho = m/n$, where m and n are positive integers, then the trajectory is n-periodic. In this case, all trajectories approach a unique n-periodic orbit, while winding around S^1 "m" times before completing one periodic orbit. If ρ is *irrational*, then the orbit is *quasi-periodic* and therefore covers the surface of a two-torus. Fig. 14 gives several important boundaries in the (α, β)-plane:

$$\text{DIV} \triangleq \{(\alpha, \beta) | \alpha = 1\}: \quad \text{divergence-free boundary}$$

$$B \triangleq \{(\alpha, \beta) | \beta = b(\alpha)\}: \quad \text{birth of the double scroll}$$

$$D \triangleq \{(\alpha, \beta) | \beta = d(\alpha)\}: \quad \text{death of the double scroll} \quad (18)$$

where $b(\cdot)$ and $d(\cdot)$ are functions that can be obtained numerically [15].

Now, in order to study the rotation number for (3), one has to find a region where the cross section of an attractor is homeomorphic to S^1 and that a homeomorphism h is indeed induced via the flow of (3) on the cross section. Since this is an extremely difficult (if not an impossible) task, we *assume* that the rotation number can be defined in following region:

$$\{(\alpha, \beta) | 1 < \alpha < b(\alpha)\}. \quad (19)$$

Fig. 15 gives a detailed global picture for the (α, β)-plane. A solid line indicates a boundary of a region where the rotation number is constant, where $1:5$ means that rotation number $\rho = 1/5$, etc. A chain line denotes the curve on which period-doubling bifurcation occurs. In order to avoid further complication of the picture, only the *onset* of the period-doubling cascade is shown. Broken lines indicate boundaries where chaos is observed. The symbol C stands for (folded torus) chaos, whereas DS stands for the double scroll. These curves are obtained by observing the trajectories via Runge-Kutta iterations. Note that there are many regions in Fig. 15 where the rotation number is equal to the same rational number. Such regions are called Arnold tongues [6].

A careful examination of Fig. 15 reveals the following empirical laws (for fixed β).

i) If $\alpha_1 > \alpha_2$ and if $\rho(\alpha_1) = m_1/n_2, \rho(\alpha_2) = m_2/n_2$, then $\rho(\alpha_1) > \rho(\alpha_2)$.

ii) There is an α_3 such that $\alpha_1 > \alpha_3 > \alpha_2$ and $\rho(\alpha_3) = (m_1 + m_2)/(n_1 + n_2), \rho(\alpha_1) > \rho(\alpha_3) > \rho(\alpha_2)$.

These empirical laws are consistent with the 1-D bifurcation diagram in Fig. 8. Fig. 16 gives the graph of ρ as a function α with $\beta = 1$. The resulting monotone-increasing function is called a devil's staircase [11] and is useful in explaining the bifurcation phenomena that resulted in Section III. The graph is obtained by observing the trajectories via Runge-Kutta iterations.

Note that the one-parameter bifurcation diagram from the previous sections is contained in this picture as a one-dimensional linear subspace ($\beta = 1$) and it can be put in a much clearer perspective in terms of this two-parameter diagram.

i) As one moves along $\beta = 1$ in the $1:5$ Arnold tongue, one hits the boundary of the $2:10$ Arnold tongue, thereby signifying a period-doubling bifurcation (Figs. 3–6(j)).

ii) When one moves to the right in the $1:6$ Arnold tongue, on the line $\beta = 1$, one does not hit the boundary of the $2:12$ Arnold tongue. This explains why we did not observe any period-doubling cascade for the period-6 attractor.

iii) As one moves to the right along the line $\beta = 1$, the circle map nature is destroyed before the system gets into the $1:6$ phase-locking. This is why we observed a sudden bifurcation of $1:6$ phase-locking into chaos (Figs. 3–6(h)). It appears that this chaotic attractor is born via an intermittency route. In order to confirm this conjecture, we computed the angle of the point on the Poincaré section $i_L = 0$, $v_{C_2} < 0$, from a reference point. Fig. 17 gives the sixth iterate $h^6(x)$ of this angular variable, at $\alpha = 13.4$, $\beta = 1$. Note that at $1:6$ phase-locking $h^6(x)$ has six stable fixed points and six unstable fixed points; hence, these six stable fixed points attract the solutions which start from any initial points. After $1:6$ phase-locking, i.e., all fixed points disappear via a *tangent bifurcation*, there are six regions called "channels" (where the unit-slope line and

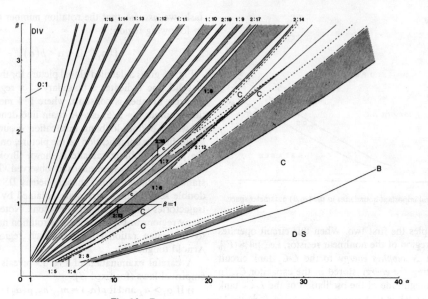

Fig. 15. Two-parameter bifurcation diagram.

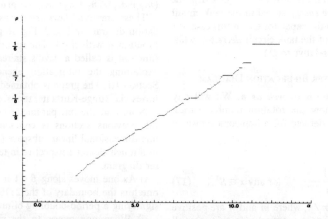

Fig. 16. The graph of ρ as a function of α at $\beta = 1$.

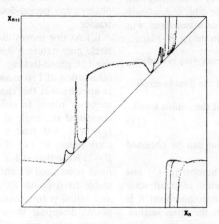

Fig. 17. Graph of $h^6(x)$ showing tangent bifurcation.

the curve of $h^6(x)$ approach each other very closely). Inside each channel, a solution behaves like a periodic orbit because $h^6(x)$ almost touches the diagonal and spends a very long period of time in the channel. Once it gets out of the channel, however, the solution behaves in an erratic manner. Finally, we remark that the 1:5 Arnold tongue overlaps with the 1:4 Arnold tongue and, hence, the right-hand boundary of the 1:5 tongue cannot be observed clearly.

ACKNOWLEDGMENT

The authors would like to thank M. Komuro of Tokyo Metropolitan University, S. Tanaka of Hitachi, K. Ayaki, T. Makise, K. Tokumasu, and T. Kuroda of Waseda University for many exciting discussions.

REFERENCES

[1] J. Stavans, F. Heslot, and A. Libchaber, "Fixed winding number and the quasi-periodic route to chaos in a convective fluid," *Phys. Rev. Lett.*, vol. 55, no. 6, pp. 596–599, Aug. 5, 1985.
[2] T. Bohr, P. Bak, and M. Hogh Jensen, "Transition to chaos by interaction of resonances in dissipative system II. Josephson junctions, charge-density waves, and standard maps," *Phys. Rev. A*, vol. 30, no. 4, pp. 1960–1969, Oct. 1984.
[3] M. Sano and Y. Sawada, "Transition from quasi-periodicity to chaos in a system of coupled nonlinear oscillator," *Phys. Lett.*, vol. 97A, no. 3, pp. 73–76, Aug. 15, 1983.
[4] V. Franceschini, "Bifurcation of tori and phase locking in a dissipative system of differential equations," *Physica*, vol. 60, pp. 285–304, 1983.
[5] J. Guckenheimer and P. Holmes, *Nonlinear Oscillations, Dynamical Systems, and Bifurcations of Vector Fields*. New York: Springer, 1983.
[6] L. Glass and R. Perez, "Fine structure phase locking," *Phys. Rev. Lett.*, vol. 48, no. 26, pp. 1772–1775, June 28, 1982.
[7] I. Shimada and T. Nagashima, "A numerical approach to ergodic problem of dissipative dynamical system," *Progr. Theor. Phys.*, vol. 61, no. 6, pp. 1605–1616, June 1979.
[8] W. F. Langford, "Numerical studies of torus bifurcations," *International Series of Numerical Mathematics*. Heidelberg/New York: Springer-Verlag, vol. 70, pp. 285–295.
[9] T. Matsumoto, L. O. Chua, and M. Komuro, "The double scroll," *IEEE Trans. Circuits Syst.*, vol. CAS-32, pp. 797–818, Aug. 1985.
[10] T. Matsumoto, L. O. Chua, and M. Komuro, "The double scroll bifurcations," *Int. J. Circuit Theory Appl.*, vol. 14, pp. 117–146, 1986.
[11] M. H. Jensen, P. Bark, and T. Bohr, "Complete devil's staircase, fractal dimension, and universality of mode-locking structure in the circle map," *Phys. Rev. Lett.*, vol. 50, no. 21, pp. 1630–1639.
[12] L. Q. Pei, F. Guo, S. X. Wu, and L. O. Chua, "Experimental confirmation of the period-adding route to chaos via nonlinear circuit," *IEEE Trans. Circuits Syst.*, vol. CAS-33, Apr. 1986.
[13] L. O. Chua, Y. Yao, and Q. Young, "Devil's staircase route to chaos via a nonlinear circuit," *Int. J. Circuit Theory Appl.*, in press.
[14] M. P. Kennedy and L. O. Chua, "Van der Pol and Chaos," *IEEE Trans. Circuits Syst.*, vol. CAS-33, pp. 974–980, Oct. 1986.
[15] L. O. Chua, M. Komuro, and T. Matsumoto, "The double scroll family, Parts I and II," *IEEE Trans. Circuits Syst.*, vol. CAS-33, pp. 1072–1118, Nov. 1986.
[16] C. Grebogi, E. Ott, and J. A. Yorke, "Crisis, sudden changes in chaotic attractors, and transient chaos," *Physica*, vol. 7D, pp. 181–200, May 1983.
[17] T. Matsumoto, L. O. Chua, and M. Komuro, "Birth and death of the double scroll," *Physica. D.*, in press.

Presently, he is Professor of Electrical Engineering, Waseda University, Tokyo, Japan. From 1977 to 1979, he was on leave at the Department of Electrical Engineering and Computer Sciences, University of California, Berkeley, CA. His interest is in nonlinear networks and fault diagnosis. He was an Overseas Associate Editor of the IEEE TRANSACTIONS ON CIRCUITS AND SYSTEMS.

Leon O. Chua (S'60–M'62–SM'70–F'74) was born in the Philippines on June 28, 1936, of Chinese nationality. He received the B.S.E.E. degree from Mapua Institute of Technology, Manila, the Philippines, in 1959, the S.M. degree from the Massachusetts Institute of Technology, Cambridge, in 1961, and the Ph.D. degree from the University of Illinois, Urbana, in 1964.

He worked for the IBM Corporation, Poughkeepsie, NY, from 1961 to 1962. He joined the Department of Electrical Engineering, Purdue University, Lafayette, IN, in 1964, as an Assistant Professor. Subsequently, he was promoted to Associate Professor in 1967, and to Professor in 1971. Immediately following this, he joined the Department of Electrical Engineering and Computer Sciences, University of California, Berkeley, where he is currently Professor of Electrical Engineering and Computer Sciences. His research interests are in the areas of general nonlinear network and system theory. He has been a consultant to various electronic industries in the areas of nonlinear network analysis, modeling, and computer-aided design. He is the author of *Introduction to Nonlinear Network Theory* (New York: McGraw-Hill, 1969) and coauthor of the book *Computer-Aided Analysis of Electronic Circuits: Algorithms and Computational Techniques* (Englewood Cliffs, NJ: Prentice-Hall, 1975). He has also published many research papers in the area of nonlinear networks and systems. He was the Guest Editor of the November 1971 Special Issue of IEEE TRANSACTION ON EDUCATION on "Applications of Computers to Electrical Education," the Editor of the IEEE TRANSACTIONS ON CIRCUITS AND SYSTEMS from 1973 to 1975, and the Guest Editor of the August and September Special Issues of the IEEE TRANSACTIONS ON CIRCUITS AND SYSTEMS on "Nonlinear Phenomena, Modeling, and Mathematics." He is presently a Deputy Editor of the *International Journal of Circuit Theory and Applications*.

Dr. Chua is a member of Eta Kappa Nu, Tau Beta Pi, and Sigma Xi. He was a member of the Administrative Committee of the IEEE Circuits and Systems Society from 1971 to 1974, and is a past President of the IEEE Circuits and Systems Society. He has been awarded four patents and is the recipient of the 1967 IEEE Browder J. Thompson Memorial Prize Award, the 1973 IEEE W. R. G. Baker Prize Award, the 1973 Best Paper Award of the IEEE Circuits and Systems Society, the Outstanding Paper Award at the 1974 Asilomar Conference on Circuits, Systems, and Computers, the 1974 Frederick Emmons Terman Award, the 1976 Miller Research Professorship from the Miller Institute, the 1982 Senior Visiting Fellowship at Cambridge University, England, the 1982/83 Alexander Humboldt Senior U.S. Scientist Award at the Technical University of Munich, W. Germany, and the 1983/84 Visiting U.S. Scientist Award at Waseda University, Tokyo, from the Japan Society for Promotion of Science. In May 1983, he received an Honorary Doctorate (Doctor honoris causa) from the Ecole Polytechnique, Federal de Lausanne, Switzerland. In July 1983, he was awarded an Honorary Professorship at the Chengdu Institute of Radio Engineering in Sichuan, the People's Republic of China. In 1984, he was awarded an Honorary Doctorate at the University of Tokushima, Japan, and the IEEE Centennial Medal. In 1985, he was awarded the Waseda University International Fellowship from Japan, the IEEE Guillemin-Cauer Prize, and the Myril B. Reed Best Paper Prize.

Takashi Matsumoto (M'71–F'85) was born in Tokyo, Japan, on March 30, 1944. He received the B.Eng. degree in electrical engineering from Waseda University, Tokyo, Japan, the M.S. degree in applied mathematics from Harvard University, Cambridge, MA, and the Dr. Eng. degree in electrical engineering from Waseda University, Tokyo, in 1966, 1969, and 1973, respectively.

Ryuji Tokunaga was born in Nagano, Japan, on August 5, 1961. He received the B.E. degree in electrical engineering from Waseda University, Tokyo, Japan, in 1985. Presently, he is a graduate student at Waseda University. His interest is in chaos in nonlinear circuit.

VI. Chaos in Noisy Systems

In this section the influence of random noise on chaotic behavior and period-doubling bifurcation is investigated.

Paper VI.1 considers the role of fluctuations on the onset and characteristics of chaotic behavior associated with period-doubling subharmonic bifurcation. It shows that the effect of noise is to produce a bifurcation gap in the set of available states.

Paper VI.2 considers the influence of noise on chaotic behavior of an acousto-optic system. Both experimental and numerical results are presented.

In Paper VI.3 the effect of noise on the forced Duffing's oscillator in the region of parameter space where different chaotic attractors coexist is investigated. It has been found that noise may lead to jumps between different basins of attraction.

The probability density function of a dissipative nonlinear system driven by stochastic and periodic forces is examined in Paper VI.4. The probability density function of noisy chaotic systems displays multiple maxima.

The impact of noise and resonant perturbation on a dynamical system exhibiting a period-doubling bifurcation is investigated in Paper VI.5. It has been noticed that a shift of the bifurcation point is proportional to the square of the amplitude of the perturbation.

Paper VI.6 introduces Lyapunov exponents in the presence of noise. These exponents are themselves random numbers with a corresponding distribution.

The effect of small addative noise on the homoclinic threshold is considered in Paper VI.7. Melnikov function is derived for a noisy system which is seen to be the Melnikov function for the corresponding noise-free system plus the correction term.

FLUCTUATIONS AND THE ONSET OF CHAOS

J.P. CRUTCHFIELD [1] and B.A. HUBERMAN
Xerox Palo Alto Research Center, Palo Alto, CA 94304, USA

Received 3 April 1980

We consider the role of fluctuations on the onset and characteristics of chaotic behavior associated with period doubling subharmonic bifurcations. By studying the problem of forced dissipative motion of an anharmonic oscillator we show that the effect of noise is to produce a bifurcation gap in the set of available states. We discuss the possible experimental observation of this gap in many systems which display turbulent behavior.

It has been recently shown that the deterministic motion of a particle in a one-dimensional anharmonic potential, in the presence of damping and a periodic driving force, can become chaotic [1]. This behavior, which appears after an infinite sequence of subharmonic bifurcations as the driving frequency is lowered, is characterized by the existence of a strange attractor in phase space and broad band noise in the power spectral density. Furthermore, it was predicted that under suitable conditions such turbulent behavior may be found in strongly anharmonic solids [2]. Since condensed matter is characterized by many-body interactions, one may ask about the effects that random fluctuating forces have on both the nature of the chaotic regime and the sequence of states that lead to it. This problem is also of relevance to the behavior of stressed fluids, where it has been suggested that strange attractors play an essential role in the onset of the turbulent regime [3]. Although there are experimental results supporting this conjecture [4–6], other investigations have emphasized the possible role of thermodynamic fluctuations directly determining the chaotic behavior [7].

With these questions in mind, we study the role of fluctuations on the onset and characteristics of chaotic behavior associated with period doubling subharmonic bifurcations. We do so by solving the problem of forced dissipative motion in an anharmonic potential with the aid of an analog computer and a white-noise generator. As we show, although the structure of the strange attractor is very stable even under the influence of large fluctuating forces, their effect on the set of available states is to produce a symmetric gap in the deterministic bifurcation sequence. The magnitude of this bifurcation gap is shown to increase with noise level. By keeping the driving frequency fixed we are also able to determine that increasing the random fluctuations induces further bifurcations, thereby lowering the threshold value for the onset of chaos. Finally, the universality of these results is tested by observing the effect of random errors on a one-dimensional map, and suggestions are made concerning the possible role of temperature in experiments that study the onset of turbulence.

Consider a particle of mass m, moving in a one-dimensional potential $V = a\eta^2/2 - b\eta^4/4$, with η the displacement from equilibrium and a and b positive constants. If the particle is acted upon by a periodic force of frequency ω_d and amplitude F, and a fluctuating force $f(t)$, with its coupling to all other degrees of freedom represented by a damping coefficient γ, its equation of motion in dimensionless units reads

$$\frac{d^2\psi}{dt^2} + \alpha\frac{d\psi}{dt} + \psi - 4\psi^3 = \Gamma \cos\left(\frac{\omega_d}{\omega_0}\right)t + f(t) \quad (1)$$

with $\psi = \eta/2\eta_0$, the particle displacement normalized to the distance between maxima in the potential (η_0

[1] Permanent address: Physics Department, University of California, Santa Cruz, CA 95064, USA.

$= (a/b)^{1/2}$, $\alpha = \gamma/(ma)^{1/2}$, $\Gamma = Fb^{1/2}/2a^{3/2}$, $\omega_0 = (a/m)^{1/2}$ and $f(t)$ a random fluctuating force such that

$$\langle f(t) \rangle = 0 \qquad (2a)$$

and

$$\langle f(0)f(t) \rangle = 2A\delta(t) \qquad (2b)$$

with A a constant proportional to the noise temperature of the system.

The range of solutions of eq. (1), in the case where $f(t) = 0$ (the deterministic limit) has been investigated earlier [1]. For values of Γ and ω_d such that the particle can go over the potential maxima, as the driving frequency is lowered, a set of bifurcations takes place in which orbits in phase space acquire periods of 2^n times the driving period, T_d. At a threshold frequency ω_{th}, a chaotic regime sets in, characterized by a strange attractor with "periodic" bands. Within this chaotic regime, as the frequency is decreased even further, another set of bifurcations takes place whereby 2^m bands of the attractor successively merge in a mirror sequence of the 2^n periodic sequence that one finds for $\omega \to \omega_{th}^+$. The final chaotic state corresponds to a single band strange attractor, beyond which there occurs an irreversible jump into a periodic regime of lower amplitude.

In order to study the effects of random fluctuations on the solutions we have just described, we solved eq. (1) using an analog computer in conjunction with a white-noise generator having a constant power spectral density over a dynamical range two orders of magnitude larger than that of the computer. Time series and power spectral densities were then obtained for different values of Γ, A and ω_d. While we found that the folding structure of the strange attractor is very stable under the effect of random forces, the bifurcation sequence that is obtained in the presence of noise differs from the one encountered in the deterministic limit.

Our results can be best summarized in the phase diagram of fig. 1, where we plot the observed set of bifurcations (or limiting set) as a function of the noise level, N, normalized to the rms amplitude of the driving term, Γ. The vertical axis denotes the possible states of the system, labeled by their periodicity $P = 2^n$, which is defined as the observed period normalized to the driving period, T_d. As can be seen, with increasing noise level a symmetric bifurcation gap appears, deplet-

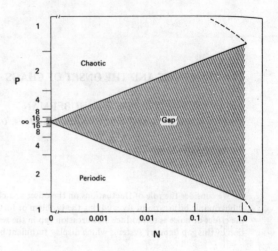

Fig. 1. The set of available states of a forced dissipative anharmonic oscillator as a function of the noise level. The vertical axis denotes the periodicity of a given state with $P = T/T_d$. The noise level is given by $N = A/\Gamma_{rms}$. The shaded area corresponds to inaccessible states.

ing states both in the chaotic and periodic phases. This set of inaccessible states is characterized by the fact that the longest periodicity which is observed before a strange attractor appears is a decreasing function of N, with the maximum number of bands which appear in the strange attractor behaving in exactly the same fashion. This gap extends over a large range of noise levels (up to $N = 1.5$), beyond which the motion either becomes unstable (i.e., $|\psi| \to \infty$; lower dashed line) or an amplitude jump takes place from the chaotic regime to a limit cycle of period 1 (upper dashed line).

We can illustrate this behavior by looking at the power spectral densities, $S(\omega)$ at fixed values of the driving frequency while increasing N. Fig. 2 shows such a sequence for $\omega_d/\omega_0 = 0.6339$, $\Gamma = 0.1175$, and $\alpha = 0.4$. Fig. 2(a) corresponds to $S(\omega)$ near the deterministic limit which, for the parameter values used, displays a limit cycle of period four. As N is increased, a transition takes place into a chaotic regime characterized by broad band noise with subharmonic content of periodicity $P = 4$ (fig. 2(b)) [‡1]. As the noise is increased even further, a new bifurcation occurs from which a new

---

[‡1] We should mention that the Poincare map corresponding to this state clearly shows a four-band strange attractor with a single fold.

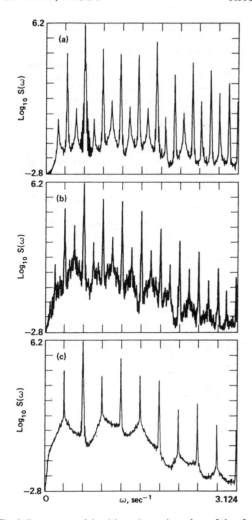

Fig. 2. Power spectral densities at increasing values of the effective noise temperature, for $\Gamma = 0.1175$, $\alpha = 0.4$, and $\omega_d = 0.6339\,\omega_0$. Fig. 2(a): $N = 10^{-4}$. Fig. 2(b): $N = 0.005$. Fig. 2(c): $N = 0.357$.

chaotic state with $P = 2$ emerges. Physically, this sequence reflects the fact that a larger effective noise temperature (and hence a larger fluctuating force) makes the particle gain enough energy so as to sample increasing nonlinearities of the potential, with a resulting motion which in the absence of noise could only occur for longer driving periods [‡2].

[‡2] In the regime of subharmonic bifurcation the dependence of response amplitude on driving frequency is almost linear.

A different set of states appears if the noise level is kept fixed while changing the driving frequency. In this case the observed states of the system correspond to vertical transitions in the phase diagram of fig. 1, with the threshold value of the driving period, T_{th}^n, at which one can no longer observe periodicities $P \geqslant 2^n$, behaving like

$$T_{th}^n = T_{th}^\infty (1 - N_n^\gamma) \qquad (3)$$

for $0 \leqslant N_n \leqslant 1$, with N_n the corresponding noise level, T_{th}^∞ the value of the driving period for which the deterministic equation undergoes a transition into the chaotic state, and γ a constant which we determined to be $\gamma = \sim 1$ for $P \geqslant 2$ [‡3].

In order to test the universality of the bifurcation gap we have just described, we have also studied the bifurcation structure of the one-dimensional map described by

$$x_{L+1} = \lambda x_L (1 - x_L) + n_L(0, \sigma^2) \qquad (4)$$

where $0 \leqslant x_L \leqslant 1$, $0 \leqslant \lambda \leqslant 4$, and n_L is a gaussian random number of zero mean and standard deviation σ. For $n_L = 0$, eq. (4) displays a set of 2^n periodic states universal to all single hump maps [8–9], with a chaotic regime characterized by 2^m bands that merge pairwise with increasing λ [10]. For $n_L \neq 0$ and a given value of σ, the effect of random errors on the stability of the limiting set is to produce a bifurcation gap analogous to the one shown in fig. 1.

The above results are of relevance to experimental studies of turbulence in condensed matter, for they show that temperature plays an important role in the observed behavior of systems belonging to this same universality class. In particular, Belyaev et al. [11], Libchaber and Maurer [12] and Gollub et al. [13] have reported that under certain conditions the transition to turbulence is preceded and followed by different finite sets of 2^n subharmonic bifurcations. It would therefore be interesting to see if temperature changes or external sources of noise in the fluids can either reduce or increase the set of observed frequencies, thus providing for a test of these ideas. In the case of solids such as superionic conductors, the expo-

---

[‡3] Using the scaling relation $(T_{th} - T_n)/(T_{th} - T_{n+1}) = \delta$ [1] this implies that the threshold noise level scales like $N_n/N_{n+1} = \delta$, with $\delta = 4.669201609 \dots$.

nential dependence on temperature of their large diffusion coefficients might provide for an easily tunable system with which to study the existence of bifurcation gaps. Last, but not least, these studies can serve as useful calibrations on the relative noise temperature of digital and analog simulations.

In concluding we would like to emphasize the wide applicability of the effects that we have reported. Beyond the experimental studies of turbulence, there exist other systems which belong to the same universality class as the anharmonic oscillator and one-dimensional maps. These systems range from the ordinary differential equations studied by Lorenz [10], Robbins [14], and Rossler [15] to partial differential equations describing chemical instabilities [16]. Since period doubling subharmonic bifurcation is a universal feature of all these models, our results provide a quantitative measure of the effect of noise on their non-linear solutions.

The authors wish to thank D. Farmer, N. Packard, and R. Shaw for helpful discussions and the use of their simulation system.

References

[1] B.A. Huberman and J.P. Crutchfield, Phys. Rev. Lett. 43 (1979) 1743.

[2] See also, C. Herring and B.A. Huberman, Appl. Phys. Lett. 36 (1980) 976.

[3] See, D. Ruelle, in Lecture notes in physics, eds. G. Dell'Antonio, S. Doplicher and G. Jona-Lasinio (Springer-Verlag, New York, 1978), Vol. 80, p. 341.

[4] G. Ahlers, Phys. Rev. Lett. 33 (1975) 1185; G. Ahlers and R.P. Behringer, Prog. Theor. Phys. (Japan) Suppl. 64 (1978) 186.

[5] J.P. Gollub and H.L. Swinney, Phys. Rev. Lett. 35 (1975) 927; P.R. Fenstermacher, H.L. Swinney, S.V. Benson and J.P. Gollub, in: Bifurcation theory in scientific disciplines, eds. D.G. Gorel and D.E. Rossler (New York Academy of Sciences, 1978).

[6] A. Libchaber and J. Maurer, J. Physique Lett. 39 (1978) L-369.

[7] G. Ahlers and R.W. Walden, preprint (1980).

[8] T. Li and J. Yorke, in: Dynamical systems, an International Symposium, ed. L. Cesari (Academic Press, New York, 1972), Vol. 2, 203.

[9] M. Feigenbaum, J. Stat. Phys. 19 (1978) 25.

[10] E.N. Lorenz, preprint (1980).

[11] Yu.N. Belyaev, A.A. Monakhov, S.A. Scherbakov and I.M. Yavorshaya, JETP Lett. 29 (1979) 295.

[12] A. Libchaber and J. Maurer, preprint (1979).

[13] J.P. Gollub, S.V. Benson and J. Steinman, preprint (1980).

[14] K.A. Robbins, SIAM J. Appl. Math. 36 (1979) 451.

[15] O.E. Rossler, Phys. Lett. 57A (1976) 397; J.P. Crutchfield, D. Farmer, N. Packard, R. Shaw, G. Jones and R.J. Donnelly, to appear in Phys. Lett.

[16] Y. Kuramoto, preprint (1980).

Noise versus chaos in acousto-optic bistability

Réal Vallée and Claude Delisle
Laboratoire de Recherches en Optique et Laser, Département de Physique, Université Laval, Québec, Québec, Canada G1K 7P4

Jacek Chrostowski
Division of Electrical Engineering, National Research Council of Canada, Ottawa, Canada K1A 0R6
(Received 2 December 1983)

We present a study of the evolution to chaos for the acousto-optic bistable device. We numerically solve the difference-differential equation describing the system which shows excellent agreement with the experiment. We analyze the influence of noise—additive and multiplicative—on the bifurcation sequence and on the onset of chaos. It is shown experimentally that both types of noise create a gap in bifurcation sequence. In addition we present a comparison between theory and experiment of the time evolution of the signal.

MS code no. AM2305   1983 PACS numbers: 42.50.+q, 42.65.−k, 05.40.+j, 42.80.−f

I. INTRODUCTION

The chaotic or turbulent behavior seen in physical, chemical, or biological systems which are governed by deterministic equations has attracted intense interest recently.[1] It has been pointed out[2] that chaotic behavior can occur in an optical bistable system which can be described by a differential-difference equation. Since then, a period-doubling route to chaos has been demonstrated in a hybrid electro-optic[3] and acousto-optic[4] device with delay in the feedback loop, all-optical passive systems,[5,6] and single-mode lasers.[7] It is well known that such dynamical systems exhibiting a continuous instability as a function of the bifurcation parameter are extremely sensitive to small perturbations, particularly at the points close to the threshold of instability.

It has been shown that added noise can lead to a truncation of the sequence of periods. Experimentally, this question has been studied in various systems[8] by applying artificial and experimentally controlled broad-band noise to the control parameter. In the case of the electro-optic device, Derstine et al. studied the influence of the shot noise of their photomultiplier which is intensity dependent, and found some departure from the theoretical model[9] based on intensity-independent noise.

In optical bistability, however, noise may appear in both forms: intensity dependent (multiplicative) and intensity independent (additive). In this paper we will consider the two sources of noise influencing our system—intensity fluctuations of the laser and bias voltage fluctuations. They appear in either of the so-named multiplicative or additive form in the difference-differential equation describing the system.

In order to see the real influence of these two kinds of noise, we simulate them by adding noisy voltage with known characteristics in different places of the loop. In Sec. II we briefly describe the acousto-optic bistability. In Sec. III we describe the model used to simulate the effect of noise. In Sec. IV we analyze the time evolution of the noisy system in the chaotic region.

II. CHAOS IN ACOUSTO-OPTIC BISTABILITY

Figure 1 shows the experimental layout of our hybrid acousto-optic bistable device. The He-Ne laser diffracted light is detected by a photodiode and the signal is delayed by an amount of $\tau_D = 5$ μsec$= 10\tau$ where τ is the response time of the system. This delay results from an intrinsic delay in the acousto-optic interaction and from the propagation time through several hundred meters of coaxial cable. The signal is fed to the rf generator (driver) which produces a voltage on the Bragg cell proportional to the feedback-signal amplitude, thus closing the loop. Further experimental details were published elsewhere.[4]

A Gaussian noise generator with a bandwidth of 5 MHz was used as a well-controlled source of noise. This noise was introduced by two methods. (1) additive—when fed into the amplifier producing a noisy offset in the loop and (2) multiplicative—when modulating the intensity of the laser by driving the second acousto-optic modulator operating in the linear mode (Fig. 1).

The experimental bifurcation sequence for both multiplicative and additive noise was obtained on the oscilloscope by means of an electronic window comparator. The results are illustrated in Fig. 2 for different noise levels. The relative width of the electronic window allowed us to obtain a well-defined branching structure while slowly varying μ (x axis). We could even observe the period-8 subharmonic at the lowest noise level. It is therefore nearly the null derivatives points of the signal that are plotted with the Z-axis input of the oscilloscope because the branches of the pitchfork bifurcations shown correspond to the regions of the most probable values of the signal.

We can easily observe (cf. Fig. 2) the progressive disappearance of the higher subharmonics with increasing noise. This general feature is discussed for the difference-equation case by Crutchfield et al.[10] In a similar manner we can say that the effect of noise in our system can be understood as a kind of dynamical average of the structure of attractors over a range of nearby parameters.

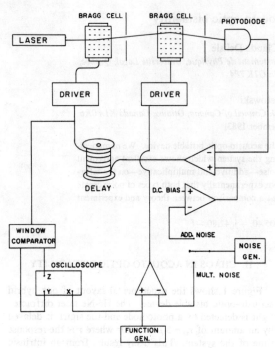

FIG. 1. Acousto-optic bistable device.

According to this view, the averaging of a periodic orbit with adjacent chaotic orbits tends to lower the transition to chaos. On the other hand, in the case of the transition from a periodic orbit to the next one, the averaging does not produce a shift of the bifurcation. In Fig. 3 we can see an enlargement of the bifurcation from the period-1 to the period-2 cycle for the following two cases: (1) when there is no noise added [Fig. 3(a)]; (2) with a multiplicative noise of amplitude $p=0.03$ [Fig. 3(b)]. It appears clearly that the bifurcation point remains globally unchanged even though a Fourier analysis of the signal reveals that the noise substantially reduces the slope of the growing of the period-2 frequency with the bifurcation parameter μ.

Moreover, we could observe, by studying the evolution of the Fourier components, that the disappearance of a periodic waveform for example, (8-P) was preceding that of its corresponding chaotic waveform (8-C) for both types of noise. Derstine et al. previously observed such a phenomenon, but for multiplicative noise only. The departure from the model developed by Crutchfield et al.,[9(a)] which predicts the simultaneous disappearance of 8-C and 8-P with increasing noise (symmetric gap), can thus be attributed to the fact that our hybrid system is described by a delay-differential equation instead of the nonlinear differential equation of the model.

III. NUMERICAL ANALYSIS OF THE STOCHASTIC EQUATION

The transient behavior of the noiseless system is given by the difference-differential equation[11]

$$\tau\frac{dX(t)}{dt}=-X(t)+\pi\{A-\mu\sin^2[X(t-\tau_D)-X_B]\}, \quad (1)$$

where X is the normalized voltage at the input of the acousto-optic driver; μ, proportional to the laser intensity, is the bifurcation parameter, and X_B and A are constants related to the voltage offset. The solution of Eq. (1) takes the form

$$X(t)=\int_{-\infty}^{t}e^{-(t-s)}F[X(s-\tau_D);\mu]ds, \quad (2)$$

where

$$F(X)=-\mu\pi\sin^2(X-X_B)+\pi A, \quad (3)$$

and in general requires numerical calculations. Solutions for the similar type of equations appearing in different examples of bistability have been presented recently.[12] Up until now, as far as we know, there was no attempt to record experimentally and compare with theory the transient behavior of the bistable system in the chaotic domain.

The deterministic evolution given by Eq. (1) is subject to at least two sources of noise in our experiment: the intensity of the laser undergoes fast δ-correlated fluctuations and the electrical part of the device also produces white noise. We model these by adding noise terms to the right-hand side of Eq. (1). The corresponding Langevin equation takes the form

$$\tau\frac{dX(t)}{dt}=-X(t)+\pi\{(A+q\xi_1)\\-(\mu+p\xi_2)\sin^2[X(t-\tau_D)-X_B]\},$$

(4)

where ξ_1, ξ_2 are Gaussian random processes with zero mean, variance one, and a correlation function given by $G(t-t')=\delta(t-t')$; p and q control the amplitudes of the multiplicative and additive noise.

If the overall delay τ_D is small compared to the response time τ, the system will end up its evolution into the steady state governed by the initial conditions and parameter μ. In this case, it shows hysteresis and bistability in three modes:[13] input intensity variation, feedback gain, and modulator-bias voltage variation.

If the response time τ becomes much faster than the delay τ_D, the time evolution can be approximated by a difference equation. It shows the famous bifurcation sequence predicted by Feigenbaum. The difference equation correctly describes most of the basic features of the bifurcation sequence.

In Fig. 4 we present the bifurcation diagram for the noisy difference equation:

$$X(t)=\pi\{(A+q\xi_1)-(\mu+p\xi_2)\sin^2[X(t-\tau_D)-X_B]\} \quad (5)$$

for different additive $q\xi_1$ and multiplicative $p\xi_2$ noise levels. Comparison with the experimental results (Fig. 2) shows good qualitative agreement, particularly in the case of additive noise.

However, the stochastic-difference equation (5) can only give us a crude approximation of the real experiment. The differential term in Eq. (1) is responsible for interest-

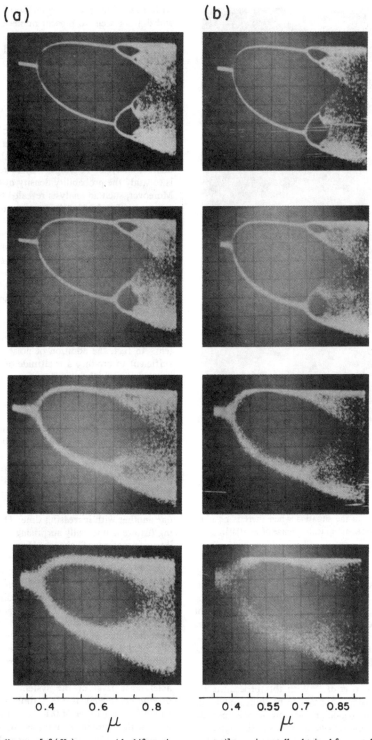

FIG. 2. Bifurcation diagram [of $\{X_n\}$ versus μ (the bifurcation parameter)] experimentally obtained for $\tau_D = 10\tau$. For the multiplicative noise the experimental error associated with the measure of the rms amplitude was of the order of 50%. For additive noise it was less than 10%. Figure (a) shows the effect of additive noise for different amplitudes q: (1) $q = 0.0004$, (2) $q = 0.001$, (3) 0.005, and (4) 0.008. Figure (b) shows the effect of multiplicative noise for different amplitudes p: (1) $p = 0$, (2) $p = 0.01$, (3) $p = 0.04$, and (4) $p = 0.15$. A residual, mainly additive, noise was always present with an amplitude of $q = 0.0004$.

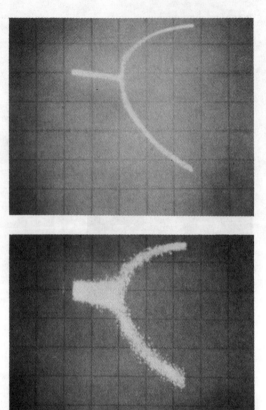

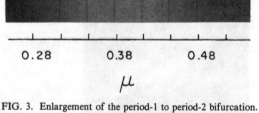

FIG. 3. Enlargement of the period-1 to period-2 bifurcation. Figure (a) corresponds to the situation when there is no noise added. Figure (b) is for a multiplicative noise of $p=0.03$.

ing features that a simple difference equation cannot explain (frequency-locked oscillations, for example). Furthermore, in order to analyze the influence of noise on chaos on a definite time scale, one needs to consider the solution of the stochastic-differential-difference equation (4).

To solve the stochastic equation (4) it would require the simulation using Monte Carlo techniques. A discrete version of Eq. (4) can be obtained following, for example, a prescription of Sancho et al.[14] With this procedure, the resulting stochastic signal appears to be quite stable and correct in the periodic region and for low noise amplitude. However, in the chaotic region, as expected, the signal began to show numerical instabilities for very low noise level, resulting from the intrinsic error (of order $\Delta^{3/2}$) in the algorithm. In order to increase the precision we used a second method based on a Runge-Kutta model with error in order three (Δ^3). The results are in good agreement with those of the preceding method for the periodic region and did not seem to present numerical instabilities in the chaotic region when adding noise. The way we have modeled the noise in our system and the particular nature of the noise considered allows us to introduce noise in a simple manner.

IV. NOISY EVOLUTION TO CHAOS

The critical point in the analysis of a noisy chaotic signal is to distinguish between the effect of noise on the chaotic state itself and its effect on the stability of the numerical analysis. A simple way to realize this distinction is to study the probability density of the calculated signal. Moreover, such an analysis revealed that, as for the difference case, the global stability of a chaotic attractor is not perturbed very much by noise. We present in Fig. 5 the effect of noise ($q=0.02$) on the probability density for the chaotic attractor at $\mu=0.78$ from which we can conclude that the chaotic signal nearly behaves the same way, statistically speaking, in presence of noise.

On the other hand, the time evolution of the signal appears to be extremely sensitive to noise. It seems, therefore, that the exact trajectory of the signal is meaningless since it is indefinitely single valued or deterministic only in the ideal case when there is no perturbation on the system. In fact, the addition of noise at a very low level is sufficient to produce a multitude of progressively diverging trajectories all originating from the same initial conditions. Moreover, even without adding noise, the only uncertainty related to the numerical analysis was enough to cause the "deterioration" of the signal. While increasing the number of calculated points, and therefore the precision of the analysis, we could observe that the divergence of the trajectories was slightly postponed.

We present at Fig. 6 the superposition of chaotic trajectories calculated with a small additive noise ($q=0.001$). This noise induces a progressive loss of the initial conditions so that all trajectories diverge more and more from one another with increasing time. Of course, this interesting feature is not really surprising since it is related to the intrinsic nature of chaos. However, one can be very surprised by the extreme sensitivity to noise of chaos for numerical and for experimental signal.

The experimental time evolution of chaos (for $\mu \approx 1$) starting from particular initial conditions was obtained by using a switching gate in the feedback loop of our system. The gate was alternately opened and closed by an external periodic signal, simultaneously triggering the oscilloscope. We could directly observe on the oscilloscope the superposition of many trajectories. The results are presented in Fig. 7 where we can see that for the intrinsic (unavoidable) noise level of the system ($q=0.004$), trajectories diverge completely inside a time interval of about $20\tau_D$ which is of the order of 0.1 msec in our system. Comparison with numerical analysis shows good qualitative agreement. The experiment also demonstrates a great dependence of trajectories on fine change of the bifurcation parameter μ. It also appears that for a very small increase of μ, one could encounter frequency-locked, period-2 chaos, and some of the harmonics predicted by

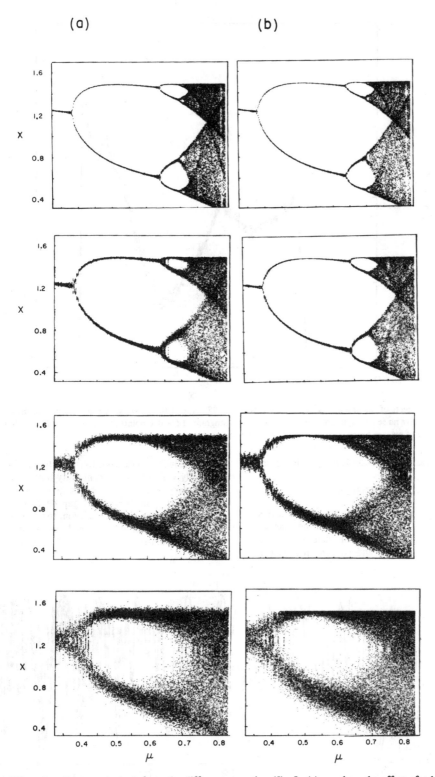

FIG. 4. Bifurcation diagram obtained from the difference equation (5). In (a) we show the effect of additive noise for (1) $q=0.0003$, (2) $q=0.0011$, (3) $q=0.0045$, and (4) $q=0.0083$. In (b) we consider the effect of multiplicative noise of amplitude: (1) $p=0.0015$, (2) $p=0.0025$, (3) $p=0.02$, and (4) $p=0.08$.

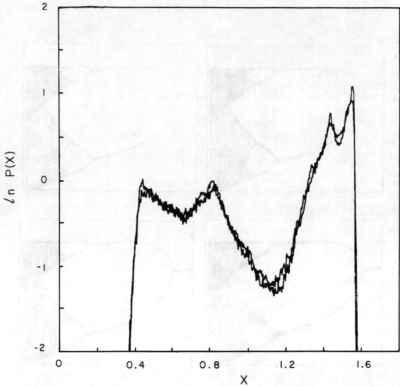

FIG. 5. Logarithmic plots of the probability density $P(X)$ for $\mu=0.78$. One of the curves shown corresponds to the noiseless case; the other is for an additive noise of amplitude $q=0.02$. $P(X)$ is a histogram of 500 000 points calculated from Eq. (4) and partitioned in 500 bins. The effect of noise is practically visible only near the peaks where it rounds off the curve.

Ikeda et al.[15] appearing and disappearing within a small interval of time.

V. CONCLUSION

We presented an experimental study of the acousto-optic bistable device in the periodic and in the chaotic regime with external noise. In the fully deterministic case, the agreement between theory and experiment is excellent. The noise was introduced into the system in two ways: by randomly varying the laser intensity (multiplicative noise) and by varying the dc offset voltage in the amplifier (additive noise). The bifurcation gap develops as the noise amplitude is increased. In addition, the "transient" chaotic signal experimentally obtained shows good agreement with the calculated one concerning the sensitivity to noise.

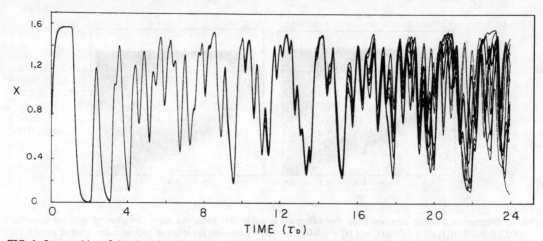

FIG. 6. Superposition of chaotic trajectories obtained numerically from Eq. (4) with $p=0$, $q=0.001$, and for $\mu=1$ and $\tau_D=10\tau$.

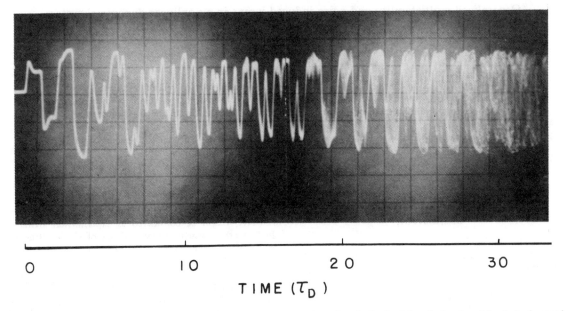

FIG. 7. Superposition of chaotic trajectories experimentally obtained, just after the feedback loop is closed, and for the background additive noise level of the system, $q = 0.0004$.

ACKNOWLEDGMENTS

The authors wish to thank P. Dubois for many helpful discussions and calculations concerning the analysis in this paper. D. Bourque is also acknowledged for his technical assistance. One of us (J.C.) would like to thank Dr. L. M. Narducci for his kind comments on our previous publication.[4] This work was supported by the Natural Sciences and Engineering Research Council of Canada.

---

[1]See, for example, *Evolution of Order and Chaos in Physics, Chemistry, and Biology*, edited by H. Haken (Springer, Berlin, 1982); *Dynamical Systems and Chaos*, edited by L. Garrido, (Springer, Berlin, 1983); *Coherence and Quantum Optics V*, edited by L. Mandel and E. Wolf (Plenum, New York, in press).

[2]K. Ikeda, H. Daido, and O. Akimoto, Phys. Rev. Lett. **45**, 709 (1980).

[3]F. A. Hopf, D. L. Kaplan, H. M. Gibbs, and R. L. Skoemaker, Phys. Rev. A **25**, 2172 (1982).

[4]J. Chrostowski, R. Vallée, and C. Delisle, Can. J. Phys. **61**, 1143 (1983).

[5]H. Nakatsuka, S. Asaka, H. Itoh, K. Ikeda, and M. Matsuoka Phys. Rev. Lett. **50**, 109 (1983).

[6]R. G. Harrison, W. J. Firth, C. A. Emshary, and I. A. Al-Saidi, Phys. Rev. Lett. **51**, 562 (1983).

[7]R. S. Gioggia and N. B. Abraham, Phys. Rev. Lett. **51**, 650 (1983).

[8]T. Kawakubo, A. Yamagita, and S. Kabashima, J. Phys. Soc. Jpn. **50**, 1451 (1981); M. W. Derstine, H. M. Gibbs, F. A. Hopf, and D. L. Kaplan, Phys. Rev. A **26**, 3720 (1982); J. Perez and C. Jeffries, Phys. Rev. B **26**, 3460 (1982).

[9](a) J. P. Crutchfield and B. A. Huberman, Phys. Lett. A **77**, 407 (1980); (b) G. Mayer-Kress and H. Haken, J. Stat. Phys. **26**, 149 (1981); J. Chrostowski, Phys. Rev. A **26**, 3023 (1982).

[10]J. P. Crutchfield, J. D. Farmer, and B. A. Huberman, Phys. Rep. **92**, 46 (1982).

[11]J. Chrostowski, C. Delisle, and R. Tremblay, Can. J. Phys. **61**, 188 (1983).

[12]K. Ikeda, K. Kondo, and O. Akimoto, Phys. Rev. Lett. **49**, 1467 (1982); M. Kitano, T. Yabuzaki, and T. Ogawa, *ibid.* **50**, 713 (1983); J. Y. Gao, J. M. Yuan, and L. M. Narducci, Opt. Commun. **44**, 201 (1983); J. Y. Gao, L. M. Narducci, L. S. Schulman, M. Squicciarini, and J. M. Yuan, Phys. Rev. A **28**, 2910 (1983).

[13]J. Chrostowski and C. Delisle, Opt. Commun. **41**, 71 (1982).

[14]J. M. Sancho, M. San Miguel, S. L. Katz, and J. D. Gunton, Phys. Rev. A **26**, 1589 (1982).

[15]K. Ikeda, K. Kondo, and O. Akimoto, Phys. Rev. Lett. **49**, 1467 (1982).

Generalized multistability and noise-induced jumps in a nonlinear dynamical system

F. T. Arecchi,* R. Badii,† and A. Politi
Istituto Nazionale di Ottica, Largo Enrico Fermi 6, 50125 Firenze, Italy
(Received 4 April 1985)

A study of the forced Duffing equation is reported, with particular reference to a region of the parameter space where five different attractors coexist. This coexistence, reported in some recent experiments, is called generalized multistability. The role of external noise in bridging the otherwise disjoint basins is explored. Noise-induced couplings are shown to be ruled by simple kinetic equations under a general assumption for the geometry of the boundaries. These kinetic equations yield low-frequency power spectra in qualitative agreement with the experimental results.

I. INTRODUCTION—THE DUFFING OSCILLATOR

In the last few years many papers have dealt with the transition from order to chaos in dissipative dynamical systems.[1] Three routes to chaos (period doubling, intermittency and quasiperiodicity) have been extensively studied as possible "scenarios"[2] chosen by a nonlinear system to eventually land into a strange attractor, and the ways a strange attractor loses its stability (crises) have been investigated.[3]

Here we study another, rather general, characteristic of dynamical systems, namely, the coexistence of many different attractors for the same control parameters. We have called this property "generalized multistability",[4,5] in order to distinguish it from the ordinary coexistence of stationary solutions. The relevance of the coexistence of infinitely many periodic unstable solutions is by now sufficiently clarified and taken as synonymous of a deterministic chaotic motion. In such cases, the role of an additional random noise is not relevant since the structure of a strange attractor is not substantially modified.

On the contrary, not much attention has been given by physicists to the possible coexistence of infinitely many periodic stable solutions, as was conjectured by Newhouse.[6] We report the core of his conjecture from Ref. 7: "... in the parameter range where the horseshoe[8] is in the process of creation, an infinite number of families of stable periodic orbits are created." The region of coexistence of these many stable orbits is a critical one, since a small noise may switch the physical system from one orbit to any other, adding a new feature to usual chaotic scenarios. Such a coexistence was numerically explored by Ueda[9] without, however, attempting to evaluate the transition rates.

A qualitative hint on the role of multiple basins of attraction is contained in some experimental observations of hydrodynamic[10,11] instabilities.

Clear evidence of generalized multistability was first shown in an electronic oscillator[12] and then in a modulated laser system.[4] In both cases, the appearance of different attractors in phase space was associated with a low-frequency spectral component due to noise-induced jumps among different attractors. Both measurements, however, might be considered as experimental artifacts. In fact other systems show evidence of single attractors made of two subregions with infrequent passages from one to the other (see, e.g., the Lorenz attractor[13]). In such cases the low-frequency tail corresponds to the sporadic passages (deterministic diffusion),[14,15] and it does not require added noise. Therefore measurements of the power spectra are insufficient to discriminate between the two phenomena, and we must in addition specify the role of noise.

The noise-induced couplings have been studied so far in a simplified model, namely a cubic recursive map, allowing for two simultaneous attractors plus a long transient bridging the two solutions.[16] Here we present a numerical study of differential systems, that is, the forced Duffing oscillator with a double-well potential,

$$\ddot{x}+\gamma\dot{x}-x+4x^3 = A\cos(\omega t) . \quad (1)$$

Many relevant features of this oscillator have been studied by Holmes and a thorough report can be found in Ref. 7.

Before entering into details, we realize already that two limit cycles can coexist for the same control parameters (A,ω), thus showing a simple example of generalized bistability. It is indeed well known that the amplitude response curve versus the frequency of the forcing term of a nonlinear oscillator is a curve bent as in Fig. 1, thus yielding a hysteresis phenomenon (bistability).[17] This means that, even inside a single potential well, two stable periodic solutions can coexist.

Dynamical systems with more than one final state have a highly complex interlacing of basins of attraction. In general the basins boundaries are not simple curves but, instead, their fractal dimension[18] is usually larger than 1, that is, there exist infinitely many points in the phase space whose neighborhoods (of radius ϵ) arbitrarily contain many points belonging to different basins of attraction for any ϵ value.

In Sec. II we present a numerical study of a limited region of the parameter space (A,ω) with particular reference to a fixed set of parameter values, where five attractors simultaneously coexist. In the Appendix the structure of a particular basin of attraction is studied in detail.

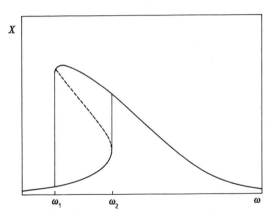

FIG. 1. Generic response curve $X(\omega)$ for a nonlinear damped oscillator vs the forcing frequency and for a fixed value of the external force. The values ω_1 and ω_2 indicate the boundaries of the hysteresis region.

In Sec. III we discuss the role of external noise in jumping from one attractor to the other, evaluating the escape times and the reinjection probabilities, thus yielding the necessary information for evaluating the power spectra.

II. THE PARAMETER SPACE

Numerical study of Eq. (1), performed for limited ranges of the external parameters A and ω, and for fixed $\gamma = 0.154$, yields many coexisting periodic islands. They correspond to attractors whose Poincaré section is made of a finite number N of points and hence they have a period N times that of the external force (Nth subharmonic).

The Poincaré section is built by plotting both x and $\dot{x}$ any time the phase of the external crosses a preassigned value. Each period N attractor arises by tangent bifurcation.[19] As the control parameters drive the system toward chaos, each of the N points of the Poincaré section generates a new one in its neighborhood, up to when we have N disconnected chaotic region, each one made of a Cantorlike set. For instance, if we start from period 3, each of the three points of the Poincaré section gives rise to a neighborhood of 2^k points at the kth bifurcation, but the three neighborhoods are still located around the initial positions and they are visited sequentially so that they can be considered as a perturbed period 3 even in the chaotic limit, where the number of points in each cluster is no longer finite. Since, starting from a period-N attractor, we can follow its N subregions up to the chaotic limit, we call "period N" that region of parameter space that includes both the strictly periodic solutions characterized by N points, as well as the chaotic ones characterized by N subregions sequentially visited. Only when, through a crisis, the N regions merge together, the attractor loses its individuality and we speak of death of the period-N attractor.

In Fig. 2 the contours of the stability regions for some of these attractors have been drawn and the respective periodicity indicated by the attached numbers.

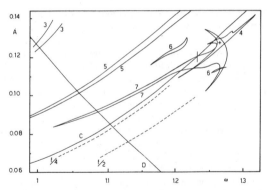

FIG. 2. Phase diagram of the Duffing equation showing the type of solutions for each pair of driving parameters (A,ω). The borderlines are the frontiers of the parameter regions corresponding to ordered motion, and the associated numbers denote the periodicity. Curves C and D show the upper limit (in the amplitude A) for the stability region of solutions confined in one valley. The vertical bar at $\omega = 1.22$ indicates the region where the escape times have been measured.

The meaning of the lines C and D is understood by referring to the two valleys of the potential of Eq. (1). Line C is a borderline above which there are no longer stable solutions confined in one valley. Below line C there is a manifold of lines, approximately parallel to it, which corresponds to a sequence of period doubling bifurcations and with mutual distances ruled by the Feigenbaum δ. These have been omitted for clarity reasons. They have already been observed experimentally (see Fig. 1 of Ref. 20) in an electronic oscillator ruled by Eq. (1); which is, however, affected by too large a noise to display the other interesting details reported in Fig. 2.

On the left of line D, there is a small limit cycle confined in one valley. This limit cycle does not undergo subharmonic bifurcations and it dies via tangent bifurcations, crossing line D.

Hence in the triangular region below the two lines C and D there is the coexistence of the small limit cycle with another one belonging to the above-mentioned Feigenbaum cascade.

The interplay between the two attractors can be appreciated if we draw the oscillator response versus the driving frequency at constant amplitude A, (moving horizontally in Fig. 2). A qualitative sketch of such a response has been already given in Fig. 1, where we see that two stable branches may coexist for $\omega_1 < \omega < \omega_2$. More precisely, at $\omega = \omega_2$ (ω_1), the smallest (largest) limit cycle disappears, yielding a point of line D (C) of Fig. 2. It is important to recall that the response curve shown in Fig. 1 is evaluated by means of perturbative techniques[17] that converge only for small amplitude solutions, when the nonlinear terms can be suitably taken into account. This may not be the case for the upper branch solution that visits highly anharmonic regions. Indeed, Fig. 1 describes only the disappearance of the large response solution, without any reference to subharmonic bifurcations, first observed in Ref. 21.

Let us return to Fig. 2 and focus our attention to the

two period-6 regions: they correspond to different attractors both extended over the two potential valleys and accompanied by the respective symmetric ones. Indeed, if $x(t)$ is a solution of Eq. (1), it is readily seen that also $-x(t)$ is a solution, with the only difference being a shift of π relative to the phase of the forcing term. Hence the symmetry properties immediately tell us that any asymmetric attractor (as the period 4 and the two period 6 ones) is accompanied by its mirrorlike image.

We have discussed a very small part of the parameter space, but it is already so rich of relevant details that the rest of the paper will be confined to discuss the phenomena occurring in pieces of Fig. 2.

In the region denoted by a cross in Fig. 2 we have the coexistence of five attractors, namely, two period 4, one period 7, and finally, two period 2. Their basins of attraction (BA) are sketched in Figs. 3—5; namely, Fig. 3(a) and 3(b) refer to the two period-2 attractors, Fig. 4 to one of the two period 4 (the other is not given for simplicity), and Fig. 5 to the single period 7. The construction of the different BA's requires, in principle, the knowledge of their boundaries, that is, of the unstable manifold of suitable saddle points. However, such curves are so interlaced that it is practically impossible to draw a globally accurate picture. Thus we have preferred to follow a more direct approach, while leaving to the Appendix partial application of the formal method.

We start from a uniform grid of 150×100 points in a

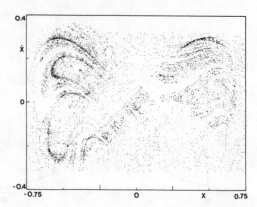

FIG. 4. Basin of attraction of one period-4 solution for the same values of A and ω as in Fig. 3.

Poincaré section with the phase of the external force put equal to 0. The x coordinates are distributed in the interval $(-0.75, 0.75)$ while the velocities stay within $(-0.4, 0.4)$. Each point of the grid, considered as the initial condition for Eq. (1), generates a trajectory that asymptotically falls in one of five coexisting attractors. Every initial point is consequently associated with the basin of attraction of the respective asymptotic solution (Figs. 3—5). Notwithstanding the mirrorlike symmetry between even-period limit cycles, their basins of attraction do not exhibit any symmetry [compare Figs. 3(a) and 3(b)]. Indeed, solutions of Eq. (1) are invariant not under the reflection $(x, \dot{x}) \rightarrow (-x, -\dot{x})$ only, but if further one shifts the phase of the forcing term by π.

Furthermore, to give a better feeling for each basin of attraction, we have reported, together with each point of the initial grid, the next two iterates on the Poincaré section. Referring, for instance, to Fig. 5, we notice seven dense regions, showing the fast contraction rate towards the seven-point attractor.

Figures 6(a) and 6(b) are magnified versions of the central parts (around the origin) of Figs. 3(a) and (5), respectively, for an x interval $(-0.09, 0.09)$ and an $\dot{x}$ interval $(-0.06, 0.06)$. The superposition of the two graphs gives an idea of the intimate interlacing of the different BA's,

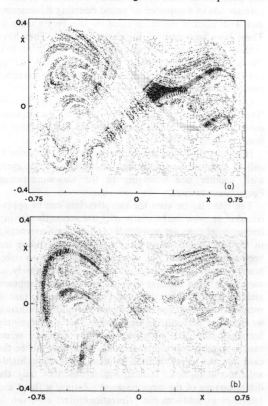

FIG. 3. Basins of attraction of the two period-2 solutions for $A = 0.117$ and $\omega = 1.17$.

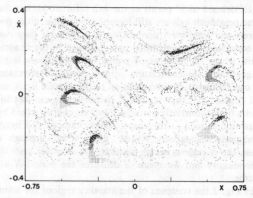

FIG. 5. Basin of attraction of the period-7 solution, taken in the same conditions as the previous figures.

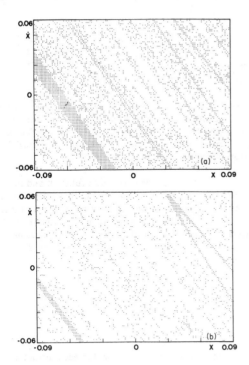

FIG. 6. Expansion of the central part of Fig. 3(a) (period-2 attractor) (a), and of Fig. 5 (period-7 attractor) (b).

thus showing how a tiny displacement in the initial condition may imply a change of the asymptotic solution, as shown recently by Grebogi et al.,[18] who have given evidence of the fractal nature of the border of a BA in a two-dimensional iteration map with two distant attractors. The same task for a differential equation is much more difficult, however a comparison between Fig. 6 and the previous Fig. 3, shows how an improvement by a factor of 8 in the definition of the initial grid does not permit a clear-cut discrimination among the BA's.

III. ROLE OF EXTERNAL NOISE—LOW-FREQUENCY SPECTRA

In the previous section we have considered the noise-free dynamical system, focusing our attention on the occurrence of different asymptotic solutions and on the structure of their BA's. Now, by adding an external white noise in Eq. (1), jumps among different attractors become possible. Such a phenomenon is particularly evident close to the marginal stability points of the attractors, when the occurrence of even very small noise spikes is sufficient to let the point leave the attractor. In the standard multistability (coexistence of different fixed points) marginal stability means that we are near a tangent bifurcation. In this case of generalized multistability another class of critical phenomena must be considered, namely the crisis.[3]

Indeed, the distance of the attractor's support from the border of its BA is equal to zero both for a tangent bifurcation and for a crisis. However, in the former case the BA itself shrinks to zero, whereas in the latter case it is the attractor that spreads up to the border of its BA.

As for the intermittency,[22] we can introduce a relevant parameter to describe a noise-induced crisis, that is, the mean escape time T from the attractor. Such a parameter has been proved[5] to depend, through a universal scaling law, on the noise-amplitude σ and on the distance from the crisis value ϵ.

$$T = \sigma^\alpha F(\epsilon/\sigma^\beta) . \qquad (2)$$

For the case of a logistic map the exponents are (Ref. 5) $\alpha = -\frac{1}{2}$ and $\beta = 1$.

Indeed, referring to the logistic map and for a Gaussian noise, the time T is

$$T = \pi\sqrt{2/\sigma} e^{\epsilon^2/\sigma^2}/D_{-3/2}(\epsilon/\sigma) , \qquad (3)$$

where D is the parabolic cylinder function.

Quite below crisis, for $\epsilon \gg \sigma$, the main dependence of T on the noise amplitude is an exponential one

$$T = \frac{\pi}{\sqrt{2}} \frac{\epsilon^{3/2}}{\sigma^2} e^{\epsilon^2/2\sigma^2} , \qquad (4)$$

very similar to the Kramers diffusion law.[23] Here, however, the physics is wholly different. Let us consider the small noise limit in order to make a sensible comparison with Kramers's approach. In our case the density of points on the attractor is generated by the deterministic equations and is barely affected by the small noise. In contrast in Kramers's problem, the probability density within the potential valleys is essentially determined by the applied noise as in any Langevin problem.

Even though the above relations (3) and (4) have been derived for discrete maps, we can reasonably assume that the same qualitative behavior has to be expected also for a differential equation. In fact for the Duffing equation (1), we have studied the dependence of the mean escape time on the noise amplitude for different values of the external force (see vertical bar in Fig. 2), moving from below to above the crisis of the period-7 attractor. Specifically, at each integration step (which was $\frac{1}{100}$ of the forcing period) $\dot{x}$ has been shifted by a random number selected from a Gaussian distribution with zero mean and rms σ. Due to the smallness of the integration step, such a procedure is a good approximation of a white Gaussian noise. The results are plotted in Fig. 7. For amplitude values below the crisis, the exponential growth clearly appears and, moreover, approaching the marginal stability point, the rate of change of T shows a slowing down that eventually leads to a saturation above the crisis value $A = A_\epsilon$. Indeed, for $A > A_c$, the attractor looses its stability and even without noise it jumps out of its previous BA. So far we have discussed the escape from an attractor. If we want to complete the description of the jumps among the attractors, we need also some information on the reinjection probabilities. The escape problem requires knowledge of the pseudo-invariant distribution of any single attractor and the distance from the border of its BA; the reinjection deals with the interlacing of the different BA's and it is clearly connected with their respective areas.

As we have seen in Sec. II, all the BA's are interlaced over infinitely small length scales, and this makes possible

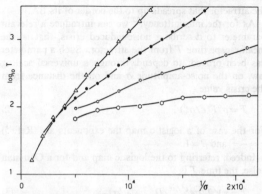

FIG. 7. Mean escape time for the period-7 region vs the inverse of the noise amplitude. All the curves refer to the same frequency, but with different A's. Namely, the symbols $\triangle$, $\bullet$, $\circ$, $\circ$ represent, respectively, $A=0.1170$, 0.1171, 0.1172, and 0.1173 ($\gtrsim A_c$).

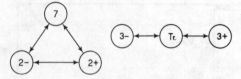

FIG. 8. Logical schemes showing the possible coupling among the attractors in two different cases: (a) Duffing equation with the three attractors (one period 7 and two period 2); (b) one-dimensional antisymmetric cubic map with two period-3 attractors plus a long transient (T).

a probability analysis of the reinjections because any attractor may be within reach (via a jump induced by a suitable noise amplitude) from a larger number of BA's that are surrounding the attractor itself.

Let us then refer to a generic situation with m simultaneously coexisting attractors and call $p_i(t)$ the instantaneous probability to be on the ith attractor. According to the previous assumption, the rate equation for p_i is

$$\dot{p}_i = -a_i p_i + \sum_{j=1}^m s_{ij} a_j p_j \;, \qquad (5)$$

where a_i is the inverse of the mean escape time $(1/T)$ from the ith attractor and s_{ij} is the jump probability from the jth attractor onto the ith one, and is grossly given by the area of the ith BA spanned from any point of the attractor by a leap of the noise amplitude order.

Two terms containing p_i are present in the right-hand side of Eq. (5). Indeed, besides $-a_i p_i$, which is the escape rate from the ith attractor, $+s_{ii}a_i p_i$ takes into account the occurrence of an immediate reinjection. This leads to a distinction between the mean escape time $T=1/a_i$ and the residence time that is increased by jumps from the attractor onto itself. Such a distinction is meaningful when the transients the system takes to "decide" which attractor to land on,[24] are very short compared to T. This is indeed the case we have analyzed ($\omega=1.22$, $A=0.114$) with two period-2 and one period-7 attractor. Whenever such transients become very long, it is still possible to describe the evolution by means of equations like Eq. (5), but the transient has to be considered as another region (like those occupied by attractors) that bridges all the multistable solutions together without any other direct coupling. An example of such a behavior has been described in Ref. 16 where, in a one-dimensional cubic map, two period-3 attractors were coupled only through a long transient.

The logical schemes referring to the two different conditions (Duffing with three attractors and cubic map) have been sketched in Fig. 8.

As shown in Ref. 16, solution of kinetic equations allows the evaluation of the correlation function and hence, under the general assumptions listed in that reference and plausible in the present case, evaluation of the stationary power spectrum. Calling $x=x(t)$ and $x'=x(t+\tau)$, the correlation function of the dynamical process is defined as the ensemble average over the joint probability distribution $p(x)p(x\,|\,x')$ of the two events, that is,

$$R(t,t+\tau) = \int dx \int dx' xx' p(x) p(x\,|\,x') \qquad (6)$$

and the averaged correlation function can be written as

$$\langle R(\tau)\rangle = \lim_{\bar{t}\to\infty} \frac{1}{2T} \int_{-\bar{t}}^{\bar{t}} R(t,t+\tau) dt \;. \qquad (7)$$

As said previously, the motions within each attractor can be taken as decorrelated from the jumps as well as decorrelated from one to another attractor. Therefore, the time average yields either the correlation function $\langle x_i x_i(\tau)\rangle$ or just $\langle x_i\rangle\langle x_j\rangle$ for $i\neq j$, where x_i coincides with $x(t)$ onto the ith attractor and is 0 elsewhere; analogously the probabilities $p(x),p(x\,|\,x')$ reduce the the jump probabilities p_i and $p_j(\tau,i)$. These latter ones are those solutions of Eq. (5) taken with the following criteria: $p_j(\tau,i)$ is the conditional probability of j at time τ, when $p_i=1$ at $\tau=0$; p_i is the asymptotic probability of i for $\tau\to\infty$, independent of the initial condition. With such assumptions, the above correlation function becomes

$$\langle R(\tau)\rangle = \sum_i \langle x_i x_j(\tau)\rangle p_i p_j(\tau,i)$$
$$+ \sum_{i\neq j} \langle x_i\rangle\langle x_j\rangle p_i p_j(\tau,i) \;. \qquad (8)$$

Neglecting the oscillating terms of $\langle x_i x_j(\tau)\rangle$, which contribute to the high-frequency spectrum, and besides a zero-frequency component, the low-frequency spectrum is made in general of $m-1$ Lorentzians, plus a background corresponding to the fast mixing within the transient region.

We have thus shown that the simultaneous coexistence of m attractors leads in general to a power spectrum made of $m-1$ Lorentzians. Symmetries in the attractors may reduce the number of independent coefficients and hence the number of Lorentzians that make the spectrum. This can be particularly relevant for fitting a limited region of the low-frequency spectrum with a power law $S(f)=f^{-\alpha}$ as discussed in Refs. 12 and 16.

We specify the above arguments with the numerical results obtained for $A=0.114$ and $\omega=1.22$. For an external noise rms equal to 2.5×10^{-4}, the mean escape times

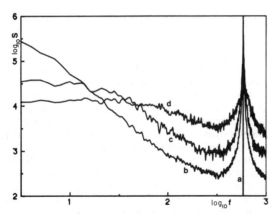

FIG. 9. Power spectrum for the Duffing oscillator for $A = 0.114$, $A = 1.22$ and for different external noise levels the following: (a) 2.5×10^{-5}, (b) 2.5×10^{-4}, (c) 5.0×10^{-4}, (d) 2.0×10^{-3}. The peak on the right corresponds to $f = 1.22/14$ and comes from the period-7 attractor.

from the period-7 attractor and the two period-2 attractors are, respectively, 158 ± 10 and 180 ± 10 (the period of the forcing term being the time unity), while the mean residence times turn out to be 413 ± 20 and 206 ± 10. The large difference between the two averages referred to the period-7 attractor yields a large value for the reinjection probability s_{77} from such attractor back to itself, which is, indeed around 62%. Therefore, since $s_{77} + 2s_{27}$ has to be 1, s_{27} turns out to be 19% while the probability s_{22} of a jump back to the period 2 is lower, namely 12%. Incidentally, $s_{2\text{-}2}$ is very close to s_{22}, even if this is not imposed by the symmetry properties. Finally, again from the normalization condition $s_{22} + s_{-22} + s_{72} = 1$, s_{72} is 76%, hence even larger than the probability of a jump from the period-7 attractor back onto itself.

All of these data contribute to determine the power spectrum S of $x(t)$ shown in Fig. 9 (see curve b). A complete characterization of the low-frequency part is, however, not yet possible, since we should also add the contribution of jumps between two different period-7 solutions (see Sec. II) and the low-frequency component of the two other attractors $(2, -2)$. A qualitative analysis of Fig. 9 shows, anyhow, that, when increasing the noise level from 2.5×10^{-5} to 2.5×10^{-4}, 5.0×10^{-4} and 2.0×10^{-3} (respectively, curves a, b, c, d), the following sequence of events occurs: in a, no jumps occur during the measurement and the solution remains in the initial attractor (namely, a period-7 one); in b, a well-defined low-frequency contribution shows up and, finally, it broadens in c, d indicating faster decay rates.

IV. CONCLUSION

We have shown how addition of noise to deterministic chaos induces a low-frequency spectral component made in general by the superposition of $m - 1$ Lorentzians, m being the number of coexisting attractors that characterize a multistable region of the parameter space. On the contrary, the high-frequency spectrum describing the decay of the correlations within each stable attractor is practically not affected by the noise.

APPENDIX: NUMERICAL CONSTRUCTION OF THE BASIN BOUNDARY OF A PERIODIC ATTRACTOR

Here we show as example, the boundaries of the period-7 attractor. The seventh iterate of the Poincaré map, is made of seven distinct fixed points. Each one of them corresponds to the same attractor except for being observed with a different phase. For simplicity we focused our attention on the lowest point on the right of Fig. 5 and reconstructed its BA in the vicinity of the point itself (see Fig. 10).

Since the period-7 attractor arises via a tangent bifurcation, it is accompanied by its unstable counterpart. This is for instance shown in Fig. 10, where S and U indicate, respectively, the stable and unstable solution.

The contour of the BA is simply defined by the stable manifold of U, and it is made of two distinct curves E and I, as it appears from Fig. 10. We now study the behavior of E and I in the regions P, Q, where they approach one another.

We start by considering a transverse section of the basin

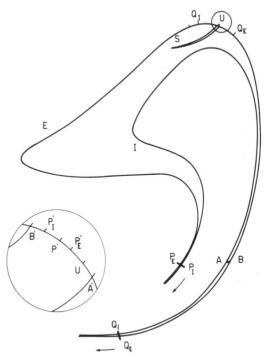

FIG. 10. Boundaries of the BA of the period-7 attractor around one of the points that make its Poincaré section, namely, the lowest one on the right of Fig. 1. S is the stable solution, while U indicates the unstable period 7 born together with the stable one via tangent bifurcation. E and I are the stable manifold of U. The expanded view around U is reported in the circle at the lower left and the role of the points P and Q is discussed in the text.

of attraction in the region P,Q. Let P_E (Q_E) and P_I (Q_I) be the end points of a section in region P (Q) belonging, respectively, to E and I. By iterating seven times P_E (Q_E) and P_I (Q_I), in order to have again the same phase, two new points P'_E (Q'_E) and P'_I (Q'_I) are generated.

If we specialize to the region P, P'_E and P'_I both fall at left of U. Moreover, when letting P_E and P_I move according to the arrow, P'_E and P'_I appear to converge towards the same point P'. Hence for continuity reasons, a point exists where E and I join together, perhaps forming a cusp.

The behavior of E and I is entirely different in the region Q. Indeed, the points Q'_E and Q'_I lie on E, but on opposite sides with respect to U and, moreover, the distance $Q'_E Q'_I$ increases when Q_E and Q_I move along the arrow. Therefore, the expansion rate $Q'_E Q'_I / Q_E Q_I$ seems to diverge because $Q'_E Q'_I$ increases as $Q_E Q_I$ decreases.

However, we can reasonably suppose that, moving forward Q_E and Q_I, the distance $Q'_E Q'_I$, starts decreasing after having reached a maximum value, thus solving the apparent paradoxical result.

This phenomenology is shared by the other six subregions of the period-7 attractor, and it seems to be a rather common feature. The phase plane $x,\dot{x}$ is, therefore, decomposable into a finite number of filamentous regions that are infinitely interlaced. This is a characteristic of complex dynamical systems that causes a high sensitivity to initial conditions in the sense of Ref. 18.

Finally, in order to show the stretching and folding properties of the seventh iterate of the Poincaré map, we have drawn the image of a transversal segment AB. Both the iterates A' and B' of A and B fall in the neighborhood of U, while the interior of AB extends from U towards S as shown in Fig. 10.

---

[*]Also at Dipartimento di Fisica, Università di Firenze, Italy.
[†]Present address: Physik Institut der Universitat, Zurich 8001, Switzerland.

[1]Due to the vast amount of papers on the subject, we report here, for simplicity, the proceedings of some recent conferences or school on this subject, namely *Chaotic Behavior in Dynamical Systems*, edited by G. Iooss and R. H. G. Helleman (North-Holland, Amsterdam 1983); *Order in Chaos*, proceedings of the Los Alamos Conference, Los Alamos, New Mexico, 1982 [Physica 7D, 3 (1983)]; N. B. Abraham, J. Gollub, and H. Swinney, Review of Haverford Workshop, Haverford, 1983 [Physica 11D, 252 (1984)].
[2]J. P. Eckmann, Rev. Mod. Phys. 58, 643 (1981).
[3]C. Grebogi, E. Ott, and J. A. Yorke, Phys. Rev. Lett. 48, 1507 (1982).
[4]F. T. Arecchi, R. Meucci, G. Puccioni, and J. Tredicce, Phys. Rev. Lett. 49, 1217 (1982).
[5]F. T. Arecchi, R. Badii, and P. Politi, Phys. Lett. A 103, 3 (1984).
[6]S. E. Newhouse, Topology 13, 9 (1974).
[7]J. Guckenheimer and P. Holmes, *Nonlinear Oscillation, Dynamical Systems, and Bifurcations of Vector Fields* (Springer, Berlin, 1983).
[8]S. Smale, *Differential and Combinatorial Topology*, edited by S. S. Cairns (Princeton University, Princeton, N.J., 1963).
[9]Y. Ueda, Report No. IPPJ-430, Nagoya University, 1979 (unpublished).
[10]J. Maurer and A. Libchaber, J. Phys. (Paris) Lett. 41, 515 (1980).
[11]M. Giglio, S. Musazzi, and U. Perini, Phys. Rev. Lett. 47, 243 (1981).
[12]F. T. Arecchi and F. Lisi, Phys. Rev. Lett. 49, 94 (1982); 50, 1328 (1983).
[13]C. Sparrow, *The Lorenz Equation* (Springer, Berlin, 1982).
[14]T. Geisel and J. Nierwetberg, Phys. Rev. Lett. 48, 7 (1982).
[15]D. D'Humieres, M. R. Beasley, B. A. Huberman, and A. Libchaber, Phys. Rev. A 26, 3483 (1982).
[16]F. T. Arecchi, R. Badii, and A. Politi, Phys. Rev. A 29, 1006 (1984).
[17]P. Hagedorn, *Nonlinear Oscillations* (Clarendon, Oxford, 1982).
[18]C. Grebogi, S. W. McDonald, E. Ott, and J. A. Yorke, Phys. Lett. A 99, 415 (1983).
[19]P. Collet and J. P. Eckmann, *Iterated Maps* (Birkhauser, Boston, 1980).
[20]F. T. Arecchi and A. Califano, Phys. Lett. A 101, 443 (1984).
[21]B. A. Huberman and J. Crutchfield, Phys. Rev. Lett. 43, 1743 (1979).
[22]Y. Pomeau and P. Manneville, Commun. Math. Phys. 74, 189 (1980).
[23]H. A. Kramers, Physica (Utrecht) 7, 284 (1940).
[24]In general, the BA chosen by the first jump does not necessarily mean that the system will asymptotically go onto the corresponding attractor. Since noise is applied at each step, the point may leave the first BA and wander over other ones, depending on the width of the BA's themselves. This wandering is what we define as "transient."

CHAOTIC DISTRIBUTION OF NON-LINEAR SYSTEMS PERTURBED BY RANDOM NOISE

T. KAPITANIAK

Institute of Applied Mechanics, Technical University of Lodz, Stefanowskiego 1/15, 90-924 Lodz, Poland

Received 10 January 1986; accepted in revised form 26 April 1986

Chaotic behaviour of the distribution function of non-linear systems like the Duffing oscillator and the non-linear pendulum perturbed by random noise is reported.

It is well known that an anharmonic system with an external periodic perturbation

$$\ddot{x} + a\dot{x} + bx + cx^3 = B \cos \omega t \quad (1)$$

can show chaotic behaviour for certain ranges of the parameters (see for example refs. [1–5]).

We consider the amplitude distribution function of the stochastic system

$$\ddot{x} + a\dot{x} + bx + cx^3 = B \cos \omega t + \eta(t), \quad (2)$$

where $\eta(t)$ is the white noise with zero mean and correlation function $\langle \eta(t), \eta(t') \rangle = D\delta(t - t')$, $D < 1$ is constant, $\langle \rangle$ indicates the ensemble average to be studied.

If information is needed only about certain properties of the system without direct reference to the distribution function, eq. (2) may be numerically integrated by a Monte Carlo simulation method [6]. As we are mainly interested in the behaviour of the distribution function, in what follows we employ the method of the Fokker–Planck–Kolmogorov (FPK) equation.

To rewrite eq. (2) in the form for which the FPK equation can be easily written [6,7] the following transformation was made: $x = x_1$, $\dot{x} = x_2$, $x_3 = B \cos \omega t$, where x_3 is the solution of the following initial-value problem:

$$\dot{x}_3 = x_4, \quad \dot{x}_4 = -\omega^2 x_3, \quad (3)$$

and $x_3(0) = B$, $x_4(0) = 0$. After the above transformation system (2) has the form

$$\dot{x}_1 = x_2, \quad \dot{x}_2 = -ax_2 - bx_1 - cx_1^3 + x_3 + \eta(t),$$
$$\dot{x}_3 = x_4, \quad \dot{x}_4 = -\omega^2 x_3. \quad (4)$$

With the distribution function denoted by $P(x_1, x_2, x_3, x_4, t \mid x_{10}, x_{20}, B, 0)$, where x_{10} and x_{20} are the initial conditions of the system (2) the Fokker–Planck–Kolmogorov equation satisfying eqs. (4) becomes

$$\frac{\partial P}{\partial t} = -\frac{\partial}{\partial x_1}[x_2 P]$$
$$- \frac{\partial}{\partial x_2}\left[(-ax_2 - bx_1 - cx_1^3 + x_3)P\right]$$
$$- \frac{\partial}{\partial x_3}[x_4 P] - \frac{\partial}{\partial x_4}[-\omega^2 x_3 P] + \frac{D}{2}\frac{\partial^2 P}{\partial x_2^2}. \quad (5)$$

This equation was solved numerically by the path-integral method, which was exactly described in ref. [8]. The main advantage of this method is the efficiency in terms of computer time compared with other methods. To verify the path-integral calculations the same equation was solved also by finite-difference methods. The amplitude distribution function was calculated from the formula

$$P(x_1, t) = \int_{-\infty}^{\infty} \int_{-\infty}^{\infty} \int_{-\infty}^{\infty} P(x_1, x_2, x_3, x_4, t \mid x_{10}, x_{20}, B, 0) \, dx_2 \, dx_3 \, dx_4. \quad (6)$$

For our numerical calculations we put in this example $a = 1.0$, $b = -10.0$, $c = 100.0$, $B = 0.1$. The initial conditions are deterministic, $x_{10} = 0.0$, $x_{20} = 0.0$ or $x_{10} = 0.1$, $x_{20} = 0.0$. The amplitude of the random noise was $D = 0.2$. Depending on the value of ω three types of distribution function can be obtained: (i) one-maximum curve $\omega = 3.2$ (fig. 1A), (ii) two-maxima curve $\omega = 3.4$ (fig. 1B) and (ii) multiple maxima chaotic curve $\omega = 3.5$ (fig. 1C). The multiple maxima curve corresponds to the value of parameters for which the system (1) shows chaotic behaviour. To characterize the chaotic behaviour of the distribution function the phase portraits were plotted for three characteristic types of the distribution functions (fig. 2).

Also the function $P(x_1, t)$ for a constant amplitude x_1 shows chaotic behaviour (fig. 3).

The chaotic type of distribution function is very sensitive to initial conditions (fig. 4).

As another example we consider the forced non-linear pendulum with external noise:

$$\ddot{x} + a\dot{x} + \lambda \sin x = B \cos \omega t + \eta(t), \qquad (7)$$

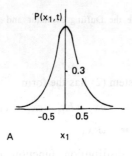
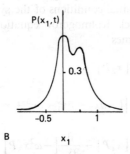
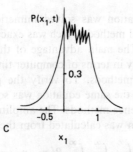

Fig. 1. Types of the amplitude distribution function for the anharmonic oscillator: $a = 1.0$, $b = -10.0$, $c = 100.0$, $B = 0.1$, $D = 0.2$, $t = 20.0$. (A) $\omega = 3.2$, (B) $\omega = 3.4$, (C) $\omega = 3.5$.

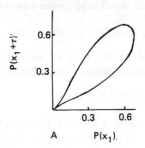
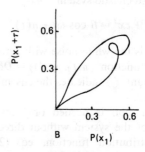
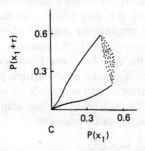

Fig. 2. Phase portraits of the amplitude distribution functions of fig. 1, $\tau = 0.2$.

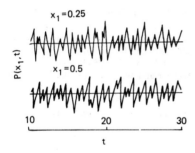

Fig. 3. $P(x_1, t)$ versus time for constant x_1.

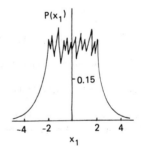

Fig. 5. Chaotic amplitude distribution function of the non-linear pendulum: $a = 1.0$, $b = 1.5$, $\lambda = 4.0$, $\omega = 0.25$, $D = 0.5$, $t = 20.0$, $x_{10} = 0$, $x_{20} = 0$.

where $\eta(t)$ is the white noise.

After the same transformation as for the anharmonic oscillator (2) the system (7) has the form

$$\dot{x}_1 = x_2, \quad \dot{x}_2 = -ax_2 - \lambda \sin x_1 + x_3 + \eta(t),$$
$$\dot{x}_3 = x_4, \quad \dot{x}_4 = -\omega^2 x_3. \qquad (8)$$

In this case the Fokker–Planck–Kolmogorov equation for the distribution function $P(x_1, x_2, x_3, x_4, t \mid x_{10}, x_{20}, B, 0)$ has the following form:

$$\frac{\partial P}{\partial t} = -\frac{\partial}{\partial x_1}[x_2 P]$$
$$-\frac{\partial}{\partial x_2}[(-ax_2 - \lambda \sin x_1 + x_3)P]$$
$$-\frac{\partial}{\partial x_3}[x_4 P] - \frac{\partial}{\partial x_4}[-\omega^2 x_3 P] + \frac{D}{2}\frac{\partial^2 P}{\partial x_2^2}. \qquad (9)$$

In this example we put for numerical calculations $a = 1.0$, $b = 1.5$, $\lambda = 4.0$, $\omega = 0.25$, and the amplitude of the random noise is $D = 0.5$. The initial conditions are deterministic $x_{10} = 0.0$, $x_{20} = 0.0$ or $x_{10} = 0.0$, $x_{20} = 0.1$.

For these parameters the system (7) without random noise shows chaotic behaviour [9]. With random noise we obtain the chaotic amplitude distribution function (6) (fig. 5).

Also the stationary state distribution functions are chaotic (fig. 6).

To summarize, it seems that the chaotic distribution functions of a non-linear system perturbed by small external white noise are characteristic for all dynamic systems which show chaotic behaviour and may provide another description of chaos.

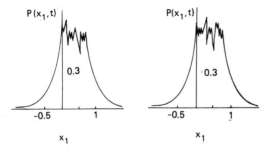

Fig. 4. Chaotic amplitude distribution function for different initial values: (left) $x_{10} = 0.0$, $x_{20} = 0.0$, (right) $x_{10} = 0.1$, $x_{20} = 0.0$.

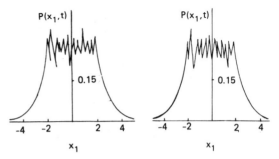

Fig. 6. Stationary state chaotic amplitude distribution functions: (left) $x_{10} = 0$, $x_{20} = 0$, (right) $x_{10} = 0$, $x_{20} = 0.1$.

References

[1] A. Kunick and W.-H. Steeb, J. Japan Phys. Soc. 54 (1985) 1220.
[2] W.-H. Steeb, W. Erig and A. Kunick, Phys. Lett. A 93 (1983) 267.
[3] B.A. Hubermann and J.P. Crutchfield, Phys. Rev. Lett. 43 (1979) 1743.
[4] Y. Ueda, Int. J. Non-linear Mech. 20 (1985) 481.
[5] S. Sato, M. Sano and Y. Sawada, Phys. Rev. A 28 (1983) 1654.
[6] D.L. Ermak and H. Buckholtz, J. Comp. Phys. 35 (1980) 169.
[7] I.I. Gichman and A.V. Skorchod, Stochastic differential equations (Springer, Berlin, 1972).
[8] G. Tagata, J. Sound Vibr. 58 (1978) 95.
[9] M.F. Wehner and W.G. Wolfer, Phys. Rev. A 27 (1983) 2663.
[10] D. D'Humieres, M. Beasley, B.A. Huberman and A. Libchaber, Phys. Rev. A 26 (1982) 3483.

Influence of perturbations on period-doubling bifurcation

H. Svensmark and M. R. Samuelsen
Physics Laboratory I, The Technical University of Denmark, DK-2800 Lyngby, Denmark
(Received 12 December 1986)

The influence of noise and resonant perturbation on a dynamical system in the vicinity of a period-doubling bifurcation is investigated. It is found that the qualitative dynamics can be revealed by simple considerations of the Poincaré map. These considerations lead to a shift of the bifurcation point which is proportional to the square of the amplitude of the perturbation. The results of this investigation are in agreement with numerical calculations for the microwave-driven Josephson junction.

I. INTRODUCTION

Dynamical systems which undergo periodic motion are of great theoretical interest and also of importance for applications. Not only the existence of periodic orbits but also the structural stability of these is important. In this paper we focus on the stability of a periodic orbit against period-doubling bifurcation in the presence of noise and a small resonant signal, a problem which has attracted considerable interest.[1-6] It is shown that the qualitative dynamics in the vicinity of a period-doubling bifurcation can be derived from very simple considerations and that these results are in qualitative agreement with previous results.[1-6] These considerations differ from the previous ones[1-3] in the way that the dynamics is investigated exclusively on the basis of the Poincaré map. Furthermore, the investigation takes its origin in the fix points in the Poincaré map and can therefore be based on linear-stability analysis. The results show that the shift of the bifurcations point is proportional to the square of the perturbating amplitude, a result which is confirmed by experiments on Josephson junctions.[7] As an example of the influence of perturbations on a period-doubling bifurcation, a numerical study of the microwave-driven Josephson junction is reported.

We consider a dissipative dynamical system modeled by the following differential equation:

$$\dot{x} = f_\mu(x,t) + a(t) + s(t), \quad (x,t) \in \mathbb{R}^n \times \mathbb{R}, \quad \mu \in I. \quad (1)$$

Here $f(\cdot,t) = f(\cdot, t+T)$ is periodic in t with period T, μ is a control parameter defined in some interval I, $a(t) = a(t+2T)$ is periodic in t with period $2T$, and $s(t)$ is a white-noise term defined by $\langle s(t)s(t+\tau)\rangle = \sigma\delta(\tau)$ and $\langle s(t)\rangle = 0$. $a(t)$ and $s(t)$ are small compared to $f(x,t)$, i.e., the perturbations are small. Let us first set these perturbations to zero. As the control parameter μ is varied, changes in the structure of the asymptotically stable solution governed by Eq. (1) may occur. In this paper it is assumed that this solution changes from a periodic orbit γ of period T to a periodic orbit γ^* of period $2T$.

II. DYNAMICS IN THE POINCARÉ MAP

A tool for investigation of the asymptotic behavior of orbits close to a closed orbit is the Poincaré map.[8] Here this map is given by a local transversal section of the orbit in phase space $\psi(x,t)$, at integer numbers of period T, $\psi(x,nT) = \psi_T^n(x)$, and is written $P(x_0) = \psi_T(x_0)$. With the aid of the Poincaré map one can derive a discrete dynamical system arising from Eq. (1),

$$x_{n+1} = P_\mu(x_n), \quad x \in \mathbb{R}^n. \quad (2)$$

A fix point of the discrete map Eq. (2) corresponds to a closed orbit of period T, and it is clear that the stability of the fix point reflects the stability of the closed orbit. A period-doubling bifurcation is associated with an eigenvalue $\lambda = -1$ at the fix point x^* of the Poincaré map Eq. (2). This means that an orbit γ^* of the orbit $\psi(x,t)$ alternates from one side of the fix point x^* to the other along the direction of the eigenvector e_λ for $\lambda = -1$. Orbits like γ^* are confined on a two-dimensional surface called the center manifold.[8] This surface can be described as a Möbius band, i.e., a band with a half twist in it (see Fig. 1). The asymptotic behavior of the orbits near a fix point, in the vicinity of a period-doubling bifurcation, can be revealed by linearization of the discrete map around the fix point. Since the asymptotic behavior in the directions of the other eigenvectors is relaxing fast, the asymptotic behavior can be described exclusively with respect to the direction of the eigenvector for $\lambda = -1$. This means that the discrete map will be one dimensional (see Fig. 1), regardless of the original dimension of the dynamical system.[8] It should be noted that this way of describing the asymptotic behavior agrees with the previous description.[1-4] Instead of the control parameter μ the eigenvalue λ of the discrete map will be used, assuming that μ and λ are linearly related in a narrow region close to the bifurcation. So linearizing Eq. (2) around the fix point, and using the mentioned assumptions, the equation describing the asymptotic behavior becomes

$$\xi_{n+1} = \lambda \xi_n, \quad \xi \in \mathbb{R}. \quad (3)$$

ξ is the deviation from the fix point in the direction of the eigenvector for $\lambda \simeq -1$, if $\lambda > -1$ the fix point is stable, but if $\lambda < -1$ the fix point has become unstable and a period-doubling bifurcation has occurred. Higher-order terms in ξ_n are not necessary since this is a linear stability analysis.

If we now look at the influence of a small resonant per-

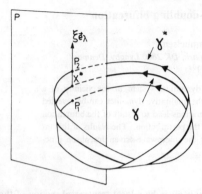

FIG. 1. Intersection of a two-dimensional manifold with the Poincaré map P. The periodic orbit γ of period T is confined to this manifold and intersects P at the fix point x^*. The vector e_λ in P is the eigenvector for the eigenvalue $\lambda = -1$. γ^* is a periodic orbit of period $2T$ which intersects P at the points p_1 and p_2.

turbation at half of the fundamental frequency [$a(t) \neq 0$ in Eq. (1)], the discrete map Eq. (2) will, of course, be changed by the perturbation. The change in the discrete map Eq. (2) is written

$$x_{n+1} = P_n^*(x_n), \quad \text{where } P_{n+2}^*(\cdot) = P_n^*(\cdot) \quad (4)$$

since the driving terms in Eq. (1) now have the periodicity $2T$. The function $P_n^*(x_n)$ is approximated in the following way:

$$P_n^*(x_n) = P(x_n) + P_n'(x_n), \quad P_n'(\cdot) = P_{n+2}'(\cdot), \quad (5)$$

where $P(x_n)$ is the unperturbed Poincaré map and $P_n'(x_n)$ is a small perturbation with the given periodicity. Linearizing around the fix point x^* gives

$$\xi_{n+1} = \lambda \xi_n + (-1)^n A_1 \xi_n + (-1)^n A_0, \quad (6)$$

where the last two terms are part of an expansion of $P_n(x)$. It is assumed that this expansion of P_n is the simplest possible with the given periodicity. In the expansion, terms such as B_0 and $B\xi_n$ are neglected since they correspond to resonant terms at the driving frequency and not on half the fundamental frequency. A_1 and A_0 are small compared with 1. Since the perturbed Poincaré map has the periodicity of 2, it is natural to seek fix points with this periodicity,

$$\xi_{n+2} = (\lambda^2 - A_1^2)\xi_n + [\lambda - (-1)^n A_1 - 1](-1)^n A_0. \quad (7)$$

Inserting the fix point ξ_n^* one gets

$$\xi_n^* = \frac{(-1)^n A_0 [\lambda - (-1)^n A_1 - 1]}{1 - (\lambda^2 - A_1^2)} \quad (8a)$$

or

$$\xi_{\text{odd}}^* - \xi_{\text{even}}^* = \frac{2 A_0 (1 - \lambda)}{1 - (\lambda^2 - A_1^2)}. \quad (8b)$$

These fix points correspond to intersection of the orbit at either even or odd numbers of the period T, that is, n ei-

ther even or odd. From this result one observes two things. First the influence of the resonant perturbation is to stabilize the system against bifurcation as seen from Eq. (7), where the squared eigenvalue is reduced by A_1^2.[3] This leads to a bifurcation shift which is proportional to A_1^2. Experiments on Josephson junctions, where the detuning is very small, confirms this result.[7] Second, as seen from Eq. (8), the small perturbation is amplified as the reduced eigenvalue tends to -1.[2]

The influence of noise alone on the asymptotic behavior can be modeled by adding a small stocastic term s_n to Eq. (3) so this becomes

$$\xi_{n+1} = \lambda \xi_n + s_n, \quad (9)$$

where s_n has the following properties; $\langle s_n \rangle = 0$ and $\langle s_n s_m \rangle = \sigma \delta_{nm}$, the bracket means ensemble averaging and δ_{nm} Kroneckers delta. s_n describes the random intersection of orbits in the Poincaré map along the direction of the eigenvector e_λ for $\lambda = -1$, caused by the noise. Equation (9) can be solved ($\lambda^n \to 0$ for $n \to \infty$),

$$\xi_{n+1} = \sum_{i=0}^{\infty} s_{n-i} \lambda^i. \quad (10)$$

The ensemble average of ξ_n is

$$\langle \xi_n \rangle = 0$$

since $\langle s_n \rangle = 0$ and the autocorrelation function is

$$\langle \xi_n \xi_{n+m} \rangle = \frac{\lambda^{|m|} \langle s_n^2 \rangle}{1 - \lambda^2}. \quad (11)$$

This result shows that the noise is amplified as the eigenvalue λ approaches -1. Since λ is negative ξ relaxes towards the fix point in an alternating fashion, centering the noise spectrum about half of the fundamental frequency.[4]

Finally, the influence of both noise and a small resonant perturbation is investigated [$a(t) \neq 0$ and $s(t) \neq 0$ in Eq. (1)]. By Eqs. (6) and (9) the equation describing the asymptotic behavior becomes

$$\xi_{n+1} = \lambda \xi_n + (-1)^n A_1 \xi_n + (-1)^n A_0 + s_n. \quad (12)$$

Using the same procedure as in Eqs. (6) and (9) the average value of ξ_n becomes as expected the same as Eq. (8a), and the autocorrelation function

$$\langle (\xi_n - \langle \xi_n \rangle)(\xi_{n+2j} - \langle \xi_n \rangle) \rangle$$
$$= \frac{\{1 + [\lambda - (-1)^n A_1]^2\}(\lambda^2 - A_1^2)^{|j|} \langle s_n^2 \rangle}{1 - (\lambda^2 - A_1^2)^2}. \quad (13)$$

It is seen that the general features are the same as when the perturbations were added separately, that is, the amplification of both the noise and the small signal and a reduction of the critical eigenvalue, i.e., a stabilization against the bifurcation. It should be noted that it is also possible to see the stabilization against the bifurcation in the case of a near-resonant signal which is different from half of the fundamental frequency. This is done in the same way as before, but now the function $P_n'(x_n)$ has a periodicity which is much larger than 2, see Eq. (5).

III. NUMERICAL CALCULATIONS

In order to illustrate the features of the above theory, the differential equation describing a microwave-driven Josephson junction is integrated numerically. The governing differential equation is[9]

$$\phi_{tt} + \alpha\phi_t + \sin\phi = A_D\sin(\omega_D t) + A_s\sin(\omega_s t) + \eta + n(t) \; . \quad (14)$$

Here ϕ is the quantum-mechanical phase difference across the junction and α is a damping parameter. A_D and A_s are the driving and resonant perturbing amplitudes normalized to the critical current of the junction I_0, ω_D, and ω_s are the corresponding frequencies normalized to the maximum plasma frequency of the junction. η is the normalized bias current and $n(t)$ is a noise current to be specified below. We note that Eq. (14) also describes a driven damped pendulum.

In the numerical calculations described here Eq. (14) is integrated using a fourth-order Runge-Kutta method with 32 points per period of the drive. 256 drive periods are integrated and the first 128 are discarded to remove transients. The noise is assumed to have a white spectrum described by the parameter $\Gamma = 2ekT/\hbar I_0$, i.e., the ratio of the thermal energy (kT) to the Josephson coupling energy $(\hbar/2e)I_0$. Typical experimental values for Josephson junctions correspond to $\Gamma = 10^{-4} - 10^{-3}$. In the calculations the white noise is constructed using a random-number generator. Throughout the calculations the following values of the parameters were used: $\alpha = 0.2$, $A_D = 0.85$, $A_s = 0$ and 0.001, $\omega_D = 1.6$ and $\omega_s = 0.8$. η and $n(t)$ were varied. The parameters were chosen so the system would get near a period-doubling bifurcation with the variation of η, here used as control parameter.

Figure 2 illustrates for $A_D = 0.85$ and $\omega_D = 1.6$ an example of the influence of noise and a small resonant signal in the vicinity of a period-doubling bifurcation. Here the amplitude of the response at the signal frequency $\omega_s = 0.8$ is plotted versus the control parameter η, which is the normalized bias current. Figure 2(a) shows the amplitude in the absence of both noise and signal $A_s = 0$ and $n(t) = 0$ (this response is in fact due to the extremely slow relaxation). In Fig. 2(b) a small signal is included ($A_s = 0.001$, $\omega_s = 0.8$) and it is seen that the small signal is amplified in the vicinity of the period-doubling bifurcation in agreement with the above theory. In Fig. 2(c) the signal is absent but now noise is included ($\Gamma = 10^{-4}$), and it is seen that the noise is amplified in the vicinity of the bifurcation. In Fig. 2(d) both noise and signal are included ($A_s = 0.001$ and $\omega_s = 0.8$) and again the amplification of the perturbations is seen.

One way of characterizing the stability of a dynamical system against a period-doubling bifurcation is to calculate the maximum Liapunov exponent.[10] When this exponent is negative the periodic orbit of period T is stable, but when the exponent tends to zero the orbit gets unstable against an orbit of period $2T$. Figure 3 illustrates the maximum Liapunov exponent $\sigma_{\max}$ versus the control parameter η. Figure 3(a) shows the maximum Liapunov exponent in the absence of a small signal and noise ($A_s = 0$ and $\Gamma = 0$) where it is seen that the exponent comes very close to zero, i.e., to a period-doubling bifurcation. In Fig. 3(b) a small signal is included ($A_s = 0.001$ and $\Gamma = 0$) and it is seen that the exponent is reduced in the vicinity of the bifurcation showing the stabilization of the system against bifurcation in agreement with the above theory. Figures 3(c) and 3(d) are similar to Figs. 3(a) and 3(b) but here noise is included ($\Gamma = 10^{-4}$). In Fig. 3(c) only minor effects of the noise are seen, whereas the presence of noise

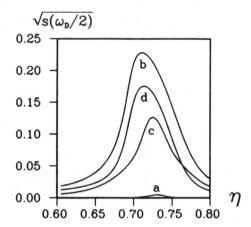

FIG. 2. The amplitude of the response at the signal frequency $\omega_s = 0.8$ for $A_D = 0.85$, $\omega_D = 1.6$ vs η. Curve a, without any perturbations $A_s = 0$ and $\Gamma = 0$. Curve b, with a small resonant perturbation, $A_s = 0.001$, $\omega_s = 0.8$, and $\Gamma = 0$. Curve c, with only noise added, $A_s = 0$ and $\Gamma = 10^{-4}$. Curve d, where both noise and a resonant perturbation are present, $A_s = 0.001$, $\omega_s = 0.8$, and $\Gamma = 10^{-4}$.

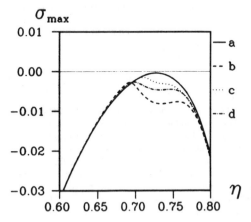

FIG. 3. Maximum Liapunov exponent $\sigma_{\max}$ vs η, for $A_D = 0.85$, $\omega_D = 1.6$. Curve a, without resonant perturbation and noise, $A_s = 0$ and $\Gamma = 0$. Curve b, with resonant perturbation and no noise $A_s = 0.001$, $\omega_s = 0.8$, and $\Gamma = 0$. Curve c, with noise but no resonant perturbation, $\Gamma = 10^{-4}$ and $A_s = 0$. Curve d, with noise and resonant perturbation, $A_s = 0.001$, $\omega_s = 0.8$, and $\Gamma = 10^{-4}$.

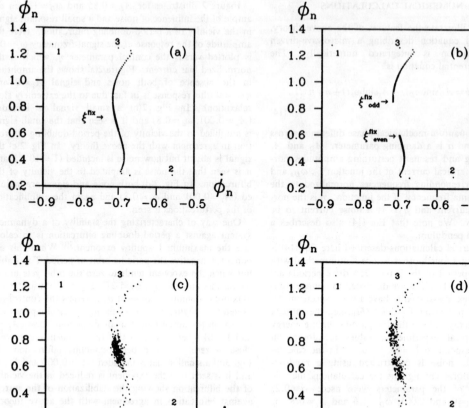

FIG. 4. Asymptotic behavior around the fix point in the Poincaré map. The numbers 1,2,3, indicate the successive intersection of orbits with the Poincaré map. $A_D = 0.85$, $\omega_D = 1.6$ and $\eta = 0.68$. (a) Without perturbations, $A_s = 0$ and $\Gamma = 0$. (b) With a resonant perturbation, $A_s = 0.001$, $\omega_s = 0.8$, and $\Gamma = 0$. (c) Only noise is present, $A_s = 0$ and $\Gamma = 10^{-4}$. (d) Both noise and a resonant perturbation are present, $A_s = 0.001$, $\omega_s = 0.8$, and $\Gamma = 10^{-4}$.

and signal in Fig. 3(d) reduces the exponent in the vicinity of bifurcation.

In the above theory it was assumed that the asymptotic behavior was confined to a line in the Poincaré map; in order to illustrate this numerically, successive intersections of orbits with the Poincaré map were calculated. Figure 4 shows the relaxation towards the fix point in the Poincaré map for $\eta = 0.68$. The relaxation is very fast in the direction perpendicular to the curve on which the essential asymptotic behavior is confined. The numbers (1,2,3,) in Fig. 4 indicate the alternating relaxation towards fix point(s) or steady state in the presence of noise. It is seen that the asymptotic behavior, with no noise added, is strictly confined to a one-dimensional curve in the Poincaré map [see Figs. 4(a) and 4(b)], and so confirming that the asymptotic behavior in the full phase space is confined on a two-dimensional manifold. In the presence of noise the asymptotic behavior is not as strictly confined to the one-dimensional curve in the Poincaré map [see Figs. 4(c) and 4(d)]. The asymptotic behavior shown in Fig. 4 agrees with the basic postulates of the present and the previous theory.[1-4]

As a result of Fig. 4, where only very small perturbations are used ($A_s = 0.001$), the dynamics close to a period-doubling bifurcation in the presence of larger perturbations (e.g., $A_s = 0.01$),[5] have to be described by a two-dimensional manifold in the full phase space, where the curvature of the manifold has to be considered (the Poincaré section is no longer a line but a curve).

IV. CONCLUSION

In conclusion it has been shown that the influence of perturbations on the qualitative dynamics of a system in the vicinity of a period-doubling bifurcation can be derived from very simple considerations on the Poincaré map. The general features of this influence are, first, an amplification of both noise and signal as the bifurcation point is approached, and second, a stabilization of the dynamical system against the bifurcation. The shift of the bifurcations point is proportional to the squared amplitude of the perturbation. Agreement between numerical studies on the microwave-driven Josephson junction and the above theory is found.

---

[1] K. Wiesenfeld and B. McNamara, Phys. Rev. Lett. **55**, 10 (1985).
[2] K. Wiesenfeld and B. McNamara, Phys. Rev. A **33**, 629 (1986).
[3] P. Bryant and K. Wiesenfeld, Phys. Rev. A **33**, 2525 (1986).
[4] K. Wiesenfeld, J. Stat. Phys. **38**, 1071 (1985).
[5] H. Svensmark, J. Bindslev Hansen, and N. F. Pedersen, Phys. Rev. A **35**, 1457 (1987).
[6] J. Heldstab, H. Thomas, T. Geisel, and G. Radons, Z. Phys. B **50**, 141 (1983).
[7] G. F. Eriksen, H. Svensmark, J. Bindslev Hansen, N. F. Pedersen, and M. R. Samuelsen (unpublished).
[8] J. Guckenheimer and P. Holmes, *Nonlinear Oscillations, Dynamical Systems and Bifurcations of Vector Fields* (Springer, New York, 1984).
[9] For a good description of Josephson junction equations see, for example, the book A Barone and G. Paterno, *Physics and Applications of the Josephson Effect* (Wiley, New York, 1982).
[10] Giancarlo Benettin, Luigi Galgani, and Jean-Marie Strelcyn, Phys. Rev. A **14**, 2338 (1976).

CHAOS IN A NOISY MECHANICAL SYSTEM WITH STRESS RELAXATION

T. KAPITANIAK

Institute of Applied Mechanics, Technical University of Lodz, Stefanowskiego 1/15, Lodz, Poland

(Received 20 February 1987, and in revised form 26 June 1987)

The influence of periodic and random external excitation on the chaotic behaviour of a self-excited system is considered. The Lyapunov exponents of the noisy system have been defined as random variables. The properties of their distribution which allow one to quantify chaos have been determined.

1. INTRODUCTION

Chaotic behaviour is known to occur in a variety of relatively simple deterministic mechanical systems [1-9]. In the present paper the influence of the random external excitation on the chaotic behaviour of the mechanical system with stress relaxation shown in Figure 1 is investigated.

The equations of motion of the mass m are

$$m\ddot{x} + \delta = F(t, \omega),$$
$$\dot{\delta} + a\delta = bx + c\dot{x}x^2 + ex + dx^2 + fx^3, \quad (1)$$

where $a = 1/\bar{a}$, $\omega \in \Omega$ and (Ω, β, μ) is a probabilistic space in which Ω, β and μ are respectively, the set of random variables, the σ-field of its Borel subsets and the probabilistic measure.

In the absence of the external excitation ($F(t, \omega) = 0$), self-excited oscillations of the system (1) occur. By calculating the maximum one-dimensional Lyapunov exponent [10-14] it has been found that the chaotic behaviour occurs when the parameters a, c and d are those corresponding to the zone indicated in Figure 2 and the parameters b, e and f satisfy the following relations: $b = \alpha - a$, $f = c/3 + \xi d$ and $e = \alpha - \gamma d$ where $\alpha \in (2 \cdot 2, 3 \cdot 9)$, $\xi \in (0 \cdot 65, 0 \cdot 9)$ and $\gamma \in (0 \cdot 9, 1 \cdot 2)$. Without loss of generality in the calculations described here $m = 1$ was taken.

Consider now the influence of a random external excitation on the chaotic behaviour of the self-excited system. The external excitation $F(t, \omega)$ is assumed to have the form

$$F(t, \omega) = A \cos \Omega_0 t + \eta(t, \omega), \quad (2)$$

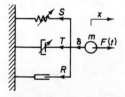

Figure 1. Model of the system: m, mass; x, displacement of the mass m; δ, internal stress; S, resistance due to the stiffness, $S = (ex + dx^2 + fx^3)\bar{a}$; T, resistance due to the damping, $T = (bx + c\dot{x}x^2)\bar{a}$; R, resistance due to stress relaxation, $R = -\bar{a}\dot{\delta}$; $F(t, \omega)$, external force.

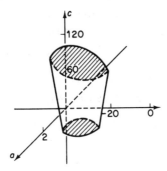

Figure 2. Chaotic zone in the parameters space.

where A and Ω_0 are constant and $\eta(t, \omega)$ is a "band-limited white noise" stochastic process with zero mean and spectral density:

$$S_0(\nu) = \begin{cases} \dfrac{\sigma^2}{\nu_{\max} - \nu_{\min}} & \text{for} \quad \nu \in [\nu_{\min}, \nu_{\max}] \\ 0 & \text{for} \quad \nu \notin [\nu_{\min}, \nu_{\max}] \end{cases}. \qquad (3)$$

σ is the variance of the process $\eta(t, \omega)$ and $[\nu_{\min}, \nu_{\max}]$ is the interval of the frequencies considered.

Recently a method in which the amplitude probability density function [15-17] has been developed to investigate the chaotic behaviour of non-linear systems perturbed by random noise. In what follows an attempt to compute the maximum one-dimensional Lyapunov exponents for such systems is described.

2. RANDOM LYAPUNOV EXPONENTS

External excitation (2) causes the equations of motion (1) to be non-autonomous and because of the random component $\eta(t, \omega)$ it will be impossible to obtain the variational equations directly and use standard methods of computation of Lyapunov exponents. Instead one can do this by the following approximation of the random function $\eta(t, \omega)$.

The stochastic process $\eta(t, \omega)$ can be approximated as a sum of harmonic components $\eta_R(t)$, where

$$\eta_R(t) = \sum_{k=1}^{K} A_k \cos(\nu_k t + \varphi_k). \qquad (4)$$

φ_k $(k = 1, 2, \ldots, K)$ are independent random variables with uniform distribution on the interval $[0, 2\pi]$. There are a few methods of calculating A_k and ν_k, such as those of Borgman [18], Rice (see reference [19]), Shinozuka [19] and Wróbel [20]. In these calculations Rice's method was used with A_k and ν_k deterministic and given by

$$A_k = \sqrt{2 S_0(\nu_k) \Delta \nu}, \quad \nu_k = (k - 0.5) \Delta \nu + \nu_{\min} + \delta \nu_k, \quad \Delta \nu = (\nu_{\max} - \nu_{\min})/K, \quad (5)$$

where $\delta \nu_k$ are uniformly distributed on the interval $[-0.1 \Delta \nu, 0.1 \Delta \nu]$. A unique realization of the process $\eta_R(t)$ is obtained by selecting the random variables φ_k and substituting them into equation (4). As a measure of the quality of these realizations the mean square error ε of the spectral density of the process $\eta(t, \omega)(S_0(\nu))$ and the spectral density of its realization $\eta_R(t)(S(\nu))$

$$\varepsilon = \frac{1}{\nu_{\max} - \nu_{\min}} \left\{ \int_{\nu_{\min}}^{\nu_{\max}} [S(\nu) - S_0(\nu)]^2 \, d\nu \right\}^{1/2} \qquad (6)$$

can be taken.

In the case of external excitation (2) and approximation (4), and after the transformation $x_1 = x$, $x_2 = \dot{x}$, $x_3 = \Omega_0 t$, $x_4 = \nu_1 t + \varphi_1, \ldots, x_{3+K} = \nu_K t + \varphi_K$, equations (1) take the form

$$\dot{x}_1 = x_2, \quad \dot{x}_2 = (1/m)a\delta + A\cos x_3 + B[\cos x_4 + \cos x_5 + \cdots + \cos x_{K+3}],$$

$$\dot{x}_3 = \Omega_0, \quad x_4 = \nu_1, \quad x_5 = \nu_2, \ldots$$

$$\dot{x}_{3+K} = \nu_{3+K}, \quad \dot{\delta} = -a\delta + bx_2 + cx_2 x_1^2 + ex_1 + dx_1^2 + fx_1^3, \tag{7}$$

with initial conditions $x_{30} = 0, x_{40} = \varphi_1, x_{50} = \varphi_2, \ldots, x_{3+K0} = \varphi_K$. The variational equations are as follows:

$$\dot{y}_1 = y_2,$$

$$\dot{y}_2 = (1/m)y_\delta + A(\sin x_3)y_3 + B[(\sin x_4)y_4 + (\sin x_5)y_5 + \cdots + (\sin x_{3+K})y_{3+K}],$$

$$\dot{y}_3 = 0, \quad \dot{y}_4 = 0, \ldots, \dot{y}_{3+K} = 0,$$

$$\dot{y}_\delta = -ay_\delta + by_2 + 2cx_1 x_2 y_1 + cx_1^2 y_2 + ey_1 + 2dx_1 y_1 + 3fx_1^2 y_1. \tag{8}$$

Without loss of generality one can put $y_3, y_4, \ldots, y_{3+K} = 1$ which simplifies the system (8).

The Lyapunov one-dimensional exponent for the noisy system (1) can be described as the following random variable:

$$\lambda(x_{10}, x_{20}, \delta_0, y_{10}, y_{20}, y_{\delta 0}, \omega) = \lim_{t \to \infty} \frac{1}{t} \ln \|y(t)\|. \tag{9}$$

The Lyapunov exponent is independent of the norm and the following one was used: $\|y\| = \sum_{n=1}^{2} |y_n|$. The largest rate of the Lyapunov exponent is selected by variation of the initial values $x_{10}, x_{20}, \delta_0, y_{10}, y_{20}, y_{\delta 0}$ with the same set of values φ_k.

3. INFLUENCE OF THE EXTERNAL EXCITATION ON THE CHAOS IN THE SELF-EXCITED SYSTEM

First, the system with only periodic deterministic excitation was investigated. For the parameter values for which an autonomous system ($F(t, \omega) = 0$) shows chaotic behaviour and for the particular values of A and Ω_0 the zone where the system is not chaotic was found; see Figure 3. In zone A the solution is periodic with a period $4\pi/\Omega_0$ and in zone B it is periodic with a period $2\pi/\Omega_0$.

In the zones A and B there are subzones A1 and B1 where the phase trajectories are double-revolving (see Figure 4) and subzones A2 and B2 with triple-revolving phase trajectories (see Figure 5). An example of the chaotic phase trajectory is shown in Figure 6.

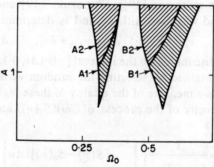

Figure 3. The influence of the periodic deterministic excitation on the chaotic behaviour: $a = 1\cdot75$, $b = 1\cdot9$, $c = 54\cdot2$, $d = -20\cdot0$, $e = 21\cdot2$, $f = 2\cdot1$.

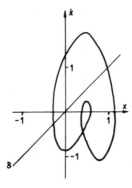

Figure 4. Double-revolving phase trajectory: $A = 1.5$, $\Omega_0 = 0.6$.

In the case of random excitation different values of the Lyapunov exponents are obtained depending on the realization of the random process $\eta(t, \omega)$. Using these values one cannot determine chaotic or regular behaviour of the system. Taking into account the probability density function of the Lyapunov exponents one can obtain two types of distribution, as shown in Figure 7.

They were obtained from 100 realizations of the process $\eta(t, \omega)$. Each realization consists of $K = 30$ harmonic components, and for all of them the mean square error of the spectral densities ε is less than 0.1.

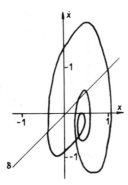

Figure 5. Triple-revolving phase trajectory: $A = 1.5$, $\Omega_0 = 0.5$.

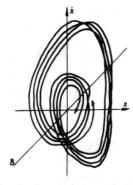

Figure 6. Chaotic phase trajectory: $A = 1.5$, $\Omega_0 = 0.32$.

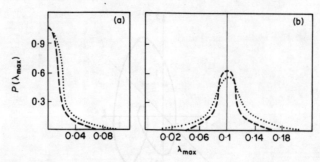

Figure 7. Distributions of random maximum one-dimensional Lyapunov exponent. (a) Regular behaviour; (b) chaotic behaviour. - - -, $\sigma = 0.05$; . . . . ., $\sigma = 0.1$.

As was checked by the method involving use of the amplitude probability density function [15–17] the first type of Lyapunov exponents distribution (see Figure 7(a)) corresponds to the regular behaviour of the system and the second one (see Figure 7(b)) to the chaotic behaviour.

In Figure 8 the zones of chaotic behaviour of the system are shown for different values of noise variance and in Figure 9 for different intervals of noise frequencies. As in the deterministic case (see again Figure 3) in the zones A and B the behaviour of the system is not chaotic. It is interesting that in the random case these non-chaotic zones increase with an increase of the noise variance.

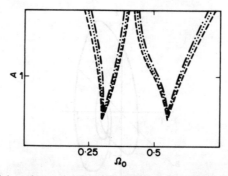

Figure 8. The influence of the noise on the chaotic behaviour of the system (1) for different values of noise variance. — · —, $\sigma = 0.05$; . . . . ., $\sigma = 0.1$; - - -, $\sigma = 0.2$.

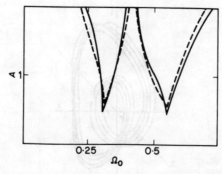

Figure 9. The influence of the noise on the chaotic behaviour of the system (1) for different values of the frequency interval. ———, $\nu_{min} = 0.4$, $\nu_{max} = 1.8$; - - -, $\nu_{min} = 0.8$, $\nu_{max} = 1.4$.

4. CONCLUSIONS

Based on the properties of the maximum one-dimensional Lyapunov exponents estimates have been made of the parameter values for which an autonomous self-excited system with stress relaxation shows chaotic behaviour.

The investigation of the influence of the periodic deterministic excitation on the chaotic behaviour of the self-excited system shows that there exist values of A and Ω_0 for which the solution is periodic with the period or double period of the external excitation.

Finally, a method has been developed of calculating Lyapunov exponents for the system with noise. Application of this method led to the finding that in the presence of external noise the chaotic zone of the system decreases.

REFERENCES

1. F. C. MOON and P. HOLMES 1979 *Journal of Sound and Vibration* **65**, 275-296; **69**, 339. A magnetoelastic strange attractor.
2. F. C. MOON 1980 *American Society of Mechanical Engineers Journal of Applied Mechanics* **47**, 638-644. Experiments on chaotic motions of a forced non-linear oscillator: strange attractors.
3. P. S. LINSAY 1981 *Physical Review Letters* **47**, 1349-1352. Period doubling and chaotic behaviour in a driven, anharmonic oscillator.
4. P. HOLMES 1982 *Journal of Sound and Vibration* **84**, 173-189. The dynamics of repeated impacts with a sinusoidally vibrating table.
5. J. AWREJCEWICZ 1986 *Journal of Sound and Vibration* **109**, 178-180. Chaos in simple mechanical systems with friction.
6. E. H. DOWELL 1982 *Journal of Sound and Vibration* **85**, 333-344. Flutter of a buckled plate as an example of chaotic motion of a deterministic autonomous system.
7. E. H. DOWELL 1984 *American Society of Mechanical Engineers Journal of Applied Mechanics* **51**, 664-673. Observation and evolution of chaos for an autonomous system.
8. R. W. LEVEN, B. POMPE, C. WILKE and B. P. KOCH 1985 *Physica* **16D**, 371-384. Experiments on periodic and chaotic motion of a parametrically forced pendulum.
9. P. J. HOLMES and F. C. MOON 1983 *American Society of Mechanical Engineers Journal of Applied Mechanics* **50**, 12-29. Strange attractors and chaos in nonlinear mechanics.
10. A. WOLF, J. B. SWIFT, H. L. SWINNEY and J. A. VASTANO 1985 *Physica* **16D**, 285-317. Determining Lyapunov exponents from a time series.
11. J. WRIGHT 1984 *Physical Review* **A29**, 2924-2927. Method for calculating a Lyapunov exponent.
12. C. FROESCHLE 1984 *Journal of Theoretical and Applied Mechanics*, Numero special, 101-132. The Lyapunov characteristic exponents and applications.
13. W.-H. STEEB, J. A. LOUW and T. KAPITANIAK 1986 *Journal of the Physical Society of Japan* **55**, 3279-3280. Chaotic behaviour of an anharmonic oscillator with two external periodic forces.
14. J. M. T. THOMPSON and H. B. STEWART 1986 *Nonlinear Dynamics and Chaos*. New York: John Wiley.
15. T. KAPITANIAK 1986 *Journal of Sound and Vibration* **107**, 177-180. A property of a stochastic response with bifurcation of a non-linear system.
16. T. KAPITANIAK 1986 *Physics Letters* **116A**, 251-254. Chaotic distribution of non-linear systems perturbed by random noise.
17. T. KAPITANIAK 1987 *Journal of Sound and Vibration.* **114**, 588-592. Quantifying chaos with an amplitude probability density function.
18. L. E. BORGMAN 1969 *Journal of Waterways and Harbours Division* **95**, 557-583. Ocean wave simulation for engineering design.
19. M. SHINOZUKA 1977 *Journal of the Acoustical Society of America* **49**, 357-367. Simulation of multivariate and multidimensional random processes.
20. J. WRÓBEL 1985 *Papers of the Technical University of Warsaw, Mechanics* no. 92. Simulation investigation of quality in non-linear stochastic machine dynamics (in Polish).

Homoclinic chaos in systems perturbed by weak Langevin noise

A. R. Bulsara
Research Branch, Naval Ocean Systems Center, San Diego, California 92152

W. C. Schieve
*Physics Department and Ilya Prigogine Center for Studies in Statistical Mechanics,
University of Texas, Austin, Texas 78712*

E. W. Jacobs
Research Branch, Naval Ocean Systems Center, San Diego, California 92152
(Received 19 May 1989)

We consider the effect of weak additive noise on the homoclinic threshold of a driven dissipative nonlinear system. A new "generalized" Melnikov function is derived for the system and is seen to be the Melnikov function for the corresponding noise-free system plus a correction term that depends on the second-order noise characteristics. The correction term is explicitly calculated for three model systems [Duffing oscillator, Josephson junction, and rf superconducting quantum interference device (SQUID)]. The effect of a distribution of dc driving terms on the chaotic attractor of a dissipative system is also examined via numerical simulation of the rf SQUID.

I. INTRODUCTION

In this paper we wish to consider the appearance of homoclinic instabilities in driven classical nonlinear systems perturbed by weak external Langevin noise. For classical dissipative noise-free systems it has been well known since Poincaré[1] that, under perturbation, the stable and unstable manifolds emanating from a hyperbolic fixed point no longer coincide, and may intersect. The resulting complicated motion is an indication of the existence of chaos and, according to an existence theorem of Smale and Moser,[2] is homeomorphic to a Markov shift map. A simple theoretical test function due to Melnikov[3-6] may be used as a test of this homoclinic instability. This function, which measures the separation between the unstable and stable solution manifolds (the system parameters that cause the function to vanish correspond to the situation in which these manifolds touch), has been applied to a number of driven nonlinear oscillators [e.g., Duffing oscillator,[4-5] Josephson junction,[7-11] and rf superconducting quantum interference device[12] (SQUID)] with the aim of determining the set of system and driving parameters that lead to the onset of the homoclinic instability.

When one considers the onset of chaos in an actual (i.e., in the real world) system, the presence of noise in the external perturbation cannot be ruled out. Recently, a number of researchers have attempted to quantify the effect of weak external Langevin-type noise (such noise manifests itself in the system dynamics as an additive term in the external perturbation) on the transition to chaos in driven nonlinear systems. Early work on this problem was carried out by Crutchfield, Farmer, and Huberman,[13] who considered the effect of noise on period doubling in a discrete system. The noise was found to introduce a gap in the bifurcation sequence, which implied a scaling behavior at the chaotic threshold, in the critical exponents. Similar work was carried out by Svensmark and Samuelson[14] on the Josephson junction; they found that, in the presence of noise and a resonant external perturbation, the bifurcation point shifted by an amount proportional to the square of the perturbation amplitude. The amplification of a small resonant periodic perturbation in the presence of noise, near the period-doubling threshold, has also been investigated by Wiesenfeld and McNamara.[15,16] Arecchi, Badii, and Politi[17] have investigated the effect of noise on the forced Duffing oscillator in the region of parameter space where different chaotic attractors coexist, finding that the noise may lead to jumps between the different basins of attraction, with the noise-induced transitions obeying simple kinetic equations. Recent work, along the same lines, has been carried out by Kautz[18] on the problem of thermally induced escape from the basin of attraction in a dc-biased Josephson junction. The average escape time has been found to increase exponentially with inverse temperature, in the low-temperature limit. Finally, Kapitaniak[19] has investigated the behavior of the probability density function (obtained from the Fokker-Planck equation) of a dissipative nonlinear system driven by random and periodic forces. He finds that, for a choice of damping and deterministic parameters such that the noise-free system is chaotic, the stationary probability density function corresponding to the noisy case exhibits multiple maxima. Further, he defines a maximal Liapounov characteristic exponent in the presence of noise. This exponent is itself a random number with a corresponding probability density function. As the noise strength increases, the mean value of this exponent approaches zero. The averaged exponent is a smoother function of the system (and driving) parameters than its noise-free analog. This indicates that the noise may actually introduce a degree of order (or smoothing) in the chaotic system; a similar conclusion was obtained earlier by Matsumoto and Tsuda,[20] who considered the effect of noise on chaotic behavior in the Belousov-Zhabotinsky reaction.

In all the work cited in the preceding paragraph, the emphasis has been on the effect of external noise on either the period-doubling bifurcations that precede the appearance of chaotic attractors, or on the chaotic attractors themselves. No mention has been made of the effect of the noise on the homoclinic threshold as defined by the Melnikov function. The first calculation of such an effect was carried out by Schieve and Petrosky.[21,22] They considered the effect of noise on the homoclinic threshold in a classical system under the influence of zero-point quantum fluctuations. A new quantum-mechanical Melnikov function defined by them was found to consist of its classical counterpart shifted by a constant quantum correction, which was found to be simply the quantum energy fluctuation on the stable and unstable manifolds. Hence the classical homoclinic threshold was suppressed by the quantum noise; the quantum Melnikov function admitted of a zero in a different region of parameter space than its classical counterpart.

In this work, we consider a classical dissipative nonlinear system driven by a deterministic perturbing term, in the presence of weak external noise. A generalized Melnikov function for the noisy system is defined in Sec. II. This function may be written down as the Melnikov function in the corresponding noise-free problem, shifted by a constant correction term which takes into account the effect of the noise. In Sec. III we describe, formally, the procedure for calculating this correction term and the calculation is demonstrated, in Sec. IV, for three model nonlinear systems (Duffing oscillator, Josephson junction, and rf SQUID). The stochastic differential equation corresponding to the dissipative nonlinear problem (specifically, the rf SQUID) in the presence of random and periodic driving terms is numerically integrated in Sec. V. The noise is introduced by allowing the *initial* values of the random force to have a Gaussian distribution with specified variance (this is equivalent to introducing a random change in the system potential for each initial value) for each integration of the stochastic differential equation. In Sec. V we also consider the probability density function corresponding to the dependent variable in the nonlinear system dynamics. This function is found to have multiple maxima, and increasing the noise level has a smoothing effect on it. We also consider the probability density function corresponding to the generalized Melnikov function of Sec. II. This calculation provides a test of the accuracy of our theoretical computations (specifically, the noise-induced shift in the homoclinic threshold) of Secs. II–IV. Corresponding to our approach, Carlson[23,24] has investigated a shift map in the presence of thermal noise by means of an analogy to the eight-vertex model of spins. He has shown that the sequences corresponding to the homoclinic points of the Cantor set are removed by the noise.

II. GENERAL CASE: EFFECT OF WEAK ADDITIVE NOISE

Let us consider a general nonlinear second-order system driven by *weak* external additive noise. The state (e.g., displacement) variable $x(t)$ describing the evolution of this system is assumed to obey the dynamic equation (the dots denote differentiation with respect to time),

$$\ddot{x} + f(x) = F(t) , \qquad (1)$$

where $F(t)$ is the random driving term which we take to be Gaussian, δ correlated with finite mean and variance σ^2

$$\langle F(t) \rangle = m, \quad \langle F(t)F(t+\tau) \rangle = \sigma^2 \delta(\tau) , \qquad (2)$$

$\delta(\tau)$ being the Dirac δ function. $f(x)$ is a nonlinear function of the dynamic variable $x(t)$. In the absence of the random force, the system (1) is conservative and represents a particle moving in a potential $U(x) \equiv \int^x f(y) dy$.

We now introduce, as perturbations, a dissipative term $-k\dot{x}$ and a deterministic (often taken to be periodic) driving term $Qg(t,t_0)$ on the right-hand side of (1). In the absence of the random force $F(t)$, the introduction of these perturbations is known to induce homoclinic behavior in the system for certain sets of values of the system and driving parameters. This behavior is characterized by a bifurcation of the separatrix (i.e., unperturbed) solution into stable and unstable solution manifolds. The small separation of these manifolds is given by the Melnikov function. When this function vanishes, the manifolds touch and above this threshold one may (for certain values of the system and driving parameters) observe chaotic behavior characterized by the appearance of strange attracting sets in phase space. The Melnikov function is given (for the noise-free system) by[3–6]

$$\Delta(t_0) = -k \int_{-\infty}^{\infty} \dot{x}_s^2(t) dt + Q \int_{-\infty}^{\infty} \dot{x}_s(t) g(t,t_0) dt . \qquad (3)$$

Here $x_s(t)$ represents the separatrix (or unperturbed) solution, i.e., the solution of (1) in the absence of any dissipative or driving terms.

In the presence of noise, the formalism of the preceding paragraph must be modified. Let us return to the noise-driven equation (1) and treat this as our "unperturbed" system. The solution $x(t)$ of (1) may be expressed as the separatrix solution plus a (small) noise-induced deviation

$$x(t) = x_s(t) + \delta x(t) , \qquad (4)$$

with a similar expression holding for the velocity $\dot{x}(t)$. Averaging over the ensemble of the random force yields for the mean displacement

$$\langle x(t) \rangle = x_s(t) + \langle \delta x(t) \rangle , \qquad (5)$$

where the angular brackets denote an average over the ensemble of $F(t)$ and a similar expression holds true for the averaged velocity $\langle \dot{x}(t) \rangle$. We now substitute the expansion (4) into (1), expand to leading order in δx, and separate out the terms that depend on the random force

$F(t)$. The result is the two equations

$$\ddot{x}_s + f(x_s) = 0 , \quad (6a)$$

$$\delta\ddot{x} - \omega^2(t)\delta x = F(t) . \quad (6b)$$

Here $\omega^2(t) \equiv -[df(x)/dx]_{x=x_s}$ is, in general, a complicated function of time due to the time dependence of the separatrix solution $x_s(t)$. Equation (6a) describes the unperturbed motion, i.e., the separatrix. Its solution may be obtained for specific forms of the function $f(x_s)$. Equation (6b) yields the stochastic component of the solution to the unperturbed problem (1). In its present form, it cannot be integrated (except numerically). We shall see, however, that the corrections to the Melnikov integral require us to compute the *averaged* quantities $\langle \delta \dot{x}(t) \rangle$ and $\langle \delta \dot{x}^2(t) \rangle$. These quantities may be computed using a procedure that will be described in Sec. III.

We now introduce a Melnikov function in the presence of the random force. This function defines the separation of the stable and unstable manifolds for each realization of the weak random force term $F(t)$ and may be written as

$$\Delta_F(t_0) = -k \int_{-\infty}^{\infty} \dot{x}^2(t) dt + Q \int_{-\infty}^{\infty} \dot{x}(t) g(t, t_0) dt , \quad (7)$$

Recalling that we have treated the noise system (1) as our unperturbed system, the analogy between (7) and the usual definition of the Melnikov function (3) is evident. One readily observes that (7) may be cast in the form

$$\Delta_F(t_0) = \Delta(t_0) + \Delta_c(t_0) , \quad (8)$$

where Δ_c is a correction term. The Melnikov function Δ_F defined in (8) is a random variable, since the correction Δ_c involves the stochastic component $\delta \dot{x}(t)$ of the velocity. One observes that the homoclinic threshold is shifted for each element of the stochastic ensemble: the zeros of (8) do not coincide with those of (3). One may first obtain an expression for its *average* value by averaging (8) over the ensemble of the random force. In this case, we obtain the averaged Melnikov function

$$\langle \Delta_F(t_0) \rangle = \Delta(t_0) + \langle \Delta_c(t_0) \rangle , \quad (9)$$

where the averaged correction is given [using (4)] by

$$\langle \Delta_c(t_0) \rangle \equiv -2k \int_{-\infty}^{\infty} \dot{x}_s(t) \langle \delta \dot{x}(t) \rangle dt$$

$$+ Q \int_{-\infty}^{\infty} g(t, t_0) \langle \delta \dot{x}(t) \rangle dt$$

$$- k \int_{-\infty}^{\infty} \langle \delta \dot{x}^2(t) \rangle dt . \quad (10)$$

The ensemble-averaged Melnikov function defined in (10) is a generalization of the usual function (3) defined in connection with the noise-free problem. It is similar to the function introduced by Schieve and Petrosky[21-22] to take into account the effects of small quantum fluctuations on the homoclinic threshold in a classical nonlinear system. The terms appearing on the right-hand side of (10) may be computed for specific model systems. In the following section we describe the calculation in general. This is followed by a computation of the correction term for specific systems. Before moving on, however, we wish to point out that a significant and formidable question exists regarding the existence of a Smale-Birkoff theorem[5] for solutions to the stochastic differential equation (1). Our work here and in the following sections suggests that under weak noise there are, following the noisy tangency, multiple crossings for each realization of the ensemble and thus a Cantor-set-like structure in this weak sense.

III. COMPUTATION OF $\langle \Delta_c(t_0) \rangle$: THE NOISE CORRECTION TO THE MELNIKOV FUNCTION

We now consider the computation of the mean value $\langle \delta \dot{x}(t) \rangle$ and the second-order correlation function $\langle \delta \dot{x}^2(t) \rangle$ appearing in (10). To do this, we employ a stochastic description in which the stochastic differential equation (6b) is replaced by a linear inhomogeneous Fokker-Planck or diffusion equation[25] for the probability density function $P(\delta x, \delta \dot{x}, t | \delta x(t_0), \delta \dot{x}(t_0), t_0)$ corresponding to the random variables $\delta x(t)$ and $\delta \dot{x}(t)$. The probability density function, as well as the second-order correlation function, may be obtained in closed form from this Fokker-Planck equation once the solutions for $\langle \delta x(t) \rangle$ and $\langle \delta \dot{x}(t) \rangle$ are known. The procedure has been outlined by van Kampen,[26] whose treatment and notation we follow in the remainder of this section.

We first consider the case when the random force term $F(t)$ has zero mean value, i.e., $m = 0$. Then, the stochastic differential equation (6b) leads one to a two-dimensional Fokker-Planck equation for the probability density function $P(\underline{y}, t)$

$$\frac{\partial P}{\partial t} = -\sum_{ij} A_{ij} \frac{\partial}{\partial y_j} y_i P + \frac{\sigma^2}{2} \sum_{ij} B_{ij} \frac{\partial^2 P}{\partial y_i \partial y_j} , \quad (11)$$

where we have defined the matrices,

$$\underline{y}(t) \equiv \begin{bmatrix} \delta x(t) \\ \delta \dot{x}(t) \end{bmatrix} ,$$

$$\underline{A}(t) \equiv \begin{bmatrix} 0 & 1 \\ \omega^2(t) & 0 \end{bmatrix} , \quad (12)$$

$$\underline{B} \equiv \begin{bmatrix} 0 & 0 \\ 0 & 1 \end{bmatrix} .$$

In order to obtain the solution of (11), we must first solve the transport equation [obtained by averaging the stochastic differential equation (6b) over the ensemble of the random force $F(t)$]

$$\frac{d^2}{dt^2} \langle \delta x(t) \rangle = \omega^2(t) \langle \delta x(t) \rangle . \quad (13)$$

In terms of the matrix $\underline{y}(t)$ defined in (12), the formal solution of (13) may be written down in the form

$$\langle \underline{y}(t) \rangle = \underline{Y}(t) \langle \underline{y}(t_0) \rangle , \quad (14)$$

$\underline{Y}(t)$ being the Green's function determined by

$$\underline{\dot{Y}}(t) = \underline{A}(t) \underline{Y}(t), \quad \underline{Y}(t_0) = 1 . \quad (15)$$

We shall return to a precise computation of the elements

of $\underline{Y}(t)$ later in this section. We define the correlation matrix $\underline{\Sigma}(t)$ by

$$\Sigma_{ij} \equiv \langle y_i y_j \rangle - \langle y_i \rangle \langle y_j \rangle \; . \tag{16}$$

It obeys the matrix differential equation

$$\underline{\dot{\Sigma}}(t) = \underline{A}\underline{\Sigma} + \underline{\Sigma}\,\underline{\tilde{A}} + \sigma^2 \underline{B} \; , \tag{17}$$

where $\underline{\tilde{A}}$ is the transpose of $\underline{A}$. The solution of (17) may be formally written down [in terms of an initial correlation matrix $\underline{\Sigma}(t_0)$] as

$$\underline{\Sigma}(t) = \underline{Y}(t)\underline{\Sigma}(t_0)\underline{\tilde{Y}}(t)$$
$$+ \sigma^2 \int_{t_0}^{t} \underline{Y}(t)\underline{Y}(t')^{-1}\underline{B}(t')\underline{\tilde{Y}}(t')^{-1}\underline{\tilde{Y}}(t)dt' \; , \tag{18}$$

and the probability density function $P(\underline{y},t)$ appearing in the Fokker-Planck equation (11) takes the form

$$P(\underline{y},t) = (4\pi^2 \det\underline{\Sigma})^{-1/2}$$
$$\times \exp[-\tfrac{1}{2}(\underline{y} - \langle \underline{y}\rangle)\underline{\Sigma}^{-1}(\underline{y} - \langle \underline{y}\rangle)] \; . \tag{19}$$

Using (16) and (18), we readily obtain

$$\langle \delta x^2(t)\rangle = \langle \delta \dot{x}(t)\rangle^2$$
$$+ [\underline{Y}(t)\underline{\Sigma}(t_0)\underline{\tilde{Y}}(t)]_{22} + \sigma^2 \int_{t_0}^{t} K_{22}(t,t')dt' \; , \tag{20}$$

where we have set $\underline{K}(t,t') \equiv \underline{Y}(t) Y(t')^{-1}\underline{B}\,\underline{\tilde{Y}}(t')^{-1}\underline{\tilde{Y}}(t)$, the kernel of the integral appearing on the right-hand side of (18). In addition to the matrix element $K_{22}(t,t')$, the expression (20) includes the initial variances defined by

$$\Sigma_{11}(t_0) = \langle \delta x^2(t_0)\rangle,$$
$$\Sigma_{12}(t_0) = \langle \delta x(t_0)\delta \dot{x}(t_0)\rangle \; ,$$
$$\Sigma_{21}(t_0) = \langle \delta \dot{x}(t_0)\delta x(t_0)\rangle \; , \tag{21}$$
$$\Sigma_{22}(t_0) = \langle \delta \dot{x}^2(t_0)\rangle \; .$$

It remains to evaluate the averaged solution $\langle \delta \dot{x}(t)\rangle$ as well as the matrix $\underline{Y}(t)$ in terms of the parameters appearing in our original stochastic differential equation (6b). We now demonstrate how this is done.

Let us return to the transport equation (13). This equation is to be solved for the average value $\langle \delta x(t)\rangle$, usually using a WKB technique [since $\omega^2(t)$ is, in general, a complicated function of time]. The precise form of the solution will depend on the model system under consideration (this will be discussed in greater detail in Sec. IV when we apply our results to three specific model systems). Let us assume, for the time being, that the function $\omega^2(t)$ admits of only one turning point (i.e., it changes sign only once) in the range $0 \leq t \leq \infty$. Then, one may write down a general solution to (13) in the form

$$\langle \delta x(t)\rangle = C_1 f_1(t) + C_2 f_2(t) \; , \tag{22}$$

where $C_{1,2}$ are integration constants (to be determined by the boundary or initial conditions). The function $f_1(t)$ is given by[27]

$$f_1(t) = S_0^{1/6}[\omega^2(t)]^{-1/4}\mathrm{Ai}[(\tfrac{3}{2}S_0)^{2/3}] \; , \tag{23}$$

where

$$S_0(t) \equiv \int_{t_c}^{t} \omega(t')dt' \; , \tag{24}$$

and Ai is the Airy function. t_c is the turning point of $\omega^2(t)$, i.e., $\omega^2(t_c) = 0$. The other part $f_2(t)$ of the solution is found, using standard techniques, to be

$$f_2(t) = f_1(t)\int_{t_0}^{t}\frac{dt'}{f_1^2(t')} \; . \tag{25}$$

Once again we emphasize that the precise behavior of these solutions (e.g., at long times) depends on the system under consideration, through the function $\omega^2(t)$ and that the equations (23)–(25) are true under the assumption that the function $\omega^2(t)$ has just one turning point in the interval $0 \leq t \leq \infty$. In such cases, the solution $f_1(t)$ generally represents the solution that converges in the $t \to \infty$ limit, whereas $f_2(t)$ diverges in this limit. In fact, one may break up the general solution (23) into distinct solutions for $t < t_c$ and $t > t_c$. These solutions are to be "patched" together in the connection region $t \approx t_c$. For this procedure to hold, i.e., for (23) to represent a unified solution for all $t > 0$, the condition[27]

$$\left|\frac{d}{dt}\omega^2(t_c)\right|^{4/3} \gg \left|\frac{d^2}{dt^2}\omega^2(t_c)\right| \; , \tag{26}$$

must be satisfied. Then the unified solution (23) may be written down in terms of three distinct solutions[27]

$$\begin{aligned}
f_1(t) &= 2\sqrt{\pi}\,\frac{(\tfrac{3}{2}S_0)^{1/6}}{|\omega^2(t)|^{1/4}}\mathrm{Ai}[(\tfrac{3}{2}S_0)^{2/3}] \approx [\omega^2(t)]^{-1/4}\exp\left\{-\int_{t_c}^{t}\omega(t')dt'\right\}, \quad t > t_c \\
&= 2\sqrt{\pi}\,\frac{(\tfrac{3}{2}S_0)^{1/6}}{|\omega^2(t)|^{1/4}}\mathrm{Ai}[-(\tfrac{3}{2}S_0)^{2/3}] \approx [-\omega^2(t)]^{-1/4}\sin\left\{\int_{0}^{t_c}[-\omega^2(t)]^{1/2}dt\right\}, \quad 0 \leq t \leq t_c \\
&\approx 2\sqrt{\pi}\left[\frac{3}{2d\omega^2(t_c)/dt}\right]^{1/6}\mathrm{Ai}\left\{\left[\tfrac{3}{2}t_c^{3/2}\left(\frac{d}{dt}\omega^2(t_c)\right)^{1/2}\right]^{2/3}\right\}, \quad t \approx t_c \; .
\end{aligned} \tag{27}$$

The constants $C_{1,2}$ are now determined. Assuming that both $\langle \delta x(t)\rangle$ and $\langle \delta \dot{x}(t)\rangle$ are initially nonzero, we readily find

$$f_2(t_0) = 0, \quad \dot{f}_2(t_0) = f_1^{-1}(t_0) \; , \tag{28a}$$

which leads to the expressions

$$C_1 = \frac{\langle \delta x(t_0)\rangle}{f_1(t_0)} \; ,$$
$$C_2 = \langle \delta \dot{x}(t_0)\rangle f_1(t_0) - \langle \delta x(t_0)\rangle \dot{f}_1(t_0) \; . \tag{28b}$$

Throughout this work we shall assume that $\langle \delta x(t_0)\rangle = 0 = \langle \delta \dot{x}(t_0)\rangle$; it then follows that $\langle \delta x(t)\rangle = 0 = \langle \delta \dot{x}(t)\rangle$ for all subsequent times t (for the $m=0$ case being currently considered). In this case, the averaged Melnikov correction term $\langle \Delta_c(t_0)\rangle$ involves only the second-order velocity correlation function $\langle \delta \dot{x}^2(t)\rangle$, in agreement with earlier calculations[21-22] on the effects of zero-point quantum fluctuations on classical chaotic systems. The above special initial conditions assume physically that, at time $t=t_0$, the *mean* displacement in the noisy unperturbed system follows the separatrix motion of the corresponding noise-free system. This condition ensures that the mean displacement follows the noise-free separatrix motion for all subsequent times as well.

The solution (22) [with $f_{1,2}(t)$ given by (23) and (25)] represents the solution of the *initial* value problem. In keeping with our fundamental assumption of weak noise, however, the problem must be solved as a *boundary* value problem with the requirement that both $\langle \delta x(t)\rangle$ and $\langle \delta \dot{x}(t)\rangle$ be finite as $t \to \infty$. In fact, we expect these quantities to follow qualitatively similar behavior to the corresponding separatrix quantities $x_s(t)$ and $\dot{x}_s(t)$, which implies that $\langle \delta \dot{x}(t_f)\rangle \to 0$ at some final time $t_f \to \infty$. Imposing these boundary conditions on the solution (22) and noting that $\dot{f}_1(t_f)/\dot{f}_2(t_f) \ll 1$ in the large t_f limit, one obtains, after some calculation, the solution of the boundary problem. This solution is diagonal,

$$\langle \delta x(t)\rangle = \frac{\langle \delta x(t_0)\rangle}{f_1(t_0)} f_1(t) ,$$
$$\langle \delta \dot{x}(t)\rangle = \frac{\langle \delta \dot{x}(t_0)\rangle}{\dot{f}_1(t_0)} \dot{f}_1(t) ,$$
(29)

and involves the implicit constraint

$$\langle \delta \dot{x}(t_0)\rangle f_1(t_0) = \langle \delta x(t_0)\rangle \dot{f}_1(t_0) , \quad (30)$$

implying that, given an arbitrary initial displacement $\langle \delta x(t_0)\rangle$, the boundary condition requires that the initial velocity perturbation be determined by the above constraint. Note that the constraint (30) is equivalent to setting the constant C_2 defined in (28b), equal to zero.

In terms of the solution (29) we readily obtain for the elements of the matrix $\underline{K}(t,t')$

$$K_{11}(t,t') = 0 = K_{12}(t,t') = K_{21}(t,t') ,$$
$$K_{22}(t,t') = \frac{\dot{f}_1^2(t)}{\dot{f}_1^2(t')} .$$
(31)

The expression (31) together with the initial condition $\langle \delta x(t_0)\rangle = 0 = \langle \delta \dot{x}(t_0)\rangle$ [note that this condition is a special case of the constraint (30)] permits one to evaluate the mean correction term $\langle \Delta_c(t_0)\rangle$ defined in (10). The correction term is seen [from (20)] to depend solely on second-order fluctuations. In addition, one readily observes from (31) that, after carrying out the t' integration in (18), the resulting term is an odd function of t. Hence this term does not contribute to the integral appearing in the third term on the right-hand side of (10) and one obtains finally the relatively simple expression

$$\langle \Delta_c \rangle = -2k \frac{\langle \delta \dot{x}^2(0)\rangle}{\dot{f}_1^2(0)} \int_0^\infty \dot{f}_1^2(t) dt , \quad (32)$$

for the correction term. The above result has been obtained from the initial variance term in (31) with t_0 set equal to zero for convenience.

Before proceeding further, we pause to comment briefly on the assumption of weak Langevin noise. Throughout this work, the noise has been assumed to be sufficiently weak that the resulting motion (in the presence of the noise, but in the absence of the dissipative and deterministic forcing terms) is not appreciably different from the separatrix motion in the noise-free case. This permits one to carry out the expansion leading to (6a) and (6b) to first order and also ensures that the system does not make excursions to neighboring wells of the potential under the influence of the noise alone (this would be the case for very strong noise). Under these conditions, one may also assume that *at initial times,* the random variables $x(0)$ and $\dot{x}(0)$ [these variables refer to the solution of (1), i.e., the right-hand side of (4)] obey a joint probability density which may be assumed to be Gibbsian:

$$P(x(0),\dot{x}(0)) \equiv N^{-1} \exp(-2E/\sigma^2) , \quad (33)$$

where N is a normalization constant and E the total energy given by $E = \frac{1}{2}\dot{x}^2(0) + U(x(0))$. One readily obtains the initial velocity correlation function from (33)

$$\langle \dot{x}^2(0)\rangle \equiv N^{-1} \int_{-\infty}^\infty \dot{x}^2(0) P(x(0),\dot{x}(0)) dx(0) d\dot{x}(0) .$$
(34)

Since $\dot{x}_s(0) \equiv 0$ one readily obtains from (4) and (34)

$$\langle \delta \dot{x}^2(0)\rangle \equiv \langle \dot{x}^2(0)\rangle = \frac{\sigma^2}{2} . \quad (35)$$

The above result enables us to express the initial variance $\langle \delta \dot{x}^2(0)\rangle$ directly in terms of the noise variance. One may now set an *approximate* upper bound on the noise strength by requiring that the variance $\langle \delta \dot{x}^2(0)\rangle$ of the initial velocity perturbation be much smaller than the depth of the potential well. This ensures that the noise variance σ^2 is also much smaller than the well depth and allows us to quantify the assumption of "weak Langevin noise."

Returning to our discussion (32) for the Melnikov correction term, we see that, on average, the homoclinic threshold has been shifted from its parameter-space location in the corresponding noise-free case by an amount proportional to the variance of the noise. Indeed, setting $\langle \Delta_F \rangle = 0$ in (9) one readily obtains the new ratio Q/k of the perturbation amplitudes corresponding to the homoclinic threshold. The result may be written in the form

$$\frac{Q}{k} = A + B\sigma^2 , \quad (36)$$

where we have defined

$$A \equiv \left[\frac{Q}{k}\right]_0 , \quad (37)$$

and

$$B \equiv \frac{C}{\int_{-\infty}^{\infty} \dot{x}_s(t) g(t,t_0) dt} , \qquad (38a)$$

$$C \equiv \frac{1}{2\dot{f}_1^2(0)} \int_{-\infty}^{\infty} \dot{f}_1^2(t) dt , \qquad (38b)$$

where we have used (32) and (35). The quantity A defined in (37) represents the threshold ratio of Q/k in the noise-free system; it is obtained by setting the right-hand side of (3) equal to zero. The second term on the right-hand side of (36) represents the correction induced by the Langevin noise. Note that this term is always positive; the noise always *elevates* the homoclinic threshold.

We now consider briefly the case when the random driving term has a finite mean value m which now appears on the right-hand side of the transport equation (13). Using the initial conditions $\langle \delta x(0) \rangle = 0 = \langle \delta \dot{x}(0) \rangle$ as before, one may write down the solution of the transport equation in the form

$$\langle \delta x(t) \rangle = mf(t) , \qquad (39)$$
$$\langle \delta \dot{x}(t) \rangle = m\dot{f}(t) ,$$

in terms of an undetermined function $f(t)$. One readily observes that, within the framework of our perturbation theory of Sec. II [the mean value m must be small and positive so that $\langle \delta x(t) \rangle \ll x_s(t)$ and $\langle \delta \dot{x}(t) \rangle \ll \dot{x}_s(t)$], the dominant contribution to the correction term (10) arises from the first two terms on the right-hand side of (10). One may verify that the absolute magnitude of the first term on the right-hand side of (10) is greater than or equal to that of the second term [for unit Q and k and for the case when $g(t,t_0)$ is an alternating periodic signal of the form to be considered in this work]. Hence the homoclinic threshold will always be elevated (or, at worst, stay the same) when the noise is assumed to have a finite mean value. In fact, one readily obtains for the threshold ratio Q/k corresponding to this case:

$$\frac{Q}{k} = \left[\int_{-\infty}^{\infty} \dot{x}_s^2(t) dt + 2m \int_{-\infty}^{\infty} \dot{x}_s(t) \dot{f}(t) dt \right] / \left[\int_{-\infty}^{\infty} \dot{x}_s(t) g(t,t_0) dt + m \int_{-\infty}^{\infty} \dot{f}(t) g(t,t_0) dt \right] ,$$

$$\approx \left[\frac{Q}{k} \right]_0 \left[1 + 2m \left[\int_{-\infty}^{\infty} \dot{x}_s(t) \dot{f}(t) dt \bigg/ \int_{-\infty}^{\infty} \dot{x}_s^2(t) dt \right] - m \left[\int_{-\infty}^{\infty} \dot{f}(t) g(t,t_0) dt \bigg/ \int_{-\infty}^{\infty} \dot{x}_s(t) g(t,t_0) dt \right] \right] . \qquad (40)$$

The preceding remarks imply that one always has $(Q/k) \geq (Q/k)_0$ so that the presence of a finite mean value in the random term can never, of its own accord, induce homoclinic behavior.

In the following section, the theory presented so far is elucidated numerically in the case of three well-known model systems. We confine ourselves to the $m=0$ case, the important point being that the Langevin noise suppresses the observation of homoclinic chaos, in qualitative agreement with the work of Carlson.[23−24]

IV. EXAMPLES

In this section, we consider three well-known nonlinear problems: the Duffing oscillator, Josephson junction, and rf SQUID. In all three cases, we assume that the system is perturbed by a dissipative term $-k\dot{x}$ and a deterministic term $g(t,t_0)$, which we take to be periodic. In the presence of weak Langevin noise, the correction term $\langle \Delta_c \rangle$ is obtained using the WKB solutions (27) and (29), as well as by direct numerical integration of the transport equation (13). The agreement between the correction terms obtained using these two procedures is seen to be extremely good. Since the calculations are qualitatively similar for all three models, we first present the salient features of each model before computing the Melnikov function in each case in the presence of weak Langevin noise having zero mean and variance σ^2.

A. Duffing oscillator

We consider the Duffing oscillator with negative stiffness (sometimes referred to as the anti-Duffing oscillator).[4−5] The dynamic equations on the separatrix are

$$\dot{x} = v \qquad (41)$$
$$\dot{v} = \beta x - \alpha x^3 .$$

The above system is seen to be homologous to a particle moving in a potential given by

$$U(x) = -\frac{\beta x^2}{2} + \frac{\alpha x^4}{4} . \qquad (42)$$

This potential (we assume $\beta = 1.0 = \alpha$ throughout this section) admits of a saddle point at $(x,v) = (0,0)$ and elliptic points at $(\pm 1, 0)$. The particular unperturbed (or separatrix) solution is found by solving (41)

$$x_s(t) = \sqrt{2} \, \text{sech} \, t . \qquad (43)$$

We now introduce weak Langevin noise. Using the procedure of Sec. II we readily set up the transport equation (13) for the mean value $\langle \delta x(t) \rangle$ with $\omega^2(t)$ given by

$$\omega^2(t) = 1 - 6 \, \text{sech}^2 t . \qquad (44)$$

This function is symmetric about $t=0$ and has a single turning point (for positive t) at $t_c = \text{sech}^{-1}(1/\sqrt{6})$.

B. Josephson junction

The separatrix response of a Josephson junction may be written in the form[28]

$$\dot{x} = v \qquad (45)$$
$$\dot{v} = -\beta \sin x ,$$

where k and β may be expressed in terms of the junction parameters and x is typically the phase of the supercurrent in the junction [it should be noted that (45) also describes the dynamics of a common pendulum in the absence of the small oscillation assumption]. Equation (45) may be derived from a potential

$$U(x) = -\beta \cos x \ . \tag{46}$$

One readily observes that this potential is periodic and has saddle points at $x = n\pi$ (n odd) with elliptic points at $x = 2n\pi$. The separatrix solution may be found by solving (45)

$$x_s(t) = 4\tan^{-1}(e^{\sqrt{\beta}t}) + \pi \ . \tag{47}$$

Introducing noise as in the preceding example, we find for this case

$$\omega^2(t) = -2\beta^{3/2}\cos x_s(t)\,\text{sech}(\sqrt{\beta}t) \ . \tag{48}$$

This function shows the same qualitative features as (44) and is not plotted. One readily obtains its turning point (for positive time) at $t_c = (1/\sqrt{\beta})\ln(\tan\tfrac{3}{8}\pi)$.

C. rf SQUID

In its simplest form, the rf SQUID consists of a single Josephson junction shorted by a superconducting loop. An external magnetic field produces a geometrical magnetic flux across the loop together with a circulating supercurrent in the loop. Hence the flux actually sensed by the SQUID is not the same as the original magnetic flux (the SQUID actually amplifies this flux because of the additional supercurrent induced in the loop by the Josephson tunneling current). The net flux is inductively coupled to a rf-driven detector cirvuit. Setting $x(t)$ equal to the flux sensed by the SQUID (in units of the universal flux quantum), the separatrix dynamics of the SQUID are given by[28]

$$\begin{aligned}\dot{x} &= v \\ \dot{v} &= -\omega_0^2 x - \beta\sin 2\pi x \ ,\end{aligned} \tag{49}$$

where, one again, the parameters β and ω_0^2 may be expressed in terms of the SQUID parameters. Homoclinic chaos in the rf SQUID has been treated in detail in Ref. 12. As with the formalism of Ref. 12, we transform to a new variable $z = 2\pi x - \pi/2$ in terms of which one may consider the system (49) from the standpoint of a particle moving in a potential,

$$U(z) = \frac{\omega_0^2}{2}\left[z + \frac{\pi}{2}\right]^2 + 2\pi\beta\omega_0^2\sin z \ . \tag{50}$$

The potential is multistable above the critical value of the nonlinearity β and is plotted in Fig. 1 for $(\beta,\omega_0) = (2.0, 1.0)$. In the remainder of this section, we shall confine our attention to the separatrix $z_1 z_2$ in Fig. 1. This renders the rf SQUID qualitatively similar to the Duffing oscillator problem discussed above (note that had we considered the separatrix $z_1 z_3$ in Fig. 1, the problem would have resembled the Josephson junction). For the case of moderate β this separatrix solution has been ob-

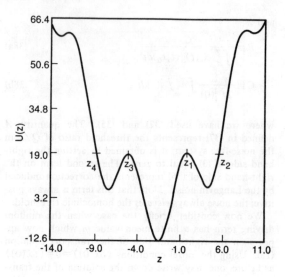

FIG. 1. Potential $U(z)$ [Eq. (50)] for the rf SQUID corresponding to $(\beta,\omega_0) \equiv (2.0, 1.0)$.

tained as[12]

$$z_s(t) = z_i - \alpha\tanh^2\zeta t \ , \tag{51}$$

where $\alpha = z_2 - z_1$ and $\zeta = (0.1482\pi\beta\omega_0^2)^{1/2}$. Introducing noise as before, we readily obtain

$$\omega^2(t) = 2\pi\beta\sin z_s(t) - 1 \ , \tag{52}$$

Again, $\omega^2(t)$ is symmetric about $t = 0$ and has a single turning point (for $t > 0$) at

$$t_c = \zeta^{-1}\tanh^{-1}\left\{\frac{[z_2 - \pi + \sin^{-1}(1/2\pi\beta)]}{\alpha}\right\}^{1/2} \ .$$

D. Results

We now assume that each of the above model systems is perturbed by a dissipative term $-k\dot{x}$ and a deterministic driving term which we take to be periodic: $g(t, t_0) \equiv Q\cos[\Omega(t + t_0)]$. Then, for the Duffing oscillator, the Melnikov function for the noise-free case is[4-5]

$$\Delta_D(t_0) = -\frac{4k}{3} + \sqrt{2}\pi\Omega Q(\sin\Omega t_0)\,\text{sech}\frac{\pi\Omega}{2} \ . \tag{53}$$

The zero of this function (taken in units of $\sin\Omega t_0$) yields the homoclinic threshold in the absence of noise, but in the presence of the dissipative and deterministic driving terms. It is plotted in Fig. 2. The curve represents the second term (53) (in units of $\sin\Omega t_0$). It is peaked at a critical frequency $\Omega_{Dc} = 2.399/\pi$. The straight line represents the absolute value of the first term. The values $(Q, k) \equiv (1.25, 2.81)$ are used in these plots. This represents the threshold case; the straight line is tangential to the curve. We now assume that weak Langevin noise of variance $\sigma^2 = 0.2$ (corresponding to an initial velocity variance $\langle\delta\dot{x}^2(0)\rangle = 0.1$) is present in the system.

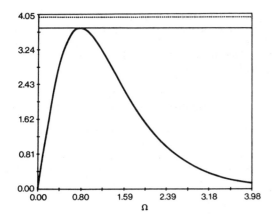

FIG. 2. Melnikov function for the Duffing oscillator. Solid curve represents the second term in (51) (in units of $\sin\Omega t_0$) for $Q=2.0$; solid line represents the first term in (53) for $k=2.81$ (threshold case). Dotted line represents the effect, on the straight line, of including noise of variance $\sigma^2=0.2$; it includes the correction $\langle \Delta_c \rangle$ calculated using the WKB approximation (29).

This leads to an elevation in the homoclinic threshold characterized by a displacement of the straight line in Fig. 2. One may also compute the displaced straight line through a direct numerical integration of the transport equation (11). The relative difference between the corrections is approximately 2.2% for this case.

The corresponding figure for the Josephson junction is readily obtained. The Melnikov function in the absence of noise is[7–10]

$$\Delta_J(t_0) = -8k\sqrt{\beta} + 2\pi Q (\cos\Omega t_0)\mathrm{sech}\left[\frac{\pi\Omega}{2\sqrt{\beta}}\right], \quad (54)$$

which we plot in Fig. 3 for $(Q,k)=(2,1.57)$ correspond-

ing to the homoclinic threshold (we set $\beta=1$ throughout). As in the preceding case, the curve represents the second term in (54) (in units of $\cos\Omega t_0$) and the straight line represents the first term. The curve is peaked at $\Omega_{Jc}=0$ because of the difference in parity of the separatrix velocity $\dot{x}_s(t)$ between this example and the Duffing oscillator. Once again, the introduction of noise with variance $\sigma^2=0.2$ leads to an elevation of the homoclinic threshold. The displaced line is plotted in the figure; its location differs from the position computed via numerical integration of the transport equation (11) by approximately 1% in this case.

Finally, we present, in Fig. 4, the Melnikov function for the rf SQUID; for the noise-free case, this has been computed in Ref. 12:

$$\Delta_S(t_0) = -\frac{16}{15}\alpha^2 \zeta k + \frac{4\pi Q \Omega^2 \sin\Omega t_0}{A_2 \sinh[\pi\Omega/(A_2\alpha)^{1/2}]}, \quad (55)$$

where $A_2 = 0.5928\pi\beta\omega_0^2$. This function, which is peaked at a frequency[12] $\Omega_{Sc}=(1.915/\pi)\sqrt{A_2\alpha}$, is plotted in Fig. 4 for $(\beta,Q,k)=(1.0,4,2.24)$; as in the preceding examples, the straight line represents the first term in (55) and the curve represents the second term (in units of $\sin\Omega t_0$). Introducing the noise term, with noise variance $\sigma^2=0.2$ as before, results in a displacement of the straight line and a shift in the homoclinic threshold; the relative error between the displaced positions of the straight line computed using direct numerical integration of the transport equation, and the WKB solution is about 1.1% for this case.

We complete the analysis by computing the threshold value of Q/k for each of these examples. This is accomplished using the general equations (37) and (38). It is instructive to evaluate this quantity at the critical frequencies Ω_{Dc}, Ω_{Jc}, and Ω_{Sc} at which the curves in Figs. 4–6 are peaked. Using the numerical parameters introduced in this section for each of our model systems we may calculate (at the critical frequency) the parameters (A,B) appearing in Eq. (36) for the new threshold value of Q/k.

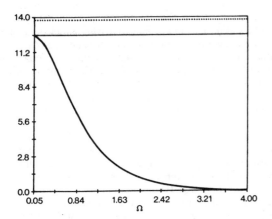

FIG. 3. Melnikov function for the Josephson junction, Eq. (54); same as Fig. 2 with $(Q,k,\sigma^2) \equiv (2.0,1.57,0.2)$.

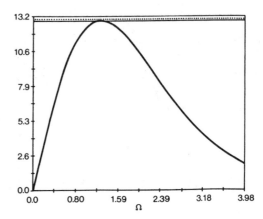

FIG. 4. Melnikov function for the rf SQUID, Eq. (55); same as Fig. 2 with $(\beta,Q,k,\sigma^2) \equiv (1.0,4.0,2.24,0.2)$.

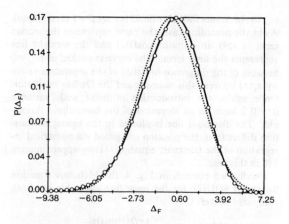

FIG. 5. Probability density function corresponding to the stochastic Melnikov function Δ_F defined in Eq. (7), for the rf SQUID. $(\beta,\omega_0^2,\Omega,Q/k,\sigma^2)\equiv(2.0,1.0,2.0768,4.734,0.8)$. Value of Q/k corresponds to the homoclinic threshold in the absence of noise. Data points represent the result of sampling the Melnikov function with 50 bins. Dotted curve is the Gaussian having the same mean and variance as the sampled data. Solid curve is obtained by assuming "perfect" Gaussian noise (see text).

We find that $(A,B)\equiv(0.71,0.13)$, $(1.27,0.31)$, and $(1.78,0.06)$ for the Duffing oscillator, Josephson junction, and rf SQUID respectively. The relatively small value of B for the rf SQUID is a consequence of our carrying out the analysis in the side well (i.e., along the separatrix z_1z_2) in Fig. 2 rather than in the much deeper center well. This small value of B manifests itself in the relatively small shift in the homoclinic threshold observed for this case (Fig. 4). In the context of the rf SQUID it is interesting to note that the quantity B decreases with increasing β: to produce a given shift in the homoclinic threshold (characterized by the ratio Q/k) requires a larger noise variance as the depth of the well is increased. One expects a similar effect to occur in the other examples (Duffing oscillator and Josephson junction) considered above, as well.

V. NUMERICAL COMPUTATION OF PROBABILITY DENSITY FUNCTIONS

In this section we consider numerical solutions of the stochastic differential equation for nonlinear dissipative systems driven by random and deterministic (periodic) forces. The stochastic differential equation for such a system may be written in the general form

$$\ddot{x}=k\dot{x}+f(x)=Q\sin\Omega t+F(t), \qquad (56)$$

where the parameters k, Q, and Ω as well as the model-dependent nonlinearity function $f(x)$ and the random force $F(t)$ have been defined in the preceding sections. The solution $x(t)$ of (56) is a random variable. Throughout this section, we will consider the effect of a random *initial value*, i.e., the force $F(t)$ is assumed to be a random variable which assumes a new realization at $t=0$ only. Thereafter, it is assumed to remain constant while the equation (56) is integrated. The integration is repeated for several different realizations of the initial value. It is apparent that each new realization of $F(t=0)$ leads to a new potential corresponding to the undriven conservative problem. By requiring the initial value $x(0)$ to correspond to the point z_2 on the potential (see, e.g., Fig. 1) for each integration of (56) we can carry out numerous integrations of this differential equation with each integration corresponding to a different realization of the initial value $x(0)$. Ultimately, the results obtained may be averaged over this ensemble of initial conditions. For the low noise variances of interest in this work, this method yields results that are expected to be fairly close to the results that might be expected if the noise term $F(t)$ was allowed to vary with time throughout the integration of (56).

A comparison of the central theoretical results of this paper to numerical simulation can be made by going back to our basic definition (7) of the stochastic Melnikov function Δ_F and noting that, for a given model system and a specified set of system and driving parameters, one may compute the quantity on the left-hand side of (7) for different realizations of the random force $F(t)$ (taken to have a specified mean and variance). The resulting set of values of Δ_F (we recall that, for each realization of the random term, Δ_F represents the separation of the stable and unstable solution manifolds of the perturbed system) follows a distribution that may be characterized by a probability density function $P(\Delta_F)$. The theory presented in this work predicts that this distribution yields a mean value $\langle\Delta_F(t_0)\rangle$ defined in Eq. (9). The result of a simulation of this probability density function appears in Fig. 5 for the rf SQUID discussed in Sec. IV. We consider, once again, the $\beta=2$ case in the side well (i.e., the separatrix z_1z_2 of Fig. 1). it is assumed that the system is initially at its noise-free homoclinic threshold $[\Delta(t_0)=0$ in (9)] corresponding to values

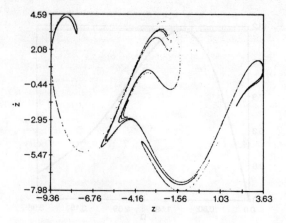

FIG. 6. Poincaré plot of $\dot{z}$ vs z for the rf SQUID with $(\beta,\omega_0^2,\Omega,k,Q)\equiv(2.0,1.0,2.0768,1.1133,10.02)$ in the absence of noise $(\sigma^2=0)$.

$(Q/k,\Omega)\equiv(4.734,2.0768)$.

The random noise term is taken to have zero mean and variance $\sigma^2=0.8$ at time $t=0$, as mentioned earlier. This initial value remains unchanged throughout the subsequent integration of (56). For each such realization of the random force, the solution $x(t)$ of the stochastic differential equation (56) and the corresponding realization Δ_F of the stochastic Melnikov function are computed. This is repeated for 75 000 realizations of the random noise term and a distribution function numerically fitted to the results. The computations have been performed on an Apollo DN3500 workstation. For this case, the theory of this paper [Eqs. (9), (10), and (32)] predicts a mean value $\langle\Delta_F\rangle=-0.110$ [since the random force has zero mean, we recall that this value depends only on the third term on the right-hand side of (10)]. In Fig. 5, we show the results of our simulation of the Melnikov function. The probability density function $P(\Delta_F)$ is plotted as a function of Δ_F. The data points represent the probability density function obtained by sorting the 75 000 realizations of Δ_F into 50 bins. This probability density function yields a mean value $\langle\Delta_F\rangle=-0.125$, which is in excellent agreement with the above-mentioned theoretical prediction for this case. The dotted curve in Fig. 5 represents the Gaussian computed with the same mean and variance as our simulated values of Δ_F. It should be noted that this mean value has been computed for the total change in Δ_F. Theoretically, the value of only the first term in Eq. (7) should change in the presence of noise having zero mean value as is the case under consideration. The accurate numerical calculation of the second term in (7) is more difficult than the calculation of the first term; such a numerical calculation yields a very small change in the second term of (7). If one assumes that the second term in (7) does not change (in accordance with the theory), then the probability density function yields a mean value $\langle\Delta_F\rangle=-0.155$. The third (solid) curve in Fig. 5 is obtained as follows: 1000 realizations of Δ_F are calculated for a set of 1000 evenly spaced forces about the zero mean. Each resulting Δ_F is assigned the corresponding probability of the force under a Gaussian distribution (i.e., perfect Gaussian noise, with given variance σ^2 is assumed). $P(\Delta_F)$ is then obtained by renormalization based on the spacing of the Δ_F. It is seen that the results obtained are in excellent agreement (the mean value of Δ_F is the same to the third decimal place) with the data points. The obvious advantage of using this procedure is that a substantially smaller amount of computer time is necessary to achieve accurate results due to the small number (1000) of realizations of F required.

The noise-induced shifts in the Melnikov function are more easily examined in light of Eq. (36). The coefficient B in (36) determines how the homoclinic threshold condition changes in the presence of noise. For the parameters under consideration, Eq. (38a) yields a value of $B=0.025$, using the theory of Sec. III. The numerical simulation (using the method employed to obtain the dotted line in Fig. 4) has been performed for $\sigma^2=0$, 0.4, 0.6, and 0.8. Each of the resulting threshold values Q/k for the four values of σ^2 falls on the same line [represented by Eq. (36)] to within 0.1%. The value of the slope is $B=0.025$ and 0.032 for the cases where, respectively, the total calculated shift, and the shift due to the change in the first term of (7) only, are considered. It is seen that these values agree quite well with the theoretical prediction.

We now return to Eq. (56) and note that, in general, one may construct[25] a two-dimensional Fokker-Planck equation for the complete probability density function $P(x,\dot{x},t|x_0,\dot{x}_0,t_0)$ corresponding to the random variable $x(t)$. The reduced equilibrium distribution function corresponding to the displacement variable x may be formally expressed in the form (up to a normalization constant)

$$P(x)=e^{-\Phi(x)}, \quad (57)$$

in terms of a generalized potential function $\Phi(x)$. In general, one cannot analytically compute the function $\Phi(x)$ except in certain approximate cases.[29,30] It has been suggested[29] that when the parameters (k,Q,Ω) in (56) are set so that the system is below its homoclinic threshold then the potential function Φ is a well-behaved differentiable function and has all the properties of a thermodynamic potential. However, when the system (56) is above its homoclinic threshold, the separatrix is no longer continuous and one obtains (as pointed out in Sec. I) an infinity of intersections of the stable and unstable solution manifolds. In this case, it has been suggested[19,29] that the potential function Φ may be nondifferentiable hence one is lead to infer that the occurrence of zeros in the Melnikov function implies that the potential Φ will be nondifferentiable. Further, Jauslin has shown[31] that any nonsingular perturbation of the form $h(x)\sin\Omega(t+t_0)$ will cause the Melnikov function to have simple zeros.

The long-time probability density function defined in (57) is now computed, numerically, for the rf SQUID. A random number generator is used to produce 40 000 realizations of the random noise $F(t=0)$ having zero mean and specified (nonzero) variance σ^2. For each of these realizations, the solution $(z,\dot{z})$ of the differential equation (56) is obtained at a fixed time t, z being the transformed variable used in our computations on the rf SQUID in Sec. IV. For each solution run (corresponding to a particular realization of the random force) we start the particle at the point z_2 in Fig. 1 with zero initial velocity. It should be noted that the points z_1 and z_2 in Fig. 1 must be recomputed for each realization of the random forcing term. We consider the system for the parameter set $(\beta,\omega_0^2,k,Q,\Omega)\equiv(2.0,1.0,1.1133,10.02,2.0768)$ corresponding to the occurrence of a chaotic attractor (these values will remain the same throughout the remainder of this work). Figure 6 shows this chaotic attractor for the noise-free case ($\sigma^2=0$). In Fig. 7 we plot the velocity $\dot{z}$ versus the displacement z for the same case ($Q=10.02$) with a finite noise term ($\sigma^2=10^{-5}$). Each point $(\dot{z},z)$ corresponds to a solution of the stochastic differential equation (56), i.e., a state of the system, for a given realization of the random force. The 40 000 solutions on this figure are all computed at $t=90.7625$ (corresponding to 30 Poincaré periods). The effect of increasing the noise ($\sigma^2=0.004$) is clearly evident in Fig. 8; the same number of points (40 000) are used in this figure, but the increased

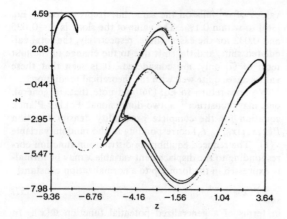

FIG. 7. $\dot{z}$ vs z obtained by solving (56) for the rf SQUID at $t=90.7625$ with $(\beta,\omega_0^2,\Omega,k,Q)$ taking on the same values as Fig. 6. The points correspond to 40 000 realizations of the random force $F(t)$ with zero mean and variance $\sigma^2=10^{-5}$.

phase space as the noise variance is increased. It must be pointed out that the maps in Figs. 7 and 8, while resembling the deterministic attractor of Fig. 6, do not display the self-replication property that is a hallmark of chaotic attractors; this is evident in Figs. 9(a) and 9(b) in which we show a magnified segment of the deterministic chaotic attractor of Fig. 6 and the corresponding segment from its noisy analog of Fig. 7 ($\sigma^2=10^{-5}$). The self-replication property of the deterministic attractor has been destroyed by the noise. In the limit of even smaller noise variances, one would expect that the phase-space mapping resembles the noise-free chaotic attractor more closely. However, the self-replication property would still be absent in this limit, albeit on a much smaller scale than in Figs. 9(a) and 9(b).

The probability density functions corresponding to the displacement z are plotted in Fig. 10 for the cases $\sigma^2=0$ and 0.04. These probability density functions are equivalent to those that would be obtained from a long-time solution of the Fokker-Planck equation corresponding to the nonlinear system (56). In each case, the proba-

noise results in a greater region of phase space being accessible to the system. We note that the deterministic attractor of Fig. 6 could have been obtained in a manner analogous to Figs. 7 and 8 if we had changed the initial condition slightly for each of the 40 000 solutions of the noise-free dynamic equation and computed each solution at the above time $t=90.7625$. In the presence of noise, the potential changes for each realization of the random force and, since we always start the system at the right endpoint z_2 (Fig. 1), we are effectively changing the initial condition for each of the 40 000 solutions. Each realization of the noise (i.e., each realization of the potential) leads to a different region of phase space, i.e., a different attractor that is accessible to the system. Accordingly, the net result in Figs. 7 and 8 is a map that is the *image* of the original (noise-free) attractor. This also explains why the system appears to traverse a greater area of

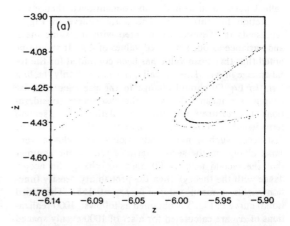

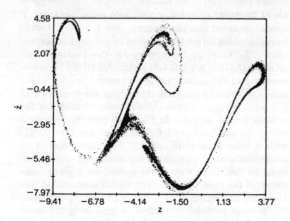

FIG. 8. Same as Fig. 7 with $\sigma^2=0.04$.

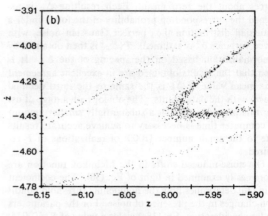

FIG. 9. (a) Magnified section of the chaotic attractor of Fig. 5; the self-replication property is evident. (b) Magnified section of the phase space plot of Fig. 7. There is no self-replication in the presence of even a small amount of noise.

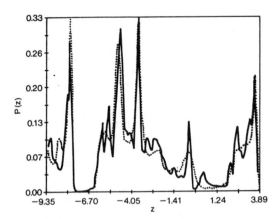

FIG. 10. Probability density function $P(z)$ (solid curve) corresponding to the displacement variable $z(t)$ of Fig. 6. The dotted curve shows the probability density function $P(z,t=90.7625)$ corresponding to the variable z of Fig. 8.

bility density function is seen to display multiple maxima, reminiscent of the nondifferentiable potentials suggested by Kapitaniak[19] and Graham and Tel.[29] The probability density function is a measure of the frequency with which each elemental area of phase space is traversed by the system. In the presence of noise, the peaks in the probability function maxima are seen to have a greater width but a smaller height than those corresponding to the noise-free case; the noise tends to "smooth" the probability density function coarse graining the deterministic "randomness" of the attractor itself. This effect is a direct consequence of the greater region of phase space that is made available to the system in the presence of noise (we recall that the same number of points, 40 000, appear in Figs. 6–8). A similar effect is evident if we consider the probability density functions corresponding to

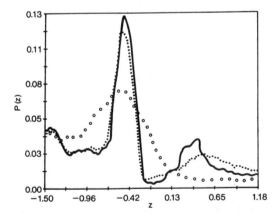

FIG. 11. Displacement probability density functions magnified over a very small range of the displacement. Solid curve represents $P(z)$ corresponding to the attractor of Fig. 6, dotted curve represents $P(z,t=90.7625)$ corresponding to the noisy case of Fig. 7, and data points represent the case of Fig. 8. The smoothing effect of the noise is evident.

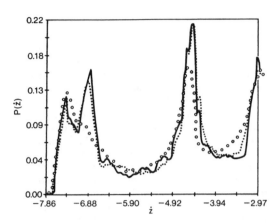

FIG. 12. Velocity probability density functions; same as in Fig. 11. The range of velocities $\dot{z}$ corresponds to the displacement range of Fig. 11.

the velocity variable. Finally, we show (in Fig. 11) a small section of the displacement probability density function $P(z)$ for the cases $\sigma^2=0$, 10^{-5}, and 0.04. The smoothing effect of the noise is evident in this figure as well as in Fig. 12 in which we plot the velocity probability density function for the range of velocities corresponding to the displacements (horizontal scale) of Fig. 11.

VI. DISCUSSION

In this work, we have presented a theory that predicts that the presence of weak Langevin noise in a dissipative nonlinear system suppresses, in the mean, homoclinic behavior that might normally be observed in the noise-free system. This result is in agreement with the work of Carlson[23-24] and Schieve and Petrosky[21-22] on the effect of quantum fluctuations on the homoclinic threshold in driven dissipative classical systems. We have introduced a generalization of the Melnikov function in the presence of the external noise. This function is stochastic, but by averaging it over the ensemble of the noise, one obtains an averaged Melnikov function that corresponds to its noise-free analog shifted by a constant correction term.

The significance of this generalized Melnikov function is worth some further discussion. It is well known[2-6] that, in a noise-free nonlinear dynamic system the zeros of the Melnikov function are associated with homoclinic tangencies and chaos via horseshoes; the Melnikov method for this case is applicable to small perturbations of the separatrix motion. In Secs. II and III above, we have assumed that the Langevin noise is weak enough that the deviations from the separatrix motion for each separate realization of the random force term $F(t)$ in (1) are very small, i.e., the second term in (4) is quite small compared to the term $x_s(t)$, which represent the separatrix solution for the noise-free case. It is further assumed that the averaged motion coincides with the separatrix of the noise-free problem, i.e., $\langle \delta x(t) \rangle = 0$ in (5) [this follows directly from our assumption of a zero mean initial condition: $\langle \delta x(t_0) \rangle = 0 = \langle \delta \dot{x}(t_0) \rangle$]. Clearly, the generalized Melnikov function, being an averaged quantity, represents a deviation from homoclinicity in the mean. The distribution of this Melnikov function (Fig. 5) obtained via numerical simulation is peaked at some mean value $\langle \Delta_F \rangle$. If this distribution were Gaussian, this mean value would correspond to the most probable value of Δ_F which, from Eq. (9), is the noise-free Melnikov function shifted by a small correction term. Figure 5 indicates that the deviation from a Gaussian distribution is small. In fact, for noise variances smaller than the value 0.8 used in this figure, the distribution $P(\Delta_F)$ approximates a Gaussian more closely. Hence, for the low-noise variances required by our perturbation-theoretic arguments of Sec. II, it is reasonable to expect that the probability of observing a homoclinic tangency *in the mean* is a maximum for the parameters which determine the zeros of the Melnikov function in the noise-free problem. In fact, in this limiting case of a very low noise, one may (assuming the distribution of Δ_F shown in Fig. 5 to be approximately Gaussian) write down an analytical expression for the probability density function that characterizes Δ_F:

$$P(\Delta_F) \approx (2\pi\sigma_F^2)^{-1/2} \exp\left[-\frac{(\Delta_F - \langle \Delta_F \rangle)^2}{2\sigma_F^2}\right],$$

σ_F^2 being the variance of Δ_F and $\langle \Delta_F \rangle$ being the mean value which is defined in Eq. (9) and may be determined directly from numerical simulations as described in Sec. V. For a given realization of the noise, the probability that Δ_F vanishes may be deduced directly from the above expression. This allows one to compute the probability that a given realization of the noise will result in a homoclinic tangency. In the limit of vanishing noise, the distribution $P(\Delta_F)$ will be very sharply peaked (approaching a δ function) at its noise-free value $\Delta(t_0)$. For finite small noise, each realization of the noise leads, in general, to a separation of the stable and unstable manifolds given by the stochastic quantity of Eq. (8); this separation is offset from its noise-free value. The comments of this paragraph, and the results embodied in Eqs. (8) and (9) might be directly verified by plotting the stable and unstable solution manifolds for a given model system (see, e.g., Refs. 4, 5, and 12) for numerous realizations of the random noise. Such a simulation would be costly, but possibly feasible.

The correction term $\langle \Delta_c \rangle$ of Eq. (9) has been estimated (within the limits of accuracy of the WKB approximation) using the procedure described in Sec. III. In this context we must point out that better accuracy may be obtained by retaining higher-order terms in the WKB expansion [the solutions (22) and (29) represent only the first-order solutions, the so-called "physical optics approximation"]. However, although the agreement between the solutions obtained using this approximation and those obtained via a direct numerical solution of the transport equation (13) may not be perfect, we find that the correction to the Melnikov function calculated using the *integral* of the square of the solution is almost the same using the two procedures. Our analysis has taken into account the case in which the Langevin noise has zero mean value as well as the case in which this mean value is finite. In the former case, the mean displacement $\langle x(t) \rangle$ as well as the associated mean velocity $\langle \dot{x}(t) \rangle$ of the noisy unperturbed system follow the separatrix motion of the corresponding noise-free system, through our choice of zero initial values in the solution of the transport equation (13). This ensures that the Melnikov correction term depends solely on the second-order statistics [in this case the initial variance $\langle \delta\dot{x}^2(0) \rangle$] of the random variable $\langle \delta\dot{x}(t) \rangle$. However, if we choose a nonzero initial condition, then the expression (10) contains contributions from all the terms on the right-hand side even if the noise term has zero mean value. The case in which the noise has a finite mean value has also been considered. In this case, the first two terms on the right-hand side of (10) contribute to the correction term (regardless of whether we assume a vanishing initial condition or not). It is seen that the presence of a finite mean value does not lower the homoclinic threshold. In this context it is worth pointing out that weak multiplicative fluctuations in the nonlinearity parameter of a driven dissipative system may, on average, *lower* the homoclinic threshold.[32]

By expressing the initial variance $\langle \delta\dot{x}^2(0) \rangle$ in terms of the variance of the noise, we have been able to quantify the assumption of weak Langevin noise [our value $\sigma^2 = 0.2$ of the external noise variance used in the numerical calculations corresponds to an initial velocity variance $\langle \delta\dot{x}^2(0) \rangle = 0.1$, which agrees with our definition of weak noise introduced in Sec. III]. The implementation of this assumption within the framework of the perturbation theory that underlies the Melnikov function, leads one to believe that the calculation of the noise-induced shift in the Melnikov function is very accurate (the correspondence with a real system is likely to improve if one assumes even weaker noise, i.e., if both m and σ^2 are reduced even further). This idea is further strengthened by the results of our numerical computations of the probability density function $P(\Delta_F)$ in Sec. V; these computations yield values of the mean $\langle \Delta_F \rangle$ that agree very well with those computed using the theory of Sec. III despite the fact that the computations of this section were carried out using a random initial value and then assuming the noise to be constant throughout the remainder of each integration of the stochastic differential equation (56). Increasing the noise variance elevates the homoclinic threshold. Since noise is present in most real experiments, one must conclude that some results of experiments carried out on deterministic nonlinear systems might well be chaotic were it not for the smoothing effect of the system noise. The numerical work of Sec. V, summarized in Figs. 6–12 displays this smoothing effect [we reiterate that these results were obtained using a random dc driving in the stochastic differential equation (56)] on the multimaximum probability density functions that characterize the chaotic regime.

ACKNOWLEDGMENTS

One of us (A.R.B.) would like to acknowledge support from the office of Naval Research under Grant No. N0001489AF00001.

[1] H. Poincaré, *New Methods of Celestial Mechanics* (Gauthier-Villars, Paris, 1899), Vol. 3, p. 391.
[2] J. Moser, *Stable and Random Motions* (Princeton University Press, Princeton, 1973).
[3] V. Melnikov, Trans. Moscow Math. Soc. **12**, 1 (1963).
[4] P. Holmes, Philos. Trans R. Soc. London Ser. A **292**, 419 (1979).
[5] J. Guckenheimer and P. Holmes, *Nonlinear Oscillations, Dynamical Systems and Bifurcations of Vector Fields* (Springer-Verlag, New York, 1983).
[6] V. Arnold, Dokl. Akad. Nauk SSSR **156**, 9 (1964) [Sov. Phys. Dokl. **5**, 581 (1964)].
[7] J. Sanders, Celest. Mech. **28**, 171 (1982).
[8] R. Kautz and R. Monaco, J. Appl. Phys. **57**, 875 (1985).
[9] R. Kautz and J. Macfarlane, Phys. Rev. A **33**, 498 (1986).
[10] Z. Genchev, Z. Ivanov, and B. Todorov, IEEE Trans. Circuits Syst. **CAS-30**, 633 (1983).
[11] V. Gubankov, S. Zyglin, K. Konstantinyan, V. Koshelets, and G. Ovsyannikov, Zh. Eksp. Teor. Fiz. **86**, 343 (1984) [Sov. Phys.—JETP **59**, 198 (1984)].
[12] W. Schieve, A. Bulsara, and E. Jacobs, Phys. Rev. A **37**, 3541 (1988).
[13] J. Cruthfield, J. D. Farmer, and B. Huberman, Phys. Rep. **92**, 45 (1982).
[14] H. Svensmark and M. Samuelson, Phys. Rev. A **36**, 2413 (1987).
[15] K. Wiesenfeld and B. McNamara, Phys. Rev. Lett. **55**, 10 (1985).
[16] K. Wiesenfield and B. McNamara, Phys. Rev. A **33**, 629 (1986).
[17] F. Arecchi, R. Badii, and A. Politi, Phys. Rev. A **32**, 402 (1985).
[18] R. Kautz, Phys. Rev. A **38**, 2066 (1988).
[19] T. Kapitaniak, *Chaos in Systems with Noise* (World Scientific, New York, 1988).
[20] K. Matsumoto and I. Tsuda, J. Stat. Phys. **31**, 87 (1983).
[21] W. Schieve and T. Petrosky, in *Quantum Optics IV*, edited by J. Harvey and D. Walls (Springer-Verlag, New York, 1986).
[22] T. Petrosky and W. Schieve, Phys. Rev. A **31**, 3907 (1985).
[23] L. Carlson, Ph.D. thesis, University of Texas, 1989.
[24] L. Carlson and W. Schieve, J. Math. Phys. (to be published).
[25] H. Risken, *The Fokker Planck Equation* (Springer-Verlag, New York, 1989).
[26] N. van Kampen, *Stochastic Processes in Physics and Chemistry* (North-Holland, Amsterdam, 1983).
[27] R. Langer, Bull. Am. Math. Soc. **40**, 545 (1934).
[28] R. Barone and G. Paterno, *Physics and Applications of the Josephson Effect* (Wiley, New York, 1982).
[29] R. Graham and T. Tel, J. Stat. Phys. **35**, 729 (1984).
[30] R. Graham and T. Tel, Phys. Rev. A **31**, 1109 (1985).
[31] H. Jauslin, J. Stat. Phys. **42**, 573 (1986).
[32] W. C. Schieve and A. R. Bulsara, Phys. Rev. A (to be published).

VII. Strange Nonchaotic Attractors

Recently a new class of attractors, i.e., strange nonchaotic has been found. An attractor which is geometrically strange (is not a finite set of points, a limit cycle or a closed curve, a smooth or piecewise smooth surface, bounded by piecewise smooth closed surface volume) but for which typical orbits have non-positive Lyapunov exponents is called strange nonchaotic.

Paper VII.1 gives some fundamental information about strange nonchaotic attractors and shows how to distinguish this class from quasiperiodic attractors basing on power spectra. It suggests the existence of strange nonchaotic attractors in phase space on a set with positive measure, i.e., they are typical for quasiperiodically forced systems.

The information dimension and Kolmogorov capacity of the strange nonchaotic attractors are discussed in Paper VII.2. It suggests that information dimension being equal to one and Kolmogorov capacity being equal to two are typical for strange nonchaotic attractors.

The possible route to chaos in quasiperiodically forced systems with strange nonchaotic attractors is presented in Paper VII.3.

QUASIPERIODICALLY FORCED DYNAMICAL SYSTEMS WITH STRANGE NONCHAOTIC ATTRACTORS

Filipe J. ROMEIRAS[a,b], Anders BONDESON[c], Edward OTT[a], Thomas M. ANTONSEN Jr.[a] and Celso GREBOGI[a]

[a] *Laboratory for Plasma and Fusion Energy Studies, University of Maryland, College Park, MD 20742, USA*
[b] *Permanent address: Centro de Electrodinâmica, Instituto Superior Técnico, 1096 Lisboa Codex, Portugal*
[c] *Institute for Electromagnetic Field Theory, Chalmers University of Technology, S 41296 Göteborg, Sweden*

Received 14 October 1986

We discuss the existence and properties of strange nonchaotic attractors of quasiperiodically forced nonlinear dynamical systems. We do this by examining a particular model differential equation, $\dot{\phi} = \cos\phi + \varepsilon\cos 2\phi + f(t)$, where f is a two-frequency quasiperiodic function of t. When $\varepsilon = 0$ the analysis of the equation is facilitated since then it can be related to the Schrödinger equation with quasiperiodic potential. We show that the equation does indeed exhibit strange nonchaotic attractors, and we consider the important question of whether these attractors are typical in the sense that they exist on a set of positive Lebesgue measure in parameter space. (The equation also exhibits two- and three-frequency quasiperiodic behavior.) We also show that the strange nonchaotic attractors have distinctive frequency spectrum; this property might make them experimentally observable.

1. Introduction

Recently attention has been given to a class of dissipative dynamical systems that typically exhibit a class of attractors that may be described as strange and nonchaotic (Grebogi et al. [1]).

Here the word strange refers to the geometrical structure of the attractor: a strange attractor is an attractor which is neither a finite set of points nor is it piecewise differentiable (that is, either a piecewise differentiable curve or surface, or a volume bounded by a piecewise differentiable closed surface).

The word chaotic refers to the dynamics of the orbits on the attractor: a chaotic attractor is one for which typical orbits have a positive Lyapunov exponent. This implies that nearby orbits diverge exponentially from one another with time and that the orbit depends sensitively on its initial conditions.

By a strange nonchaotic attractor we therefore mean an attractor which seems to be geometrically strange but for which typical nearby orbits do not diverge exponentially with time.

An example of a strange nonchaotic attractor is exhibited by the one-dimensional quadratic map, $x_{n+1} = C - x_n^2$, at the point of accumulation of period doublings. The attractor is a Cantor set of dimension $\simeq 0.538$ and the Lyapunov exponent for a typical orbit is zero (Grassberger [2]). Attractors of the same type occur at the infinite number of C values representing points of accumulation of period doublings of orbits of period $2^N P$ corresponding to the infinite number of periodic windows occurring as C is varied. Nevertheless these strange nonchaotic attractors only occupy a set of C values of zero Lebesgue measure. That is, if we were to pick a C value at random, the probability of that value yielding a strange nonchaotic attractor would be zero. In this sense we say that this type of attractor is not typical of the quadratic map.

As another example, consider the circle-map, $\theta_{n+1} = \Omega + \theta_n + k\sin\theta_n \pmod{2\pi}$, for parameter

Reprinted with permission from Physica, Vol. 26D, pp. 277–294
Copyright (1987), Elsevier Science Publishers B.V.

values $k = 1$ (the critical case) and Ω chosen to give irrational winding number. In this case the Lyapunov exponent is zero, and the density of orbit points $\rho(\theta)$ is zero on a dense set of $\theta \in [0, 2\pi]$. Due to this behavior of $\rho(\theta)$ one might call this nonchaotic attractor strange. Again, however, the measure in parameter space (k, Ω) where this occurs is zero, and hence this type of attractor is not typical for the circle map*.

On the other hand, examples of systems where strange nonchaotic attractors are typical, in the sense of occupying a set of positive Lebesgue measure in the parameter space, were given by Grebogi et al. [1]. They examined a particular class of maps of the general form

$$x_{n+1} = g(x_n, \theta_n), \tag{1a}$$
$$\theta_{n+1} = \theta_n + 2\pi\omega [\mod 2\pi], \tag{1b}$$

where g is a 2π-periodic function of its second argument and ω is an irrational number**. In the paper of Grebogi et al. θ_n was always taken to be a scalar while two cases were considered for x_n, g: in case (1) x_n, g were scalars and eq. (1a) was taken to be of the form

$$x_{n+1} = 2\lambda (\tanh x_n) \cos \theta_n;$$

in case (ii) $x_n = (u_n, v_n)$ and $g = (g_1, g_2)$ were two-dimensional vectors, and eq. (1a) was taken to be of the form

$$\begin{bmatrix} u_{n+1} \\ v_{n+1} \end{bmatrix} = \frac{\lambda}{1 + u_n^2 + v_n^2} \begin{bmatrix} 1 & 0 \\ 0 & \gamma \end{bmatrix}$$
$$\times \begin{bmatrix} \cos \theta_n & \sin \theta_n \\ -\sin \theta_n & \cos \theta_n \end{bmatrix} \begin{bmatrix} u_n \\ v_n \end{bmatrix}.$$

In case (i) it was shown that the map exhibits a strange nonchaotic attractor with one negative Lyapunov exponent (the other, corresponding to eq. (1b), is trivially zero) for all λ in the range $|\lambda| > 1$. In case (ii) it was shown that the attractor is also strange with two negative Lyapunov exponents (again the third is trivially zero) for all λ above some critical value $\lambda_c(\gamma)$. The important result is the existence of strange nonchaotic attractors on a set of positive measure in parameter space.

Maps of the form (1) may be obtained from two-frequency quasiperiodically forced nonlinear systems (see section 2). It might therefore be suspected that such nonlinear systems can exhibit strange nonchaotic attractors for a positive measure of the parameter space. Off hand, however, it is not at all clear that this suspicion will be fulfilled, since the functions $g(x_n, \theta_n)$ (eq. (1a)) used by Grebogi et al. [1] are highly artificial and were specifically constructed to demonstrate the possibility of strange nonchaotic attractors with positive measure in parameter space. Thus we cannot decide, on the basis of the work by Grebogi et al. [1], what the situation will be for quasiperiodically forced systems that are likely to arise in applications.

It is the aim of the present paper to discuss the existence and the properties of strange nonchaotic attractors of dynamical systems forced at two incommensurate frequencies by examining a specific model given by the non-autonomous first order differential equation

$$\frac{d\phi}{dt} = \cos \phi + \varepsilon \cos 2\phi + f(t), \tag{2a}$$

where f is two-frequency quasiperiodic (that is, $f(t) = \hat{f}(\omega_1 t, \omega_2 t)$, where $\hat{f}$ is 2π-periodic in both its arguments and ω_1/ω_2 is irrational) and ε is a variable parameter. In our study f was actually taken to be of the form

$$f(t) = K + V(\cos \omega_1 t + \cos \omega_2 t), \tag{2b}$$

where $\omega_1 = \frac{1}{2}(\sqrt{5} - 1)$ and $\omega_2 = 1$ were kept fixed while K, V were allowed to vary. Equations of this type are perhaps the simplest ordinary differential equations yielding a strange attractor.

---

*For a third example, also with zero measure in parameter space, occurring for a system of the form of eqs. (1a, b), below, see fig. 6 of Sethna and Siggia [3].

**Equations of this form have also been investigated from other points of view by Sethna and Siggia [3] and by Kaneko [4].

Specifically we are primarily interested in (i) the question of parameter space measure (typicality), posed above, and (ii) in elucidating possible power spectral signatures of these attractors. (We believe that the signature we shall discuss may be a useful diagnostic in experiments.)

In the special case $\varepsilon = 0$, eq. (2) can be regarded as a simplified form of the well-known equation of the damped forced pendulum

$$\frac{d^2\phi}{dt^2} + \nu \frac{d\phi}{dt} + \Omega^2 \sin\phi = f(t), \tag{3}$$

corresponding to strong damping and forcing (so that the inertial term $d^2\phi/dt^2$ can be neglected). Eq. (3) has also been extensively used as a model of the current driven Josephson junction (see Gwinn and Westervelt [5], Kautz and MacFarlane [6], and references therein). The existence of strange nonchaotic attractors for eq. (3) will be considered in a separate publication (Romeiras and Ott [7]).

Again in the case $\varepsilon = 0$, eq. (2) can be related by the so-called Prüfer transformation of the dependent variable $\phi(t) \to \psi(x)$, defined by

$$e^{i\phi} = \frac{\psi'/\psi + ic}{\psi'/\psi - ic}.$$

(see Johnson and Moser [8]) with the change of independent variable $t \to x$, defined by

$$x = \frac{1}{2c}\left[t + \int_0^t f(E)\,dE\right],$$

to the linear "time independent" Schrödinger equation

$$\psi'' + k^2(x)\psi = 0, \quad \text{with } k^2 = c^2 \frac{f-1}{f+1}.$$

Here the prime denotes differentiation with respect to x and c is an arbitrary constant. Since f is quasiperiodic in t, k^2 will be quasiperiodic in x. Thus the theory of the Schrödinger equation with quasiperiodic potential (see the reviews by Simon [9] and Souillard [10]) can be used to aid in understanding the behavior of the solutions of eq. (2). This has been done by Bondeson et al. [11] where some of the results presented in this paper were first briefly described.

The fact that the case $\varepsilon = 0$ allows a reduction to the Schrödinger equation, although convenient for analysis, also means that we are dealing with a special case whose properties might differ qualitatively from those typical of equations of the general form

$$\frac{d\phi}{dt} = g_\eta(\phi, t), \tag{4}$$

where $g_\eta(\phi, t)$ is periodic in ϕ and two-frequency quasiperiodic in t, and $\eta = (\eta_1, \eta_2, \ldots, \eta_N)$ denotes the parameters of the system. Thus we also investigate the $\varepsilon \neq 0$ case, since the Prüfer transformation does not apply, and hence more generally applicable qualitative behavior should occur*. Indeed, as shown in sections 3 and 4, the $\varepsilon \neq 0$ case has many more resonances than the $\varepsilon = 0$ case**.

The plan of this paper is as follows. In section 2 we present the detailed results of the numerical study of eq. (2) in the case $\varepsilon = 0$. In sections 3 and 4 we discuss the case $\varepsilon \neq 0$. Finally in section 5 we summarize the main conclusions of the study. For $\varepsilon = 0$ we find that there are indeed strange nonchaotic attractors for a positive measure of parameters. Thus such attractors are typical for this system. There is, however, an important difference as compared to the situations occurring in Grebogi et al. [1]). Namely, parameter values corresponding to strange nonchaotic attractors of eq. (2) with $\varepsilon = 0$ do not occur on an interval, but rather lie on a Cantor set of positive measure. In the case $\varepsilon \neq 0$, our numerical results strongly suggest that the

---

*In the ε term, in place of $\cos(2\phi)$ we could have used some other smooth 2π-periodic function of ϕ.

**For $\varepsilon = 0$ we shall see that the winding number plateaus (resonances) occur at $W = l\omega_1 + m\omega_2$, whereas for $\varepsilon \neq 0$ they occur at $W = (l/n)\omega_1 + (m/n)\omega_2$, where W is the winding numbers and l, m, n are integers. The latter case is to be expected in the general context of equations of the form (4), while the former case is specific to eq. (2) with $\varepsilon = 0$.

same situation also applies, although we have no direct analytical support (the Schrödinger equation correspondence only applies for $\varepsilon = 0$). Thus we believe that strange nonchaotic attractors are typical for quasiperiodically forced systems of the form (4), where by "typical" we mean that they occur on a set of positive measure in parameter space.

2. The case $\varepsilon = 0$

2.1. Characterization of attractors

Before starting with the presentation and discussion of the numerical results we introduce the main quantities used to characterize the attractors, namely, the Lyapunov characteristic exponent, the winding number, the surface of section plot and the frequency spectrum.

The *Lyapunov characteristic exponent*, Λ, for an orbit $\phi(t)$ of eq. (4) is defined by [12]

$$\Lambda = \lim_{T \to \infty} \frac{1}{T} \ln \frac{|\phi^{(1)}(T)|}{|\phi^{(1)}(0)|},$$

where $\phi^{(1)}$ denotes the solution of the equation of first variation

$$\frac{d\phi^{(1)}}{dt} = A(t)\phi^{(1)}, \quad A(t) = \frac{\partial g_\eta}{\partial \phi}(\phi(t), t). \quad (5)$$

By using the explicit form of the solution of eq. (5) we can also write

$$\Lambda = \lim_{T \to \infty} \frac{1}{T} \int_0^T A(t) \, dt.$$

The Lyapunov exponent represents the mean exponential rate of divergence of two initially close trajectories*.

---
*The nonautonomous eq. (4) can be written as an autonomous system of three first order equations of the form $d\phi/dt = \hat{g}_\eta(\phi, \psi_1, \psi_2)$, $d\psi_1/dt = \omega_1$, $d\psi_2/dt = \omega_2$. In this case there are three Lyapunov exponents. However, two of them, corresponding to the second and third equations, are zero; the third is Λ given above.

The *winding number*, W, is defined by

$$W = \lim_{T \to \infty} \frac{\phi(T) - \phi(0)}{T}.$$

It represents the mean angular frequency of the solution.

The *surface of section plot* is obtained by strobing the solution of eq. (4) at times

$$t_n = \frac{2\pi}{\omega_2} n + t_0, \quad (6)$$

where n is an integer, and plotting

$$\phi_n = \phi(t_n)[\mathrm{mod}\, 2\pi], \quad (7)$$

versus

$$\theta_n = \omega_1 t_n [\mathrm{mod}\, 2\pi]. \quad (8)$$

The solution of eq. (4) thus generates a discrete time map

$$\phi_{n+1} = G(\phi_n, \theta_n), \quad (9a)$$
$$\theta_{n+1} = \theta_n + 2\pi\omega [\mathrm{mod}\, 2\pi], \quad (9b)$$

where $\omega = \omega_1/\omega_2$. The function G must be invertible for ϕ since eq. (4) can also be solved going backward in time.

The *frequency spectrum* was obtained in the following way: (1) From the sequence $\{\phi_n\}_{n=0, M-1}$ obtain the new sequence $\{s_n\}_{n=0, M-1}$. $s_n = h_n P(\phi_n)$, where P is some smooth 2π-periodic function (which was actually taken to be $P(\phi) = \cos \phi$) and $h_n = \frac{1}{2}(1 - \cos 2\pi n/M)$; the multiplication by h_n corresponds to the so-called Hanning's method of leakage reduction and is a means of smoothing out spurious spectral features introduced by the finite duration of the time series. (ii) Calculate the discrete Fourier transform $\{S_k\}_{k=0, M-1}$ of the sequence $\{s_n\}_{n=0, M-1}$ defined by

$$S_k = \sum_{n=0}^{M-1} s_n \exp\left(-i\frac{2\pi}{M} kn\right), \quad k = 0, M-1,$$
$$(10)$$

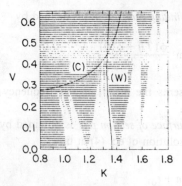

Fig. 1. Diagram of the K-V-plane showing regions where $\Lambda < 0$ (hatched) or $\Lambda = 0$ (blank). The curve denoted by (C) is the critical curve. The curve denoted by (W) is the curve of constant winding number $W = 0.9277\ldots$ along which the orbits of fig. 8 and the corresponding spectra of fig. 9 were calculated [$\varepsilon = 0.0$].

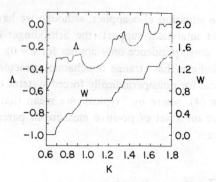

Fig. 2. Curves of the Lyapunov exponent (Λ) and the winding number (W) versus K at $V = 0.55$ [$\varepsilon = 0.0$].

by using a Fast Fourier Transform algorithm. (See Brigham [13] as a general reference on these topics and Powell and Percival [14].)

The differential system (4)–(5) was integrated by using a 4th order Runge-Kutta method with 32 time steps per period of the $\cos t$ driver. The number of driver periods, N, was taken between 2×10^3 and 4×10^5, depending on the circumstances. For the FFT algorithm $M = 2^{16}$ points were used. The computations were carried out on a CRAY X-MP machine; by using the vector mode operation (see Petersen [15]), when possible, a 15 fold increase in speed was achieved.

In the remainder of this section we present and discuss the numerical results for eq. (2) with $\varepsilon = 0$.

2.2. Lyapunov exponents and winding numbers

Fig. 1 shows a diagram of the K-V-plane giving regions where Λ is negative (hatched) or zero (blank). The criterion for negative Lyapunov exponent used in this figure is $\Lambda < -10^{-4}$. The diagram was obtained by taking a grid with 201 values of K and 66 values of V; the integration of the differential system was taken over a variable number of driver periods going from $N = 2 \times 10^3$ for most cases up to $N = 16 \times 10^3$ for the more slowly converging ones. The diagram exhibits a structure similar to the Arnold tongues of the circle map (see, for example, Devaney [16], p. 111).

Fig. 2 shows curves of the Lyapunov exponent (Λ) and the winding number (W) as functions of K at a fixed value of V ($V = 0.55$). It was obtained by taking 241 values of K and integrating over $N = 10^4$ driver periods. The curve of W versus K is a "devil's staircase": a continuous non-decreasing curve with a dense set of open intervals on which W is constant and given by

$$W = l\omega_1 + m\omega_2, \qquad (11)$$

where l, m are integers. Between these plateaus there is a Cantor set, generally of positive measure, on which W increases with K. These results follow from the general character of fig. 2 and the correspondence with the Schrödinger equation with quasiperiodic potential [11]. For small K in fig. 2 (i.e., above curve (C) in fig. 1), Λ is negative on both the Cantor set and the plateaus, while for large K, Λ is zero on the Cantor set and negative on the plateaus. The regions where eq. (11) holds appear in fig. 1 as the narrow tongues emerging at small V. The set of points where the tongues touch each other appear to lie on a smooth (but non-unique) critical curve which separates the region where Λ is always negative from the region where Λ is either negative or zero. An approximation to such a critical curve is indicated as curve (C) in fig. 1.

Table 1

Case	Winding number	Lyapunov exponent	Type of attractor	Analogy with solutions of Schrödinger eq.	Fig.
A	$W \neq l\omega_1 + m\omega_2$	$\Lambda = 0$	3-freq. quasiperiodic	extended	3a
B	$W = l\omega_1 + m\omega_2$	$\Lambda < 0$	2-freq. quasiperiodic	stop band	3b
C	$W \neq l\omega_1 + m\omega_2$	$\Lambda < 0$	strange nonchaotic	localized	3c

2.3. Surface of section characterization

The three distinct combinations of winding numbers (either satisfying eq. (11) or not) and Lyapunov exponents (either negative or zero) give rise to surface of section plots with qualitatively different characteristics – see table I. In terms of the analogy with the Schrödinger equation [11], cases A, B, and C correspond to extended, stop band and Anderson localized solutions, respectively (see Aubry and Andre [17], as well as refs. [9, 10], for a discussion of Anderson localization in incommensurate lattices).

In case A, the three frequencies W, ω_1 and ω_2 are typically irrationally related, and eq. (2) will exhibit three-frequency quasiperiodic behavior. A typical orbit apparently generates a smooth density of points densely filling the surface of section. This is illustrated in fig. 3a ($V = 0.55$, $K = 1.54$, $N = 2 \times 10^5$).

In case B, the frequency W is rationally related to ω_1 and ω_2 and eq. (2) will exhibit two-frequency quasiperiodic behavior. The attracting orbit in the surface of section lies on a smooth single-valued curve, $\phi = F(\theta)$, which wraps l times around the torus in ϕ for each time it wraps once around in θ. This is illustrated in fig. 3b ($V = 0.55$, $K = 1.39$, $N = 10^5$). Note that in this example eq. (11) is satisfied with $l = 0$, $m = 1$, hence $W = 1$; as is clear from the figure the curve does not wrap around the torus in the ϕ direction.

The behavior of the attractors in cases A and B can be explained by the following argument. Let us suppose that ϕ is a three-frequency quasiperi-

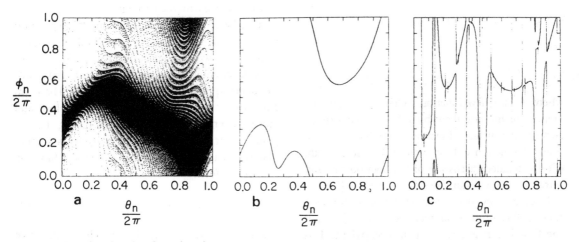

Fig. 3. Surface of section plot of a) a three-frequency quasiperiodic attractor ($V = 0.55$, $K = 1.54$); b) a two-frequency quasiperiodic attractor ($V = 0.55$, $K = 1.39$); c) a strange nonchaotic attractor ($V = 0.55$, $K = 1.34$) [$\varepsilon = 0.0$].

odic function of t,

$$\phi(t) = \hat{\phi}(\omega_1 t, \omega_2 t, Wt), \tag{12}$$

where $\hat{\phi}$ is 2π-periodic in each of its arguments and ω_1, ω_2, W are irrationally related. By strobing ϕ at times t_n given by (6) we obtain

$$\phi_n \equiv \phi(t_n) = \hat{\phi}(\omega_1 t_n, \omega_2 t_n, Wt_n),$$

or, using the periodicity of $\hat{\phi}$ in its second argument

$$\phi_n = \hat{\phi}(\omega_1 t_n, \omega_2 t_0, Wt_n). \tag{13}$$

If we now introduce the variable θ_n defined by (8) this expression can be written in the form

$$\phi_n = \hat{\hat{\phi}}\left(\theta_n, \frac{W}{\omega_1}\theta_n\right),$$

where $\hat{\hat{\phi}}$ is a new function which is 2π-periodic in both arguments. As the θ-map (eq. (9b)) is ergodic and W/ω_1 is irrational this expression will generate an orbit that will densely fill up the whole surface of section. In the particular case of two-frequency quasiperiodic behavior we have $W = l\omega_1 + m\omega_2$ and therefore by substituting into (13) and using the periodicity of $\hat{\phi}$ in its third argument we obtain

$$\phi_n = \hat{\phi}(\omega_1 t_n, \omega_2 t_0, l\omega_1 t_n + m\omega_2 t_0),$$

or

$$\phi_n = F(\theta_n),$$

where F is a new function which is 2π-periodic. Hence, in this case the orbit will generate a single curve in the surface of section. Note that, since the original function $\hat{\phi}$ is smooth, F is also smooth. (For a detailed discussion of quasiperiodicity and attractors on a N-torus see Grebogi et al. [18].)

In case C the attractor is geometrically strange: there still is a functional relationship $\phi = F(\theta)$ but the function F is discontinuous everywhere. This can be verified in the following way: (i) to verify the existence of the relationship $\phi = F(\theta)$ we initialize a large number of points at a single initial θ value but with different initial ϕ values and find that after a large number of iterates N, say, all the orbits are attracted to a single value ϕ_N; (ii) that $\phi = F(\theta)$ cannot be a continuous curve follows from the fact that the winding number is irrationally related to ω_1, ω_2; (iii) finally, that $\phi = F(\theta)$ is discontinuous everywhere follows from the fact that the θ map is ergodic. An example of a strange nonchaotic attractor is given in fig. 3c ($V = 0.55$, $K = 1.34$, $N = 2 \times 10^5$; the corresponding Lyapunov exponent is $\Lambda \simeq -0.07167$).

2.4. Frequency spectra

In figs. 4(a, b, c) we have plotted the frequency spectra of the orbits which correspond to the surface of section plots of figs. 3(a, b, c), respectively. We have plotted $|S_k/\max_{k=0,M-1}(S_k)|$, $k = 0, M/2$, versus k/M; the remaining components of the discrete Fourier transform satisfy the symmetry relation, $S_k = S^*_{M-k}$, $k = M/2 + 1, M - 1$, where * denotes the complex conjugate. The figures show that the spectrum of the two-frequency quasiperiodic attractor (fig. 4b) is concentrated at a discrete set of frequencies while the spectra of both the three-frequency quasiperiodic (fig. 4a) and the strange attractor (fig. 4c) have a much richer harmonic content.

In order to explain the behavior of the spectra it is convenient to consider separately two comparisons: (i) two-frequency quasiperiodic versus three-frequency quasiperiodic; and (ii) two-frequency quasiperiodic versus strange. The comparisons (i) and (ii) correspond to below and above the critical line, respectively (cf. fig. 1).

In the first comparison we start by assuming that ϕ is a three-frequency quasiperiodic function of t of the form

$$\phi(t) = \hat{\phi}(\omega_1 t, \omega_2 t, Wt), \tag{12}$$

where $\hat{\phi}$ is smooth and 2π-periodic in its three arguments. Then the function s defined by $s(t) =$

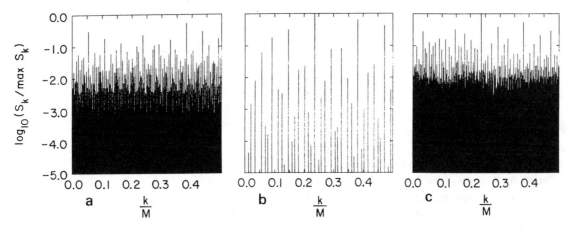

Fig. 4. Frequency spectrum of the attractors of a) fig. 3a; b) fig. 3b; c) fig. 3c.

$P(\phi(t))$, where P is smooth and 2π-periodic, can be expanded in the multiple Fourier series

$$s(t) = \sum_{p,q,r=-\infty}^{+\infty} C_{pqr} \exp\left[\mathrm{i}(p\omega_1 + q\omega_2 + rW)t\right].$$

By strobing s at times t_n given by (6) we now obtain

$$s_n \equiv s(t_n)$$
$$= \sum_{p,q,r} C_{pqr} \exp\left[\mathrm{i}(p\omega_1 + q\omega_2 + rW)t_0 + \mathrm{i}(p\omega_1 + qW)\frac{2\pi}{\omega_2}n\right],$$

or, formally performing the r sum,

$$s_n = \sum_{p,q} \bar{C}_{pq} \exp\left[\mathrm{i}(p\omega_1 + qW)\frac{2\pi}{\omega_2}n\right]. \quad (14)$$

The discrete Fourier transform of the sequence $\{s_n\}$ is given by

$$S_k = \sum_{p,q} \bar{C}_{pq} \sum_{n=0}^{M-1} \exp\left[\mathrm{i}2\pi n\left(\frac{p\omega_1 + qW}{\omega_2} - \frac{k}{M}\right)\right].$$

Thus, as expected, the spectrum of the three-frequency quasiperiodic attractor will be peaked at the frequencies that satisfy $k/M = (p\omega_1 + qW)/\omega_2$, modulo 1, where W is irrationally related to both ω_1 and ω_2. In the particular case of two-frequency quasiperiodic behavior we have $W = l\omega_1 + m\omega_2$ and (14) can be simplified to

$$s_n = \sum_p \bar{\bar{C}}_p \exp\left(\mathrm{i}2\pi n p \frac{\omega_1}{\omega_2}\right).$$

The corresponding discrete Fourier transform is

$$S_k = \sum_p \bar{\bar{C}}_p \sum_{n=0}^{M-1} \exp\left[\mathrm{i}2\pi n\left(p\frac{\omega_1}{\omega_2} - \frac{k}{M}\right)\right].$$

The spectrum is therefore peaked at the frequencies $k/M = p\omega_1/\omega_2$, modulo 1.

Moving now to the second comparison (two-frequency quasiperiodic versus strange) we start by recalling that for both types of attractors there is a functional relationship $\phi = F(\theta)$ which defines either a smooth or a strange curve. The same applies to the curve defined by $\tilde{s}(\theta) = P(F(\theta))$, where P is 2π-periodic. The function $\tilde{s}$ can be expanded in a Fourier series

$$\tilde{s}(\theta) = \sum_{p=-\infty}^{+\infty} \tilde{C}_p \exp(\mathrm{i}p\theta). \quad (15)$$

From here we obtain the sequence $\{\tilde{s}_n = \tilde{s}(\theta_n)\}$

and the corresponding discrete Fourier transform

$$\tilde{S}_k = \sum_p \tilde{C}_p \exp(\mathrm{i} p\omega_1 t_0)$$

$$\times \sum_{n=0}^{M-1} \exp\left[\mathrm{i} 2\pi n\left(p\frac{\omega_1}{\omega_2} - \frac{k}{M}\right)\right].$$

For both types of attractors the spectra are peaked at the frequencies $k/M = p\omega_1/\omega_2$, modulo 1. The difference between the two comes from the very different smoothness of the function $\phi = F(\theta)$ and hence of the asymptotic behavior as $|p| \to \infty$ of the Fourier coefficients. For the smooth F we expect that $\tilde{C}_p \sim \exp(-\nu|p|)$, as $|p| \to \infty$ (ν is a constant). For the strange F we expect a much slower decay of the Fourier coefficients; if we assume that the discontinuities are of the form $|\theta - \theta_0|^{-\beta}$, as $\theta \to \theta_0$, with $0 < \beta < 1$, then $\tilde{C}_p \sim |p|^{\beta-1}$, as $|p| \to \infty$.

In order to obtain a more quantitative characterization of the spectra of the three types of attractors we introduce a spectral distribution $\mathcal{N}(\sigma)$ defined as the number of spectral components larger than some value σ. From the behavior of the Fourier coefficients of the series (15) we expect that this function will behave like

$$\mathcal{N}(\sigma) \sim \log\frac{1}{\sigma}$$

for the two-frequency quasiperiodic attractor and like

$$\mathcal{N}(\sigma) \sim \sigma^{-\alpha}, \qquad (16)$$

for the strange attractor, with $\alpha^{-1} = 1 - \beta$ (these results can be obtained by solving $\sigma \sim e^{-\nu\mathcal{N}}$ and $\sigma \sim \mathcal{N}^{-1/\alpha}$ for $\mathcal{N}$, respectively). In the case of the three-frequency quasiperiodic attractor the function $\hat{\phi}$ introduced in (12) is smooth and therefore the Fourier coefficients $\overline{C}_{pq}$ which appear in (14) should behave like $\overline{C}_{pq} \sim \exp[-\nu(p^2 + q^2)^{1/2}]$ as $p^2 + q^2 \to \infty$; hence

$$\mathcal{N}(\sigma) \sim \left(\log\frac{1}{\sigma}\right)^2,$$

for the three-frequency quasiperiodic attractor. In

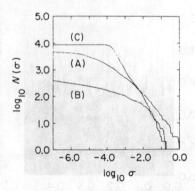

Fig. 5. Spectral distributions of the attractors of (A) fig. 3a; (B) fig. 3b; (C) fig. 3c.

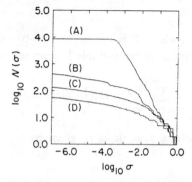

Fig. 6. Variation of the spectral distribution across the transition from two-frequency quasiperiodic to strange behavior. The curves are for $V = 0.55$ and the following values of K: (A) 1.346; (B) 1.347; (C) 1.35; (D) 1.39 [$\varepsilon = 0.0$].

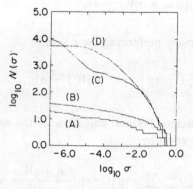

Fig. 7. Variation of the spectral distribution across the transition from two-frequency quasi-periodic to three-frequency quasiperiodic behavior. The curves are for $V = 0.55$ and the following values of K: (A) 1.49; (B) 1.536; (C) 1.537; (D) 1.54 [$\varepsilon = 0.0$].

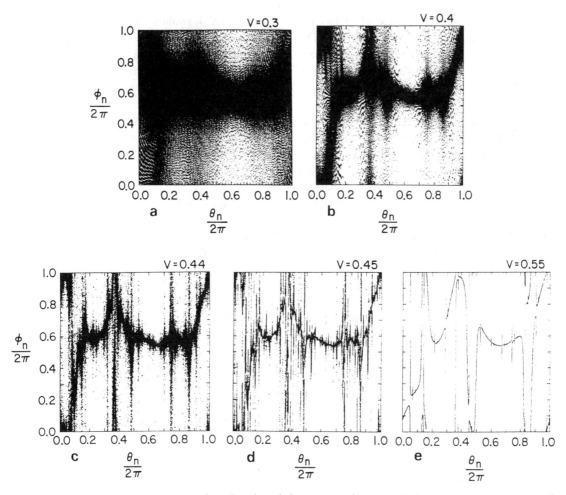

Fig. 8. Surface of section plots of attractors for points situated along a curve of constant winding number $W = 0.9277...$ on the K-V-plane: a) $V = 0.30$, $K = 1.331752$; b) $V = 0.40$, $K = 1.329461$; c) $V = 0.44$, $K = 1.331199$; d) $V = 0.45$, $K = 1.331790$; e) $V = 0.55$, $K = 1.34$. Figs. 8a, 8b correspond to three-frequency quasiperiodic attractors while figs. 8c, 8d, 8e correspond to strange attractors [$\varepsilon = 0.0$].

fig. 5 we have plotted the spectral distributions for the three attractors of fig. 3. The results seem to confirm the above asymptotic predictions. In the case of the strange nonchaotic attractors this and other plots of $\mathcal{N}(\sigma)$ indicate that $1 < \alpha < 2$. Two remarks are in order at this point: (i) the levelling out shown by the curves (A) and (C) for small σ is due to the use of a finite number of points in the calculation of the transform; increasing M causes the levelling out to occur at smaller σ; (ii) the use of the previously mentioned Hanning smoothing is quite important in obtaining meaningful results for $\mathcal{N}(\sigma)$ with time series of reasonable length.

The spectral distribution $\mathcal{N}(\sigma)$ should vary continuously with the parameters. In order to illustrate this we have plotted $\mathcal{N}(\sigma)$ for a constant value of V ($V = 0.55$) and different values of K located near the edges of a plateau ($W = 1.0$). Fig. 6 illustrates the transition from two-frequency quasiperiodic (curves (D), (C), (B)) to strange behavior (curve (A)). We see that the change of $\mathcal{N}(\sigma)$ across the transition point (from (B) to (A)) is much greater than the change of $\mathcal{N}(\sigma)$ for points on the plateau (for example (B) and (D)) even though the change from (B) to (A) corresponds to a much smaller variation in K (1 part in 1346 for the first case; 43 parts in 1347 for the second). Fig. 7 illustrates the transition from two-frequency quasiperiodic (curves (A), (B)) to three-frequency quasiperiodic behavior (curves (C), (D)). Again the change of $\mathcal{N}(\sigma)$ across the transition point (from (B) to (C)) is much greater and more rapid than the change corresponding to variations of K within the plateau (cf. figure caption for K values).

In order to illustrate the important transition from three-frequency quasiperiodic to strange behavior we have followed the evolution of the attractors along a curve of constant winding number in the KV-plane (the curve used, $W = 0.9277\ldots$, is plotted as curve (W) in fig. 1). As V is increased the three-frequency quasiperiodic attractor undergoes a transition to a strange attractor at some critical value $V = V_c(W)$. From the evidence of the numerical results (both the surface of section plots and the spectra) the transition seems to occur somewhere in the interval $0.42 < V < 0.44$. [Note that the Lyapunov exponent does not help much in finding the transition very accurately as it is always very small over the interval $0.42 < V < 0.44$ ($|\Lambda| < 10^{-5}$).]

In figs. 8(a, b, c, d, e) we have plotted the surface of section plots of the attractors for several pairs of (K, V) values along the constant winding number curve. In fig. 9 we have plotted the spectral distributions of some of these attractors. One point that is worth emphasizing is that the exponent introduced in eq. (16) has a value close to 2 near the transition point which then rapidly decreases to 1 as one moves away from this point. A heuristic argument for why α approaches 1 is given in ref. 11.

3. The case $\varepsilon \neq 0$

Eq. (2) with $\varepsilon = 0$ is special in the sense that it is related to the linear Schrödinger equation with quasiperiodic potential by Prüfer's transformation. An interesting question to be asked is to what extent do equations of the more general form (4) exhibit behavior similar to that of eq. (2) with $\varepsilon = 0$. In order to answer this question we have considered eq. (2) with $\varepsilon \neq 0$. In all the numerical experiments reported here we have taken $\varepsilon = 0.2$.

Fig. 10 shows a diagram of the KV-plane giving regions where Λ is negative (hatched) or zero (blank): the criterion for negative Lyapunov exponent is $\Lambda < -10^{-4}$. Fig. 11 shows Λ and W versus K at a fixed value of V. The numerical procedure used to obtain these results for $\varepsilon = 0.2$ is the same as used for the $\varepsilon = 0$ case, except that the grid in the K-V-plane was taken with 241 values of K.

These figures are qualitatively similar to those found in the case $\varepsilon = 0$. An important difference is the more detailed structure corresponding to the appearance of a larger number of tongues and plateaus where the winding number is fixed. This is due to the fact that for $\varepsilon \neq 0$ the plateaus occur at winding numbers

$$W = \frac{l}{n}\omega_1 + \frac{m}{n}\omega_2, \tag{17}$$

where l, m, and n are integers, whereas for $\varepsilon = 0$ the plateaus occur at $W = l\omega_1 + m\omega_2$, cf. eq. (11). This will be demonstrated analytically using perturbation theory in section 4.

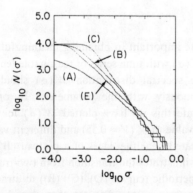

Fig. 9. Spectral distributions of the attractors of (A) fig. 3a; (B) fig. 3b; (C) fig. 3c; (E) fig. 3e.

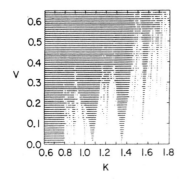

Fig. 10. Diagram of the K-V-plane showing regions where $\Lambda < 0$ (hatched) or $\Lambda = 0.0$ (blank) [$\varepsilon = 0.2$].

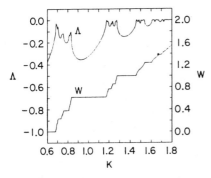

Fig. 11. Curves of the Lyapunov exponent (Λ) and the winding number (W) versus K at $V = 0.55$ [$\varepsilon = 0.2$].

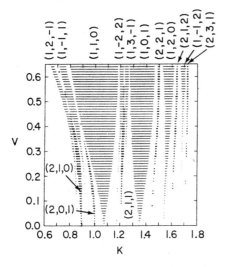

Fig. 12. Diagram of the K-V-plane showing the most important resonances. The resonances are identified by the triplet (n, l, m) (cf. eq. (17)) [$\varepsilon = 0.2$].

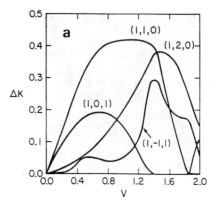

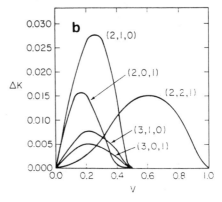

Fig. 13. Curves giving the width of several resonances as a function of V: a) $n = 1$; b) $n = 2, 3$ [$\varepsilon = 0.2$].

In order to better illustrate the appearance of these resonances for $n \neq 1$ we have plotted in fig. 12 a diagram of the K-V-plane showing the position of the most prominent resonances for $n = 1, 2$, identified by the triplets (n, l, m). One interesting feature of some of the $n = 2$ resonances is that their width reaches a maximum at an intermediate value of V and then decreases. Actually this behavior is also exhibited by the other resonances, although for larger values of V. This can be seen in fig. 13(a, b) where we have plotted the width of several resonances as a function of V. This figure also shows that at least one of the resonances, $(1, 0, 1)$, reappears again for larger values of V.

Again the system exhibits three different types of attractors: three-frequency quasiperiodic, two-frequency quasiperiodic and strange nonchaotic.

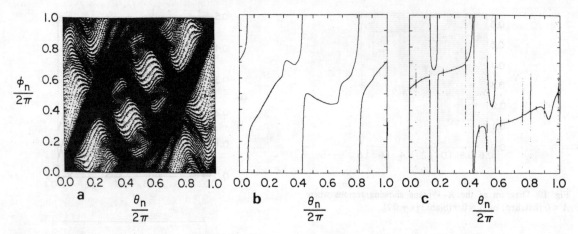

Fig. 14. Surface of section plots of a) a three-frequency quasiperiodic attractor ($V = 0.55$, $K = 1.53$); b) a two-frequency quasi-periodic attractor ($V = 0.55$, $K = 1.25$); and c) a strange nonchaotic attractor ($V = 0.55$, $K = 0.83$) [$\varepsilon = 0.2$].

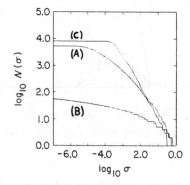

Fig. 15. Spectral distributions of the attractors of (A) fig. 14a; (B) fig. 14b; (C) fig. 14c.

Table I, giving the correspondence between the different types of attractors, the Lyapunov exponent and the winding number for eq. (2) with $\varepsilon = 0$, is valid in the more general case, except that eq. (11) has to be replaced by eq. (17) and the analogy to the Schrödinger equation does not apply. In figs. 14(a, b, c) we have plotted surface of section plots for each type of attractor.

In fig. 15 we have plotted the spectral distributions of the attractors to which refer the plots of fig. 14. These distributions seem to follow the same laws of variation obtained in the $\varepsilon = 0$ case.

One direct consequence of eq. (17) is that the two-frequency quasiperiodic attractor is now multivalued in the surface of section; the degree of multiplicity is given by n, provided that l, n and m, n are taken to be relatively prime integers. An example of a two-frequency quasiperiodic attractor with $n = 1$ was given in fig. 14b. An example of an attractor with $n = 2$ is given in fig. 16.

Figs. 10 and 11 seem to indicate that the measure of the Cantor set where strange nonchaotic attractors occur is reduced in relation to the case $\varepsilon = 0$. Is this measure still positive or is it zero? In order to obtain relevant evidence on this question we have performed the following numerical experiment. For $\varepsilon = 0.2$ we have taken the set of K values

$$\{ K^{(i)} = 0.685 + 0.001(i-1), \quad i = 1, 146 \},$$

which lie between the two widest plateaus $W = 0.0$ and $W = \omega$ (the points $K = 0.684$ and $K = 0.831$ are already on these plateaus, respectively), and for each of these values we calculated the winding numbers of the orbits with parameters

$$K^{(i)} - \Delta, \quad K^{(i)}, \quad K^{(i)} + \Delta,$$

by integrating eq. (2) over $N = 10^5$ driver periods,

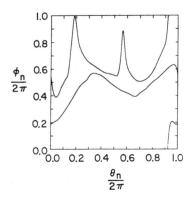

Fig. 16. Surface of section plot of a multivalued ($n = 2$) two-frequency quasiperiodic attractor ($V = 0.55$, $K = 1.225$) [$\varepsilon = 0.2$].

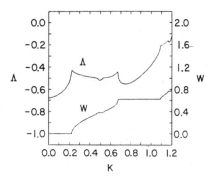

Fig. 17. Curves of the Lyapunov exponent (Λ) and the winding number (W) versus K at $V = 1.0$ [$\varepsilon = 0.2$].

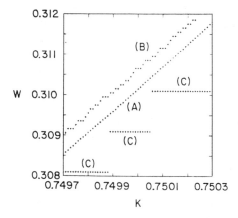

Fig. 18. Curves of W vs. K in the neighborhood of $K = 0.75$ for $V = 0.55$ and the following values of N: (A) 10^5; (B) 10^4; (C) 10^3. Note that the curves (B) and (C) were shifted by ± 0.0005 in the vertical direction for clarity [$\varepsilon = 0.2$].

with $\Delta = 10^{-5}$. When at least two of these three winding numbers are equal we say that K is on a plateau while if they are different we say that K is on the Cantor set. By proceeding in this way we found that $36/146 \simeq 24.7\%$ of points are on the Cantor set. Repeating this study for other small Δ values yields similar results; in particular for $\Delta = 4 \times 10^{-5}$ and 16×10^{-5} we obtain $36/146 \simeq 24.7\%$ and $38/146 \simeq 26.0\%$, respectively. Thus we believe that $\simeq 25\%$ of the measure of the interval $K \in [0.685, 0.830]$ is on the Cantor set where eq. (17) is not satisfied. This result seems to indicate that, although the measure of the Cantor set is reduced as ε becomes different from zero, it remains positive. Two observations are in order. The first has to do with what we mean by equal and different winding numbers; we found that the distinction between the two cases is always very sharp; for points on the plateaus the difference between the calculated winding numbers is always less than 10^{-8} while for points on the Cantor set it is always larger than 5×10^{-5}. The second observation is related to the necessity of integrating the differential equation over a large number of driver periods; if this number is not large enough, spurious plateaus, at a distance in W of approximately $1/N$ from each other, tend to appear; this numerical phenomenon is illustrated in fig. 18 where we have plotted W versus K in the neighborhood of $K = 0.75$ (which is a point on the Cantor set) for different values of N.

Our conclusions regarding the measure of the Cantor set where strange nonchaotic attractors occur are based on the results obtained at $V = 0.55$. If we increase V then this measure also seems to increase; that is, it is easier to find strange nonchaotic attractors for larger values of V. This can be seen by comparing fig. 17, which shows curves of Λ and W versus K at $V = 1.0$, with fig. 11 (note that the horizontal scale is the same in both figures, although the K values are taken to be different so that the interval between the two plateaus at $W=0$ and $W=\omega$ is included in both figures).

Towards the end of the paper by Bondeson et al. [11] some preliminary calculations on the $\varepsilon \neq 0$ case were mentioned and, on their basis, it was speculated that "the measure of the Cantor set with strange attractors seems to be zero or at least is very small". As indicated by our discussion above, we now believe that this is not so; the previous statement may have been due to the presence of apparent, but not real, plateaus which occur as a result of the finite-length-orbit-effect illustrated in fig. 18. Nevertheless, while our present numerical results are strongly suggestive, in the absence of rigorous analytical results, there is room for uncertainty concerning this issue.

To conclude this section, we reiterate the main conclusions implied by our $\varepsilon \neq 0$ numerical experiment: (i) the strange nonchaotic attractors are characterized by $\mathcal{N}(\sigma) \sim \sigma^{-\alpha}$; and (ii) the strange nonchaotic attractors apparently exist on a Cantor set of positive measure in parameter space. These conclusions are the same as those for $\varepsilon = 0$ (section 2). There are, however, other aspects, namely the greater profusion of resonances for $\varepsilon \neq 0$, for which the cases $\varepsilon = 0$ and $\varepsilon \neq 0$ are fundamentally different from each other. In the next section this will be demonstrated analytically using perturbation theory.

4. Perturbation theory

In this section we use perturbation theory to study the phase-locked solutions of eq. (2) for small values of the parameter V. Actually our perturbation analysis is slightly more general and applies to differential equations of the form

$$\frac{d\phi}{dt} = h(\phi) + K + \mu v(t), \qquad (18)$$

where h is 2π-periodic and v is two-frequency quasiperiodic with incommensurate frequencies ω_1 and ω_2. μ is an ordering parameter.

In order to solve (18) for small μ we assume that

$$\phi = \phi_0 + \mu \phi_1, \quad K = K_0 + \mu K_1,$$

and introduce a new independent variable

$$\tau = t - t_0(\mu t),$$

where t_0 is a slowly varying function of t. Substituting this ansatz into (18) and equating the coefficients of equal powers of μ we obtain, to $\mathcal{O}(\mu^0)$,

$$\frac{d\phi_0}{d\tau} = h(\phi_0) + K_0, \qquad (19)$$

and to $\mathcal{O}(\mu^1)$,

$$\frac{d\phi_1}{d\tau} - h'(\phi_0)\phi_1 = t_0'(\mu t)\frac{d\phi_0}{d\tau} + K_1 + v(\tau + t_0). \qquad (20)$$

The solution of (19) is given implicitly by

$$\tau = \int_0^{\phi_0(\tau)} \frac{d\phi}{h(\phi) + K_0},$$

assuming that $h(\phi) + K_0 > 0$. This solution can be expressed in the form

$$\phi_0(\tau) = W_0 \tau + \phi_{01}(W_0 \tau),$$

where ϕ_{01} is 2π-periodic and $W_0 = 2\pi/T_0$ with

$$T_0(K_0) = \int_0^{2\pi} \frac{d\phi}{h(\phi) + K_0}.$$

The function t_0 will be chosen in such way that the solution of (20) is not secular. With this choice the winding number W of the solution $\phi(t)$ of (18) is given by

$$W = \lim_{t \to \infty} \frac{\phi(t)}{t} = W_0 \left(1 - \lim_{t \to \infty} \frac{t_0(\mu t)}{t}\right). \qquad (21)$$

The locked solutions will therefore occur if it is further possible to choose $t_0(\mu t)$ such that $\lim_{t \to \infty} t_0(\mu t)/t = 0$.

Eq. (20) can be rewritten in the form

$$\frac{d}{d\tau}[\phi_1 I(\tau)] = t_0^1(\mu t) + I(\tau)[K_1 + v(\tau + t_0)], \tag{22}$$

where

$$I(\tau) = [h(\phi_0(\tau)) + K_0]^{-1}.$$

In order to proceed let us assume that the integrating factor I has the Fourier series expansion

$$I(\tau) = \sum_{n=-\infty}^{+\infty} I_n e^{inW_0\tau}, \tag{23}$$

while the forcing function v has the double Fourier series expansion

$$v(t) = \sum_{l,m=-\infty}^{+\infty} V_{l,m} e^{i(l\omega_1 + m\omega_2)t}. \tag{24}$$

Without loss of generality we take $V_{0,0} = 0$ (any non-zero $V_{0,0}$ can always be included in K_0). Substituting (23) and (24) into (22) we obtain, after some slight rearrangement,

$$\frac{d}{d\tau}[\phi_1 I(\tau)] = t_0'(\mu t) + K_1 I_0 + K_1 \sum_{\substack{n=-\infty \\ n \neq 0}}^{+\infty} I_n e^{inW_0\tau}$$

$$+ \sum_{n,l,m=-\infty}^{+\infty} I_n V_{l,m} e^{i(l\omega_1 + m\omega_2)t_0} e^{i(nW_0 + l\omega_1 + m\omega_2)\tau}. \tag{25}$$

From now on two cases have to be considered separately.

As the first case we assume that $nW_0 + l\omega_1 + m\omega_2 \neq 0$, for all n, l, m. Then the condition for the solution of (25) not to have a secular term is

$$t_0'(\mu t) + K_1 I_0 = 0,$$

which, on integration, gives

$$t_0(\mu t) = -K_1 I_0 \mu t + t_{00},$$

where t_{00} is a constant. If this condition is satisfied the solution of (25) will have the form

$$\phi_1(\tau) = \phi_{11}(W_0\tau) + \phi_{12}(W_0\tau, \omega_1\tau, \omega_2\tau),$$

where ϕ_{11} and ϕ_{12} are 2π-periodic functions of their arguments. Putting these results together we see that the solution of (18) is, except for the secular term $W_0(t - t_0)$, three-frequency quasiperiodic. The corresponding winding number is

$$W = W_0(1 + \mu K_1 I_0).$$

Noting that I_0, the average value of the integrating factor, can be expressed in terms of W_0 by

$$I_0 = \frac{1}{W_0(K_0)} \frac{dW_0(K_0)}{dK_0},$$

this expression simply indicates that for small variations μK_1 around K_0 the winding number is given by the first two terms of a Taylor expansion.

As the second case we assume there are integer triplets $(\hat{n}, \hat{l}, \hat{m})$ for which the resonance condition

$$\hat{n}W_0 + \hat{l}\omega_1 + \hat{m}\omega_2 = 0, \tag{26}$$

is satisfied. Then the condition for the solution of (25) not to have a secular term is

$$t_0'(\mu t) + K_1 I_0 + R(W_0 t_0) = 0, \tag{27}$$

where

$$R(W_0 t_0) = \sum_{\substack{\hat{n},\hat{l},\hat{m}=-\infty \\ \hat{n}W_0 + \hat{l}\omega_1 + \hat{m}\omega_2 = 0}}^{+\infty} I_{\hat{n}} V_{\hat{l},\hat{m}} e^{-i\hat{n}W_0 t_0},$$

is the resonant part of $I(\tau)v(\tau + t_0)$. Eq. (27) is a first order differential equation for t_0; as R is 2π-periodic its solution will satisfy $\lim_{t \to \infty} t_0(\mu t) = L < \infty$ provided

$$\min R(W_0 t_0) < -K_1 I_0 < \max R(W_0 t_0), \tag{28}$$

where L is such that

$$I_0 K_1 + R(W_0 L) = 0.$$

From these results we conclude that in this second case (18) will have two-frequency quasiperiodic solutions with winding number $W = W_0$, where W_0 satisfies (26), provided that the parameter K remains sufficiently close to K_0 so that (28) is satisfied. Note that condition (28) implies that the width (in K space) of the phase locked regions scales as the magnitude of $R(W_0 t_0)$.

The perturbation results that we have just described are valid for (18) with arbitrary 2π-periodic h and two-frequency quasiperiodic v. Let us now analyse the two special cases (i) $h(\phi) = \cos\phi$, for which (18) can be reduced to the Schrödinger equation, and (ii) $v(t) = K(\cos\omega_1 t + \cos\omega_2 t)$, as considered in the numerical work described in the present paper.

In case (i) it can be shown that the integrating factor for (20) has the form

$$I(\tau) = \frac{1}{W_0^2} \left[K_0 + \sin(W_0 \tau + \varphi) \right].$$

where $W_0(K_0) = (K_0^2 - 1)^{1/2}$ and φ is a constant. That is, I is monochromatic with frequency W_0. If we follow through the perturbation analysis we verify that in this special case the resonance condition (26) takes the form

$$W_0 + \hat{l}\omega_1 + \hat{m}\omega_2 = 0,$$

in agreement with our results of section 2.

In case (ii), if we again follow through the analysis, we verify that the resonance condition can only take one of the two particular forms

$$\hat{n} W_0 \pm \omega_1 = 0, \quad \hat{n} W_0 \pm \omega_2 = 0.$$

That is, to $\mathcal{O}(\mu^1)$, only resonances of one of these two forms can occur. Note, however, that resonances of the general form (26) can still be found in this case by taking the perturbation theory to $\mathcal{O}(\mu^{\hat{l}+\hat{m}})$.

5. Conclusions

We have discussed the existence and properties of strange nonchaotic attractors exhibited by the first order ordinary differential equation with two-frequency quasiperiodic forcing, eq. (2). The following are the most important conclusions of our study:

i) For a fixed value of V the curve giving the winding number W as a function of K is a "devil's staircase": a continuous, non-decreasing curve with a dense set of open intervals on which the winding number is constant; between these intervals there is a Cantor set of apparently positive Lebesgue measure on which the winding number increases with K.

ii) On the intervals the Lyapunov exponent is always negative while on the Cantor set it is either negative (above the critical curve) or zero (below the critical curve).

iii) On the intervals the equation exhibits two-frequency quasi-periodic attractors. On the Cantor set the equation exhibits either three-frequency quasiperiodic attractors (when $\Lambda = 0$) or strange nonchaotic attractors (when $\Lambda < 0$).

iv) The frequency spectra of the three types of attractors have in general clearly distinctive characteristics. If we introduce a spectral distribution $\mathcal{N}(\sigma)$ giving the number of spectral components larger than some value σ, then we have: $\mathcal{N}(\sigma) \sim \sigma^{-\alpha}$, $1 < \alpha < 2$, for strange nonchaotic attractors; $\mathcal{N}(\sigma) \sim \log(1/\sigma)$, for two-frequency quasiperiodic attractors; $\mathcal{N}(\sigma) \sim \log^2(1/\sigma)$, for three-frequency quasiperiodic attractors.

v) The equation we have studied can be considered as a strong damping model of the driven pendulum and Josephson junction. Our present results, namely those related to the form of the frequency spectrum of the attractors, indicate that it should be possible in experiments with these physical devices to identify the strange nonchaotic attractors via an $\mathcal{N}(\sigma)$ diagnostic. This is even made more plausible by the fact that these attractors seem to exist on a set of positive measure.

Finally we note that the qualitative conclusions numerically obtained for eq. (2) are expected to hold for general first order quasiperiodically forced equations of the form $d\phi/dt = g_\eta(\phi, t)$, where η represents parameters and the explicit t dependence of g_η is quasiperiodic. In particular, our results suggest that the existence of strange nonchaotic attractors on a positive measure in parameter space should apply. [We emphasize, however, that for $\varepsilon \neq 0$ our evidence for the existence of strange nonchaotic attractors on a set of positive measure is purely numerical. A rigorous proof confirming (or refuting) the numerical evidence remains a challenging problem for future study.]

Acknowledgements

This work was supported by the U.S. Department of Energy, Office of Basic Energy Sciences (Applied Mathematics Program) and the Office of Naval Research. Filipe Romeiras was also supported by the Portuguese Instituto Nacional de Investigação Científica during his sabbatical leave from the Instituto Superior Técnico of the Universidade Técnica de Lisboa.

References

[1] C. Grebogi, E. Ott, S. Pelikan and J.A. Yorke, Physica 13D (1984) 261.
[2] P. Grassberger, J. Stat. Phys. 26 (1981) 173.
[3] J.P. Sethna and E.D. Siggia, Physica 11D (1984) 193.
[4] K. Kaneko, Prog. Theor. Phys. 71 (1984) 282.
[5] E.G. Gwinn and R.M. Westervelt, Phys. Rev. A 33 (1986) 4143.
[6] R.L. Kautz and J.C. MacFarlane, Phys. Rev. A 33 (1986) 498.
[7] F.J. Romeiras and E. Ott, Phys. Rev. A (1987) to be published.
[8] R. Johnson and J. Moser, Commun. Math. Phys. 84 (1982) 403.
[9] B. Simon, Adv. Appl. Math. 3 (1982) 463.
[10] B. Souillard, Phys. Rep. 103 (1984) 41.
[11] A. Bondeson, E. Ott and T.M. Antonsen Jr., Phys. Rev. Lett. 55 (1985) 2103.
[12] G. Benettin, L. Galgani, A. Giorgilli and J.-M. Strelcyn, Meccanica 15 (1980) 21.
[13] E.O. Brigham, The fast Fourier transform (Prentice-Hall, Englewood Cliffs, NJ, 1974).
[14] G.E. Powell and I.C. Percival, J. Phys. A: Math. Gen. 12 (1979) 2053.
[15] W.P. Petersen, Commun. ACM 26 (1983) 1008.
[16] R.L. Devaney, An introduction to chaotic dynamical systems (Benjamin/Cummings, Menlo Park, CA, 1986).
[17] S. Aubry and G. Andre, Ann. Isr. Phys. Soc. 3 (1980) 133.
[18] C. Grebogi, E. Ott and J.A. Yorke, Physica 15D (1985) 354.

DIMENSIONS OF STRANGE NONCHAOTIC ATTRACTORS

Mingzhou DING [1], Celso GREBOGI and Edward OTT [1,2]
Laboratory for Plasma Research, University of Maryland, College Park, MD 20742, USA

Received 3 October 1988; accepted for publication 15 March 1989
Communicated by A.P. Fordy

Strange nonchaotic attractors in two-dimensional maps exhibit zero Lyapunov exponent along one direction and negative along the other. Evidence is presented indicating that the capacity dimension of these attractors is two while their information dimension is one.

1. Introduction

It was shown in refs. [1-4] that a typical feature of two frequency quasiperiodically forced dynamical systems is the existence of strange nonchaotic attractors. These attractors are not a finite set of points, or a smooth curve or surface, or a volume bounded by a piece-wise smooth surface. Hence, they are strange geometrically. On the other hand, a typical orbit on a strange nonchaotic attractor has nonpositive Lyapunov exponents. Therefore, they are nonchaotic in the sense that they do not exhibit sensitive dependence on initial conditions. It was also shown in refs. [1-4] that strange nonchaotic attractors can display distinct dynamical and Fourier spectral properties, and that they appear to be typical for quasiperiodically forced systems. Here by "typical" we mean that these attractors occur on a set of positive measure in the parameter space. That is, if we pick a point at random in the parameter space the system has nonzero probability to yield such an attractor.

In this paper we discuss the metric properties of strange nonchaotic attractors in phase space. Two important quantities for characterizing such properties are the capacity (box-counting) dimension D_0

[1] Department of Physics and Astronomy, University of Maryland, College Park, MD 20742, USA.
[2] Department of Electrical Engineering, University of Maryland, College Park, MD 20742, USA.

and information D_1 (see ref. [5] for more details). These two dimensions can be defined in the following way. Consider a given compact set S and assume that to cover set S we need $N(r)$ cubes of edge length r and that for each cube i we have a probability $p_i(r)$ associated with it. (For the case of attractors p_i is identified as the fraction of time a typical orbit spends in the ith cube.) Then the capacity dimension D_0 of S is defined as

$$D_0 = \lim_{r \to 0} \frac{\log N(r)}{|\log r|}, \qquad (1)$$

and the information dimension D_1 is defined as

$$D_1 = \lim_{r \to 0} \frac{H(r)}{\log r}, \qquad (2)$$

where

$$H(r) = \sum_{i=1}^{N(r)} p_i(r) \log p_i(r).$$

The dynamical systems we study in this paper are systems which are driven at two incommensurate frequencies. As mentioned at the beginning of the paper such systems can exhibit strange nonchaotic attractors. A typical example of such systems is the following pendulum equation [3]:

$$\frac{d^2\phi(t)}{dt^2} + \nu \frac{d\phi(t)}{dt} + g \sin \phi(t)$$
$$= F + G \sin(\omega_1 t + \alpha_1) + H \sin(\omega_2 t + \alpha_2),$$

where ω_1 and ω_2 are incommensurate. If we sample the system at time intervals corresponding to one of the driving frequencies, say, $\omega_1 t_n = 2\pi n$, we obtain a discrete map for the variables ϕ and $\theta = \omega_2 t$. The form of this discrete map is

$$\phi_{n+1} = F(\phi_n, \theta_n), \qquad (3)$$

$$\theta_{n+1} = [\theta_n + 2\pi\omega], \qquad (4)$$

where $\omega = \omega_1/\omega_2$ is irrational and we henceforth use the square bracket to indicate that modulo 2π of the enclosed expression is taken. Past works [1–4] show that strange nonchaotic attractors realized in the θ–ϕ phase space exhibit zero Lyapunov exponent along the θ direction and negative Lyapunov exponent along the ϕ direction. Thus the Kaplan–Yorke formula [6] (which gives D_1 in terms of Lyapunov exponents) predicts that their information dimension is unity, and this is consistent with our numerical results. In this paper, however, we show that there is evidence indicating that their capacity (box-counting) dimension is much larger than one, and, in fact, we believe that it is two.

Instead of studying general two-frequency quasiperiodically forced systems, we focus on two dynamical systems of the form of (3) and (4) with different F functions. For the first system [1] we take

$$F(x_n, \theta_n) = 2a \tanh x_n \cos \theta_n$$

and obtain

$$x_{n+1} = 2a \tanh x_n \cos \theta_n, \qquad (5)$$

$$\theta_{n+1} = [\theta_n + 2\pi\omega], \qquad (6)$$

where a is a parameter and ω is irrational. It was shown in ref. [1] that the attractors of the system (5)–(6) are strange and nonchaotic for $|a| > 1$. For the second system we take

$$F(\phi_n, \theta_n) = [\phi_n + 2\pi K + V \sin \phi_n + C \cos \theta_n]$$

and obtain the following quasiperiodically forced circle map [4],

$$\phi_{n+1} = [\phi_n + 2\pi K + V \sin \phi_n + C \cos \theta_n], \qquad (7)$$

$$\theta_{n+1} = [\theta_n + 2\pi\omega], \qquad (8)$$

where K, V and C are three parameters and ω is irrational. It was shown in ref. [4] that the transition from quasiperiodicity to chaos is mediated by the existence of strange nonchaotic attractors for this system. We find that the first system allows a fairly convincing heuristic analytical treatment supporting our anticipation that $D_0 = 2$. The second system, however, has properties that we believe are perhaps more representative of systems encountered in practice.

The organization of this paper is the following. In section 2 we present an analytical analysis of the capacity dimension D_0 for the system of (5) and (6). In section 3 we discuss our numerical experiments on D_0 and D_1 for both systems. In section 4 we summarize the conclusions of this paper.

2. Analytical results for D_0

In this section we study the system of (5) and (6) and we define an attractor as the closure of the limit set of a typical orbit as time tends to $+\infty$. Consider the area: $0 < \theta < 2\pi$, $-\infty < x < \infty$ and apply eqs. (5) and (6) once to this area, we obtain the shaded area in fig. 1. It is shown in ref. [1] that for $|a| > 1$ the attractor is strange and that the orbit $x_n = 0$, $\theta_n = [\theta_0 + 2\pi n\omega]$ is unstable, but the line $x = 0$ is included in the attractor. Thus the attractor is the limit set obtained as we keep on mapping the shaded region [1]. To estimate the capacity dimension D_0 for this attractor, we take the following heuristic approach. Fig. 1 shows a strip of width 2ϵ symmetrically located with respect to the θ-axis. There are four

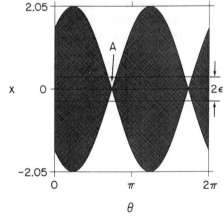

Fig. 1. The first image of the area: $0 \leq \theta \leq 2\pi$, $-\infty < x < \infty$ under the map (5)–(6).

identical blank wedges in the ϵ-strip. The successive applications of eqs. (5) and (6) map the tips of these four original wedges along the θ-axis ergodically. At each step, the number of wedges double and also the four original wedges widen. Thus the remaining shaded area inside the ϵ-strip is shrinking. We argue that the widening of the original wedges soon approaches a limiting slope (this is confirmed numerically). If ϵ is small we can approximate the edges of the wedges by straight lines, and the slope of limiting wedges, λ, is a monotonically increasing function of the parameter a. In what follows we proceed as if we started with four limiting wedges. At each step of the iteration, their images delete some shaded area from the ϵ-strip. We wish to show in what follows that the remaining area is still positive. If this is so then $D_0 = 2$ follows from the definition (1).

We now consider the effect on the shaded area of the ϵ-strip of the repeated iterations of the map. Noting the symmetry we only need to consider one wedge, say, wedge A. If ϵ is sufficiently small the edges of the wedges inside the strip are approximately straight lines. From the above discussion, the slope of the edges of wedge A is $\rho_0 = \lambda$, and the area below $x = \epsilon$ is ϵ^2/ρ_0; the slope of the edges of the first image of wedge A is $\rho_1 = \lambda |\lambda \cos(\pi/2 + 2\pi\omega)|$, and the area below $x = \epsilon$ is ϵ^2/ρ_1; ...; the slope of the edges of the nth image of A is

$$\rho_n = \lambda \left| \prod_{k=1}^{n} \lambda \cos(\pi/2 + 2\pi k\omega) \right|,$$

and the area below $x = \epsilon$ is ϵ^2/ρ_n.

We shall ignore the possibility that a wedge and one of its subsequent images overlap, and thus what we really get is an upperbound on the total area deleted from the ϵ-strip. This upperbound B constructed by the above procedure can be written explicitly as

$$B = \frac{\epsilon^2}{\lambda}(1 + \Sigma), \qquad (9)$$

where

$$\Sigma \equiv \sum_{n=1}^{\infty} \left| \prod_{k=1}^{n} \cos(\pi/2 + 2\pi k\omega) \right|^{-1} \lambda^{-n}. \qquad (10)$$

We wish to prove that the sum Σ converges. If this is so, then, since ϵ is small, the deleted area which is proportional to ϵ^2 by (9) is less than the area of the ϵ-strip which is $4\pi\epsilon$. Hence we would have $D_0 = 2$.

For many different combinations of λ and irrational ω we have tested the convergence of Σ numerically and found that it only depends on the parameter λ. There appears to be a critical $\lambda_c \approx 2.1$. For $\lambda > \lambda_c$ the sum Σ converges for all irrational ω that we tested, while it apparently diverges for $\lambda < \lambda_c$. To prove the convergence of Σ analytically, we assume that ω belongs to a special class of irrational numbers refered to as c-constant type irrational numbers [7]. In this class any irrational number ω has the following property: For any integers m and n,

$$|n\omega - m| \geq c/n,$$

where c is an constant for all irrational numbers in this class.

Consider the nth term

$$\Sigma_n = \left| \prod_{k=1}^{n} \cos(\pi/2 + 2\pi k\omega) \right|^{-1} \lambda^{-n}$$

in the sum Σ.

If the argument of the cos-function, $[\pi/2 + 2\pi k\omega]$, is very close to $\pi/2$ or $3\pi/2$, the cosine is very small and Σ_n might consequently be large. However, notice that $[\pi/2 + 2\pi k\omega]$ for $k = 1, 2, 3, ..., n$ fill the interval $[0, 2\pi]$ ergodically as n tends to ∞. So we can expect that only a small fraction of the k values yield arguments close to $\pi/2$ or $3\pi/2$. For large n, λ^n is also large, these small cos-values might typically not cause the series Σ to diverge. In what follows we quantify the above ideas.

Fig. 2 shows the function $|\cos\theta|$. Consider the distribution of $[\pi/2 + 2\pi k\omega]$ on the interval $[0, 2\pi]$ for $k = 1, 2, 3, ..., n$. Assume there are n_1 points in region 1, n_2 points in region 2, n_3 in region 3, n_4 in region 4, and $n_1 + n_2 + n_3 + n_4 = n$.

Let us see how close any of these n points can get to $\pi/2$. Since $[\pi/2 + 2\pi k\omega] = \pi/2 + 2\pi k\omega - 2\pi m$, where m is an integer, the distance between $[\pi/2 + 2\pi k\omega]$ and $\pi/2$ is

$$|\pi/2 + 2\pi k\omega - 2\pi m - \pi/2| \geq \frac{2\pi c}{k} \geq \frac{2\pi c}{n}. \qquad (11)$$

Similarly, the distance between $[\pi/2 + 2\pi k\omega]$ and $3\pi/2$ is

169

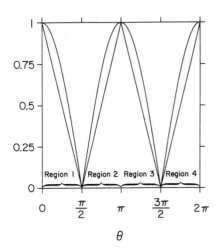

Fig. 2. Functions $|\cos\theta|$, $(2/\pi)|\theta-\pi/2|$, and $(2/\pi)|\theta-3\pi/2|$.

$$|\pi/2+2\pi k\omega-2\pi m-3\pi/2| \geq \frac{2\pi c}{4k} \geq \frac{2\pi c}{4n}. \quad (12)$$

The distance between any two points $[\pi/2+2\pi k_1\omega]$ and $[\pi/2+2\pi k_2\omega]$ is

$$|\pi/2+2\pi k_1\omega-2\pi m_1-\pi/2-2\pi k_2\omega+2\pi m_2|$$

$$\geq \frac{2\pi c}{|k_1-k_2|} \geq \frac{2\pi c}{n}. \quad (13)$$

Now consider the regions 1 and 2 in fig. 2 (because they share the same singular point $\theta=\pi/2$). Since $|\cos\theta| \geq (2/\pi)|\theta-\pi/2|$ in these two regions (the piece-wise linear curve in fig. 3), the inequalities (11)–(13) yield

$$\left|\prod_{k_i \in \text{region } j}^{n_j} \cos(\pi/2+2\pi k_i\omega)\right|\lambda^{n_j}$$

$$> (2/\pi)^{n_j}(2\pi c/n)^{n_j}n_j!\lambda^{n_j}$$

$$= (4\lambda c/n)^{n_j}n_j!, \quad (14)$$

where $j=1, 2$. Stirling's formula for $n!$ is [8]

$$n! = (2\pi n)^{1/2} n^n e^{-n+\theta/12n} \quad (0<\theta<1)$$

$$\geq (2\pi n)^{1/2} n^n e^{-n}.$$

For n large enough, the points are approximately evenly distributed on the four equal intervals. Thus

$$n_j/n < \tfrac{1}{5}, \quad j=1, 2, 3, 4.$$

Then substituting the above information in (14) we have

$$\left|\prod_{k_i \in \text{region } j}^{n_j} \cos(\pi/2+2\pi k_i\omega)\right|\lambda^{n_j}$$

$$> (4\lambda c/5e)^{n_j}(2\pi n_j)^{1/2}, \quad (15)$$

where $j=1, 2$. Along the same line of argument, we obtain a similar inequality for regions 3 and 4,

$$\left|\prod_{k_i \in \text{region } j}^{n_j} \cos(\pi/2+2\pi k_i\omega)\right|\lambda^{n_j}$$

$$> (\lambda c/5e)^{n_j}(2\pi n_j)^{1/2}, \quad (16)$$

where $j=3, 4$. Combining eqs. (15) and (16) we have

$$\left|\prod_{k=1}^{n} \cos(\pi/2+2\pi k\omega)\right|\lambda^{n}$$

$$> (\lambda c/5e)^n 4^{n_1+n_2}(2\pi)^2(n_1 n_2 n_3 n_4)^{1/2}$$

$$> (\lambda c/5e)^n. \quad (17)$$

Or

$$\Sigma_n < (5e/\lambda c)^n.$$

If $5e/\lambda c < 1$ or $\lambda > 5e/c$, then the summation of the terms $(5e/\lambda c)^n$ is a convergent geometric series, and thus Σ converges. Hence $D_0 = 2$.

As to the information dimension D_1, since both systems we consider in this paper are two-dimensional maps and the strange nonchaotic attractors exhibited by them have zero Lyapunov exponent along one direction and negative along the other, the Kaplan–Yorke formula [6] predicts that their information dimension is unity.

3. Numerical results for D_0 and D_1

In this section we present our numerical calculation on the dimensions D_0 and D_1 for the two systems (5)–(6) and (7)–(8). In all our numerical experiments we use the reciprocal of the golden mean for ω (cf. eqs. (6) and (8)), namely, $\omega=(\sqrt{5}-1)/2$. The first example we study is generated by (5) and (6) for $a=1.025$. The phase space plot of this strange nonchaotic attractor is given in fig. 3 where

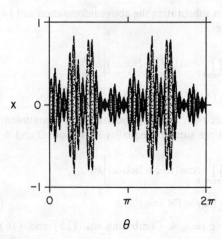

Fig. 3. The strange nonchaotic attractor generated by (5) and (6) for $a=1.025$.

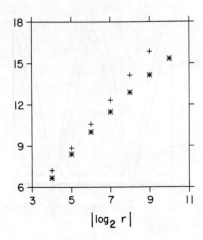

Fig. 4. Plots of $\log_2 N(r)$ versus $|\log_2 r|$ (plus signs) and $|\sum_{i=1}^{N(r)} p_i(r) \log p_i(r)|/\log 2$ versus $|\log_2 r|$ (stars) for the attractor shown in fig. 3.

40000 points are plotted after discarding the initial 4000 iterates (to eliminate the effect of transients). The box sizes we used to cover the attractor vary from $r=2^{-4}$ to $r=2^{-10}$. For every r we count the minimum number of boxes $N(r)$ that are needed to cover the whole attractor and compute

$$H(r) = \sum_{i=1}^{N(r)} p_i(r) \log p_i(r).$$

Fig. 4 shows the plots of $\log N(r)$ versus $|\log r|$ (plus signs) and $|\sum_{i=1}^{N(r)} p_i(r) \log p_i(r)|$ versus $|\log r|$ (stars) for the total of 10^9 iterates. Using the last three plus points and the last three star points yields a capacity dimension D_0 of approximately 1.8, and an information dimension D_1 of approximately 1.2. These values are quite uncertain since they are based on the last three points of their respective plots in fig. 4. In particular, the slopes calculated using all of the plotted points yield a smaller value for D_0 and a larger value for D_1, indicating an opposite curvature tendency in the range tested. It would thus be desirable to significantly extend the range of $\log r$. To do so would require many more iterates of the map. As it was the present calculation takes hours of CPU time on a Cray-2 supercomputer and large memory requirement. Nevertheless, the result presented above (fig. 4) indicates that D_1 is significantly less than D_0, in agreement with our conjecture that $D_0=2$ and $D_1=1$. Note that, in fig. 4, we neglect the last plus

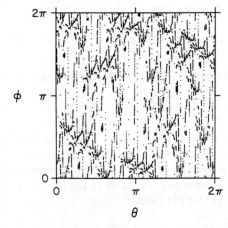

Fig. 5. The strange nonchaotic attractors generated by (7) and (8) for $C=1.2$, $V=0.95$ and $K=0.2841$.

point corresponding to $r=2^{-10}$. The reason is because if we add more orbit points to the total of 10^9 iterates the number of boxes corresponding to this box size still increases, while all the other plus points are saturated before we reach the total of 10^9 iterates.

The second example of strange nonchaotic attractors we study is generated by (7) and (8), the quasiperiodically forced circle map, for $C=1.2$, $V=0.95$ and $K=0.2841$. The phase space plot of this attractor is given in fig. 5 where 40000 points are plotted after discarding the initial 4000 points. The sizes of

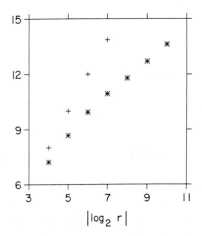

Fig. 6. Plots of $\log_2 N(r)$ versus $|\log_2 r|$ (plus signs) and $|\sum_{i=1}^{N(r)} p_i(r) \log p_i(r)|/\log 2$ versus $|\log_2 r|$ (stars) for the attractor shown in fig. 5.

boxes we use to cover the attractor range from $r=2^{-4}$ to $r=2^{-10}$. Fig. 6 shows the plots of $\log N(r)$ versus $|\log r|$ (plus signs) and $|\sum_{i=1}^{N(r)} p_i(r) \log p_i(r)|$ versus $|\log r|$ (stars). As we can see four plus points lie on a pretty good straight line corresponding to $D_0 = 1.95$. Our ability to obtain plus points for small r values was limited by lack of more reliable orbit points. The last four star points also lie on a good straight line corresponding to $D_1 \approx 0.9$. The first three star points correspond to large box sizes and they are apparently not in the scaling zone. Again our numerical results, while far from conclusive, do not contradict our conjecture that $D_0 = 2$ and $D_1 = 1$.

4. Conclusions

In this paper we discuss the characterization of strange nonchaotic attractors in phase space using their capacity dimension D_0 and information dimension D_1. Based on analytical arguments of section 2 we conjecture that $D_0 = 2$ for these attractors, and based on the Kaplan–Yorke formula [6] we conjecture that $D_1 = 1$. Our numerical experiments performed on two examples of strange nonchaotic attractors generated by two different systems support these conjectures.

Acknowledgement

This work was supported by the U.S. Department of Energy (Basic Energy Science) and the Office of Naval Research (Physics division).

References

[1] C. Grebogi, E. Ott, S. Pelikan and J.A. Yorke, Physica D 13 (1984) 261.
[2] A. Bondeson, E. Ott and T.M. Antonsen, Phys. Rev. Lett. 55 (1985) 2103;
F.J. Romeiras, A. Bondeson, E. Ott, T.M. Antonsen and C. Grebogi, Physica D 26 (1987) 277.
[3] F.J. Romeiras and E. Ott, Phys. Rev. A 35 (1987) 4404.
[4] M. Ding, C. Grebogi and E. Ott, to be published.
[5] J.D. Farmer, E. Ott and J.A. Yorke, Physica D 7 (1983) 153.
[6] J. Kaplan and J.A. Yorke, Functional differential equations and the approximation of fixed points (Springer, Berlin, 1978) p. 288.
[7] J. Belissard, R. Lima and D. Testard, Commun. Math. Phys. 88 (1983) 207.
[8] M. Abramovitz and R.A. Stegun, Handbook of mathematical functions (National Bureau of Standards, Washington, 1964) p. 257.

J. Phys. A: Math. Gen. **23** (1990) L383-L387. Printed in the UK

LETTER TO THE EDITOR

Route to chaos via strange non-chaotic attractors

T Kapitaniak†§, E Ponce†∥ and J Wojewoda‡§

† Department of Applied Mathematical Studies and Centre of Nonlinear Studies, University of Leeds, Leeds LS2 9JT, UK
‡ Division of Dynamics and Control, University of Strathclyde, James Weir Building, 75 Montrose St, Glasgow G1 1XJ, UK

Received 28 December 1989

Abstract. The route to chaos in quasiperiodically forced systems is investigated. It has been found that chaotic behaviour is obtained after breaking of three-frequency torus, but strange non-chaotic attractors are present before three-frequency quasiperiodic behaviour occurs.

Ruelle and Takens [1] suggested that strange attractors could arise after a finite sequence of Hopf bifurcations. Later it was specified by Newhouse, Ruelle and Takens [2] that after three bifurcations strange attractors could arise.

Recently it was found that there are two types of strange attractors [3-6].

The word 'strange' refers to the geometrical structure of the attractor and an attractor which is not:

a finite set of points;
a limit cycle (closed curve);
a smooth (piecewise smooth) surface;
bounded by a piecewise smooth closed surface volume;

is called a strange attractor. An attractor is chaotic if at least one Lyapunov exponent is positive (typically nearby orbits diverge exponentially with time). From what was said above, one finds that a strange non-chaotic attractor is an attractor which is geometrically strange, but for which typical orbits have non-positive Lyapunov exponents.

Ding et al [4] suggested that the route to chaos from two-frequency quasiperiodicity on a T^2 torus is via three-frequency quasiperiodicity on a T^3 torus and strange non-chaotic attractors. In this letter we investigate the Ruelle-Takens-Newhouse route to chaos in the systems with two-frequency quasiperiodic forcing, i.e. from a T^2 torus to chaos. The aim of this work is to show that the route

two-frequency quasiperiodicity → strange non-chaotic attractors

→ three-frequency quasiperiodicity

→ strange chaotic attractors   (1)

is possible and that the T^2 torus breaks before creation of the T^3 torus.

First consider van der Pol's oscillator:

$$\ddot{x} + d(x^2-1)\dot{x} + x = a \cos \omega t \cos \Omega t \qquad (2)$$

§ Permanent address: Institute of Applied Mechanics, Technical University of Lodz, Stefanowskiego 1/15, 90-924 Lodz, Poland.
∥ Permanent address: Department of Applied Mathematics, University of Sevilla, Avda. Reina Mercedes, 41012 Sevilla, Spain.

© 1990 IOP Publishing Ltd

L384   *Letter to the Editor*

where a, d, ω, Ω are constants. We considered $a = d = 5.0$, $\Omega = \sqrt{2} + 1.05$ and $\omega \in [0, 0.01]$. Equation (2) has four-dimensional phase space:

$$(x, \dot{x}, \Theta_1 = \omega t, \Theta_2 = \Omega t) \in R^2 \times S^1 \times S^1.$$

We can reduce the study of (2) to the study on an associated three-dimensional Poincaré map obtained by defining a three-dimensional cross section to a four-dimensional phase space by fixing the phase of one of the angular variables and allowing the remaining three variables that start on the cross section to evolve in time under the effect of the flow generated by (2) until they return to the cross section. If we fix the phase Θ_2, the Poincaré map is defined as a set:

$$M(t_0) = \{(x(t_n), \dot{x}(t_n), \Theta_1(t_n)) | t_n = 2\Pi n/\Omega + t_0, n = 1, 2, \ldots\}$$

where t_0 is initial time. To describe the surface of the Poincaré map we plot

$$x(t_n) \text{ against } \dot{x}(t_n).$$

Alternative surfaces can be obtained by plotting $x(t_n)$ against $\Theta_1(t_n)$, and $\dot{x}(t_n)$ against $\Theta_1(t_n)$ mod 2Π. Of course, to characterise the attractor we also used Lyapunov exponents given by:

$$\lambda = \lim_{t \to \infty} \{(1/t) \ln[d(t)/d(t_0)]\}$$

where $d = \sqrt{|y^2 + \dot{y}^2|}$ and y denotes the solution of the equation variational to (2).

The winding number for orbit $x(t)$ of (2) defined by the limit

$$w = \lim_{t \to \infty} \{(\alpha(t) - \alpha(t_0))/t\}$$

where $(x, \dot{x}) = (r \cos \alpha, r \sin \alpha)$ is another quantity.

The plot of the Lyapunov exponent against ω has been shown in figure 1 (the largest non-zero exponent has been taken). For two-frequency quasiperiodic behaviour we have a negative Lyapunov exponent and winding number fulfilling the relation:

$$w = (1/n)\omega + (m/n)\Omega \tag{3}$$

where l, m, n are integers. With further decrease of ω, the Lyapunov exponent is still negative but the winding number does not satisfy relation (3) and we have the example of a strange non-chaotic attractor. When the winding number does not satisfy relation (3) and the Lyapunov exponent is zero, three-frequency quasiperiodic behaviour is present (point F_3 in figure 1). No evidence of three-frequency quasiperiodic behaviour has been found in the transition from two-frequency quasiperiodic behaviour to the strange non-chaotic attractor. This type of behaviour is present on the boundary between strange non-chaotic behaviour and chaos. In figure 2(a)–(c) the Poincaré map is shown for two-frequency quasiperiodic, strange non-chaotic and strange chaotic behaviour. In the case of both strange attractors the Poincaré map has the same structure. In figure 2(a) we observe that the strange non-chaotic attractors do not exist on the T^2 torus as this figure shows that it is broken. We investigated 1200 different attractors of system (2) and in all examples we observed the same sequence (1) and the same properties of the Poincaré map as described in figure 2.

Finally consider the map:

$$\begin{aligned} \Phi_{n+1} &= [\Phi_n + 2\Pi K + V \sin \Phi_n + C \cos \Theta_n] \\ \Theta_{n+1} &= [\Theta_n + 2\Pi \omega] \end{aligned} \tag{4}$$

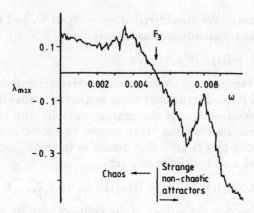

Figure 1. The largest non-zero Lyapunov exponent of equation (2) against ω; $a = d = 5$, $\Omega = \sqrt{2} + 1.05$.

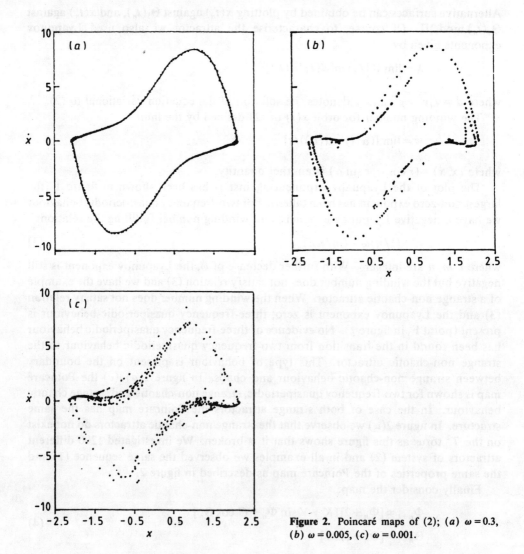

Figure 2. Poincaré maps of (2); (a) $\omega = 0.3$, (b) $\omega = 0.005$, (c) $\omega = 0.001$.

L386 *Letter to the Editor*

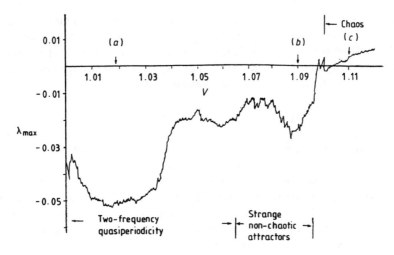

Figure 3. Lyapunov exponent (5) against V for map (4).

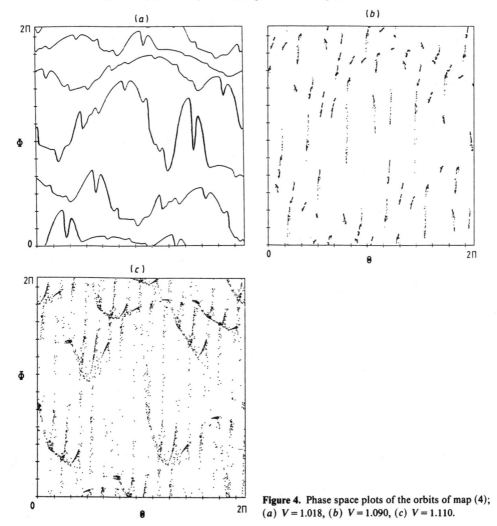

Figure 4. Phase space plots of the orbits of map (4); (a) $V = 1.018$, (b) $V = 1.090$, (c) $V = 1.110$.

where K, V, C and ω are constants, ω is irrational (in our numerical experiments the golden mean $\omega = (\sqrt{5}-1)/2$ has been taken) and the square brackets indicate that modulo 2Π of the expression is taken. In [4] it was shown that the transition from quasiperiodicity to chaos leads through the region where strange non-chaotic attractors are present. In figure 3 the plot of Lyapunov exponent:

$$\lambda = \lim_{n \to \infty} (1/n) \sum_{k=1}^{n} \ln|1 + V \cos \Phi_k| \tag{5}$$

is shown against V. Up to $V \approx 1.067$ condition (3) is fulfilled, but for larger values of V we observe the transition to strange non-chaotic attractors. Strange non-chaotic attractors occur again before three-frequency quasiperiodicity. In figure $4(a)-(c)$ phase space plots of the orbits corresponding to two-frequency quasiperiodicity, strange non-chaotic and chaotic behaviours are shown for the V values indicated in figure 3.

To summarise, in this letter we show that the possible route to chaos in two-frequency quasiperiodically forced systems is as follows: two-frequency quasiperiodicity → strange non-chaotic attractors → three-frequency quasiperiodicity → chaos.

References

[1] Ruelle D and Takens F 1971 *Commun. Math. Phys.* **20** 167
[2] Newhouse S, Ruelle D and Takens F 1978 *Commun. Math. Phys.* **64** 35
[3] Romieras F and Ott E 1987 *Phys. Rev.* A **35** 4404
[4] Ding M, Grebogi C and Ott E 1989 *Phys. Rev.* A **39** 2593
[5] Ding M, Grebogi C and Ott E 1989 *Phys. Lett.* **137A** 167
[6] Kapitaniak T and Wojewoda J 1990 Strange nonchaotic attractors of quasiperiodically forced van der Pol's oscillator *J. Sound and Vibration* in press

VIII. Spatial Chaos

The papers in this section show that chaotic behavior can be observed not only in time but also in space.

Paper VIII.1 is an attempt to define chaotic behavior in finite boundary value problems. A symbolic dynamics technique is used to show how the limiting chaos dominates the behavior of the finite boundary value problem.

The connection between generalized bifurcation into turbulence on one side and soliton and localized buckling of shells on the other side is conjectured in Paper VIII.2. It is pointed out that approximate homoclinic and heteroclinic solitons can be perturbed to produce spatial chaos.

Mathematical analysis of spatially complex behavior of elastic system is presented in Paper VIII.3 where the spatial aspects of equilibrium states exhibited by infinitely long rods, buckled by loads applied at their ends are analyzed. The analysis uses the Hamiltonian structure of equilibrium equations and Melnikov theory.

Classical static-dynamic analogies are used to demonstrate spatial chaos and localization of deformations in the elastica in Paper VIII.4.

Purely spatial chaos of loop and envelope solitons localization in long elastic strings is considered in Paper VIII.5. Possible connections to some problems in molecular biology are discussed.

Spatiotemporal dynamics of a one-dimensional chain of forced oscillators is investigated in Paper VIII.6. It describes phenomena characteristic for continuous space-time systems.

Chaos as a Limit in a Boundary Value Problem

Claus Kahlert and Otto E. Rössler

Institute for Physical and Theoretical Chemistry, University of Tübingen

Z. Naturforsch. **39 a**, 1200−1203 (1984); received November 8, 1984

A piecewise-linear, 3-variable autonomous O.D.E. of C^0 type, known to describe constant-shape travelling waves in one-dimensional reaction-diffusion media of Rinzel-Keller type, is numerically shown to possess a chaotic attractor in state space. An analytical method proving the possibility of chaos is outlined and a set of parameters yielding Shil'nikov chaos indicated. A symbolic dynamics technique can be used to show how the limiting chaos dominates the behavior even of the finite boundary value problem.

Reaction-diffusion equations occur in many disciplines [1, 2]. Piecewise-linear systems are especially amenable to analysis. A variant to the Rinzel-Keller equation of nerve conduction [3] can be written as

$$\frac{\partial}{\partial t} u = \frac{\partial^2}{\partial x^2} u + \mu[-u + v - \beta + \theta(u - \delta)],$$

$$\frac{\partial}{\partial t} v = -\varepsilon u + v, \qquad (1)$$

where $\theta(a) = 1$ if $a > 0$ and zero otherwise; δ is the threshold parameter.

Focusing on those solutions to (1), which are wave-trains of constant shape, one arrives at the following 3-variable ordinary differential equation (see [4, 5]):

$$u' = w,$$
$$v' = \frac{\varepsilon u - v}{c},$$
$$w' = \mu[u - v + \beta - \theta(u - \delta)] - cw. \qquad (2)$$

Here $' = d/ds$ whereby s represents the wave variable, $x - ct$.

As compared to (1), (2) contains an additional free parameter, the travelling speed c. It has to be chosen such that the corresponding solution of (2) becomes a solution of (1) − with the latter subjected to the specific boundary conditions assumed.

In an earlier note [5], (2) was shown to produce finite wave-trains of arbitrary finite pulse number (and pulse arrangement) if either cyclic or natural boundary conditions are assumed. This result was obtained by an analytical matching method. The connection to chaos theory (Smale [6] basic sets) was not evident at the time. In the following, even the possibility of manifest chaos will be demonstrated.

In Fig. 1, a chaotic attractor of (2) is presented. The flow can be classified as an example of "screw-type-chaos" (cf. [7]). A second example of a chaotic attractor is shown in Fig. 2. A 1-D projection of a 2-D cross section through the chaotic flow in a separating plane is also given (Fig. 2c). One-dimensional maps with a shape of this type generate chaos (cf. [8]). The (mathematically alone existing) attractor could nevertheless be an embedded limit cycle of high periodicity that is numerically inaccessible [8]. The presence of a narrow second dimension in the real map does not affect these conclusions (cf. [9]).

Equation (2), being piecewise linear, is also amenable to analytical methods, however. In the technique of "Poincaré half maps" [10], each half system is analyzed separately. Each half map is defined, from the "entry region" in the separating plane $u = \delta$ of one half system, into the corresponding "exit region". The two half maps "flush" (that is, the entry region of one is the other's exit region and vice versa) because (2) has unique trajectories. The resulting combined overall map (two coupled implicit algebraic equations [10]) is an admissible "ordinary" Poincaré map.

In the examples of chaos found so far in (2) − Figs. 1 and 2 −, either half system contains a saddle focus in state space whose 1-D manifold is unstable (contracting eigenvector) while the 2-D manifold is

---

Reprint requests to Dr. C. Kahlert, Institut für Physikalische und Theoretische Chemie der Universität Tübingen, Auf der Morgenstelle 8, D-7400 Tübingen.

C. Kahlert and O. E. Rössler · Chaos as a Limit in a Boundary Value Problem

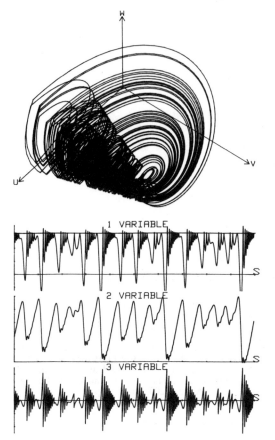

Fig. 1. Screw-type chaos in (2). Numerical solution of the initial value problem using a standard Runge-Kutta-Merson integration routine with step error 10^{-8}. Parameters chosen: $\beta = 1$, $\delta = 0.01$, $\varepsilon = 150$, $\mu = 10$, $c = 3.5$. Initial conditions: $u(0) = 0.009979$, $v(0) = 0.9285$, $w(0) = 0.06406$. Axes: -0.004 to 0.012 for u, 0.86 to 1.04 for v, -0.175 to 0.175 for w, 0 to 100 for s (upper trace), 0 to 30 (lower traces). The horizontal line in the first time plot indicates the threshold value $u = \delta$.

stable (containing the expanding focus). For the parameters chosen, only one of the two half systems possesses a "real" steady state of this type; the other has its steady state in the region of definition of the first one ("virtual" steady state) [5].

In the domain of the first half map one always finds a "distortion" (branch cut) that ends in a logarithmic spiral around a singular point [11]. This distortion generates a "splitting" in the half map [10]. The other half map in general does not undo this splitting. All one needs for chaos to occur is that part of an image of a neighborhood of the distortion is mapped back into the same neighborhood. In that case the overall Poincaré map contains at least one Smale [6] horseshoe. Smale horseshoes, in turn, imply "nontrivial basic sets" [6] and hence infinitely many periodic trajectories of differing periods — implying chaos (non manifest at least). If both half maps are non-expanding (non-positive divergence in (2)), manifest chaos is possible [11].

A special case within the above scenario is also analytically accessible. A trajectory that is "homoclinic" with respect to the saddle focus in state space is generated whenever the line of intersection of the two-dimensional manifold of the first half system with the separating plane is mapped back, by the second half map, in such a way that its image hits precisely the intersection point of the one-dimensional manifold of the first half system with the separating plane. Such homoclinicity to a saddle focus in state space was shown by Shil'nikov [12] to imply the formation of infinitely many Smale horseshoes (under a mild condition of the eigenvalues). One particular set of parameters for (2) yielding a situation that fulfills the conditions of Shil'nikov's theorem is as follows: $\beta = 1$, $\delta = 0.01$, $\varepsilon = 150$, $\mu = 10$, $c = 4.30523810140\ldots$ (cf. [5]). Note that the Shil'nikov case corresponds to the innermost part of the spiral distortion (mentioned above) being mapped back upon itself.

Any such "ordinary" homoclinic trajectory corresponds to a single-pulse wave train on the infinite linear fiber in (1). Similarly, if not the first but the second image of the straight line mentioned above hits the one-dimensional manifold as described, the corresponding homoclinic situation implies a double-pulse wave-train, and so forth up to arbitrarily long finite wave trains (see the examples, up to order 10, in [5]). Cf. also [13], where similar results were predicted for the Hodgkin-Huxley equation.

The connection between this finite wave-train case and the infinite case is nontrivial, however. Finite pulse sequences by definition cannot represent wave chaos. A symbolic dynamics method can be used to make the relationship more transparent. The excitable medium is first divided into "distance quanta" [5, 14]. Then, for any solution the presence or absence of a supra-threshold ($u_{\max} > \delta$) pulse in each bin can be symbolized by a "1" or "0", respectively. The number N of realized com-

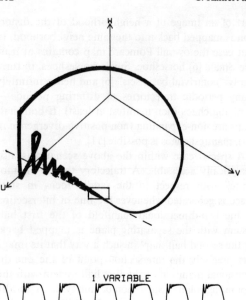

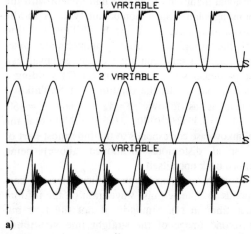

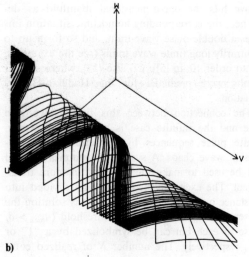

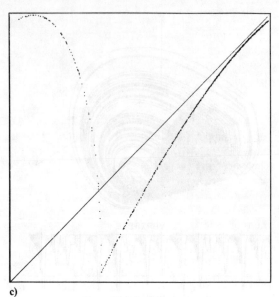

Fig. 2. A second type of chaos in (2). Parameters: $\beta = 1$, $\delta = 0.01$, $\varepsilon = 150$, $\mu = 10$, $c = 9.0$. (a) Phase space plot for 100 s-units. Initial conditions: $u(0) = 0.01000$, $v(0) = 0.9985$, $w(0) = -0.003726$. Axes: -0.001 to 0.011 for u, 0.96 to 1.03 for v, -0.05 to 0.05 for w, 0 to 15 for s. (b) Blow up. Axes: 0.0099 to 0.01 for u, 1.003 to 1.008 for v, -0.0025 to 0.0025 for w. (c) Approximate 1-D map of the interval $(1.003, 1.008)$ of the dashed line in the $u = \delta$ plane indicated in (a). About 2200 points (up to $s = 5000$) are shown.

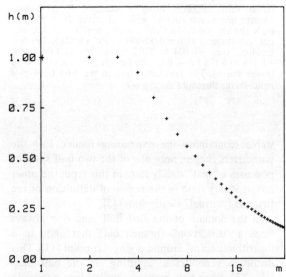

Fig. 3. Plot of $h(m)$ for the 10-pulse wave train indicated in the text.

binations of 0's and 1's as a function of the length, m, of a "running window", can be obtained for this solution.

In this way, a quantity h can be arrived at that is defined as follows:

$$h(m) := \frac{1}{m} \log_2 N(m). \qquad (3)$$

As an example, take the wave train of 10 pulses as shown in Fig. 16c [5]. It possesses the following symbolic representation: "101111100110011", with zeros in front and behind. A plot of $h(m)$ for this sequence is shown in Figure 3.

The point is that if we had taken a longer wave train of comparable local complexity, we would have obtained essentially the same graph once more. Only the left-hand portion of the graph (with unit height in the present case) would be elongated. (For comparable-complexity wave-trains, this elongation is proportional to the logarithm of wave-train length.) The "tail" (which in the case of Fig. 3 starts at $m = 4$) would have essentially the same form again, going asymptotically to zero due to the $1/m$ factor in (3).

Mathematically, the straight (or wiggly, in other cases) "plateau" in the left-hand portion of the graph of Fig. 3 has no name. However, if it had infinite length (because we had analyzed a complex infinite-length wave-train) then its value would be the topological entropy of the corresponding infinite wave-train and, by implication, of the corresponding chaotic regime (cf. [15]).

We think that the presence of this potentially growing plateau ("pre-entropy") can be used to differentiate between asymptotically chaotic and not asymptotically chaotic boundary value problems within the present class – also and even if only finite lengths are accessible. This conjecture opens the possibility that even a finite boundary value problem may be shown to be "related to chaos".

Finally the stability problem is worth mentioning. A wave-train in (1), being described by a trajectory of (2), will in general not share the stability properties of the latter. Nevertheless for the chaotic attractor of Fig. 2, the original P.D.E. can be expected to possess an attracting set of chaotic waves. (Small perturbations of a wave-train would lead back to this set, but not necessarily to the same element.) This is because the wave velocity c in Fig. 2a ($c = 9.0$) is more than twice the value of that of a single unstable pulse ($c = 4.3...$), for the same parameters, in (1). (See the above "Shil'nikov set".) Although it is known (cf. [4, 5]) that multiple unstable pulses may travel somewhat faster than single ones, a speed increase by more than 100% appears highly unlikely. Therefore, the wave train described indirectly by Fig. 2 probably consists of stable pulses. It hereby goes without saying that such "temporal stability" of the individual components of a wave train does not necessarily imply "spatial stability" of the wave train [16].

To sum up, wave-train chaos exists in 1-D excitable media. The infinite wave-train case is embedded in an exploding scenario of possibilities in which the limiting case (chaos) makes its presence felt even in the finite boundary value problem already.

C. K. acknowledges support by a grant from the Studienstiftung. O. E. R. thanks Yoshiki Kuramoto, Norman Packard and Rob Shaw for discussions. Both thank John Rinzel for his help with the manuscript.

[1] Arthur T. Winfree, The Geometry of Biological Time, Springer-Verlag, New York 1980.
[2] Paul C. Fife, Mathematical Aspects of Reacting and Diffusing Systems, Springer-Verlag, New York 1979.
[3] John Rinzel and Joseph B. Keller, Biophys. J. **1**, 445 (1973).
[4] John A. Feroe, SIAM J. Appl. Math. **42**, 235 (1982).
[5] Claus Kahlert and Otto E. Rössler, Z. Naturforsch. **38a**, 648 (1983).
[6] Stephen Smale, Bull. Amer. Math. Soc. **73**, 747 (1967).
[7] Otto E. Rössler, Bull. Math. Biol. **39**, 275 (1977).
[8] Pierre Collet and Jean-Pierre Eckmann, Iterated Maps on the Interval as Dynamical Systems, Birkhäuser, Boston 1980.
[9] Igor Gumowski and Christian Mira, Dynamique Chaotique, Transformations Ponctuelles, Transition Ordre-Disordre, Cépadues Editions, Toulouse 1980.
[10] Bernhard Uehleke and Otto E. Rössler, Z. Naturforsch. **38a**, 1107 (1983); Z. Naturforsch. **39a**, 342 (1984).
[11] Claus Kahlert, in preparation.
[12] L. P. Shil'nikov, Soviet Math. Docl. **6**, 163 (1965).
[13] Gail A. Carpenter, SIAM J. Appl. Math. **36**, 334 (1979).
[14] John W. Evans, Neil Fenichel, and John A. Feroe, SIAM J. Appl. Math. **42**, 219 (1982).
[15] James P. Crutchfield and Norman Packard, Physica **D7**, 201 (1983).
[16] John Rinzel, Biophys. J. **15**, 975 (1975).

On the Connection between Statical and Dynamical Chaos

M. S. El Naschie and S. Al Athel

Z. Naturforsch. **44a**, 645–650 (1989); received March 18, 1989

> The paper discusses the interrelationship between statical chaos and dynamical initial value problems. It is pointed out that approximate homoclinic and heteroclinic solitons can be perturbed to produce spacial asymptotic chaos in some buckled structural elastic systems which constitute strictly speaking boundary value problems.

Introduction

In marked contrast to initial value problems of dynamics, boundary value problems seem to preclude the possibility of chaos. Superficially the argument seems to be as compelling as it is simple. For chaos to take place, infinite time is needed [1]. Boundary value problems are by definition finite and consequently elastostatical, and structural systems cannot display chaos. Nevertheless experience teaches us that quasistatical systems with finite dimensions can sometimes display great complexity [2]. To understand this apparent contradiction, we need only to extend the notion of time to that of time-like co-ordinates. The well known statical-dynamical analogy of Kirschoff provides the frame work for such an extension [3, 4]. In what follows we will attempt to show that by virtue of this analogy and the use of a kind of stretched spacial co-ordinate, boundary value problems could exhibit a considerable complexity and tend asymptotically toward quasistatical chaotic states [5–8].

1. The Forced Pendulum – An Initial Value Problem

1.1. Parametric Excitation

The equation of motion of a mathematical pendulum which is forced periodically in the vertical direction at the pivot is easily found from the equilibrium condition to be

$$\ddot{\phi} - (\Omega^2/\omega^2) \sin \phi = (a/l) \sin \tau \sin \varphi,$$

where ϕ is the angle of rotation, l the length of the pendulum, ω the frequency of excitation, $\Omega = \sqrt{g/l}$ the natural frequency, g the earth acceleration, a the amplitude of excitation, $\tau = \omega t$ the non-dimensional time and $(*) = d()/d\tau$. Since the corresponding homogeneous differential equation describes a homoclinic orbit [7, 8] for the initial conditions $\phi = \pi$ and $\dot{\phi} = 0$, then following Poincaré's homoclinic criterion one expects the forcing to initiate a chaotic motion near the hyperbolic saddle, i.e. the inverted dynamical unstable position. The next step is to use the most direct and simplest characterization of possible chaos, namely the Poincaré map method. Since we are dealing with an initial value problem, all what is needed is a simple numerical integration of the equations using a marching technique. Some of the Poincaré maps obtained for system parameters

$$\Omega^2/\omega^2 = 1; \qquad a/l = 0.01 \quad \text{and}$$
$$\Omega^2/\omega^2 = 0.0272222; \quad a/l = 0.15,$$

and various initial conditions are shown in Figure 1. They are quite comprehensive and show, may be for the first time for this very simple conservative problem, the behaviour anticipated by the celebrated KAM theorem [7, 8].

1.2. Simple Periodic-moment Excitation

In anticipation of the use of the statical-dynamical analogy in studying the elastica, it is essential to consider once more the preceding problem. However, this time the system will be excited externally by a simple periodic moment, so that the corresponding differential equation would simplify to

$$\ddot{\phi} - (\Omega^2/\omega^2) \sin \phi = a \sin \omega t.$$

Again following Poincarés homoclinic criterion, the system may be shown to be susceptible to chaotic motion, which is easily shown to be the case using the Poincaré map method in conjunction with an appropriate numerical technique. The relevant Poincaré

Reprint requests to Prof. Dr. M. S. El Naschie, 51 Pitt Place, Church Street, Epsom, Surrey KT17 4PY, England, U.K.

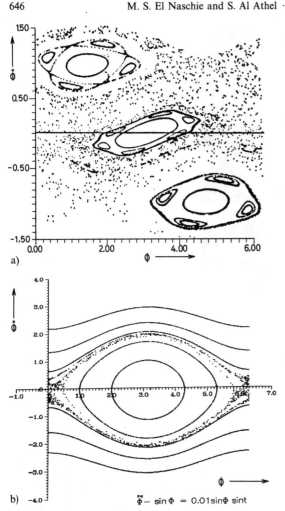

Fig. 1. Poincaré maps with various initial conditions for the parametrically excited pendulum in case of
(a) $\Omega_2^2/\omega_2^2 = 0.0272222$ and $a/l = 0.15$,
(b) $\Omega^2/\omega = 1$ and $a/l = 0.01$.

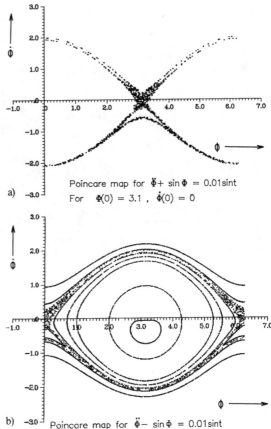

Fig. 2. Poincaré maps for the pendulum with periodic moment excitation and system values $\Omega^2/\omega^2 = 1$ and $a/l = 0.01$ in case of
(a) $\phi(0) = 3.1$ and $\dot\phi(0) = 0$,
(b) fifteen different initial values.

maps are shown in Figure 2. We may note at this stage that the shape of the maps was found to be extremely sensitive towards the step size of numerical integration and great care must be exercised to avoid wrong conclusions.

2. The Elastica – A Boundary Value Problem

Next we consider the differential equation of elastica [9]

$$\phi'' - \lambda^2 \sin\phi = 0$$

in terms of the angle of rotation ϕ, where $\lambda = \sqrt{P/\alpha}$, P is the axial load, α the bending stiffness, $(') = d(\ )/ds$, and s is the arch length. We see clearly that in this form the differential equation is mathematically identical to that of the unforced pendulum and that s plays the same role as the time in dynamics. Subsequently we suppose that the strut middleaxis possesses an initial sinusoidal crookedness. Taking only the first order effect of this shape imperfection, one finds

$$\alpha(\phi + a\sin\omega s)'' - P\sin\phi = 0.$$

That means

$$\phi'' - \lambda^2 \sin\phi = a\omega^2 \sin\omega s.$$

Differentiating with respect to the time-like spacial co-ordinate ωs, where ω is the frequency of axial im-

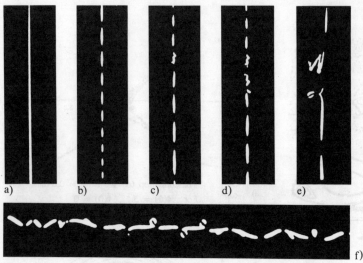

Fig. 3. Initiation of soliton-like asymptotic chaos in very long metal steel band with 3 D spacial imperfection. Note that in these simple physical experiments homoclinic as well as heteroclinic points are now possible. Exact mathematical conditions for the appearance of each one of these points may be established. – (a) The perfect band. – (b) The imperfect band. – (c) The first appearance of loops. – (d), (e), (f) Subsequent quasi random soliton loops. Note the similarity to the soliton found in plasma by Ichikawa et al. [21].

perfection, one finds

$$\ddot{\phi} - (\lambda^2/\omega^2) \sin \phi = a \sin \omega s,$$

where $(\dot{\ }) = d(\ )/d\omega s$ and a is the amplitude of axial shape imperfection. Now, that we do not need to consider more than the first order imperfection is theoretically founded on Koiter's theory of initial post-buckling [9–11] and in a wider mathematical context on catastrophe theory [11]. It is then important to realize that the differential equation of sinusoidally imperfect elastica, and the pendulum which is excited externally by a periodic moment, are mathematically identical. The only difference is that the first is a boundary value problem whilst the second is an initial value problem.

3. Chaos in Initial and Boundary Value Problems

A boundary value problem has two major consequences from the present discussion point of view. The first is regarding its finiteness, which precludes chaos by definition, as previously indicated. Second it cannot be integrated numerically using a direct marching technique. However, the first point could be bypassed by the physical assumption of infinitely long elastica, so that chaos cannot be excluded a priori. The second point is subsequently resolved automatically since the assumption of an infinite spacial domain and the analogy to the dynamical pendulum problem allows us to use a direct marching technique, exactly as in the case of an initial value problem. Consequently, the dynamical chaos of the moment excited pendulum near to the inverted position can be reinterpreted now in the spacial domain as a random formation of strophoid-like loops which were described for the first time by Leonard Euler in the appendix to his famous treatise on the calculus of variations (see Fig. 10, Tabular IV of [12]). These loops in turn can be interpreted within two extremely important fields of current mathematical research. First they may be viewed as a spacial-statical homoclinic connection formed by a rod of infinite length undergoing exceedingly large deformation in the limiting case when the angle tends towards $\phi_0 = \pi$ and a single loop is formed in the rod, a situation which corresponds in the dynamical analogy to the pendulum starting close to the inverted position of unstable equilibrium and making just one revolution as explained in great detail in the classical treatise of Love [3]. The second interpretation of these loops is that of so called solitons, which are in general localized wave solutions of permanent form [13, 14]. It is spacial imperfection which perturbs these solitary homoclinic loops into chaos so that they may appear at random in the theoretically infinitely long elastica (see Figs. 4 and 5). Nevertheless, long as elastica may be they are of course of finite length and consequently we

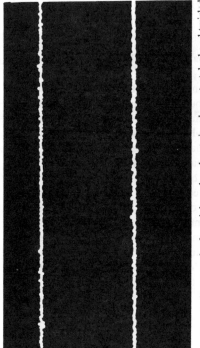

Fig. 5a. A photograph of spatial random knots in a long stretched and twirled rubber band as the stretching force is gradually released.

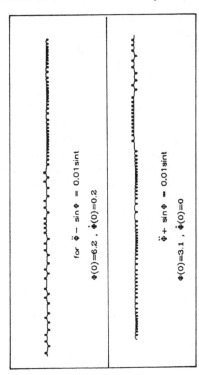

Fig. 5b. Result of numerical integration of the infinitely long periodically imperfect elastica.

for $\ddot{\Phi} - \sin\Phi = 0.01\sin t$

$\Phi(0) = 6.2$, $\dot{\Phi}(0) = 0.2$

$\ddot{\Phi} + \sin\Phi = 0.01\sin t$

$\Phi(0) = 3.1$, $\dot{\Phi}(0) = 0$

Fig. 4a. A photograph of some random soliton loops in a compressed very long elastic metal tape with periodic torsional imperfection.

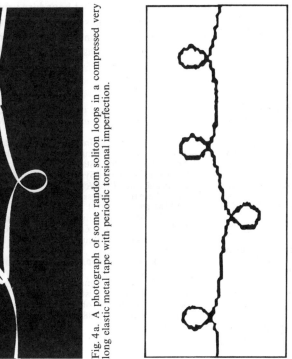

Fig. 4b. Result of numerical integration of the periodically imperfect planar elastica ($\ddot{\phi} - \sin\phi = 0.01 \sin t$; $\phi(0) = 6.2$; $\dot{\phi}(0) = 0.2$).

may anticipate in general only a complex behaviour which is tending in theory towards a limiting chaos, but never truly reaches it. The consequence of such behaviour seems to have been foreseen for the first time by O. E. Roessler and his associates in a completely different context [5]. Such chaos may thus be termed asymptotic chaos. It is noteworthy, however, that we do not need the exceedingly large deflection in order to obtain the solitary loops if we remove the constraint of two dimensions and admit a third one. This fact may be readily demonstrated by a very simple experiment using a long fibreoptic string or a steel tape measure. Giving the tape a spacial imperfection, a small compression is sufficient to produce several such approximate homoclinic or heteroclinic points at random in the tape. In fact, the statical dynamical analogy of Kirchoff was extended long ago in 3 D by Hess [3], and the existence of spacial homoclinic and heteroclinic points [15] can be established. These are in turn the source of asymptotic chaos in 3 D elastica. Figure 3 shows the experimentally observed progressively complex appearance of an approximate homoclinic soliton in a long steel band*. This way many dynamical chaos problems [16–20] may be reinterpreted spacially. In Figs. 4 and 5 comparison is made between numerical integration and simple illustrative experiments of spacial asymptotic chaos.

4. Localized Damped Buckling Forms as Solitons

In conclusion it is important to remark that localized damped buckling forms also represent another form of a solitary homoclinic situation of the spacial-statical type which may exist in a boundary value problem and may also be perturbed by spacial imperfection into chaos [6–8]. The solitary solution acts thus in a sense as a sort of stretched co-ordinate, necessary to accommodate the infinity which chaos requires. The connection to buckling of shell-like structures is a more or less obvious application of this conjecture. A modified version of the inverse scattering method can be used to prove the existence of localized envelope soliton solutions for shell-like structures, but this will be given elsewhere.

5. Conclusion

The statical counterparts of Poincaré's homoclinic orbits can be perturbed by shape imperfection to produce a random sequence of localized soliton-like deformations. It is shown that boundary value problems can display asymptotic spacial chaos in the sense of Roessler, which is relevant to a deep understanding of localized buckling of elastic and shell-like structures.

---

* Heteroclinic points are now easily formed when the third dimension of the real physical space is admitted. Our model is thus a mechanical realization of soliton and chaos simultaneously.

---

[1] Hao Bai-Lin, Chaos, World Scientific, Singapore 1985.
[2] M. S. El Naschie, High speed Deformation of Shells, High Velocity Deformation of Solids (K. Kawata, ed.), Springer, New York 1977, p. 363.
[3] A. E. H. Love, A Treatise on the Mathematical Theory of Elasticity, Dover Edition, Cambridge University Press, Cambridge 1944.
[4] P. J. Holmes and J. E. Marsden, Horseshoes and Arnold Diffusion for Hamiltonian Systems on Lie Groups, Indiana University Mathematical Journal **32**, 273 (1988).
[5] C. Kahlert and O. E. Roessler, Z. Naturforsch. **39a**, 1200 (1984).
[6] M. S. El Naschie, Chaos and Generalized Bifurcation in Science and Engineering. Current Advances in Mechanical Design and Production (Y. H. Kabil and M. E. Said, eds.), Pergamon Press, London 1988, p. 389.
[7] M. S. El Naschie, J. Appl. Mathem. Mech. (ZAMM) **69**, T367 (1989). Also paper presented at GAMM, Vienna 5–7 April, 1988.
[8] M. S. El Naschie, Order chaos and generalized bifurcation. Submitted to J. Engineering Sci., K.S. University in May (1987). To appear in **14**, No. 2, p. 437.

[9] M. S. El Naschie, The initial postbuckling of an extensional ring under external pressure. Int. J. Mech. Sci. **17**, 387 (1975).
[10] J. M. T. Thompson, Instabilities and Catastrophes in Science and Engineering, John Wiley, Chichester 1982.
[11] M. S. El Naschie, Stability Bifurcation and Catastrophes in Nonlinear Systems. Current Advances in Mechanical Design Production (Y. H. Kabil and M. E. Said, eds.), Pergamon Press, London 1988, p. 407. Also ZAMM **69**, 378 (1989).
[12] L. Euler, Methodus Inveniendi Lineas curvas Maximi Minimive properietate Gaudentes (Appendix, De Curvis elasticis), Marcum Michaelem Bousquet, Lausanne and Geneva 1944.
[13] P. G. Drazin, Soliton, Cambridge University Press, Cambridge 1983.
[14] M. S. El Naschie, Generalized Bifurcation and Local Buckling as Deterministic Chaos, Euro Mech 242, Application of Chaos Concepts to Mechanical Systems, September 26–29, 1988, Univeristy of Wuppertal, Abstracts, p. 23.

[15] H. Troger, On point mapping for mechanical system possessing homoclinic and heteroclinic points. J. Appl. Mech. **46**, 468 (1979).
[16] C. S. Hsu, Cell to Cell Mapping, Springer, New York 1987.
[17] A. Kunick and W. H. Steeb, Chaos in dynamischen Systemen, Wissenschaftsverlag, Mannheim 1986.
[18] G. Schmidt and A. Tondl, Nonlinear Vibration, Cambridge University Press, Cambridge 1986.
[19] A. B. Pippard, Response and Stability, Cambridge University Press, Cambridge 1985.
[20] F. Moon, Chaotic Vibration, John Wiley, New York 1987.
[21] Y. H. Ichikawa and N. Yajima, Recent developments of soliton research in plasma physics. In: Statistical physics and chaos in fusion plasma (C. W. Harton and L. W. Reichl, eds.). John Wiley, New York 1984, pp. 79–90.

Spatially Complex Equilibria of Buckled Rods

ALEXANDER MIELKE & PHILIP HOLMES

1. Introduction

In this paper we study the spatial aspects of equilibrium states exhibited by infinitely or arbitrarily long rods, buckled by loads applied at their ends. Our methods exploit the Hamiltonian structure of the equilibrium equations and use regular perturbation methods based on completely integrable cases (MELNIKOV theory). We obtain a qualitative description of classes of solutions close to such limiting cases, which correspond to geometrical symmetries and the vanishing of certain stress components. Our results imply that there exist spatially irregular or chaotic equilibrium states for rods under the appropriate load conditions.

KIRCHHOFF [1859] was apparently the first to remark the analogy between the equilibrium equations of a rod loaded at its end and the equations of motion of a heavy rigid body pivoted at a fixed point. In this analogy the arclength along the axis of the rod plays the rôle of a time-like coordinate. LARMOR [1884] subsequently extended the analogy to rods with initial curvature and twist; cf. LOVE [1927, §§ 259–264]. The discussions in KIRCHHOFF and LOVE assume linear constitutive relations and the equations thus obtained are precisely analogous to the rigid body equations. However, as we show in the next section, the analogy extends to general nonlinear hyperelastic materials. The (non-canonical) Hamiltonian structure of the rigid body equations is preserved, while the quadratic Hamiltonian is replaced by a general function. This structure, and the existence of certain integrals, derive from underlying symmetries and group structures in the problem. While the derivation of the rod equations is relatively well known (cf. ANTMAN & KENNEY [1981], ANTMAN [1984], MIELKE [1987]), the Hamiltonian structure is not normally emphasized and so we outline it in Section 2. Here we mean the Hamiltonian structure of the static problem with the arclength as timelike variable, in contrast to the dynamic problem which is a partial differential equation with time and arclength as independent variables. See KRISHNAPRASAD, MARSDEN, & SIMO [1986] for the Hamiltonian structure in that case.

While there is an elegant non-canonical formulation for the three degree of freedom rigid body equations (cf. HOLMES & MARSDEN [1983]) we find it more convenient here to work with a reduced two degree of freedom system in canonical coordinates. In general this system is expected to be non integrable; in fact our main results prove this to be true near certain limiting cases. However, in the two cases of zero resultant force and of circular symmetry, an additional integral can be found and the equations solved completely. The former case corresponds to the absence of gravitational forces (moments) and the latter to the well known 'Lagrange' top (GOLDSTEIN [1980, Chapter 5]). In both cases smooth manifolds of heteroclinic or homoclinic orbits exist, and using the perturbative techniques of MELNIKOV [1963] in the Hamiltonian context of HOLMES & MARSDEN [1982, 1983], we are able to prove that these manifolds break to give transverse homoclinic orbits in the presence of small resultant forces or asymmetries. Then, by arguments familiar in dynamical systems (SMALE [1963, 1967], GUCKENHEIMER & HOLMES [1988, Ch. 4–5]), it follows that spatially chaotic equilibrium states occur.

This paper is organized as follows. In Section 2 we outline the derivation of the equilibrium equations and discuss the Hamiltonian structure. In Section 3 we perform our first reduction, obtaining a canonical two degree of freedom Hamiltonian system. The main results are given in Section 4, together with a discussion of their physical implications, including a rough description of the spatial shapes exhibited by rods in such 'chaotic' states. The remainder of the paper is devoted to proofs of the two main theorems of Section 4. Section 5 contains a brief outline of the second reduction and application of MELNIKOV's method to the resulting periodically perturbed single degree of freedom system. A detailed treatment of this material is contained in HOLMES & MARSDEN [1982, 1983]. The two main theorems are then proved in Section 6 and 7 by computation of Melnikov functions for appropriate limiting cases. Computational details are relegated to the Appendix. In order to make explicit computations we restrict ourselves to stresses which are sufficiently small in magnitude, so that linear elasticity dominates. The geometric nonlinearities are, however, unrestricted.

2. The Rod Equations in Hamiltonian Form

The model of a rod treated in this paper takes the following form (*cf.* ANTMAN [1984], KRISHNAPRASAD, MARSDEN, & SIMO [1986]). We consider a prismatic, elastic body with reference configuration $\Omega = \mathbb{R} \times \Sigma \in \mathbb{R}^3$ or $\bar{\Omega} = I \times \Sigma \in \mathbb{R}^3$, where $t \in \mathbb{R}$ or $t \in I \subset \mathbb{R}$ denotes the axial variable and $x = (x_1, x_2) \in \Sigma$, the cross-section Σ being a bounded domain in $\mathbb{R}^2$. We assume the deformations of the rod have the approximate form $\varphi: \Omega, \bar{\Omega} \to \mathbb{R}^3$, with

$$\varphi(t, x) = r(t) + R(t) \begin{pmatrix} x \\ 0 \end{pmatrix}. \tag{2.1}$$

Here the $r(t) \in \mathbb{R}^3$ denotes the position of the deformed axis and $R(t) \in \mathrm{SO}(3)$ specifies the position of the rigidly transformed cross-section. See Figure 1. Note that $d_3 = R(t) e_3$, the local coordinate axis perpendicular to the transformed cross-section, need not be tangent to $r'(t)$. Thus we allow for shear deformations.

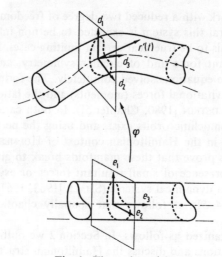

Fig. 1. The rod model.

However, the local coordinate system $\{d_1, d_2, d_3\}$, $d_i = R(t) e_i$, is still orthogonal so that $d_3(t) = d_1(t) \times d_2(t)$, as in ANTMAN's formulation.

There are differing conventions for definition of strains. We adopt the following. We set

$$v = R^T r' - e_3, \quad \Omega = R^T R' \qquad (2.2\,\text{a, b})$$

so that $\Omega = -\Omega^T \in \text{so}(3)$, and finally define u by

$$u \times a = \Omega a \qquad (2.2\,\text{c})$$

for arbitrary vectors $a \in \mathbb{R}^3$. Since u and Ω are related by (2.2c), we sometimes write $\Omega(u)$ subsequently. Thus $u = u(t)$, $v = v(t)$ are both 3-vectors; they are the strains in body coordinates. Specifically, u_1 and u_2 represent bending in the d_2, d_3 and d_1, d_3 planes respectively and u_3 is torsion ($=$ bending in d_1, d_2); v_1 and v_2 are shears in the d_1, d_2 directions and v_3 is extension in the d_3 direction; cf. ANTMAN & KENNEY [1981, § 2].

The elastic properties of the rod are described by a strain energy function $W(u, v): \mathbb{R}^6 \to \mathbb{R}$, which is assumed to have a nondegenerate minimum at $(u, v) = (0, 0)$, corresponding to the undeformed state $r(t) = te_3$, $R(t) = I$. The stresses in body coordinates are then given by

$$m = \frac{\partial W}{\partial u}, \quad n = \frac{\partial W}{\partial v}, \qquad (2.3)$$

the components of m being bending and torsion moments and of n, shear and extension forces.

The equilibrium equations for a rod loaded only at its ends are easily expressed in terms of the force and moment vectors F, M in spatial coordinates (e_1, e_2, e_3). They simply express the fact that F and M are constants. Thus, via (2.1), in body coordinates, we have

$$F = Rn = \text{const},\tag{2.4a}$$

$$M = Rm + r \times Rn = \text{const}.\tag{2.4b}$$

Differentiating (2.4a), we have $R'n + Rn' = 0$ or

$$n' = -R^T R' n = -\Omega n = n \times u.\tag{2.5}$$

Differentiating (2.4b) and applying R^T from the left yields $R^T R' m + m' + R^T(r' \times Rn + r \times (Rn)') = 0$ or, in view of (2.4a),

$$R^T R' m + m' + R^T r' \times n = 0.\tag{2.6}$$

Using (2.2) equation (2.6) becomes

$$m' = -\Omega m + n \times (e_3 + v) = m \times u + n \times (e_3 + v).\tag{2.7}$$

Equations (2.5) and (2.7) involve the stresses and the strains, but, as W has a nondegenerate minimum at $u = v = 0$, it is locally convex and hence the relations (2.3) are locally invertible. We may therefore write the inverses

$$u = u(m, n), \quad v = v(m, n)\tag{2.8}$$

and there furthermore exists a real valued function $H(m, n): \mathbb{R}^6 \to \mathbb{R}$ such that

$$u = \frac{\partial H}{\partial m}, \quad e_3 + v = \frac{\partial H}{\partial n}.\tag{2.9}$$

H and W are related via a Legendre transform; in the dynamical analogy, W plays the rôle of a Lagrangian and H of the Hamiltonian (*cf.* ARNOLD [1978]).

We can now write the equilibrium equations (2.5), (2.7) entirely in terms of the stresses as a non-canonical Hamiltonian system

$$\binom{m}{n}' = J(m, n)\, \nabla H(m, n),\tag{2.10}$$

where

$$J = -J^T = \begin{pmatrix} \Omega(m) & \Omega(n) \\ \Omega(n) & 0 \end{pmatrix}, \quad \Omega(a)\, b = a \times b.\tag{2.11}$$

Alternatively (2.10) can be written in terms of the Lie-Poisson bracket defined in HOLMES & MARSDEN [1983, eqns. (3.17–18)] for the heavy rigid body. In that case the Hamiltonian is simply $H = \tfrac{1}{2} \left(\dfrac{m_i^2}{I_i} \right) + \text{Mgl}\, n_3$, but we note that the symplectic structure permits formulation of equations for general Hamiltonian functions.

Regardless of the precise form of H, the group structure implies that the two functions $I_1 = |n|^2$ and $I_2 = m \cdot n$ are constants of motion for (2.10), along with the Hamiltonian H itself, as one can readily verify. These are the only quantities derivable from $F = \text{const}$ and $M = \text{const}$, that are invariant under motions in $SO(3) \times \mathbb{R}^3$; *i.e.*, invariant under changes in R and r. (Note, $F \to RF$,

$M \to RM + r \times RF$). In the rigid body analogy I_1 corresponds to conservation of magnitude of the gravity vector and I_2 to conservation of the component of angular momentum in the direction of the gravity vector.

Remark. In the above discussion it is not clear how this special Hamiltonian structure in (2.11) arises. Here we point out briefly how it can be deduced using the methods for Hamiltonian systems on Lie groups as discussed in ABRAHAM & MARSDEN [1978, Ch. 4].

In our case the Lie group is the Euclidean group $SO(3) \times \mathbb{R}^3$, *i.e.* the group of rigid transformations with multiplication $(R_2, r_2)(R_1, r_1) = (R_2 R_1, R_2 r_1 + r_2)$. The Lagrangian L is given by $L(R, r, R', r') = W(R^T R', R^T r' - e_3)$. Since W is invariant under the action of G, it is more convenient to use the "body coordinates" (basis (d_1, d_2, d_3)) than the "spatial coordinates" (with basis (e_1, e_2, e_3)), *i.e.* u and $v + e_3$ are the strains in body coordinates. Now L can be written as a function $\tilde{L} \colon G \times g \to \mathbb{R}$; $(R, r, \Omega(u), v + e_3) \to W(u, v)$, where $g = T_e G$ is the Lie algebra of G with Lie bracket $[(\Omega_1, v_2), (\Omega_2, v_2)] = (\Omega_1 \Omega_2 - \Omega_2 \Omega_1, \Omega_1 v_2 - \Omega_2 v_1)$.

Similarly the corresponding Hamiltonian H is defined on $G \times g^*$ (g^* = dual of g) rather than on the cotangent bundle T^*G. The variables in g^*, being conjugate to the variables in G, are exactly the stresses m and n in body coordinates. Of course $H = H(m, n)$ is given as in (2.9).

The price for working in body coordinates must be paid when the Hamiltonian structure of $G \times g^*$ is calculated, since "inertial effects" appear. Using Theorem 4.4.1 of ABRAHAM & MARSDEN [1978], we obtain the symplectic form

$$\omega_{(R, r, \Gamma(m), n)}((R'_1, r'_1, \Omega_1, n_1), (R'_2, r'_2, \Omega_2, n_2))$$

$$= R^T R'_1 : \Omega_2 - R^T R'_2 : \Omega_1 + R^T r'_1 \cdot n_2 - R^T r'_2 \cdot n_1$$

$$+ \Gamma(m) : (R^T R'_1 R^T R'_2 - R^T R'_2 R^T R'_1) + n \cdot (R^T R'_1 R^T r'_2 - R^T R'_2 R^T r'_1)$$

where $\Omega_1 : \Omega_2 = \operatorname{tr}(\Omega_1^T \Omega_2)$ and $\Gamma(m) b = m \times b$.

The associated vector field X_H is defined by

$$\omega_{(R, r, \Omega, u)}(X_H, (R'_2, r'_2, \Omega_2, n_2)) = \frac{\partial H}{\partial R} : R'_2 + \frac{\partial H}{\partial r} r'_2 + \frac{\partial H}{\partial \Gamma} : \Omega_2 + \frac{\partial H}{\partial n} \cdot n_2.$$

A direct calculation results in

$$R' = R \frac{\partial H}{\partial \Gamma}, \quad r' = R \frac{\partial H}{\partial n}, \quad n' = \frac{\partial H}{\partial \Gamma} n,$$

$$\Gamma' = \Gamma \frac{\partial H}{\partial \Gamma} - \frac{\partial H}{\partial \Gamma} \Gamma + \tfrac{1}{2}\left(n \otimes \frac{\partial H}{\partial n} - \frac{\partial H}{\partial n} \otimes n\right),$$

(2.12)

which are exactly the equations established above, when Γ and m are identified. In addition, we remark that the invariants $F = Rn$ and $M = Rm + r \times Rn$ are exactly the momentum mappings obtained from the invariance of L under the action of G (ABRAHAM & MARSDEN [1978, Theorem 2.12]).

In the analysis below we find it convenient to fix $I_1 = a^2$, $I_2 = ab$ and work with the reduced Hamiltonian system restricted to this (family of) four manifold(s). In Section 3 we show that a symplectic structure exists such that this restriction is a canonical two degree of freedom system. However, the limiting cases which we study below are best described initially in terms of the non-canonical coordinates (m, n).

The set $n = 0$ is clearly an invariant manifold for (2.10); restricted to this manifold the equation becomes

$$m' = m \times u(m, 0), \tag{2.13}$$

which has the additional integral $I_3 = |m|^2$. From MIELKE [1987] we know that, without loss of generality, we can take u to have the form

$$u(m, 0) = (\alpha_1 m_1, \alpha_2 m_2, \alpha_3 m_3) + \mathcal{O}(|m|^2) \tag{2.14}$$

where $0 < \alpha_1 \leq \alpha_2$ and $\alpha_3 > \frac{1}{2}(\alpha_1 + \alpha_2)$. Hence there are two generic cases: $\alpha_1 < \alpha_2 < \alpha_3$ and $\alpha_1 < \alpha_3 < \alpha_2$. We will also consider the special case $\alpha_1 = \alpha_2 < \alpha_3$, which occurs in the case of certain cross-sectional symmetries. The phase portraits on spheres $I_3 =$ constant, sufficiently small, are then qualitatively identical to those for the Euler equations for the gravity free rigid body (GOLDSTEIN [1980, Ch. 5]). Figure 2 shows the case $\alpha_1 < \alpha_2 < \alpha_3$. Note the heteroclinic orbits connecting the hyperbolic fixed points at $m_2 = \pm I_2$, $m_1 = m_3 = 0$. Solutions can be written down explicitly in terms of elliptic functions (WHITTAKER [1937, § 69]) and the heteroclinic orbits are hyperbolic functions (*cf.* HOLMES & MARSDEN [1983]).

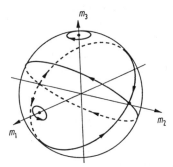

Fig. 2. Heteroclinic cycles for the pure bending case.

This limit corresponds to a rod in pure bending and torsion with zero shear and extension. The fixed points $m_2 = \pm I_2$ correspond to circularly coiled rods: one of the classic planar equilibria found by EULER [1744].

In the case $\alpha_1 < \alpha_3 < \alpha_2$ the phase portraits are similar to Figure 2, but the heteroclinic orbits now connect the fixed points $m = (0, 0, \pm m_3)$. When $\alpha_1 = \alpha_2 < \alpha_3$ the portrait degenerates to a family of circles parallel to the equatorial circle $m_1^2 + m_2^2 =$ const., $m_3 = 0$ and m_3 is an additional constant of motion (see below).

The existence of these integrable limits suggests that we consider a rod with small shear and extension. Letting

$$n = \varepsilon \bar{n}, \quad |\bar{n}| = 1 \tag{2.15}$$

and dropping the bars, we put (2.11) into the form

$$\begin{aligned} m' &= m \times u(m, \varepsilon n) + \varepsilon n \times (e_3 + \mathcal{O}(\varepsilon)), \\ n' &= n \times u(m, \varepsilon n), \end{aligned} \tag{2.16}$$

An analysis of this system lies behind our first main result, Theorem 4.1.

A second important case is provided by rods whose cross-sections are symmetric with respect to the dihedral group D_N with $N \geq 3$. (Thus, Σ is invariant under rotation by $2\pi/N$). In that case the two constants α_1 and α_2, which appear at leading order in the function $u(m, 0)$ of (2.14) are necessarily equal. If we truncate the Hamiltonian at quadratic terms, then equation (2.16) admits the additional integral m_3, which results from the rotational symmetry of the rod at that order. This corresponds to the integrable Lagrange top, m_3 being the component of angular momentum along the axis of symmetry, and $1/\alpha_1$, $1/\alpha_2$ being the (equal) moments of inertia. Our second main result, Theorem 4.2, concerns this case. Here the perturbation parameter is the scaling parameter δ, with $|n| = \delta^2$ and $m = \delta \hat{m}$. As $\delta \to 0$ we approach the rotationally symmetric Hamiltonian $\hat{H}(\hat{m}, \hat{n}) = \frac{1}{2}(\alpha_i \hat{m}_i^2) + \hat{n}_3$ with $\alpha_1 = \alpha_2 (= \alpha)$. The term of lowest order that breaks circular symmetry while respecting D_N-symmetry is then δ^{N-2} Re $(\hat{m}_1 + i\hat{m}_2)^N$.

The fact that H contains terms of higher order in the nonlinear elastic case makes the present analysis more subtle than that of the dynamical analogue carried out by HOLMES & MARSDEN; for example delicate scaling arguments, carried out in Section 3, are necessary to prove Theorems 4.1 and 4.2.

Solution of the Hamiltonian system (2.10) does not specify the spatial state of the rod. Equipped with the stresses $m(t)$ and $n(t)$ as functions of arclength $t \in \mathbb{R}$ (respectively I) we must then integrate the relations (2.2a, b)

$$r' = R(v(m, n) + e_3) \tag{2.17a}$$

$$R' = R\Omega(u(m, n)), \tag{2.17b}$$

using the functions $u = \dfrac{\partial H}{\partial m}$ and $v = \dfrac{\partial H}{\partial n} - e_3$ from (2.9). This yields $r = r(t)$, $R = R(t)$ and hence, via (2.1), the deformation $\varphi(t, x)$.

3. First Reduction: A Symplectic Transformation

The Hamiltonian system (2.10) has a degenerate symplectic structure and is, in fact, a parameterized family of two degree of freedom systems. To reveal this explicitly and make elementary calculations possible, we wish to reduce via the momentum mapping which defines the invariants $I_1 = |n|^2$ and $I_2 = m \cdot n$ (ABRAHAM & MARSDEN [1978, § 4.3]).

Let the reduced phase space be denoted

$$P_{a,b} = \{(m, n) \in \mathbb{R}^6 \mid |n|^2 = a^2, \, m \cdot n = ab\} \tag{3.0}$$

for real constants $a \geq 0$, $b \in \mathbb{R}$. We are especially concerned with limiting cases in which additional integrals $|m|$ and m_3 arise. This suggests that we pick a coordinate system in $P_{a,b}$ which has $|m|$ and m_3 as two of the new variables. We define these variables (r, s, σ, ϱ) by

$$m = \begin{pmatrix} \tilde{r} \cos \sigma \\ \tilde{r} \sin \sigma \\ s \end{pmatrix}, \tag{3.1a}$$

$$n = \frac{ab}{r^2} \begin{pmatrix} \tilde{r} \cos \sigma \\ \tilde{r} \sin \sigma \\ s \end{pmatrix} + a \frac{\bar{r}}{r^2} \begin{pmatrix} s \cos \sigma \\ s \sin \sigma \\ -\tilde{r} \end{pmatrix} \cos \varrho + a \frac{\bar{r}}{r} \begin{pmatrix} -\sin \sigma \\ \cos \sigma \\ 0 \end{pmatrix} \sin \varrho, \tag{3.1b}$$

where $\tilde{r} = \sqrt{r^2 - s^2}$ and $\bar{r} = \sqrt{r^2 - b^2}$. Thus $m_3 = s$, $|m| = r$ and it can be verified that $|n|^2 = a^2$ and $m \cdot n = ab$, so that the point $(m, n) = (m(r, s, \sigma), n(r, s, \sigma, \varrho))$ lies in $P_{a,b}$ as required.

To obtain the symplectic form $\overline{\omega}$ on the reduced space $P_{a,b}$ in (r, s, σ, ϱ) coordinates, we need the transformation matrix

$$G = \frac{\partial(r, s, \sigma, \varrho)}{\partial(m, n)}, \tag{3.2}$$

where elements are computed by inversion of equations (3.1 a, b). The variables r, s and σ are easily obtained in terms of m, n:

$$r = \sqrt{m_1^2 + m_2^2 + m_3^2}, \quad s = m_3, \quad \sigma = \arctan(m_2/m_1) \tag{3.3a}$$

and ϱ is implicitly defined via

$$r^2 n_3 = abs - a\bar{r}\tilde{r} \cos \varrho \Rightarrow \varrho = \bar{\varrho}(r, s, n_3). \tag{3.3b}$$

Thus we have

$$G = \begin{bmatrix} \dfrac{\tilde{r}}{r} \cos \sigma & \dfrac{\tilde{r}}{r} \sin \sigma & 0 & 0 & 0 & 0 \\ 0 & 0 & 1 & 0 & 0 & 0 \\ -\dfrac{1}{\tilde{r}} \sin \sigma & \dfrac{1}{\tilde{r}} \cos \sigma & 0 & 0 & 0 & 0 \\ \dfrac{\tilde{r}}{r} \varrho_1 \cos \sigma & \dfrac{\tilde{r}}{r} \varrho_1 \sin \sigma & \dfrac{s}{r} \varrho_1 + \varrho_2 & 0 & 0 & \varrho_3 \end{bmatrix}, \tag{3.4}$$

where

$$\varrho_1 = \frac{\partial \bar{\varrho}}{\partial r}, \, \varrho_2 = \frac{\partial \bar{\varrho}}{\partial s} \quad \text{and} \quad \varrho_3 = \frac{\partial \bar{\varrho}}{\partial n_3} = \frac{r^2}{a\bar{r}\tilde{r} \sin \varrho} \tag{3.5}$$

are obtained from (3.3b). We shall not require the explicit expressions for ϱ_1 and ϱ_2.

Writing the matrix J of (2.12) explicitly in m, n coordinates, we have

$$J = \begin{bmatrix} 0 & -m_3 & m_2 & 0 & -n_3 & n_2 \\ m_3 & 0 & -m_1 & n_3 & 0 & -n_1 \\ -m_2 & m_1 & 0 & -n_2 & n_1 & 0 \\ 0 & -n_3 & n_2 & 0 & 0 & 0 \\ n_3 & 0 & -n_1 & 0 & 0 & 0 \\ -n_2 & n_1 & 0 & 0 & 0 & 0 \end{bmatrix}. \tag{3.5}$$

The new symplectic form $\bar{\omega}$ is then given by the matrix

$$J_{ab} = GJG^T, \tag{3.7}$$

and an elementary calculation using (3.4)–(3.6) and (3.1) yields

$$J_{a,b} = \begin{bmatrix} 0 & 0 & 0 & 1 \\ 0 & 0 & 1 & 0 \\ 0 & -1 & 0 & 0 \\ -1 & 0 & 0 & 0 \end{bmatrix}. \tag{3.8}$$

Thus the reduced Hamiltonian system can be written

$$\dot{x} = J_{a,b} \nabla_x H_{a,b}(x) \tag{3.9}$$

where $x = (r, s, \sigma, \varrho)^T$ and $H_{a,b}(r, s, \sigma, \varrho) = H(m(r, s, \sigma), n(r, s, \sigma, \varrho))$. The explicit form of $J_{a,b}$ implies that (3.9) is canonical with (s, σ) and (r, ϱ) as conjugate pairs of variables.

To appreciate that this reduction process simplifies the problem, we consider the following special cases. For a rotationally symmetric rod we know that $H(m, n)$ is of the form

$$H(m, n) = \mathcal{H}(m_1^2 + m_2^2, m_3, n_1^2 + n_2^2, n_3, m_1 n_1 + m_2 n_2, m_1 n_2 - m_2 n_1) \tag{3.10}$$

(cf. ANTMAN & KENNEY [1981]). Now $H_{a,b}$ is given by

$$H_{a,b}(r, s, \delta, \varrho) = \mathcal{H}\left(r^2 - s^2, s, a^2\left(1 - \frac{(bs - \tilde{r}\bar{r}\cos\varrho)^2}{r^4}\right), a\frac{bs - \tilde{r}\bar{r}\cos\varrho}{r^2},\right.$$
$$\left.\frac{a}{r^2}(b\tilde{r}^2 + \bar{r}^2 s \cos\varrho), \frac{a\bar{r}\tilde{r}}{r}\sin\varrho\right) \tag{3.11}$$

and is therefore independent of σ. Thus σ is a cyclic variable and s a constant of the motion.

On the other hand the case $a = 0$ $(n = 0)$ leads to

$$H_{0,b}(r, s, \sigma, \varrho) = H(\bar{r}\cos\sigma, \tilde{r}\cos\sigma, s, 0, 0, 0), \tag{3.12}$$

and here ϱ is cyclic and r a constant of the motion.

To perform explicit calculations we restrict our analysis to very small m and n. Then only the lowest order terms of $H(m, n) = \frac{1}{2} \alpha_i m_i^2 + n_3 + \mathcal{O}(|m|^3, |n||m|, |n|^2)$ are relevant. We use the scaling $m = \delta \hat{m}$, $n = \varepsilon \delta^2 \hat{n}$ with $|\hat{n}| = 1$ and define

$$\hat{H}(\hat{m}, \hat{n}, \varepsilon, \delta) = \frac{1}{\delta^2} H(\delta \hat{m}, \varepsilon \delta^2 \hat{n}) = \frac{1}{2} \alpha_i \hat{m}_i^2 + \varepsilon \hat{n}_3 + \mathcal{O}(\delta). \tag{3.13}$$

The corresponding $(a, b, r, s, \sigma, \varrho)$-scaling is

$$a = \varepsilon \delta^2 \hat{a}, \quad b = \varepsilon \delta^3 \hat{b}, \quad r = \delta \hat{r}, \quad s = \delta \hat{s}, \quad (\sigma = \hat{\sigma}, \varrho = \hat{\varrho}), \tag{3.14}$$

where $\hat{a} = 1$. Henceforth we need the (r, s, σ, ϱ)-coordinates in the scaled version only. Thus we drop the hats on (r, s, σ, ϱ) and b but retain them on $(\hat{m}, \hat{n})$. We define

$$\tilde{H}_b(r, s, \sigma, \varrho, \varepsilon, \delta) = \frac{1}{\delta^2} H_{\varepsilon \delta^2, \varepsilon \delta^3 b}(\delta r, \delta s, \sigma, \varrho)$$

$$= \frac{1}{2} (\alpha_1 \bar{r}^2 \cos^2 \sigma + \alpha_2 \bar{r}^2 \sin^2 \sigma + \alpha_3 s^2) \tag{3.15}$$

$$+ \frac{\varepsilon}{r^2} (bs - \bar{r}r \cos \varrho) + \mathcal{O}(\delta).$$

Observe that the limit $\delta = 0$ is identical to the heavy rigid body if $\varepsilon \neq 0$ and note that ε is not necessarily small; this is important in our analysis of the D_N-symmetric rod.

In both cases the canonical structure of the reduced system explicitly reveals the symmetries implicit in the original non-canonical structure.

The case $\alpha_1 < \alpha_3 < \alpha_2$ requires a slightly different coordinate system. In that case the unperturbed $(\varepsilon = 0)$ behavior is essentially the identical to that of $\alpha_1 < \alpha_2 < \alpha_3$ with m_2, m_3 and n_2, n_3 interchanged. Instead of (3.1a, b), we define an analogous coordinate system by interchanging m_2 and m_3 in definition (3.1a) and n_2 and n_3 in (3.1b). Everything goes through in the same manner as previously, except that the Hamiltonian (3.15) is replaced by

$$\tilde{H}_b = \frac{1}{2} (\alpha_1 \bar{r}^2 \cos^2 \sigma + \alpha_3 \bar{r}^2 \sin^2 \sigma + \alpha_2 s^2)$$

$$+ \frac{\varepsilon a}{r^2} ((b\tilde{r} + \bar{r}s \cos \varrho) \sin \sigma + \bar{r}r \cos \sigma \sin \varrho) + \mathcal{O}(\delta). \tag{3.16}$$

In the rotationally symmetric case we use the same scaling but with $\varepsilon = 1$. Here, since $\alpha_1 = \alpha_2$, for $\varepsilon = 0$ the invariant spheres $|m| = \delta$ are filled with periodic orbits lying in the planes $m_3 = $ const. To get a homoclinic solution we fix $\varepsilon = 1$ and treat the limit $\delta = 0$, which is exactly the case treated in HOLMES & MARSDEN [1983].

4. The Main Results: Equilibrium States of Buckled Rods

We state our main result in terms of the two-dimensional systems on the reduced phase spaces $P_{a,b} = \{(m, n) \mid |n| = a, m \cdot n = ab\}$:

Theorem 4.1. *Let* $H(m, 0) = \frac{1}{2}\alpha_i m_i^2 + \mathcal{O}(|m_i|^3)$. *We then have the following cases*:

I. $0 < \alpha_1 < \alpha_2 < \alpha_3$. *For all sufficiently small* ε, $\delta > 0$ *and all* $b \in (-1, 1)$ *there exists a pair of periodic solutions* $(m, n) = (0, \pm\delta, 0; 0) + \mathcal{O}(\delta^2)$ *on* $P_{\varepsilon\delta^2, \varepsilon\delta^3 b} \cap \{H = \frac{1}{2}\alpha_2 \delta^2\}$ *which are connected by transverse heteroclinic cycles.*

II. $0 < \alpha_1 < \alpha_3 < \alpha_2$. *For all sufficiently small* ε, $\delta > 0$ *and all b satisfying*

$$b^2 < \frac{\pi^2 \alpha_3^2}{\pi^2 \alpha_3^2 + 4(\alpha_2 - \alpha_3)(\alpha_3 - \alpha_1) \cosh^2\left(\frac{\pi \alpha_3}{2\sqrt{(\alpha_2 - \alpha_3)(\alpha_3 - \alpha_1)}}\right)} \quad (4.1)$$

there exists a pair of periodic solutions $(m, n) = (0, 0, \pm\delta; 0) + \mathcal{O}(\delta^2)$ *on* $P_{\varepsilon\delta^2, \varepsilon\delta^3 b} \cap \{H = \frac{1}{2}\alpha_3 \delta^2\}$ *which are connected by transverse heteroclinic cycles.*

The second result concerns rods which are symmetric with respect to rotations through the angle $2\pi/N$ about their axes in the reference configuration. In the (r, s, σ, ϱ)-coordinates this corresponds exactly to the transformation $\sigma \to \sigma + 2\pi/N$. Moreover we assume that the cross-section has also a reflectional symmetry; and without loss of generality let the d_2-axis be the symmetry axis. Then the Hamiltonian $H = H(m, n)$ is invariant under the reflections $S_1 : (m, n) \to (m_1, -m_2, -m_3, -n_1, n_2, n_3)$ and $S_3(m, n) \to (m_1, m_2, -m_3, -n_1, -n_2, n_3)$ which correspond to reflection of material points in the rod with respect to the d_2, d_3-plane and d_1, d_2-plane respectively. Observe that only the composition $S = S_1 S_3$ maps $P_{a,b}$ onto itself if $b \neq 0$. Since $S : (r, s, \sigma, \varrho) \to (r, s, -\sigma, -\varrho)$ the Hamiltonian $\tilde{H}_b$ satisfies

$$\tilde{H}_b(r, s, \sigma, \varrho, \varepsilon, \delta) = \tilde{H}_b(r, s, \sigma + 2\pi/N, \varrho, \varepsilon, \delta) = \tilde{H}_b(r, s, -\sigma, -\varrho, \varepsilon, \delta). \quad (4.2)$$

To find the lowest order at which the first nontrivial D_N-symmetric term can occur we have to appeal to the methods in invariant theory (*cf.* BUZANO, GEYMONAT, & POSTON [1985]). In our case the corresponding polynomials, being D_N-invariant but not rotationally symmetric, are of the form $\text{Re}(m_1 + im_2)^k (n_1 + in_2)^{N-k}$ where $k = 0, 1, \ldots, N$. However, since our scaling is of the form $m = \delta\hat{m}$, $n = \delta^2\hat{n}$, (recall $\varepsilon = 1$ here) the lowest order term is $\text{Re}(m_1 + im_2)^N$ with order $\mathcal{O}(\delta^N)$.

For $N \geq 3$ and under the assumption that $H = H(m, n)$ is $N + 1$ times continuously differentiable, we conclude that the scaled Hamiltonian has the form

$$\hat{H}_b(r, s, \sigma, \varrho, \varepsilon, \delta) = \hat{H}_b^*(r, s, \varrho, \varepsilon, \delta) + c\,\delta^{N-2}(r^2 - s^2)^{N/2} \cos N\sigma + \mathcal{O}(\delta^{N-1}).$$

$$(4.3)$$

Observe that the quadratic terms are σ-independent, i.e. $H(m, n) = \frac{1}{2}(\alpha(m_1^2 + m_2^2) + \alpha_3 m_3^2) + \mathcal{O}(|m|^3)$ and thus $\alpha_1 = \alpha_2 = \alpha < \alpha_3$. Without loss of generality we restrict our attention to the case $\varepsilon = 1$.

We can now state the second main result:

Theorem 4.2. *Let the Hamiltonian satisfy* (4.2) *with* $N \geq 3$ *and assume that c in* (4.3) *is nonzero. Then, for all sufficiently small* $\delta > 0$ *and all b with* $0 < b^2 < 4/\alpha$

there exists a periodic solution $m = \delta b e_3 + \mathcal{O}(\delta^2)$, $n = \delta^2 e_3 + \mathcal{O}(\delta^3)$ on $P_{\delta^2, \delta^3 b} \wedge \left\{ H = \delta^2 \left(1 + \frac{\alpha_3}{2} b^2 \right) \right\}$ which possesses transverse homoclinic orbits.

To appreciate the physical implications of these results, we must anticipate some of the material outlined in Section 5. The theorems are proved by perturbation arguments involving solutions lying close to homoclinic or heteroclinic orbits to hyperbolic saddle points of the unperturbed systems. The existence of transverse homoclinic or heteroclinic points then implies, via the Smale-Birkhoff homoclinic theorem (GUCKENHEIMER & HOLMES [1983, § 5.3]), that there exist solutions which remain in a neighborhood of the unperturbed homoclinic or heteroclinic orbits and which are chaotic in the following sense. There are two (or more) disjoint closed sets in the phase space which the solutions pass through in any prescribed sequence. The sets can be chosen to lie in any neighborhood of the saddle point(s); in the event of a set of multiple homoclinic orbits or heteroclinic cycles, this implies that the perturbed solution passes near different members of the set in any order. The consequence is that nonperiodic orbits near the transverse homoclinic or heteroclinic orbits are, to first order, quasi-random superpositions of single, unperturbed homoclinic or heteroclinic orbits. Thus, to understand typical global structures of perturbed orbits, and the spatial equilibrium states to which they correspond, we must first consider the spatial states corresponding to unperturbed orbits.

As we observed at the end of Section 2, to obtain equilibrium shapes from the stresses $(m(t), n(t))$, we must integrate equations (2.17a, b) and recall the original spatial description of the deformed rod of equation (2.1). We will concentrate on the implications for the vector $r(t)$ describing the position of the axis of the deformed rod. To arrange the discussion clearly we will only deal with the scaling limit $\delta = 0$. Thus we scale $r(t)$ and $R(t)$ in the following way

$$r(t) = \frac{1}{\delta} \hat{r}(\delta t), \quad R(t) = \hat{R}(\delta t). \tag{4.4}$$

For $\delta = 0$ we are left with

$$\hat{r}' = \hat{R} e_3, \quad \hat{R}' = \hat{R} \begin{bmatrix} 0 & -\alpha_3 \hat{m}_3 & \alpha_2 \hat{m}_2 \\ \alpha_3 \hat{m}_3 & 0 & -\alpha_1 \hat{m}_1 \\ -\alpha_2 \hat{m}_2 & \alpha_1 \hat{m}_1 & 0 \end{bmatrix} \tag{4.5a, b}$$

if $\hat{m}(t)$ is already known.

In the case $0 < \alpha_1 < \alpha_2 < \alpha_3$ the four heteroclinic solutions on the manifold $|\hat{m}| = 1$, $|\hat{n}| = 0$ are given by

$$\begin{bmatrix} \hat{m}_1(t) \\ \hat{m}_2(t) \\ \hat{m}_3(t) \end{bmatrix} = \begin{bmatrix} K_1 \sqrt{-a_1/a_2} \operatorname{sech}(\sqrt{a_1 a_3}\, t) \\ -K_1 K_2 \tanh(\sqrt{a_1 a_3}\, t) \\ K_2 \sqrt{-a_3/a_2} \operatorname{sech}(\sqrt{a_1 a_3}\, t) \end{bmatrix} \tag{4.6}$$

where $K_1, K_2 \in \{-1, 1\}$, $a_1 = \alpha_3 - \alpha_2 > 0$, $a_2 = \alpha_1 - \alpha_3 < 0$, and $a_3 = \alpha_2 - \alpha_1 > 0$. It will be convenient to express the rotation matrix $\hat{R}$ in terms of the

classical Euler angles ψ, θ, φ:
$$\hat{R} = D_\psi \bar{D}_\theta D_\varphi, \tag{4.7a}$$
where
$$D_\psi = \begin{bmatrix} \cos\psi & -\sin\psi & 0 \\ \sin\psi & \cos\psi & 0 \\ 0 & 0 & 1 \end{bmatrix}, \quad \bar{D}_\theta = \begin{bmatrix} 1 & 0 & 0 \\ 0 & \cos\theta & -\sin\theta \\ 0 & \sin\theta & \cos\theta \end{bmatrix}. \tag{4.7b}$$

Since $F = \hat{R}\hat{n} = 0$ for the heteroclinic orbits, we have $M = \hat{R}\hat{m} = $ const. We are free to orient our spatial coordinates and, therefore, we pick $M = e_3$ since this simplifies the calculations. Use of the definition (4.7) yields then
$$\hat{m} = \hat{R}^T e_3 = \begin{bmatrix} \sin\theta \sin\varphi \\ \sin\theta \cos\varphi \\ \cos\theta \end{bmatrix}. \tag{4.8}$$

Comparing this expression with (4.6), we find
$$\cos\theta = \hat{m}_3, \quad \varphi = \arctan(\hat{m}_1/\hat{m}_2). \tag{4.9}$$

We will also require ψ. From (4.5b) and (4.7) we derive, after some calculation, the relation
$$\psi' \sin\theta = \alpha_1 \hat{m}_1 \sin\varphi + \alpha_2 \hat{m}_2 \cos\varphi, \tag{4.10}$$
or, with (4.8) and (4.9),
$$\psi' = \alpha_1 \sin^2\varphi + \alpha_2 \cos^2\varphi = \frac{\alpha_1 \hat{m}_1^2 + \alpha_2 \hat{m}_2^2}{\hat{m}_1^2 + \hat{m}_2^2}. \tag{4.11}$$

Assuming without loss of generality that $\psi(0) = 0$, by direct integration we obtain
$$\psi(t) = \alpha_2 t - \arctan(\sqrt{a_3/a_1} \tanh(\sqrt{a_1 a_3}\, t)). \tag{4.12}$$

We are now in a position to compute the equilibrium shape in terms of the displacement vector $\hat{r}(t)$ by integrating (4.5a). Using (4.7) we obtain
$$\hat{r} = \hat{r}(0) + \int_0^t \begin{bmatrix} \sin\theta \sin\psi \\ -\sin\theta \cos\psi \\ \cos\theta \end{bmatrix} dt. \tag{4.13}$$

Realizing that $s = \hat{m}_3 = \cos\theta$, $\sin\theta = \sqrt{1-s^2}$, and $\psi(t) = -\varrho(t)$ (cf. (6.2)), we use the relations (A.5a, b) to obtain
$$\hat{r}_1(t) = \hat{r}_1(0) - \frac{1}{\alpha_2}\sin\theta\cos\psi + \frac{\sqrt{-a_1 a_2}}{\alpha_2}\int_0^t \cos^2\theta \sin\alpha_2 t\, dt,$$
$$\hat{r}_2(t) = \hat{r}_2(0) + \frac{1}{\alpha_2}\sin\theta\sin\psi + \frac{\sqrt{-a_1 a_2}}{\alpha_2}\int_0^t \cos^2\theta \cos\alpha_2 t\, dt, \tag{4.14}$$
$$\hat{r}_3(t) = \hat{r}_3(0) + K_2 \frac{2}{\sqrt{-a_1 a_2}}\arctan(\tanh(\sqrt{a_1 a_3}\, t/2)).$$

To understand the geometric shape of these solutions we first look at the limits $t \to \pm\infty$. Since $\cos\theta(t) \to 0$ and $\psi(t) - \alpha_2 t \to \pm A_1$ with $A_1 = \arctan(\sqrt{a_3/a_1})$, we have for $t \to \pm\infty$

$$\hat{r}(t) = \hat{r}(0) + \begin{bmatrix} A_2 \\ \pm A_3 \\ \pm K_2 A_4 \end{bmatrix} + \frac{1}{\alpha_2} \begin{bmatrix} \cos(\alpha_2 t \pm A_1) \\ \sin(\alpha_2 t \pm A_1) \\ 0 \end{bmatrix} + \mathcal{O}(e^{-\sqrt{a_1 a_3}|t|}), \quad (4.15)$$

where A_2 is some constant and A_3 and A_4 are given by

$$A_3 = \frac{\pi}{2\sqrt{-a_1 a_2}} \operatorname{cosech}\left(\frac{\pi\alpha_2}{2\sqrt{a_1 a_3}}\right) < \frac{1}{\alpha_2}, \quad A_4 = \pi/2\sqrt{-a_1 a_2}. \quad (4.16)$$

The oscillatory behavior for large t corresponds to the fact that the saddle points $\hat{m} = (0_1, \pm 1, 0)$, $\hat{n} = 0$, represent rods in pure bending, coiled onto themselves in one plane, here parallel to the 1, 2-plane. The full rod configuration, obtained by numerical integration of (4.14) from $t = -10$ to 10, is sketched in Figure 3.

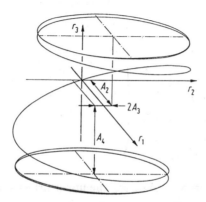

Fig. 3. Sketch of the rod configuration corresponding to the unperturbed heteroclinic orbits for $0 < \alpha_1 = 2 - 1/\sqrt{3} < \alpha_2 = 2 < \alpha_3 = 2 + \sqrt{3}$.

As Figure 2 indicates, on the sphere $|\hat{m}| = 1$ there are four heteroclinic solutions. Yet we obtain only two different geometric shapes ($K_2 = \pm 1$). The difference between the cases $K_1 = +1$ and $K_1 = -1$ in (4.6) is obtained by changing φ to $\varphi + \pi$, i.e. the centerline stays the same but the rod is first rotated by π around its centerline and then bent into the same configuration.

To obtain an idea of what the chaotic equilibrium states of such a rod may look like, we remind the reader that such solutions always remain near one of the perturbed heteroclinic solutions. These are themselves close ($\mathcal{O}(\varepsilon)$) to the unperturbed solutions. Hence we may think of the chaotic states as fairly arbitrary combinations of "elements", each of which is, up to $\mathcal{O}(\varepsilon)$, an unperturbed solution taken over a large but finite arclength. Of course we have to satisfy two conditions. First there is a "dynamical" condition that near a saddle point there are only three choices for the continuing solution: either it remains there or leaves in the neighborhood of one of the two branches of the unstable manifold. Second, there is the overall compatibility of the equilibrium to consider. We must arrange each "elementary" configuration in $\mathbb{R}^3$ in such a way that the resultants F and M are the same as for the other elements.

The two other cases, $0 < \alpha_1 < \alpha_3 < \alpha_2$ and $0 < \alpha_1 = \alpha_2 < \alpha_3$, lead in a similar fashion to the corresponding shapes of heteroclinic or homoclinic solutions in the unperturbed case. For $0 < \alpha_1 < \alpha_3 < \alpha_2$ the four heteroclinic solutions are

$$\hat{m}(t) = \begin{bmatrix} K_1 \sqrt{-a_1/a_3} \, \text{sech}(\sqrt{a_1 a_2}\, t) \\ K_2 \sqrt{-a_2/a_3} \, \text{sech}(\sqrt{a_1 a_2}\, t) \\ K_1 K_2 \tanh(\sqrt{a_1 a_2}\, t) \end{bmatrix}. \tag{4.17}$$

From (4.8) we obtain $\cos \theta = K_1 K_2 \tanh(\sqrt{a_1 a_2}\, t)$ and $\varphi = K_1 K_2 \arctan \sqrt{a_1/a_2}$. Hence, (4.11) yields $\psi(t) = \alpha_3 t$ and (4.13) gives

$$\hat{r}(t) = \hat{r}(0) + \int_0^t \begin{bmatrix} \text{sech}(\sqrt{a_1 a_2}\, t) \sin \alpha_3 t \\ -\text{sech}(\sqrt{a_1 a_2}\, t) \cos \alpha_3 t \\ K_1 K_2 \tanh(\sqrt{a_1 a_2}\, t) \end{bmatrix} dt.$$

In the limit $t \to \pm\infty$ $\hat{r}(t)$ behaves as follows

$$\hat{r}(t) = \hat{r}(0) + \begin{bmatrix} B_1 \\ \pm B_2 \\ \pm B_3 \end{bmatrix} + K_1 K_2 \begin{bmatrix} 0 \\ 0 \\ t \end{bmatrix} + \mathcal{O}(e^{-\sqrt{a_1 a_2}|t|}) \tag{4.18}$$

where $B_2 = \dfrac{\pi}{2\sqrt{a_1 a_2}} \, \text{sech}\left(\dfrac{\pi \alpha_3}{2\sqrt{a_1 a_2}}\right)$.

In the case $0 < \alpha_1 = \alpha_2 = \alpha < \alpha_3$ the unperturbed solution is a homoclinic solution defined by

$$r(t) = \sqrt{b^2 + \gamma^2 \, \text{sech}^2\left(\frac{\alpha \gamma}{2} t\right)}, \quad s = b,$$

$$\sigma(t) = \left(\frac{\alpha}{2} - \alpha_3\right) bt, \quad \cos \varrho = \frac{\alpha}{2} r^2 - 1, \tag{4.19}$$

where $\gamma^2 = \dfrac{4}{\alpha} - b^2$. Now $F = R\hat{n}$ is different from zero, and we may assume $F = e_3$. In a manner similar to (4.8) we obtain by using (3.1)

$$\hat{n} = \begin{bmatrix} \dfrac{b\bar{r}}{r^2} \cos \sigma (1 + \cos \varrho) - \dfrac{\bar{r}}{r} \sin \sigma \sin \varrho \\ \dfrac{b\bar{r}}{r^2} \sin \sigma (1 + \cos \varrho) + \dfrac{\bar{r}}{r} \cos \sigma \sin \varrho \\ \dfrac{b^2}{r^2} - \dfrac{\bar{r}^2}{r^2} \cos \varrho \end{bmatrix} = \begin{bmatrix} \sin \theta \, \sin \varphi \\ \sin \theta \, \cos \varphi \\ \cos \theta \end{bmatrix}.$$

Hence, (4.19) gives us

$$\cos\theta = 1 - \frac{\alpha\gamma^2}{2}\operatorname{sech}^2\left(\frac{\alpha\gamma}{2}t\right).$$

Using (4.13) again, we find that the foregoing implies for $\hat{r}(t)$ the limit behavior

$$\hat{r}(t) = \begin{bmatrix} \pm C_1 \\ C_2 \\ \pm C_3 \end{bmatrix} + \begin{bmatrix} 0 \\ 0 \\ t \end{bmatrix} + \mathcal{O}\left(e^{-\frac{\alpha\gamma}{2}|t|}\right) \qquad (4.20)$$

for $t \to \pm\infty$. In these two cases the asymptotic states (corresponding to the saddle points in (m, n)-space) are straight rods; in (4.18) in pure torsion and in (4.20) in mixed torsion and tension.

5. Second Reduction: Melnikov's Method

In this section we outline the method of MELNIKOV [1963], as generalized and adapted to the analysis of Hamiltonian systems having two degrees of freedom by HOLMES & MARSDEN [1982, 1983]. We first outline the reduction procedure (cf. BIRKHOFF [1927, Ch. 8], WHITTAKER [1937, Ch. 12], ARNOLD [1978, § 45B]).

We start with a Hamiltonian of the form

$$H_\varepsilon = H_0(q, p, I) + \varepsilon H_1(q, p, \theta, I, \varepsilon), \qquad (5.1)$$

where (q, p) are conjugate variables and (I, θ) are conjugate variables in action-angle form, so that H_1 is 2π-periodic in θ. For $\varepsilon = 0$ θ is a cyclic variable and Hamilton's equations are completely integrable. We suppose $H_1(q, p, \theta, I, 0)$ is bounded on bounded sets and that H_0 and H_1 are sufficiently differentiable for the power series manipulations which follow (C^3 will suffice).

Our specific assumptions on the unperturbed Hamiltonian $H_0(q, p, I)$ are that
(1) The system

$$\dot{q} = \partial H_0/\partial p, \quad \dot{p} = -\partial H_0/\partial q \qquad (5.2)$$

possesses a homoclinic orbit $\bar{x}_h = (\bar{q}(t - t_0; h), \bar{p}(t - t_0; h))$ to a hyperbolic fixed point $x_0 = (q_0, p_0)$ for each total energy $H_0 = h$ in some interval $J \subset \mathbb{R}$. Note that $\bar{x}_h$ depends on h via the action $I = I_h$ corresponding to the homoclinic orbit and total energy: $H_0(\bar{x}_h, I_h) = h$.

(2) For $h \in J$ and $(q, p) = \bar{x}_h(t - t_0)$, the frequency

$$\Omega_0 = \frac{\partial H_0}{\partial I} = \Omega_0(\bar{x}_h(t - t_0), I_h) \qquad (5.3)$$

of the unperturbed system $H_0(\bar{x}_h, I_h) = h$ satisfies $|\Omega_0| \geq \delta > 0$.

Under these hypotheses, we may invert the equation

$$H_\varepsilon(q, p, \theta, I) = h \in J \qquad (5.4)$$

in a neighborhood of the unperturbed homoclinic orbit and solve for the action I_h as a function of $(q, p, \theta; h)$:

$$I_h = \mathscr{I}_\varepsilon(q, p, \theta; h) = \mathscr{I}_0 + \varepsilon \mathscr{I}_1 + \mathcal{O}(\varepsilon^2). \tag{5.5}$$

Moreover, hypothesis (2) implies that the equation $d\theta/dt = \Omega_\varepsilon = \Omega_0 + \varepsilon \dfrac{\partial H_1}{\partial I}$ can be inverted for small ε, and hence that "real" time t can be replaced by the angle θ to give

$$\frac{dq}{d\theta} = \frac{dq}{dt} \bigg/ \frac{d\theta}{dt} = \Omega_\varepsilon^{-1} \frac{\partial H_\varepsilon}{\partial p}, \quad \frac{dp}{d\theta} = -\Omega_\varepsilon^{-1} \frac{\partial H_\varepsilon}{\partial q}. \tag{5.6}$$

Implicit differentiation of $H_\varepsilon = h$ yields

$$\frac{\partial H_\varepsilon}{\partial q} + \Omega_\varepsilon \frac{\partial \mathscr{I}_\varepsilon}{\partial q} = 0 = \frac{\partial H_\varepsilon}{\partial p} + \Omega_\varepsilon \frac{\partial \mathscr{I}_\varepsilon}{\partial p}, \tag{5.7}$$

so that, from (5.6)–(5.7), we obtain the reduced equation

$$q' = -\frac{\partial \mathscr{I}_\varepsilon}{\partial p}(q, p, \theta; h), \quad p' = \frac{\partial \mathscr{I}_\varepsilon}{\partial p}(q, p, \theta; h) \tag{5.8}$$

on each energy surface $H_\varepsilon = h \in J$. Here $(\ )'$ denotes $\dfrac{d}{d\theta}(\ )$, differentiation with respect to the angle variable, which plays the rôle of the new time.

The series expansion (5.5) for $\mathscr{I}_\varepsilon$ can be computed directly by expanding $H_0 + \varepsilon H_1 = h$ with I replaced by $\mathscr{I}_0 + \varepsilon \mathscr{I}_1 + \ldots$ One obtains

$$\mathscr{I}_0 = \mathscr{I}_0(q, p; h) = H_0(q, p)^{-1}(h), \tag{5.9a}$$

$$\mathscr{I}_1 = \mathscr{I}_1(q, p, \theta; h) = \frac{-H_1(q, p, \theta, H_0^{-1}(q, p)(h))}{\Omega_0(q, p, H_0^{-1}(q, p)(h))}, \tag{5.9b}$$

where $H_0(q, p)^{-1}(h)$ denotes inversion of H_0 with respect to the variable I. Thus (5.8) takes the form of a periodic perturbation of an integrable Hamiltonian system: specifically, from (5.7) and the expansion (5.5), (5.9), we have

$$\begin{aligned} q' &= -\frac{\partial \mathscr{I}_0}{\partial p} - \varepsilon \frac{\partial \mathscr{I}_1}{\partial p} \\ &\qquad\qquad\qquad + \mathcal{O}(\varepsilon^2), \\ p' &= \frac{\partial \mathscr{I}_0}{\partial q} + \varepsilon \frac{\partial \mathscr{I}_1}{\partial q} \end{aligned} \tag{5.10}$$

and the unperturbed vector field $\left(-\dfrac{\partial \mathscr{I}_0}{\partial p}, \dfrac{\partial \mathscr{I}_0}{\partial q}\right)$ is simply $\Omega_0^{-1} \left(\dfrac{\partial H_0}{\partial p}, -\dfrac{\partial H_0}{\partial q}\right)$, a scaled version of the unperturbed field of the original problem, restricted to (q, p) space. Thus hypothesis (1) implies that, for $\varepsilon = 0$, (5.10) has a homoclinic orbit to a hyperbolic fixed point.

The standard MELNIKOV method as developed in GUCKENHEIMER & HOLMES [1953, § 4.5] can be applied directly to (5.10). We have

Proposition 5.1. *Let $\bar{x}_h = (\bar{q}_h(\theta - \theta_0), \bar{p}_h(\theta - \theta_0))$ denote the homoclinic orbit to the fixed point p_0 of the unperturbed Hamiltonian system $\mathscr{I}_0(q, p; h)$ in the energy surface $H_\varepsilon = h$ and define the Melnikov function*

$$M_h(\theta_0) = \int_{-\infty}^{\infty} \{\mathscr{I}_0, \mathscr{I}_1\} (\bar{q}_h(\theta), \bar{p}_h(\theta), \theta + \theta_0) \, d\theta. \tag{5.11}$$

Then for $\varepsilon \neq 0$ sufficiently small, if $M_h(\theta_0)$ has simple zeros, the stable and unstable manifolds of the perturbed fixed point $p_\varepsilon = p_0 + \mathcal{O}(\varepsilon)$ of the Poincaré map corresponding to (5.8) intersect transversely for the perturbed system $\mathscr{I}_\varepsilon$. If $M_h(\theta_0)$ is bounded away from zero, then the manifolds do not intersect.

Proof. See GREENSPAN & HOLMES [1982] or GUCKENHEIMER & HOLMES [1983]. The main ideas and original proof are due to MELNIKOV [1963] (*cf.* ARNOLD [1964]).

It is unnecessary to compute $\mathscr{I}_0$ and $\mathscr{I}_1$ explicitly, for we have

Lemma 5.2. (HOLMES & MARSDEN [1983])

$$M_h(\theta_0) = \int_{-\infty}^{\infty} \left\{H_0, \frac{H_1}{\Omega_0}\right\}_{(q,p)} (\bar{q}_h(t), \bar{p}_h(t), I_h, \bar{\theta}(t) + \theta_0) \, dt, \tag{5.12}$$

where $\{\cdot, \cdot\}_{(q,p)}$ denotes that only the variables (q, p) are used in the bracket evaluation. I_h is the (constant) action given by $H_0(\bar{q}, \bar{p}, I_h) = h$ and $\bar{\theta}(t) = \int_0^t \Omega(\bar{q}_h(s), \bar{p}_h(s), I_h) \, ds$.

Proof. Consider the equation

$$H_0(q, p, \mathscr{I}_0 + \varepsilon \mathscr{I}_1) + \varepsilon H_1(q, p, \theta, \mathscr{I}_0 + \varepsilon \mathscr{I}_1) = h + \mathcal{O}(\varepsilon^2);$$

this implies that

$$\frac{\partial H_0}{\partial q} = -\Omega_0 \frac{\partial \mathscr{I}_0}{\partial q} \frac{\partial H_0}{\partial p} = -\Omega_0 \frac{\partial \mathscr{I}_0}{\partial p} \tag{5.13}$$

as well as $\mathscr{I}_1 = -H_1/\Omega_0$, as in (5.9b). Therefore

$$\{\mathscr{I}_0, \mathscr{I}_1\} = \frac{\partial \mathscr{I}_0}{\partial q} \frac{\partial \mathscr{I}_1}{\partial p} - \frac{\partial \mathscr{I}_0}{\partial p} \frac{\partial \mathscr{I}_1}{\partial q}$$

$$= -\frac{1}{\Omega_0} \frac{\partial H_0}{\partial q} \frac{\partial}{\partial p}(H_1/\Omega_0) + \frac{1}{\Omega_0} \frac{\partial H_0}{\partial p} \frac{\partial}{\partial q}(-H_1/\Omega_0)$$

$$= \frac{1}{\Omega_0} \left\{H_0, \frac{H_1}{\Omega_0}\right\}_{(q,p)}. \tag{5.14}$$

Since $d\theta = \Omega \, dt$, substitution of (5.14) in (5.11) yields (5.12). □

Proposition 5.1 and Lemma 5.2 together yield:

Theorem 5.3. *If $H_\varepsilon = H_0 + \varepsilon H_1$ satisfies hypotheses (1) and (2) and the Melnikov function $M_h(\theta_0)$ of (5.12) has simple zeros, then, for $\varepsilon \neq 0$ sufficiently small, there exist transverse homoclinic orbits to a hyperbolic periodic orbit on the energy surface $H_\varepsilon = h$.*

The Smale-Birkhoff homoclinic theorem (SMALE [1963], [1967], GUCKENHEIMER & HOLMES [1983]) then implies

Corollary 5.4. *The Poincaré map associated with H_ε on the level set $H_\varepsilon^{-1}(h)$ has a hyperbolic, non-wandering Cantor set Ω_h on which the map is conjugate to a subshift of finite type.*

As MOSER [1973] shows, this in turn implies

Corollary 5.5. *H_ε possesses no analytic integrals of motion independent of the total energy H_ε itself.*

In the first two situations treated in this paper, rather than a homoclinic orbit to a fixed point we have a cycle of four heteroclinic orbits connecting a pair of saddle points (*cf.* Figure 2). The transverse homoclinic orbits of Theorem 5.3 become transverse heteroclinic cycles, but otherwise it and the conclusions of the corollaries stand unchanged. See HOLMES & MARSDEN [1983, Figure 5] for an impression of the structure of such cycles.

6. Proof of Theorem 4.1

To prove the theorem we compute Melnikov functions for suitably scaled versions of the Hamiltonian. Specifically, for case I, $\alpha_1 < \alpha_2 < \alpha_3$, we take (3.15) and for case II, $\alpha_1 < \alpha_3 < \alpha_2$, (3.16). In particular we restrict the computations to the limit case $\delta = 0$; this suffices since the dependence on δ is continuous. The conclusions of the theorem then follow upon application of Theorem 5.3. Certain computational details are relegated to the Appendix.

We remark that the unperturbed Hamiltonians $H_0(r, s, \sigma)$ differ only in transposition of α_2 and α_3. It therefore suffices to consider the unperturbed heteroclinic solutions only the first case. The unperturbed Hamilton's equations may be written

$$\dot{s} = \frac{\bar{r}^2}{2} \beta'(\sigma),$$

$$\dot{\sigma} = s(\beta(\sigma) - \alpha_3), \qquad (6.1)$$

$$\dot{r} = 0,$$

$$\dot{\varrho} = -r\beta(\sigma) \stackrel{\text{def}}{=} \Omega(r, \sigma)$$

where $\beta(\sigma) = \alpha_1 \cos^2 \sigma + \alpha_2 \sin^2 \sigma$ and we recall $\tilde{r}^2 = r^2 - s^2$. Without loss of generality we fix the unperturbed "momentum" $r = 1$. in which case one of the four heteroclinic orbits connecting the fixed points $(s, \sigma) = (0, \pm\pi/2)$ takes the form

$$s = \sqrt{-a_3/a_2} \operatorname{sech}(-\sqrt{a_1 a_3} \, t),$$
$$\sigma = \arctan(\sqrt{-a_2/a_1} \sinh(-\sqrt{a_1 a_3} \, t)), \quad (6.2)$$
$$\varrho = -[\alpha_2 t + \arctan(\sqrt{a_3/a_1} \tanh(-\sqrt{a_1 a_3} \, t))],$$

where $a_1 = \alpha_3 - \alpha_2 > 0$, $a_2 = \alpha_1 - \alpha_3 < 0$, $a_3 = \alpha_2 - \alpha_1 > 0$ and $a_1 + a_2 + a_3 = 0$. Moreover $\sigma(t) \to \pm\pi/2$ and $\varrho(t) \to \pm\infty$ as $t \to \pm\infty$ and the orbits lie on the level set $H_0 = \alpha_2/2$. The other heteroclinic orbits are obtained by appropriate sign changes (*cf.* Figure 2).

To apply Lemma 5.2, we must compute

$$M(\varrho_0) = \int_{-\infty}^{\infty} \left\{ H_0, \frac{H_1}{\Omega} \right\}_{(s,\sigma)} (s_h(t), \sigma_h(t), r_h = 1, \varrho_h(t) + \varrho_0) \, dt.$$

The Poisson bracket is given by

$$\frac{\partial H_0}{\partial s} \frac{\partial (H_1/\Omega)}{\partial \sigma} - \frac{\partial H_0}{\partial \sigma} \frac{\partial (H_1/\Omega)}{\partial s}$$

$$= s(\alpha_3 - \beta(\sigma))(b_s - \tilde{r}r \cos(\varrho + \varrho_0)) \frac{\partial \beta/\partial \sigma}{r^2 \beta^2(\sigma)} + \frac{\tilde{r}^3 \partial \beta/\partial \sigma}{2r^2 \beta(\sigma)} \left(b + \frac{\tilde{r}s}{\tilde{r}} \cos(\varrho + \varrho_0) \right).$$
$$(6.3)$$

Using (6.2), we see that the ϱ_0-independent part of (6.3) is odd in t and therefore vanishes in integration to yield

$$M(\varrho_0) = \int_{-\infty}^{\infty} \frac{\tilde{r}r \, \partial \beta/\partial \sigma}{2\beta^2(\sigma)} s(3\beta(\sigma) - 2\alpha_3) \cos(\varrho + \varrho_0) \, dt$$

$$= \sqrt{1 - b^2} \left(\int_{-\infty}^{\infty} \frac{s\dot{s}}{\beta^2(\sigma) \sqrt{1 - s^2}} (2\alpha_3 - 3\beta(\sigma)) \sin \varrho \, dt \right) \sin \varrho_0, \quad (6.4)$$

where we have again used the odd/even properties of (6.2) and the differential relation $\partial \beta/\partial \sigma = 2\dot{s}/\tilde{r}^2$, as well as setting $r = 1$. Equation (6.4) is evaluated in the Appendix to yield

$$M(\varrho_0) = \frac{\sqrt{1 - b^2} \, \pi}{\sqrt{(\alpha_3 - \alpha_2)(\alpha_3 - \alpha_1)}} \operatorname{cosech}\left(\frac{\alpha_2 \pi}{2\sqrt{(\alpha_3 - \alpha_2)(\alpha_2 - \alpha_1)}} \right) \sin \varrho_0. \quad (6.5)$$

This function has simple zeros for all $\alpha_1 < \alpha_2 < \alpha_3$ and $|b| < 1$, and the proof in case I is complete.

In case II we obtain the Poisson bracket from (3.16). As before, the Melnikov function has a constant (ϱ_0-independent) part and a part periodic in ϱ_0. However, here the constant part does not vanish identically. After using the properties of the unperturbed solutions (6.2) and interchanging the rôles of α_2 and α_3, so that now $\beta(\sigma) = \alpha_1 \cos^2 \sigma + \alpha_3 \sin^2 \sigma$, we obtain

$$M(\varrho_0) = b \int_{-\infty}^{\infty} \frac{\tilde{r}s}{\beta(\sigma)} \left[(\alpha_2 - \beta) \left(\frac{\partial \beta/\partial \sigma}{\beta(\sigma)} \sin \sigma - \cos \sigma \right) - \frac{\partial \beta/\partial \sigma}{2\beta(\sigma)} \sin \sigma \right] dt$$

$$+ \sqrt{1 - b^2} \left(\int_{-\infty}^{\infty} \frac{s(\alpha_2 - \beta(\sigma))}{\beta(\sigma)} \left[\left(\frac{\partial \beta/\partial \sigma}{\beta(\sigma)} \sin \sigma - \cos \sigma \right) s \cos \varrho \right. \right. \quad (6.6)$$

$$\left. + \left(\frac{\partial \beta/\partial \sigma}{\beta(\sigma)} \cos \sigma + \sin \sigma \right) \sin \varrho \right] dt + \int_{-\infty}^{\infty} \frac{\tilde{r}^2 \, \partial \beta/\partial \sigma}{2\beta(\sigma)} \sin \sigma \cos \varrho \, dt \right) \cos \varrho_0.$$

These integrals are evaluated in the Appendix to give

$$M(\varrho_0) = -\frac{2b}{\alpha_3} - \frac{\sqrt{1 - b^2} \, \pi}{\sqrt{(\alpha_2 - \alpha_3)(\alpha_3 - \alpha_1)}} \operatorname{sech} \left(\frac{\alpha_3 \pi}{2\sqrt{(\alpha_2 - \alpha_3)(\alpha_3 - \alpha_1)}} \right) \cos \varrho_0.$$

(6.7)

We conclude that, if

$$b^2 < \frac{\alpha_3^2 \pi^2}{\alpha_3^2 \pi^2 + 4(\alpha_2 - \alpha_3)(\alpha_3 - \alpha_1) \cosh^2 \left(\frac{\alpha_3 \pi}{2\sqrt{(\alpha_2 - \alpha_3)(\alpha_3 - \alpha_1)}} \right)}, \quad (6.8)$$

then $M(\varrho_0)$ has simple zeros and thus that transverse homoclinic orbits exist for ε, δ sufficiently small. The proof of Theorem 4.1 is complete. □

Remarks on Case II: $\alpha_1 < \alpha_3 < \alpha_2$.

In this case the Melnikov function has a constant part and, if $|b|$ is sufficiently close to 1, so that $|b| < 1$ but (6.8) is violated, then $M(\varrho_0)$ has no zeros. Thus, by

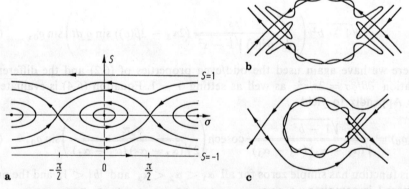

Fig. 4a–c. Hyperbolic fixed points and invariant manifolds in the (σ, s) cross section, $r = 1$. (a) Unperturbed problem, $\alpha_1 < \alpha_2 < \alpha_3$ and $\alpha_1 < \alpha_3 < \alpha_2$. (b) Perturbed problem, $\alpha_1 < \alpha_2 < \alpha_3$ and $\alpha_1 < \alpha_3 < \alpha_2$ if b^2 satisfies (6.8). (c) A possible situation if $\alpha_1 < \alpha_3 < \alpha_2$ and b^2 fails to satisfy (6.8) so that $M(\varrho_0)$ has no zero.

Proposition 5.1, the stable and unstable manifolds the two perturbed fixed points do not intersect. We observe that this does not necessarily imply that no transverse homoclinic orbits exist, since the stable and unstable manifolds of a single fixed point might still intersect, as indicated in Figure 4(c). However, the perturbation calculations give no information on this.

7. Proof of Theorem 4.2

As is shown in Section 4 the scaled Hamiltonian of a D_N-symmetric rod has the form

$$\tilde{H}_b(r, s, \sigma, \varrho, 1, \delta) = \tilde{H}_0(r, s, \varrho) + \delta \tilde{H}_s(r, s, \varrho, \delta) \qquad (7.1)$$
$$+ c\, \delta^{N-2}\, \bar{r}^N \cos N\sigma + \mathcal{O}(\delta^{N-1}),$$

where ε is chosen equal to 1 and

$$\tilde{H}_0(r, s, \varrho) = \tfrac{1}{2}(\alpha \bar{r}^2 + \alpha_3 s^2) + \frac{1}{r^2}(bs - \bar{r}r \cos \varrho). \qquad (7.2)$$

The rotationally symmetric part $\delta \tilde{H}_s$ is even in ϱ (cf. (4.2)) and allows for nonlinear terms of lower order ($\mathcal{O}(\delta^M)$, $1 \leq M \leq N-2$) than the first D_N-symmetric term $\delta^{N-2} \bar{r}^N \cos N\sigma$.

The unperturbed equations are

$$\dot{r} = \frac{\bar{r}r}{r^2} \sin \varrho,$$

$$\dot{\varrho} = -\alpha r + \frac{2bs}{r^3} + \frac{\cos \varrho}{r^3}\left(\frac{b^2 \bar{r}}{\bar{r}} + \frac{s^2 \bar{r}}{\bar{r}}\right), \qquad (7.3)$$

$$\dot{s} = 0,$$

$$\dot{\sigma} = (\alpha - \alpha_3) s - \frac{b}{r^2} - \frac{s\bar{r}}{r^2 \bar{r}} \cos \varrho \overset{\text{def}}{=} \Omega(r, s, \varrho).$$

Here σ is the cyclic variable. We restrict our attention to the special case $s = b$ with $0 < b^2 < \dfrac{4}{\alpha}$, for which the phase portrait in the (r, ϱ)-plane appears as in Figure 5. The line $r = b$ is degenerate in the (m, n) coordinate system: for $\bar{r} = \sqrt{r^2 - b^2} = 0$, (m, n) is independent of ϱ (cf. (3.1b)). Thus the heteroclinic solution lying on the level set

$$\tfrac{1}{2}(\alpha(r^2 - b^2) + \alpha_3 b^2) + \frac{b^2 - (r^2 - b^2)\cos \varrho}{r^2} = \frac{\alpha_3}{2} b^2 + 1 \qquad (7.4)$$

connecting the fixed points $(r, \varrho) = \left(b, \arccos\left(\dfrac{\alpha b^2}{2} - 1\right)\right)$ is a homoclinic orbit to the saddle point $(m, n) = (0, 0, b; 0, 0, 1)$ in the original coordinates. The condition $b^2 < \dfrac{4}{\alpha}$, necessary for the saddle to exist, corresponds to the angular momentum condition for a "slow top" in the classical Lagrange analysis (*cf.* GOLDSTEIN [1980, Ch. 5]).

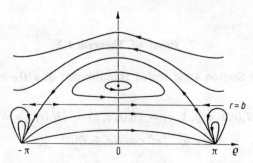

Fig. 5. Unperturbed (r, ϱ) phase plane for equation (7.3) with $s = b$.

We shall require an explicit expression for the homoclinic solution as a function of r, on the set $s = b$, as well as an implicit relation:

$$\tilde{r}(t) = \bar{r}(t) = \sqrt{r^2(t) - b^2} = \gamma \operatorname{sech}\left(\frac{\alpha\gamma}{2}t\right), \tag{7.5a}$$

$$r^2 = \frac{2}{\alpha}(1 + \cos\varrho), \tag{7.5b}$$

where $\gamma^2 = \dfrac{4}{\alpha} - b^2 > 0$.

To apply the Melnikov method we must compute the Poisson bracket

$$\left\{H_0, \frac{H_1}{\Omega}\right\} = \frac{\partial H_0}{\partial r}\left(\frac{1}{\Omega}\frac{\partial H_0}{\partial \varrho} - \frac{H_1}{\Omega^2}\frac{\partial \Omega}{\partial \varrho}\right) - \frac{\partial H_0}{\partial \varrho}\left(\frac{1}{\Omega}\frac{\partial H_0}{\partial r} - \frac{H_1}{\Omega^2}\frac{\partial \Omega}{\partial r}\right). \tag{7.6}$$

From (7.3) and (7.5b) we see that, on the homoclinic orbit, $\Omega(r, s = b, \varrho)$ is constant and takes the value

$$\Omega = b\left(\frac{\alpha}{2} - \alpha_3\right), \tag{7.7}$$

which is nonzero since $\alpha_3 > \alpha$.

The perturbation Hamiltonian H_1 divides into two parts: δH_s and $c\,\delta^{N-2}\,\tilde{r}^N \cos N\sigma$. Since H_0, H_s and Ω are even in ϱ, $\dfrac{\partial H_0}{\partial \varrho}$, $\dfrac{\partial H_0}{\partial \varrho}$ and $\dfrac{\partial \Omega}{\partial \varrho}$ are odd in ϱ and $\left\{H_0, \dfrac{\partial H_s}{\Omega}\right\}$ is also odd in ϱ. From this fact and the evenness of r, $\bar{r}$, and $\tilde{r}$ in t we deduce that this part of the Poisson bracket is odd in t and so does not contribute to the Melnikov integral. Thus we need only compute the part involving

$c\delta^{N-2} \bar{r}^N \cos N\sigma$. Using (7.6a, b), we obtain

$$\frac{\partial H_0}{\partial r} = \frac{\alpha \bar{r}^2}{r}, \qquad \frac{\partial H_0}{\partial \varrho} = \frac{\bar{r}^2}{r^2} \sin \varrho,$$

$$\frac{\partial (H_1/\Omega)}{\partial r} = \left(\frac{N r \bar{r}^{N-2}}{\Omega} - \frac{\alpha b \bar{r}^N}{\Omega^2 r}\right) \cos N\sigma, \qquad \frac{\partial (H_1/\Omega)}{\partial \varrho} = -\frac{c \bar{r}^N}{\Omega^2 r^2} \sin \varrho \cos N\sigma. \quad (7.8)$$

The Melnikov integral therefore reduces to

$$M(\sigma_0) = \int_{-\infty}^{\infty} \left(-\frac{c N \bar{r}^N}{\Omega r} \sin \varrho \cos (N(\sigma + \sigma_0))\right) dt, \quad (7.9)$$

where $\sigma(t) = \Omega t$, since Ω is constant.

Using (7.8) and the fact that $r, \bar{r}$ are even in t while $\sin \varrho$ is odd and $\dot{r} = (\bar{r}^2/r^2) \sin \varrho$, we may rewrite (7.9) as follows:

$$M(\varrho_0) = c\left(\int_{-\infty}^{\infty} \frac{\sin (N(\Omega t))}{\Omega} N r \bar{r}^{N-2} \dot{r} \, dt\right) \sin (N\sigma_0).$$

Integrating by parts and substituting from (7.5), we obtain

$$M(\varrho_0) = c\left(\int_{-\infty}^{\infty} -N \cos (N\Omega t) \bar{r}^N(t) \, dt\right) \sin (N\sigma_0)$$

$$= -N c \gamma^N \left(\int_{-\infty}^{\infty} \operatorname{sech}^N\left(\frac{\alpha \gamma}{2} t\right) \cos (N\Omega t) \, dt\right) \sin (N\sigma_0) \quad (7.10)$$

$$= \frac{-2 N c \gamma^{N-1}}{\alpha} \left(\int_{-\infty}^{\infty} \operatorname{sech}^N (\tau) \cos \left(\frac{2 N \Omega}{\alpha \gamma} \tau\right) d\tau\right) \sin (N\sigma_0).$$

Writing the integral of (7.10) as $I_N(\omega)$, $\omega = \frac{2N\Omega}{\alpha\gamma} = \frac{bN}{\gamma}\left(1 - \frac{2\alpha_3}{\alpha}\right)$, by integrating by parts twice we deduce the recurrence relation

$$I_{N+2}(\omega) = \frac{\omega^2 + N^2}{N(N+1)} I_N(\omega), \quad N \geq 0. \quad (7.11)$$

This relation, together with the evaluations of I_1 and I_2 by the calculus of residues, namely

$$I_1(\omega) = \pi \operatorname{sech}\left(\frac{\pi \omega}{2}\right),$$
$$I_2(\omega) = \pi \omega \operatorname{cosech}\left(\frac{\pi \omega}{2}\right), \quad (7.12)$$

guarantees that $I_N(\omega) \neq 0$ for all $N \geq 1$ and $\alpha, \gamma \neq 0$. We conclude that the Melnikov function (7.10) has simple zeroes under the hypotheses of the theorem. Applying Theorem 5.3 completes the proof. $\square$

Remark. We have assumed $N \geq 3$, so that the explicit σ-dependent term in the Hamiltonian appears at higher order. In the case $N = 2$ the relevant term is $\tilde{c} r^2 \cos 2\sigma$, which occurs at the same order as the rotationally symmetric part H_0.

In this case we have

$$H = \tfrac{1}{2}(\bar{r}^2(\alpha + 2c\cos 2\sigma) + \alpha_3 s^2) + \frac{bs - \bar{r}\tilde{r}\cos\varrho}{r^2}. \tag{7.14}$$

Here the problem reduces to the perturbed Lagrange top considered by HOLMES & MARSDEN [1983]. Using the fact that $\beta(\sigma) = \alpha_1\cos^2\sigma + \alpha_2\sin^2\alpha = \left(\dfrac{\alpha_1 + \alpha_2}{2}\right) + \left(\dfrac{\alpha_1 - \alpha_2}{2}\right)\cos 2\sigma$, we see explicitly that the condition $c \neq 0$ corresponds to inequality of the (inverse)-moments of inertia, $\alpha_1 \neq \alpha_2$.

Appendix. Computation of Melnikov Functions for Theorem 4.1

We give the necessary calculations to evaluate the first order approximations of the Melnikov functions in equations (6.4) and (6.6).

In the case $0 < \alpha_1 < \alpha_2 < \alpha_3$ the relevant parts of the Hamiltonian are

$$H_0 = \tfrac{1}{2}((r^2 - s^2)\beta(\sigma) + \alpha_3 s^2),$$

$$\frac{H_1}{\Omega} = \frac{\bar{r}\tilde{r}}{r^3\beta(\sigma)}\cos\varrho - \frac{bs}{r^3\beta(\sigma)}. \tag{A.1}$$

From (6.2) we deduce that the heteroclinic solution with $r = 1$ satisfies

$$s(t) = \sqrt{-a_3/a_2}\, S(t), \quad \beta(\sigma) = \frac{\alpha_2 - \alpha_3 s^2}{1 - s^2}, \tag{A.2a + b}$$

$$\sin\sigma(t) = -\frac{T(t)}{\sqrt{1 - s^2}}, \quad \cos\sigma(t) = \sqrt{-a_1/a_2}\,\frac{S(t)}{\sqrt{1 - s^2}}, \tag{A.2c + d}$$

$$\sin\varrho(t) = \sqrt{-a_1/a_2}\,\frac{1}{\sqrt{1 - s^2}}(\sqrt{a_3/a_1}\, T(t)\cos\alpha_2 t - \sin\alpha_2 t) \tag{A.2e}$$

$$\cos\varrho(t) = \sqrt{-a_1/a_2}\,\frac{1}{\sqrt{1 - s^2}}(\cos\alpha_2 t + \sqrt{a_3/a_1}\, T(t)\sin\alpha_2 t) \tag{A.2f}$$

where $a_1 = \alpha_3 - \alpha_2, a_2 = \alpha_1 - \alpha_3, a_3 = \alpha_2 - \alpha_1$ and

$$S(t) = \operatorname{sech}(\sqrt{a_1 a_3}\, t), \quad T(t) = \tanh(\sqrt{a_1 a_3}\, t).$$

We further need the relations

$$d\varrho = \dot{\varrho}\, dt = -\beta\, dt, \quad d\sigma = \dot{\sigma}\, dt = -s(\alpha_3 - \beta)\, dt, \tag{A.3a + b}$$

$$\dot{s} = \frac{\partial H}{\partial \sigma} = \tfrac{1}{2}\bar{r}^2\frac{\partial \beta}{\partial \sigma} = \tfrac{1}{2}(1 - s^2)\frac{\partial \beta}{\partial \sigma}, \tag{A.4}$$

$$\frac{d}{dt}(\sqrt{1 - s^2}\sin\varrho) = -\alpha_2\sqrt{1 - s^2}\cos\varrho + a_3\sqrt{-a_1/a_2}\, S^2(t)\cos\alpha_2 t, \tag{A.5a}$$

$$\frac{d}{dt}(\sqrt{1 - s^2}\cos\varrho) = \alpha_2\sqrt{1 - s^2}\sin\varrho + a_3\sqrt{-a_1/a_2}\, S^2(t)\sin\alpha_2 t. \tag{A.5b}$$

The Melnikov integral then yields

$$M(\varrho_0) = \int_R \left(\frac{\partial H_0}{\partial s} \frac{\partial (H_1/\Omega)}{\partial \sigma} - \frac{\partial H_0}{\partial \sigma} \frac{\partial (H_1/\Omega)}{\partial s} \right) dt$$

$$= \int_R \left\{ s(\alpha_3 - \beta) \left[bs - \sqrt{1-b^2} \sqrt{1-s^2} \cos(\varrho + \varrho_0) \right] \right.$$

$$\left. + \tfrac{1}{2} (1-s^2) \beta \left[b + \frac{\sqrt{1-b^2}}{\sqrt{1-s^2}} s \cos(\varrho + \varrho_0) \right] \right\} \frac{\frac{\partial \beta}{\partial \sigma}}{\beta^2} dt.$$

Since $s, \beta \cos \varrho$ are even in t and $\sin \varrho, \frac{\partial \tau}{\partial \sigma}$ are odd this integral simplifies, after use of (A.4), to

$$M(\varrho_0) = b \cdot 0 + \sqrt{1-b^2} I^* \sin \varrho_0,$$

$$I^* = \int_R \frac{s \dot{s}}{\sqrt{1-s^2} \beta^2} (2\alpha_3 - 3\beta) \sin \varrho \, dt,$$

as in equation (6.4). Expressing β in terms of s via (A.2b), we have

$$I^* = \int_R g(s) \dot{s} \sin \varrho \, dt, \tag{A.6}$$

where

$$g(s) = \frac{s\sqrt{1-s^2}}{(\alpha_2 - \alpha_3 s^2)^2} (2\alpha_3(1-s^2) - 3(\alpha_2 - \alpha_3 s^2))$$

and

$$G(s) = \int g(s) \, ds = \frac{(1-s^2)^{3/2}}{\alpha_2 - \alpha_3 s^2} = \frac{\sqrt{1-s^2}}{\beta}.$$

Hence, upon integration by parts, we have

$$\int g(s) \dot{s} \sin \varrho \, dt = G(s) \sin \varrho - \int G(s) \dot{\varrho} \cos \varrho \, dt$$

$$= \frac{\sqrt{1-s^2}}{\beta} \sin \varrho + \int \sqrt{1-s^2} \cos \varrho \, dt$$

$$= \frac{\sqrt{1-s^2}}{\beta} \sin \varrho - \frac{\sqrt{1-s^2}}{\alpha_2} \sin \varrho + \frac{a_3}{\alpha_2} \sqrt{\frac{a_1}{-a_2}} \int S^2(t) \cos \alpha_2 t \, dt.$$

The last equality is a consequence of (A.5a). Now

$$I^* = \left(\frac{(1-s^2)}{\alpha_2 - \alpha_3 s^2} - \frac{1}{\alpha_2} \right) \sqrt{1-s^2} \sin \varrho \Big|_{-\infty}^{\infty} + \frac{a_3}{\alpha_2} \sqrt{\frac{a_1}{-a_2}} \int_R S^2(t) \cos \alpha_2 t \, dt,$$

so that the boundary terms vanish and, letting $\tau = \sqrt{a_1 a_3}\, t$, we are left whith

$$I^* = \frac{1}{\alpha_2} \sqrt{\frac{a_3}{-a_2}} \int_R \operatorname{sech}^2(\tau) \cos\left(\frac{\alpha_2}{\sqrt{a_1 a_3}} \tau\right) d\tau$$

$$= \frac{\pi}{\sqrt{-a_1 a_2}} \operatorname{cosech}\left(\frac{\pi \alpha_2}{2\sqrt{a_1 a_3}}\right), \tag{A.7}$$

from which we obtain (6.5). The final integral is evaluated by the method of residues.

For the case $0 < \alpha_1 < \alpha_3 < \alpha_2$, as indicated in Section 4, we use the (r, s, σ, ϱ)-coordinates for the vectors $(m_1, m_3, m_2)^T$ and $(n_1, n_3, n_2)^T$ rather than for m and n. Thus, the unperturbed solutions are given by the formulae (6.2), (A.2) and (A.3), as above, but with α_2 and α_3 interchanged and the heteroclinic orbits now connect the points $m = (0, 0, \pm 1)$. The perturbation Hamiltonian, H_1, however is quite different. From (3.16) we have

$$H_0 = \tfrac{1}{2}((r^2 - s^2)\beta(\sigma) + \alpha_2 s^2),$$

$$\frac{H_1}{\Omega} = -\frac{b\bar{r} \sin \sigma}{r^3 \beta} - \frac{\bar{r}}{r^3 \beta}(s \sin \sigma \cos \varrho + r \cos \sigma \sin \varrho)$$

where now $\beta(\sigma) = \alpha_1 \cos^2 \sigma + \alpha_3 \sin^2 \sigma$.

The Melnikov function is

$$M(\varrho_0) = \int_{-\infty}^{\infty} \left(\frac{\partial H_0}{\partial s} \frac{\partial (H_1/\Omega)}{\partial \sigma} - \frac{\partial H_0}{\partial \sigma} \frac{\partial (H_1/\Omega)}{\partial s}\right) dt.$$

With $r = 1$ and after omitting the odd terms in the integrand, we are left with

$$M(\varrho_0) = b I_1 + \sqrt{1 - b^2}\, (I_2 - I_3) \cos \varrho_0, \tag{A.8}$$

where I_1, I_2 and I_3 are the integrals defined implicitly by comparison of (A.8) and (6.6). For the first, we have

$$I_1 = \int_{-\infty}^{\infty} \sqrt{1 - s^2} \left[(\alpha_2 - \beta(\sigma)) s \frac{d}{d\sigma}\left(-\frac{\sin \sigma}{\beta}\right) - \frac{s}{2\beta(\sigma)} \frac{\partial \beta}{\partial \sigma} \sin \sigma\right] dt. \tag{A.9}$$

Using $\dot{\sigma} = s(\beta - \alpha_2)$ from (6.1) (with $\alpha_3 \leftrightarrow \alpha_2$) and $\beta'(\sigma) = 2\dot{s}/\bar{r}^2$, we may rewrite (A.9) and obtain

$$I_1 = \int_{-\pi/2}^{\pi/2} \sqrt{1 - s^2} \frac{d}{d\sigma}\left(\frac{-\sin \sigma}{\beta(\sigma)}\right) d\sigma - \int_{-\infty}^{\infty} \frac{s \sin \sigma}{\beta(\sigma)\sqrt{1 - s^2}} \dot{s}\, dt$$

$$= \sqrt{1 - s^2}\left(\frac{-\sin \sigma}{\beta(\sigma)}\right)\bigg|_{-\pi/2}^{\pi/2} - \int_{-\pi/2}^{\pi/2} \frac{s \sin \sigma}{\beta(\sigma)\sqrt{1 - s^2}} \frac{ds}{d\sigma} d\sigma + \int_{-\pi/2}^{\pi/2} \frac{s \sin \sigma}{\beta(\sigma)\sqrt{1 - s^2}} \frac{ds}{d\sigma} d\sigma$$

$$= -\frac{2}{\alpha_3}, \tag{A.10}$$

since the two last integrals cancel one another.

In the ϱ_0-dependent part of (A.8) we have

$$I_2 = \int_{-\infty}^{\infty} s(\alpha_2 - \beta(\sigma)) \left[s \cos \varrho \frac{d}{d\sigma}\left(-\frac{\sin \sigma}{\beta(\sigma)}\right) + \sin \varrho \frac{d}{d\sigma}\left(-\frac{\cos \sigma}{\beta(\sigma)}\right) \right] dt, \quad \text{(A.11a)}$$

$$I_3 = \int_{-\infty}^{\infty} \frac{(1-s^2)}{2\beta(\sigma)} \frac{\partial \beta}{\partial \sigma} \sin \sigma \cos \varrho \, dt. \quad \text{(A.11b)}$$

Using $s(\beta - \alpha_2) = \dot{\sigma}$ again we have, in terms of $d\sigma$,

$$I_2 = \int_{-\pi/2}^{\pi/2} \left[s \cos \varrho \frac{d}{d\sigma}\left(-\frac{\sin \sigma}{\beta(\sigma)}\right) + \sin \varrho \frac{d}{d\sigma}\left(-\frac{\cos \sigma}{\beta(\sigma)}\right) \right] d\sigma,$$

or, after integration by parts and transformation back to dt:

$$I_2 = \int_{-\pi/2}^{\pi/2} \left[\sin \sigma \frac{d}{d\sigma}(s \cos \varrho) + \cos \sigma \frac{d}{d\sigma}(\sin \varrho) \right] \frac{d\sigma}{\beta(\sigma)}$$

$$= \int_{-\infty}^{\infty} \left[\sin \sigma \frac{d}{dt}(s \cos \varrho) + \cos \sigma \frac{d}{dt}(\sin \varrho) \right] \frac{dt}{\beta(\sigma)}. \quad \text{(A.12)}$$

Note that, in integration by parts, the boundary terms vanish since s and $\cos \sigma \to 0$ as $\sigma \to \pm \pi/2$. In I_3 we use (A.4) to obtain

$$I_3 = \int_{-\infty}^{\infty} \dot{s} \sin \sigma \cos \varrho \frac{dt}{\beta}. \quad \text{(A.13)}$$

Thus, from (A.12) and (A.13), we have, after cancellation and using $\dot{\varrho} = -\beta(\sigma)$:

$$I_2 - I_3 = \int_{-\infty}^{\infty} [s \sin \sigma \sin \varrho - \cos \sigma \cos \varrho] \, dt. \quad \text{(A.14)}$$

Finally, we use the relations (A.2) (with $\alpha_2 \leftrightarrow \alpha_3$, so that $a_1 \to -a_1$, $a_2 \to -a_3$, $a_3 \to -a_2$) and the fact that $S^2 + T^2 = 1$ to reduce (A.14) to the form

$$I_2 - I_3 = -\int_{-\infty}^{\infty} S(t) \cos \alpha_3 t \, dt$$

$$= -\frac{1}{\sqrt{a_1 a_2}} \int_{-\infty}^{\infty} \text{sech } \tau \cos\left(\frac{\alpha_2 \tau}{\sqrt{a_1 a_2}}\right) d\tau$$

$$= -\frac{\pi}{\sqrt{a_1 a_2}} \text{sech}\left(\frac{\alpha_2 \pi}{2\sqrt{a_1 a_2}}\right). \quad \text{(A.15)}$$

Thus from (A.8), (A.10) and (A.15), we have

$$M(\varrho_0) = -\frac{2b}{\alpha_3} - \frac{\sqrt{1-b^2}\,\pi}{\sqrt{a_1 a_2}} \text{sech}\left(\frac{\alpha_2 \pi}{2\sqrt{a_1 a_2}}\right) \cos \varrho_0, \quad \text{(A.16)}$$

which gives (6.7).

Acknowledgment. The research reported here was supported by ARO under grant DAAG 29-85-C-0018 (Mathematical Sciences Institute), NSF under MSM 84-02069 and AFOSR under 84-0051.

References

R. ABRAHAM & J. E. MARSDEN [1978] *Foundations of Mechanics*, 2nd edition, Benjamin/Cummings, Reading, MA.

S. S. ANTMAN [1984] Large lateral buckling of nonlinearly elastic beams, *Arch. Rational Mech. Anal.* **84**, 293–305.

S. S. ANTMAN & C. S. KENNEY [1981] Large buckled states of nonlinearly elastic rods under torsion, thrust and gravity, *Arch. Rational Mech. Anal.* **76**, 289–338.

V. I. ARNOLD [1964] Instability of dynamical systems with several degrees of freedom. *Sov. Math. Dokl.* **5**, 581–585.

V. I. ARNOLD [1978] *Mathematical Methods of Classical Mechanics* Springer Verlag, N.Y., Heidelberg, Berlin (Russian original, Moscow, 1974).

G. D. BIRKHOFF [1927] *Dynamical Systems*, A.M.S. Publications, Providence, R.I.

E. BUZANO, G. GEYMONAT, & T. POSTON [1985] Post-buckling behavior of a nonlinearly hyperelastic thin rod with cross section invariant under the dihedral group D_n, *Arch. Rational Mech. Anal.* **89**, 307–388.

L. EULER [1744] Additamentum I de Curvis Elasticis, Methodus Inveniendi Lineas Curvas Maximi Minimivi Proprietate Gaudentes, Lausannae in *Opera Omnia I* **24**, Füssli, Zürich [1960], pp. 231–297.

B. D. GREENSPAN & P. J. HOLMES [1983] Homoclinic orbits, subharmonics, and global bifurcations in forced oscillations, in *Nonlinear Dynamics and Turbulence* (editors: B. Barenblatt, G. Iooss & D. D. Joseph) Pitman, London. Chapter 10, pp. 172–214.

H. GOLDSTEIN [1980] *Classical Mechanics*, Addison-Wesley, Reading, MA.

J. GUCKENHEIMER & P. J. HOLMES [1983] *Nonlinear Oscillations, Dynamical Systems and Bifurcations of Vector Fields*, Springer Verlag, New York (corrected second printing, 1986).

P. J. HOLMES & J. E. MARSDEN [1982] Horseshoes in perturbations of Hamiltonian systems with two degrees of freedom, *Comm. Math. Phys.* **82**, 523–544.

P. J. HOLMES & J. E. MARSDEN [1983] Horseshoes and Arnold diffusion for Hamiltonian system on Lie groups, *Indiana University Math. J.* **32**, pp. 273–310.

G. KIRCHHOFF [1859] Über das Gleichgewicht und die Bewegung eines unendlich dünnen elastischen Stabes, *Journal für Mathematik (Crelle)* **56**, 285–313.

P. S. KRISHNAPRASAD, J. E. MARSDEN & J. C. SIMO [1986], The Hamiltonian structure of nonlinear elasticity: the convective representation of solids, rods and plates, (preprint).

J. LARMOR [1884] On the direct application of the principle of least action to the dynamics of solid and fluid systems, and analogous elastic problems, *Proc. London Math. Society* **15**, 170–184.

A. E. H. LOVE [1927] *A Treatise on the Mathematical Theory of Elasticity* Cambridge University Press (4th edition).

V. K. MELNIKOV [1963] On the stability of the center for time period perturbations, *Trans. Moscow Math. Soc.* **12**, 1–57.

A. MIELKE [1987] Saint-Venant's problem and semi-inverse solutions in nonlinear elasticity, MSI Tech. Rep. '87-25, Arch. Rational Mech. Anal., to appear.

J. MOSER [1973] *Stable and Random Motions in Dynamical Systems*, Princeton University Press, Princeton, N.J.

S. Smale [1963] Diffeomorphisms with many periodic points, in *Differential and Combinational Topology*, ed. S. S. Cairns, pp. 63–80, Princeton University Press, Princeton, N.J.

S. Smale [1967] Differentiable Dynamical Systems, *Bull. Amer. Math. Soc.* 73, 747–817.

E. T. Whittaker [1937] *A treatise on the Analytical Dynamics of Particles and Rigid Bodies* (4th edition) Cambridge University Press, Cambridge.

Mathematisches Institut A
Universität Stuttgart
Federal Republic of Germany

and

Mathematical Sciences Institute and
Departments of Theoretical & Applied Mechanics
and Mathematics
Cornell University
Ithaca, New York

(Received July 20, 1987)

SPATIAL CHAOS AND LOCALIZATION PHENOMENA IN NONLINEAR ELASTICITY

J.M.T. THOMPSON and L.N. VIRGIN

Department of Civil Engineering, University College London, Gower Street, London WC1E 6BT, UK

Received 30 september 1987; revised manuscript received 20 November 1987; accepted for publication 23 November 1987
Communicated by A.P. Fordy

Classical static–dynamic analogies are invoked to demonstrate spatial chaos and localization of deformations in the *elastica* of a post-buckled strut. Some conjectures are then made relating homoclinic events in the dynamic analogy of a strut on a nonlinear elastic foundation to the spatial localization of the buckling pattern.

1. Introduction

Nonlinear dynamics, including for example the new theory of chaotic motions in deterministic systems, has undergone a major revolution in recent years. High-speed digital computation and powerful topological phase-space concepts have been mutually-stimulating ingredients of a world-wide explosion of activity.

It could well be that these geometrical concepts and techniques are ripe for a fruitful re-interpretation into the spatial domain, with the space coordinate s replacing the time t as the independent variable. Our preliminary studies reported here show that this could lead to major new insights into the onset of spatial chaos, and perhaps more importantly the triggering of spatial localization.

In this short note, we look first at the well-known static–dynamic analogy between the large spatial deformations of elastic rods and the global dynamics of rigid pendular bodies. This we feel could be invaluable in the numerous deformation problems of elastic "lines" associated for example with molecular chains, biological hairs and filaments, textiles, optical fibres, magnetic tapes, wires and oil pipelines. We present an example of spatial chaos and localization in the planar deformations of an elastic rod, and finally make some speculations about the role of homoclinic events in the localization of structural buckling modes.

2. The planar elastica

It is well-known that the motions of a rigid, undamped pendulum passing close to its inverted unstable saddle equilibrium become chaotic when the pendulum is driven by a deterministic periodic forcing [1]. There is, moreover, a classical analogy due to Kirchhoff [2,3] between the planar pendulum described by the angular displacement θ as a function of the time t, and the large static deflections of an axially loaded elastic column described by the centre-line rotation θ as a function of the arc length s. So if the centre-line of the column has a small initial sinusoidal out-of-straightness of magnitude F the equation of the elastic line becomes

$$EI \frac{d^2\theta}{ds^2} + P \sin \theta = F \sin \omega s,$$

which corresponds precisely to the sinusoidally driven pendulum equation

$$mL \frac{d^2\theta}{dt^2} + mg \sin \theta = F \sin \omega t.$$

Here EI is the bending stiffness of the elastic strut subjected to the axial load P: m is the mass of the pendulum bob, L is the length of its light rigid arm, and g is the gravitational constant. In each equation F is a measure of the perturbation amplitude, and ω is a measure of its spatial or temporal frequency.

For $F=0$, some sample phase trajectories for the unforced pendulum, and the corresponding spatial

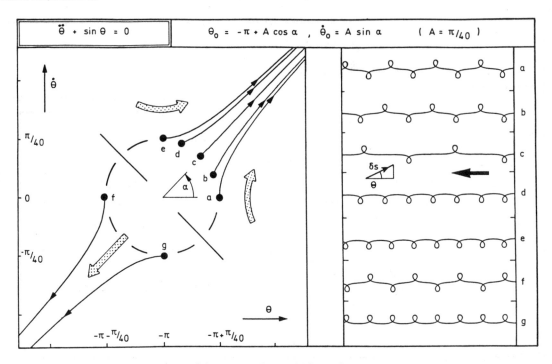

Fig. 1. Sample phase-space motions and the corresponding spatial forms of a perfect elastic rod in tension with $EI = P = 1$.

forms of the perfect strut, are shown in fig. 1. Here we can infer that the homoclinic saddle-connection of the pendulum, during which it could be said to "swing in infinite time from the fully inverted position and back", corresponds to a long elastic beam in tension with a single *localized* loop at its centre, as shown schematically in fig. 2. This tensile state of the strut is clearly physically stable for deformations constrained to the plane of the paper, and can be thought of as a grossly post-buckled configuration of a compressed strut in which the loaded ends have passed through one another. Now it is perturbation of the homoclinic connection that leads to chaos in the driven pendulum, and a sample numerical Runge–Kutta integration with small F is shown in fig. 2. This is plotted in three alternative ways, as $\theta(t)$ for the pendulum, as $\theta(s)$ for the beam, and as a Poincaré section $(\dot{\theta}, \theta)$.

With the starting conditions chosen to ensure that the motion passes repeatedley close to the inverted configuration, the pendulum experiences a choatic motion with random sequences of oscillation and tumbling. Easily recognized sequences of continuous tumbling are labelled A, B, C, and D in the large time-history.

The corresponding spatial deflected form of the elastic strut is shown in the lower plot of fig. 2. Here we have a chaotic spatial sequence of random looping, with the tumbling sequences A to D of the pendulum clearly identifiable as sequences of one-sided looping.

The Poincaré section shows the values of angular displacement and velocity sampled stroboscopically at the forcing frequency, when the time t is a multiple of the forcing period $T = 2\pi/\omega$. The points are scattered randomly around the homoclinic separatrix of the unforced equation, and we can recall that in this undamped hamiltonian problem we do not expect the characteristic *attractors* familiar in dissipative systems [4].

3. The spatial elastica

For the three-dimensional deformations of an elastic rod or wire, there is a generalized kinetic anal-

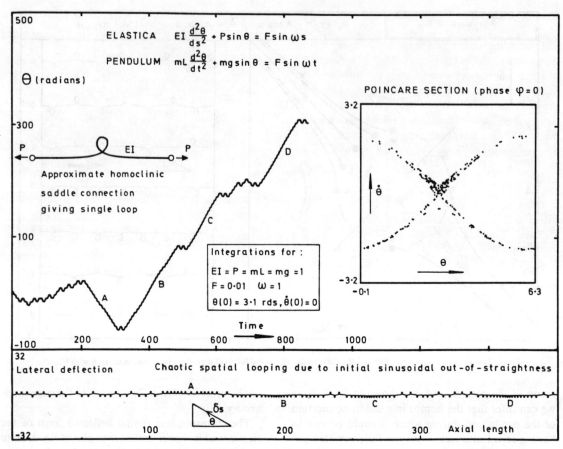

Fig. 2. Chaotic pendulum motion and the analogous spatial form of an imperfect elastic rod in tension, showing a random sequence of localized loops. Inset is the corresponding Poincaré section.

ogy introduced by Kirchhoff and Clebsch [2,3,5]. Here the elasticity equations of a thin prismatic bar that is bent and twisted through arbitrarily large deformations by a force, a couple and a torque applied to each end, correspond to the equations of motion of a rigid pendular body turning about a fixed point. So we can again expect spatial *localization phenomena* corresponding to invariant manifolds of the unstable inverted equilibrium state of the rigid body: and the localized form will be nudged into chaos by small perturbations. An example of a spatially localized three-dimensional form of an elastic strip is shown in fig. 3, derived from a photograph of a length of cine-film.

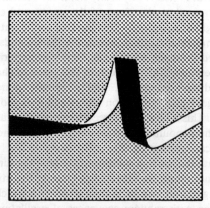

Fig. 3. A three-dimensional spatial localization of a thin elastic strip.

4. Local buckling as a homoclinic event

The elastica problem just described was devised to throw some light on the important *localized buckling* of structural components [6]. Here much attention has been focused on the response of an axially compressed beam on an elastic foundation, equivalent to the rotationally-symmetric deformations of a cylindrical shell. The linear buckling deformations of such a beam are governed by the well-known equation

$$\text{EI}\frac{d^4y}{dx^4} + P\frac{d^2y}{dx^2} + ky = 0,$$

where $y(x)$ is the lateral deflection, EI is the bending stiffness, P is the axial compressive load and k is the foundation stiffness (7). The critical buckling load of an infinitely long beam is given by

$$P^C = 2\sqrt{k\text{EI}}.$$

We suggest that it might be useful to examine, using the geometrical phase-space methods of dynamical systems theory, the *dynamical analogy* of this elasticity equation obtained by simply replacing the axial coordinate x by the time t. Writing the dynamic problem as a set of four coupled first-order differential equations in the normal way, it is easy to establish that the divergence of an ensemble of trajectories in the four-dimensional phase space is identically zero: so the dynamical analogy has no dissipation, and is hamiltonian in nature.

Now for $P < P^C$ the general solution of the beam equation is [7]

$$y = a e^{\beta x} \sin(\alpha x + c) + b e^{-\beta x} \sin(\alpha x + d),$$

where α and β are known constants, and a, b, c and d are the four arbitrary constants dependent on the boundary conditions. A typical form of this solution, for a long beam under low P, is shown in fig. 4a, where the effects of end-loads (lateral forces W, and bending moments M) decay rapidly, leaving the central section of the beam essentially straight.

In the dynamic analogy, the equilibrium solution $y(t) = 0$ is clearly a *saddle* with a two-dimensional spiralling *outset* (unstable manifold) given by the first term, and a two-dimensional spiralling *inset* (stable manifold) given by the second term. The form of fig. 4a thus represents motion initiated close to the inset,

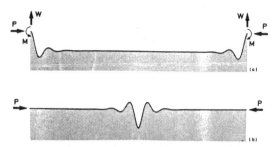

Fig. 4. Spatial forms of a strut on a nonlinear elastic foundation. Here (a) shows decaying end disturbances at low P, while (b) shows a localized buckling mode at high P.

leading to a close encounter with the saddle, followed by divergence close to the outset.

As P is increased through P^C the saddle becomes a *centre*, so we have the rather paradoxical result that as the long beam becomes statistically unstable, the analogous dynamical system becomes stable. In fact as P is *decreased* through P^C the dynamical system undergoes what aeronautical engineers would call an undamped flutter instability.

The linear problem that we have outlined so far is of course trivial, but much interest in structural engineering focuses on the strut on a *nonlinear* elastic foundation, for which the above elasticity equation would have an additional term which we might here take as $-cy^3$. This makes the dynamics potentially more instructive, with additional equilibrium solutions at $y = \pm\sqrt{k/c}$. The local linear features of these new equilibrium solutions are readily established from the appropriate characteristic equations, while the full four-dimensional nonlinear phase portrait between the three equilibria will clearly hold the key to the spatial localization phenomena that are of concern to the engineer.

With such a softening nonlinearity, a localized buckle pattern similar to that sketched in fig. 4b is often observed in practice, once the compression approaches P^C. This must clearly correspond dynamically to a homoclinic orbit, since in the limit of an infinitely long beam, the outset of $y = 0$ is seen to return as the inset.

Now since the dynamic analogy has an inverted flutter bifurcation (in a four-dimensional phase space), we might try to illustrate schematically what may happen by an adaptation of the two-dimen-

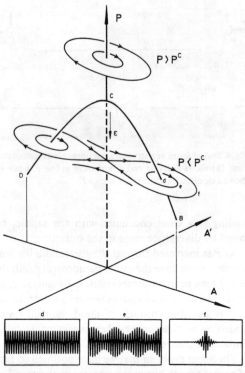

Fig. 5. Phase-space portraits of an inverted super-critical pitch-fork bifurcation, providing a schematic illustration of the homoclinic orbits that might govern spatial localization phenomena. In double-scale analysis of amplitude and phase modulation the envelope-amplitude A is governed by $4A'' - A + A^3 = 0$.

sional portraits of an inverted super-critical pitch-fork bifurcation (inverted stable cusp). The result is shown in fig. 5, where the secondary path DCB which is an equilibrium path for the pitch-fork must here be interpreted as a trace of periodic solutions bifurcating from C. At $P < P^C$ the spatial localization corresponds to the homoclinic orbit leaving and returning to the phase-space origin. So we can conjecture that the amplitude of any local buckling mode that might be induced in a beam by small imperfections or dynamic disturbances will tend to zero as P is increased to P^C.

As with the elastica, we can expect a perturbation of a localized solution to yield chaos. So for our axially compressed beam on a nonlinear foundation we would expect to observe a randomly spaced sequence of localizations due to a regular sinusoidal spatial imperfection.

5. Concluding remarks

Much more work clearly needs to be done to fully develop and extend the analogies presented here. An interesting extension of the perfect elastica is, for example, to consider self-weight loading which has the effect of replacing P by a term proportional to time, giving a non-autonomous dynamical analogy with a three-dimensional phase space. The role of structured *boundary conditions*, which we have largely ignored here, also needs to be carefully examined, since they allow only a *sub-set of the dynamics orbits* to emerge as physically realizable spatiale equilibrium states.

The ideas presented here arose from stimulating discussions with Giles Hunt [8] and his doctoral student Helen Bolt from Imperial College, and a comprehensive joint paper is now being prepared [9].

We might finally observe that to overcome the analytical and visualization problems inherent in four-dimensional phase space, it is tempting to use techniques of amplitude and phase modulation, following for example the double-scale expansions of Lange and Newell [10] and Potier-Ferry [11]. Indeed, Potier-Ferry's figure 9.8 showing the two-dimensional phase portrait of the modulation amplitude A versus its derivative A' corresponds, after appropriate scaling for the distance ϵ from P^C, to our $P < P^C$ plane in fig. 5. The forms of his solutions are sketched at the foot of fig. 5.

Note added

The authors are grateful to one of the referees for drawing their attention to the paper of Holmes and Marsden [12]. On page 298 of this article devoted to the analogous problem of rigid body dynamics they make a passing remark about the possibility of an arbitrary sequence of loops in the elastica. In a current paper, which has just come to our attention, Mielke and Holmes [13] establish the hamiltonian structure of the spatial elastica equations for a general nonlinear hyperelastic material and prove the existence of spatial chaos on the basis of a Melnikov analysis (our classical elastica analogies do of course relate to *linear* elastic rods undergoing large nonlinear geometrical deformations). These studies reinforce our view that the elastica can be a useful proving ground for ideas directed towards the more important buckling localizations that are our main concern here.

References

[1] A.J. Lichtenberg and M.A. Lieberman, Regular and stochastic motion (Springer, Berlin, 1983).
[2] G.R. Kirchhoff, On the equilibrium and the movements of an infinitely thin bar, Crelles Journal 56 (1859).
[3] R. Frisch-Fay, Flexible bars (Butterworths, London, 1962).
[4] J.M.T. Thompson and H.B. Stewart, Nonlinear dynamics and chaos (Wiley, New York, 1986).
[5] A. Clebsch, Theorie der Elasticität fester Körper (Leipzig, 1862).
[6] V. Tvergaard and A. Needleman, On the development of localized buckling patterns, in: Collapse: the buckling of structures in theory and practice, eds. J.M.T. Thompson and G.W. Hunt (Cambridge Univ. Press, Cambridge, 1983).
[7] M. Hetényi, Beams on elastic foundation (Univ. of Michigan Press, Ann Arbor, 1946).
[8] G.W. Hunt, Bifurcations of structural components, to be published.
[9] G.W. Hunt, H.M. Bolt and J.M.T. Thompson, Structural localization phenomena and the dynamical phase-space analogy, to be published.
[10] C.G. Lange and A.C. Newell, SIAM J. Appl. Math. 21 (1971) 605
[11] M. Potier-Ferry, Amplitude modulation, phase modulation and localization of buckling patterns, in: Collapse: the buckling of structures in theory and practice, eds. J.M.T. Thompson and G.W. Hunt (Cambridge Univ. Press, Cambridge, 1983).
[12] P.J. Holmes and J.E. Marsden, Indiana Univ. Math. J. 32 (1983) 273.
[13] A. Mielke and P.J. Holmes, Spatially complex equilibria of buckled rods, Technical Report '87-56, Mathematical Sciences Institute, Cornell University (July 1987), to be published in Arch. Rat. Mech. Anal.

Soliton chaos models for mechanical and biological elastic chains

M.S. El Naschie
King Abdul Aziz City for Science and Technology, Riyadh, Saudi Arabia

and

T. Kapitaniak [1]
Department of Applied Mathematical Studies and Centre for Nonlinear Studies, University of Leeds, Leeds LS2 9JT, UK

Received 27 March 1990; revised manuscript received 26 April 1990; accepted for publication 3 May 1990
Communicated by A.P. Fordy

The possibility of purely spatial chaos of loop and envelope soliton localization in long elastic strings is considered. Possible connections to some problems in molecular biology are discussed.

1. Introduction and preliminary remarks

There has been an increased interest in complexity and spatial chaos in recent years [1–5]. In particular, since the discovery of loop solitons and their interaction [6,7] as well as the possibility of a purely spatial chaos [8], there is a renewed interest in the Euler elastica and its applications [9].

El Naschie established the connection between the loop soliton and the Milke–Holmes chaotic elastica using a dynamical version of the Euler elastica [5,9]. He also drew attention to the possibility of interpreting the instability waves in curved compressed thin material surfaces (i.e. shells) as envelope soliton turbulence [9,10]. Thompson and Virgin were the first to publish a numerical confirmation of the theoretical results of Milke and Holmes using an elementary but neat model [11].

There is some intriguing likeness, at least a purely visual one, between the elastica configurations and protein transformations. The primary structure of protein [12], which is made up of long linear chains of covalently linked amino acids, for instance, resembles the periodic instability waves of compressed elastica. Depending on certain geometrical parameters, a long elastica inside a long circular pipe would lose stability when compressed and form a helical structure [13]. The secondary structure also looks very close to the soliton loops of the chaotic elastica. Finally, the tertiary structure looks very much like the strongly coiled elastica anticipated theoretically in ref. [5] and confirmed numerically in refs. [9,13].

Protein chains are not rigid. Similar to the elastica, they are flexible. This elasticity is essential in understanding protein deformation [12,14]. Not unlike DNA, soliton chaos of the elastica also conveys a definite code when translated into symbolic dynamics.

In the present work we study numerically what we may term spatial strange attractors in the elastica. This might be of interest, since strange attractors are regarded occasionally as generators of information.

Due to the analogy between the Hamiltonian of the elastica and the Hamiltonian of a circular elastic ring under external pressure [15], the similarity may be extended to circular DNA. In fact in some elementary demonstration using a long twirled and stretched elastic band, we observe not only the spatial complexity and pseudo-random loops shown in refs. [5,9], but we can more frequently and very

---

[1] Permanent address: Institute of Applied Mechanics, Technical University of Lodz, Stefanowskiego 1/15, 90-924 Lodz, Poland.

clearly observe supercoiling in the elastic band very similar to that of DNA. It seems also that spatial chaos generated by periodic fluctuation in the elastica may fit well into a known analogy between polymer chains and Brownian motion and this in turn is another connection to fractals.

We may recall as explained in detail in refs. [5,9] that for a soliton loop to form in the planar elastica, very large deformation is needed first, then we must assume that the ends can pass without obstruction through each other, which is of course physically impossible. If the planar two-dimensional constraint is removed however, the loop soliton forms in three dimensions without the need for a large deflection. This might be related to another observation in protein. There the primary structures can be considered planar, however, the secondary structures must be taken as three-dimensional [14]. Nevertheless, if the lateral movement of the elastica is restrained in some way, for instance through electromagnetic forces as in electric conductors [16] or by elastic forces as in axisymmetrical deformation of beams attached laterally to the elastic medium [9], then there can be a possibility for another type of soliton in two dimensions and without very large deflection. This is the envelope soliton well known from the solution of the nonlinear Schrödinger equation [17]. In the present work we give numerical confirmation for the conjecture made in refs. [5,10] that elastic material surfaces, such as shells, exhibit under certain conditions purely spatial and statical soliton chaos.

In all the problems considered here we study the influence of band-limited white noise on the randomness of the soliton [18,19], and we show that this spatial chaos may be eliminated or reduced by adding noise. Needless to say a phenomenon indistinguishable from spatial chaos can only be observed in the unforced system by adding band-limited white noise.

The idea of using soliton to model DNA is of course not new at all. It has been considered in some pioneering work by Davydov and Kislukla [20]. Highly interesting results were reported in numerous excellent papers by Scott [21].

These researchers go of course far deeper into the real and far more difficult problems of molecular biology. We on the other hand are familiar only with the global logic of molecular biology and are not in a position to comment in depth on the exact nature of the analogy suggested here.

Nevertheless we hope that our detailed knowledge of the elastica and statical chaos may be of some value, however limited, to the specialist who may be able to draw a clearer picture. In addition we hope that this work clearly shows that soliton and chaos are not contradictory and even essential as noted in a different context by Ueda and Noguchi [22].

2. The dynamical elastica – loop soliton

Consider the following nonlinear differential equation which describes the dynamical behaviour of the elastica,

$$W_{tt} + W_{xx} + 2\varepsilon [W_{xx}(1+W_x^2)^{-3/2}]_{xx} = 0, \qquad (1)$$

where $(\ )_x = d/dx$ and $(\ )_t = d/dt$. Here W is the nondimensional displacement, $\varepsilon = \alpha/2PA$, α is the bending stiffness. A is the cross-sectional area of the elastica, P is the axial force, x is the axial coordinate and t is the time.

Introducing the stretched coordinates

$$x_1 = x+t \quad \text{and} \quad t_1 = \varepsilon t,$$

noting that when a loop forms in the elastica, then compression is reversed into tension [13] and using ϕ and s as coordinate system where ϕ is the slope of the central line and s is the arch length, our PDE reduces to

$$\dot{\phi} + \cos\phi \, (\sec\phi\,\phi_{ss})_s = 0, \qquad (2)$$

where a dot denotes d/dt and $(\ )_s = d/ds$. Using the inverse scattering transformation, this equation can be shown to possess a loop soliton solution [6,7,26–28]. Some elementary experimental demonstrations of these loops were reported in ref. [13]. The time independent version of the last equation is nothing but the familar nonlinear ODE of the Euler elastica,

$$\phi_{ss} + \sin\phi = 0. \qquad (3)$$

Now we perturb this equation by first adding periodic spatial forcing (imperfection),

$$\phi_{ss} + \sin\phi = a\sin\omega s, \qquad (4)$$

and then adding band-limited white noise perturbation to it:

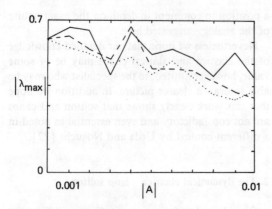

Fig. 1. The most probable value of the distribution of maximum Lyapunov exponents $|\lambda_{max}|$ for eq. (5) versus noise intensity A: $a=0.01$, $\omega=1$, $\phi(0)=2$, $\dot\phi(0)=0$; (----) $\nu_{min}=0.5$, $\nu_{max}=3.5$; (·····) $\nu_{min}=0.5$, $\nu_{max}=1.5$; (---) $\nu_{min}=2.5$, $\nu_{max}=3.5$.

$$\phi_{ss}+\sin\phi=a\sin\omega s+A\sum_{i=1}^{N}\sin(\nu_i s+\gamma_i), \qquad (5)$$

where ν_i and γ_i are random variables.

The second component of eq. (5) is an approximation of a band-limited white noise with a spectral density

$$S(\nu)=\sigma/(\nu_{max}-\nu_{min}), \quad \nu\in[\nu_{min},\nu_{max}],$$
$$=0 \qquad \nu\notin[\nu_{min},\nu_{max}], \qquad (6)$$

σ is constant, ν_{min} and ν_{max} are the band frequencies of the noise. γ_i are independent random variables with uniform distribution on the interval $[0, 2\pi]$, A and ν_i are given by

$$A=\sqrt{2\sigma}/N, \quad \nu_i=(i-0.5)\Delta\nu+\nu_{min},$$
$$\Delta\nu=(\nu_{max}-\nu_{min})/N. \qquad (7)$$

Some of the obtained numerical results for $a=0.01$,

$0.00\leqslant A\leqslant 0.01$ and three different perturbation frequency bands $0.5\leqslant\nu_i\leqslant 3.5$, $0.5\leqslant\nu_i\leqslant 1.5$ and $2.5\leqslant\nu_i\leqslant 3.5$, $N=300$ are shown in figs. 1 and 2. Two different representations are used: A spatial plot which shows the actual form which the infinite elastica should take and also the plot of the most probable value of the maximum Lyapunov exponent distribution over a number of noise realizations $|\lambda_{max}|$ (100 realizations of noise for different γ_i have been considered) [18,19] versus noise intensity A. As has been shown in refs. [18,19] a positive value of $|\lambda_{max}|$ indicates a chaotic stochastic process while a nonpositive one is characteristic for a regular stochastic process. The results agree qualitatively with some experimental demonstration reported in refs. [5,9].

To conclude this part we consider the influence of positive damping as well as what might appear as somewhat artificial spatial forcing. This forcing arises however in a natural way in the parametric forcing of the corresponding damped pendulum problem. Thus we study the following equation of the elastica,

$$\phi_{ss}+\delta\phi_s+b\sin\phi$$
$$=a\sin\omega s\sin\phi+A\sum_{i=1}^{N}\cos(\nu_i s+\gamma_i) \qquad (8)$$

for different parametric values as well as initial conditions.

For the deterministic and noise perturbed nonlinear dynamics we plot the corresponding spatial strange attractor in the region of the strange attractor [23]. Fig. 3 shows clearly the immense richness of information which these looping patterns can produce for infinitely long s. This may be relevant to some problems in molecular biology.

From the plot of the most probable value of the Lyapunov exponent distribution $|\lambda_{max}|$ shown in fig. 4 we can observe that for some value of noise inten-

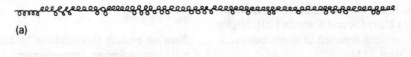

Fig. 2. Examples of spatial plot for eq. (5): (a) $A=0.0015$; (b) $A=0.0080$.

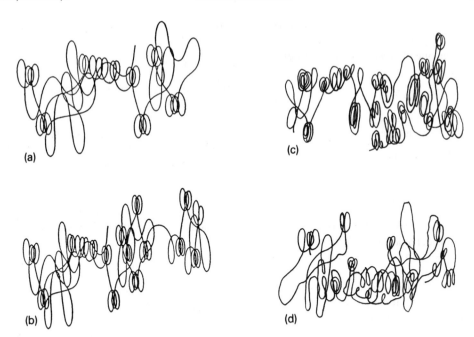

Fig. 3. Spatial strange attractors for eq. (8), $\delta=0.15$, $b=1$, $a=0.94$; (a) $\omega=1.56$, $A=0$; (b) $\omega=1.58$, $A=0$; (c) $\omega=1.56$, $A=0.1$; (d) $\omega=1.58$, $A=0.15$.

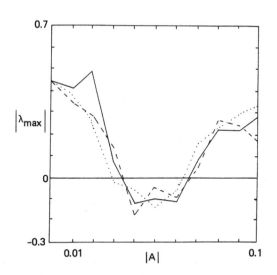

Fig. 4. The most probable value of the distribution of maximum Lyapunov exponents – $|\lambda_{max}|$ for eq. (8) versus noise intensity A, $a=0.15$, $\delta=0.01$, $\omega=1$, $b=0.0272222$, $\phi(0)=6$, $\dot\phi(0)=0$; (----) $\nu_{min}=0.5$, $\nu_{max}=3.5$; (·····) $\nu_{min}=0.5$, $\nu_{max}=1.5$; (---) $\nu_{min}=2.5$, $\nu_{max}=3.5$.

sity A, $|\lambda_{max}|$ is negative, which implies elimination of chaos (transition from chaotic to regular stochastic process) in the sense of symbolic dynamics as shown in the spatial plot of fig. 5b. The shape of

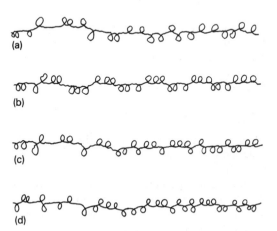

Fig. 5. Examples of spatial plot for eq. (8): (a) $A=0.02$; (b) $A=0.05$; (c) $A=0.08$; (d) $a=0$, $N=2$, $\nu_1=1$, $\nu_2=\sqrt{2}/10$, $A=0.1$.

loops in the spatial plot is not the same because of random forcing, but if we indicate the upper loop as 1 and the lower one as 0 we obtain a periodic sequence of symbols:

000111000111000111000111... ,

while for the chaotic spatial plots of figs. 5a and 5c we have aperiodic sequences:

000111100001001000110001...

and

000111011100111010011100... .

Another interesting type of behaviour can be observed if we consider a particular form of eq. (5) by taking $a=0$, $N=2$ and ν_1 and ν_2 to be incommensurable. In this case we can observe the behaviour presented in fig. 5d which seems to be chaotic. it is chaotic in the sense that it has an aperiodic sequence in the symbolic representation:

011100100111000111010111010100... ,

but we have no sensitive dependence on the initial conditions as the Lyapunov exponents are negative. This type of spatial strange behaviour is related to the so-called strange nonchaotic attractors [23–25].

3. Instability waves in an elastic structure – envelope soliton

Consider the following nonlinear partial differential equation which may be used to describe the propagation of buckling waves in an elastic medium such as the axisymmetrical deformation of an axially compressed cylindrical shell,

$$\alpha W'''' + \sigma W'' + c_1 W - c_2 W^2 + \rho \ddot{W} = 0 . \quad (9)$$

For a radial strain obeying a logarithmic law, this equation was used in refs. [9,13] to study the instability waves due to buckling.

Now depending on the number of slow spaces and slow time, different reduced differential equations for the complex amplitude of deflection A may be obtained. For instance, using

$$x = x_0, \quad x_1 = \varepsilon x_0, \quad x_2 = \varepsilon^2 x ,$$
$$t = t_0, \quad t_1 = \varepsilon t_0, \quad t_2 = \varepsilon^2 t_1 ,$$

one finds the following Ginzburg–Landau type equation [17]:

$$\alpha_1 A'' - \alpha_2 A + i\alpha_3 A' + i\alpha_4 \dot{A} + \alpha_5 A|A|^2 = 0 , \quad (10)$$

where $i=\sqrt{-1}$, an accent denotes d/dx and a dot denotes d/dt.

On the other hand the PDE may be drastically reduced to an ODE by reducing stretching to only $x_1 = \varepsilon x$. This leads to the following stationary nonlinear Schrödinger equation [17],

$$\alpha_1 A'' - \alpha_2 A + \alpha_5 A|A|^2 = 0 . \quad (11)$$

This equation is easily integrated by elementary methods and gives the soliton solution

$$A = \frac{6}{\sqrt{19}} \text{sech}(\tfrac{19}{12} x_1) , \quad (12)$$

for $\alpha = c_1 = c_2 = c_3 = r = 1$. The homoclinicity of this solution may be established easily as shown in ref. [9].

An optimum choice of the number of slow spaces and slow times which restores the dynamical character of the problem we have, however, when we take

$$x_1 = \varepsilon x_0, \quad t_1 = \varepsilon t_0, \quad t_2 = \varepsilon^2 t_1 .$$

This leads to the nonlinear Schrödinger equation

$$\alpha_1 A'' - \alpha_2 A + i\alpha_4 \dot{A} + \alpha_5 A|A|^2 = 0 , \quad (13)$$

with the well-known solution [17]

$$A(x, t)|_{t=t_0} = a \,\text{sech}\, bx \cos cx , \quad (14)$$

where a, b and c are constant.

Either way we expect spatial forcing to yield spatial envelope soliton chaos. Thus we consider first the periodically forced equation

$$A'' + k_1 A' - k_2 A + k_3 A^3 = k_4 \cos k_5 s . \quad (15)$$

The results of the numerical integrations for different parameter values which are: $k_1 = 0.01$, $k_2 = 0.25$, $k_3 = 19/(4 \times 18)$, $k_5 = 1$ and different values of k_4 are shown in fig. 6. They fully confirm the expectations expressed earlier in refs. [5,9,13].

Subsequently the forcing by band-limited white noise,

$$A'' + k_1 A' - k_2 A + k_3 A^3 = \bar{A} \sum_{i=1}^{N} \cos(\nu_i s + \gamma_i) , \quad (16)$$

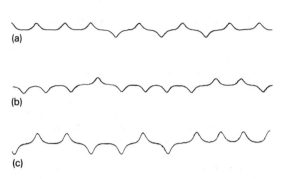

Fig. 6. Examples of spatial plot for eq. (15): $k_1=0.01$, $k_2=0.25$, $k_3=19/(4\times 18)$, $k_5=1$, $A(0)=1.37649$, $\dot{A}(0)=0$; (a) $k_4=0.001$; (b) $k_4=0.002$; (c) $k_4=0.003$.

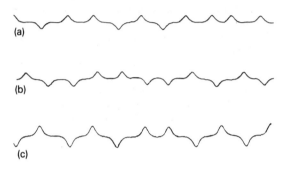

Fig. 7. Examples of spatial plot for eq. (16): k_1–k_3 as in fig. 6: (a) $\bar{A}=0.001$, $|\lambda_{max}|=0.06$; (b) $\bar{A}=0.002$, $|\lambda_{max}|=0.07$; (c) $\bar{A}=0.003$, $|\lambda_{max}|=-0.03$.

is considered. In fig. 7 we show a spatial plot of the same system as in fig. 6 this time, however, with band-limited white noise forcing.

In all the above examples the soliton strange behaviour would thus have a similar symbolic dynamic representation. However, in many cases we have no sensitive dependence on initial conditions – fig. 7c.

4. Conclusions

Based on the symbolic dynamics of a single spatial plot, it is not easy if at all possible to distinguish between chaos, strange nonchaotic behaviour and random behaviour. However, based on the distribution of the maximum Lyapunov exponents, a distinction can be made between chaotic and nonchaotic "strange" behaviour.

There seems to be some likeness between the deformation of the elastica and DNA chains. The deformation in an elastic band however is in principle reversible while the transformation from DNA to RNA was never observed to be reversible.

In terms of the mechanics of deformable bodies DNA chains act as if they had in-locked internal compression inside them, a kind of pre-stressing with a very weak elastic bond, which is checked by the bending and axial stiffness of the silhouette of the chain. When through chemical reactions this stiffness and the bond are eroded, collapse follows. This is not very much unlike the coiling of a long twirled and stretched rubber band when the stretching forces are gradually released. If this outrageously elementary mechanical picture is anywhere near correct, then it is of course extremely unlikely that an increase in the stiffness could ever restore the original situation and if the analogy holds, then there can be no RNA to DNA transformation.

Of course there is still the possibility known from materials with memory which may regain the original form by an influx of energy.

We hope to have shown clearly through the numerical results that spatial chaos may help in understanding complexity. The role of random perturbation in eliminating spatial chaos sheds light on the therapeutic effect of vibration in the medical treatment of bone disorder. Based upon previous dynamical observations [18,19] this effect is fully expected although it should be regarded as counter-intuitive that a type of spatial forcing which on its own produces stochasticity should eliminate another type of chaos where intuition may suggest that more complicated behaviour is expected. The work also stresses the view expressed probably for the first time by Ueda that soliton and chaos should not be regarded as contradictory.

Finally we may note that spatial damping may be thought of as a kind of nonconservative force similar to that known in dynamical stability as follower forces [29].

References

[1] S. Wolfram, Nature 311 (1984) 419.

[2] B.M. Herbs and W.-H. Steeb, Z. Naturforsch 43a (1988) 727.
[3] P. Reichert and R. Schilling, Phys. Rev. B 25 (1984) 917.
[4] H.G. Purwins, G. Klempt and J. Berkemeier, Festkörperprobleme 27 (1987) 27.
[5] M.S. El Naschie and S. Al Athel, Z. Naturforsch. 44a (1989) 645.
[6] Y.H. Ichikawa and N. Yajima, in: Statistical physics and chaos in fusion plasmas, eds. C.W. Haston and L.E. Reichl (Wiley, New York, 1984) pp. 79–90.
[7] K. Konno and A. Jeffrey, in: Advances in nonlinear waves Vol. 1, ed. L. Debnath (Pitman, London, 1984) pp. 162–183.
[8] A. Mielke and P. Holmes, Arch. Rat. Mech. Anal. 101 (1988) 319.
[9] M.S. El Naschie, J. Phys. Soc. Japan 58 (1989) 4310.
[10] M.S. El Naschie, K.S.U. J. Eng. Sci. 14 (1988) 437.
[11] J.M.T. Thompson and L.N. Virgin, Phys. Lett. A 126 (1988) 491.
[12] H. Frauenfelder, in: Emerging syntheses in science, ed. D.Pines (Addison-Wesley, Reading, 1988) pp. 155–164.
[13] M.S. El Naschie, S. Al Athel and A.C. Walker, in: Proc. IUTAM Symp. on Nonlinear dynamics in engineering, ed. W. Schiehlen (Springer, Berlin, 1990) pp. 67–74.
[14] P. Sheeler and D. Bianchi, Cell and molecular biology (Wiley, New York, 1980).
[15] M.S. El Naschie, Int. J. Mech. Sci. 17 (1975) 387.
[16] N. Dolbin and A. Morozov, Zh. Prikl. Met. Tekh. Fiz. 3 (1966) 97.
[17] P.G. Drazin and R.S. Johnson, Solitons (Cambridge Univ. Press, Cambridge, 1989).
[18] T. Kapitaniak, Chaos in systems with noise (World Scientific, Singapore, 1988).
[19] T. Kapitaniak, J. Sound Vib. 123 (1988) 391.
[20] A.S. Davydov and N.I. Kislukla, Phys. Stat. Sol. (b) 59 (1973) 465.
[21] A.C. Scott, in: Emerging syntheses in science, ed. D. Pines (Addison-Wesley, Reading, 1988) pp. 133–152.
[22] Y. Ueda and A. Noguchi, J. Phys. Soc. Japan 52 (1983) 713.
[23] F.J. Romeiras and E. Ott, Phys. Rev. A 35 (1987) 4404.
[24] T. Kapitaniak, E. Ponce and J. Wojewoda, J. Phys. A 23 (1990) L383.
[25] T. Kapitaniak, Chaotic oscillations in mechanical systems (Manchester Univ. Press, Manchester, 1990), in press.
[26] K. Konno, Y.H. Ichikawa and M. Wadati, J. Phys. Soc. Japan 50 (1981) 1025.
[27] T. Shimizu, K. Sawada and M. Wadati, J. Phys. Soc. Japan 52 (1983) 36.
[28] Y.H. Ichikawa, K. Konno and M. Wadati, in: Long-time prediction in dynamics, eds. C.W. Horton Jr. et al. (Wiley-Interscience, New York, 1983) pp. 345–369.
[29] M.S. El Naschie, Stress, stability and chaos in structural engineering (McGraw-Hill, New York, 1990).

Spatiotemporal dynamics in a dispersively coupled chain of nonlinear oscillators

David K. Umberger,* Celso Grebogi, Edward Ott,[†] and Bedros Afeyan[‡]
Laboratory for Plasma Research, University of Maryland, College Park, Maryland 20742
(Received 17 October 1988)

> A one-dimensional chain of forced nonlinear oscillators is investigated. This model exhibits typical behavior in periodically forced, spatially extended, nonlinear systems. At low driving amplitudes characteristic domainlike structure appears accompanied by simple asymptotic time dependence. Before reaching its final state, however, the chain behaves chaotically. The chaotic transients appear as intermittent bursts mainly concentrated at the domain walls. At higher driving, the chaotic transient becomes longer and longer until the time dependence apparently corresponds to sustained chaos with the chain state characterized by the absence of domainlike spatial structure.

I. INTRODUCTION

In the past, most of the work on chaotic dynamics has been concentrated on the temporal behavior of low-dimensional systems. Many physical systems of interest, however, as fluid flows, require the study of very high-dimensional systems which have intricate spatial and temporal evolution properties. Models which might reveal, therefore, some of the fundamental properties of spatially extended nonlinear systems are of great interest. One such model is proposed and studied in this work.

For low-dimensional systems, the Poincaré surface of section technique transforms continuous time systems (flows) into discrete time systems (maps). Furthermore, simple maps not necessarily derived from Poincaré surfaces of section (e.g., the quadratic map, the Hénon map, etc.) have served as useful models displaying typical temporal behavior of low-dimensional dynamical systems. This has motivated the suggestion that spatiotemporal phenomena, characteristic of continuous space-time systems (such as described by partial differential equations) might be modeled by systems that are discrete in both space and time.[1,2] Work along these lines has revealed that a class of coupled map lattices can exhibit spatial domain like structures (kink-antikink patterns, for example). Furthermore, it has been shown that these systems can exhibit highly nontrivial behavior such as spatial intermittency, period doubling of kink-antikink patterns, the coexistence of laminar and chaotic regions, and so forth.

It is, however, an open question whether phenomena in coupled map lattice systems are really indicative of typical phenomena in systems that are continuous in time and space. An intermediate approach, which might be used to bridge the gap between continuous space-time systems and discrete time systems, is a system that is continuous in time and discrete in space. The systems considered here consist of simple second-order ordinary differential equations defined on a one-dimensional lattice of oscillators with nearest-neighbor coupling. A mechanical realization of the system we consider is illustrated in Fig. 1. In the uncoupled limit the basic ordinary differential equation (ODE) describing the dynamics of each oscillator is chosen to be the forced Duffing equation.[3] This equation is known to display a rich variety of dynamical behavior: from periodic to chaotic solutions, multiple attractors, and fractal basin boundaries.

For the specific parameter values we use to investigate our coupled ODE lattice system, we observe two regimes, one at relatively low forcing and one with a somewhat higher forcing. In the first case we observe the formation of spatial domains where groups of neighboring lattice points are in similar states. Furthermore, these low forcing cases evolve to a state which is periodic with the same period as the driving, i.e., a *fixed point* of the time one stroboscopic map (Poincaré). Even though the final state is a simple one, the transient dynamics[4] through which the system passes is rich in structure. In particular, we observe *extremely* intermittent behavior wherein the time evolution of a given oscillator alternates irregularly between slowly evolving periods and periods of rapid chaotic oscillation. These oscillation patterns propagate along the chain. These spatially structured chaotic transients appear to be very typical in this system. In the strongly forced case, the system seems to evolve as sustained chaotic final state. Here, the spatial patterns are irregular with no apparent domain structure. In common with the lower forcing case, the tendency for extremely intermittent temporal behavior persists, but now it is sustained rather than transient.

In Sec. II we introduce and discuss the specific system studied. In Sec. III we present numerical results and their interpretations. Section IV presents conclusions.

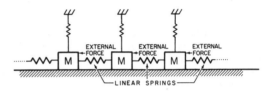

FIG. 1. Illustration of the unforced coupled ODE lattice chain.

II. COUPLED ORDINARY DIFFERENTIAL EQUATION LATTICE SYSTEM

The chain of nonlinear oscillators whose behavior we investigate is a collection of N forced Duffing oscillators coupled via a linear dispersive term whose equations of motion are

$$\ddot{x}_i(t) = -\gamma \dot{x}_i(t) + \frac{\sigma}{2} x_i(t)[1 - x_i^2(t)] + f \cos(\omega_D t) + \varepsilon D_2[x_i(t)] , \quad (1)$$

where $i = 1, \ldots, N$. Dots denote differentiating with respect to the time t, and D_2 is the second spatial differencing operator

$$D_2[x_i] \equiv x_{i+1} - 2x_i + x_{i-1} . \quad (2)$$

The variable $x_i(t)$ is the displacement at time t of the ith oscillator from its point of unstable equilibrium in the absence of damping, driving, and coupling. The parameters γ, σ, f, ω_D, and ε are, respectively, the damping strength, the strength of the restoring force, the amplitude and frequency of the external driving, and the coupling strength. We use periodic boundary conditions $x_0 = x_N, x_1 = x_{N+1}$.

Taking a Poincaré surface of section of the system, Eq. (1) results in a coupled map lattice. The section is obtained by examining the system's state after each driving period $\tau = 2\pi/\omega_D$. Letting $v_i(t) \equiv \dot{x}_i(t)$ denote the velocity of the ith oscillator at time t, the state of the system at time $t = n\tau$ with n an integer is given by the $2N$-tuple $\xi_n \equiv (x_1(n\tau), \ldots, x_N(n\tau), v_1(n\tau), \ldots, v_N(n\tau))$. Then, the sequence of states ξ_k, for $k = 0, 1, \ldots$, is generated by a deterministic, discrete-time dissipative map $\hat{T}$, which takes ξ_k into ξ_{k+1}. That is, $\xi_{k+1} = \hat{T}(\xi_k)$. The map $\hat{T}$ differs from the most commonly studied coupled map lattice systems[1,2] in three important respects. First, $\hat{T}$ is invertible and, hence, it mimics a flow. Second, it takes two variables per site to specify its state rather than one, as is the case in coupled logistic maps, for example. Third, and most important, as our system is integrated over one driving period τ, the motion of any fixed oscillator can, depending on the system parameters, be influenced by oscillators quite far away. The range of interaction depends on, among other things, the coupling strength and the damping. Hence, the map $\hat{T}$ typically exhibits long-range coupling, even though the system described by Eq. (1) has a nearest-neighbor coupling. On the other hand, except for a few notable examples, previous work on coupled map lattices were strictly limited to nearest-neighbor coupling.[2]

To show how a coupled set of ordinary differential equations can bridge the gap between coupled map lattices and partial differential equations, we seek to obtain the continuous limit of a dispersively coupled Duffing chain of oscillators. The continuum limit of Eq. (1) can be shown to be

$$\frac{\partial^2 \psi}{\partial t^2} = -\gamma \frac{\partial \psi}{\partial t} + \frac{\sigma}{2} \psi(1 - \psi^2) + f \cos(\omega_D t) + \varepsilon \frac{\partial^2 \psi}{\partial z^2} . \quad (3)$$

The correspondence between Eqs. (1) and (3) is seen readily by making the heuristic substitutions $i \to z$, $x_i(t) \to \psi(z,t)$, and $D_i \to \partial^2/\partial z^2$. We expect that as N is increased in Eq. (1), its solutions approach solutions of Eq. (3) more and more closely. Linearizing Eq. (3) about the stable equilibria $\psi_0 = \pm 1$ and assuming the usual harmonic space-time dependence of the form $\psi \sim \exp(-i\omega t + ikz)$, yields the dispersion relation $\omega^2 = \varepsilon k^2 + \sigma - i\omega\gamma$. Thus $\sqrt{\varepsilon}$ is proportional to the group velocity of the wave, γ results in linear damping of the wave, while σ can be interpreted as a susceptibility of the medium. Setting $\sigma = 0$ corresponds to waves propagating in vacuum, where there is no dispersion. In the material medium, σ causes the waves to be dispersive. Thus the coupled lattice ODE system (1) might be expected to model typical characteristic phenomena to be expected in forced, spatially extended nonlinear wave systems with dispersion and damping.

Returning to Eq. (1), we note that when the coupling is turned off, i.e., $\varepsilon = 0$, the system becomes a collection of independent oscillators each having the equation of motion

$$\ddot{x}_i(t) = -\gamma \dot{x}_i(t) + \frac{\sigma}{2} x_i(t)[1 - x_i^2(t)] + f \cos\omega_D t . \quad (4)$$

This is the equation of motion of a particle moving in a double-well potential when damping and driving are added. The well has points of stable equilibrium at $x_i = \pm 1$ and a point of unstable equilibrium at $x_i = 0$. The system equation (4) has been studied extensively,[3] and it has been shown numerically that it is capable of various types of behavior such as chaos and "final-state sensitivity."[5] The latter results from the coexistence of several attractors whose basins of attraction have fractal boundaries.[6] It is only natural to wonder, therefore, what types of behavior are possible in a lattice dynamical system whose local dynamics are so rich.

Some intuition from low-dimensional dynamics can be brought to bear on this question in the weak-coupling limit. In this limit, we expect that the motion of a single oscillator in the chain is qualitatively similar to a system obtained by adding noise to Eq. (4). For example, it is often found that attractors are stable under small noisy perturbations. Thus, if f, ω_D, σ, and γ are chosen so that Eq. (4) yields motion on some attractor, the corresponding noisy motion will occur on a slightly "fuzzed out" version of that attractor. We expect then that an oscillator of Eq. (1) also behaves in this way when the coupling is weak; the attractor of the full system will be, roughly, the direct product of N strange attractors of Eq. (4).

In cases where Eq. (4) has multiple attractors whose basins vary in size, the effect of adding noise is that of "washing out" the smaller basins of attraction; the external noise tends to "knock" an orbit originally in such a basin out into a basin of larger measure. If a smoothly varying initial condition is used in the corresponding, weakly coupled chain, different oscillators can tend to the different attractors of Eq. (4). The effect of the coupling will be to send oscillators originally intended for attractors with small basins into the attractors having the larger basins. Thus, we expect the coupling to "smooth out" the fine-scale structure of the basin of the local dynamics. This would imply that fine-scale structures of partial differential equation systems would be wiped out on a length scale determined by the coupling strength; the finer-scale structures of the basins would be eliminated.

Another important issue associated with the chain when the local dynamics has multiple attractors is that of domain formation. Recall that in such cases the effect of using smoothly varying initial conditions is to send different oscillators to different attractors when the coupling is small. Groups of nearby oscillators in the chain whose initial conditions lie in the same basin all tend towards the same attractor. Other adjacent groups tend towards different attractors, forming domains. Walls separating such domains are associated with oscillators whose initial conditions are close to basin boundaries.

III. NUMERICAL EXPERIMENTS

In this section we present the results of experiments performed by numerically integrating Eq. (1). The experiments examine the transient as well as the asymptotic behavior of the solutions of Eq. (1). The experiments are of two types. In the first type, the parameters of the system (ε, γ, σ, f, ω_D, and N) are fixed, and the initial condition ξ_0 is varied. The second type of experiment involves increasing the driving strength f with the remaining parameters and initial conditions held fixed. All the simulations were done with a fifth-order Runge-Kutta algorithm with variable step size[7] and chain length $N = 256$.

We begin by examining the chain when the parameters γ, f, ω_D, and σ have values such that the local dynamics have several coexisting attractors. One such situation occurs when $\gamma = 0.15$, $f = 0.10$, $\omega_D = 0.833$, and $\sigma = 1$. (We fix $\sigma = 1$ throughout this work.) For these parameter values, the single Duffing oscillator has[3] two periodic attractors with a fractal basin boundary separating their basins. As f is increased there is transition to chaotic motion at $f_c = 0.12$. We investigate here the properties of the chain in both these regimes starting with the case where each single oscillator undergoes periodic motion. Using $\varepsilon = 0$ (i.e., no coupling) and the initial condition

$$x_i(0) = a + 0.45 \sin\left[\frac{2\pi i}{N}\right], \quad (5a)$$

$$v_i(0) = 0, \quad (5b)$$

with the offset $a = 0.003$, we find that under $\hat{T}$ some of the oscillators asymptote to fixed points, while others asymptote to period-6 orbits. This is depicted in Fig. 2 where the displacements of the oscillators are plotted against lattice position for six successive iterates of $\mathbf{T}$ for $1025 \leq n \leq 1031$. (We choose n large so that the transient has died down.) Evidently, there are four coexisting attractors, an *up* (positive displacement) fixed point, a *down* (negative displacement) fixed point, and a pair of period-6 limit cycles, one up and one down. The particular attractor that a given oscillator tends to depends on its initial condition. Since a smoothly varying (in space) initial condition was used, the lattice breaks up into domains determined by groups of nearby oscillators asymptoting

to the same attractor. Experiments and simulations on coupled cells of chemical oscillators have shown rhythm splitting behavior[8–10] similar to that observed here (viz., part of the chain exhibits a fixed point state while other parts exhibit periodic time dependence).

As the coupling ε is increased the period-6 motion disappears, and the displacement as a function of lattice position smooths out. This is shown in Fig. 3 for $\varepsilon = 1.0$, where the initial condition and the parameters of the local dynamics are the same as for Fig. 2. Note that, unlike Fig. 2, the state shown in Fig. 3 is a fixed point of $\hat{T}$. That is, the same state is observed at all times t separated by $2\pi/\omega_D$ (for Fig. 2 the attractor is a fixed point of $\hat{T}^6$). The smoothing of the displacement as a function of the lattice position seen in Fig. 3 is expected as ε is increased since the dispersive coupling term increases in importance. Note that in Fig. 3, x_i versus i still has two basic domains corresponding, respectively, to up and down states. Within each domain, however, there is a sinusoidal oscillation which has a wavelength of about 14 lattice sites. Based on the total lattice length of 256 points, this corresponds to a mode number of $m = \frac{256}{14} \simeq 18$.

When the offset a in Eq. (5a) is varied (holding the system parameters fixed as for Fig. 3), it is found that, for different values of a, the system asymptotes to a large number of different final states, most of them fixed points of $\hat{T}$. For example, Fig. 4 is a plot of x_i versus i at $n = 2048$ for $a = 0.00301$. This initial condition again results in a fixed point final state with two domains and an $m = 18$ oscillation, but the down domain has broadened considerably from that of Fig. 3. Not only does this system possess many attractors, but the actual asymptotic motion which results apparently depends sensitively on the choice of the initial condition. This final-state sensitivity[5,6] is expected in systems which possess fractal basin boundaries.[4–6]

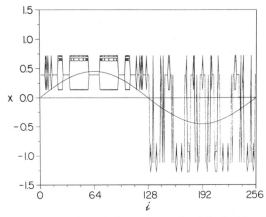

FIG. 2. Plot of the displacement x vs the chain site i for six consecutive iterates ($1025 \leq n \leq 1031$) in the asymptotic state. Superimposed, also shown is the sinusoidal profile for the initial condition. The parameters are $\gamma = 0.15$, $f = 0.10$, $\omega_D = 0.833$, $\varepsilon = 0$, and $a = 0.003$.

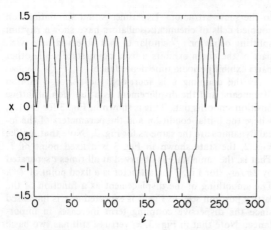

FIG. 3. Same plot as in Fig. 2 but now $\varepsilon = 1.0$, and $n = 2048$.

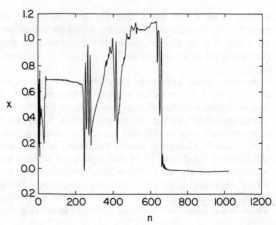

FIG. 5. Plot of displacement x vs time n for the oscillator $i = 61$ and $1 \leq n \leq 1024$. The parameters are the same as in Fig. 4.

One might expect that the sensitivity to final state described above would be reflected in the transient behavior of the chain as well. In Fig. 5 we plot the displacement of the $i = 61$ oscillator versus time between $n = 1$ and $n = 1024$ for the case of Fig. 4. Note that the time series shows intervals of roughly constant displacement interrupted by large-excursion chaotic bursts. Apparently, this oscillator tries to lock onto a fixed point, but is disrupted from doing so by disturbances originating from other portions of the lattice and is occasionally thrown in the fractal basin boundary region where it exhibits chaotic motion. At around $n = 650$, this locking-bursting behavior dies down and the oscillator settles into its final, fixed displacement state. In Fig. 6 we show a plot of x_i versus n for $i = 120$ of the same chain as in Fig. 5. This oscillator shows relatively wild chaotic behavior which also ceases at about $n = 650$. Comparing Figs. 5 and 6 between $n = 1$ and 200, we see that during the interval when oscillator 61 is almost static, oscillator 120 is experiencing relatively fast, large amplitude changes in the displacement. This is an indication that the details of the transient behavior which occurs depends on the position in the chain.

We illustrate this with another fixed point example where $a = 0.003$ (as in Fig. 3) is used, but where the driving is slightly stronger, $f = 0.11$ instead of the value $f = 0.10$ in Fig. 3. Figure 7 shows the chain displacements as a function of lattice position superimposed from $n = 2024$ to $n = 2048$. The system again settles to a fixed point. Figure 8 is the same as Fig. 7 except that the displacements are superimposed for $600 \leq n \leq 661$. Note that the region near the left domain wall undergoes a great deal of motion while other parts of the chain, including the region near the right domain wall, is relative-

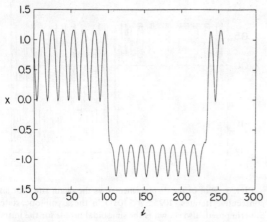

FIG. 4. Same plot as in Fig. 3 but for $a = 0.00301$ instead of $a = 0.00300$.

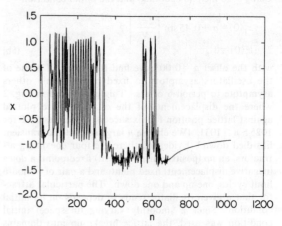

FIG. 6. Same as Fig. 5 but for the oscillator $i = 120$.

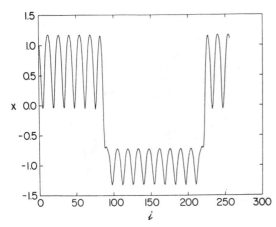

FIG. 7. Superposition of plots of x vs i for $2024 \leq n \leq 2048$. This plot is for stronger driving $f = 0.11$ with $a = 0.003$.

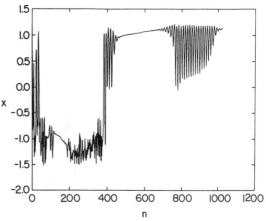

FIG. 9. Plot of x vs n for the oscillator $i = 226$ and $1 \leq n \leq 1024$. The parameters are the same as in Fig. 7.

ly static. In Fig. 9 x_i versus n is shown for $i = 226$ and $1 \leq n \leq 1024$. This oscillator is in the stable zone of Fig. 8 and undergoes relatively *static* behavior during the time interval $500 \leq n \leq 700$. This contrasts greatly with the behavior of oscillator 101, which is well in the *unstable* zone. This oscillator shows rather large excursions during the same time interval, as shown in Fig. 10. Another feature is that when Figs. 9 and 10 are compared, we see that at $n \simeq 700$ the $i = 226$ oscillator starts to undergo rapid changes just as oscillator 101 becomes roughly static. This is depicted in Fig. 11 which shows x_i versus i for $750 \leq n \leq 800$.

We speculate that the transient behavior[4] discussed in the previous examples arises from the existence of many attractors in our chain and fractal basin boundaries separating the various basins. The gross features of the spatial structure of the final state of the examples discussed so far are qualitatively similar in that there are two domains separated by sharp kinklike walls with a spatial oscillation of 14 lattice sites ($m = 18$) in each domain. The main differences between the final states are the sizes of the domains and the positions of the domain walls. We have seen that groups of nearby oscillators can lock temporarily into an $m = 18$, *transient structure* that lasts through hundreds of iterations of $\hat{T}$. Thus, it appears that the confusion experienced by an orbit as to which attractor to asymptote to shows up locally in the development of highly structured transient regions. For this model we have found that the domain walls appear to have rather complex dynamics. Standard continuous unforced models that show domain formation do not exhibit such structure. We speculate that the complex be-

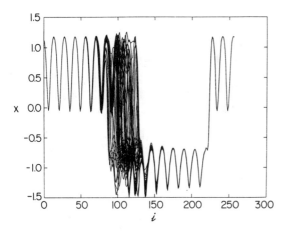

FIG. 8. Same as Fig. 7 but for $600 \leq n \leq 661$.

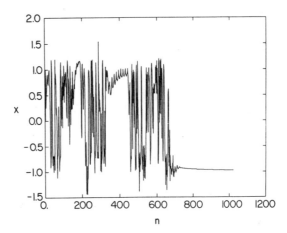

FIG. 10. Same as Fig. 9 but for the oscillator $i = 101$.

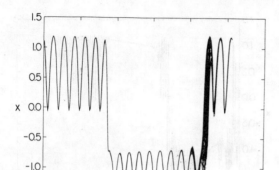

FIG. 11. Same as Fig. 7 but for $750 \leq n \leq 800$.

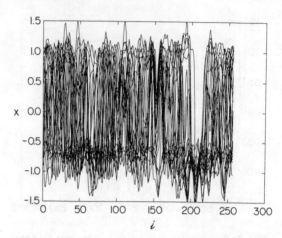

FIG. 13. Plot of x vs i showing the superposition of iterates $2024 \leq n \leq 2048$ for the same parameters as Fig. 12.

havior we observe is due to the forcing coupled with the Duffing nonlinearity rather than with the discrete-space aspect of our system.

The observation of spatio-temporal chaotic bursts during the transient phase leads us to look into the types of behavior possible when the asymptotic behavior is not as simple as fixed points. With this in mind, we examine the effects of increased driving. The local dynamics is known to make a transition to chaos around $f_c = 0.12$ when $\gamma = 0.15$, $\omega_D = 0.833$, so keeping $\varepsilon = 1.0$, we increase f.

Figure 12 shows a disordered chain state at $n = 2048$ when the driving is set to $f = f_c = 0.12$ [the initial condition is that of Eqs. (5) with $a = 0.003$]. Note the absence of the $m = 18$ structures. To get an indication of the temporal behavior of this system, Fig. 13 shows a superposition of 25 iterates of $\hat{T}$ for $2024 \leq n \leq 2048$. The time series for oscillator 1 and its power spectrum, all for the time interval $1025 \leq n \leq 2048$, are shown in Figs. 14 and 15, respectively. This behavior persists past $n = 8192$.

Thus at $f = 0.12$, the chain state is highly disordered with a chaotic temporal evolution. The transition to this behavior takes place at $f \simeq 0.112$. At $f = 0.112$, the chain displacements look qualitatively the same as in Fig. 12 (see also Fig. 13), as shown in Fig. 16. However, on examination of the time series, there are similarities to what was seen in the fixed point cases, namely, that certain oscillators attempt to lock onto a local fixed point attractor and that these locking intervals are interrupted by chaotic bursts. This is seen in Fig. 17, where $x_i(n)$ is plotted for $i = 1$ from $n = 1024$ to $n = 2048$. Note the relatively stable (low-amplitude) motion from $1200 < n < 1700$. A superposition of chain displacements from $n = 1304$ to $n = 1324$ (Fig. 18) reveals the temporary formation of an orderly stable structure analogous to the $m = 18$ structure of the fixed point experiment. Oscillator 1 is in this region which accounts for its relatively small amplitude motion.

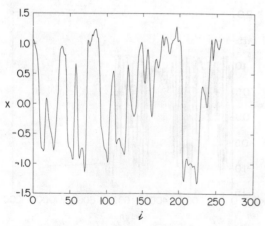

FIG. 12. The chain state x vs i for $f = 0.12$ and $n = 2048$.

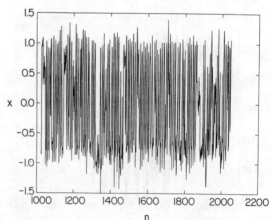

FIG. 14. Plot of x vs n for oscillator $i = 1$ and $1025 \leq n \leq 2048$ and for the same parameters as in Fig. 12.

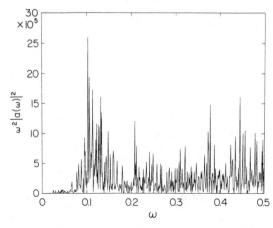

FIG. 15. Power spectrum of the trajectory shown in Fig. 14. $a(\omega)$ is the Fourier amplitude and the frequency ω is in units of ω_D.

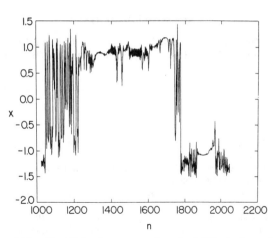

FIG. 17. Plot of x vs n for $f=0.112$, $i=1$, and $1024 \leq n \leq 2048$.

This type of transition to sustained chaos is phenomenologically similar to transitions to chaos in low-dimensional dynamical systems which undergo crises.[4] In particular, if we look at a dynamical variable as a function of time, for different values of the parameter, we observe that chaotic transients become longer and longer until we obtain sustained chaos at the crisis value of the parameter.

IV. CONCLUSION

In this paper we have numerically investigated a one-dimensional lattice system of dissipative, forced, nonlinear ordinary differential equations. The continuum limit of the this system is a nonlinear driven dissipative wave equation. In the absence of forcing, small-amplitude linear waves on the continuum system display both dispersion and damping. The parameter space of this system is very large, and so our investigation of it has not been exhaustive. The central, most interesting feature revealed in our numerical investigations was the *extreme* intermittency of the temporal behavior of the system. We believe that this is a typical feature of forced, nonlinear, spatially extended systems with wave propagation, dispersion, and damping.

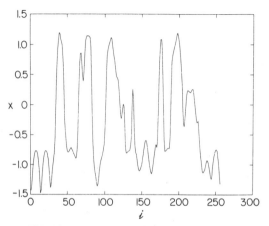

FIG. 16. Plot of x vs i for $f=0.112$ and $n=2048$.

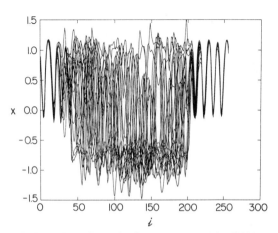

FIG. 18. Plot of x vs i for $f=0.112$ and $1304 \leq n \leq 1324$.

ACKNOWLEDGMENTS

This work was supported by the U.S. Department of Energy (Office of Basic Energy Sciences) and the Office of Naval Research (Physics). We would like to thank V. Englisch and K. Kaneko for useful discussions.

---

*Permanent address: Nordisk Atomfysik, Blegdamsvej 17, DK-2100 Copenhagen, Denmark.
†Also Department of Electrical Engineering and Department of Physics and Astronomy.
‡Permanent address: Laboratory for Laser Energetics, University of Rochester, Rochester, NY 14620.

[1] I. Waller and R. Kapral, Phys. Rev. A **30**, 2047 (1984); K. Kaneko, Progr. Theor. Phys. **72**, 480 (1984); **74**, 1033 (1985); T. Yamada and H. Fujisaka, ibid., **72**, 885 (1984); Phys. Lett. **124A**, 421 (1987); R. Kapral, Phys. Rev. A **31**, 3868 (1985); H. Fujisaka and T. Yamada, Progr. Theor. Phys. **74**, 918 (1985); K. Kaneko, Physica **23D**, 436 (1986); R. Kapral and G.-L. Oppo, ibid. **23D**, 455 (1986); R. J. Deissler and K. Kaneko, Phys. Lett. **119A**, 397 (1987); F. Kaspar and H. G. Schuster, ibid. **113A**, 451 (1986); Phys. Rev. A **36**, 842 (1987).
[2] J. D. Keeler and J. D. Farmer, Physica **23D**, 415 (1988) also included longer-range coupling in their system of coupled logistic maps. There are also examples of cellular automata studies which have longer-range coupling.
[3] For example, F. C. Moon and G. -X. Li, Phys. Rev. Lett. **55**, 1439 (1985).
[4] C. Grebogi, E. Ott, and J. A. Yorke, Physica **7D**, 181 (1983).
[5] C. Grebogi, S. W. McDonald, E. Ott, and J. A. Yorke, Phys. Lett. A **99**, 415 (1983).
[6] S. W. McDonald, C. Grebogi, E. Ott, and J. A. Yorke, Physica **17D**, 125 (1985).
[7] W. H. Press, B. P. Flannery, S. A. Teukolsky, and W. T. Vetterling, *Numerical Recipes* (Cambridge University Press, New York, 1986), Chap. 15.
[8] H. Fujii and Y. Sawada, J. Chem. Phys. **69**, 3830 (1978).
[9] M. Marek and I. Stuchl, Biophys. Chem. **3**, 241 (1975).
[10] I. Stuchl and M. Marek, J. Chem. Phys. **77**, 2956 (1982).

IX. Fractal Basin Boundaries

In nonlinear systems it is possible for more than one attractor to exist. The range of values of initial conditions for which the solution tends towards a given attractor is called a basin of attraction. In case of two or more attractors, the transition from one basin of attraction to another is called a basin boundary. In many systems it has been shown that this boundary has got a fractal structure. This proves that even when the system is not chaotic we cannot predict its behavior, as small changes of system parameters can move the system to another attractor.

Fundamental properties of fractal basin boundaries are presented in Paper IX.1. A number of examples are given.

Connection between fractal structure of basin boundary and the appearance of homoclinic orbits in the Poincaré map calculated by Melnikov method is shown in Paper IX.2.

Examples of smooth and fractal basin boundaries of an impact oscillator are given in Paper IX.3.

The erosion of smooth and fractal basins of attractions is discussed in Paper IX.4. Integrity measures which allow to estimate the safety of engineering systems are introduced.

FRACTAL BASIN BOUNDARIES

Steven W. McDONALD,[a] Celso GREBOGI,[a] Edward OTT[a,b] and James A. YORKE[c]

Received 9 November 1984
Revised manuscript received 13 April 1985

Basin boundaries for dynamical systems can be either smooth or fractal. This paper investigates fractal basin boundaries. One practical consequence of such boundaries is that they can lead to great difficulty in predicting to which attractor a system eventually goes. The structure of fractal basin boundaries can be classified as being either locally connected or locally disconnected. Examples and discussion of both types of structures are given, and it appears that fractal basin boundaries should be common in typical dynamical systems. Lyapunov numbers and the dimension for the measure generated by inverse orbits are also discussed.

1. Introduction

Much of the interest in nonlinear dynamical systems has focused on the existence of periodic, quasiperiodic, and chaotic attractors, and the investigation of how these arise [1]. It is also important to recognize, however, that the analysis of a typical dissipative dynamical system may be complicated by the fact that initial conditions in different regions of phase space may generate orbits which exhibit different time-asymptotic behavior. That is, it is possible, and even common, that at fixed values of system parameters, more than one attractor may be present. The set of initial conditions (more precisely, the closure of this set) which eventually approach each particular attractor is called its basin of attraction. In this paper, we will be interested in the variety and structure of the boundaries which separate basins of attraction.

In order to illustrate the concepts of coexisting attractors, basins of attraction and basin

[a] Laboratory for Plasma and Fusion Energy Studies, University of Maryland, College Park, MD 20742, USA.
[b] Department of Electrical Engineering and Department of Physics and Astronomy, University of Maryland, College Park, MD 20742, USA.
[c] Department of Mathematics and Institute for Physical Sciences and Technology, University of Maryland, College Park, MD 20742, USA.

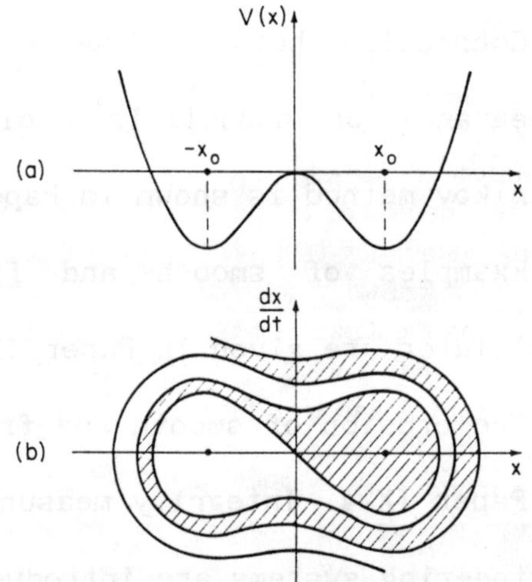

Fig. 1. (a) Potential $V(x)$ for a point particle moving in one dimension. With friction, almost every initial condition eventually comes to rest at one of the equilibrium points, x_0 or $-x_0$. (b) Phase (velocity-position) space for the system in (a). The basin of attraction for x_0 (crosshatched) is separated from the basin of attraction for $-x_0$ (blank) by a smooth basin boundary curve.

boundaries, consider the simple case of a point particle moving under the influence of friction in a potential $V(x)$ as shown in fig. 1a. For almost any initial condition, the orbit will eventually come to rest at either of the two stable fixed points at $x = \pm x_0$. Fig. 1b schematically depicts the phase

Reprinted with permission from Phys. Lett., Vol. 17D, pp. 125–153
Copyright (1985), Elsevier Science Publishers B.V.

space of the system and the basins of attraction of these two fixed point attractors. An initial condition chosen in the crosshatched region eventually comes to rest at $x = x_0$, while any initial condition in the blank region tends to $x = -x_0$. The boundary separating these basins is the smooth curve passing through the origin. Points on the boundary do not tend to any attractor and so must be mapped to other boundary points: the basin boundary is an invariant set under the system action. In the example of fig. 1, initial conditions on the boundary approach the unstable fixed point at the origin (*not* an attractor); i.e., the boundary is the stable manifold of an unstable orbit (although this will not always be the case).

Another property of the boundary in fig. 1 is that it is a smooth curve. It is a main point of this paper that a basin boundary need not be a smooth curve or surface. Indeed, for a wide variety of systems it is common for boundaries to exhibit a fractal structure and to be characterized by a noninteger dimension.

The importance of studying the structure of basin boundaries is illustrated by the following example. Consider the simple two-dimensional phase space diagram schematically depicted in fig. 2. There are two possible final states, or attractors, denoted by A and B. The region to the left (right) of the basin boundary Σ is the basin of attraction for attractor A (or B, respectively). Let us consider an initial condition and measure its coordinates. Now suppose that this measurement has an uncertainty ε in the sense that the actual initial condition might be anywhere in a disc of radius ε centered at the measured value. In fig. 2, points 1 and 2 represent two such measured initial conditions. While the orbit generated by initial condition 1 is definitely attracted to B, initial condition 2 is uncertain in that it may be attracted to either A or B. Now assume that initial conditions are chosen randomly with uniform distribution in the rectangular region shown in fig. 2. We consider the fraction $f(\varepsilon)$ of initial conditions which are uncertain as to which attractor is approached when there is an initial error ε. For the simple case of fig. 2, initial conditions within a strip of width 2ε centered on the boundary are uncertain; thus, $f(\varepsilon)$ is proportional to ε. In section 2, however, we demonstrate that systems with fractal boundaries are more sensitive to initial uncertainty and can obey

$$f \sim \varepsilon^\alpha, \qquad (1.1)$$

where α is *less than one*. We call α the *uncertainty exponent* and we say that these systems possess *final state sensitivity* [2]. We believe that many typical dynamical systems exhibit this behavior. In cases where the uncertainty exponent α is significantly less than unity, a substantial reduction in the error in the initial condition, ε, produces only a relatively small decrease in the uncertainty of the final state as measured by f. Furthermore, we show that the uncertainty exponent α is the difference between the dimension of the phase space and the "capacity dimension" of the basin boundary. This is explained in section 2 (cf. eq. (2.5) for a definition of capacity dimension). The increased sensitivity of final states to initial condition error when $\alpha < 1$ provides an important motivation for the study of fractal basin boundaries.

Another reason for interest in fractal basin boundaries is that as a system parameter is varied, a chaotic attractor can be suddenly destroyed in a collision with the basin boundary (we have called such events *crises* [3]); for values of the parameter beyond the crisis point, long chaotic transients occur [4]. Variation of the parameter in the opposite direction produces the creation of a chaotic attractor (a "route to chaos").

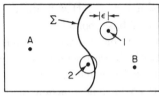

Fig. 2. A schematic region of phase space divided by the basin boundary Σ into basins of attraction for the two attractors A and B. Points 1 and 2 represent two initial conditions with uncertainty ε.

Several standard dissipative dynamical systems possess fractal basin boundaries. The work of Cartwright and Littlewood [5], Levinson [6], and Levi [7] can be shown to imply that the forced Van der Pol oscillator exhibits a fractal boundary due to the existence of a "horseshoe" in the dynamics. A similar analysis by Kaplan and Yorke [8] of the Lorenz system [9] of ordinary differential equations can also be shown (section 4.2) to imply the existence of a fractal basin boundary in a parameter regime below that at which the creation of the strange attractor occurs. Finally, we show that the simple one-dimensional logistic map possesses a fractal boundary in the period three regime [10] (or rather the third iterate of the map does). These examples are discussed in section 4, and it is shown that in these cases the basin boundary exhibits a fractal Cantor set structure. The basin structure of a simple two-dimensional mapping which models this behavior is shown in fig. 3a. Here there are two fixed point attractors ($A^\pm$); initial conditions in the dark region are attracted to A^+, while the blank region is the basin for A^-. The fine-scale complexity of this type of fractal boundary is further revealed under magnification in fig. 3b. (Note that the graininess in fig. 3 is due to finite resolution; i.e., fig. 3b should be regarded as consisting of an infinite number of dark and blank strips.)

The basin boundary in fig. 3 is not a continuous curve. In contrast, we observe in section 4 that fractal boundaries in two-dimensional maps can be continuous curves. This type of boundary is also "Cantor-like": a typical smooth curve crossing the boundary intersects it in an uncountable set that contains no segments. See section 3 for further discussion and classification of different types of fractal basin boundaries.

Perhaps the best-known examples [11] of boundaries which are fractal curves are found in two-dimensional dynamics of the form $z \to F(z)$, where F is an analytic function of the complex variable $z = x + iy$. Fig. 4a shows the complex plane divided into the two basins of the analytic map given by $F(z) = z^2 + 0.9z \exp(2\pi i \Omega)$, with

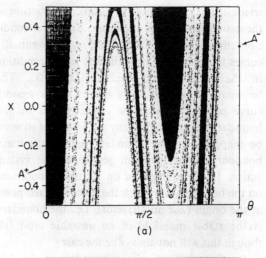

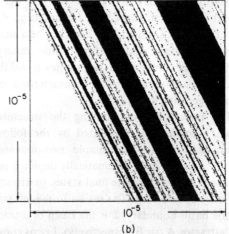

Fig. 3. (a) Basins of attraction for the two attractors A^+ (dark region) and A^- (blank region). This two-dimensional system is governed by the mapping in eqs. (4.1). (b) Magnification by a factor of 10^5 of the region in (a) given by $1.92200 \le \theta \le 1.92201$ and $-0.50000 \le x \le -0.49999$.

$\Omega = (\sqrt{5} + 1)/2$, the golden mean. Here, orbits originating in the interior blank region are attracted to the fixed point at the origin, whereas the dark region is the set of points which escape to infinity (the basin for the point at infinity). The boundary separating these regions is quite complicated and has "snowflake" structure on arbitrarily small scale, as suggested by the magnification in fig. 4b. Such boundaries can be classed as "quasi-circles" (see sections 3 and 5 for a defini-

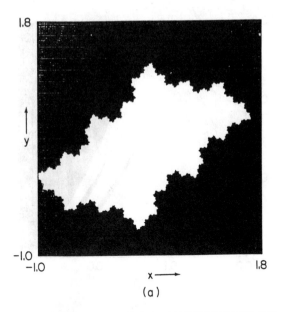

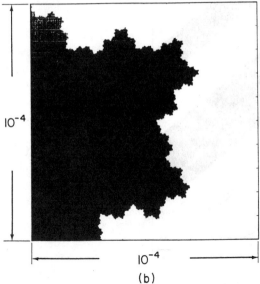

Fig. 4. (a) Basins of attraction for the fixed point at the origin (blank region) and the point at infinity (dark region). The mapping is given by the analytic function $z_{n+1} = z_n^2 + 0.9 z_n \exp(2\pi i \Omega)$, where $z \equiv x + iy$ and Ω is the golden mean. (b) Magnification by a factor of 10^4 of the region in (a) $0.72809 \le x \le 0.72819$, $0.02209 \le y \le 0.02219$.

tion and discussion of quasicircles), fractal curves that have a moderate degree of regularity.

Complex analytic maps are a very restricted class and possess special features not typical of two-dimensional maps in general. In particular, as a consequence of their Cauchy–Riemann structure they do not have chaotic attractors (the two Lyapunov numbers are equal). In section 5, therefore, we shall also study more general maps of the plane that provide more suitable models of typical nonlinear physical systems (possibly of higher dimension). These more general maps yield basin boundaries which are curves but lack the quasicircle property. An example of this more general set of maps is shown in fig. 5a. The blank region consists of initial conditions which are attracted to a chaotic attractor (which is also shown) and again, the dark region is the basin of attraction for the point at infinity. The basin boundary has a rather complicated structure, as figs 5(b) and 5(c) show, and the numerical studies of this paper indicate that it is indeed a fractal set. This magnification, however, reveals a fractal character quite different from that exhibited by the boundary of the analytic map; here, the boundary appears "stretched" or "striated" in contrast to the "snowflake" appearance of fig. 4. For the class of two-dimensional maps that we study, basin boundaries can be either smooth or fractal. When they are fractal, they almost always exhibit a "stretched" structure similar to fig. 5.

In section 6 we discuss the measure generated by inverse orbits on the basin boundary, the Lyapunov numbers for this measure, and the dimension of the measure. Section 7 concludes the paper with a summary of our main results.

2. Final state sensitivity

Even in the absence of a detailed description of the dynamics of the systems shown in figs. 1–5, one can readily understand the effect that initial condition error has on the ability to predict which final state will be approached by a particular trajectory. If an initial condition specified with error ε is within that distance of the basin boundary, the attractor to which it will be attracted cannot be predicted with certainty. In the

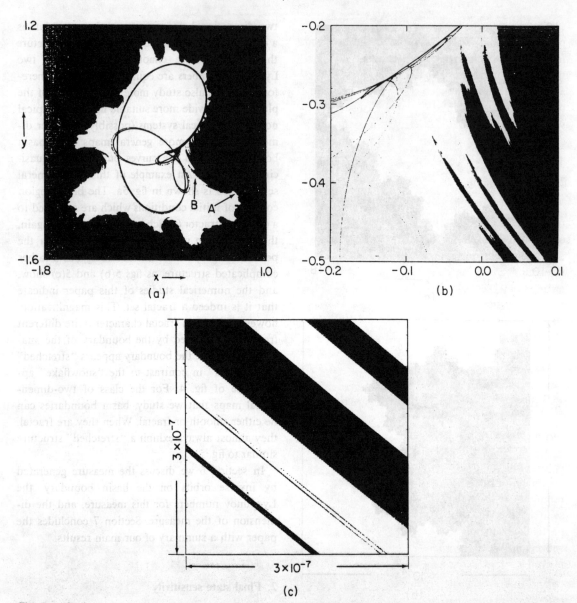

Fig. 5. (a) Basin structure of the mapping given in eqs. (5.15). The chaotic attractor attracts all initial conditions in the blank region while the dark region is the basin for orbits which escape to infinity. An unstable fixed point is located on the basin boundary at A. (b) Magnification of a region near B in (a). (c) Magnification to a scale of 10^{-7} of the region $-0.5316359 \leq x \leq -0.5316356$, $-1.1584159 \leq y \leq -1.1584156$ showing locally striated structure of the boundary.

present section we shall develop this notion of *final state sensitivity* in terms of the fraction of phase space consisting of initial conditions which are uncertain (in the above sense) when specified with error ε. We focus not on the actual size of the uncertain fraction of phase space, but rather on the way in which this fraction scales as the initial condition error is reduced. Furthermore, since the location and structure of a basin boundary depends on the system parameters, uncertainty in system parameter values could affect the ability to predict the final state from an initial condition, independent of the degree of precision with which the initial condition is chosen. Therefore, we shall also investigate parameter sensitivity: the scaling of the uncertain fraction of parameter space with variation in the parameter error, when the initial condition is fixed.

2.1. Initial condition uncertainty

In the schematic example shown in fig. 2 with two attractors (A and B) and a simple smooth one-dimensional basin boundary separating the finite volume of phase space, the uncertain region for initial conditions with error ε is simply determined by thickening the basin boundary by the amount ε. Any initial condition in this strip can change from the basin of A to that of B (or vice versa) if perturbed by an amount ε or less. The area of this region is proportional to the error ε; as ε is reduced, the "uncertain fraction" of phase space satisfies

$$f(\varepsilon) \sim \varepsilon. \tag{2.1}$$

For $f(\varepsilon)$ to be well defined in cases where the phase space of the system is infinite, we shall restrict initial conditions to lie in some fixed finite subregion of phase space which contains the basin boundary (e.g., the region pictured in fig. 3a). In such a case, the actual magnitude of $f(\varepsilon)$ at a particular value of ε will depend on the choice of subregion. We are primarily interested in the scaling of $f(\varepsilon)$ as ε becomes small (e.g., eq. (2.1)), however, and we believe that this behavior is independent of the subregion selected. (More precisely, the exponent α in eq. (1.1) is independent of the subregion.)

We now consider the phase space region shown in fig. 3 and examine the dependence of the "uncertain fraction" of phase space $f(\varepsilon)$ on the error ε. This is accomplished by selecting 8192 random initial conditions over the region of fig. 3a and iterating each to determine its final state (i.e., to ascertain the basin in which each is located). Each

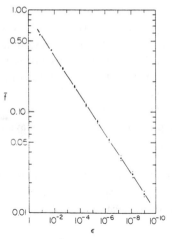

Fig. 6. Log-log plot of $\tilde{f}$ versus ε for the phase space shown in fig. 3.

initial condition (θ_0, x_0) is then perturbed in the horizontal direction by $\pm \varepsilon$ to produce two perturbed initial conditions $(\theta_0 \pm \varepsilon, x_0)$. All 16.384 perturbed initial conditions are iterated to determine their basins, and this data is compared to that for the unperturbed initial conditions. If either of the two perturbed initial conditions associated with a particular unperturbed initial condition is in a basin different from the unperturbed one, we say that the initial condition is uncertain under the error ε. We record the fraction $\tilde{f}(\varepsilon)$ of these initial conditions as the error ε is varied: we expect this fraction of uncertain initial conditions to be proportional to the uncertain fraction $f(\varepsilon)$ of phase space volume. See subsection 2.3 for further discussion of this.

The variation of $\tilde{f}(\varepsilon)$ for this system with decreasing error ε is plotted in fig. 6. The error bars at each data point were computed on the basis that occurrences of uncertain initial conditions are random events, and hence the error in their number $N(\varepsilon)$ at a particular value of ε is $\sqrt{N(\varepsilon)}$. Here, the log-log plot indicates a power law behavior with exponent 0.2,

$$\tilde{f}(\varepsilon) \approx 0.9 \varepsilon^{0.2}. \tag{2.2}$$

The consequences of such a relationship are remarkable. For an error of 0.125 in the initial conditions (about 3% of the width of the phase space shown in fig. 3a), the final states of approximately 59% of the initial conditions could not be predicted with certainty. Reducing the error by about a factor of 60 to 0.002 results in a decrease of $\bar{f}$ to 26%, only about a factor of two increase in one's confidence in the ability to predict.

We expect such a power law behavior,

$$f(\varepsilon) \sim \varepsilon^{\alpha}, \qquad (2.3)$$

to be common in dissipative dynamical systems. The uncertainty exponent α is related to the dimension of the basin boundary,

$$\alpha = D - d. \qquad (2.4)$$

Here, D is the dimension of the phase space and d is the dimension of the basin boundary. We use the capacity definition of dimension [12]

$$d = \lim_{\delta \to 0} \frac{\ln N(\delta)}{\ln(1/\delta)}, \qquad (2.5)$$

where $N(\delta)$ is the minimum number of D-dimensional cubes of side δ required to completely cover the basin boundary. This definition simply expresses the scaling of $N(\delta)$ with the cube size as δ is decreased; for example, we see that the power law

$$N(\delta) \sim \delta^{-d} \qquad (2.6)$$

satisfies (2.5). That (2.3) and (2.4) result from (2.5) can be seen heuristically as follows: setting the cube edge δ equal to the initial condition error ε, the volume of the uncertain region of phase space will be of the order of the total volume of all $N(\varepsilon)$ D-dimensional cubes of side ε required to cover the boundary. Since the volume of one of these cubes is ε^D, this uncertain volume is of the order $\varepsilon^D N(\varepsilon)$. With (2.6) for $N(\varepsilon)$, we estimate the uncertain phase space volume to be of the order of $\varepsilon^D N(\varepsilon) \sim \varepsilon^{D-d}$, which gives (2.3) and (2.4). A more precise statement is given by the following rigorous result.

Theorem. The uncertain fraction f of a finite region of a D-dimensional phase space associated with initial condition error ε obeys

$$\lim_{\varepsilon \to 0} \frac{\ln f(\varepsilon)}{\ln \varepsilon} = \alpha, \qquad (2.7)$$

if and only if the basin boundary has capacity dimension $d = D - \alpha$.

A proof of this theorem is provided at the end of this section.

The expression (2.3) reduces to the linear relation (2.1) in cases where the basin boundary is a smooth curve or surface so that its dimension is one less than that of the phase space, $d = D - 1$. In general, since the basin boundary divides the phase space, its dimension d must satisfy $d \geq D - 1$. Thus, the existence of fractional dimension boundaries allows for values $0 < \alpha \leq 1$. From (2.4), one concludes that the result (2.2) for the mapping in fig. 3 indicates an experimentally measured fractal basin boundary dimension of approximately 1.8. For the analytic mapping shown in fig. 4 and discussed in section 5.1, the experimental measurement of $\bar{f}(\varepsilon)$ as described above again yields the relationship (2.3) with a value of $\alpha \approx 0.69$; by (2.4), this indicates a basin boundary dimension of about 1.3. The more general quadratic map in fig. 5 produces a similar value $\alpha \approx 0.7$ for a similar dimension of about 1.3, despite the quite different appearance of the basin boundary.

2.2. *Parameter sensitivity*

In addition to affecting the specification of an initial condition, error may also be present in the selection of the parameters of a system. A small error in a system parameter might alter the location or structure of a basin boundary so that a fixed initial condition shifts from one basin to another. In a finite region of parameter space, the fraction of parameter values which will produce such a shift for a given initial condition when perturbed by a parameter error δ is the uncertain fraction $f_p(\delta)$ of parameter space. Here we investigate the scaling of $f_p(\delta)$ at small δ for the map shown in fig. 3. As discussed in section 4.1,

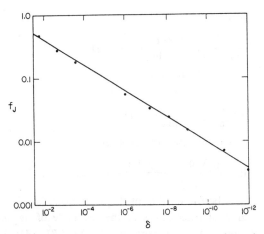

Fig. 7. Log-log plot of f_J versus parameter error δ for the phase space of fig. 3 (eqs. (4.1)).

the parameter we shall vary in this system is denoted J_0 (cf. eqs. (4.1)).

We consider the sensitivity of the basin boundary to variations of the system parameter J_0. We select a single initial condition $(\theta_0, x_0) = (1.05, 0.3)$ with a random choice of J_0 in the interval $0.1 \le J_0 \le 0.3$ and determine which attractor ($A^\pm$) is approached by this orbit. Then J_0 is perturbed to $J_0 \pm \delta$ and the same initial condition is iterated for both perturbed maps. If either of the parameter perturbations has shifted the basin boundary so that the fixed initial condition has changed from the original basin to the other, we say that this particular unperturbed choice of J_0 is uncertain under the error δ. This experiment is repeated for 12,800 random choices of J_0 in the same interval (all with the same single initial condition) for values of the parameter error δ between $10^{-3} \ge \delta \ge 10^{-12}$. The fraction $f_J(\delta)$ of the uncertain parameter values (in the sense given above) in the parameter interval is plotted as a function of the parameter error δ in fig. 7.

The result of fig. 7 again indicates a power law relationship between the uncertain fraction of parameter space and the parameter error, $f_J(\delta) \sim \delta^{0.2}$. (While it is possible that the dimension of a basin boundary could vary widely over a parameter range, the dimension of the boundary in this system remains close to 1.8 in the parameter interval examined.)

2.3. *Remarks*

1) While the power law behavior (2.2) of $f(\varepsilon)$ observed for the map in fig. 3 satisfies the limiting relationship (2.7), there are other forms of $f(\varepsilon)$ at small ε that would also yield the same value of α. For example, the behavior $f(\varepsilon) \sim \varepsilon^\alpha \ln \varepsilon$ also satisfies (2.7). Indeed, an example given in the appendix shows that one can have a smooth basin boundary ($\alpha = 1$) with $f(\varepsilon) \sim \varepsilon \ln(1/\varepsilon)$ for small ε.

2) Imagine that we attempt to initialize the system at a point η in the region of interest (e.g., the rectangle of fig. 3a). If η has an uncertainty ε, then the actual initial condition is $\bar\eta = \eta + \varepsilon\Delta$, where $\varepsilon\Delta$ is a random number representing the error. The quantity Δ might be thought of as being generated from some probability density $p(\Delta)$. For example, without loss of generality p can be chosen so that $\int \Delta^2 p(\Delta)\, d^D\Delta \equiv 1$, in which case ε is the variance of the error $\varepsilon\Delta$. Now choose η randomly in the region of interest and use η to determine the attractor to which the system tends. Since the actual initial condition is $\bar\eta$ rather than η, the prediction of the final state might in fact be incorrect. Let $\tilde{f}(\varepsilon)$ denote the probability of this error. Note that $\tilde{f}(\varepsilon)$ and $f(\varepsilon)$ are different: $\tilde{f}(\varepsilon)$ is the probability of making an incorrect prediction, while $f(\varepsilon)$ is the probability of *being able* to make an incorrect prediction if $|\Delta| < 1$. The quantity $\tilde{f}(\varepsilon)$ is measurable by performing repeated experiments. For example, our calculated quantity $\tilde{f}(\varepsilon)$ for the numerical experiments of fig. 3a is an approximation to $2\tilde{f}(\varepsilon)$ for the case $\Delta = (\Delta_x, \Delta_y)$, $p(\Delta) = \delta(\Delta_y)[\delta(\Delta_x - 1) + \delta(\Delta_x + 1)]/2$, where here $\delta(\Delta_x)$ is the standard delta-function. The quantity $f(\varepsilon)$ is, however, in principle much harder to approximate in a numerical experiment.

Pelikan [13] has proven that $\tilde{f}(\varepsilon) \sim f(\varepsilon)$ under the following restrictions: (i) p = constant in $|\Delta| < 1$ and zero otherwise; and (ii) the map is strictly expanding on the basin boundary. The restriction (ii), for example, applies to certain one-dimensional maps and to Julia sets (e.g., fig. 4), but not to our example in fig. 3. In the general case, we believe that the following statement should hold.

Conjecture. For typical dynamical systems

$$\lim_{\varepsilon \to 0} \frac{\ln \tilde{f}(\varepsilon)}{\ln f(\varepsilon)} = 1.$$

2.4. Proof of theorem (2.7)

Let $B(\varepsilon, A)$ be the set of points within ε of a closed bounded set A and assume A has capacity dimension d. Cover phase space with a cubic grid with cube edge length ε. Each point x of A is in a box of the grid. Any point y within ε of A lies within ε of some point x in A. Therefore y lies in one of 3^D boxes which are the original cube or a cube touching the original cube (see fig. 8). Hence, $B(\varepsilon, A)$ can be covered using no more than $3^D N(\varepsilon)$ cubes where $N(\varepsilon)$ is the number of ε cubes needed to cover A. Thus the volume of the set $B(\varepsilon, A)$ satisfies

$$\text{Vol}(B(\varepsilon, A)) \leq 3^D \varepsilon^D N(\varepsilon). \qquad (2.8)$$

Now we cover A using cubes from a grid with edge lengths $\varepsilon/D^{1/2}$. We choose that size grid since any two points within such a cube are within ε of each other. Thus every point in every cube used in the cover is within ε of A, and so lies in $B(\varepsilon, A)$. Such a cover has $N(\varepsilon/D^{1/2})$ cubes, so

$$\text{Vol}(B(\varepsilon, A)) \geq (\varepsilon/D^{1/2})^D N(\varepsilon/D^{1/2}). \qquad (2.9)$$

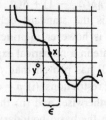

Fig. 8. Schematic representation of a basin boundary set A in a phase space covered with cubes of edge ε.

Thus,

$$\frac{\ln D^{-D/2}}{\ln \varepsilon} + \frac{\ln N(\varepsilon/D^{1/2})}{\ln \varepsilon} + D$$
$$\leq \frac{\ln[\text{Vol}(B(\varepsilon, A))]}{\ln \varepsilon}$$
$$\leq \frac{\ln 3^D}{\ln \varepsilon} + \frac{\ln N(\varepsilon)}{\ln \varepsilon} + D.$$

From the definition of d, eq. (2.5),

$$\lim_{\varepsilon \to 0} \frac{\ln[\text{Vol}(B(\varepsilon, A))]}{\ln \varepsilon} = D - d.$$

Let A be the portion of the basin boundary in the region of interest (as in the rectangle of fig. 3a). Since $f(\varepsilon)$ is proportional to $\text{Vol}(B(\varepsilon, A))$, eq. (2.7) follows.

3. Classification of fractal basin boundaries

3.1. Classification

In figs. 3, 4, and 5 of section 1 we have given several examples of fractal basin boundaries. The boundaries in these figures appear to have quite different structure, and indeed the types of basin boundaries which they exemplify are fundamentally different. In this section we offer a classification of fractal basin boundaries, ordered below in terms of increasing degree of regularity of the boundary:

(i) boundaries which are *locally disconnected* (e.g., fig. 3);

(ii) boundaries which are *locally connected* but are not quasicircles (e.g., fig. 5);

(iii) boundaries which are *quasicircles* (e.g., fig. 4).

3.2. Definitions

A closed set is *disconnected* if it can be split into two parts A and B such that

$$\min|\alpha - \beta| > 0$$

for α in A and β in B. If this cannot be done then the set is *connected*.

A set is *locally connected* if, given any point η in the set and any sufficiently small ε, then there exists a $\delta(\varepsilon, \eta) \leq \varepsilon$ with the following property:

P) For every point ξ in the set satisfying $|\eta - \xi| \leq \delta(\varepsilon, \eta)$, there is a connected subset of the original set containing η and ξ and lying wholly in the ε-ball centered at η.

If the property P is not satisfied for every η in the set, then we say that the boundary is *locally disconnected*.

If the property P can be satisfied with $\delta = \kappa \varepsilon$ for some constant κ independent of η, then the set is a *quasicircle*.

From our definition, it follows that the boundary is locally connected if it is a continuous curve or surface. A continuous surface lying in a D-dimensional phase space is a surface which is parametrically representable as $x = g(s)$, where s is a $D-1$-dimensional parameter vector, x is a point on the boundary, and g is a continuous D-dimensional vector function of s.

If the basin boundary is locally connected and bounded, then one can show that $\delta(\varepsilon, \eta)$ can be chosen independent of η, $\delta(\varepsilon, \eta) = \delta(\varepsilon)$.

3.3. Examples

In order to give the reader a feel for these definitions we list below several illustrative examples applying these definitions.

Example 1. A simple smooth curve such as the basin boundary pictured in fig. 1 is connected (but not fractal).

Example 2. The logistic map, $x_{n+1} = rx_n(1 - x_n)$, has two attractors for $0 < r < 4$. One of these attractors is $x = -\infty$. The other is located in the region $1 > x > 0$ and may be either periodic or chaotic depending on r. The basin boundary for these two attractors are the two points $x = 0$ and $x = 1$. To apply the definition, identify η in the definition with one of these two points, say $x = 0$, and take $\varepsilon < 1$. Then $\xi = \eta = 0$, and property P is clearly satisfied. The boundary is locally connected.

Example 3. In section 5 we shall consider a mapping of a cylinder for which the basin boundary can be determined analytically and is given by

$$y = B(x) = -\sum_{j=1}^{\infty} \lambda_y^{-j} \cos(2\pi \lambda_x^{j-1} x), \quad (3.1)$$

where λ_x is an integer, $\lambda_x > \lambda_y > 1$ and $-1 < x < 1$. This basin boundary is shown in fig. 19. The sum (3.1) converges absolutely and uniformly for $\lambda_y > 1$ so that $B(x)$ is a continuous curve; therefore, the basin boundary is locally connected. As such, (3.1) can be formally differentiated term by term,

$$\frac{dB}{dx} = \frac{2\pi}{\lambda_x} \sum_{j=1}^{\infty} \left(\frac{\lambda_x}{\lambda_y}\right)^j \sin(2\pi \lambda_x^{j-1} x). \quad (3.2)$$

Since $\lambda_x > \lambda_y > 1$, we see that (3.2) diverges. Thus, the boundary $y = B(x)$ is a continuous but nowhere differentiable curve. Furthermore, the boundary has infinite length and can be shown [14] to have capacity dimension $d = 2 - (\ln \lambda_y / \ln \lambda_x)$.

We now show heuristically that this basin boundary is not a quasicircle. Let us focus on the behavior of (3.1) near $x = 0$, where $B(x)$ is schematically depicted in fig. 9. As prescribed in the definition of local connectedness, we select the point $\eta \equiv (x_*, B(x_*))$ on the boundary to be the center of an ε-ball. In order to apply the property P, we ask the following question: if we construct a ball of radius δ around η (as shown in fig. 9) which includes the boundary point $\xi_* \equiv (-x_*, B(-x_*))$, how large must ε be so that a connected piece of the boundary containing η and ξ_* lies within the ε-ball centered at η? We shall show that as x_* becomes smaller, the required radius ε diminishes as $\delta(\varepsilon) \sim \varepsilon^\gamma$, with $\gamma > 1$.

From (3.1) we note that $B(x_*) = B(-x_*)$, so that the distance between ξ_* and η is simply the horizontal displacement $\delta = 2x_*$. Referring to fig. 9, we see that for η and ξ_* to lie on a connected piece of the boundary contained entirely within an ε-ball centered at η, we must choose ε large enough to include the boundary point $m \equiv (0, B(0))$. Therefore, $\varepsilon(x_*) \geq |m - \eta| \approx |B(x_*) - B(0)|$, for $\varepsilon \gg x_*$ (since $\delta \sim \varepsilon^\gamma$, $\gamma > 1$, and x_* is small). Now we note that (3.1) implies $B(x/\lambda_x) = \lambda_y^{-1}[B(x) - \cos(2\pi x/\lambda_x)]$, so that $B(0) = \lambda_y^{-1}[B(0) - 1]$. Com-

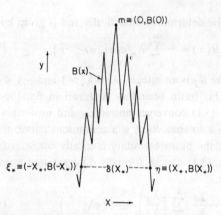

Fig. 9. Schematic representation of the basin boundary $B(x)$ of fig. 19 near $x = 0$. Choosing the point η on the boundary, the radius of an ε-ball around η must be $\varepsilon = |m - \eta|$ so as to include a connected piece of $B(x)$ connecting η to the point ξ_* a distance δ away.

bining these two relations, we see that $\varepsilon(x_*/\lambda_x)$ is

$$\begin{aligned}\varepsilon(x_*/\lambda_x) &\approx |B(x_*/\lambda_x) - B(0)| \\ &\approx \lambda_y^{-1}|B(x_*) - B(0) + 1 \\ &\quad - \cos(2\pi x_*/\lambda_x)| \\ &\approx \lambda_y^{-1}|B(x_*) - B(0)| \\ &\approx \varepsilon(x_*)/\lambda_y,\end{aligned}$$

where we have neglected the contribution $1 - \cos(2\pi x_*/\lambda_x)$ for small x_*. Thus we see that as x_* is diminished by a factor of λ_x, the required radius of the ε-ball decreases by a factor of λ_y. Thus, $\varepsilon(x_*) \sim x_*^{1/\gamma}$, with $\gamma = (\ln\lambda_x/\ln\lambda_y) > 1$. Combining this with $\delta(x_*) \sim x_*$, we have

$$\delta(\varepsilon) \leq \mathcal{O}(\varepsilon^\gamma).$$

While we have derived this result by considering a region in the vicinity of $x = 0$, we emphasize that the same behavior must occur near every $x = m/\lambda_x^n$ for all positive integers m and n. Thus, since $\gamma > 1$, to satisfy property P, $\delta(\varepsilon)$ must shrink to zero faster than linearly with ε as $\varepsilon \to 0$; we cannot satisfy property P with $\delta(\varepsilon) = \kappa\varepsilon$. We conclude that (3.1) is an example of a fractal basin boundary that is connected but is not a quasicircle. Due to the similar appearance upon magnification of the basin boundary of fig. 5 and that in fig. 19, we believe that the boundary in fig. 5 is also con- nected but not a quasicircle (cf. section 5 for further discussion).

Example 4. By definition the largest connected subset of a Cantor set is a single point; yet in any neighborhood of a point in a Cantor set there are an infinite number of other points in the Cantor set. Thus a Cantor set is locally disconnected. In the next section we discuss a one-dimensional map model of the Lorenz system of differential equations; for this map the basin boundary is a Cantor set. In addition, the magnification of the basin boundary shown in fig. 3b is essentially a Cantor set of parallel line segments (see section 4), and thus also is an example of a disconnected basin boundary.

3.4. Discussion

In the following section we discuss several dynamical systems with locally disconnected fractal basin boundaries. Following that, section 5 is devoted to a discussion of locally connected basin boundaries. Locally connected basin boundaries arising from an analytic map of a single complex variable, $z \to F(z)$, where $z = x + iy$ and F is analytic, have been studied for a long time. Such maps yield quasicircle basin boundaries [11]. However, such maps are also special, and we believe that the quasicircle property should not be expected to occur in *typical* dynamical systems. To our knowledge, our discussion of section 5 is the first (with the exception of our preliminary note, ref. 4) to address locally connected fractal basin boundaries that are not quasicircles.

From the examples of section 5 it appears that locally connected fractal basin boundaries require a minimum system dimensionality for their existence. In particular, for smooth invertible maps, the dimension must be at least three, corresponding to four for continuous time systems (flows). For smooth noninvertible maps the dimensionality is evidently required to be at least two. In contrast, like chaotic attractors, locally disconnected fractal basin boundaries can occur in smooth one-dimen-

4. Locally disconnected basin boundaries

The development of the present understanding of chaotic motion in dynamical systems has relied on the investigation of several standard physical models. In particular, the behavior of the forced van der Pol oscillator in certain parameter regimes was studied by Cartwright and Littlewood [5] who suggested the presence of chaotic (although nonattracting) orbits as well as an infinity of periodic orbits (also nonattracting). The ensuing work of Levinson [6] on a class of chaotic orbits in this system led to the geometric interpretation of Smale [15] and his introduction of the horseshoe map. This general construction has since influenced the analysis of chaotic behavior in many systems, and its specific application to the van der Pol oscillator is described in the work of Levi [7]. Levi points out, in passing, that the van der Pol system possesses a basin boundary that is a Cantor set and Flaherty and Hoppensteadt [16] comment briefly that Levinson's solutions are orbits on this boundary. However, there is virtually no literature on fractal basin boundaries, and the important practical implications of this type of boundary structure, which we discuss in this paper, have not been studied.

The basin boundary for the case studied in refs. 5–7 and 16 is an example of a locally disconnected fractal basin boundary. In this section we shall give a detailed discussion of several other examples of locally disconnected fractal basin boundaries. The general impression emerging from the discussion of this section is that such boundaries may be commonly encountered in practical situations. We begin in section 4.1 with an examination of the basin boundary (shown in fig. 3) which arises in a model similar to (but simpler than) that used by Levi to describe the qualitative behavior of the van der Pol oscillator. While the van der Pol equation has two attracting orbits of periods n and $n+2$ for some n, our model has two attracting fixed points.

4.1. A model annulus map

Consider a map on the annulus $\mathscr{A}$ in the plane as shown in fig. 10. We use polar coordinates (r, θ) with θ measured from the horizontal axis as shown. Phase points in $\mathscr{A}$ at time $t=0$ are mapped into the thinner convoluted annulus $\mathscr{A}'$ at $t=1$. Thus we are considering a discrete time map (which might arise via Poincaré surface of section [1] from a continuous time system). The effect of this map M on the region between $\theta = 0$ and $\theta = \pi$ is to squeeze the annulus radially, stretch it in length, and to fold the region near $\theta = \pm\pi/2$ as shown. The points $A^{\pm}$ at $\theta = 0, \pi$ are attracting fixed points of the map. The map M is taken to be symmetric under reflection across the horizontal axis ($\theta \to 2\pi - \theta$). An explicit map with these properties will be given later.

The angular aspect of the two-dimensional map M may be simulated by the one-dimensional map $M_1(\theta)$ which is shown in fig. 11. In addition to the stable fixed points $A^{\pm}$ at $\theta = 0, \pi$, there are three unstable fixed points: S_+ at $\theta = \theta_+$, S_- at $\theta = \theta_-$, and S_0 at $\theta = \pi/2$. In the full two-dimensional geometry of the annulus, these points are located at the crosses in fig. 10; they are within the image annulus $\mathscr{A}'$ and continue to be within all successive images of the original annulus $\mathscr{A}$. Thus, $S_{\pm}$

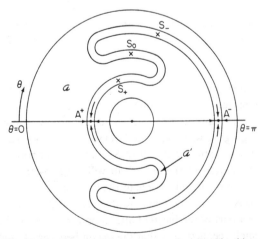

Fig. 10. The mapping M maps the annulus $\mathscr{A}$ to the thinner convoluted annulus $\mathscr{A}'$ such that there are two attracting fixed points $A^{\pm}$ and saddle points S_+, S_-, and S_0.

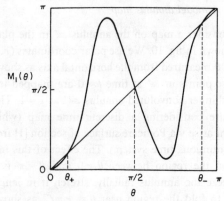

Fig. 11. The angular behavior $M_1(\theta)$ of the annulus map M of fig. 10.

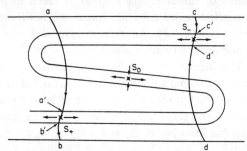

Fig. 12. Schematic illustration of the convolution of $\mathscr{A}'$ near $\theta = \pi/2$.

and S_0 are actually saddles due to the radial contraction under M.

The action of M in the convoluted region near $\theta = \pi/2$ is schematically illustrated in fig. 12, where the arc has been straightened out. Here, the saddles $S_\pm$ and S_0 are shown with their local contracting and stretching directions. The extension $\overline{ab}$ of the contracting direction of S_+ is the part of the stable manifold of S_+ which maps to $\overline{a'b'}$ under one iteration of M; similarly, $\overline{cd}$ is the piece of the stable manifold of S_- which maps to $\overline{c'd'}$. Comparing fig. 12 with fig. 10 and in view of the properties of the stable manifolds of S_+ and S_-, all points in the region to the left of $\overline{ab}$ will map toward the attractor A^+; all points to the right of $\overline{cd}$ will tend to A^-. Thus, the region to the left (right) of $\overline{ab}(\overline{cd})$ is a part of the basin of attraction for $A^+(A^-)$. We shall see, however, that the region between $\overline{ab}$ and $\overline{cd}$ contains more pieces of each of these basins in an extremely intermingled order.

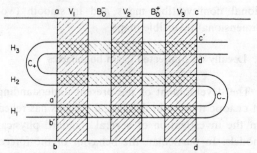

Fig. 13. Further simplification of the region in fig. 12, identifying the action of M on particular subregions of the rectangle abdc: $M(V_i) = H_i$ for $i = 1, 2, 3$, and $M(B_0^\pm) = C_\pm$.

The region of the annulus near $\theta = \pi/2$ in fig. 12 is further schematized in fig. 13, where $\overline{ab}$ and $\overline{cd}$ have been straightened. Again, the action of the map M on the rectangle abdc is to squeeze the region vertically, stretch it out horizontally (to more than triple its original length), and then form the "s" shape with $\overline{ab} \to \overline{a'b'}$ and $\overline{cd} \to \overline{c'd'}$. Thus, the crosshatched region V_1 at the left end of abdc is mapped to the lowest horizontal strip H_1 (also crosshatched). Similarly, V_2 maps to H_2 and V_3 to H_3. The region to the left of $\overline{ab}$ is a piece of the basin of attraction of the attractor A^+; the region to the right of $\overline{cd}$ is a piece of the basin for A^-.

The vertical strip B_0^- between V_1 and V_2 is the preimage of the arc C_- which connects H_1 and H_2; B_0^+ maps to the arc C_+. Since the arc $C_+(C_-)$ is to the left of $\overline{ab}$ (right of $\overline{cd}$), its future iterates tend to the attractor $A^+(A^-)$. Therefore, we see that the vertical strip $B_0^+(B_0^-)$ is a piece of the basin of attraction of $A^+(A^-)$. The basins of the two attractors are now seen to be "tangled", with the regions B_0^+ and B_0^- in "inverted" order (i.e., B_0^+ is closer to A^- than is B_0^-). The future of the three remaining strips (V_1, V_2, and V_3) is as yet undetermined; we define the union of these strips to be the set $\beta_1 \equiv V_1 \cup V_2 \cup V_3$.

We now consider the future of the horizontal strip $H_1 = M(V_1)$. The segment $H_1 \cap B_0^- \equiv H_1^-$ will eventually tend to the attractor A^- since it is in the basin of attraction for A^-. Similarly, the segment $H_1 \cap B_0^+ \equiv H_1^+$ will tend to A^+. As H_1 is the image of V_1 with $\overline{ab} \to \overline{a'b'}$, one may now subdivide V_1 into thinner vertical strips: these are the preimages of $H_1^\pm$ and the remaining three sections of H_1, $H_{1i} \equiv H_1 \cap V_i$ ($i = 1, 2, 3$). This is shown in fig. 14, where the strip V_1 is partitioned into five thinner strips. The strips V_{1i} map to H_{1i}; these are separated by the strips $B_1^\pm$ which map to $H_1^\pm$.

Since H_1^+ eventually maps to A^+, its preimage B_1^+ must be a piece of the basin of attraction for A^+. Similarly, B_1^- is a piece of the basin of attraction for A^-. The other two vertical strips, V_1 and V_2, may be similarly partitioned and so contain pieces $B_i^\pm$ ($i = 2, 3$) of the basin of $A^\pm$. The alternating order of these pieces indicates an even higher degree of tangling than previously noted. At this stage there are nine vertical strips V_{ij} ($i, j = 1, 2, 3$) with undetermined futures. The union of these strips with two indices is defined as the set $\beta_2 \equiv \bigcup_{i,j} V_{ij}$.

Indeed, this structure continues when one considers the future $M(V_{ij})$ of V_{ij}. For illustration, consider $M(V_{11}) \equiv H_{11}$. Since $H_{11}^\pm \equiv H_{11} \cap B_1^\pm$ will eventually map to $A^\pm$, the vertical strip V_{11} is further divided into the 5 substrips $B_{11}^\pm$ and V_{11i} such that $H_{11}^\pm = M(B_{11}^\pm)$ and $H_{11i} = M(V_{11i})$ ($i = 1, 2, 3$). We see that at the nth stage of this process a vertical strip $V_{j_1 j_2 \ldots j_n}$ ($j_i = 1, 2, 3$) can be further identified to contain substrips $B_{j_1 j_2 \ldots j_n}^\pm$, which are pieces of the basins of $A^\pm$, separated by a new series of substrips $V_{j_1 j_2 \ldots j_{n+1}}$. At each step n we define the union of the 3^n vertical strips with undetermined future $V_{j_1 j_2 \ldots j_n}$ to be the set

$$\beta_n \equiv \bigcup_{j_1, j_2, \ldots, j_n} V_{j_1, j_2, \ldots, j_n}.$$

This process of division is similar to that found in the standard horseshoe construction [17] and produces a Cantor set. As $n \to \infty$, the basin boundary β is given by

$$\beta \equiv \lim_{n \to \infty} \bigcap_n \beta_n.$$

Thus the part of the basin boundary lying in the rectangle abcd of fig. 13 is the product of a Cantor set (running horizontally) and an interval (running vertically). Hence the boundary is locally disconnected (cf., example 4 of section 3.3).

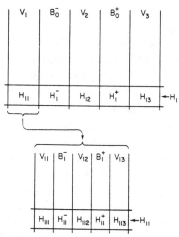

Fig. 14. Simplifying fig. 13, the top picture shows only the vertical strips and the lowest horizontal strip $H_1 = M(V_1)$. The bottom picture shows the partitioning of V_1 into the five substrips which are preimages of the five pieces of H_1: $M(V_{1i}) = H_{1i}$ for $i = 1, 2, 3$ and $M(B_1^\mp) = H_1^\mp$. Also identified are the pieces involved in the next level of the construction: the intersections of the substrips of V_1 with H_{11}.

We now give a specific example of a map which has the essential geometric properties outlined above,

$$\theta_{n+1} = \theta_n + a \sin 2\theta_n - b \sin 4\theta_n - x_n \sin \theta_n, \quad (4.1a)$$
$$x_{n+1} = -J_0 \cos \theta_n, \quad (4.1b)$$

where x may be thought of as the radial distance from the center of the annulus. The angles θ and $\theta + 2\pi$ are identified as equivalent and the desired reflection symmetry for $\theta \to 2\pi - \theta$ is present. At fixed x_n, the graph of θ_{n+1} versus θ_n is similar to that in fig. 11 and is adjustable with the two parameters a and b. The x-dependent part of (4.1a) and the form of (4.1b) were chosen in a simple way so that the map is everywhere contracting for $J_0 < 1$ (the Jacobian determinant is $J_0 \sin^2 \theta$).

Like fig. 10, this map has two fixed points $A^+ = (\theta_+, x_+) = (0, -J_0)$ and $A^- = (\theta_-, x_-) = (\pi, J_0)$ which are attracting for values of a, b, and J_0 such that $|1 + 2a - 4b + J_0| < 1$. The parameters must also be adjusted for these two fixed points to be the only attractors present. Our choice of $a = 1.32$, $b = 0.9$, and $J_0 = 0.3$ seems to satisfy these requirements.

We have examined the basin structure of (4.1) with the stated choice of parameters by considering a grid of 256×256 initial conditions over the region $0 \leq \theta \leq \pi$ and $-0.5 \leq x \leq 0.5$. Each initial condition was iterated 100 times, which is long enough to ensure that all these orbits come within a distance of 10^{-3} from one of the two attractors. The basins for A^+ and A^- are shown in fig. 3, which was constructed by placing a dot at the location of each initial condition whose trajectory approached A^+ at $(0, -0.3)$. The solid vertical dark region to the left of $\theta \approx 0.3$ corresponds to the region to the left of $\overline{ab}$ in fig. 13; the blank region between $2.8 \leq \theta \leq \pi$ corresponds to the region to the right of $\overline{cd}$. This correspondence with figs. 13 and 14 can be continued: the blank parabolic strip at $0.8 \leq \theta \leq 1.4$ is B_0^-, the inverted dark parabolic region at $1.8 \leq \theta \leq 2.4$ is B_0^+, the blank strip at $\theta \approx 0.5$ is B_1^- and the somewhat thinner dark strip a $\theta \approx 0.7$ is B_1^+, etc. The slight tilt of the strips and the parabolic shapes are due to the x-dependence of the map which was, for the most part, ignored in the schematic figs. 12 and 13.

The richness of this basin structure is further revealed upon magnification of the small region $1.92200 \leq \theta \leq 1.92201$, $-0.50000 \leq x \leq -0.49999$ (a scale of 10^{-5}) as shown in fig. 3b. The basins are evidently intertwined to an extreme degree and the fractal Cantor set nature of the basin boundary is indicated.

4.2. Other examples of locally disconnected boundaries

The preceding discussion of the existence of basin boundaries which are locally disconnected was illustrated by a two-dimensional mapping (4.1) selected to exhibit the properties of an annulus map developed in figs. 10–14. There are, however, more familiar systems which possess this type of basin boundary structure. There has also been recent experimental evidence implying the existence of fractal basin boundaries in an actual physical system. Here we briefly describe three additional examples in which fractal basin boundaries occur.

1) The one-dimensional logistic map

The basin structure of a one-dimensional map with N coexisting attractors sometimes consists of N collections of one-dimensional line segments (basins) separated by a basin boundary composed of a finite set of points. In other examples, the basin boundary is fractal, a Cantor set of fractional dimension. The uncertainty exponent α observed in the power law behavior (1.1) described in section 2 could be used to determine the dimension (2.4) of the basin boundary. An example of a fractal basin boundary is provided by the logistic map, here written in the form

$$x_{n+1} = F_\lambda(x_n) = \lambda - x_n^2, \quad |x| \leq 2. \quad (4.2)$$

In the parameter range $1.75 \leq \lambda \leq 1.79$, the attractor is a periodic orbit with period three (cf. refs. 10, 18, and 19 for results concerning this regime). In order to construct a one-dimensional map with more than one attractor, we consider the third iterate $F_\lambda^{(3)}$ of (4.2): each element of the period three attractor is a fixed point attractor of $F_\lambda^{(3)}$ with its own basin of attraction. In terms of the original map, each basin is the set of initial conditions which eventually progress around the period three orbit in the same phase. Thus, when considering the original map F_λ we speak of "final phase sensitivity", while for $F_\lambda^{(3)}$ we refer to final state sensitivity as before.

The final state sensitivity experiment was performed as described in section 2 for the third iterate map $F_\lambda^{(3)}$ with $\lambda = 1.75$. The fraction $\bar{f}(\varepsilon)$ of 12,800 initial conditions which changed from one basin to another when perturbed by an error ε is expected to be proportional to the uncertain volume (length) of phase space; the behavior of $\bar{f}(\varepsilon)$ as ε is reduced is shown in fig. 15. Again, a power law behavior $\bar{f}(\varepsilon) \sim \varepsilon^{0.03}$ is evident, with an uncertainty exponent which implies a basin boundary dimension (2.4) of approximately 0.97.

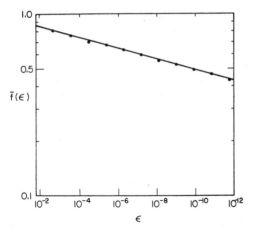

Fig. 15. Log–log plot of $\bar{f}$ versus ε for the third iterate of the logistic map in the period three regime.

2) *Lorenz system*

The Lorenz system [9] of three ordinary differential equations (a truncated model of fluid convection) progresses through several regimes of motion as a parameter r is increased. For values of r in the range $0 < r < 1$, there is a single fixed point attractor at the origin of the three-dimensional phase space representing a stationary fluid without flow. As r is increased through unity, this fixed point becomes unstable and two separate fixed point attractors appear (representing left- and right-handed convective rolls); these are stable in the regime $1 < r < r_c$. A strange attractor appears at a parameter value r_* slightly less than r_c; in this parameter range $r_* < r < r_c$, the two fixed point attractors coexist with the strange attractor. Increasing r through the critical value r_c brings the instability of the two fixed points, and for $r > r_c$ the only attractor is the strange attractor. We concentrate on the parameter range $r < r_*$ in which there are only the two coexisting fixed point attractors.

We consider the three-dimensional phase space in rectangular coordinates. The two fixed point attractors A and B lie in a plane P defined by $z = r - 1$ (in Lorenz's notation) as shown in fig. 16a. We follow Kaplan and Yorke [8] and consider the mapping ϕ on P induced by the repeated downward ($\dot{z} < 0$) intersections of a trajectory in the three-dimensional space with this plane.

In the parameter range $1 < r < r_*$, the origin $(0, 0, 0)$ is an unstable fixed point possessing a two-dimensional stable manifold S_0 and a one-dimensional unstable manifold U_0. The curve L in

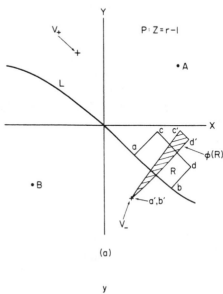

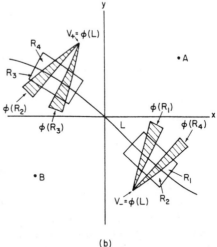

Fig. 16. (a) The flow of the Lorenz system in a three-dimensional phase space induces the mapping ϕ on the plane P (defined as constant z). This qualitative illustration of ϕ shows how points in a region R on P near the stable manifold L map across L and converge near the unstable manifold V_-. (b) More complete illustration of the action of ϕ on regions R_i near L (on both sides).

fig. 16a represents the first intersection of the stable manifold S_0 with P. The points $V_\pm$ are the first downward intersections of U_0 with P, following U_0 in both directions from the origin. Thus, ϕ is not defined on L, and ϕ is discontinuous at L.

Numerical investigation of the orbits for r slightly less than r_* reveal the nature of ϕ, as illustrated in fig. 16a by $R \to \phi(R)$. In general, points far from L on either side uniformly approach the attractor (A or B) on the same side (cd $\to$ c'd'). A region near L, however, will map to the other side of L when r is sufficiently close to r_* (abdc $\to$ a'b'd'c'). This characteristic of the map is due to a corresponding feature of the flow in this regime: orbits starting near L on P are drawn down toward the origin and then are pushed out and travel near the unstable manifold U_0, crossing over [8] the curve L on the next downward return to the plane P, intersecting P near V_+ or V_-. The schematic triangular shape of the region $\phi(R)$ results from the fact that the closer points in R are to the curve L, the more they tend to be mapped to the same point in $\phi(R)$ (ab $\to$ a'b'). The area of $\phi(R)$ is also much smaller than that of R due to the strong contraction experienced by the flow for the standard parameters of the Lorenz system.

The flow admits inversion symmetry in the plane P so that, following Kaplan and Yorke [8], we show in fig. 16b a more complete (yet qualitative) picture of the mapping ϕ near L. Here, the boundaries of the regions R_i ($i = 1, \ldots, 4$) which coincide with L map to the pointed tips of the thin triangles shown in fig. 16b. This picture of ϕ, called a "broken horseshoe" in ref. 8, can be used to show that the boundary separating the basins of attraction of A and B is a Cantor set.

As in the previous development in section 4.1, figs. 12–14, of a Cantor set boundary, we redraw fig. 16b as a "reduced broken horseshoe" in fig. 17. That is, we imagine trimming the area of the upper portion of R_1 until the upper boundary of $\phi(R_1)$ coincides with the upper boundary of R_1. Similarly, we trim $R_{2,3,4}$ until we obtain fig. 17a. Comparison of this picture with that of fig. 16b reveals that the crosshatched regions of $\phi(R_1)$ and

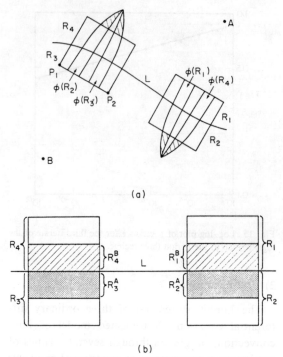

Fig. 17. (a) Reduced broken horseshoe of the map ϕ. The cross-hatched portions of $\phi(R_1)$ and $\phi(R_4)$ will eventually map to attractor B (if R_2 was extended further from L, the image of that portion of R_2 would map to between $\overline{p_1 p_2}$ and attractor B (cf. $\phi(R_2)$). (b) Simplified illustration of the regions R_i, showing only the subregions which map to the cross-hatched regions of $\phi(R_i)$ in (a). For example, R_1^B maps to the pointed tip of $\phi(R_1)$ and thence toward B; R_1^B is a piece of the basin of B.

$\phi(R_4)$ shown in fig. 17a will map to some region between the segment $\overline{p_1 p_2}$ of R_3 and the attractor B. Thus, under successive applications of ϕ, these two crosshatched regions will approach B; the same argument implies that the crosshatched pieces of $\phi(R_2)$ and $\phi(R_3)$ (c.f. fig. 17a) will eventually converge to A. This allows us to identify the subregions R_1^B and R_4^B of R_1 and R_4 which are in the basin of B and the subregions R_2^A and R_3^A of R_2 and R_3 which are in the basin of A. These regions are shown as the hatched and shaded strips in the even more schematic fig. 17b. The basin structure in the blank strips of R_1 through R_4 is as yet undetermined.

Fig. 18a shows just the two regions R_1 and R_2, separated by the curve L. The hatched strip R_1^B in

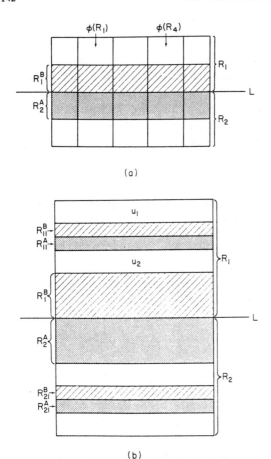

Fig. 18. (a) Illustration of the intersections of R_1 and R_2 with $\phi(R_1)$ and $\phi(R_4)$ (vertical strips, cf. fig. 17a)). (b) Further division of the horizontal blank portions of R_1 and R_2 of (a). The blank region u_1 maps to the blank portion of $\phi(R_1) \cap R_1$. $R_{11}^B \to R_1^B$ (and thence toward B), $R_{11}^A \to R_2^A$ (and thence toward A) and u_2 maps to the blank portion of $\phi(R_1) \cap R_2$. Thus $R_{11}^B(R_{11}^A)$ is in the basin of B(A) and u_1 and u_2 are yet undetermined. Similar arguments apply for the subdivision of R_2.

R_1 has been determined to be a piece of the basin of B, and R_2^A in R_2 is a piece of the basin for A. Now the vertical strips representing the intersection of $\phi(R_1)$ and $\phi(R_4)$ with R_1 and R_2 are also shown. From the definition of R_1^B, we see that points lying in the intersections $\phi(R_1) \cap R_1^B$ and $\phi(R_4) \cap R_1^B$ will eventually approach B, while the region $\phi(R_1) \cap R_2^A$ and $\phi(R_4) \cap R_2^A$ will approach A. Thus, by considering $\phi^{-1}(\phi(R_1))$ we can further deduce that the blank strip of R_1 must at least have the basin structure as shown in fig. 18b. The strip R_{11}^B maps into R_1^B and so is a piece of the basin of attraction for B; similarly, R_{11}^A is a piece of the basin for A. Analysis of the implications of fig. 18a for the region R_4 (due to the intersection of $\phi(R_4)$ with R_1^B and R_2^A) results in a similar partitioning of R_4. Indeed, all four original regions can be analyzed in this manner and can be shown to have the alternating basin structure illustrated in fig. 18b. Furthermore, consideration of the intersections $\phi(R_1) \cap R_{11}^B$ and $\phi(R_1) \cap R_{11}^A$ leads to a similar partitioning of the blank strip u_1; the blank strip u_2 is partitioned by considering $\phi(R_1) \cap R_{21}^B$ and $\phi(R_1) \cap R_{21}^A$. As in the development of section 4.1, this continued process results in a standard horseshoe construction; that is, the remaining undetermined "blank strips" at each stage produce a basin boundary which, in the limit, forms a Cantor set. Almost every point of R_1 and R_2 will eventually approach either A or B, the exceptions being the points on the Cantor set of lines that are mapped into $R_1 \cup R_2$ by all iterates of the map. An indirect effect of the existence of this Cantor set boundary in the parameter regime r slightly less than r_* is provided by the observation of long chaotic transients in numerical investigations, specifically those of Yorke and Yorke [20].

3) *The experiments of Bergé and Dubois* [21]

These authors have performed experiments on the Bénard instability in a low aspect ratio rectangular cell for high Rayleigh number and high Prandtl number. They observe that the system can have multiple attractors, and that a rather long chaotic transient exists before the system settles into one of the attractors. Since the system evolution during the transient phase depends strongly on initial conditions, it is to be expected that the final state will also, and that $\alpha \ll 1$ in (2.3) will apply. At somewhat lower Rayleigh number ($450 \gtrsim r > 200$), different stable attractors still simultaneously coexist but long chaotic transients do not occur. Even in this range it is probable that final state sensitivity will occur.

5. Locally connected basin boundaries

In this section we discuss basin boundaries which are continuous curves or surfaces, but are fractal in that they possess a noninteger capacity dimension and are nowhere differentiable. We shall focus on two-dimensional maps because these systems are relatively simple, computationally rapid to evolve in time, and because they offer a fairly clear visualization of the properties of basin boundaries. We first discuss the tangent map and its Lyapunov numbers.

For a general two-dimensional map M, the tangent map TM (the Jacobian matrix of M) maps infinitesimal displacements $(\delta x_n, \delta y_n)$ near (x_n, y_n) to $(\delta x_{n+1}, \delta y_{n+1})$ near (x_{n+1}, y_{n+1}). We now define the Lyapunov number of an orbit. Consider a forward orbit O^+ given by the sequence $(x_0, y_0; x_1, y_1; \ldots; x_N, y_N)$. By the chain rule for partial differentiation, $TM^{(N)}(x_0, y_0) = TM(x_N, y_N) \cdot TM(x_{N-1}, y_{N-1}) \cdot \ldots \cdot TM(x_0, y_0)$. Let $l_1^+(N)$ and $l_2^+(N)$ be the eigenvalues of $TM^{(N)}(x_0, y_0)$, with $|l_1^+| \geq |l_2^+|$. The Lyapunov numbers for the initial condition (x_0, y_0) are

$$\mu_{1,2}^+ \equiv \lim_{N \to \infty} |l_{1,2}^+(N)|^{1/N} \quad (5.1)$$

and the Lyapunov exponents are $h_{1,2}^+ = \ln \mu_{1,2}^+$. In practice, these limits appear to exist and can be estimated easily in numerical computations. In this paper we shall also be interested in Lyapunov numbers for inverse orbits. That is, we start at (x_0, y_0) and consider a sequence of preimages $(x_0, y_0; x_{-1}, y_{-1}; \ldots; x_{-N}, y_{-N})$; we denote this backwards orbit O^-. Infinitesimal displacements $(\delta x_0, \delta y_0)$ near (x_0, y_0) are mapped backwards by the N-composed linear map $TM^{(-N)}(x_0, y_0) = TM^{-1}(x_{-N}, y_{-N}) \cdot TM^{-1}(x_{-N+1}, y_{-N+1}) \cdot \ldots \cdot TM^{-1}(x_0, y_0)$, where TM^{-1} is the inverse of the matrix TM. Denoting the eigenvalues of $TM^{(-N)}$ by $l_1^-(N)$ and $l_2^-(N)$, with $|l_1^-| \leq |l_2^-|$, the Lyapunov numbers associated with O^- are

$$\mu_{1,2}^- \equiv \lim_{N \to \infty} |l_{1,2}^-|^{-1/N}. \quad (5.2)$$

The backward orbit O^- from (x_0, y_0) is in general not uniquely determined by (x_0, y_0) because the map M may not have a unique inverse. One must remember, therefore, that the backwards Lyapunov numbers $\mu_{1,2}^-$ refer to a particular backwards orbit; a different backwards orbit from (x_0, y_0) may possess a different set of backwards Lyapunov numbers.

5.1. Analytic maps

A particular class of two-dimensional maps can be constructed by identifying the plane $\mathbb{R}^2$ with the one-dimensional complex plane $\mathbb{C}^1$ and considering mappings $z_{n+1} = F(z_n)$, where F is an analytic function of the single complex variable $z = x + \mathrm{i} y$. Such a mapping can be resolved to give a two-dimensional map with

$$x_{n-1} = f(x_n, y_n) = \mathrm{Re}\, F(x_n + \mathrm{i} y_n),$$
$$y_{n-1} = g(x_n, y_n) = \mathrm{Im}\, F(x_n + \mathrm{i} y_n).$$

For brevity, we refer to two-dimensional maps which satisfy this condition as *analytic maps*. Several properties of these maps greatly facilitate their analysis, and therefore they have been the subject of much study [11]. We emphasize, however, that this is a very special class of two-dimensional maps because the functions $f(x, y)$ and $g(x, y)$ must obey the Cauchy–Riemann conditions in these cases.

The attractors in analytic maps may be fixed points or periodic orbits. A particular analytic map can possess more than one attractor with basins of attraction separated by continuous curves. Despite being continuous, a basin boundary in an analytic map may be nowhere differentiable and may possess a fractal dimension. For our purposes the main facts that we should be aware of concerning analytic maps are (1) they do not exhibit strange attractors; (2) their basin boundaries (Julia sets) are typically fractal curves which exhibit two-dimensional structure on arbitrarily small scales; and (3) their basin boundaries are quasicircles. An example of such a basin boundary is shown in fig. 4.

5.2. An explicit construction of a locally connected fractal boundary

In this subsection we consider a two-dimensional map for which an explicit expression for the basin boundary can be derived,

$$x_{n+1} = \lambda_x x_n \quad (\text{mod } 1), \tag{5.3a}$$

$$y_{n+1} = \lambda_y y_n + \cos 2\pi x_n, \tag{5.3b}$$

Here λ_x is an integer, x is restricted to the unit interval, but y is allowed to take on all values. The case $\lambda_x = 2$ and $|\lambda_y| < 1$ has been studied by Kaplan and Yorke [22] and has been shown to possess a chaotic attractor which attracts almost all initial conditions. In the following, however, we study the quite different case [4] $\lambda_x > \lambda_y > 1$.

The linearized (or tangent) map

$$\text{TM}(x, y) = \begin{pmatrix} \lambda_x & 0 \\ -2\pi \sin 2\pi x & \lambda_y \end{pmatrix} \tag{5.4}$$

has eigenvalues λ_x and λ_y, which are both greater than one. Thus, since local displacements are always expanded independent of location or initial direction, there are no attractors for this map at finite y. Furthermore, initial conditions with large $|y_0|$ rapidly approach $\pm \infty$ as $\lambda_y^n y_0$, depending on the sign of y. Since no finite attractors exist, we see that the two points at infinity ($y = +\infty$ and $y = -\infty$) are the only attractors.

A picture of the basins for each of these attractors is displayed in fig. 19 for the case $\lambda_x = 3$, $\lambda_y = 1.5$. The dark region is the basin B_- for the attractor at $y = -\infty$ and the blank region is the basin B_+ for $y = +\infty$. The magnification in fig. 19b reveals that linear structure is present on a very small scale; indeed, similar pictures are obtained with repeated magnification of the boundary region. The linear, or "striated" structure on all scales is quite different from the "two-dimensional" or "snowflake" form exhibited by the complex analytic maps of the previous subsection.

We now show that the basin boundary (fig. 19) is a continuous curve of the form $y = B(x)$. In order to proceed, we make use of the fact that since the basin boundary repels trajectories (they all eventually approach $\pm \infty$), it is an attractor for inverse orbits. The map (5.3) is noninvertible;

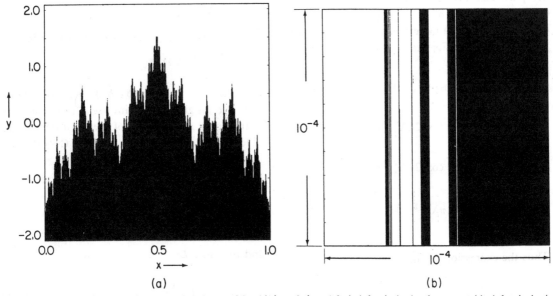

Fig. 19. (a) Basin structure of the system given in eqs. (5.3) with $\lambda_x = 3$, $\lambda_y = 1.5$: dark for the basin of $y = -\infty$, blank for the basin of $y = +\infty$. (b) Magnification of the region $0.38317 \leq x \leq 0.38327$, $0.4450 \leq y \leq 0.4451$ revealing striated boundary structure.

there are λ_x (an integer) inverses of each point x_n: $x_{n-1}^{(j)} = (j + x_n)/\lambda_x$, $j = 0, \ldots, \lambda_x - 1$. Once a particular inverse of x_n is chosen, the inverse y_{n-1} is determined from (5.3b) as

$$y_{n-1} = \frac{1}{\lambda_y}(y_n - \cos 2\pi x_{n-1}). \tag{5.5}$$

Each of the inverse sequences $\{(x_n, y_n), (x_{n-1}, y_{n-1})\ldots\}$ must approach the basin boundary.

The construction of the basin boundary now proceeds as follows: First, we select an initial value x_0 at which we wish to determine the corresponding y value for the boundary, $y_0 = B(x_0)$. Noting that the evolution (5.3a) of the x-coordinate is independent of y, we iterate x_0 N times to produce the one-dimensional x-orbit $\{x_0, x_1, \ldots, x_N\}$, where $x_n = \lambda_x^n x_0$ (mod 1). Now choose an arbitrary value of y_N to pair with the final point x_N in order to determine the point (x_N, y_N) in the plane. There are λ_x^N orbits which result in (x_N, y_N) after N steps; one of these must begin with initial value $x = x_0$ and proceed along the x-orbit already constructed. Thus, we can form the particular inverse sequence which begins at (x_N, y_N) and ends at (x_0, y_0) by using repeated applications of (5.5) to determine the y_n, since all the x_n are known. For example, we have

$$y_{N-1} = \frac{1}{\lambda_y}(y_N - \cos 2\pi x_{N-1})$$
$$= \frac{1}{\lambda_y} y_N - \frac{1}{\lambda_y} \cos(2\pi \lambda_x^{N-1} x_0),$$
$$y_{N-2} = \frac{1}{\lambda_y^2} y_N - \frac{1}{\lambda_y^2} \cos(2\pi \lambda_x^{N-1} x_0)$$
$$- \frac{1}{\lambda_y} \cos(2\pi \lambda_x^{N-2} x_0). \tag{5.6}$$

Thus the expression for y_0 is

$$y_0 = \frac{y_N}{\lambda_y^N} - \sum_{j=1}^{N} \lambda_y^{-j} \cos(2\pi \lambda_x^{(j-1)} x_0), \tag{5.7}$$

and, letting $N \to \infty$, we see that the basin boundary is

$$y = B(x) = -\sum_{j=1}^{\infty} \lambda_y^{-j} \cos(2\pi \lambda_x^{j-1} x). \tag{5.8}$$

As discussed in section 3.3, this expression implies that the basin boundary $B(x)$ is a continuous, nowhere differentiable curve (but not a quasicircle). For a check on our numerical procedure for measuring the uncertainty exponent α, we have applied it to the case shown in fig. 19 ($\lambda_x = 3$, $\lambda_y = 1.5$). The resulting value $\alpha \approx 0.38$ compares well with that given by the expression for the capacity dimension (see section 3.3): $\alpha = D - d = 2 - d = (\ln \lambda_y / \ln \lambda_x) = 0.37\ldots$ in this case.

We have presented now two examples of locally connected fractal basin boundaries. The typical boundary for a quadratic analytic map (see fig. 4) was described as possessing a "snowflake" structure, whereas the fractal boundary for the map (5.3) exhibits a "striated" or stretched shape. In order to explain the difference between the boundary of fig. 19 and that of fig. 4, we consider the composed linear map

$$\delta x_{-n} = TM^{-1}(x_{-n+1})$$
$$\cdot TM^{-1}(x_{-n+2}) \cdot \ldots \cdot TM^{-1}(x_0) \cdot \delta x_0$$
$$\equiv L^{(-n)}(x_0) \cdot \delta x_0, \tag{5.9}$$

where $x = (x, y)$, $\delta x = (\delta x, \delta y)$, and each matrix $TM^{-1}(x_j)$ is given by the inverse of (5.4) evaluated at the point x_j along any of the λ_x^n inverse orbits of length n generated from x_0. From (5.4) we have

$$TM^{-1}(x_j) = \begin{pmatrix} \lambda_x^{-1} & 0 \\ \dfrac{2\pi \sin 2\pi x_j}{\lambda_x \lambda_y} & \lambda_y^{-1} \end{pmatrix} \tag{5.10}$$

and therefore $L^{(-n)}$ is of the form

$$L^{(-n)}(x_0) = \begin{pmatrix} \lambda_x^{-n} & 0 \\ A_n(x_0) & \lambda_y^{-n} \end{pmatrix}. \tag{5.11}$$

The eigenvalues of $L^{(-n)}$ are $(\lambda_x^{-n}, \lambda_y^{-n})$. Therefore, the Lyapunov numbers for any inverse sequence of (5.3), as defined in (5.2), are

$$\mu_1^- = \lambda_x > \mu_2^- = \lambda_y > 1. \quad (5.12)$$

Due to the inequality of the two eigenvalues the action of $L^{(-n)}$ on a small disk at x_0 produces a highly eccentric ellipse with length-to-width ratio of the order of $(\lambda_x/\lambda_y)^n$. This results in the structure of fig. 19. This is in contrast to the case of analytic maps for which the Cauchy–Riemann structure implies that an initial disk remains a disk upon iteration (implying that the two Lyapunov numbers of the map are equal). This is manifested in the "snowflake" structure of the Julia set in fig. 4.

5.3. Nonanalytic noninvertible quadratic maps on the plane

The previous examples of fractal curve basin boundaries have arisen in very special cases of two-dimensional maps. The class of analytic maps is special because of the Cauchy–Riemann relations. The map given in eq. (5.3) is also special in that the eigenvalues of the tangent map are independent of x and y, and one can construct the explicit form of the basin boundary. Thus, in order to investigate the properties of basin boundaries one might expect to exist in typical dissipative dynamical systems, we consider in this subsection more general maps of the plane, which have none of the special qualities discussed above.

We shall restrict our attention to quadratic maps of the plane

$$x_{n+1} = a_{xx}x_n^2 + a_{xy}x_ny_n + a_{yy}y_n^2 + a_xx_n \\ + a_yy_n + a_c, \quad (5.13)$$
$$y_{n+1} = b_{xx}x_n^2 + b_{xy}x_ny_n + b_{yy}y_n^2 + b_xx_n + b_yy_n + b_c.$$

In general, a map of this form is noninvertible and can therefore be argued to exhibit properties found in systems of higher dimension. Examples of (5.13), such as the Hénon map (which is invertible), have been studied and have been shown to possess chaotic attractors. Also, since analytic maps are a subset of the class (5.13), one might expect general quadratic maps to exhibit fractal boundaries. Thus, we expect that, in general, (5.13) can have both fractal basin boundaries and chaotic attractors. Our results lead us to believe that striated fractal basin boundaries are "common" while the snowflake structure fractal basin boundaries found in analytic maps are not. By this we mean that, if coefficients of (5.13) are chosen randomly from some ensemble with a continuous probability density, then the probability of obtaining a map with a striated fractal basin boundary would be positive, while the probability of a map with a snowflake structure fractal basin boundary would be zero. (We also find smooth basin boundaries to be common.)

Surveying the twelve-dimensional parameter space of (5.13) in search of fractal boundaries and chaotic attractors is clearly an inordinately large task. Therefore, we begin with an analytic map

$$x_{n+1} = x_n^2 - y_n^2 + \tfrac{1}{2}x_n - \tfrac{1}{2}y_n, \\ y_{n+1} = 2x_ny_n + \tfrac{1}{2}x_n + \tfrac{1}{2}y_n. \quad (5.14)$$

This map has an attracting fixed point at the origin and an attractor at infinity; the basin boundary shown in fig. 20 is fractal, though somewhat less spectacular than that in fig. 4. We then alter this map to be nonanalytic by changing the coefficients of the linear terms and adding constant terms until a map with a chaotic attractor is found. Such a map is given by

$$x_{n+1} = x_n^2 - y_n^2 + x_n - 0.297y_n + 0.048, \\ y_{n+1} = 2x_ny_n + x_n - 0.6y_n. \quad (5.15)$$

The structure of this map, shown in fig. 5, is studied in more detail in ref. 23. Again, initial conditions in the dark region escape to infinity; in this case, however, orbits in the interior blank region approach a chaotic attractor which is also

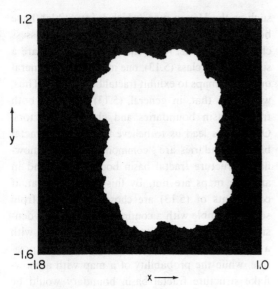

Fig. 20. Basin structure for the analytic map given in eqs. (5.14): dark for the basin of infinity, blank for the basin of the fixed point at the origin.

displayed. This attractor (here constructed from a single initial condition) contains self-intersections due to the noninvertibility of (5.15).

On the scale of fig. 5a, the basin boundary appears to be characterized by roughly self-similar whorls repeated with varying size and distortion around the boundary at intervals which seem to accumulate near the point labeled A in the lower right-hand corner. This is the location of an unstable fixed point. The tangent map at this fixed point has complex conjugate eigenvalues of magnitude greater than one. Thus, in the vicinity of this fixed point the map is expanding and rotating; this behavior of the boundary near the fixed point is revealed under repeated magnification. This structure replicated around the boundary has long spiral arms (see fig. 5b). As the coefficient a_y in (5.15) is varied, the attractor is observed to collide with these arms in a "crisis" [3, 23], destroying the attractor and its basin.

The microscopic structure at a representative location on the boundary is shown in fig. 5c. This magnification by a factor of 10^7 exposes local linear striations similar to that found for eq. (5.3)

(see fig. 19). Further magnification of any of the interfaces between dark and blank in fig. 5c produces a similar picture of striations on smaller scale, although the pattern of transversal spacing may not be the same. Apparently, the magnification of almost any small region around the boundary will reveal such structure; the degree of magnification required before this structure is apparent, however, may depend on the location. Thus, the "rotating" structures on large scale are ultimately replaced on finer scale by striations indicating local stretching. An interpretation connecting these observations would be that the local leaves or layers are just the flattened extensions of many greatly stretched spiral arms in the neighborhood of some small whorl. This is roughly displayed on a large scale in the structure of the object labelled B emanating from the spiral fixed point at A in fig. 5a.

The result to be emphasized here is that a non-special choice of parameters for a two-dimensional quadratic map can produce either a smooth boundary or a boundary with a fractal character of local striations as opposed to the "snowflake" pattern of analytic maps. The uncertainty exponent α for fig. 5 is approximately 0.7. This would indicate a fractal dimension of 1.3, using (2.4).

The "striations" of the boundary in this example are reminiscent of the simple model for this behavior provided by (5.3) of the previous subsection. In that case, the linear structure was interpreted in terms of the Lyapunov numbers of the inverse mapping. That model is simple, however, in that not only can one calculate this effect, but the stretching occurs in the same direction (y) independent of location on the boundary and the Lyapunov numbers are independent of inverse path. Here, no analytic calculation of Lyapunov numbers is possible. Thus, a numerical computation of inverse orbits and corresponding Lyapunov numbers (which now will depend on the inverse sequence chosen) is required to investigate this connection. This will be discussed in section 6.

In addition to this single example (5.15), we have surveyed many other noninvertible quadratic maps of the class (5.13). We restricted our attention to those maps for which the origin is a fixed point, $a_c = b_c = 0$. Furthermore, we considered only the cases in which the origin is unstable so that either some other finite attractor exists of almost all orbits escape to infinity. For each case considered in this class, we examined the orbits generated by 63 initial conditions in the neighborhood of the origin; if all of these did not tend to infinity, we concluded that for this case a finite attractor must exist. Sampling 6400 maps in this way, we found that roughly 17% of the cases satisfied this condition for possessing multiple attractors. From this collection of approximately 1000 maps, we selected several random examples for further investigation of their basin boundaries. On this basin, our findings can be summarized as follows:

1) In general, a quadratic noninvertible map of the plane may possess multiple attractors with a basin boundary which can either be a smooth curve or a fractal.

2) In the case of a fractal curve basin boundary, the local fractal structure generally exhibits linear striations (as opposed to the "snowflake" pattern of analytic maps).

Thus we believe that small-scale, striated structure in fractal basin boundaries should commonly occur, but that the small-scale, snowflake, two-dimensional structure of complex analytic maps is very special and should not occur in typical situations.

6. Lyapunov numbers and dimension for the measure on fractal basin boundaries

6.1. *Lyapunov numbers for inverse orbits*

Here we discuss the Lyapunov numbers as applied to the examples of locally-connected basin boundaries in two-dimensional maps considered in the previous section. In particular, we consider the Lyapunov numbers for an orbit generated by inverse images of an initial point in the vicinity of the basin boundary (such points are attracted to the boundary). Since the maps which we discuss in this section do not have a unique inverse, we need a rule for choosing which preimage of a point is to be on the inverse orbit. Here we shall choose the preimage at random with equal probability for each of the possible preimages. In our numerical calculations of Lyapunov numbers, we do this by making use of a random number generator. An orbit produced in this way will generate an asymptotic measure on the basin boundary. We conjecture that this measure is the same for almost all initial conditions in any neighborhood of the boundary for which almost all inverse orbits are attracted to the boundary. We call the measure generated by random choices of preimages the *natural measure* for the basin boundary.

Fig. 21 depicts the result of plotting 200,000 inverse iterates for the map of fig. 5 of the previous section. The distribution of points on the boundary appears to be extremely nonuniform, emphasizing certain regions very heavily, while leaving others comparatively unexplored. We numerically obtain the smallest Lyapunov number by finding the largest eigenvalue of the composition of the inverse tangent map along the inverse orbit (cf. eq. (5.2)). Due to the finite precision of numerical computations, the largest Lyapunov number μ_1^- is usually inaccurately given by the smallest eigenvalue of the composed inverse tangent map. Therefore, in order to compute it we take products

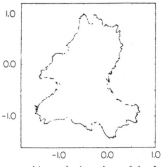

Fig. 21. Inverse orbit on the boundary of fig. 5 (eqs. (5.15)) with 200,000 iterations. At each stage of the inverse iteration, the choice between two possible preimages was made at random.

of the determinant of the individual tangent maps along the inverse orbit to obtain the determinant of the N-times composed inverse tangent map. Since this determinant is equal to the product of the eigenvalues of the composed matrix, and since we know the largest eigenvalue, we thus determine the smallest eigenvalue.

We emphasize that the Lyapunov numbers that we obtain in this way are Lyapunov numbers appropriate to the natural measure generated by the inverse orbit. That is, the stretching properties of those regions of the boundary with larger measures are more strongly weighted in the calculation of the Lyapunov numbers than are those regions with smaller measures.

For the three examples discussed in section 5 we have the following results for the Lyapunov numbers:

1) For the analytic map of fig. 4, a numerical calculation of the Lyapunov numbers (both are equal for analytic maps) yields $\mu_{1,2}^- = 2$ to within numerical accuracy (± 0.01).

2) For the map (5.3), it is easy to show that $\mu_1^- = \lambda_x$ and $\mu_2^- = \lambda_y$.

3) For the map of fig. 5 we numerically obtain $\mu_1^- = 2.15$ and $\mu_2^- = 2.09$.

6.2. *Dimension of the measure for fractal basin boundaries*

For the case of chaotic *attractors*, a formula for the dimension of the measure on the attractor has been conjectured [12, 22] and gives correct results in examples where the dimension of the measure is known. For a discussion and definitions of the dimension of the measure see Farmer, Ott, and Yorke [12]. In this subsection we develop an analogous conjecture for the dimension of the natural measure generated by inverse orbits on a fractal basin boundary. For definiteness in this discussion, we assume that almost all points on the boundary have two preimages. This assumption is, in fact, satisfied by our examples of figs. 4 and 21, and is also satisfied by eqs. (5.3) if $\lambda_x = 2$. The heuristic arguments leading to our expression for the dimension of the natural measure of these basin boundaries (eq. (6.1)) are similar to those for the previous development for the case of chaotic attractors; the main difference is that here we need to account for the fact that there are two preimages for points on the boundary.

Say we have an initial set of N_0 small squares of side ε_0 covering the boundary. Now apply the inverse map n times to each square on the boundary. Each square will be mapped to 2^n small parallelograms, typically with length of the order of $\varepsilon_0/(\mu_2^-)^n$ and width of the order of $\varepsilon_0/(\mu_1^-)^n$ (we assume that both μ_1^- and μ_2^- are larger than one). Since the boundary is invariant, the collection of these $2^n N_0$ parallelograms also covers the basin boundary. We can typically cover each parallelogram with of the order of $(\mu_1^-/\mu_2^-)^n$ small boxes of side $\varepsilon_n \equiv \varepsilon_0/(\mu_1^-)^n$. Thus, if we assume that it takes $N(\varepsilon) \sim \varepsilon^{-d}$ boxes of side ε to cover the boundary, then

$$N(\varepsilon_0)/N(\varepsilon_n) = \left[2(\mu_1^-/\mu_2^-)\right]^{-n}$$
$$= (\varepsilon_n/\varepsilon_0)^d = (\mu_1^-)^{-nd}.$$

Thus taking logarithms, we obtain

$$d_L = 1 + \left[\ln(2/\mu_2^-)\right]/\ln \mu_1^-, \qquad (6.1)$$

where the subscript L denotes that (6.1) is a prediction based on the Lyapunov numbers, and we call d_L the Lyapunov dimension of the boundary. Since the Lyapunov numbers reflect behavior for points on the boundary weighted by the natural measure on the boundary, we conjecture that d_L is equal to the dimension of the natural measure of the boundary (which is usually not the same as the capacity dimension). See ref. 12 for further discussion.

For the case of the analytic map of fig. 4, $\mu_1^- = \mu_2^- = 2$ yields $d_L = 1$. This is in agreement with the rigorous theoretical result of Manning [24] for Julia sets of polynomial analytic maps. Note, however, that the capacity dimension as measured by the final state sensitivity exponent α is approximately $1.3 > 1$. For fig. 4, this may be viewed as a reflection of the fact that the measure generated by randomly chosen inverse iterates is highly concentrated on a set of lower dimension than the capacity of the boundary. The situation is very similar for the case of the boundary shown in fig. 5 for a nonanalytic quadratic map. In that case our numerically calculated Lyapunov numbers yield $d_L \approx 0.94$, while the final state sensitivity exponent yields a capacity dimension of about 1.3. For the case of the boundary for the map (5.3) with $\lambda_x = 2$, (6.1) yields

$$d_L = 2 - \frac{\ln \lambda_y}{\ln 2},$$

in agreement with the rigorous result of Kaplan, Mallet-Paret and Yorke [14].

7. Conclusions

Basin boundaries for typical dynamical systems can be expected to be either smooth or fractal. In this paper we have investigated fractal basin boundaries. Our main conclusions are as follows:

1) Fractal basin boundaries can strongly affect the ability of predicting to which attractor a system eventually goes. In particular, a substantial reduction in the error in the initial condition can, under some circumstances, lead to only modest increase in predictive ability (section 2).

2) Fractal basin boundaries can be classified as being either locally disconnected or locally connected (section 3).

3) Examples of locally disconnected basin boundaries resulting from horseshoe type dynamics appear to be common (section 4).

4) Locally connected basin boundaries may or may not be quasicircles (section 3). Quasicircle basin boundaries, as exemplified by analytic mappings of a single complex variable, however, appear to be very special and are not to be expected in general (section 5). As a result, fine scale structure of locally connected boundaries should, most commonly, exhibit a one-dimensional (or striated) appearance (e.g., fig. 5), rather than a two-dimensional (or snowflake) appearance (e.g., fig. 4).

5) The dimension of the measure generated by inverse orbits on a fractal basin boundary and its possible connection with the associated Lyapunov numbers has been discussed (section 6).

Acknowledgements

This work was supported by the Air Force Office of Scientific Research, the Department of Energy (Office of Scientific Computing), the National Science Foundation, and the Office of Naval Research.

Appendix A

$\alpha = 1$ *Basin boundaries with $f(\varepsilon)$ not proportional to ε*

In this appendix we consider basin boundaries for which $\alpha = 1$ according to our definition (2.7), but for which $f(\varepsilon) \cong k\varepsilon$ for small ε does not hold. In particular, consider the example in fig. 22a. This figure shows two basins of attraction (the dark and blank regions) for the Hénon map at parameter values such that the map has two distinct strange attractors. The map corresponding to this figure is given by $x_{n+1} = 1 - 1.0807 x_n^2 + y_n$, $y_{n+1} = 0.3 \, x_n$. Fig. 22b shows a magnification of the boxed region in fig. 22a and fig. 22c is a further magnification of the boxed region in fig. 22b. The replication evident on comparison of figs. 22b and 22c is *not* a consequence of fractal structure, but rather of the fact that an unstable saddle fixed point is located in the center of these figures. Indeed, the basin boundary here is a smooth curve with dimension one.

To understand how the structure illustrated in figs. 22 arises, consider the stable and unstable manifolds of the fixed point (FP), as shown in fig. 23. The stable manifold (shown as a dashed line in

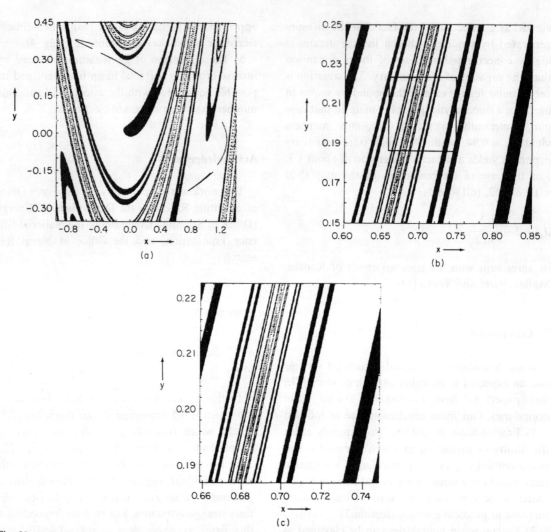

Fig. 22. (a) Basin structure of the Hénon map at parameter values for which both a four-piece chaotic attractor (shown, with blank basin) and a six-piece chaotic attractor (not shown, dark basin) exist. Figs. 22a–c adapted from the second of ref. [3]. (b) Magnification of rectangle in (a). The vertical bands accumulate (transversally, along the unstable manifold) at the piece of the stable manifold (cf. fig. 22) running vertically just left of center in the small rectangle; indeed, the fixed point FP in fig. 25 is in this small rectangle. (c) Magnification of small rectangle in (b) by a factor λ_u^2.

fig. 23) is in the direction along the bands in fig. 22, while the unstable manifold cuts across them. In particular, the bands accumulate at the piece of the unstable manifold shown dashed in fig. 23. The stable and unstable eigenvalues of the Jacobian matrix at the fixed point are $\lambda_s = 0.17931\ldots$ and $\lambda_u = -1.6731\ldots$. Now consider the application of the inverse mapping to fig. 22b. Bands crossing the unstable manifold will be drawn in toward the fixed point, contracted in width (by approximately $|\lambda_u|^{-1}$), flipped to the other side of the fixed point (because λ_u is negative), and stretched out along the stable manifold (by λ_s^{-1}). In fact, fig. 22c is a magnification of fig. 22b by precisely the amount λ_u^2 (corresponding to two applications of the inverse map).

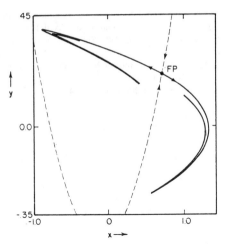

Fig. 23. Pieces of the stable (dashed) and unstable (solid) manifolds of the unstable fixed point (FP) in the Hénon map. Adapted from the second of ref. [3].

We now consider fig. 22b and attempt to deduce how the uncertain fraction of this region behaves for small initial error ε. First we note that the phase space in fig. 22b that is within ε of the dashed portion of the stable manifold of FP shown in fig. 23 is uncertain. We now wish to estimate the volume of the uncertain phase space that is outside this inner strip. Since, for small ε this volume will turn out to be larger than the volume of the strip itself, it is the outside region which will determine $f(\varepsilon)$. In order to estimate the number of boundary lines outside the inner strip, we first note that all of the boundary lines outside the inner strip in fig. 22b can be regarded as being generated (via inverse mapping) from the four outermost boundary lines. Say we consider these four outermost lines and apply the inverse map to them until they first fall within the inner strip of width ε. Call the number of iterates necessary to do this r. If w is the x-width of fig. 22b, in order of magnitude we have

$$\varepsilon/w \sim |\lambda_u|^{-r}. \text{ Thus,}$$

$$r \cong K_1 \ln(1/\varepsilon),$$

for small ε, where K_1 is a constant. Since by construction these boundary lines that are outside the inner strip are all separated by distances of the order of ε or greater, we have that for small ε, $f(\varepsilon)$ is proportional to $r\varepsilon$ or

$$f(\varepsilon) \cong K_2 \varepsilon \ln(1/\varepsilon). \tag{A.1}$$

Other examples also exist. In particular, if one considers the final phase sensitivity (cf. section 4.2-1) of the period two orbit for the logistic map (in the parameter range below that at which period doublings accumulate), then one finds $f(\varepsilon) \cong K_3 \ln(1/\varepsilon)$. As in the case of figs. 22, the reason for this behavior involves a period one unstable fixed point.

References

[1] E. Ott, Rev. Mod. Phys. 53 (1981) 655; and references therein.
[2] C. Grebogi, S.W. McDonald, E. Ott and J.A. Yorke, Phys. Lett. 99A (1983) 415.
[3] C. Grebogi, E. Ott and J.A. Yorke, Phys. Rev. Lett. 48 (1982) 1507; Physica 7D (1983) 181.
[4] C. Grebogi, E. Ott and J.A. Yorke, Phys. Rev. Lett. 50 (1983) 935.
[5] M.L. Cartwright and J.E. Littlewood, Ann. Math. 54 (1951) 1; J. London Math. Soc. 20 (1945) 180.
[6] N. Levinson, Ann. Math. 50 (1949) 127.
[7] M. Levi, Mem. Amer. Math. Soc. 32 (1981) 244.
[8] J.L. Kaplan and J.A. Yorke, Comm. Math. Phys. 67 (1979) 93.
[9] E.N. Lorenz, J. Atmos. Sci. 20 (1963) 130.
[10] T.-Y. Li and J.A. Yorke, Am. Math. Mon. 82 (1975) 985.
[11] G. Julia, J. Math. Pure Appl. 4 (1918) 47. P. Fatou, Bull. Soc. Math. France 47 (1919) 271. B.B. Mandelbrot, Ann. N.Y. Acad. Sci. 357 (1980) 249.
[12] J.D. Farmer, E. Ott and J.A. Yorke, Physica 7D (1983) 153.
[13] S. Pelikan, "A Dynamical Meaning of Fractal Dimension," U. Minn. IMA Preprint #73 (1984).
[14] J.L. Kaplan, J. Mallet-Paret, and J.A. Yorke, Ergod. Th.& Dyn. Sys. 4 (1984) 261.
[15] S. Smale, Bull. Amer. Math. Soc. 73 (1967) 747.
[16] J.E. Flaherty and F.C. Hoppensteadt, Stud. Appl. Math. 58 (1978) 5.

[17] J. Guckenheimer and P. Holmes, Nonlinear Oscillations, Dynamical Systems, and Bifurcation of Vector Fields, Appl. Math. Sci. vol. 42 (Springer, New York, 1983).
[18] S. Smale and R.F. Williams, J. Math. Biol. 3 (1976) 1.
[19] G. Pianigiani and J.A. Yorke, Transactions Amer. Math. Soc. 252 (1979) 351.
[20] J.A. Yorke and E.D. Yorke, J. Stat. Phys. 21 (1979) 263.
[21] P. Bergé and M. Dubois, Phys. Lett. 93A (1983) 365.
[22] J.L. Kaplan and J.A. Yorke, in: Functional Differential Equations and Approximation of Fixed Points, Lecture Notes in Math. vol. 730 (Springer, New York, 1979) p. 228.
[23] S. McDonald, C. Grebogi, E. Ott and J.A. Yorke, Phys. Lett. 107A (1985) 51.
[24] A. Manning, Ann. Math. 119 (1984) 425.

Fractal Basin Boundaries and Homoclinic Orbits for Periodic Motion in a Two-Well Potential

F. C. Moon and G.-X. Li

Department of Theoretical and Applied Mechanics, College of Engineering, Cornell University, Ithaca, New York 14850

(Received 2 July 1985)

A fractal-looking basin boundary for forced periodic motions of a particle in a two-well potential is observed in numerical simulation. The fractal structure seems to be correlated with the appearance of homoclinic orbits in the Poincaré map as calculated by Holmes using the method of Melnikov. Below this critical forcing amplitude the basin boundary appears to be smooth and nonfractal. This example raises questions about predictability in nonchaotic dynamics of nonlinear systems.

PACS numbers: 03.20.+i, 05.40.+j, 46.30.−i

In a series of papers and reports, Yorke and co-workers[1-3] presented numerical evidence for fractal boundaries between basins of attraction for nonchaotic attractors. In their study they looked at two-dimensional maps with multiple nonchaotic attractors. In a recent lecture, Yorke suggested that equations representing flows of dynamical systems with multiple nonchaotic attractors, i.e., equilibrium points, limit cycles, periodic orbits, etc., might possess fractal boundaries between the two or more basins of attraction. This would imply that for small uncertainty in the initial conditions near this boundary absolute predictability might be impossible even if a solution is proved to exist and is unique. In a recent Letter[4] fractal basin boundaries have in fact been found for the forced pendulum.

In this Letter we have applied these ideas to the problem of forced motions of a particle in a two-well potential, and relate the appearance of fractal basin boundaries to the appearance of homoclinic orbits in the Poincaré map as examined by Holmes[5] in an earlier study. The two-well potential describes the motion of a buckled elastic beam or an electron in a plasma.[5-8] The governing equation under study is

$$\ddot{x} + \gamma \dot{x} - \tfrac{1}{2} x(1-x^2) = f \cos\omega t. \quad (1)$$

The importance of this model is that the chaotic and nonchaotic dynamics have been analyzed in great detail by Holmes[5] as well as by numerical simulation and by analog computer. The results of this work have been verified in experiments by Moon[6] for a buckled elastic beam. Because of the close agreement between theory, experiment, and numerical simulation, we have confidence that the numerical results presented in this paper reflect the actual properties of the dynamical system (1) and the physical systems it claims to model.

In his theoretical study of (1), Holmes used the method of Melnikov to derive a necessary criterion for chaotic motion based on the existence of homoclinic orbits in the Poincaré map when $f > f_c$, where

$$f_c = (\gamma\sqrt{2}/3\pi\omega)\cosh(\pi\omega/\sqrt{2}). \quad (2)$$

This criterion gives the condition for the intersection of stable and unstable manifolds associated with the saddle point of the Poincaré map ($\omega t = 2\pi n$; n is an integer).

It is the thesis of this paper that this criterion is a necessary condition for the appearance of fractal basin boundaries between two periodic attractors and that unpredictability in the presence of uncertainties in initial conditions may be a property of the two-well potential even when the attractors are not chaotic.

Experiments on the forced motion of a buckled beam have shown that below some critical $f \equiv f_c^*$ ($f_c^* > f_c$) the motion is periodic and above f_c^* the motion may be chaotic. Fractal properties of the boundary between periodic and chaotic motions for this equation have recently been studied experimentally,[7] and the fractal dimension of the Duffing-Holmes attractor has recently been calculated by the authors.[8]

For the frequency $\omega = 0.833$ and damping $\gamma = 0.15$ this equation shows chaotic behavior for $f \gtrsim 0.159$ when a fourth-order Runge-Kutta numerical integration is used while the critical Holmes value from (2) is $f_c = 0.088$. In the present study we looked at the *nonchaotic regime* $0.05 < f < 0.1$. For long times, only periodic motions are possible, an orbit about the right or the left equilibrium position $x = \pm 1$, $\dot{x} = 0$. We then explored the initial-condition space $(x_0, \dot{x}_0 \equiv v_0)$ to find the basin of attraction of these two attractors.

The criterion used to determine the long-term state of the orbit was first to ignore the initial transient, equivalent to five driving periods, and then wait until the trajectory has made five orbits about either the left or the right equilibrium position $x = \pm 1$ by looking at the long-time average of $x(t)$. If the orbit went to the right attractor we plotted a symbol; if it went to the left attractor we left a blank. In this way the edge of the dark symbols represents the basic boundary.

The numerical results are plotted in Figs. 1–5. For low values of f ($\omega = 0.833$), $f \sim 0.05$, the boundary looks smooth as shown in Fig. 1. However, for $f = 0.1$, which is greater than the Holmes critical value of 0.088, the coarse-scale boundary shows some

© 1985 The American Physical Society

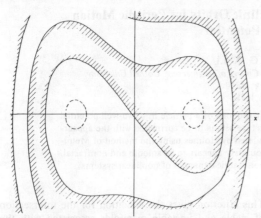

FIG. 1. Smooth basin boundary (solid lines) for periodic motion about equilibrium points $x = \pm 1$, $\dot{x} = 0$ (dotted line) for forcing amplitude $f = 0.05$ and frequency $\omega = 0.833$ (damping $\gamma = 0.15$).

fingers or whiskers indicating possible fractal behavior as shown in Fig. 2 for 400×400 initial conditions. To confirm this we performed a sequence of successive enlargements of smaller and smaller regions of phase space near the suspected fractal boundary as shown in Fig. 3. Each enlargement consisted of 10^4 initial conditions (100×100) and each showed finer and finer structure indicating a possible fractal boundary. A composite photograph with 400×400 initial conditions is shown in Fig. 3.

All of the numerical results were obtained with use of a VAX 750 computer. Most of the data were run with a Runge-Kutta solver with a step size of 0.25. However, the data in Fig. 2 were also run with a 0.1 step size and the results were almost identical. Further, a dozen or more individual points from Fig. 2 were selected at random near the fractal-looking boundary and run for a long time to make sure that the criterion for left or right periodic attractor was operating correctly. In all cases the long-time orbits were periodic. Most were period-one orbits, but a few were period three as judged by use of Poincaré maps.

In a paper on the forced Duffing equation (1), Holmes[5] showed that the chaotic motions were preceded by the appearance of an infinite set of homoclinic orbits in the Poincaré map of the periodically forced system for a critical value of the forcing amplitude f. These orbits occurred when the unstable and stable manifolds, emanating from the saddle point at the origin, intersected.

For low damping, however, $\gamma < 0.2$, the first author has shown[6] in experiments on vibrations of a buckled elastic beam that the condition (2) was only a necessary one for chaotic vibrations of a buckled beam (i.e., $f_c < f_c^*$). Thus there lies a region in the parameter space (f, ω, γ) where homoclinic orbits exist but chaos is not likely (i.e., for $f_c < f < f_c^*$). We conjecture that Holmes's criterion (2) may give the critical value of "f" for a fractal basin boundary between two nonchaotic periodic attractors about the left or right equilibrium points.

First, this conjecture is supported by the data in Fig. 4. Here the results of Runge-Kutta simulation are compared with the Holmes criterion (2). The circles indicate that smooth, nonfractal looking boundaries were obtained similar to Fig. 1. The star data points indicate the appearance of a nonsmooth boundary. Because of the costs and computer time involved only one frequency ($\omega = 0.833$) was explored on a finer

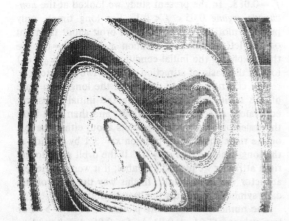

FIG. 2. Fractal-looking basin boundary for forcing amplitude $f = 0.1$ ($\omega = 0.833$, $\gamma = 0.15$) calculated from 160×10^3 initial conditions in the domain $-2.4 \leq x_0 \leq 2.4$, $-1.2 \leq v_0 \leq 1.2$.

FIG. 3. Composite photograph of finer scale enlargement of Fig. 2 basin boundary for 160×10^3 initial conditions, $f = 0.1$, $\omega = 0.833$, $\gamma = 0.15$, $-0.375 \leq x_0 \leq -0.275$, $0.045 \leq v_0 \leq 0.075$.

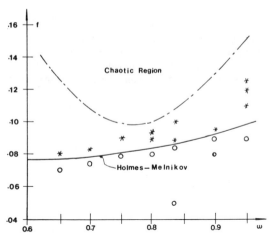

FIG. 4. Comparison of the Holmes criterion for homoclinic orbits with numerical evidence for smooth and nonsmooth basic boundaries. The lower bound of the chaotic region is shown by the dashed curve.

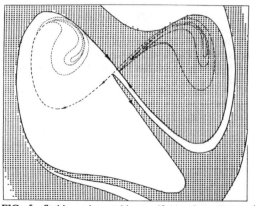

FIG. 5. Stable and unstable manifolds of the Poincaré map superimposed on the basin boundary for forcing amplitude at the Holmes critical value ($f = 0.0856$, $\omega = 0.8$, $\gamma = 0.15$).

scale to ascertain the self-similar, fractal nature of the boundary. The dashed line represents a lower bound on the chaos criterion.

Second, there is numerical evidence that the appearance of fractal-looking structure in the basin boundary is coincident with the intersection of stable and unstable manifolds of the Poincaré map as shown in Fig. 5. When $f < f_c$, one can argue that the stable manifold of the Poincaré map and the basin boundary are coincident. [This conclusion was also found for an approximate two-dimensional cubic map associated with (1) by Yamaguchi and Mishima.[9]] Using Holmes's results we show in Fig. 5 the saddle point of the Poincaré map calculated from (1). It is evident from Fig. 5 that at the Holmes criterion, the stable manifold develops a fold or finger which touches the unstable manifold shown as a dotted curve. Thus it appears that the criterion for homoclinic orbits in the forced two-well potential problem is coincident with the change from a smooth to an irregular and perhaps fractal basin boundary.

Yorke and co-workers have shown that the fraction ϕ of uncertain initial conditions in the phase space as a function of the radius of the sphere of uncertainty ϵ in initial conditions has the following relation:

$$\phi \sim \epsilon^{D-d}, \qquad (3)$$

where D is the dimension of the phase space and d is the fractal dimension of the basin boundary. For example, for $d = 1.5$, an uncertainty in initial conditions of $\epsilon = 0.01$ (compared to order one) yields an uncertainty fraction of 10%. For $\epsilon = 0.05$, $\phi = 22$%.

Fractal properties of chaotic motions have been of course the subject of great interest in the past decade. Such behavior is associated with a sensitivity to initial conditions and a loss of information about the motion as time proceeds. The importance of the conjectures of Yorke and co-workers is that a larger class of nonlinear phenomena may suffer from inherent unpredictability than was previously thought. This includes transient and periodic as well as chaotic problems. This discovery is ironic in the age of the supercomputer in which numerical simulation through finite-element, finite-difference, and other CAD/CAM software promises to increase our analysis and prediction capability of the physical world.

A sensitivity of numerical simulation predictions of nonlinear phenomena has been known anecdotally in the numerical prediction industry. Recently a study by Symonds[10] has appeared in the literature which is somewhat related to the problem in this paper. There he tried to predict the end-state behavior of the transient excitation of an elastic-plastic beam arch. The end states involve periodic oscillations about two possible buckled positions of the beam arch. Thirteen different investigators ran the same problem with different numerical codes and obtained different answers. Such inability to obtain consistent results from numerical codes may be the consequence of fractal basic boundaries in either the initial-condition space or the parameter space for the nonlinear problem.

This work was supported in part by a grant from the U.S. Air Force Office of Scientific Research, Mathematical Sciences and Aerospace Division.

---

[1]C. Grebogi, E. Ott, and J. A. Yorke, Phys. Rev. Lett. **50**, 935 (1983).

[2] C. Grebogi, S. W. McDonald, E. Ott, and J. A. Yorke, Phys. Lett. **99A**, 415 (1983).
[3] S. W. McDonald, C. Grebogi, E. Ott, and J. A. Yorke, Laboratory for Plasma and Fusion Energy Studies, University of Maryland, Report No. UMLPF 84-017, 1984 (unpublished).
[4] E. G. Gwinn and R. M. Westervelt, Phys. Rev. Lett. **54**, 1613 (1985).
[5] P. J. Holmes, Philos. Trans. Roy. Soc. London **292**, 419 (1979).
[6] F. C. Moon, J. Appl. Mech. **47**, 638 (1980).
[7] F. C. Moon, Phys. Rev. Lett. **53**, 962 (1984).
[8] F. C. Moon and G. X. Li, to be published.
[9] Y. Yamaguchi and N. Mishima, Phys. Lett. **109A**, 1613 (1985).
[10] P. Symonds, to be published.

FRACTAL BASIN BOUNDARIES OF AN IMPACTING PARTICLE

Heikki M. ISOMÄKI [1], Juhani VON BOEHM and Raimo RÄTY
Helsinki University of Technology, SF-02150 Espoo 15, Finland

Received 24 June 1987; revised manuscript received 23 September 1987; accepted for publication 30 October 1987
Communicated by A.R. Bishop

The basins and the basin boundaries of the dissipative sinusoidally driven particle moving and impacting in the harmonic potential are studied in the range where the winding number (the impact/period ratio) forms an incomplete Farey-organised devil's staircase. The general shapes of the basins of the (periodic) devil's staircase attractors are the same. The boundaries between the basins of the devil's staircase attractors and the two other periodic attractors exhibit fractal structure. The fractal dimension, calculated from large regions of the phase space, does not depend on the subregion indicating that the boundaries are statistically self-similar and have a unique fractal dimension. The Farey-organisation affects the fractal dimension considerably. The calculated large fractal dimension, typically ≈ 1.7, indicates long transients and high sensitivity to noise although all the attractors are periodic in the devil's staircase range.

Dynamical systems with multiple attractors (final states) are common in nonlinear physics. Multiple attractors imply multiple basins, which are the sets of initial conditions approaching particular attractors when time approaches infinity. The boundaries of the basins can be either smooth or fractal corresponding to integer or noninteger fractal dimension D, respectively. If the basin boundary is of fractal type, much more effort is needed in an accurate calculation of the motion: even if the attractors were periodic, small uncertainties in the initial conditions near the fractal boundary make the prediction of the final motion impossible [1–8]. Fractal basin boundaries have been found in several important physical systems [1–8] but systematic studies with the calculated D from large regions of the phase space are still rare. The widely studied complex maps may also exhibit fractal basin boundaries (Julia sets) [9,10] that, however, fully differ from those found in typical physical systems [1,8].

The purpose of this paper is to study systematically the dependence of the fractal basin boundaries on the system parameter in a realistic physical system, the dissipative sinusoidally driven particle moving and impacting in the harmonic potential

[1] Also at the Academy of Finland.

[8,11,12] in the range of a Farey organised phase-locking (the devil's staircase) [9,13]. The final state sensitivity with respect to initial conditions is studied by calculating the fractal dimension D of the basin boundary from large regions of the phase space. The possible multi-dimensionality of the basin boundaries, found recently in the motion of the kicked double rotor [3], is studied by using magnifications of the phase space. This is related to the important question of how many possible dimensions a basin boundary can have [5].

The fully general dimensionless equations of motions of the dissipative sinusoidally driven particle impacting at $x=0$ in the harmonic potential are as follows,

$$\ddot{x}(t)+c\dot{x}(t)+x(t)=\cos(\omega t), \quad x>0,$$
$$\dot{x}(t_i^+) = -e\dot{x}(t_i^-), \quad x(t_i)=0. \quad (1)$$

In the calculation the damping coefficient c and the coefficient of restitution e are fixed to the values $c=0.4$ and $e=1.0$. The angular frequency ω serves as a system parameter. At the impact moments t_i at the origin the particle velocity changes discontinuously from $\dot{x}(t_i^-)$ to $\dot{x}(t_i^+)$ according to Newton's rule of impact. In the following (i, p) denotes an attractor with i impacts during the period p given in the units $2\pi/\omega$. The number of the rebounds within the unit of time is the winding number $q=i/p$ which is associated with the phase-locking phenomenon [13]. q tends to decrease with increasing ω. We consider the motion only for $\infty > q \gtrsim 1/2$ because the motions are expected to be qualitatively similar for $1 > q > 1/2$ and $1/n > q > 1/(n+1)$, $n=2, 3, \ldots$.

The overall motions of the impacting particle are outlined in the following (the details will be found in ref. [14]). For $\infty > q \gtrsim 1$, q forms a harmless staircase [15]. With $q=1$ we find the period-doubling bifurcations leading to chaos [16,17]. The calculated Feigenbaum universality number for the period-doubling sequence is $\delta=4.672$, in close agreement with the theoretical number $\delta=4.699$ [17]. In the chaotic region, close to the Feigenbaum accumulation point, the attractors exhibit universality, i.e., the order of the low-periodic windows obeys MSS-sequence [18] and the calculated universality numbers $4.611 < \delta < 4.679$, $\theta = 0.179$ and $\rho = 0.879, 0.882$ agree closely with the theoretical numbers for one-dimensional unimodal maps: 4.699 [17], 0.187 [19] and 0.892 [20], respectively. For $1 \gtrsim q \gtrsim 3/4$ the motions are mainly chaotic and irregular exhibiting crises [21]. We find a period-halving sequence at $q=3/4$ leading to the attractor (3, 4). For $3/4 > q > 3/5$ we find an incomplete devil's staircase (DS) [9,15] which coexists with the attractors with $q=7/10$ and, at the high-ω side, also with the attractors with $q=1/2$ (fig. 1a). (The motions beyond $q=1/2$ are expected to be qualitatively similar to those beyond $q=1$.) The behaviour of the basin boundaries is studied in the DS range where the motion is very rich due to the rapidly changing phase-locking phenomenon.

The general forms of the attractors (i, p) and thus the q's of DS satisfy the Farey relationship [13]

$$(i, p) = (3m+3n, 4m+5n) \triangleq m^*(3, 4) + n^*(3, 5),$$

$$q = \frac{i}{p} = \frac{3}{4+w}, \quad (2)$$

where w is the ratio of the number n of the attractors (3, 5) to the total number $m+n$ of attractors (3,4) and (3, 5); $m, n \geq 0$. We find that usually the largest stability region (the longest q-step) between the attractors with $q_1=i_1/p_1$ and $q_2=i_2/p_2$ occurs for the concatenated attractor with $q=(i_1+i_2)/(p_1+p_2)$.

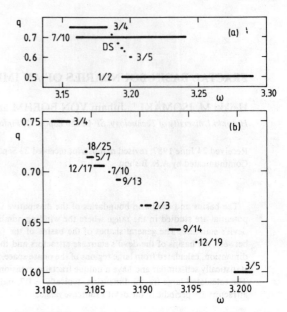

Fig. 1. Winding number q (the impact/period ratio). ω is the angular frequency of the driving force. DS denotes the devil's staircase magnified in (b). In (b) only the steps larger than $\Delta\omega \approx 0.0001$ are shown. The fractions denote the reduced q. The step $q=1/2$ in (a) starts in the range of $q=9/14$ in (b) and continues to the right. The step $q=3/4$ in (b) continues to the left.

Due to this impact-period-adding property [12] w locks at rational numbers between 0/1 and 1/1 and forms a Farey sequence [13] where the adjacent fractions a/b and c/d satisfy the unimodularity condition $|ad-bc|=1$. The magnified DS with the q-steps larger than $\Delta\omega \approx 0.0001$ are presented in fig. 1b. The w's corresponding to these q's are unimodular Farey numbers (see eq. (2)) $w=0/1$, 1/6, 1/5, 1/4, 2/7, 1/3, 1/2, 2/3, 3/4, 1/1. Improving the resolution by diminishing $\Delta\omega$ we find further q-steps that again satisfy the Farey construction. Hence the Farey organisation describes the behaviour of the phase-locking in DS when ω is changed.

We present in figs. 2a, 2c and 2b the $(\varphi, \dot{x}_0)$ basins at the ω's where the basic attractors (3, 4) and (3, 5) and their first concatenated attractor (6, 9), respectively, exist in DS (the corresponding attractors are presented in fig. 3 of ref. 3 of ref. [11]). Fig. 2d shows the basins beyond DS where the attractors (1, 2) and (14, 20), the first bifurcation of the attractor (7, 10), coexist. The phase of the driving force $\varphi = \omega t_i$ and the velocity of the particle $\dot{x}_0 = \dot{x}(t_i^+)$, both given

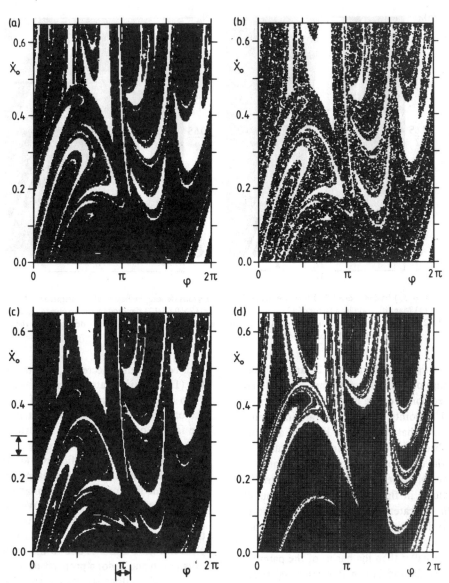

Fig. 2. $(\varphi, \dot{x}_0)$-basins at (a) $\omega = 3.18$; (b) $\omega = 3.19$; (c) $\omega = 3.20$; (d) $\omega = 3.24$. The white, black and grey regions denote the basins of the attractors of the upper q-branch ($q=7/10$), DS ($3/4 \geqslant q \geqslant 3/5$) and the lower q-branch ($q=1/2$) (see fig. 1), respectively. The stable attractors are (a) (7, 10), (3, 4); (b) (7, 10), (6, 9); (c) (7, 10), (3, 5), (1, 2); (d) (14, 20), (1, 2). φ and $\varphi + 2\pi$ are equivalent and the (periodic) fixed points of all the stable attractors are included in (a)–(d) (10, 13, 11 and 15 points, respectively). $0 \leqslant \varphi \leqslant 2\pi$, $0 \leqslant \dot{x}_0 \leqslant 0.65$.

at the impact $x(t_i) = 0$, are chosen as the initial conditions for the basin plots because they are the most natural and convenient ones to use due to the piecewise linearity of the motion, see eq. (1). In all basin plots (figs. 2 and 3) the grid in the $(\varphi, \dot{x}_0)$ plane is 200×260 and the white, black and grey regions denote the basins of the attractors of the upper q-branch ($q=7/10$), DS ($3/4 \geqslant q \geqslant 3/5$) and the lower q-branch ($q=1/2$) in fig. 1a, respectively (in these plots all the three attractors are simultaneously stable only in figs.

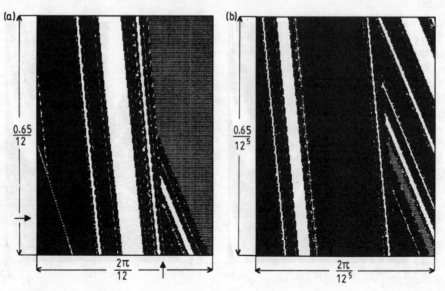

Fig. 3. Magnified (φ, $\dot{x}_0$) basins at $\omega = 3.2$. The white, black and grey regions denote the basins of the attractors (7, 10), (3, 5) and (1, 2), respectively. (a) Magnification by a factor of 12 of the rectangle ($2\pi/12 \times 0.65/12$) denoted by the arrows in fig. 2c, $2.95 \lesssim \varphi \lesssim 3.474$, $0.26 \lesssim \dot{x}_0 \lesssim 0.3142$. Magnification by a factor of 12^5 of the rectangle ($2\pi/12^5 \times 0.65/12^5$) denoted by the arrows in (a), $3.3277052 \lesssim \varphi \lesssim 3.3277305$, $0.26838517 \lesssim \dot{x}_0 \lesssim 0.26838778$.

2c and 3). In figs. 2a–2d φ and $\varphi + 2\pi$ are equivalent and the range of $\dot{x}_0$ is so chosen that all stable (periodic) fixed points of the coexisting attractors (10, 13, 11 and 15 points, respectively) are included.

Fig. 2 shows that the $q = 7/10$ basin of attraction serves as a background that is sliced by the other basins when the corresponding attractors become stable. The general shapes of the DS basins of attraction (black patterns in figs. 2a–2c) are similar but the basin of the concatenated attractor (fig. 2b), probably due to the larger impact- and period-numbers, is more entangled by the $q = 7/10$ basin of attraction and is thus smaller than the basins of the parent attractors (figs. 2a and 2c) in this restricted region. (This is also a possible reason for the discontinuity of the step $q = 2/3$: the high-periodic attractors in the gap are not found numerically due to the narrow and highly entangled basins.) When ω is increased the attractor (1, 2) becomes stable and cuts its basin (grey pattern in fig. 2c) mainly out of the (3, 5) basin of attraction. When ω is further increased the last attractor of DS (3, 5) becomes unstable and its basin discontinuously disappears, mainly due to the occupation by the (7, 10) basin of attraction, which in turn continuously decreases, due to the increase of the (1, 2) basin of attraction (fig. 2d).

To explore the fine structure of the boundary of the basins we have magnified nested subregions by the factors of 12^m, $m = 1, 2, \ldots, 5$ starting from the rectangle denoted by arrows in fig. 2c. The 12 and $12^5 = 248\,832$ times magnifications are shown in figs. 3a and 3b, respectively. These magnifications and all the intermediate ones exhibit similar parallel striped [1] structure (see also ref. [8]). We conjecture that this would continue similarly in further magnifications. The graininess of the boundaries in the basin plots in figs. 2 and 3 is due to the finite grid in the calculation and is not a property of the basin boundary. The ubiquitous stripes are straight and regular which rules out the convergence errors. Hence we conclude that the basin boundaries in figs. 2 and 3 are genuinely of a fractal Cantor type [1]. The infinitely narrow disconnected stripes, forming transversally a Cantor set of the dimension $D - 1$, are due to the Smale-horseshoe-type Poincaré map developed by the homoclinic intersections of the stable and unstable manifolds of a fixed saddle point [2,7]. The horseshoe is not an attractor – although it can cause long chaotic transients in the motion – and therefore the fractal basin boundary is found despite the fact that the attractors in figs. 2 and 3 are *periodic*. We conjecture that the fractal basin boundaries are found in the whole DS range (e.g., figs. 2a–2c and 3) and in the close neighbourhood of DS (e.g., fig. 2d).

The existence of the homoclinic intersections has recently been used as a precursor to chaos for the driven pendulum and related oscillators [22]. However, for certain systems, like the Hénon map, the homoclinic intersections imply the boundary crisis which on the contrary destroys the chaotic attractor [4,21]. The first order Melnikov's method [23] is usually used for determining criteria for the existence of homoclinic intersections [22] which recently have also been used for describing the appearance of the fractal basin boundaries for the periodic attractors [7]. Melnikov's method is, however, directly inapplicable for the impact oscillator because the phase space associated with its unperturbed equation of motion (without the damping and sinusoidal forcing terms in eq. (1)) does not have any unstable saddle point.

To characterize quantitatively the basin boundaries we have calculated their fractal dimension D using the uncertainty-exponent method by Grebogi, McDonald, Ott and Yorke [1]. If a fractal set is covered by the identical two-dimensional pieces, linearly related in scale r to a certain starting cover, their number n and total area h, both relative to this starting cover, are r-dependent and satisfy the following equations [1,9]

$$nr^D = 1, \quad h = nr^2 \quad (r \to 0). \tag{3}$$

To get D we first calculate the uncertainty fraction $f \sim h$, $r \to 0$ [1]. f is the relative area of the uncertain phase plane which consists of the initial conditions $(\varphi, \dot{x}_0)$ converging to an attractor other than that found with the perturbed initial condition. D can be then calculated using the uncertainty exponent α [1] derived from eq. (3)

$$f \sim h = r^{2-D},$$

$$\alpha = \lim_{r \to 0} \frac{\ln(f)}{\ln(r)} = 2 - D. \tag{4}$$

f is calculated by perturbing φ by $\pm r\Delta\varphi$, where $\Delta\varphi$ is 0.05 times the horizontal width of the basin plot (see, e.g., figs. 2 and 3) and $r = 1, 0.9, ..., 0.1$. All the

Table 1
The fractal dimension D of the basin boundaries at $\omega = 3.2$ (3 coexistent attractors: (7, 10), (3, 5), (1, 2)). m denotes the exponent of the magnification 12^m. In both columns $m = 0$ corresponds to the same starting region of fig. 2c.

m	D	D in ref. [8]
0	1.67	1.67
1	1.65	1.65
2	1.65	1.57
3	1.62	1.68
4	1.62	1.60
5	1.65	1.70

$200 \times 260 = 52000$ grid points are considered as initial conditions in the calculation.

We have calculated D's for the nested subregions at $\omega = 3.2$ with the basins of the attractors (1, 2), (3, 5) and (7, 10), starting from the region in fig. 2c. The result is shown in table 1 (the magnifications 12^m, $m = 0, 1$ and 5 correspond to figs. 2c, 3a and 3b, respectively). The close agreement found here and in ref. [8] indicates that D seems to be independent of the selected subregion although the basin plots differ considerably (compare, e.g., figs. 2c, 3a, 3b). Hence in this sense the basin boundary exhibits statistical self-similarity with the fractal dimension $D \approx 1.65$ at $\omega = 3.2$. In the DS range we have not found such a multi-dimensionality as was recently found for the kicked double rotor [3]. This unique dimension for the basin boundary is in agreement with the recent calculation for the forced damped pendulum [5]. Close to the basin boundary the uncertain fraction of the phase plane f satisfies the nonlinear scaling equation with respect to an initial condition error ϵ/β (cf. eq. (4))

$$f(\epsilon/\beta) = f(\epsilon)/\beta^{2-D}. \tag{5}$$

Therefore a slow decrease of f is found even in substantially improved measurements, e.g., the reduction of ϵ by a factor of $\beta = 10^6$ decreases f only by a factor of about 10^2 for $D \approx 1.65$. Hence the large value of D indicates long transients and high sensitivity to noise although all the attractors are *periodic* in the DS range.

In fig. 4 D is presented in the DS range where the three attractors (1, 2), (3, 5) of DS (7, 10) are stable and at the ends of the step $q = 3/5$ where the attractor (3, 5) has become unstable (high-periodic q-

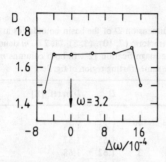

Fig. 4. Fractal dimension D in the range of the step $q=3/5$ of DS and at the ends of the step. $\Delta\omega = \omega - 3.2$. In the calculation the region of the phase plane is the same as in fig. 2.

steps of DS having extremely narrow basins may, however, appear here). In the calculation the ranges of $\dot{x}_0$ and φ are the same as in fig. 2 at all ω's. In fig. 4 D attains two non-integer values: D is very close to ≈ 1.7 in the range of the step $q=3/5$ but falls abruptly to ≈ 1.5 at both ends of the step. Hence the Farey organisation affects the fractal dimension considerably. We have found that in the range of the step $q=3/5$ always all the three basins are involved in the striped structure and we have not found different fractal dimensions in different regions (cf. refs. [3] and [5]).

In conclusion, the basins and the basin boundaries of the dissipative sinusoidally driven particle moving and impacting in the harmonic potential are studied in the range where the winding number q forms an incomplete Farey-organised devil's staircase. The $q=7/10$ basin of attraction serves as a background sliced by the basins of the devil's staircase attractors and the attractors of the branch $q=1/2$. The general shapes of the basins of the devil's staircase attractors are the same. The basins of the concatenated attractors are more entangled by the $q=7/10$ basin of attraction than the basins of their parent attractors. The basin boundaries exhibit fractal structure. The fractal dimension, calculated from large regions of the phase space, does not depend on the subregion indicating that the boundaries are statistically self-similar and have a unique fractal dimension. Hence no multidimensionality (different fractal dimensions in different regions) was found either. The Farey-organisation (the appearance of the concatenated q-steps) affects the fractal dimension considerably whereas the fractal dimension remains approximately constant in the range of each q-step. The calculated large fractal dimension, typically ≈ 1.7, indicates long transients and high sensitivity to noise although all the attractors are *periodic* in the devil's staircase range.

We thank T. Aalto and M.A. Ranta for the amicable collaboration, L. Malmi for his skillful assistance in computer graphics and F.H. Ling, C. Grebogi and F.C. Moon for sending their preprints in refs. [2], [5] and [22], respectively. We are grateful to the staffs of the computer centers of the Finnish State VTKK, Helsinki University of Technology TEKOLA and the University of London ULCC for their expert help with the computations. H.M.I. thanks T. Kotera, F. Peterka, G. Schmidt and A. Tondl for illuminating discussions.

References

[1] C. Grebogi, E. Ott and J.A. Yorke, Phys. Rev. Lett. 50 (1983) 935; 51 (1983) 942;
C. Grebogi, S.W. McDonald, E. Ott and J.A. Yorke, Phys. Lett. A 99 (1983) 415;
S.W. McDonald, C. Grebogi, E. Ott and J.A. Yorke, Physica D 17 (1985) 125.

[2] C. Grebogi, E. Ott and J.A. Yorke, Phys. Rev. Lett. 56 (1986) 1011; Physica D 24 (1987) 243;
Lj. Kocarev, Phys. Lett. A 121 (1987) 274;
F.H. Ling, On fractal attracting basin boundaries and their consequences, preprint, Jiao Tong University (1987).

[3] C. Grebogi, E. Kostelich, E. Ott and J.A. Yorke, Phys. Lett. A 118 (1986) 448; 120 (1987) 497; Physica D 25 (1987) 347.

[4] C. Grebogi, E. Ott and J.A. Yorke, Physica D 7 (1983) 181.

[5] C. Grebogi, H.E. Nusse, E. Ott and J.A. Yorke, Basic sets: sets that determine the dimensions of basin boundaries, preprint, University of Maryland (1987).

[6] A. Arneodo, P. Coullet, C. Tresser, A. Libchaber, J. Maurer and D. d'Humières, Physica D 6 (1983) 385;
S. Takesue and K. Kaneko, Prog. Theor. Phys. 71 (1984) 35;
Y. Yamaguchi and N. Mishima, Phys. Lett. A 104 (1984) 179; 109 (1985) 196;
O. Decroly and A. Goldbeter, Phys. Lett. A 105 (1984) 259;
R.G. Holt and I.B. Schwartz, Phys. Lett. A 105 (1984) 327;
I.B. Schwartz, Phys. Lett. A 106 (1984) 339; J. Math. Biol. 21 (1985) 347;
E.G. Gwinn and R.M. Westervelt, Phys. Rev. Lett. 54 (1985) 1613;
M. Iansiti, Q. Hu, R.M. Westervelt and M. Tinkham, Phys. Rev. Lett. 55 (1985) 746;

489

F.T. Arecchi, R. Badii and A. Politi, Phys. Rev. A 32 (1985) 402;
M. Napiórkowski, Phys. Lett. A 113 (1985) 111;
J.M. Aguirregabiria and J.R. Etxebarria, Phys. Lett. A 122 (1987) 241.

[7] F.C. Moon and G.-X. Li, Phys. Rev. Lett. 55 (1985) 1439;
E.G. Gwinn and R.M. Westervelt, Phys. Rev. A 33 (1986) 4143; Physica D 23 (1986) 396;
J.M.T. Thompson, S.R. Bishop and L.M. Leung, Phys. Lett. A 121 (1987) 116.

[8] H.M. Isomäki, J. von Boehm and R. Räty, in: Nonlinear science: theory and applications (Manchester Univ. Press, Manchester, 1987), to be published.

[9] B.B. Mandelbrot, Fractals – form, chance and dimension (Freeman, San Francisco, 1977).

[10] H.-O. Peitgen and P.H. Richter, in: Advances in solid state physics, Vol. XXV, ed. P. Grosse (Vieweg, Braunschweig, 1985) p. 55; J. Non-Equilib. Thermodyn. 11 (1986) 243;
H.E. Benzinger, S.A. Burns and J.I. Palmore, Phys. Lett. A 119 (1987) 441.

[11] H.M. Isomäki, J. von Boehm and R. Räty, Phys. Lett. A 107 (1985) 343.

[12] H.M. Isomäki, J. von Boehm and R. Räty, in: Advances in solid state physics, Vol. XXV, ed. P. Grosse (Vieweg, Braunschweig, 1985) p. 83.

[13] J. Farey, Philos. Mag. J. (London) 47 (1816) 385;
T. Allen, Physica D 6 (1983) 305;
J. Maselko and H.L. Swinney, Phys. Scr. T9 (1985) 35; Phys. Rev. Lett. 55 (1985) 2366.

[14] H.M. Isomäki, J. von Boehm and R. Räty, in: Proc. XIth Int. Conf. on Nonlinear oscillations ICNO-XI (Hungarian Academy of Sciences, Budapest, 1987), to be published; to be published.

[15] J. Villain and M.B. Gordon, J. Phys. 13 (1980) 3117.

[16] R.M. May, Nature 261 (1976) 459;
S. Grossmann and S. Thomae, Z. Naturforsch. 32a (1977) 1353;
C. Tresser and P. Coullet, C.R. Acad. Sci. A 287 (1978) 577.

[17] M.J. Feigenbaum, J. Stat. Phys. 19 (1978) 25; 21 (1979) 669.

[18] N. Metropolis, M.L. Stein and P.R. Stein, J. Comb. Theory (A) 15 (1973) 25.

[19] E.N. Lorenz, Ann. N.Y. Acad. Sci. 357 (1980) 282.

[20] J.A. Ketoja and J. Kurkijärvi, Phys. Rev. A 33 (1986) 2846;
J.A. Ketoja, Period-doubling university in multidimensional dissipative and conservative systems, Report TKK-F-B 109 (Helsinki Univ. of Technology, Dept. of Technical Phys., 1987) p. 1.

[21] C. Grebogi, E. Ott and J.A. Yorke, Phys. Rev. Lett. 48 (1982) 1507.

[22] M. Bartuccelli, P.L. Christiansen, N.F. Pedersen and M.P. Soerensen, Phys. Rev. B 33 (1986) 4686;
R.L. Kautz and J.C. Macfarlane, Phys. Rev. A 33 (1986) 498;
F.C. Moon, J. Cusumano and P.J. Holmes, Physica D 24 (1987) 383;
G. Cicogna and F. Papoff, Europhys. Lett. 3 (1987) 963;
G. Cicogna, Phys. Lett. A 121 (1987) 403.

[23] V.K. Melnikov, Trans. Moscow Math. Soc. 12 (1963) 1.

INTEGRITY MEASURES QUANTIFYING THE EROSION OF SMOOTH AND FRACTAL BASINS OF ATTRACTION

M. S. SOLIMAN AND J. M. T. THOMPSON

Department of Civil Engineering, University College London, Gower Street, London WC1E 6BT

(*Received 22 December 1988, and in revised form 30 March 1989*)

The phase portraits of a dissipative dynamical system are characterized by the existence of one or more stable attractors, which typically include point equilibria, periodic oscillations (harmonic and subharmonic), quasi-periodic solutions and chaotic attractors. Each attractor is embedded in its own domain or basin of attraction, bounded by a separatrix associated with an unstable saddle solution. Under the variation of a control parameter, as the attractors move and bifurcate, the basins also undergo corresponding changes and metamorphoses. These changes in size and shape are usually continuous but can be discontinuous as when an attractor vanishes, along with its basin, at a saddle-node bifurcation. Associated with the homoclinic tangling of the invariant manifolds of the saddle solution, basin boundaries can also change in nature from smooth to fractal. In this paper, the escape of a driven oscillator from a cubic potential well, as an archetypal example, is used to explore the engineering significance of the basin erosions that occur under increased forcing. Various measures of engineering integrity of the constrained attractor are introduced: a global measure assesses the overall basin area; a local measure assesses the distance from the attractor to the basin boundary; and a velocity measure is related to the size of impulse that could be sustained without failure. Since engineering systems may be subjected to pulse loads of finite duration, attention is given to both the absolute and transient basin boundaries. The significant erosion of these at homoclinic tangencies is particularly highlighted in the present study, the fractal basins having a severely reduced integrity under all three criteria.

1. INTRODUCTION

Much work has been done on how point, periodic, quasi-periodic and chaotic attractors are created, changed and destroyed as a system parameter is varied. The mechanisms include the well-known local bifurcations, together with subharmonic cascades, intermittencies, crises, etc. In addition it is important to recognize that in typical systems several attractors often coexist at fixed parameter values. This has lead to much interest in basins of attraction, and how they too undergo changes and metamorphoses. These basin boundary metamorphoses include smooth-fractal metamorphosis, fractal-fractal metamorphosis, etc.

Recent papers by Grebogi, Ott and Yorke [1-3] on these events have indicated that the structures of basin boundaries are determined by accessible saddle orbits which lie on the basin boundary, and that changes occur when these saddle orbits become inaccessible. This can occur as a result of a homoclinic tangle of the stable and unstable manifolds of a saddle orbit on the basin boundary.

In this paper we investigate, on the macroscopic (global) level, how the size of the basin of attraction changes with a control parameter. As an illustrative example we shall consider the problem of the sinusoidally forced motions of a particle in a single potential well, $V = x^2/2 - x^3/3$. This system is chosen because the escape of a dynamical system is a recurrent theme in physics and engineering, and this particular potential is the universal generic *metastable* form that is always encountered as mechanical system approaches a static fold or limit point. The roll response of a vessel, with the inevitable asymmetries of off-centre cargo and lateral wind loading, can for example be modelled by an equation of this type [4].

We thus consider the mechanical oscillator with the single generalized co-ordinate x described by the equation

$$\ddot{x} + \beta \dot{x} + x - x^2 = F \sin \omega t \qquad (\dot{x} \equiv y), \tag{1}$$

where β, ω, F are system parameters and a dot denotes differentiation with respect to time. We focus attention throughout on phase $\phi = 180°$, and achieve this by the simple expendient of immediately replacing F by $-F$ in equation (1). Firstly, we shall consider how the basin of attraction is affected by the appearance of a homoclinic tangency (predicted by the method of Melnikov). In order to illustrate this, we shall consider the erosion of the basin of attraction of the main sequence of attractors from the fundamental state $F = x = y = 0$ to the final blue sky event, at $F = F^x$, whether the latter be a simple cyclic fold or a boundary crisis.

Secondly, the nature (smooth/fractal) of the basin of attraction will be considered, and escape from a simple cyclic fold will be used to illustrate, that, if the invariant manifolds of a saddle cycle have already homoclinically tangled at some value F^T, they must "detangle" before any saddle-node bifurcation that involves this saddle cycle. In addition, we emphasize that for a saddle node bifurcation the area of the absolute basin of attraction remains finite at the escape value, F^x. This area will then drop to zero when the relevant attractor loses its stability catastrophically at the final bifurcation, which results in the inevitable jump to failure.

Thirdly, we shall consider the transient escape behaviour. This is important because systems are not always subjected to constant forcing for long periods. This is especially true in the field of marine and naval technology where offshore structures are often subjected to a steady train of waves for relatively short periods of time, making the short term response of considerable significance. Transient basin diagrams [5] may be used to show how the transient basins of the attractor at infinity change as a system parameter is varied both before and after the blue sky catastrophe. Long transient behaviour is common after a saddle-node fold as examined by Van Damme and Walkering [6].

Finally, we shall consider how, if the local stability of an attractor under small but finite disturbances is to be examined, numerous factors other than the size and nature (smooth/fractal) of its basin of attraction must be investigated. These include how an attractor would respond to impact loading, noise, etc. We consider the sensitivity of an attractor subjected to impact loading, together with an alternative criterion of local integrity based on a suitably defined minimum distance from the attractor to the basin boundary.

2. TRANSIENT BASINS AND INTEGRITY MEASURES

In studying transient times one must acknowledge the fact that a system does not usually reach an attractor in finite time. One must therefore introduce a suitably defined neighbourhood of the attractor. For a point attractor in a Poincaré section an obvious neighbourhood would be a small disc centred on the attractor.

In the present work we are only concerned with transients, from a local window of starts, that lead to escape from a potential well with x tending to infinity; that is to say the transients leading to the *attractor at infinity*. As our criterion of nearness to this we specify $x > 20$, based on the experience that for our equation a computer usually "crashes" due to overflow shortly after x passes 20.

We can therefore define a transient basin, $\mathscr{A}_\tau$, as the set of all starting points that reach $x = 20$ in $t < \tau$. It is also convenient to define the constraint basin, $\mathscr{C}_\tau$, as the set of all points that reach $x = 20$ in $t > \tau$ (or not at all). The transient boundary, $\mathscr{B}_\tau$, between these two areas is then defined as the set of all points that reach $x = 20$ in precisely time τ (allowing us to write $\mathscr{A}_\tau$ as simply the *complement* of $\mathscr{C}_\tau \cup \mathscr{B}_\tau$). The normally defined, *absolute* basin of attraction is clearly $\mathscr{A}_\infty$, obtained by letting τ tend to infinity.

We are not particularly concerned here with the fate of the constrained, non-escaping trajectories. Most will be attracted to a well defined attractor (or one of a pair of attractors, in the region of hysteresis) lying on the main sequence of stable attractors (harmonic, subharmonic and chaotic) originating from the equilibrium state $F = x = y = 0$ under slowly incremented F. A small number may, however, be attracted to the large number of unexplored but highly localized competing attractors that are invariably encountered in problems of this type, examples being the S^3 and S^6 subharmonics associated with the saddle-node cascade scenarios [7].

To quantify the erosion of the basin of constraint we introduce three measures which might serve to assess the engineering integrity of the main attractor.

The first concerns the area of $\mathscr{C}_\tau$ within a prescribed window. Using a grid of N starts, we write the proportion that fall within $\mathscr{C}_\tau$ as G_τ. We shall in fact measure τ in number of forcing cycles, m, and then write the proportion as G_m. This *global integrity measure* (*GIM*) is, conveniently, independent of the finite attractors onto which the constrained motions settle.

For a given point attractor within $\mathscr{C}_\infty$ a useful deterministic measure of its integrity would seem to be the minimum distance, L_τ, in the (x, y) Poincaré section, from the attractor to the transient basin boundary $\mathscr{B}_\tau$. This gives us our second *local integrity measure* (*LIM*), L_m, written again in terms of forcing cycles.

Our third measure, based on the concept that a mechanical oscillator might be subjected to an impulse, in which it could be thought to experience an instantaneous step change in velocity, involves the minimum distance in the direction of $+y$ or $-y$. The minimum distance in the Poincaré section from the point attractor to the boundary $\mathscr{B}_\tau$ in the direction of positive y is written as I^+, and in the direction of $-y$ as I^-. With either a positive or negative sense, we thus have the *impulsive integrity measure* (*IIM*) denoted by I_m.

A fourth, *stochastic integrity measure* (*SIM*) can be defined in terms of the mean escape time when the attractor is subjected to white noise of prescribed intensity. This could well prove to be a particularly useful measure of integrity, and will be the subject of future investigations.

3. NUMERICAL SIMULATIONS

The transient basin diagrams (Figures 3-7 of section 5) were obtained by performing a fourth order Runge-Kutta numerical algorithm on equation (1), comparison with a variety of alternative numerical integration schemes having confirmed the accuracy of this approach. Here 100×100 initial conditions were chosen in the form of a grid and integrations were continued until either the escaping criterion was satisfied (arbitrarily chosen as $x > 20$), or the maximum allowable number of forcing periods was reached. For obvious reasons of computational economy, the allowable number was here taken as $m = 16$. All the times to escape were stored with their corresponding initial conditions, and transient basin diagrams were plotted by assigning a different shade to assigned transient intervals. Global integrity curves, $G_m(F, \omega, \beta)$ versus F, can also be drawn from the stored data.

4. CHOICE OF FREQUENCIES

In order to understand how a homoclinic tangency would affect the size of a basin of attraction, several values of frequency, ω, involving different routes to escape from the $F = x = y = 0$ fundamental state, were chosen and compared. The values of ω were chosen from a consideration of Figures 1(a) and 1(b). These bifurcation diagrams were obtained by plotting curves, in (F, ω) space for fixed $\beta(=0\cdot1)$, at which a homoclinic tangency, a period-doubling flip and a saddle-node fold bifurcation occur [7].

The equation of the Melnikov homoclinic tangency line was obtained by applying the method of Melnikov [4]. The condition for a homoclinic tangency between the stable and unstable manifolds of the saddle cycle close to the $x = 1$ hilltop was found to be

$$F^M = \beta \sinh(\pi\omega)/5\pi\omega^2. \qquad (2)$$

This is, of course, only an *approximation* to the true homoclinic tangency, F^T, and gives misleading results at large F and ω, as we shall see. For this reason we have *sketched* a dashed line in Figure 1(a) on which we expect the *actual* homoclinic tangency to occur. The fold and flip bifurcation lines were obtained by using a bifucation following routine, the flip being the first period-double from the fundamental $n = 1$ to a subharmonic $n = 2$ attractor. The right-hand insert of Figure 1(b) shows four typical traces of the steady state responses at constant ω, with F plotted against the stroboscopically sampled x.

The values of ω were chosen as follows: $\omega = 3\cdot4$ at which there exists an $n = 1$ periodic attractor from the $F = 0$ fundamental state all the way to the final saddle-node blue sky event; $\omega = 2\cdot4$ at which there also exists an $n = 1$ periodic attractor until the final escape (in this case a homoclinic tangency occurs at some value $F^T(\approx F^M)$; $\omega = 1\cdot0$ and $0\cdot85$ are similar in that in both cases there exists an $n = 1$ periodic attractor—during which there is homoclinic tangency—which has a period-doubling cascade leading to a boundary crisis (at the latter there exists a hysteresis fold before the homoclinic tangency); $\omega = 0\cdot65$, where there exists an $n = 1$ fundamental steady state in which escape occurs at a simple fold (in addition, there coexists an $n = 1$ periodic attractor with a period-doubling cascade leading to a chaotic attractor).

5. EROSION OF THE BASIN OF ATTRACTION

5.1. RESULTS FOR $\omega = 3\cdot4$

The first case considered was that of $\omega = 3\cdot4$. Here there exists a monotonic trace of fundamental ($n = 1$) harmonic oscillations originating at $F = x = y = 0$ and terminating in a simple cyclic fold at $F^x \approx 8\cdot11$. For the sake of this presentation it is useful to assume that if trajectories do not escape within the maximum allowable time (16 forcing cycles) they will remain constrained, and hence that the $\mathcal{B}_{16}$ basin boundary represents the absolute boundary between the attractor at infinity and the period-one attractor ($\mathcal{B}_{16} \approx \mathcal{B}_\infty$), and $\mathcal{C}_{16}$ represents the corresponding absolute basin of attraction ($\mathcal{C}_{16} \approx \mathcal{C}_\infty$).

The erosion of the $\mathcal{C}_{16}$ basin can be seen quantitatively in Figure 2. The proportion of initial conditions constrained decreases in a smooth manner from $F = 0$ to $F = 8\cdot0$; none are constrained at $F = 9\cdot0$ as there only exists the attractor at infinity.

No homoclinic tangency was expected to occur on the basis of the Melnikov analysis (see Figure 1) and thus it was expected that the basin of attraction would remain smooth throughout. Indeed, as can be seen from Figure 3 (transient basin diagrams for $\omega = 3\cdot4$),

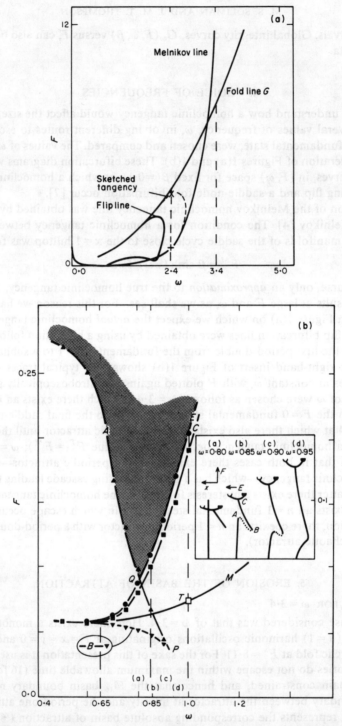

Figure 1. (a) Bifurcation diagram for $\beta = 0.1$ showing the saddle-node fold G and flip C (first period-double from the fundamental $n = 1$ solution) curves, as well as the Melnikov line in the (F, ω) control space. Also shown is a sketched line showing the form that we expect the actual homoclinic tangency line to follow. The vertical dashed lines indicate the frequency values examined in this report. (b) Blow-up of Figure 1(a). Inset are four typical traces of the steady state responses at constant ω, with F plotted against x as sampled stroboscopically at phase $\phi = 0$. The vertical dashed lines indicate the frequency values examined in this report. Shaded area, escape.

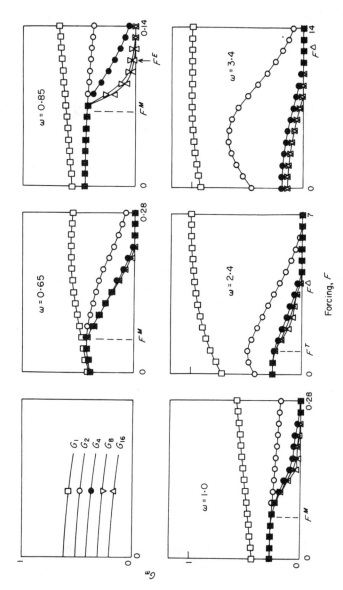

Figure 2. Global integrity measure curves for the given frequencies.

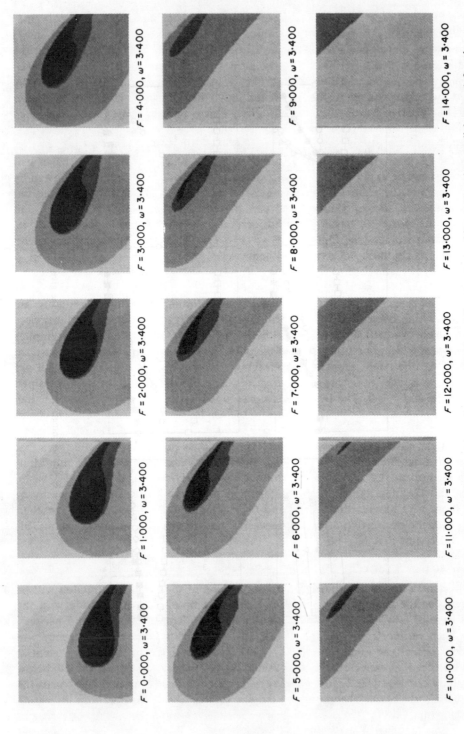

Figure 3. Transient basin diagrams. Each shade represents how many forcing cycles to escape: white, $t < 1$ cycle; very light grey, 1–2 cycles; grey, 2–4 cycles; dark grey, 4–8 cycles; very dark grey, 8–16 cycles; black, $t > 16$ cycles. For each diagram, $\beta = 0.1$ and $\phi = 180°$. Control parameters: $\omega = 3.4$ in the window $-1.5 < x < 1.5$, $-1.5 < y < 3.0$.

although the $\mathscr{C}_{16}$ basin changes in size, position and shape as F is gradually increased from $F = 0$ to $F = F^x$, it seems to remain smooth. It is also important to discuss the behaviour of the transient boundaries and their corresponding transient basins. As can be seen in Figure 3 each different shade represents the basin of attraction for a particular transient-length interval. It must be pointed out, however, that the range of initial conditions were chosen such that the overall behaviour could be observed, and at the same time enough detail preserved for the purpose of this study. As can be seen each basin seems to be smooth and once again as the value of the forcing amplitude is gradually increased the basins change in size. However, unlike the $\mathscr{C}_{16}$ basin which vanished after F^x ($\approx 8 \cdot 11$) the other transient basins still exist. This is perhaps most clearly illustrated by looking at the $\mathscr{C}_4$ basin. At $F = 8 \cdot 0$ the $\mathscr{C}_4$ basin is reminiscent of the $\mathscr{C}_{16}$ basin at lower values of F. However, as F is increased to $9 \cdot 0$ the $\mathscr{C}_{16}$ basin has disappeared whereas the $\mathscr{C}_4$ basin still exists in approximately the place where the $\mathscr{C}_{16}$ basin had resided. This can be explained by the fact that a saddle-node bifurcation is followed by long transients in the region of the extinguished basin [6]. It must also be pointed out that this phenomenon seems not only to take place at the saddle-node bifurcation, but also when part of the $\mathscr{C}_{16}$ basin is eroded; here also the eroded region generates orbits with long escape times (see Figure 3, $F = 6 \cdot 0$ to $7 \cdot 0$). This is indeed what one would expect from the continuity of dynamical behaviour.

The sizes of the transient basins may be seen quantitatively in Figure 2. Each curve represents the proportion of initial conditions being constrained, G_m, for m cycles (as denoted by the specimen diagram). For low values of forcing most initial conditions are constrained within one cycle. This is partly due to the fact that all the simulations were started at phase $\phi = 180°$ (by replacing F by $-F$ as previously mentioned). This implies that the system is being first pushed *into* the potential well, tending initially to constrain the system. As one might expect, as the forcing increases a little, a greater number are constrained; as is indeed the case when practically all the initial conditions are constrained for one cycle at $F = 7 \cdot 0$. The proportion constrained by two cycles initially rises then falls. Here we must point out that this is true for the window of initial conditions considered and might not have been the case if a different range had been chosen, due to the specific geometry of the basins. From the transient basin diagrams of Figure 3 it can be seen that there is a considerable shift in the $\mathscr{C}_2$ basin and this may account for the rise and fall of G_2. The erosion of the $\mathscr{C}_4$ basin seems to follow closely the trend of the $\mathscr{C}_{16}$ basin. However, the G_4 integrity curve clearly illustrates that there still exists the $\mathscr{C}_4$ basin for $F > F^x$. It can also be seen that the G_8 and G_{16} curves almost lie on top of one another, implying that the proportion escaping between 8 and 16 cycles is relatively small. This reinforces the view that $G_{16} \approx G_\infty$.

5.2. RESULTS FOR $\omega = 2 \cdot 4$

The effect of a homoclinic tangency on the erosion of the basin of attraction will be considered for the case of $\omega = 2 \cdot 4$. Here a direct comparison can be made with the previous case of $\omega = 3 \cdot 4$, as the routes to escape were identical (i.e., an $n = 1$ periodic attractor with escape from a simple fold); however, here a homoclinic tangency was expected to occur (see Figure 1) at $F^M \approx 1 \cdot 05$, well before the final blue sky event at $F^x \approx 4 \cdot 13$. Indeed, as expected, the $\mathscr{C}_{16}$ basin of attraction seems to remain smooth below the Melnikov criterion (see Figure 4). However, once the forcing amplitude exceeds the Melnikov limit, the basins become more complex with whisker-like projections indicating fractal behaviour [8-11]. This smooth-fractal basin boundary metamorphosis is due to the homoclinic tangling of the stable and unstable manifolds of the saddle periodic orbit on the basin boundary as Grebogi *et al.* [1-3] have indicated. As the forcing amplitude

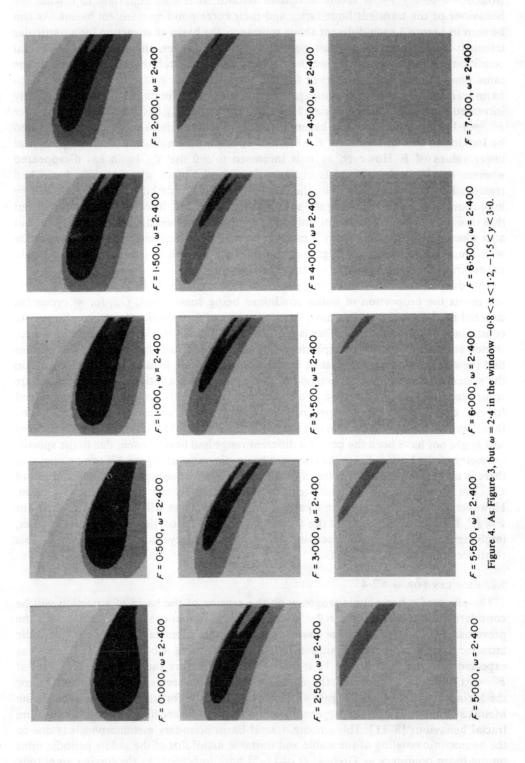

Figure 4. As Figure 3, but $\omega = 2.4$

is further increased the basin becomes more complex in appearance as new tongue-like projections gradually erode more and more of the basin. However, at $F = 3.5$, although a considerable part of the basin has been eroded, the structure seems to become less complex in appearance. In fact, at $F = 4.0$ the basin seems once again to have become smooth. This is indeed the case, as we shall examine later. Indeed, it is clear from centre manifold considerations that on approaching a saddle-node bifurcation there can be no homoclinic tangling of the saddle manifolds. So at this ω value one must have a detangling before F^x: i.e., a smooth-fractal metamorphosis as well as a fractal-smooth metamorphosis must take place.

The effect of the homoclinic tangency on the size of the basin of attraction can be seen quantitatively in Figure 2. For $F < F^T$ ($\approx F^M$) as was seen previously, a small gradual shrinkage of the $\mathcal{C}_{16}$ basin takes place. After the homoclinic tangency, the catchment region is very rapidly eroded, reducing by about 50% from $F = 1.0$ to $F = 2.0$. Furthermore, it is important to note that for $F < F^M$ the G_{16}, G_8 and G_4 integrity curves seem to be coincident, while after the tangency they diverge appreciably. This, as will be more clearly illustrated later on, is due to the emergence of the finger-like projections and chaotic transients.

5.3. RESULTS FOR $\omega = 1.0$ AND $\omega = 0.85$

In this study, $\omega = 1.0$ and 0.85 were also considered and the transient basin diagrams are reproduced in Figures 5 and 6. In both cases an $n = 1$ periodic attractor terminates in a Feigenbaum cascade, leading to a chaotic folding band attractor, and finally escape is triggered by the blue sky disappearance of the attractor at $F^x \approx 0.220$ and $F^x \approx 0.109$ respectively. The essential difference is that the latter exhibits a region of hysteresis. In addition to seeing how the homoclinic tangency will affect the size of the basin of attraction in this particular problem, the effect of a hysteresis loop will be examined as well as the transient behaviour after a chaotic blue sky event.

It is perhaps in the case of $\omega = 0.85$, from all the cases studied, that the effect of the homoclinic tangency is most clearly illustrated. This is due to the fact that there seems to be only a small movement in the position of the basin of attraction and hence the overall picture is more easily examined. As seen in the $\omega = 0.85$ global integrity curves there is little or no change in the size of the transient basins before the homoclinic tangle. It is important to point out that, apart from the G_1 curve, the curves here seem to be almost coincident, implying that a relatively small proportion escape between 2 and 16 cycles. However, as the Melnikov criterion is exceeded there is a dramatic erosion in the $\mathcal{C}_{16}$ basin with a not so dramatic erosion of the $\mathcal{C}_8$, $\mathcal{C}_4$ and $\mathcal{C}_2$ basins, as indicated by the divergence of the integrity curves. This can be clearly observed in the transient basin diagrams of Figure 6: for $F = 0.0$ to 0.07 there is little or no change in the transient basins; however at $F = 0.07$ a small finger-like projection appears, whereupon at $F = 0.08$ there is a sudden shrinking of the $\mathcal{C}_{16}$ basin and a complete change in the appearance of the picture. It can also be noted that although the total areas of the $\mathcal{C}_8$, $\mathcal{C}_4$ and $\mathcal{C}_2$ basins decrease after the homoclinic tangency, there is an increase in the proportion of initial conditions which escape between 2-4, 4-8 and 8-16 cycles. Correspondingly, the $\mathcal{C}_{16}$ basin will depart more significantly from the $\mathcal{C}_\infty$ basin. This due to the development of the homoclinic tangle and the emergence of the finger-like projections, causing chaotic transients as an escape sequence maps from one finger to the next. A detailed illustration of this mapping sequence has been given by Thompson [7]. As F is increased further towards F^x, there is a continuation of the erosion, although not so dramatic, of the $\mathcal{C}_{16}$, $\mathcal{C}_8$, $\mathcal{C}_4$, $\mathcal{C}_2$ and $\mathcal{C}_1$ basins. At F^x (≈ 0.109) there is a boundary crisis of the chaotic attractor and an instantaneous destruction of the $\mathcal{C}_\infty$ basin of attraction, analogous to

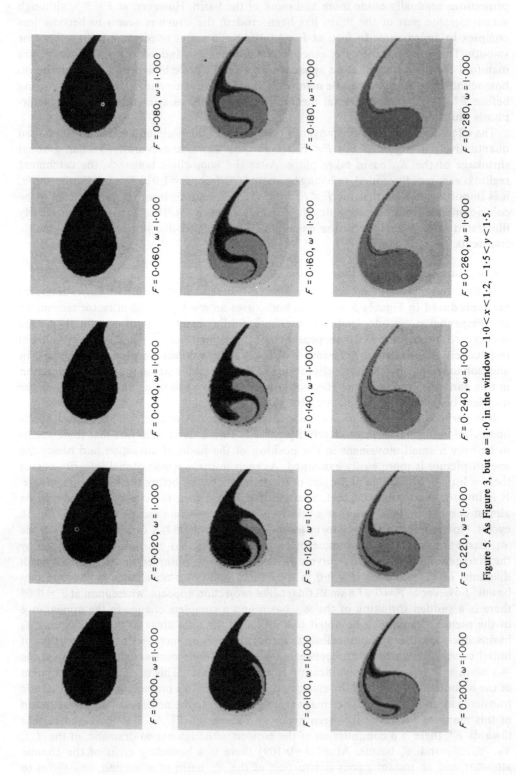

Figure 5. As Figure 3, but $\omega = 1\cdot 0$ in the window $-1\cdot 0 < x < 1\cdot 2$, $-1\cdot 5 < y < 1\cdot 5$.

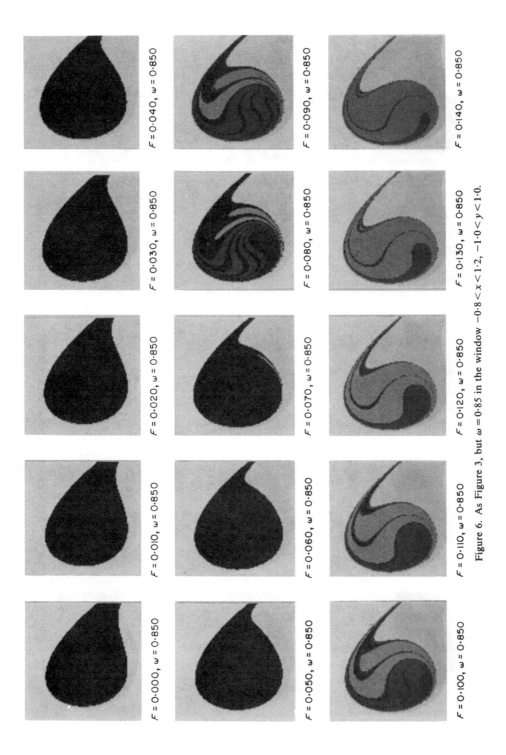

Figure 6. As Figure 3, but $\omega = 0.85$ in the window $-0.8 < x < 1.2$, $-1.0 < y < 1.0$.

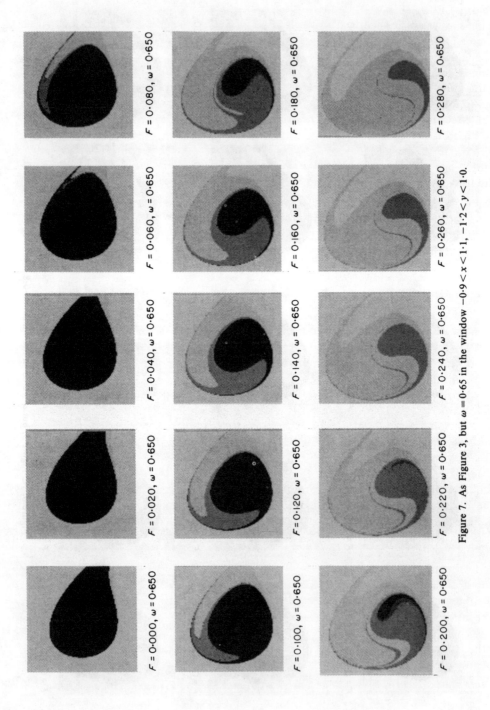

Figure 7. As Figure 3, but $\omega = 0.65$ in the window $-0.9 < x < 1.1$, $-1.2 < y < 1.0$.

that elucidated by Abraham and Stewart for the van der Pol oscillator [12]. However, there still exist well-defined transient basins after the blue sky catastrophe, and on the macroscopic level there seems to be no immediate change in the $\mathscr{C}_{16}$, $\mathscr{C}_8$, $\mathscr{C}_4$, $\mathscr{C}_2$, and $\mathscr{C}_1$ basins. Even at $F = 0.110$ there exists a $\mathscr{C}_{16}$ basin, as seen in the transient basin diagrams as well as in the integrity curve, G_{16}. This transient chaotic dynamical behaviour beyond a crisis, as described by Gwinn and Westervelt [13], is associated with the folding in the phase space. At a crisis the collision with an accessible saddle (here D^6 as demonstrated by Thompson [7]) destroys the $\mathscr{C}_\infty$ basin, although the "metastable" basin remains. The scaling of chaotic transients beyond the boundary crisis has been shown [7] to follow the exponential laws of Grebogi et al. [14].

The effect of the hysteresis can be considered by comparing the results for $\omega = 0.85$ with those of $\omega = 1.0$ where the diagrams (Figures 5 and 6) and integrity curves are seen to behave in a similar fashion. Looking more closely at the region of hysteresis for $\omega = 0.85$, one can see that there exists a resonant hysteresis in the region between two cyclic folds at $F \approx 0.05$ and $F \approx 0.07$; F^M is at 0.065. From inspection of the $\omega = 0.85$ integrity curves there seems to be little or no change in the size of the $\mathscr{C}_{16}$ basin (or in any of the basins) during the region of hysteresis. This implies that the total non-escaping region is unaffected by the hysteresis loop, although the size of the individual basins for the two coexisting $n = 1$ attractors is continually changing over this range. This is because the size and nature of the total absolute non-escaping boundary is determined by the inset of the hilltop saddle while the separatrix between the two coexisting $n = 1$ basins is determined by the resonant saddle between the coexisting attractors.

5.4. RESULTS FOR $\omega = 0.65$

The final value chosen was $\omega = 0.65$. Here there exists a monotonic trace of the fundamental ($n = 1$) solution terminating in escape from a simple cyclic fold at F^A. Also present, although not encountered in a natural loading sequence from $F = x = y = 0$, is an $n = 1$ periodic attractor created by the saddle-node at F^B which period doubles to a chaotic attractor. The range of F in which the fold B, flip C and crisis occur is extremely small and coincides roughly with the Melnikov value $F^M = 0.057$ [4]. For the transient basin diagrams for $\omega = 0.65$ (Figure 7) it can be seen that as the forcing amplitude increases from $F = 0$ the $\mathscr{C}_{16}$ basin increases in size as well as remaining smooth until the Melnikov criterion is reached. Here a fractal appearance is observed although it is not entirely clear. As F is further increased, the gradual erosion of the $\mathscr{C}_{16}$ basin is (unlike in the cases of $\omega = 0.85$ and 1.0) by one finger-like projection, which grows in size encircling the $\mathscr{C}_{16}$ basin. In fact, the $\mathscr{C}_{16}$ basin seems once again to have become smooth, and its erosion is in a similar fashion to that at $\omega = 2.4$: and in both cases termination of the $n = 1$ attractor is by escape from a simple fold, with long transients after the saddle node bifurcation. Here, however, the saddle of the saddle-node bifurcation is not the hilltop cycle, but the resonant saddle created at fold B.

The integrity curves confirm these deductions. After a slight increase in the size of the $\mathscr{C}_{16}$ basin, it starts to diminish after F^M. However, the G_{16}, G_8 and G_4 curves seem to remain roughly coincident, indicating the lack of the finger-like projections. This suggests that the basin of the attraction might have become smooth before the final escape. The sequence of events at $\omega = 0.65$ is indeed not entirely clear, and warrants further study.

6. TANGLING AND DETANGLING

The effect of the homoclinic tangency on the erosion of a basin of attraction has already been considered. It is also useful to know whether a basin is smooth or fractal. The

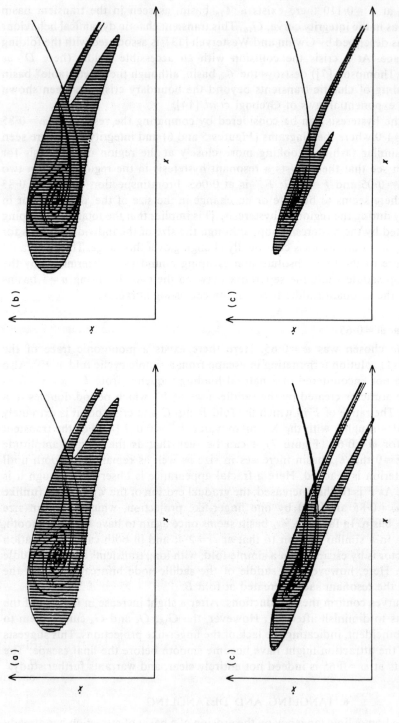

Figure 8. Tangling and detangling of the invariant manifold of the hilltop saddle cycle. (1) invariant manifold analysis (curved lines) and (2) grid of starts for the basin of attraction (dots forming vertical hatching). $\beta = 0.1$, $\phi = 180°$, $\omega = 2.4$. (a) $F < F^T$, $F = 1.0$ in the window $-1.0 < x < 1.5$, $-1.5 < y < 1.5$; (b) $F > F^T$, $F = 1.1$ in the window $-1.0 < x < 1.5$, $-1.5 < y < 1.5$; (c) $F < F^D$, $F = 3.5$ in the window $-0.2 < x < 1.3$, $0.5 < y < 2.5$; (d) $F > F^D$, $F = 3.8$ in the window $-0.2 < x < 1.3$, $0.5 < y < 2.5$.

analytical method of Melnikov can be used to predict, at which forcing amplitude, F^M, the appearance of a homoclinic tangency, F^T, takes place. However, as discussed in section 5.2, there exists a value of forcing at which the basin seems to have become smooth, indicating a "detangling" of the stable and unstable manifolds. In this section we show that this is indeed the case; critical forcing amplitudes, in which a smooth–fractal metamorphosis or a fractal–smooth metamorphosis takes place can be defined.

As an illustration of this we shall consider the case of $\omega = 2 \cdot 4$ where escape occurs from a simple fold at $F^x \approx 4 \cdot 13$. Here the homoclinic tangency between the stable and unstable manifolds predicted by Melnikov's method was at $F^M \approx 1 \cdot 05$. We deduce that at some forcing amplitude F^D, such that $F^T < F^D < F^x$, there must be a detangling of the stable and unstable manifolds before the final saddle-node bifurcation. This must happen, as just before a saddle-node bifurcation there can be no crossing of the stable and unstable manifolds, as is apparent from the centre manifold concepts [8]. Figure 8 shows the stable and unstable manifolds of the hilltop saddle cycle superimposed on the basin of attraction for several forcing amplitudes. These were obtained by plotting orbits in the Poincaré sections at phase $\phi = 180°$, backwards in time from a ladder of starts along the ingoing eigenvector of the saddle and forward in time from the outgoing eigenvector. The accuracy of the Melnikov criterion can be seen in Figures 8(a) and 8(b). For $F < F^M$ the stable and unstable manifolds do not cross; as the forcing just exceeds the Melnikov criterion ($F > F^M$) the stable manifold develops a finger which crosses the unstable manifold, as shown in Figure 8(b). As F is further increased, as seen in Figure 4 of the transient basin diagrams for $\omega = 2 \cdot 4$, it can be seen that the basin becomes more complex in character until about $F = 4 \cdot 0$ where the basin seems to have become smooth once again. Indeed, this is confirmed in Figures 8(c) and 8(d) where at $F = 3 \cdot 5$ there is a tangling of the manifolds and at $F = 3 \cdot 8$ there is not. These results imply that approximately for $0 < F < 1 \cdot 1$ no tangling occurs (hence the basin is smooth), for $1 \cdot 1 < F < 3 \cdot 65$ a tangling occurs and the basin is therefore fractal, and for $3 \cdot 65 < F < F^x$ there is again no tangling of the manifolds (smooth basin). Figure 1(a) shows a sketched line of the form that we expect the actual homoclinic tangency line to follow. The two crosses are values at which actual homoclinic tangencies were found.

Although this would be the simplest case in which this type of behaviour would occur for our particular oscillator, for other frequencies with complicated routes to escape, this scenario would be common. For example, in the case of $\omega = 0 \cdot 85$, in which $F^M \approx 0 \cdot 065$ and escape occurs from a chaotic blue sky event at $F^x \approx 0 \cdot 109$, there exists [7] a reversed period doubling cascade and chaos at $F \approx 0 \cdot 70$ leading to a second "escape" from a simple fold, and thus a "detangling" should once again occur. It can thus be deduced that although the Melnikov criterion, F^M, quite accurately predicts the homoclinic tangency, F^T, over a certain range of parameter values, there exists another critical value, F^D, in which a detangling takes place as shown in Figure 1(a). Relevant recent work on fractal distributions and the stable manifold has been done by Vazquez *et al.* [15].

7. LOCAL INTEGRITY OF THE ATTRACTOR

If one were to consider the "local stability" (but not infinitesimally local) of a particular attractor, numerous factors other than the size (section 5) and the nature (section 6) of its basin of attraction must be considered. These include the response to impact loading and external noise, position of the attractor within the basin, etc. Some specific examples are as follows. (a) Often a system settled on a particular attractor may experience a nearly instantaneous change of velocity due to an impact loading, perhaps causing the system to jump from one attractor to another. (b) Basins of attraction are often finely divided, as in the case of a fractal, where the separation between the attractor and the boundary can be small; the addition of external noise can easily push trajectories across these boundaries, as in the case discussed by Gwinn and Westervelt [13] of noise-induced intermittency of a driven damped pendulum. (c) Often real dynamical systems do not settle down to "true" periodic, subharmonic or chaotic motion; this can be due to a small random disturbance (noise) of either mechanical, thermal or electrical origin. In the case of a period one oscillation the long-term behaviour, viewed through Poincaré sections, would then appear to be a scatter of dots around this "attractor".

7.1. LOCAL INTEGRITY MEASURE

In this section we shall consider how close the attractor is to the basin boundary, the distance in the stroboscopically sampled ($t = 2i\pi/\omega$, $i = 1, 2, 3, \ldots$) Poincaré section (x, y) offering one measure of the engineering robustness of the attractor. Local integrity curves, similar to those of the global integrity are drawn, as shown in Figure 9, in this case the abscissa being the distance, L_m, to the transient basin boundary, $\mathcal{B}_m$. They were obtained by using the stored data and the co-ordinates of the main sequence attractor and measuring the minimum distance from the attractor to the $\mathcal{B}_m$ basin boundary. From the previous considerations of section 3 it will be useful in our discussion to speak as if $L_{16} \approx L_\infty$, the distance from the attractor to the absolute basin boundary.

On first glance the results seem to be in close agreement with the global integrity curves. As one would expect, the shrinking of the basin generates a reduction in the distance between the attractor and the basin boundary. This can be clearly seen in the case of $\omega = 3 \cdot 4$ where, as F is increased, the basin shrinks, resulting in the reduction of the local integrity measure, L_{16}, between the $n = 1$ periodic attractor and the $\mathcal{B}_{16}$ basin boundary. This is also the trend for the L_8, L_4, L_2 and L_1 local integrity curves.

The effect of a homoclinic tangency on the local integrity measure can be seen in the case of $\omega = 2 \cdot 4$. Here, the homoclinic tangency causes a dramatic erosion of the $\mathcal{C}_{16}$ basin, resulting in the reduction of L_{16}. Once again (as in the case of the global integrity curves), the less dramatic reduction of L_8, L_4, L_2, and L_1 after the homoclinic tangency gives rise to a divergence of the local integrity curves.

However, for $\omega = 1 \cdot 0$, although initially there is no erosion of the basin of attraction, as seen from the global integrity curves, before the Melnikov criterion, there is a continuous reduction in all the local integrity curves. This implies that as F is gradually increased successive attractors become nearer to the various transient basin boundaries. In addition, it can also be seen that the homoclinic tangency does not significantly change the trend of the local integrity curves as most of the erosion of the basin takes place outside the vicinity of the attractor. It must also be pointed out that, unlike the previous cases of $\omega = 3 \cdot 4$ and $2 \cdot 4$, where at a fixed forcing amplitude the L_{16}, L_8, L_4, L_2 and L_1 integrity measures were significantly different, here they seem to be coincident; this implies that in the vicinity of the attractor the distance between $\mathcal{B}_{16}$ and the other transient basin boundaries is extremely small, and hence a slight change in the starting conditions could mean either no escape or escape within one forcing cycle. This can be clearly seen in Figure 5.

In the case of $\omega = 0 \cdot 85$, the results are as expected; however, there is a sudden reduction in the local integrity just after $F \approx 0 \cdot 07$. This is not due to the sudden erosion of the basin, but due to the hysteresis jump in which the new attractor is considerably closer to the basin boundary. In this connection we should emphasize that the attractor chosen in this study is always the one that would be observed *physically* under the slow increase of F from zero; for the values of F chosen, this attractor is either an $n = 1$ harmonic or an $n = 2$ subharmonic, and in the latter case the *minimum* of two distances was chosen.

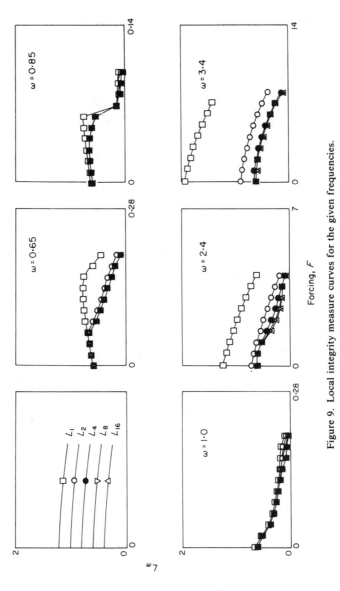

Figure 9. Local integrity measure curves for the given frequencies.

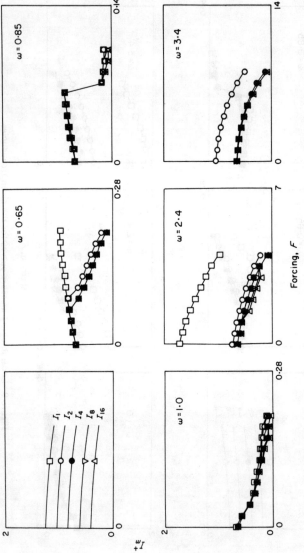

Figure 10. Impulsive integrity measure curves for the given frequencies (positive impact).

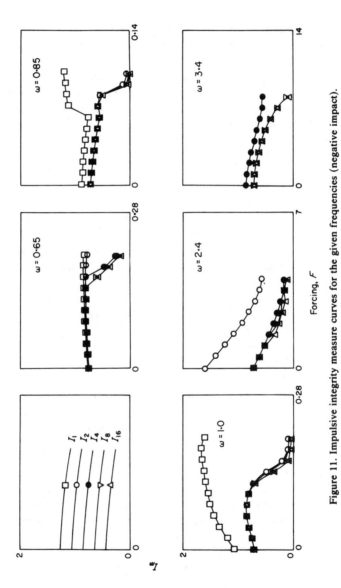

Figure 11. Impulsive integrity measure curves for the given frequencies (negative impact).

Once again, the homoclinic tangency does not seem to cause a dramatic change in the local integrity curves. For $\omega = 0.65$, the size of the basin of attraction grows as F increases for $F < F^M$, and drops dramatically for $F > F^M$. This also seems to be the trend of the local integrity curves.

7.2. IMPACT LOADING

In this section we will consider the response of a system subjected to an instantaneous change of velocity. In Figures 10 and 11 it is shown how large a positive or negative impact, measured here by Δy, would be needed before the various transient basin boundaries are reached. These results were obtained by performing a line of starts in the $(+/-)$ $\dot{x}$ direction from the relevant attractor. The required impact, I_m, to cause the attractor to cross the $\mathcal{B}_m$ basin could thus be obtained.

For $\omega = 3.4$ and fixed F, the required impact to cross successive transient basin boundaries increases; the I_1^+ and I_1^- are not shown as the impacts required to cross $\mathcal{B}_1$ were larger than those considered in this study. As F is increased the required impact (both positive and negative) to cause the attractor to escape reduces in magnitude (where $I_{16} \approx I_\infty$); this is due to the erosion of the basin of attraction as seen in Figure 2. Just before the saddle-node fold, the value of $I_m^\pm$, is relatively small, indicating that the attractor is extremely sensitive at such loading.

In the case of $\omega = 2.4$ the effect of a homoclinic tangency on the impulsive integrity measure was found to follow a similar trend to those of the local and global integrity curves. After the tangency there is a sudden reduction in $I_{16}^\pm$ as well as a less dramatic reduction of $I_8^\pm$, $I_4^\pm$, $I_2^\pm$ and $I_1^\pm$, as indicated by the divergence of the impulsive integrity curves. This is due to the appearance of the whisker-like projections making the attractor more sensitive to such loading.

For $\omega = 1.0$, as F is increased, successive attractors get closer and closer to the "northern" basin boundary. This has the effect of reducing the required positive impact to reach failure but increasing the magnitude of negative impact required to cause the stable attractor to escape. However, as the Melnikov criterion is exceeded, as well as there being a sudden reduction in the size of the basins, there is a dramatic reduction of I_m^- but hardly any change in I_m^+. This can be explained by the fact that most of the erosion takes place in the negative $\dot{x}$ direction from the attractor.

For $\omega = 0.85$, the hysteresis jump that would be *physically* observed under the slow increase of F, causes the *relevant* attractor to jump much closer to the basin boundary: the positive impact to cause failure is thus *instantaneously* reduced, while the negative impact is instantaneously increased. The curves seem to be unaffected by the homoclinic tangency ($F^T = 0.065$). In fact there seems to be a critical value ($F \approx 0.08$) at which a sudden drop in I_m^- occurs, and it can thus be deduced that the erosion of the basin by the finger-like projections has taken place in the vicinity of the attractor and caused it to be considerably more sensitive to such an impact. These observations also hold for the case of $\omega = 0.65$.

8. CONCLUSIONS

In this work we have investigated how on the macroscopic level, the size and nature (smooth/fractal) of basins of attraction change as a system parameter is varied: namely, as the forcing amplitude of our driven oscillator is increased from the $F = 0$ fundamental state to and beyond a bifurcation or a crisis. Our main conclusions are as follows.

(A) The area of the basin of attraction, until its final destruction at F^x, exhibits no discontinuous jump in size except at the final bifurcation. However, the appearance of a homoclinic tangency of the stable and unstable manifolds, resulting in a fractal basin, dramatically enhances the erosion of the basin. In addition, the erosion of both smooth and fractal basins resulted in a region (residue) of long transient orbits; trajectories initiated in the vicinity of the attractor, but not in its basin, would remain near it for long periods.

(b) The size of the basin at F^x, although finite, was extremely small for damping level $\beta = 0.1$. Beyond a crisis this basin was destroyed and long transient behaviour was observed. At a saddle-node bifurcation the transient behaviour was roughly periodic [6], and at a boundary crisis the transients were chaotic [13]. However, on the macroscopic level, no significant change in the transient basins was observed just before and beyond a crisis, where there exists a "metastable" basin of attraction.

(c) A region of hysteresis does not effect the erosion of the total non-escaping basin of attraction, although the sizes of the competing basins are changing continuously [16].

(d) The Melnikov criterion, F^M, accurately predicted the appearance of a homoclinic tangency, F^T, for a certain range of frequencies. However, for forcing levels above this criterion a "detangling" at F^D was observed. Hence, a more realistic representation of regimes where the basin was smooth or fractal could be defined as seen in Figure 1(a).

(e) Global integrity curves are a useful tool when investigating basin boundary metamorphoses; however, it must be pointed out that the size of a basin is just one of the numerous factors which must be considered when analyzing the finite stability of an attractor. For example, whether the basin is smooth or fractal, and the position of the attractor within the basin, must be considered.

Impulsive and local integrity curves can be used, together with global integrity curves to give an indication of the local stability of an attractor, and hence an estimation of its response to external forces such as an impact loading.

(1) A hysteresis jump or rapid movement of the attractor, as a system parameter is varied, can cause the attractor to move much closer to the boundary without any change in the size of the overall basin: this can result in a sudden, instantaneous and substantial reduction in the *impulsive* and *local* integrity measures.

(2) Attractors lying in fractal basins of attraction can be extremely sensitive to impact loading or noise which can cause a trajectory to move from one attractor to another.

(3) On approaching a boundary crisis or a saddle-node fold, the attractor was extremely sensitive to impact loading as its basin of attraction was often very small. One should notice here, that, unlike the global integrity measure G_∞, the local integrity measure, L_∞, drops *smoothly to zero* at a saddle-node bifurcation.

(4) An important aspect, that we have not explored in the present work, is that of the phase which we took rather arbitrarily as $\phi = 180°$: further work is needed to examine the significance of varying this angle.

A comprehensive presentation of the underlying bifurcations of equation (1) has been given by Thompson [7], and some valuable analytical studies have been made by Virgin [17].

ACKNOWLEDGMENTS

The second author would like to acknowledge the award of a five-year Senior Fellowship by the Science and Engineering Research Council of Great Britain, during the tenure of which this investigation has been made.

REFERENCES

1. C. GREBOGI, E. OTT and J. A. YORKE 1983 *Physica* **7D**, 181-200. Crises, sudden changes in chaotic attractors, and transient chaos.

2. C. GREBOGI, E. OTT and J. A. YORKE 1986 *Physical Review Letters* **56**, 1011–1014. Metamorphoses of basin boundaries in nonlinear dynamical systems.
3. C. GREBOGI, E. OTT and J. A. YORKE 1987 *Physica* **24D**, 243–262. Basin boundary metamorphoses: changes in accessible boundary orbits.
4. J. M. T. THOMPSON, S. R. BISHOP and L. M. LEUNG 1987 *Physics Letters* **121A**, 116–120. Fractal basins and chaotic bifurcations prior to escape from a potential well.
5. C. PEZESHKI and E. H. DOWELL 1987 *Journal of Sound and Vibration* **117**, 219–232. An examination of initial condition maps for the sinusoidally excited buckled beam modeled by the Duffing's equation.
6. R. VAN DAMME and T. P. VALKERING 1987 *Journal of Physics A: Math. Gen.* **20**, 4161–4171. Transient periodic behaviour related to a saddle-node bifurcation.
7. J. M. T. THOMPSON 1989 *Proceedings of the Royal Society, London* **A421**, 195–225. Chaotic phenomena triggering the escape from a potential well.
8. J. M. T. THOMPSON and H. B. STEWART 1986 *Nonlinear Dynamics and Chaos*. Chichester: John Wiley.
9. F. C. MOON and G. X. LI 1985 *Physical Review Letters* **55**, 1439–1442. Fractal basin boundaries and homoclinic orbits for periodic motion in a two-well potential.
10. F. C. MOON 1987 *Chaotic Vibrations: an Introduction for Applied Scientists and Engineers*. New York: John Wiley.
11. S. W. MCDONALD, C. GREBOGI, E. OTT and J. A. YORKE 1985 *Physica* **17D**, 125–153. Fractal basin boundaries.
12. R. H. ABRAHAM and H. B. STEWART 1986 *Physica* **21D**, 394–400. A chaotic blue sky catastrophe in forced relaxation oscillations.
13. E. G. GWINN and R. M. WESTERVELT 1986 *Physica* **23D**, 396–401. Horseshoes in the driven, damped pendulum.
14. C. GREBOGI, E. OTT and J. A. YORKE 1986 *Physical Review Letters* **57**, 1284–1287. Critical exponent of chaotic transients in nonlinear dynamical systems.
15. E. C. VAZQUEZ, W. H. JEFFERYS and A. SIVARAMAKRISHNAN 1987 *Physica* **29D**, 84–94. Fractal distributions in conservative systems: direct observations of the stable manifold.
16. J. M. T. THOMPSON and Y. UEDA 1989 *Dynamics and Stability of Systems* (in press). Basin boundary metamorphoses in the canonical escape equation.
17. L. N. VIRGIN 1988 *Journal of Sound and Vibration* **126**, 157–165. On the harmonic response of an oscillator with unsymmetric restoring force.

REFERENCES

Papers included in this selection are marked with *

Abarbanel, H. D. I. (1983). Universality and strange attractors in internal-wave dynamics. J. Fluid Mech., **135**, 407-434.

Abraham, N. B. (1983). A new focus on laser instabilities and chaos. Laser Focus, **19**, 73-81.

Abraham, N. B., Gollub, J. P., and Swinney, H. L. (1984). Meeting Report: Testing nonlinear dynamics. Physica, **11D**, 252-264.

Abraham, R. H. (1985). Choostrophes, intermittency, and noise. In Chaos, Fractals and Dynamics, P. Fisher and W. R. Smith (eds). Dekker: New York.

Abraham, R. H. (1986). Complex Dynamical Systems: Selected Papers. Aerial Press: Santa Cruz, CA.

Abraham, R. H., Kocak, H., and Smith, W. R. (1985). Chaos and intermittency in an endocrine system model. In Chaos Fractals and Dynamics, P. Fisher and W. R. Smith (eds). Dekker: New York.

Abraham, R. N., and Shaw, C. D. (1982). Dynamics: The Geometry of Behaviour, Part One, Periodic Behaviour. Aerial Press: Santa Cruz, Ca.

Abraham, R. H., and Shaw, C. D. (1983). Dynamics: The Geometry of Behaviour, Part Two, Chaotic Behaviour. Aerial Press: Santa Cruz, Ca.

Abraham, R. H., and Shaw, C. D. (1985). Dynamics: The Geometry of Behaviour, Part Three, Global Behaviour. Aerial Press: Santa Cruz, Ca.

Afraimovich, V. S., Bykov, V. V., and Shilnikov, L. P. (1977). On the origin and structure of the Lorenz attractor. Sov. Phys. Dokl., **22**,

253-255.

Aizawa, Y., and Ueza, T. (1982). Global aspects of the dissipative dynamical systems. II. Periodic and chaotic responses in the forced Lorenz system. Prog. Theor. Phys., **68**, 1864-1879.

Alexander, N. A. (1989). Production of computational portraits of bounded invariant manifolds. J. Sound Vib., **135**, 63-77.

Ananthakrishna, G., and Valsakumar, M. C. (1983). Chaotic flow in a model for repeated yielding. Phys. Lett., **95A**, 69-71.

Andereck, C. D., Dickman, R., and Swinney, H. L. (1983). New flows in a circular Couette system with co-rotating cylinders. Phys. Fluids, **26**, 1395-1401.

Andropov, A.A., Vitt, E. A., and Khaiken, S. E. (1966) Theory of Oscillators. Pergamon Press: Oxford.

Arecchi, F. T., and Califano, A. (1984). Low-frequency hopping phenomena in nonlinear systems with many attractors. Phys. Lett., **101A**, 443-446.

Arecchi, F. T., Meucci, R., Puccioni, G., and Tredicce, J. (1982). Experimental evidence of subharmonic bifurcations, multistability, and turbulence in a Q-switches gas laser. Phys. Rev. Lett., **49**, 1217-1220.

*Arecchi, F. T., Badii, R., and Politi, A. (1985). Generalized multistability and noise-induced jumps in a nonlinear dynamical system. Phys. Rev., **A32**, 402-408.

Aref, H. (1983). Integrable, chaotic, and turbulent vortex motion in two-dimensional flows. Ann. Rev. Fluid Mech., **15**, 345-389.

Ariaratnam, S. T., Xie, W. C., and Vrscay, E. R. (1989). Chaotic motion under parametric excitation, Dyn. Stab. Sys., **4**, 111-130.

Arneodo, A., Coullet, P., and Tresser, C. (1981a). A possible new mechanism for the onset of turbulence. Phys. Lett., **81A**, 197-201.

Arneodo, A., Coullet, P., and Tresser, C. (1981b). Possible new strange attractors with spiral structure. Commun. Math. Phys., **79**, 573-579.

Arneodo, A., Coullet, P., Tresser, C., Libchaber, A., Maurer, L., and D'Humieres, D. (1983). On the observation of an uncompleted cascade in a Rayleigh-Bernard experiment. Physica, **6D**, 385-392.

Arnold, V. I. (1965). Small denominators. I. Mappings of the circumference onto itself. Am. Math. Soc. Transl., Ser. 2, **46**, 213-284.

Arnold, V. I. (1977). Loss of stability of self-oscillations close to resonance and versal deformations of equivalent vector fields. Functional Anal. Appl., **11**, 85-92.

Arrowsmith, D. K., and Place, C. M. (1990). An Introduction to Dynamical Systems. Cambridge University Press, Cambridge.

Ashwin, P., King, G. P., and Swift, J. W. (1990). Three identical oscillators with symmetric coupling. Nonlinearity, **3**, 585-602.

Aswin, P. (1990). Symmetric chaos in systems of three and four oscillators. Nonlinearity, **3**, 603-618.

Atmanspacher, H. and Scheingraber, H. (1987). A fundamental link between system theory and statistical mechanics. Found. Phys., **17**, 939-963.

Atmanspacher, H., Scheingraber, H., and Voges, W. (1988). Global scaling properties of chaotic attractor reconstructed from time series, Phys. Rev., **A37**, 1314-1322.

Awrejcewicz, J. (1989). Two kinds of evolution of strange attractors for the example of a particular non-linear oscillator. ZAMP, **40**, 375-386.

Badii, R., Broggi, G., Derighetti, B., and Ravani, M. (1988). Dimension increase in filtered chaotic signals. Phys. Rev. Lett.,**66**, 979-982.

Baesens, C., and Nicolis, G. (1983). Complex bifurcations in a periodically forced normal form. Z. Phys. B, **52**, 345-354.

Bak, P., Bohr, T., and Jensen, M. H. (1985). Mode-locking and the transition to chaos in dissipative systems. Phys. Scr., **T9**, 50-58.

Baker, G. L., and Gollub, J. P. (1990). Chaotic Dynamics. Cambridge University Press, Cambridge.

Bapat, C. N., and Sankar, S. (1986). Periodic and chaotic motions of a mass-spring system under harmonic force. J. Sound Vib., **108**, 533-536.

Bartissol, P., and Chua, L. O. (1988). The double hook. IEEE Trans. Circuits Syst., **CAS-35**, 1512-1522.

Battelino, P. M., Grebogi, C., Ott, E., and Yorke, A. J. (1989). Chaotic attractors on a 3-torus, and torus break-up. Physica, **39D**, 299-314.

Benettin, G., Galgani, L., Giorgilli, A., and Stralcyn, J.-M. (1980). Lyapunov characteristic exponents for smooth dynamical systems and for hamiltonian systems: a method for computing all of them. Meccanica, **15**, 9-12.

Ben-Jacob, E., Goldhirsch, I., Imry, Y., and Fishman, S. (1982). Intermittent chaos in Josephson junctions. Phys. Rev. Lett., **49**, 1599 - 1602.

Benjamin, T. B., and Mullin, T. (1981). Anomalous modes in the Taylor experiment. Proc. R. Soc. Lond. A, **377**, 1-26, 27-43.

Berge, P. Dubois, M., Manneville, P., and Pomeau, Y. (1980). Intermittency in Rayleigh-Bernard convection. J. Phys. Lett., **41**, L341

- L345.

Berge, P. Pomeau, Y., and Vidal, Ch. (1984). L'Ordre dans le Chaos. Hermann: Paris.

Berry, M. V. (1981). Regularity and chaos in classical mechanics, illustrated by three deformations of a circular 'billiard'. Eur. J. Phys., **2**, 91-102.

Berry, M. V., Percival, J. C., and Weiss, N. O. (eds) (1987). Dynamical Chaos. Princeton University Press: Princeton, NJ.

Bier, M., and Bountis, C. (1984). Remerging Feigenbaum trees in dynamical systems. Phys. Lett., **104A**, 239-244.

Birkhoff, G., and Rota, G.-C. (1978). Ordinary differential equations, Blaishell: Waltham, MA.

Birkhoff, G. D. (1913). Proof of Poincare's geometric theorem. Trans. Am. Math. Soc., **14**, 14-22.

Birkhoff, G. D. (1927). Dynamical systems. American Mathematical Society: Providence, RI.

Birkhoff, G. D. (1950). Collected Mathematical Papers. American Mathematical Society: Providence: RI.

Bowen, R. (1978). On Axiom A diffeomorphisms (CBMS Regional Conference Series in Mathematics, vol. 35). American Mathematical Society, Providence: RI.

Bowen, R. and Ruelle, D. (1985). The ergodic theory of Axiom A flows. Invent. Math., **79**, 181-202.

Brandstater, A., Swift, J., Swinney, H. L., Wolf, A., Farmer, J. D., Jen, E., and Crutchfield, P. J. (1983). Low-dimensional chaos in a hydrodynamic system. Phys. Rev. Lett., **51**, 1442-1445.

Brorson, S. D., Dewey, D., and Linsay, P. S. (1983). Self-replicating attractor of a driven semiconductor oscillator. Phys. Rev., **28**, 1201-1203.

*Brunsden, V., and Holmes, P. J. (1987). Power spectra of strange attractors near homoclinic orbits. Phys. Rev. Lett., **58**, 1699-1702.

Brunsden, V., Cottrel, J., and Holmes, P. J. (1988). Power spectra of chaotic vibrations of a buckled beam. Preprint Cornell University.

Bullard, E. (1978). The disk dynamo. In Topics in Nonlinear Mechanics, S. Jorna (ed.), pp. 373-389. American Institute of Physics: New York.

*Bulsara, A. R., Schieve, W. C., and Jacobs, E. W. (1990). Homoclinic chaos in systems perturbed by weak Langevin noise. Phys. Rev., **A41**, 668-681.

Campbell, D. K., and Rose, H. A. (eds) (1983). Order in chaos. Physica, **7D**.

Carr, J., and Eilbeck, J. C. (1984). One-dimensional approximation for a quadratic Ikeda map. Phys. Lett., **104A**, 59-62.

*Cartwright, M. L., and Littlewood, J. E. (1945). On non-linear differential equations of the second order. I. The equation $\ddot{y} - k(1-y^2)\dot{y} + y = b\lambda k\cos(\lambda t + a)$, k large. J. Lond. Math. Soc., **20**, 180-189.

Casati, G., Ford, J., Vivaldi, F., and Visscher, W. M. (1984). One-dimensional classical many-body system having a normal thermal conductivity. Phys. Rev. Lett., **52**, 1861-1864.

Chen, Q., and Ott, E. (1990). Cross-sections of chaotic attractors. Phys. Lett., **147A**, 450-454.

Chirikov, B. V. (1979) A universal instability of many dimensional oscillator systems. Phys. Rep., **52**, 263-379.

Chow, S. N., and Hale, J. K. (1982). Methods of Bifurcation Theory. Springer-Verlag: New York.

Chua, L. O., Komuro, M., and Matsumoto, T. (1986). The double scroll family. IEEE Trans. Circuits Syst., **CAS-33**, 1073-1118.

Chua, L. O., and Lin, G. N. (1990). Canonical realization of Chua's circuits family. IEEE Trans. Circuits Syst., **CAS-37**, 885-902.

Collet, P. and Eckmann, J.-P. (1980b). Iterated Maps on the intervals as dynamical systems. Birkhauser: Boston.

Collet, P., Eckmann, J.-P., and Koch, H. (1981). Period doubling bifurcations for families of maps on R^n. J. Stat. Phys., **76**, 211-254.

Collet, P., Eckmann, J.-P., and Lanford, O. E. (1980). Universal properties of maps on an interval. Commun. Math. Phys., **76**, 211-254.

Coppersmith, N. K. (1980). Nonlinear behaviour in rail vehicle dynamics. In New Approaches to Nonlinear Problems in Dynamics, P. J. Holmes (ed.) pp. 173-194. SIAM: Philadelphia.

Coullet, P., Tresser, C., and Arneodo, A. (1979). Transition to stochasticity for a class of forced oscillators. Phys. Lett., **72A**, 268 - 270.

Coullet, P., and Tresser, C. (1978). Iterations d'endomorphismas et groupe de renormalization. J. de Phys., **39**, Coll. C5-C25.

Chistiansen, P.L., and Parmentier, R.D. (eds) (1989). Structure, Coherence and Chaos in Dynamical Systems. Manchester University Press: Manchester.

*Crutchfield, J. P. and Huberman, B. A. (1980). Fluctuations and the onset of chaos. Phys. Lett., **77A**, 407-410.

Crutchfield, J. P., Farmer, J. D., and Huberman, B. A. (1982). Fluctuations and simple chaotic dynamics. Phys. Rep., **92**, 45-82.

Curry, J. (1978). A generalized Lorenz system. Commun. Math. Phys., **60**, 193-204.

Curry, J. (1979). On the Henon transformation. Commun. Math. Phys., **68**, 129-140.

Curry, J. H., and Johnson, J. R. (1982). On the rate of approach to homoclinic tangency. Phys. Lett., **92A**, 217-220.

Cvitanovic, P. (ed.) (1984). Universality in Chaos. Adam Hingler: Bristol.

Day, R. H. (1982). Irregular growth cycles. Am. Econ. Rev., **74**, 406-414.

Day, R. H. (1983). The emergence of chaos from classical economic growth. Q. J. Econ., 201-213.

Derrida, B., Gervois, A., and Pomeau, Y. (1979). Universal metric properties of bifurcations of endomorphisms. J. Phys. A., **12**, 269-296.

Devaney, R. L. (1986). An Introduction to Chaotic Dynamical Systems. Benjamin/Cummings : Merlo Park.

Devaney, R. L. (1977). Blue sky catastrophes in reversible and Hamiltonian systems. Indiana Univ. Math. J., **26**, 247-263.

*D'Humieres, D., Beasley, M. R., Huberman, B. A., and Libchaber, A. (1982). Chaotic states and routes to chaos in the forced pendulum. Phys. Rev., **A26**, 3483-3496.

Diener, M. (1984). The canard unchained, or how fast/slow dynamical systems bifurcate. Math. Intell., **6**, 38-49.

*Ding, M., Grebogi, C., and Ott, E. (1989). Dimensions of strange nonchaotic attractors. Phys. Lett., **137A**, 167-172.

Ding, M., Grebogi, C., and Ott, E. (1989). Evolution of attractors in quasiperiodically forced systems: from quasiperiodic to strange nonchaotic to chaotic. Phys. Rev., **A39**, 2593-2598.

Dowell, E. H. (1980). Nonlinear aeroelasticity. In New Approaches to Nonlinear Problems in Dynamics, P. J. Holmes (ed.), pp. 147-172. SIAM: Philadelphia.

Dowell, E. H. (1982). Flutter of a buckled plate as an example of chaotic motion of deterministic autonomous system. J. Sound Vib., **85**, 333-334.

Dowell, E. H. (1984). Observations and evolution of chaos for an autonomous system. ASME J. Appl. Mech., **51**, 664-673.

Dowell, E. H., and Pezeshki, C. (1988). On necessary and sufficient conditions for chaos to occur in Duffing's equation: an heuristic approach. J. Sound. Vib. , **121**, 195-200.

*Dowell, E. H., and Pereshki, C. (1986). On the understanding of chaos in Duffing's equation including a comparison with experiment. ASME J. Appl. Mech., **53**, 5-9.

Dressler, U. (1988). Symmetry property of the Lyapunov spectra of a class of dissipative dynamical systems with viscoutic damping, Phys. Rev., **A38**, 2103-2109.

Duffing, G. (1918). Erzwungene Schwingungen bei Veranderlicher Eigenfrequenz. Vieweg: Braunschweig.

Eckmann, J.-P., and Ruelle, D. (1985). Ergodic theory of chaos and strange attractors. Rev. Mod. Phys., **57**, 617-656.

Eilbeck, J. C. (1984). The sine-Gordon equation - from solitons to chaos. Bull. Inst. Math. Appl., **20**, 77-81.

El Naschie, M. S., Al Athel, S., and Walker, A. C. (1989). Localized buckling as statical homoclinic soliton and spacial complexity. Proceedings UTAM Symposium on Nonlinear Dynamics in Engineering. Springer-Verlag: New York. to appear.

*El Naschie, M. S., and Al Athel, S. (1989). On the connection between statical and dynamical chaos. Z. Naturforsch., **44a**, 645-650.

El Naschie, M. S. (1988). Generalized bifurcation and shell buckling as spacial statical chaos. ZAMM, **69**, T367-T377.

El Naschie, M. S. (1988). Order, chaos and generalized bifurcation. K. S. U. J. Eng. Sci., **14**, 437-444.

El Naschie, M. S. (1990). Stress, Stability and Chaos. McGraw-Hill, London.

*El Naschie, M. S., and Kapitaniak, T. (1990). Soliton chaos models for mechanical and biological elastic chains. Phys. Lett., **147A**, 275-281.

Elvey, J. S. N. (1983). On the elimination of de-stabilizing motions of articulated mooring towers under steady sea conditions. IMA J. Appl. Math., **31**, 235-252.

Epstein, I. R. (1983). Oscillations and chaos in chemical systems. Physica, **7D**, 47-56.

Eschenazi, E., Solari, H. G., and Gilmore, R. (1989). Basin of attraction in driven dynamical systems, Phys. Rev., **A39**, 2609-2627.

Everson, R. M., (1986). Chaotic dynamics of a bouncing ball. Physica, **19D**, 355-383.

Farmer, J. D. (1982). Chaotic attractors of an infinite-dimensional dynamical system. Physica, **4D**, 366-393.

Farmer, J. D., Ott, E., and Yorke, J. A. (1983). The dimension of chaotic attractors. Physica, **7D**, 153-180.

Farmer, J. D., Hart, J., and Weidman, P. (1982). A phase space analysis of baroclinic flow. Phys. Lett., **91A**, 22-24.

Fauve, S., Laroche, C., Libchaber, A., and Perrin, B. (1984). Chaotic phases and magnetic order in a convective fluid. Phys. Rev. Lett., **52**, 1774-1777.

Fauve, S., Laroche, C., and Perrin, B. (1985). Competing instabilities in a rotating layer of mercury heated from below. Phys. Rev. Lett., **55**, 208-210.

Feeny, B. F., and Moon, F. C. (1989). Autocorrelation on symbolic dynamics for a chaotic dry-friction oscillator. Phys. Lett., **141A**, 397-400.

Feigenbaum, M. J. (1978). Quantitative universality for a class of nonlinear transformations. J. Stat. Phys., **19**, 25-52.

Feigenbaum, M. J. (1980). The onset spectrum of turbulence. Phys. Lett., **76A**, 375-379.

Feigenbaum, M. J. (1983). Universal behaviour in nonlinear systems. Physica, **7D**, 16-39.

Feigenbaum, M. J. (1979). The onset spectrum of turbulence. Phys. Lett. **74A**, 375-378.

Feigenbaum, M. J., Kadanoff, L. P., and Shenker, S. J. (1982). Quasiperiodicity in dissipative systems: a renormalization group analysis. Physica, **5D**, 370-386.

Feingold, M., and Peres, A. (1983). Regular and chaotic motion of coupled rotators. Physica, **9D**, 433-438.

Feit, S.D. (1978). Characteristic exponents and strange attractors. Commun. Math. Phys., **61**, 249-260.

Fenstermacher, P. R., Swinney, H. L., and Gollub, J. P. (1979). Dynamical instabilities and the transformation to chaotic Taylor vortex flow. J. Fluid Mech., **94**, 103-128.

Flaherty, J. E., and Hoppensteadt, F. C. (1978). Frequency entrainment of a forced van der Pol's oscillator. Stud. Appl. Math., **18**, 5-15.

Flashner, H., and Hsu, C. S. (1983). A study of nonlinear periodic systems via the point mapping methods. Int. J. Numer. Meth. Eng., **19**, 185-215.

Ford, J. (1978). A picture book of stochasticity. In Topics in Nonlinear Dynamics, S. Jorna (ed.), pp. 121-146. American Institute of Physics: New York.

Ford, J. (1983). How random is a coin toss? Phys. Today, **36**, 40-48.

Fowler, A. C., and McGuinness, M. J. (1982a). A description of the Lorentz attractor at high Prandtl number. Physica, **5D**, 149-182.

Fowler, A. C., and McGuinness, M. J. (1982b). Hysteresis in the Lorenz equations. Phys. Lett., **92A**, 103-106.

Fraser, A. M., and Swinney, H. L. (1986). Using mutual information to find independent coordinates for strange attractors. Phys. Rev., **A33**, 198-214.

Frederickson, P., Kaplan, J.L., Yorke, E. D., and Yorke J. A. (1983). The Liapunov dimension of strange attractors. J. Differ. Eq., **49**, 185-207.

Froehling, H., Crutchfield, J. P., Farmer, D., Packard, N. H., Shaw, R. (1980). On determining the dimension of chaotic flow. Physica, **3D**, 605-617.

Fujisaka, H., and Yamada, T. (1983). Stability theory of synchronized motion in coupled-oscillator systems. Prog. Theor. Phys., **69**, 32-47, **70**, 1240-1248; **72**, 885-894.

Garrido, L. (ed.) (1983). Dynamical Systems and Chaos (Springer Lecture Notes in Physics, vol. 179). Springer-Verlag: Berlin.

Gaspard, P., and Nicolis, G. (1983). What can we learn from homoclinic orbits in chaotic dynamics? J. Stat. Phys., **31**, 499-518.

Gavrilov, N. K., and Shilnikov, L. P. (1972-73). On three-dimensional dynamical systems close to systems with a structurally unstable homoclinic curve. Math USSR Sb., **88**, 467-485; **90**, 139-156.

Gibson, G., and Jeffries, C. (1984). Observation of period doubling and chaos in spinwave instabilities in yttrium iron garnet. Phys. Rev., **29**, 811-818.

Giglio, M., Musazzi, S., and Perini, U. (1981). Transition to chaotic behaviour via a reproducible sequence of period-doubling bifurcations. Phys. Rev. Lett., **47**, 243-246.

Gilpin, M. E. (1979). Spiral chaos in a predator-prey model. Am. Naturalist, **113**, 306-308.

Gioggia, R. S., and Abraham, N. H. (1983). Routes to chaotic output from a single-mode, DC-excited laser. Phys. Rev. Lett., **51**, 650-653.

Glass, L. and Perez, R. (1982). The fine structure of phase locking. Phys. Rev. Lett., **48**, 1772-1775.

Gleick, J. (1987). Chaos: making a new science. Viking, New York.

Gollub, J. P., and Benson, S. V. (1980). Many routes to turbulent convection. J. Fluid Mech., **100**, 449-470.

Gollub, J. P., and Swinney, H. L. (1975). Onset of turbulence in a rotating fluid. Phys. Rev. Lett., **35**, 927-930.

Golubitsky, M., and Schaeffer, D. (1985). Singularity and Groups in Bifurcation Theory. Springer-Verlag: New York.

Golubitsky, M., and Stewart, I. (1985). Hopf bifurcation in the presence of symmetry. Arch. Rat. Mech. Anal., **87**, 107-165.

Gorman, M., Widmann, P. J., and Robbins, K. A. (1984). Chaotic flow regimes in a convection loop. Phys. Rev. Lett., **52**, 2241-2244.

*Grabec, I. (1986). Chaos generated by the cutting process. Phys. Lett.,

117A, 384-386.

Grabec, I. (1987). Dynamics of cutting process. Preprint University of Ljubjana.

Graham, R. (1976). Onset of self-pulsing in laser and the Lorenz model. Phys. Lett., **58A**, 440-442.

Grassberger, P., and Procaccia, I. (1983). Measuring the strangeness of strange attractors. Physica, **9D**, 189-208.

Greborgi, C., Ott, E., and Yorke, J. A. (1982). Chaotic attractors in crisis. Phys. Rev. Lett., **48**, 1507-1510.

Greborgi, C., Ott, E., and Yorke, J. A. (1983a). Are three-frequency quasiperiodic orbits to be expected in typical nonlinear dynamical systems? Phys. Rev. Lett., **51**, 339-342.

Greborgi, C., Ott, E., and Yorke, J. A. (1983b). Crises, sudden changes in chaotic attractors and transient chaos. Physica, **7D**, 181-200.

Greenspan, B. D., and Holmes, P. J. (1984). Repeated resonance and homoclinic bifurcation in a periodically forced family of oscillators. SIAM J. Math. Anal., **15**, 69-97.

Grossmann, S., and Thomae, S. (1977). Invariant distributions and stationary correlation functions of the one-dimensional discrete process. Z. Naturforsch., **32a**, 1353-1358.

Guckenheimer, J. (1973). Bifurcation and catastrophe. In Dynamical Systems, M.M. Peixoto (ed.). Academic Press: New York.

Guckenheimer, J. (1976). A strange strange attractor. In The Hopf Bifurcation and Its Applications, J. E. Marseden and M. McCracken (eds), pp. 368-381. Springer-Verlag: New York.

Guckenheimer, J. (1977). On the bifurcation of maps of the interval. Invent. Math., **39**, 165-178.

Guckenheimer, J. (1979). Sensitive dependence on initial conditions for one-dimensional maps. Commun. Math. Phys., **70**, 133-160.

Guckenheimer, J. (1980). Symbolic dynamics and relaxation oscillations. Physica, **1D**, 227-235.

Guckemheimer, J., and Buzyna, G. (1983). Dimension measurements for geostrophic turbulence. Phys. Rev. Lett., **51**, 1438-1441.

Guckenheimer, J., and Holmes, P.J. (1983). Nonlinear Oscillations, Dynamical Systems, and Bifurcations of Vector fields. Springer-Verlag: New York.

Guckenheimer, J. and Williams, R. F. (1979). Structural stability of Lorenz attractors. Publ. Math. IHES, **50**, 59-72.

Guevara, M. R., Glass, L., and Shier, A. (1981). Phase locking, period-doubling bifurcations, and irregular dynamics in periodically stimulated cariac cells. Science, **214**, 1350-1352.

Gurel, O., and Rossler, O. E. (eds) (1979). Bifurcation Theory and Applications in Scientific Disciplines (Annals of the New York Academy of Sciences, vol. 316). New York Academy of Science: New York.

Haken, H. (1975). Analogy between higher instabilities in fluids and lasers. Phys. Lett. **53A**, 77-78.

Hao, B.-L. (ed.) (1984). Chaos. World Scientific Publ.: Singapore. 2nd edition (1990).

Hao, B.-L. (1990). Elementary Symbolic Dynamics and Chaos in Dissipative Systems. World Scientific: Singapore.

Hao, B.-L. (ed.) (1988). Directions in Chaos, vol. 1, (1989) vol. 2. (1990) vol. 3.

Harrison, R. G., Firth, W. J., Emshary, C. A., and Al-Saidi, I. A. (1983). Observation of period doubling in an all-optical resonator containing NH_3 gas. Phys. Rev. Lett., **51**, 562-565.

Harth, E. (1983). Order and chaos in neural systems: an approach to the dynamics of higher brain functions. IEEE Trans. Syst. Man. Cybern., **SNC-13**, 782-789.

Hassard, B. D., Kazarinoff, N. D., and Wan, Y.-H. (1981). Theory and Applications of Hopf Bifurcation. Cambridge University Press: Cambridge.

Hayashi, C. (1964). Nonlinear Oscillations in Physical Systems. McGraw-Hill: New York.

Hayashi, C. (1980). The method of mapping with reference to the doubly asymptotic structure of invariant curve. Int. J. Nonlinear Mech., **15**, 341-348.

Hayashi, H., Ishizuka, S., and Hirakawa, K. (1983). Transition to chaos via intermittency in the Onchidium pacemaker neuron. Phys. Lett., **98A**, 474-476.

Held, G. H., Jeffries, C., and Haller, E. E. (1984). Observation of chaotic behaviour in an electron-hole plasma in Ge. Phys. Rev. Lett., **52**, 1037-1040.

Helleman, R. H. G. (1979). Exact results for some linear and nonlinear beam-beam effects. In Nonlinear Dynamics and the Beam-Beam Interaction, M. Month and J. C. Herrara (eds.), pp. 236-256.

Helleman, R. H. G. (ed.) (1980). Nonlinear Dynamics (annals of the New York Academy of Science, vol. 357). New York Academy of Science: New York.

Henon, M. (1976). A two-dimensional mapping with strange attractor. Commun. Math. Phys., **50**, 69-77.

Henon, M., and Pomeau, Y. (1975). Two strange attractors with a simple structure. In Turbulence and Navier-Stokes Equation, (Springer Lecture Notes in Mathematics, vol. 668). Springer-Verlag: New York.

Henon, M., and Heiles, C. (1964). The applicability of the third intergal of motion, some numerical experiments. Astron. J., **69**, 73-78.

Herbst, B. M., and Steeb, W.-H. (1988). Parametrically driven one-dimensional Sine-Gordon equation and chaos. Z. Naturforsch., **43a**, 727-733.

Herzel, H., Eberlig, W., and Schulmeister, Th. (1987). Nonuniform chaotic dynamics and effect of noise in biochemical systems. Z. Narurf., **42a**, 136-142.

Hirsch, J. E., Huberman, B. A., and Scalapino, D. J. (1982). Theory of intermittency. Phys. Rev., **25A**, 519-532.

Hirsch, M. W. (1984). The dynamical systems approach to differential equations. Bull. Am. Math. Soc., **11**, 1-64.

Holden, A. V. (ed.) (1986). Chaos. Manchester University Press: Manchester.

*Holmes, C., and Holmes, P. J. (1981). Second order averaging and bifurcations to subharmonics in Duffing's equation. J. Sound Vib., **78**, 161-174.

Holmes, P. J. (1977). Bifurcations to divergence and flutter in flow-induced oscillations: a finite-dimensional analysis, J. Sound Vib., **53**, 471-503.

Holmes, P. J. (1979). A nonlinear oscillator with a strange attractor. Phil. Trans. R. Soc. Lond. A, **292**, 419-448.

Holmes, P. J. (1980). Unfolding a degenerate nonlinear oscillator. In Nonlinear Dynamics, R. H. G. Helleman (ed.), pp. 473-488. New York Academy of Science: New York.

Holmes, P. J. (1982). The dynamics of repeated impacts with a sinusoidally vibrating table. J. Sound Vib., **84**, 173-189.

Holmes, P. J., and Marsden, J. E. (1978). Bifurcations to divergence and flutter in flow-induced oscillations: an infinite-dimensional analysis.

Automatica, **14**, 367-384.

*Holmes, P. J., and Marsden, J. E. (1981). A partial differential equation with infinitely many periodic orbits: chaotic oscillations of a forced beam. Arch. Rat. Mech. Anal., **76**, 135-165.

Holmes, P. J. and Moon, F. C. (1983). Strange attractors and chaos in nonlinear mechanics. ASME J. Appl. Mech., **50**, 1021-1032.

Holmes, P. J., and Rand, D. A. (1976). The bifurcations of Duffing's equation: an application of catastrophe theory. J. Sound Vib., **44**, 237-253.

Holmes, P. J., and Rand, D. A. (1978). Bifurcations of the forced van der Pol's oscillator. Q. Appl. Math., **35**, 495-509.

Holmes, P. J., and Rand, D. A. (1980). Phase portraits and bifurcations of the nonlinear oscillator $x + (a+\gamma x^2)x + \delta x^3 = 0$. Int. J. Nonlinear Mech., **15**, 449-458.

Holmes, P. J., and Whitley, D. C. (1983). On the attracting set for Duffing's equation. II. A geometrical model for moderate force and damping. Physica, **7D**, 11-123.

Holmes, P. J., and Whitley, D. C. (1984). Bifurcations of one- and two-dimensional maps. Phil. Trans. R. Soc. Lond. A, **311**, 43-102; Erratum, **312**, 601-602.

Hopf, E. (1942). Abzweigung einer periodischen Losung van einer stationaren Losung eines Differentialsystems. Ber. Math.-Phys. Klasse Sachs. Akad. Wiss. Leipzig, **94**, 1-22. English translation in Marsden and McCracken (1976).

*Hsieh, S.-R., and Shaw, S. W. (1990). The stability of modes at rest in a chaotic system. J. Sound Vib., **138**, 421-431.

Hsu, C.S. (1987). Cell-to-Cell Mapping. Springer-Verlag: New York.

Hsu, C. S. (1980a). A theory of index for point mapping dynamical systems. ASME J. Appl. Mech., **47**, 185-190.

Hsu, C. S. (1980b). A theory of cell-to-cell mapping for nonlinear dynamical systems. ASME J. Appl. Mech., **47**, 931-939.

Huang, J.C., Kao, Y.H., Wang, C.S., and Gou, Y.S. (1989). Bifurcation structure of the Duffing oscillator with asymmetric potential well. Phys. Lett., **136A**, 131-138.

Huberman, B., and Crutchfield, J. P. (1979). Chaotic states of anharmonic systems in periodic fields. Phys. Rev. Lett., **43**, 1743-1747.

Huberman, B. A., Crutchfield, J. P., and Packard, N. H. (1980). Noise phenomena in Josephson junctions. Appl. Phys. Lett., **37**, 750-752.

Hudson, J. L., Hart, M., and Marinko, D. (1979). An experimental study of multiple peak periodic and nonperiodic oscillations in the Belousov-Zhabotinskii reaction. J. Chem. Phys., **71**, 1601-1606.

Hudson, J. L., and Rossler, O. E. (1985). Chaos and complex oscillations in strered chemical reactors. In Dynamics of Nonlinear Systems, V. Hlavacek (ed.). Gordon and Breach: New York.

Hurt, G. W. (1981). An algorithm for the nonlinear analysis of compound bifurcation. Phil. Trans. R. Soc. Lond. A, **300**, 443-471.

Ikeda, K., Daido, H., and Akimoto, O. (1980). Optical turbulence: chaotic behaviour of transmitted light from a ring cavity. Phys. Rev. Lett., **45**, 709-712.

Infeld, E., and Rowlands, G. (1990). Nonlinear Waves, Solitons and Chaos. Cambridge University Press, Cambridge.

Iooss, G., and Joseph, D. D. (1980). Elementary Stability and Bifurcation Theory. Springer-Verlag: New York.

Iooss, G., Helleman, R., and Stora, R. (eds) (1983). Chaotic Behaviour of Deterministic Systems. North-Holland: Amsterdam.

Iooss, G., and Langford, W. F. (1980). Conjectures on the route to turbulence via bifurcation. In Nonlinear Dynamics, R. H. G. Helleman (ed.) pp. 489-505. New York Academy of Science: New York.

*Isomäki, H. M., von Boehm, J., and Räty, R. (1988). Fractal basin boundaries of an impacting particle. Phys. Lett., **126A**, 484-490.

Isomäki, H. M., von Boehm, J., and Räty, R. (1989). Chaos and the structure of the basin boundaries of a dissipative oscillator. In Structure, Coherence and Chaos in Dynamical Systems, P. L. Christiansen, R. D. Parmentier (eds), Manchester University Press: Manchester.

*Isomäki, H. M., von Boehm, J., and Räty, R. (1985). Devil's attractors and chaos of a driven impact oscillator. Phys. Lett., **107A**, 343-346.

Jeffries, C., and Perez, J. (1983). Observation of a Pomeau-Manneville intermittent route to chaos in a nonlinear oscillator. Phys. Rev., **26A**, 2117-2122.

Jensen, M. H., Bak, P., and Bohr, T. (1983a). Complete Devil's staircase fractal dimension and universality of mode-locking structures. Phys. Rev. Lett., **50**, 1637-1640.

*Jensen, M. H., Bak. P., Christiansen, P. and Bohr, T. (1983b). Josephson junctions and circle maps. Solid State Commun., **51**, 231-234.

Jensen, M. H., Bak, P., and Bohr, T. (1984). Transition to chaos by interacting resonanses in dissipative systems, I and II. Phys. Rev. **A30**, 1960-1980.

Kadanoff, L. P. (1983). Roads to chaos. Phys. Today, **36**, 46-53.

*Kahlert, C., and Rössler, O. E. (1984). Chaos as a limit in a boundary value problem. Z. Naturforsch., **39a**, 1200-1203.

Kaneko, K. (1984). Supercritical behaviour of disordered orbits of a circle map. Prog. Theor. Phys., **72**, 1089-1103.

Kaneko, K. (1986). Collapse of Tori and Origin of Chaos. World

Scientific: Singapore.

Kaneko, K. (1989). Spatiotemporal chaos in one- and two-dimensional coupled map lattice. Physica, **D37**, 60-82.

Kaneko, K. (1989). Chaotic but regular posi-nega switch among coded attractors by cluster-size variation. Phys. Rev. Lett., **63**, 219-223.

*Kapitaniak, T. (1986). Chaotic distribution of non-linear systems perturbed by random noise. Phys. Lett., **116A**, 251-254.

Kapitaniak, T., Awrejcewicz, J. and Steeb, W.-H. (1987). Chaotic behaviour of anharmonic oscillator with almost periodic excitation. J. Phys. A., **20**, L355-L358.

Kapitaniak, T. (1987a). Quantifying chaos with amplitude probability function. J. Sound Vib., **114**, 588-592.

Kapitaniak, T. (1987b). Chaotic behaviour of anharmonic oscillators with time delay. J. Phys. Soc. Jpn., **56**, 1951-1954.

Kapitaniak, T. (1988a). Combined bifurcations and its transition to chaos in the non-linear oscillator with two external periodic forces. J. Sound Vib., **121**.

*Kapitaniak, T. (1988b). Chaos in a noisy mechanical system with stress relaxation. J. Sound Vib., **123**, 391-396.

Kapitaniak, T. (1988c). Chaos in Systems with Noise. World Scientific: Singapore, (second edition 1990).

*Kapitaniak, T. (1990). Analytical condition for chaos behaviour of the Duffing oscillator. Phys. Lett., **144A**, 322-324.

Kapitaniak, T., and Wojewoda, J. (1990). Strange nonchaotic attractors of quasiperiodically forced van der Pol's Oscillator. J. Sound Vib., **138**, 162-169.

*Kapitaniak, T., Ponce, E., and Wojewoda, J. (1990). Route to chaos via strange non-chaotic attractors. J. Phys. **A23**, L383-L387.

Kaplan, J. L., and Yorke, J. A. (1979a). Chaotic behaviour of multidimensional difference equations. In Functional Differential Equations and Approximation of Fixed Points, H. O. Peitgen and H. O. Walther (eds), pp. 228-237 (Springer Lecture Notes in Mathematics, vol.730). Springer-Verlag: New York.

Kaplan, J. L., and Yorke, J. A. (1979b). Perturbulence, a regime observed in a fluid flow model of Lorenz. Commun. Math. Phys., **67**, 93-108.

Katok, A. B. (1980). Lyapunov exponents, entropy and periodic points for diffeomorphisms. Publ. Math. IHES, **51**, 137-174.

Kawakami, H. (1990). Bifurcation Phenomena in Nonlinear Systems and Theory of Dynamical Systems. World Scientific: Singapore.

Keolian, R., Turkevich, L. A., Putterman, S. J., Rudnick, I., and Rudnick, J. A. (1981). Subharmonic sequences in the Faraday experiment: departures from period doubling. Phys. Rev. Lett., **47**, 1133-1136.

King, G., and Swinney, H. L. (1983). Limits of stability and irregular flow patterns in wavy vortex flow. Phys. Rev., **A27**, 1240-1243.

Klinker, T., Meyer-Ilse, W. and Lautenborn, W. (1984). Period doubling and chaotic behaviour in a driven Toda oscillator. Phys. Lett., **101A**, 371-375.

Koch, B. P., Leven, R. W., Pompe, B., and Wilke, C. (1983). Experimental evidence for chaotic behaviour of a parametrically forced pendulum. Phys. Lett., **96A**, 219-224.

*Koch, B. P., and Leven, R. W. (1985). Subharmonic and homoclinic bifurcations in a parametrically forced pendulum. Physica, **16D**, 1-13.

Kotic, K. L. (1990). Lectures on Dynamical Systems, Structural Stability and Their Applications. World Scientific: Singapore.

Kocak, H. (1986). Differential and Difference Systems in Mechanics and

Physics. Springer-Verlag: New York.

Kunick, A., and Steeb, W.-H. (1985). Coupled chaotic oscillator. J. Phys. Soc. Jpn., **54**, 1220-1233.

Kunick, A., and Steeb, W.-H. (1987). Chaos in system with limit cycle. Int. J. Nonlinear Mech., **22**.

Kuramoto, Y. (1978). Diffusion-induced chaos in reaction systems. Prog. Theor. Phys. (Suppl.), **64**, 346-367.

Landau, L. D. (1944). On the problem of turbulence. C. R. Acad. Sci. URSS, **44**, 311-318.

Lanford, O. E. (1982a). The strange attractor theory of turbulence. Ann. Rev. Fluid Mech., **14**, 347-364.

Lanford, O. E. (1982b). A computer assised proof of the Feigenbaum conjecture. Bull. Am. Math. Soc., **6**, 427-434.

Lanford, O. E. (1985). A numerical study of the likelihood of phase locking. Physica, **14D**, 403-408.

Lauterborn, W., and Cramer, E. (1981). Subharmonic route to chaos observed in acoustic. Phys. Rev. Lett., **47**, 1445-1448.

Lean, G. H. (1971). Subharmonic motions of moored ship subjected to wave action. R. Inst. Naval Archit. Lond. Suppl. Pap., **113**, 387-399.

Lee, K. K. (1990). Lectures on Dynamical Systems, Structural Stability and Their Application. World Scientific: Singapore.

Leipnik, R. B., and Newton, T. A. (1981). Double strange attractor in rigid body motion with linear feedback control. Phys. Lett., **86A**, 63-67.

Levi, M. (1981). Qualitative analysis of the periodically forced relaxation oscillations. Mem. Am. Math. Soc., **214**, 1-47.

Levi, M., Hoppensteadt, F., and Miranker, W. (1978). Dynamics of the Josephson junction. Q. Appl. Math., **35**, 167-198.

Li, T. Y., and Yorke, J. A. (1975). Period three implies chaos. Am. Math. Monthly, **82**, 985-992.

Li, G. X., and Moon, F. C. (1990). Criteria for chaos of a three-well potential oscillator with homoclinic and heteroclinic orbits. J. Sound Vib., **136**, 17-34.

Libchaber, A., Fauve, S., and Laroche, C. (1983). Two-parameter study of the routes to chaos. Physica, **7D**, 73-84.

Libchaber, A., Laroche, C., and Fauve, S. (1982). Period doubling cascade in mercury: a quantitative measurement. J. Phys. Lett., **43**, L211-L216.

Libchaber, A., and Maurer, J. (1982). A Rayleigh-Bernard experiment: helium in a small box. In Nonlinear Phenomena at Phase Transitions and Instabilities, T. Riste (ed.), pp. 259-286. Plenum: New York.

Lichtenberg, A. J., and Lieberman, M. A. (1982). Regular and Stochastic Motion. Springer-Verlag: New York.

Linsay, P. (1981). Period doubling and chaotic behaviour in a driven anharmonic oscillator. Phys. Rev. Lett., **47**, 1349-1352.

Linsay, P., and Cumming, A. W. (1989). Three-frequency quasiperiodicity, phase locking and the onset of chaos. Physica, **40D**, 196-217.

Lorenz, E. N. (1963). Deterministic nonperiodic flow. J. Atoms. Sci., **20**, 130-141.

Lorenz, E. N. (1964). The problem of deducing the climate from the governing equations. Tellus, **16**, 1-11.

Lorenz, E. N. (1980). Noisy periodicity and reverse bifurcation. In Nonlinear Dynamics, R. H. G. Helleman (ed.), pp. 282-291. New York Academy of Science: New York.

Lorenz, E. N. (1984). The local structure of chaotic attractor in four dimensions. Physica, **13D**, 90-104.

MacDonald, A. H., and Plischke, M. (1983). Study of the driven damped pendulum: application to Josephson junctions and change-density-wave systems. Phys. Rev., **27B**, 201-211.

MacKay, R. S., and Tresser, C. (1984). Transition to chaos for two-frequency systems. J. Phys. Lett., **45**, L741-L746.

McLaughlin, J. B., and Orszag, S. A. (1982). Transition from periodic to chaotic thermal convection. J. Fluid Mech., **122**, 123-142.

Malraison, B., Atten, P., Berge, P., and Dubois, M. (1983). Dimension of strange attractors: an experimental determination for the chaotic regime of two convective systems. J. Phys. Lett., **44**, L897-L902.

Maldelbrot, B. (1983). The Fractal Geometry of Nature. W. H. Freeman: San Francisco.

Mankin, J. C., and Hudson, J. L. (1984). Oscillatory and chaotic behaviour of a forced exothermic chemical reaction. Chem. Eng. Sci., **39**, 1807-1814.

Marcus, P. S. (1981). Effects of truncation in modal representations of thermal convection. J. Fluid Mech., **103**, 241-256.

Marsden, J. E., and McCracken, M. (1976). The Hopf Bifurcation and Its Applications. Springer-Verlag: New York.

Marzec, C. J., and Spiegel, E. A. (1980). Ordinary differential equations with strange attractors. SIAM J. Appl. Math., **38**, 403-421.

*Matsumoto, T., Chua, L. O., and Tokunaga, R. (1987). Chaos via torus breakdown. IEEE Trans. Circuits Syst., **CAS-34,** 240-253

Maurer, J., and Libchaber, A. (1979). Rayleigh-Benard experiment in liquid helium; frequency loking and the onset of turbulence. J. Phys. Lett., **40**, L419-L423.

Maurer, J., and Libchaber, A. (1980). Effect of the Prandtl number on the onset of turbulence in liquid helium. J. Phys. Lett., **41**,

L515-L518.

May, R. M. (1976). Simple mathematical models with very complicated dynamics. Nature, **261**, 459-467.

May, R. M. (1979). Bifurcations and dynamic complexity in egological systems. In Bifurcation Theory and Applications in Scientific Disciplines, O. Gurel and O. E. Rossler (eds), pp. 517-529.

Mayer-Kress, G., and Haken, H. (1981). Intermittent behaviour of the logistic system. Phys. Lett., **82A**, 151-155.

*McDonald, S. W., Grebogi, C., Ott, E., and Yorke, J.A. (1985). Fractal basin boundaries, Physica, **17D**, 125-153.

Melnikov, V. K. (1963). On the stability of the center for time periodic perturbations. Trans. Moscov Math. Soc., **12**, 1-57.

*Mielke, A., and Holmes, P. J. (1988). Spatially complex equilibria for buckled rods. Arch. Rat. Mech. Anal., **101**, 319-348.

Miles, J. (1988). Directly forced oscillations of an inverted pendulum. Phys. Lett., **133A**, 295-297.

*Miles, J. (1988). Resonance and symmetry breaking for the pendulum. Physica, **31D**, 252-268.

Milnor, J. (1985). On the concept of attractor. Commun. Math. Phys., **99**, 177-195.

Miracky, R. F., Clarke, J., and Koch, R. H. (1983). Chaotic noise observed in a resistively shunted self-resonant Josephson tunnel junction. Phys. Rev. Lett., **50**, 856-859.

*Moon, F. C. (1980). Experiments on chaotic motions of a forced nonlinear oscillator: strange attractors. ASME J. Appl. Mech., **47**, 638-644.

Moon, F. C. (1984). Fractal boundary for chaos in a two-state mechanical oscillator. Phys. Rev. Lett., **53**, 962-964.

Moon, F. C. (1887). Chaotic Vibration. Wiley: New York.

Moon, F. C., and Holmes, P. J. (1979). A magnetoelastic strange attractor. J. Sound Vib., **65**, 285-296; Errata, **69**, 339.

Moon, F. C., and Shaw, S. W. (1983). Chaotic vibrations of a beam with nonlinear boundary conditions. Int. J. Nonlinear Mech., **18**, 465-477.

*Moon, F. C., and Li, G.-X., (1985). Fractal basin boundaries and homoclinic orbits for periodic motion in a two-well potential. Phys. Rev. Lett., **55**, 1439-1442.

Moon, F. C., Cusumano, J., and Holmes, P. J. (1987). Evidence for homoclinic orbits as a precursor to chaos in a magnetic pendulum. Physica, **24D**, 383-390.

Morimoto, Y. (1989). Transition phenomena in two interacting van der Pol oscillators. Phys. Lett., **141A**, 407-411.

Nakatsuka, H., Asaka, S., Itoh, H., Ikeda, K., and Matsuoka, M. (1983). Observation of bifurcation to chaos in an all-optical bistable system. Phys. Rev. Lett., **50**, 109-112.

Nauenberg, M., and Rudnick, J. (1981). Universality and power spectrum at the onset of chaos. Phys. Rev., **27B**, 493-498.

Nayfeh, A. H., and Mook, D. T. (1979). Nonlinear Oscillations. Wiley: New York.

Newhouse, S. E., Ruelle, D., and Takens, F. (1978). Occurrence of strange axiom A attractors near quasiperiodic flows on T^m, $m \geq 3$. Commun. Math. Phys., **64**, 35-40.

Nicolis, G., and Prigogine, I. (1977). Self-Organization in Non-Equilibrium Systems. From Dissipative Structures to Order Through Fluctuations. Wiley: New York.

Nicolis, J. C. (1990). Chaos and Information Processing. World Scientific: Singapore.

Orszag, S. A., and McLaughlin, J. B. (1980). Evidence that random behaviour is genetic for nonlinear differential equations. Physica, **1D**, 68-79.

Ostlund, S., Rand, D., Sethna, J., and Siggia, E. (1983). Universal properties of the transition from quasiperiodicity to chaos in dissipative systems. Physica, **8D**, 303-342.

Ott, E. (1981). Strange attractors and chaotic motions of dynamical systems. Rev. Mod. Phys., **53**, 655-671.

Ottino, J. M. (1989). The Kinematics of mixing: Stretching, Chaos and Transport. Cambridge University Press, Cambridge.

*Packard, N. H., Crutchfield, J. P., Farmer, J. D., and Shaw, R. S. (1980). Geometry from a time series. Phys. Rev. Lett., **45**, 712-716.

Parkinson, G. V., and Smith, J. D. (1964). The square prism as an aeroelastic nonlinear oscillator. Q. J. Mech. Appl. Math., **17**, 225-239.

*Parlitz, U., and Lauterborn, W. (1985). Superstructure in the bifurcation set of the Duffing's equation $\ddot{x} + d\dot{x} + x + x^3 = f \cos(\omega t)$. Phys. Lett., **107A**, 351-355.

Parlitz, U., and Lauterborn, W. (1986). Resonances and torsion numbers of driven dissipative nonlinear oscillator. Z. Naturforsch., **41a**, 605-614.

*Parlitz, U., and Lauterborn, W. (1987). Period-doubling cascades and devil's staircases of the driven van der Pol oscillator. Phys. Rev., **A36**, 1428-1434.

Peckham, B. B. (1990). The necessity of the Hopf bifurcation for periodically forced oscillators. Nonlinearity, **3**, 261-280.

Perez, J., and Jeffries, C. (1982). Direct observation of a tangent bifurcation in a nonlinear oscillator. Phys. Lett., **92A**, 82-84.

Pesin, J. B. (1977). Charatacteristic Lyapunov exponents and smooth ergodic theory. Russ. Math. Surv., **32**, 55-114.

Pezeshki, C., and Dowell, E. H. (1987). An examination of initial condition maps for the sinusoidally excited buckled beam modelled by the Duffing's equation. J. Sound Vib., **117**, 219-232.

Pippard, B. (1982). Instability and chaos: physical models of everyday life. Interdisc. Sci. Rev., **7**, 92-101.

Pismen, L. M. (1982). Bifurcation sequences in a third-order system with a folded slow manifold. Phys. Lett., **89A**, 59-62.

*Poddar, B., Moon, F. C., and Mukherjee, S. (1988). Chaotic motion of an elastic-plastic beam. ASME J. Appl. Mech., **55**, 185-189.

Poincare, H. (1880-90). Memoire sur les courbes definies par les equations differentielles I-VI, Oeuvre I. Gauthier-Villars: Paris.

Poicare, H. (1890). Sur les equations de la dynamique et le probleme de trois cops. Acta Math., **13**, 1-270.

Poincare, H. (1899). Les Methodes Nouvelles de la Mecanique Celeste, vols. 1-3. Gauthier-Villars: Paris.

Pomeau, Y. (1983). The intermittent transition to turbulence. In Nonlinear Dynamics and Turbulence, G. I. Barenblatt, G. Iooss, and D. D. Joseph (eds). Pitman: London.

Pomeau, Y. and Manneville, P. (1980). Intermittent transition to turbulence in dissipative dynamical systems. Commun. Math. Phys., **74**, 189-197.

Pomeau, Y., Roux, J. C., Rossi, A., Bachelart, S., and Vidal, C. (1981). Intermittent behaviour in the Belousov-Zhabotinsky reation. J. Phys. Lett., **42**, L271-L273.

Preston, C. (1983). Iterates of Maps on an Interval (Springer Lecture Notes In Mathematics, vol. 999). Springer-Verlag: New York.

Prigogine, I. (1980). From Being to Becoming: Time and Complexity in the Physical Sciences. W. H. Freeman: San Fransisco.

Pustylnikov, L. D.(1978). Stable and oscillating motions in non-autonomous dynamical systems. Trans. Moscow Math. Soc., **14**, 1-101.

Rand, D. A. (1978). The topological classification of Lorenz attractors. Math. Proc. Camb. Phil. Soc., **83**, 451-460.

Rand, D. A., Oslund, S., Sethna, J., and Siggia, E. (1982). A universal transition from quasi-periodicity to chaos in dissipative systems. Phys. Rev. Lett., **49**, 132-135.

Rand, D. A., and Young, L. S. (eds) (1981). Dynamical Systems and Turbulence (Springer Lecture Notes in Mathematics, vol. 898). Springer-Verlag: New York.

Rand, R. H., and Amburaster, D. (1987). Perturbation Methods, Bifurcation Theory and Computer Algebra. Springer-Verlag: New York.

Raty. R., van Boehm, J., and Isomaki, H. M. (1984). Absence of inversion-symmetric limit cycles of even periods and the chaotic motion of Duffing's oscillator. Phys. Lett., **103A**, 289-292.

Rollins, R. W., and Hunt, E. R. (1984). Intermittent transient chaos at interior crises in the diode resonator. Phys. Rev., **29A**, 3327-3334.

Romeiras, F., and Ott, E. (1987). Strange nonchaotic attractors of the damped pendulum with quasiperiodic forcing. Phys. Rev., **35A**, 4404-4413.

*Romeiras, F., Bondeson, A., Ott, E., Antonsen, T. M. Jr. and Grebogi, C. (1987). Quasiperiodically forced dynamical systems with strange nonchaotic attractors, Physica, **26D**, 277-294.

Romeiras, F., Grebogi, C., and Ott, E. (1988). Critical exponents for power spectra scaling at mergings of chaotic bands. Phys. Rev., **38A**, 463-468.

Rosen, R. (1970). Dynamical Systems Theory in Biology. Wiley-Interscience: New York.

Rossler, O. E. (1976a). An equation for continuous chaos. Phys. Lett., **57A**, 397-398.

Rossler, O. E. (1976b). Different types of chaos in two simple differential equations. Z. Naturforsch., **31a**, 1664-1670.

Rossler, O. E. (1977). Horseshoe-map chaos in the Lorenz equation. Phys. Lett., **60A**, 392-394.

Rossler, O. E. (1979). Continuous chaos-four prototype equations. In Bifurcation Theory and Applications in Scientific Disciplines, O. Gurel and O.E. Rossler (eds), pp. 376-392. New York Academy of Sciences: New York.

Rosşler, O. E. (1981). The gluing together principle and chaos. In Nonlinear Problems of Analysis in Geometry and Mechanics, M. Atteia, D. Bancel, and I. Gumowski (eds), pp.50-56. Pitman: Boston.

Rossler, O. E. (1983). The chaotic hierarchy. Z. Naturforsch., **38a**, 788-801.

Roux, J.-C. (1983). Experimental studies of bifurcations leading to chaos in the Belousov-Zhabotinsky reations. Physica, **7D**, 57-68.

Roux, J.-C., Simoyi, R. H., and Swinney, H. L. (1982). Observation of a strange attractor. Physica, **8D**, 257-266.

Ruelle, D. (1979). Sensitivity dependence on initial conditions and turbulent behaviour of dynamical systems. In Bifurcation Theory and Applications in Scientific Disciplines, O. Gurel and O. E. Rossler (eds), pp. 408-446. New York Academy of Science: New York.

*Ruelle, D. (1980). Strange attractors. Math. Intelligencer, **2**, 126-137.

Ruelle, D. (1981). Small random perturbations of dynamical systems and the definition of attractors. Commun. Math. Phys., **82**, 137-151.

Ruelle, D. and Takens, F. (1971). On the nature of turbulence. Commun. Math. Phys., **20**, 167-192, **23**, 343-344.

Ruelle, D. (1989). Elements of Differentiable Dynamics and Bifurcation Theory. Academic Press: London.

Rudowski, J., and Szemplinska-Stupnicka, W. (1987). On an approximate criterion for chaotic motion in a model of a buckled beam.

Ingenieur-Archiv., **57**, 243-255.

Russell, D. A., and Ott, E. (1981). Chaotic (strange) and periodic behaviour in instability saturation by the oscillating two-stream instability. Phys. Fluids, **24**, 1976-1988.

Salam, F. M. A., Marsden, J. E., and Varaiya, P. P. (1983). Chaos and Arnold diffusion in dynamical systems. IEEE Trans. Circuits Syst., **CAS-30**, 697-708.

*Salam, F. M. A. (1987). The Melnikov technique for highly dissipative systems. SIAM J. Appl. Math., **47**, 232-243.

Salam, F. M. A., and Levi, M. L. (eds) (1988). Dynamical Systems Approaches to Nonlinear Problems in Systems and Circuits. SIAM: Philapelphia.

Sano, M., and Sawada, Y. (1983). Transition from quasi-periodicity to chaos in a system of coupled nonlinear oscillators. Phys. Lett., **97A**, 73-76.

*Sato, S., Sano, M., and Sawada, Y. (1983). Universal scaling property in bifurcation structure of Duffing's and of generalized Duffing's equations. Phys. Rev., **A28**, 1654-1658.

Schecter, S. (1987). The saddle-node separatrix-loop bifurcation. SIAM J. Math. Anal., **18**, 1142-1156.

Schecter, S. (1987). Melnikov's method at a saddle-node and the dynamics of the forced Josephson junction. SIAM J. Math. Anal., **18**, 1699-1715.

Scheurle, J. (1986). Chaos in quasiperiodically forced systems. ZAMP, **37**, 12-28.

Schuster, H. G. (1984). Deterministic Chaos. Physik-Verlag: Weinheim, (second edition 1988).

Scott, S. K. (1987). Oscillations in simple models of chemical systems. Accounts of Chemical Research, **20**, 186-191.

Seelig, F. F. (1980-83). Unrestricted harmonic balance. Z. Naturforsch., **35a**, 1054-1061, **38a**, 636-640, **38a**, 729-735.

Shaw, R. (1981). Strange attractors, chaotic behaviour, and information flow. Z. Naturforsch., **36a**, 80-112.

Shaw, R. (1984). The Dripping Faucet as a Model Chaotic System. Aerial Press: Santa Cruz, CA.

Shaw, S. W., and Holmes, P. J. (1983a). A periodically forced impact oscillator with large dissipation. ASME J. Appl. Mech., **50**, 849-857.

*Shaw, S. W., and Holmes, P. J. (1983b). A periodically forced piecewise linear oscillator. J. Sound Vib., **90**, 129-155.

Shaw, S. W., and Holmes, P. J. (1983c). Periodically forced linear oscillator with impacts: chaos and long-period motions. Phys. Rev. Lett., **51**, 623-626.

Shenker, S. J. (1982). Scaling behaviour in a map of a circle onto itself: empirical results. Physica, **5D**, 405-411.

Shilnikov, L. P. (1976). Theory of the bifurcation of dynamical systems and dengerous boundaries. Sov. Phys. Dokl., **20**, 674-676.

Shimada, J., and Hagashima, T. (1979). A numerical approach to ergodic problem of dissipative systems. Prog. Theor. Phys., **61**, 1605-1616.

Shtern, V. N. (1983). Attractor dimension for the generalized Baker's transformation. Phys. Lett., **99A**, 268-270.

Shtern, V. N., and Shumova, L. V. (1984). Metamorphoses of preturbulence. Phys. Lett., **103A**, 167-170.

Simo, C. (1979). On the Henon-Pomeau attractor. J. Stat. Phys., **21**, 465-494.

Sing, P., and Joseph, D. D. (1989). Autoregressive methods for chaos on binary sequences for the Lorenz attractor. Phys. Lett., **135A**, 247-252.

Smale, S. (1967). Differentiable dynamical systems. Bull. Am. Math.

Soc., **73**, 747-817.

*Soliman, M. S., and Thompson, J. M. T. (1989). Integrity measures quantifying the erosion of smooth and fractal basins of attraction. J. Sound Vib., **135**, 453-475.

Sparrow, C. (1982). The Lorenz Equations. Springer-Verlag: New York.

Sri Namachchivaya, N., and Hilton, H. (1987). Chaotic motion of a forced nonlinear oscillator. Canadiam Math. Soc., Conference Proceedings, vol.8, 561-577.

Stavans, J., Heslot, F., and Libchaber, (1985). Fixed winding number and the quasiperiodic route to chaos in a convective fluid. Phys. Rev. Lett., **55**, 596-599.

Steeb, W.-H., and Louw, J. A. (1986). Chaos and Quantum Chaos. World Scientific: Singapore.

Steeb, W.-H., Louw, J. A., and Kapitaniak, T. (1987). Chaotic behaviour of an anharmonic oscillator with two external periodic forces. J. Phys. Soc. Jpn., **55**, 3279-3281.

Steeb, W.-H., Huang, J. C., and Gou, Y. S. (1989). A comment on the chaotic behaviour of van der Pol's equations with an external periodic excitation. Z. Naturforsch., **44a**, 932-935.

Stone, E. (1990). Power spectra of the stochastically forced Duffing oscillator. Phys. Lett., **148A**, 434-442.

Stone, E., and Holmes, P. (1989). Noise induced intermittency in a model of a turbulent boundary layer. Physica, **37D**, 20-32.

Stoker, J. J. (1950). Nonlinear Vibrations. Wiley: New York.

Stoker, J. J. (1980). Periodic forced vibrations of systems of relaxation oscillators. Commun. Math. Phys., **33**, 215-240.

*Svensmark, H., and Samuelsen, M. R. (1987). Influence of perturbations on period-doubling bifurcation. Phys. Rev., **A36**, 2413-2417.

Swinney, H. L. and Gollub, J. P. (eds) (1981). Hydrodynamic Instabilities and the Transition to Turbulence. Springer-Verlag: New York.

Swinney, H. L. and Roux, J. C. (1984). Chemical chaos. In Nonequilibrium Dynamics in Chemical Systems, C. vidal (ed.). Springer-Verlag: New York.

Szemplinska-Stupnicka, W. (1988). The refined approximate criterion for chaos in a two-state mechanical oscillator. Ingenieur-Archiv, **58**, 554 - 566.

Szemplinska-Stupnicka, W. (1987). Secondary resonanses and approximate models of transition to chaotic motion in nonlinear oscillators. J. Sound Vib., **117**, 155-172.

*Szemplińska-Stupnicka, W. (1988). The refined approximate criterion for chaos in a two-stage mechanical oscillator. Ingenieur-Archiv, **58**, 354-366.

Takayasu, H.(1990). Fractals in Physical Sciences. Manchester University Press: Manchester.

Takens, F., (1974). Forced oscillations and bifurcations. Comm. Math. Inst. Rijkuniv. Utrecht, **3**, 1-59.

Takens, F. (1980). Detecting strange attractors in turbulence. In Dynamical Systems and Turbulence, D. A. Rand and L. -S. Young (eds) (Springer Lecture Notes in Mathematics, vol. 898). Springer-Verlag: New York.

Taki, M. (1987). Melnikov's method for nonperiodic perturbations and the bifurcations in a Josephson junction. Phys. Rev, **35B**, 3267-3270.

*Takimoto, N. and Yamashida, H. (1987). The variational approach to the theory of subharmonic bifurcations. Physica, **26D**, 251-276.

Tang. T., Dowell, E. H. (1988). Numerical simulations of periodic and chaotic responses in stable Duffing's system. Int. J. Nonlinear Mech., **22**, 401-425.

Tang, D. M., Dowell, E. H. (1988). On the threshold force for chaotic motions for a forced buckled beam. ASME J. Appl. Mech., **55**, 190-196.

Tavakol, R. K., and Tworkowski, A. S. (1984). On the occurrence of quasiperiodic motion on three tori. Phys. Lett., **100A**, 65-67.

Teman, R. (1988). Infinite-Dimensional Dynamical Systems in Mechanics and Physics. Springer-Verlag: New York.

Testa, J., Perez, J. and Jeffries, C. (1982). Evidence of universal chaotic behaviour of a driven nonlinear oscillator. Phys. Rev. Lett., **48**, 714-717.

Thom, R. (1975). Structural Stability and Morphogenesis. W. A. Benjamin: Reading, MA.

Thompson, J. M. T. (1982). Stability and Catastrophes in Science and Engineering. Wiley: Chichester.

Thompson, J. M. T. (1979). Stability predictions through a succession of folds. Phil. Trans. R. Soc. Lond. A, **292**, 1-23.

*Thompson, J. M. T. (1983). Complex dynamics of compliant off-shore structures. Proc. R. Soc. Lond. A, **387**, 407-427.

Thompson, J. M. T., Bokaian, A. R., and Ghaffari, R. (1983). Subharmonic resonances and chaotic motions of a bilinear oscillator. IMA J. Appl. Math., **31**, 207-234.

Thompson, J. M. T., Bokaian, A. R., and Ghaffari, R. (1984). Subharmonic and chaotic motions of compliant offshore structures and articulated mooring towers. ASME J. Energy Resources Tech., **106**, 191-198.

Thompson, J. M. T., and Elvey, J. S. N. (1984). Elimination of subharmonic resonances of compliant marine structures. Int. J. Mech. Sci., **26**, 419-425.

*Thompson, J. M. T., and Ghaffari, R. (1982). Chaos after period-doubling bifurcations in the resonance of an impact oscillator. Phys. Lett., **91A**, 5-8.

Thompson, J. M. T., and Ghaffari, R. (1983). Chaotic dynamics of an impact oscillator. Phys. Rev., **27A**, 1741-1743.

Thompson, J. M. T., and Stewart, H. B. (1984). Folding and mixing in the Birkhoff-Shaw chaotic attractor. Phys. Lett., **103A**, 229-231.

Thompson, J. M. T., and Stewart, H. B. (1986). Nonlinear Dynamics and Chaos. Wiley: Chichester.

Thompson, J. M. T., and Thompson, R. J. (1980). Numerical experiments with a strange attractor. Bull. Inst. Math. Appl., **16**, 150-154.

Thompson, J. M. T., and Virgin, L. N. (1986). Predicting a jump to resonance using transient maps and beats. Int. J. Nonlinear Mech, **21**, 234-245.

*Thompson, J. M. T., and Virgin, L. N. (1988). Spatial chaos and localization phenomena in nonlinear elasticity. Phys. Lett., **126A**, 491-496.

Thompson, J. M. T., Bishop, S. R., and Leung, L. M. (1987). Fractal basin and chaotic bifurcations prior to escape from a potential well. Phys. Lett., **121A**, 116-120.

Tomita, K. (1982). Chaotic response of nonlinear oscillators. Phys. Rep., **86**, 113-167.

Tresser, C. (1983). Nouveaux types de transitions vers une entropie topologique positive. C. R. Acad. Sci. Paris, **296**, 729-732.

Troger, H. (1979). On point mapping for mechanical system possessing homoclinic and heteroclinic points. ASME J. Appl. Mech., **46**, 468-476.

*Tseng, W.-Y., and Dugundji, J. (1971). Nonlinear vibrations of a buckled beam under harmonic excitation. ASME J. Appl. Mech., **38**, 467-476.

Tsuda, I. (1981). Self-similarity in the Beloussov-Zhabotinsky reaction.

Phys. Lett., **85A**, 4-8.

*Ueda, Y. (1979). Random phenomena resulting from nonlinearity in the system described by Duffing's equation. Int. J. Non-Linear Mechanics, **20**, 481-491.

Ueda, Y. (1980a). Steady motions exhibited by Duffing'sequation: a picture book of regular and chaotic motions. In New Approaches to Nonlinear Problems in Dynamics, P. J. Holmes (ed.) pp. 311-322. SIAM: Philadelpia.

Ueda, Y. (1980b). Explosion of strange attractors exhibited by Duffing's equation. In Nonlinear Dynamics, R. H. G. Helleman (ed.), pp. 422-434. New York Academy of Science: New York.

Ueda, Y. (1986). Survey of strange attractors and chaotically transitional phenomena in the system governed by Duffing's equation. In Complex and Distributed Systems: Analysis, Simulation and Control, S. G. Tzafestas and P. Borne (eds), Elsevier: Amsterdam.

*Ueda, Y., and Akamatsu, N. (1980). Chaotically transitional phenomena in the forced negative-resistance oscillator. IEEE Trans. Circuits Syst., **CAS-28**, 217-224.

Ueda, Y., Akamatsu, N., and Hayashi, C. (1973). Computer simulation of nonlinear ordinary differential equations and non-periodic oscillations Trans. IECE Jpn., **56A**, 218-255.

*Umberger, D. K., Grebogi, C., Ott, E., and Afeyan, B. (1989). Spatiotemporal dynamics in a dispersively coupled chain of nonlinear oscillators. Phys. Rev., **A39**, 4835-4842.

*Vallée, R., Delisle, C., and Chrostowski, J. (1984). Noise versus chaos in acousto-optic bistability. Phys. Rev., **A30**, 336-342.

Van der Pol, B. (1926). On relaxation-oscillations. Phil. Mag. (7), **2**, 978-992.

Van der Pol, B. (1927). Forced oscillations in a circuit with nonlinear resistance (reception with reactive triode). Phil. Mag. (7), **3**, 65-80.

*Van der Pol, B., and Van der Mark, J. (1927). Frequency demultiplication. Nature, **120**, 363-364.

Van der Pol, B. and Van der Mark, J. (1928). The heart beat considered as a relaxation oscillation, and an electrical model of the heart. Phil. Mag. (7), **6**, 763-775.

Wiesenfield, K. (1985). Virtual Hopf phenomenon: a new precursor of period-doubling bifurcations. Phys. Rev., **32A**, 1744-1751.

Wiesenfeld, K., and McNamara, B. (1986). Small-signal amplification in bifurcating dynamical systems. Phys. Rev., **A33**, 629-642.

Wiggins, S. (1987). Chaos in the quasiperiodically forced Duffing's oscillator. Phys. Lett., **124A**, 386-392.

Wiggins, S. (1989). Global Bifurcation and Chaos. Springer-Verlag: New York.

Wiggins, S. (1990). Introduction to Applied Nonlinear Systems ans Chaos. Springer, New York.

Williams, R. F. (1977). The structure of Lorenz attractors. In Turbulence Seminar Berkeley 1976/77, P. Bernard and T. Ratiu (eds), pp. 94-112. Springer-Verlag: New York.

*Wolf, A., Switft, J. B., Swinney, H. L. and Vastano, J. A. (1985). Determining Lyapunov exponents from a time series. Physica, **16D**, 285-317.

*Wu, S. (1987). Chua's circuit family. IEEE Proc., **75**, 8, 1022-1032.

Yagasaki, K. (1990). Second-order averaging and chaos in quasiperiodically forced weakly nonlinear oscillators. Physica, **44D**, 445-458.

Yamaguchi, Y., and Sakai, K. (1983). New type of 'crisis' showing histeresis. Phys. Rev., **A27**, 2755-2762.

Yeh, W. J., He, D. R., and Kao, Y. H. (1984). Fractal dimension and self

-similarity of the devil's staircasein a Josephson-junction simulator. Phys. Rev. Lett., **52**, 480-483.

Yeh, W. J., and Kao, Y. H. (1983). Intermittency in Josephson junctions. Appl. Phys. Lett.,**42**, 299-301.

Ysfanskii, S. L., and Beresnevich, V. I. (1983). Influence of elastic-characteristics asymmetry an the vibrations of nonlinear system. Sov. Mach. Sci., **4**, 28-32.

*Zak, M. (1984). Deterministic representation of chaos in classical dynamics. Phys. Lett., **107A**, 125-128.

Zaslawsky, G. M. (1978). The simplest case of a strange attractor. Phys. Lett., **69A**, 145-147.

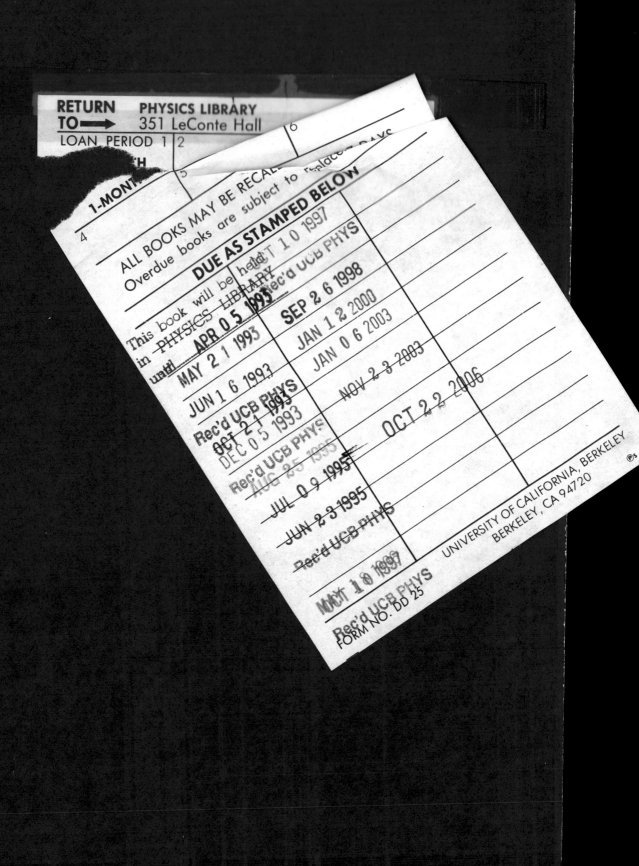